Liquid Rocket Engine Combustion Instability

Liquid Rocket Engine Combustion Instability

Edited by
Vigor Yang
William E. Anderson
Pennsylvania State University
University Park, Pennsylvania

Volume 169
PROGRESS IN ASTRONAUTICS AND AERONAUTICS

Paul Zarchan, Editor-in-Chief
Charles Stark Draper Laboratory
Cambridge, Massachusetts

Published by the American Institute of Aeronautics and Astronautics, Inc.
370 L'Enfant Promenade, SW, Washington, DC 20024-2518

ISBN 1-56347-183-3

Table of Contents

Part 4. Stability Testing Methodology

Preface

Combustion instabilities have long been recognized as one of the most difficult problems in the development of liquid propellant rocket engines. Although their existence probably predated even the invention of the V-2 rockets during World War II, the ability to measure high-frequency pressure oscillations provided the first indication to rocket developers in the early 1950s that the source of many rocket failures could be attributed to combustion instability. In spite of the relatively long history of combustion instability, including a number of spectacular failures, and the resultant large body of work that has been devoted to the area, there is still a lack of fundamental knowledge regarding its prediction and the specific physical mechanisms that bring upon its onset. This lack of understanding has resulted in severe penalties in terms of escalating the cost and time for development of new systems and of the excessive margins that have been built into propulsion systems to avoid the occurrence of instability.

This volume is the result of a concerted effort by a core group of workers active in the field who believed there was a distinct need for more effective communication and an organized effort to deal with the technological problems associated with liquid rocket engine combustion instability. The seeds of this effort were planted in the 1980s with the realization that the hardware-intensive technological approaches used in the 1960s to develop stable rocket engines would not be viable alternatives in an economically sensitive era. Specific justification for a renewed effort in solving these technological problems included the planned development of a new heavy launch vehicle in the United States, the self-destruction of an Ariane rocket in 1980, and a number of rocket development efforts in countries that were beginning or expanding their ability to launch military and commercial satellites. In the United States, the Air Force Phillips Laboratory, the NASA Lewis Research Center, the Air Force Office of Scientific Research, and the Propulsion Engineering Research Center at Pennsylvania State University were instrumental in providing a stimulus for renewed activity in combustion stability research that forms the basis for much of the material presented in this volume.

The last effort undertaken in the United States to assemble a reference text on liquid rocket engine combustion instability was concluded in 1972, when works from a series of NASA- and Air Force-sponsored research programs resulted in the publication of NASA SP-194, *Liquid Propellant Rocket Combustion Instability.* NASA SP-194 took three years to plan and another five years to write and publish; it still serves as the foremost reference text on combustion instability. In the 20-odd years since the publication of NASA SP-194, improved computational and experimental techniques along with a number of development experiences with engines world-wide have offered many opportunities for advancement of the state-of-the-art. However, subsequent published works were scattered in journals, conference proceedings, and contract reports. Furthermore, there were tendencies for members of the liquid rocket engine combustion stability community to present their work at meetings that had restricted access; to withhold the work from public dissemination to maintain a technological edge; or to present or publish sporadically. Thus, the liquid rocket engine community was virtually starved for an updated source of technical information related to combustion instability. This series volume was designed to serve as an extension to NASA SP-194 and to be used as a reference by engineers and scientists in the field of liquid rocket propulsion.

The volume had its immediate origin in the First International Symposium on Liquid Rocket Engine Combustion Instability, which was hosted by the Propulsion Engineering Research Center at Pennsylvania State University in January 1993. This meeting was a unique and landmark event, and was attended by over 50 of the world's leading experts in the field. Representatives from China, France, Germany, Korea, The Netherlands, Russia, Taiwan, Ukraine, and the United States were present. Especially noteworthy were the presentations of work from the Commonwealth of Independent States and from China that previously had been unknown in the West. This symposium ostensibly represented one of the first open technical exchanges with these formerly closed communities in this critical propulsion technology area.

This volume has been designed foremost to be a useful reference. Although it is tailored specifically to liquid rocket combustion instability, much of the material can also be applied to other types of liquid-fueled propulsion systems such as ramjets, gas turbine engines, diesel engines, and liquid-propellant guns. Since no book on unsteady combustion of liquid-fueled systems has been published as a research monograph in the past 20 years, with the exception of a few distantly related volumes on solid propellant combustion, workers in these areas may also find the volume useful and relevant.

A total of 21 chapters are included. Each chapter was prepared by one or more experts in the subject area, and represents contributions from academia, government laboratories, and industry. The authors were asked to orient their discussion toward four specific goals: provide an overview of the subject area, especially advancements made since the publication of NASA SP-194 in 1972; summarize the state-of-the-art knowledge; provide details of their methodology; and identify future research needs. The volume consists of the following four major subject areas: 1) engine phenomenology and case studies; 2) fundamental mechanisms of combustion instability; 3) combustion instability analysis; and 4) stability testing methodology.

Finally, we would like to express our deep gratitude toward the individuals and groups whose help has made the publication of this volume possible and its preparation a rewarding experience. Of course, the authors collectively have contributed years to this volume. A number of individuals contributed to the volume by reviewing the submitted work and making their suggestions for improving the book; a list of reviewers follows. Enabling financial support was provided by NASA, the Chungnam National University of South Korea, Aerojet, and the Propulsion Engineering Research Center and the College of Engineering at Pennsylvania State University. Jerry Thompson was instrumental in organizing the contributions from the Commonwealth of Independent States.

We would also like to thank Mark Fisher at NASA Marshall Space Flight Center for his help in the preparation of final drawings, Penny Barr and Mary Newby at the Propulsion Engineering Research Center for their help in preparing the manuscripts and in interacting with the authors and the publisher, and Christine Kalmin and Heather Brennan at AIAA for their professional and friendly help in publishing the volume.

The individuals who gave their time and expertise toward reviewing the chapters in this volume deserve special recognition. Their names appear below.

P. Combs	R. Ingebo	D. Preclik
C. Chen	J. Ito	R. Priem
F. Culick	A. Karagozian	F. Reardon
I. Gökalp	S. Margolis	R. Reitz
M. Habiballah	W. Mayer	J. Shuen
U. Hedge	M. Micci	W. Strahle
A. Hersh	J. Muss	F. Williams
R. Hewitt	T. Nguyen	H. Yang
J. Hulka	S. Pal	M. Zaller
E. Hurlbert	J. Pieper	

Vigor Yang
William E. Anderson
July 1995

Chapter 1

Overview of Combustion Instabilities in Liquid-Propellant Rocket Engines

Fred E. C. Culick*
California Institute of Technology, Pasadena, California 91125
and
Vigor Yang†
Pennsylvania State University, University Park, Pennsylvania 16802

I. Introduction

COMBUSTION instabilities were discovered in solid- and liquid-propellant rocket engines at about the same time in the late 1930s. Since then, unstable oscillations have occurred in most, if not practically all, new development programs. Indeed, because of the high density of energy release in a volume having relatively low losses, conditions normally favor excitation and sustenance of oscillations in any combustion chamber intended for a propulsion system.

Figure 1 is an abbreviated chronology of some major events and features of the subject during the past 50 years. In one form or another, combustion instabilities have been under continuous study during all of that period. In time, however, the emphasis naturally has shifted, depending on what sort of full-scale systems experienced difficulties. During World War II in the United States, it seems that virtually all work in this subject was concerned with elimination of high-frequency resonant burning (the term used at the time) in small tactical solid rocket motors. The common treatment was usually a form of passive control, involving installation of baffles, resonance rods, or some other modification of geometry. Since then, the need to solve problems of instabilities in solid rockets has continued for motors of all sizes. Much of the basic understanding that has been gained is applicable to liquid rockets, despite the obvious differences in the systems.

Although work on combustion instabilities in liquid rockets began in the early 1940s, significant progress was neither achieved nor required until after World War II with the development of large intercontinental ballistic missiles (ICBMs). During the 1960s, the needs of the Apollo program motivated a large amount of work on instabilities, rendered particularly important because of the astronauts

---

*Professor of Mechanical Engineering and Jet Propulsion.

†Professor of Mechanical Engineering.

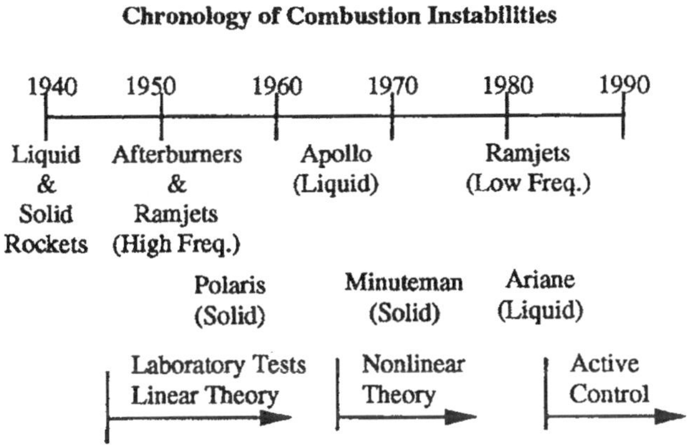

Fig. 1 Chronology of combustion instability.

onboard the rocket. That experience formed part of the basis for developing the Space Shuttle main engine. Practically no progress was made in the United States from the early 1970s to the middle 1980s, whereupon interest rose again with the prospects for developing the advanced launch system (ALS), which was subsequently canceled. In France, a small but significant program has been continuously supported since 1981 as part of the Ariane program. Chapters 4, 10, and 20 of this volume give most researchers in the West their first extensive acquaintance with the considerable body of work accomplished over many years in the former Soviet Union.

Combustion instabilities in jet-engine afterburners have been troublesome since the invention of these devices in the late 1940s. The types of problems encountered at that time still persist: high-frequency transverse modes, similar to the oscillations often found in liquid rockets. With the development of high-bypass engines, however, conditions became favorable for the entire length of the engine to participate in the motions, leading to the presence of longitudinal vibrations with lower frequencies that are much more difficult to treat simply by making changes in geometry.

Similarly, although earlier liquid-fueled ramjets had instabilities mainly in the high-frequency range, more compact designs in the late 1970s led to longitudinal oscillations having lower frequencies in the range of a few hundred hertz. Those problems received a great deal of attention during the 1980s, and several concentrated research programs have produced a quite good basic understanding of the causes and possible means of treatment.

Solid rockets also have exhibited instabilities over wide ranges of frequency, from the low frequencies arising in large ICBMs and strap-on booster rockets to very high frequencies in tactical solid rockets. Research on the problem in solid rockets has continued from the early 1950s to the present, at varying levels of intensity. Although in comparison with liquid rockets details of the problem (notably the basic causes) are significantly different in solid rockets, much of what has been learned about the general aspects of the problem is directly applicable to liquid rockets.

Indeed, quite generally, progress made in any one of these systems is relevant, and often importantly so, to understanding and treating combustion instabilities in other sorts of systems. This review of instabilities in liquid-propellant rocket engines is brief, but nevertheless we shall often refer to results achieved in other systems as well, especially to encourage workers in the field to be aware of, if not conversant with, combustion instabilities in all types of propulsion systems.

To begin to understand the essential characteristics of combustion instabilities, it is best first to distinguish linear and nonlinear behavior. Linear behavior presents only one general problem, linear instability, which received widespread attention during the 1950s and 1960s; see, e.g., the monograph by Crocco and Cheng[1] and the comprehensive compilation of works edited by Harrje and Reardon.[2] Any disturbance may be synthesized as an infinite series of harmonic motions. An approximate analysis developed over many years (see Culick[3] and Culick and Yang[4]) allows one to use classical acoustic modes as the terms in the series and to compute the perturbations of the complex wave number for each mode due to various contributing processes in a combustion chamber. The real part of the wave number gives the frequency shift, and the imaginary part gives the growth (or decay) constant associated with each mode. Vanishment of the imaginary part determines the formal condition for linear stability, whose dependence on the parameters characterizing the system is then known.

Two basic nonlinear problems arise when dealing with combustion instabilities: determining the conditions for the existence and stability of limit cycles for a linearly unstable system and finding the conditions under which a linearly stable system may become unstable to a sufficiently large disturbance. In the language of modern dynamical systems theory, these two problems are identified as supercritical and subcritical bifurcations, respectively. The term bifurcation refers to the characteristic that the character of the steady behavior of the system suffers a qualitative change abruptly as a parameter of the system is varied continuously. This may at first seem an unnecessarily formal description of the phenomenon. In fact, the framework provided by the approximate analysis and application of some of the ideas of dynamical systems theory forms a widely useful and convenient basis for understanding combustion instabilities.

Figure 2 is a schematic diagram suggesting the commonly accepted point of view of combustion instabilities. A measurement of the pressure at a fixed location shows, roughly, a time dependence similar to that for the displacement of a simple mechanical oscillator. Hence, it is natural to suppose that the fluctuation of pressure

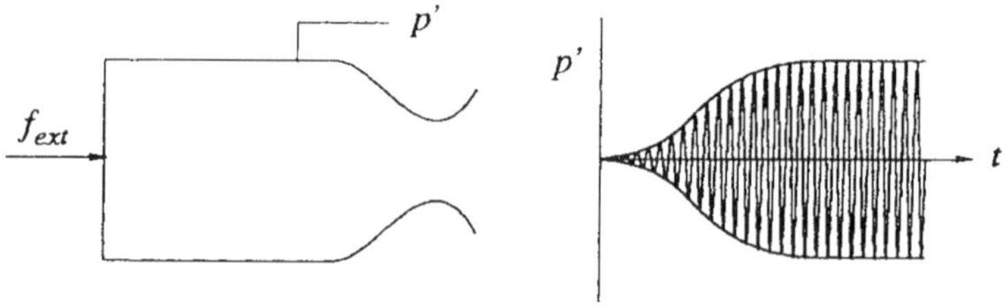

$$\frac{d^2 p'}{dt^2} + 2\alpha \frac{dp'}{dt} + \omega_0^2 p' = f(t) = f_{int} + f_{ext}$$

$$p' = g(f) \Rightarrow f = g^{-1}(p')$$

Fig. 2 Pressure oscillations in a combustion chamber.

from its mean value satisfies the oscillator equation with damping constant α, natural frequency ω_0, and forcing $f(t)$,

$$\frac{\mathrm{d}^2 p'}{\mathrm{d}t^2} + 2\alpha\frac{\mathrm{d}p'}{\mathrm{d}t} + \omega_0^2 p' = f(t) = f_{\mathrm{int}} + f_{\mathrm{ext}} \tag{1}$$

Usually there are no significant external forces acting on the flow in a combustion chamber: it is a good approximation to neglect f_{ext}. Then the internal force represents those processes not typically accounted for in α and ω_0, notably combustion and, in some formulations, nonlinear behavior.

For the moment, we confine attention to linear behavior exhibited as the exponentially growing amplitude in Fig. 2. It is convenient to assume that the terms $2\alpha\,\mathrm{d}p'/\mathrm{d}t + \omega_0^2 p'$ contain all linear processes except those associated with combustion. Thus, the internal forcing f_{int} contains the mechanisms tending to excite the instability. Although a priori determination of f_{int} is not always feasible, the following heuristic form may be assumed as a physically useful representation:

$$f_{\mathrm{int}} = q(p') = 2\alpha_c\frac{\mathrm{d}p'}{\mathrm{d}t} + \omega_c^2 p' \tag{2}$$

With a view to later remarks on active control of combustion instabilities, we interpret the preceding model with the block diagram drawn in Fig. 3. The equation of motion, Eq. (1), is abbreviated by the relationship $p' = g(f)$ between the input (sum of internal and external forcing) and the output. The internal forcing (i.e., combustion response) depicted in Fig. 3 appears as a feedback path. Because the system is linear, the Laplace transform may be applied to represent the system in the frequency domain. The transfer function relating the pressure to the external force is

$$\frac{P(s)}{F_e(s)} = \frac{G(s)}{1 - Q(s)G(s)} \tag{3}$$

[Note that the force f_{int} is fed back positively to correctly represent the system defined by Eq. (1); hence, the minus sign, rather than the more familiar plus sign, appears in the denominator of Eq. (3).] The natural motions, possible when no external force is acting, are defined by equating the denominator of the transfer function to zero,

$$Q(s)G(s) - 1 = 0 \tag{4}$$

For the simple example here, with Eqs. (1) and (2) we find

$$G(s) = \frac{1}{s^2 + 2\alpha s + \omega_0^2} \qquad \text{and} \qquad Q(s) = 2\alpha_c s + \omega_c^2 \tag{5}$$

Substitution in Eq. (4) gives

$$s^2 + 2(\alpha - \alpha_c)s + \left(\omega_0^2 - \omega_c^2\right) = 0 \tag{6}$$

The roots of this equation are

$$s = -(\alpha - \alpha_c) \pm i\sqrt{\left(\omega_0^2 - \omega_c^2\right) - (\alpha - \alpha_c)^2} \tag{7}$$

Hence, as earlier argued, s has a possible positive real part, and the natural motion is unstable if $\alpha_c > \alpha$, assuming both α and α_c to be positive.

The condition $\alpha - \alpha_c = 0$ defines neutral stability; motions neither grow nor decay. When $\alpha - \alpha_c$ is expressed in terms of the parameters characterizing the system, the locus $\alpha - \alpha_c = 0$ in the space of parameters is called the stability boundary. In the literature of combustion instabilities, the stability boundary is normally computed in a quite different fashion but, as we shall see later, the results are equivalent.

A continuum, such as the gaseous environment inside a combustion chamber, is basically an infinite degree of freedom system. For time-harmonic behavior of the (linear) unsteady motions, however, the chamber boundary conditions allow solutions only at discrete modal frequencies. Each mode can basically be treated as a one-degree-of-freedom oscillator, demonstrating the utility of Eq. (1). It is well established that combustion response (i.e., the coefficients α_c and ω_c) varies with frequency. Therefore, the degree to which combustion destabilizes a mode depends on the frequency of the mode itself. If the energy gain due to combustion response is greater than energy damping, then an instability of the particular mode results. Mathematically, this condition is equivalent to having a positive real part of an eigenvalue in Eq. (7).

Although practical situations are profoundly more complex than the simplified system in Fig. 3, the preceding remarks summarize the basis for treating, and if possible eliminating, combustion instabilities in practice. Beginning with an unstable mode, there are clearly several ways to achieve the desired result that the modes should be stable. First, damping might be increased with the addition of attenuating devices, such as small resonant cavities or acoustic liners, a common practice in liquid-propellant rocket engines. This tactic will cause a small frequency shift, but a second method of stabilization involves changes of geometry to displace the resonant peaks sufficiently in frequency so that no peak is lower than the driving curve. That is one consequence of using baffles: breaking the chamber into effectively smaller compartments introduces new internal boundary conditions, causing the normal modes to occur at lower frequencies. On the other side of the energy balance, the strategy consists in reducing the driving in the vicinity of the unstable resonance. That is, generally, a more time-consuming and

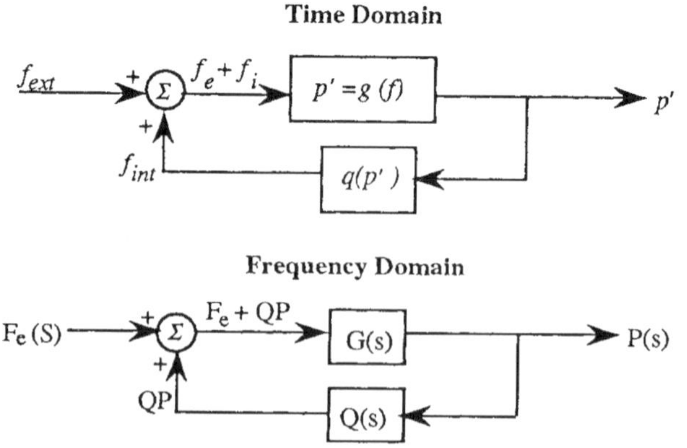

Fig. 3 Feedback loop of unsteady motions in a combustion chamber.

expensive procedure but is commonly a necessary approach when it is not possible to introduce sufficient damping to stabilize the system. Work to change injector design or the entry conditions of the injected propellants, in some sense never thoroughly understood, produces favorable effects on the driving curve.

The remainder of this chapter is divided into sections broadly summarizing the subjects covered in this volume. Following descriptions of several cases of instabilities, we briefly discuss the important fundamental processes generally recognized to be most significant in respect to combustion instabilities in liquid-propellant rocket engines. Numerical simulations and analysis are barely more than touched upon in Secs. IV and V. The crucially important practical matter of stability rating is reviewed in Sec. VI. A curious feature of the subject is that whereas data, but no theory, are available for stability rating (essentially the problem of nonlinear instability), theoretical results and relatively little data are available for linear stability. Thus, it is fair to state that research on combustion instabilities in liquid-propellant rocket engines has produced mainly formal theoretical results and a useful way of interpreting observations, but practically nothing of value for basic comparisons of theory and experiment. The most difficult problems of behavior found in operational systems remain unsolved by theory.

Section VII covers the very important subject of control. Only passive control has been applied to practical systems. Active control offers some promise for the future, but only after very substantial progress.

II. Two Examples of Combustion Instabilities in Operational Engines

In the preceding section, we have emphasized that distinguishing linear and nonlinear behavior produces one way of classifying combustion instabilities, particularly important for theoretical work. The distinctions between linear and nonlinear oscillations have also strongly influenced the planning and interpretation of experimental work. In some respects, a more immediately useful scheme to begin imposing order on the wide variety of observations is based on the propellants used. There are three main classes of propellants: liquid oxygen (LOX) with a hydrocarbon fuel, liquid oxygen and liquid hydrogen (commonly referred to as cryogenic propellants), and storable propellants, the most common being nitrogen tetroxide as the oxidizer and monomethylhydrazine (MMH) or A-50 [a mixture of 50% hydrazine, N_2H_4, and 50% unsymmetrical dimethylhydrazine (UDMH)] as the fuel. Since several chapters of this volume address many examples of combustion instabilities, we shall discuss briefly only two cases, both taken from the Apollo program: the first stage F-1 engine, which used propellants LOX/RP-1; and the lunar module descent engine (LMDE), which, of course, used storable propellants.

A. Combustion Instabilities in the F-1 Engine

The problem of instabilities in the F-1 engine was serious for more than seven years during development of the engine. Recently, Oefelein and Yang[5] presented a comprehensive review of the subject, the first such work based on a thorough study of the original reports and memoranda. Here, we shall give only a cursory discussion and a partial summary.

The F-1 engine was qualified for manned flight in 1966 to satisfy heavy-lift requirements of the Apollo program, but its origins can be traced back to several engines developed in the 1950s by Rocketdyne, principally the E-1; the Atlas

MA-2; and a derivative of the engine used in the Thor/Jupiter vehicle, the H-1, itself later used in Saturn I. Of the 44 tests of the first F-1 design carried out from January 1959 to May 1960, 20 experienced combustion instabilities with peak amplitudes greater than or comparable to the average pressure. This was contrary to predictions based on experimental and theoretical work at the time, and relatively poor instrumentation apparently provided little new useful information. The oscillations caused serious erosion and burning of the injector face, indicating the presence of large radial and tangential motions.

Distinctions between linear and nonlinear instabilities were recognized during the period 1960–1962 and, at least implicitly, helped guide the development of tests. In the literature of that time, the term self-triggering was often used to refer to spontaneous or linear instability, and dynamic instability meant that a linearly stable system was nonlinearly unstable to a sufficiently large disturbance. An important conclusion established by the early tests (1959–1962) and confirmed throughout subsequent work was that baffles were necessary to provide dynamic stability.

To circumvent the instability problems, a program called Project First was established to develop a stable F-1 engine. Project First lasted for four years and progressed in three steps: preliminary flight rating tests (PFRT), flight rating tests (FRT), and flight qualification tests. These stages ran from October 1962 to June 1963, June 1963 to January 1965, and January 1965 to September 1966, respectively, with the bulk of the work concerned with improving the injectors and the baffles. During this period, there were 207 full-scale tests with 11 injectors in PFRT, 422 tests with 46 injectors in FRT, and 703 tests with 51 injectors in the qualification program—a total of 1332 tests with 108 injectors. The F-1 engine from its beginning to flight was subject to more than 2000 full-scale tests, probably the most intensive (and expensive) program ever devoted primarily to solving a problem of combustion instabilities.

Figure 4 shows examples of the 5U injector with and without baffles. Iterations with respect to this pattern evolved into the flight qualified injector. Without baffles, the engine was prone to instabilities, a characteristic consistent with previous experience with engines using LOX/RP-1 as propellants. With a suitable combination of baffle configuration and injector design, the engine not only did not exhibit self-excited oscillations (linear instabilities) but was also stable to finite disturbances (nonlinear stability). Pressure records in the figure illustrate that behavior.

The report by Oefelein and Yang[5] provides details concerning the sequence of design changes made during the four years of Project First. What matters here are the main conclusions. For at least two reasons, longitudinal or organ-pipe modes tended to be more stable than tangential modes. First, the exhaust nozzle provides considerably more damping for longitudinal than for tangential modes; second, the physical and chemical processes near the injector face—containing practically all of the mechanisms producing instabilities—are generally more sensitive to velocity fluctuations parallel to the injector face than to unsteady motions normal to the face. Moreover, the velocity fluctuation near the head end is relatively small in a longitudinal wave, vanishing at a rigid end surface.

Several mechanisms for exciting instabilities are evidently active in the engine. The dominant ones seem to have been coupling between oscillations in the chamber and unsteady motions within the injection elements (injection coupling), periodic pulsed combustion of excess liquid-propellant accumulations on

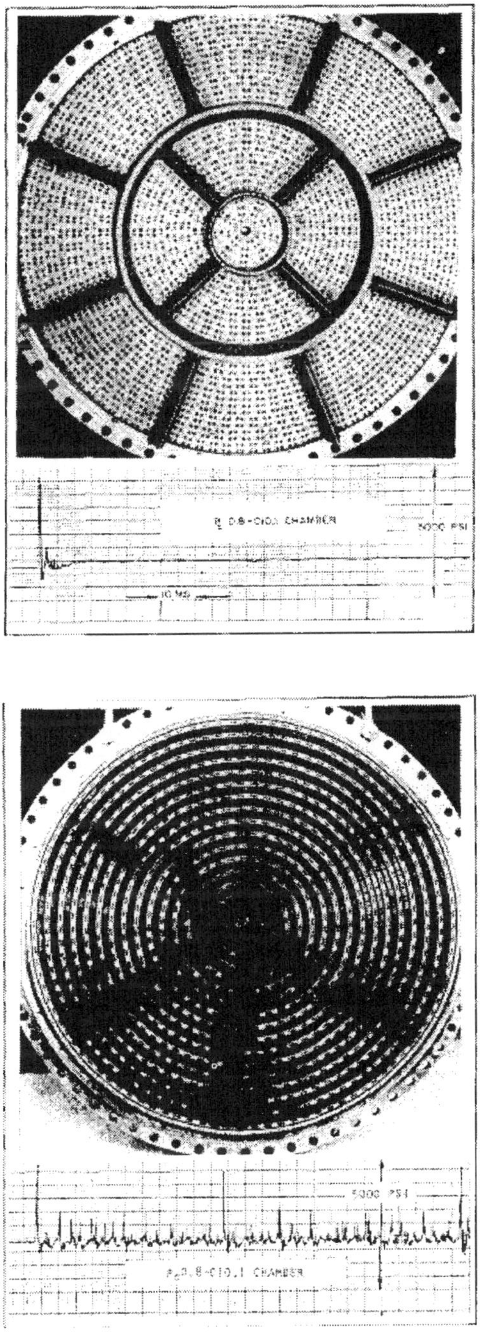

Fig. 4 **Importance of baffles with respect to dynamic stability of F-1 engines.**

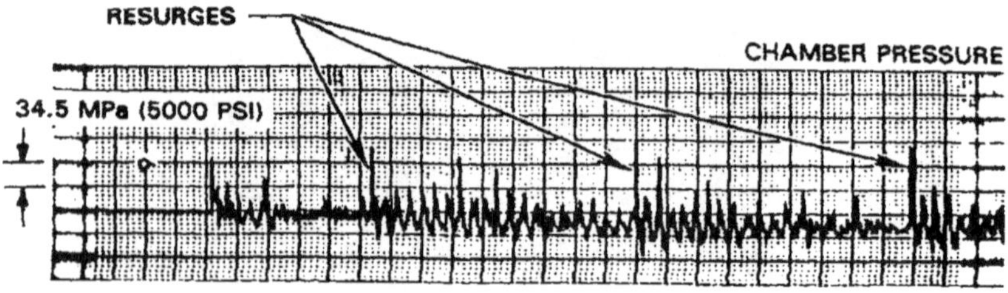

Fig. 5 Pressure trace exhibiting resurge phenomenon.

boundary surfaces accompanying film cooling, and transverse displacements of the injected fuel and oxidizer jets when exposed to oscillations of velocity parallel to the injector face. The first tangential mode was the troublesome oscillation, eventually subdued by a combination of modifications of the injector design and use of baffles extending downstream from the injector face.

Primary mechanisms responsible for instabilities in the F-1 engine were apparently located within three regions near the injector face and a fourth region associated with film cooling of surfaces. The last was identified with the dominant cause of a phenomenon called resurging, frequently observed following explosion of a bomb for rating dynamic stability. Figure 5 reproduces pressure traces showing resurging observed during the PFRT program. The dominant cause was pulsed combustion of liquid fuel detached from the liquid layer produced with the film cooling of the chamber. A secondary cause may have been rapid combustion of droplets at supercritical pressures. In any event, resurging was essentially eliminated by determining the best conditions of film cooling, the thickness of the film being the main parameter.

Whereas the mechanism associated with film cooling is largely specific to the injector design, other causes near the injector face are of much more general significance and remain the subject of research. The three regions referred to are 1) that nearest the injector face, containing the spray fans and all processes generating liquid drops of fuel, 2) the region of fuel vaporization extending roughly 10 in. downstream of the face; and 3) the region farther downstream in which both fuel and oxidizer are gaseous and where most of the combustion occurs. In the region closest to the injector, the dominant processes involve the dynamics of the liquid jets and the spray fans formed by impinging jets. Within this region, which extended roughly 3 in. from the injector face, formation of fuel droplets is nearly complete, and LOX drops were almost entirely vaporized. Note that the baffles finally used extended 3 in. downstream from the face, which had a diameter of 40 in. Hence, the baffles effectively shadowed the region in which interactions between jets and the processes forming the liquid drops are most important. A secondary consequence, whose importance was not quantified in the F-1 program, was the conversion of tangential fluctuations to longitudinal motions that are then damped by the action of the exhaust nozzle.[6] In region 2, fuel drops suffer breakup due to aerodynamic forces and are vaporized. It is here that vaporization is encouraged by the high relative velocity between gaseous oxygen and the liquid fuel drops. If the gaseous fuel and oxygen are not uniformly mixed as a result of the processes in regions 1 and 2, interactions with oscillations in the flow can cause fluctuations of the mixture ratio and, hence, of the burning rate. Interactions in region 3 highlighted the basic importance of the location of the combustion zone. Tests with design changes to produce larger fuel drops and higher relative velocity between

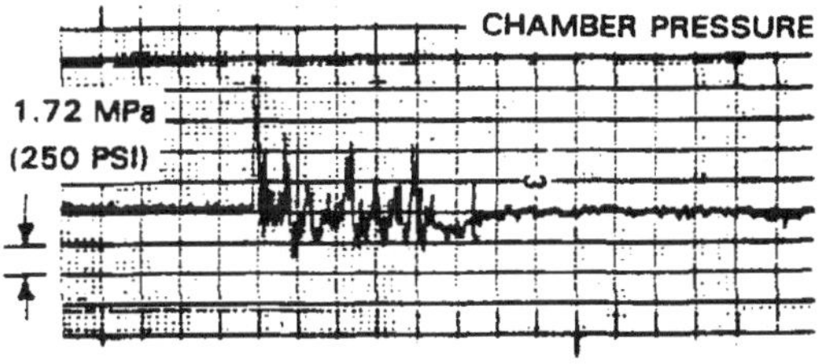

Fig. 6 Pressure trace exhibiting damping characteristics of FRT injector of F-1 engine.

the drops and the gaseous oxygen showed that the zone of vigorous burning moved downstream and improved stability. The final injector produced a specific impulse of 265.4 s and characteristic velocity efficiency equal to 93.8%, confirmed with 703 full-scale engine tests. Figure 6 shows test results typical of the behavior following bombing.

B. Combustion Stability of the Lunar Module Descent Engine

The Apollo LMDE operated with the storable propellants, N_2O_4 oxidizer and A-50 fuel. Owing to the stringent requirements for descent to landing on the moon, the engine had to be throttleable (ultimately over the range of 1000–9850 pounds thrust), with a maximum single firing time of 1000 s but with three starts required. Essentially, no combustion instability could be tolerated in the entire range of operation. The TR 201 engine,[7,8] conceived first by Elverum at the NASA Jet Propulsion Laboratory and later developed and produced by TRW, successfully met all of the requirements and remains one of the most remarkable liquid rocket engines ever used.

The general configuration of the engine and a sketch of the pintle injector are given in Chapter 5. Fuel is injected as a cylindrical tubular sheet along the axis, and oxidizer is injected radially as a number of jets near the end of the movable sleeve. On first acquaintance, one might expect this configuration to exhibit poor performance, but that is not the case. The specific impulse exceeds 300 s above 3000 pounds thrust. What is equally remarkable is that even during severe tests of dynamic stability by bombing, the engine "has never experienced a case of sustained high-frequency instability," a statement referring to the LMDE operated within its design envelope. The chief reason for the stability rests on the relative distributions of acoustic pressure and combustion energy release. Because of the way in which the liquid fuel and oxidizer are injected, the energy release tends to be most concentrated in an annular region between the axis and the chamber boundary and roughly midway between the head end and the nozzle. The central axial location is unfavorable to exciting the fundamental longitudinal mode (and, in fact, all odd modes), which is additionally highly damped by the exhaust nozzle. Moreover, the intermediate radial position has the least destabilizing influence on the first radial and first tangential modes.

In view of the remarks in the Introduction, one cannot expect an engine to be absolutely stable under all possible conditions. The TR 201 did show transient oscillations, particularly when rapidly throttled or at full thrust with a fuel-rich mixture or under some conditions of partial failure. Moreover, the envelope for stable operation changes when different propellants are used. It is, however, fair to state that the engine is stable over a wide range of conditions for which it is qualified. The method of bombing to assess dynamic stability will be discussed briefly in Sec. VI.

III. Fundamental Processes

The purpose of the injection system and combustion chamber in a liquid rocket engine is to accomplish controllably the conversion of liquid propellants to product gases at high temperature and pressure. Thrust is then produced by expansion through the exhaust nozzle, transforming potential and thermal energy into kinetic energy. Although the characteristics of the flow in the nozzle are significant in determining the overall efficiency of the engine, we are concerned here chiefly with unsteady processes upstream of the nozzle entrance. The only significant contribution of the nozzle to combustion instabilities arises from its position, setting the downstream boundary conditions on waves in the chamber; its most important influence is attenuating longitudinal oscillations.

Figure 7 is a broad summary of the various elementary processes. It is convenient to display them in serial form, roughly corresponding to the sequence of events from propellant supply to exhaust. It is important, however, to realize that several different processes may take place simultaneously in a given region of space.

Several chapters in this volume provide good summaries of past work in this subject, the present state of understanding, and reports of research in progress. There is, therefore, no need here to dwell on details. We shall rather try to clarify the context within which the elementary processes may act as mechanisms for exciting and sustaining combustion instabilities. The discussion here, and in any consideration of fundamental processes, cannot be complete in a practical sense;

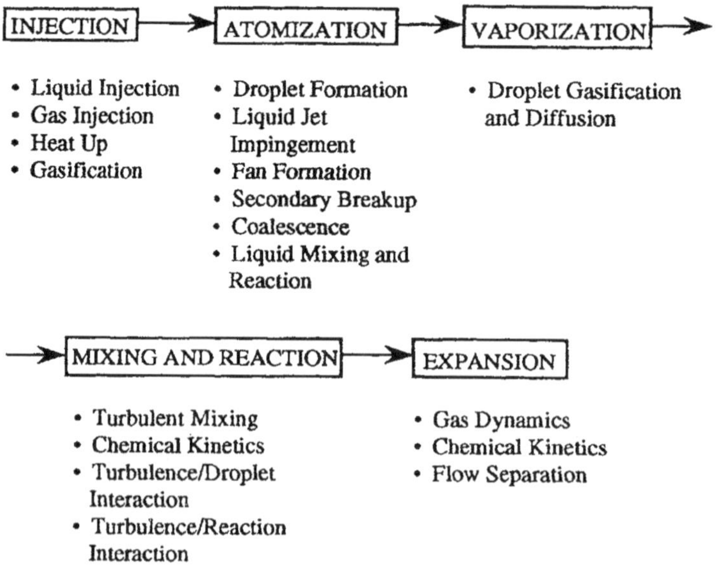

Fig. 7 Physicochemical processes involved in liquid rocket thrust chambers.

for a given engine, there are inevitably characteristics peculiar to the design. For example, we have seen in the case of the F-1 engine that accumulation of liquid fuel during film cooling of the chamber was responsible for an unacceptable spiking unsteady motion.

To bring some order to the possible importance of the many processes shown in Fig. 7, it is helpful to overlay a classification according to injector types.[9] For that purpose, we lump all possibilities into three classes distinguished by the configurations of the liquids when they are discharged by the injection elements into the chamber: 1) impinging jets, 2) concentric jets, and 3) jet/sheet (noncoaxial). The last class arises notably in the lunar descent module engine. Little is known about the fundamental processes involved in this type of pintle injector, although there is research in progress.

We have already discussed the F-1 engine using injectors based on impinging jets; other examples include the lunar ascent engine, several launch vehicles, and the Space Shuttle orbital maneuvering engine. Notable applications of concentric injectors include the RL-10 and the Space Shuttle main engine. In this volume, problems of impinging jets are discussed in Chapter 8 by Anderson and Santoro; concentric or coaxial injection processes are covered in Chapter 6 by Vingert et al., which includes an excellent review, and in Chapter 7 by Zaller and Klem. There seems to be no work which provides a basis for assessing quantitatively the relative stability of the various configurations. In this regard, there has been a suggestion that the Soviets regarded injectors with impinging jets as more prone to causing combustion instabilities than those using coaxial jets, even with hydrocarbon fuels. We have been unable to verify that conclusion.

Figure 8, taken from Ref. 10, is a photograph showing one type of situation for a case of impinging liquid jets. Immediately following impingement, a spray fan or sheet is formed. Either because of disturbances already growing in the jets or because the sheet is itself unstable (possibly for both reasons), the sheet breaks up. There are several forms of the breakup but, in any event, a cloud of liquid drops or droplets is formed having a fairly broad distribution of size and velocity. Because of the surrounding atmosphere of moving gases, the sheet and the drops are subject to shear forces that encourage breakup of both. In a combustion chamber, these processes take place in an atmosphere consisting of combustion product gases and reactant gases as well.

The greater part of what is known about the events in which liquid jets are transformed to drops has come from tests with liquids at room temperature and with no combustion. Whereas it is true that such tests are necessary, valuable, and do produce a great deal of useful information, one must never forget that the situation may be drastically different when combustion occurs. Heat transfer upstream and pressure disturbances generated by combustion of individual drops may have strong effects on the jets and spray sheets before and during breakup. For example, with hypergolic propellants, blow-a-part of the reactive streams and popping or intermittent explosion of liquid may be sufficiently vigorous to excite instabilities.

Whatever the environment, the first step in the transformation of liquid jets to drops must involve some sort of instability. For the case of impinging jets, the jets themselves may be unstable, or the spray fan is unstable. Concentric jets are unstable at their interface. The basic processes responsible for the instabilities are characterized by various dimensionless groups that provide scaling laws (see Chapters 6 and 9 for particularly good discussions of scaling).

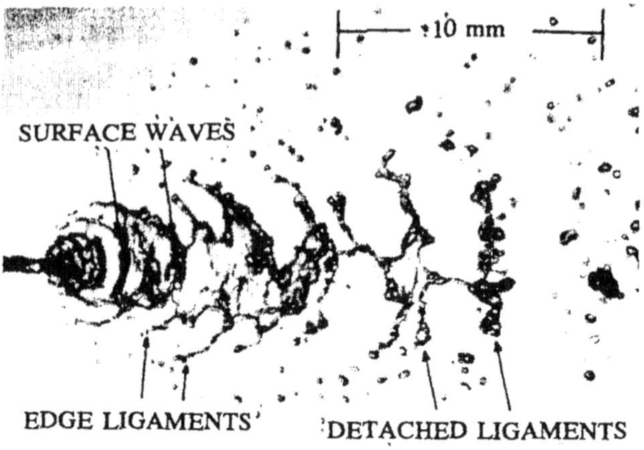

Fig. 8 Typical spray formed by two turbulent impinging water jets.

Determining the distributions of drop size and velocity theoretically is an extremely difficult problem, more so than the matter of incipient instability. In general, the influences of aerodynamic forces on the breakup of jets and the secondary breakup of drops is so complicated as to be beyond realistic analysis.

Combustion occurs in the gaseous phase and must be preceded by mixing. In a LOX/hydrocarbon system, such as in the F-1, vaporization of the liquid oxygen occurs much more rapidly than that of the liquid fuel. Hence, a significant part of the spray combustion involves vaporization and subsequent combustion of the fuel in an oxygen-rich atmosphere. It appears that some of the consequences of detailed design changes on the injector elements may be explained ultimately by the large difference in vaporization rates. That characteristic must also have much to do with differences in the stability of concentric injectors compared with impinging jets when the same propellant combination is used.

Whereas burning of single drops has been actively studied for many decades and is quite well understood (see Chap. 11 by Sirignano et al. for details), spray combustion, particularly for dense clouds of interacting drops under conditions found in a rocket combustion chamber, is poorly understood. Only ad hoc modeling is available for unsteady combustion of sprays. In addition to the purely fluid-mechanical and chemical problems, serious further complications arise in combustion chambers operated at high pressure, namely, those arising from behavior in the supercritical regime. This is not a new subject, but thorough analysis for conditions appropriate to combustion instabilities in liquid rocket engines is recent and unfinished.[11–14] The fundamental origin of the special behavior accompanying combustion near the critical point is the rapid variations of thermophysical properties. For a given substance or mixture of substances, the critical properties are well defined. However, the special characteristics of behavior at the critical point itself extend, albeit attenuated, into some region around the critical point. Hence, it is clearly unnecessary for the chamber conditions to be precisely at the critical state to cause unusual phenomena during combustion.

To be specific, consider a liquid drop of oxygen injected as liquid (and hence having a temperature below the critical point) into a hot fuel-rich environment—the simplest case being hot hydrogen. Suppose that the chamber pressure is greater than the critical value for some mixture of the fuel and oxidizer. As the drop follows its trajectory, its temperature rises, oxygen vaporizes, and gaseous hydrogen diffuses inward. Somewhere between the center of the drop and the environment far away, the temperature and pressure may, if conditions are suitable, assume the critical values for the local mixture ratio of oxygen and hydrogen. For a spherically symmetric flow, the critical point is reached on a spherical surface. Locally, the sharp variations of properties with temperature will produce drastic changes in the flowfield at and near that surface. Globally, those effects have a substantial influence on the vaporization rate.

If oscillations of pressure and temperature occur in the environment, then it is possible, even though the average chamber pressure may be far from critical, that critical conditions are reached in some point of the drop under unsteady conditions. Then transient fluctuations can be amplified by the large fluctuations of thermodynamic properties. The result is a new possibility for coupling between acoustical motions in the chamber and droplet burning. The most thorough analysis of droplet vaporization and combustion under supercritical conditions has been covered in recent publications in Refs. 11–14. As the preceding remarks suggest, calculations in which supercritical behavior is ignored are seriously in error when the chamber pressure is greater than the critical states of the injected propellants.

IV. Analysis and Numerical Simulations

As in many fields of engineering, the rapidly increasing power of computing resources continues to motivate increasingly elaborate calculations of unsteady behavior in combustion chambers. The term numerical simulation refers to numerical solution to the governing equations of motion, comprising the fundamental conservation equations and models of all physical processes included in the formulation. There are two sorts of inaccuracies in numerical simulation: computational errors and those arising from approximations in the equations. The numerical analysis itself cannot be perfect, although for most applications contemporary methods have practically negligible errors in their range of validity. Modeling the physical processes is necessarily imperfect and is by far the main reason for inaccurate results. Thus, even with the remarkable progress in computing power, it is still not realistic to attempt to solve fully three-dimensional, time-dependent problems for combustion chambers; both the modeling of combustion processes and available computing power are inadequate.

Other than the unavoidable limitations just discussed and the expenses involved, the main disadvantage of numerical simulation is that each case computed is just that: one case. Understanding global behavior and trends over the possible ranges of parameters defining a given problem can be extremely expensive and time consuming and often does not lead to clear, simple rules useful for interpreting experimental data and for practical design. Hence, there will always be a need for formal analysis, which must in any event be approximate.

An approximate analysis must be founded on the same governing equations as a numerical simulation, but commonly with some simplifications in models of the physical processes, to help gain solutions. Hence for a given problem, numerical simulations should be at least as accurate as analytical results. The great advantages of an approximate analysis are the possibility for deducing general rules and the opportunity to investigate theoretically the global behavior of a complicated system. For the subject of combustion instabilities, the most important use of the second characteristic is in understanding nonlinear behavior.

Thus, the entire subject of combustion instabilities may be viewed as divided into three parts: experiments and full-scale tests, analysis and theory, and numerical simulations, sometimes also viewed as a part of computational fluid dynamics (CFD). The case of the F-1 engine discussed in Sec. II is an (perhaps, the) extreme example of how expensive a problem of combustion instabilities can be in a development program. Probably the chief practical purpose of both analysis and numerical simulations is to provide a sound basis for reducing the amount of experimentation and testing required. It will never be possible to eliminate testing entirely, but theoretical work can achieve a great deal by narrowing the ranges of parameters that must be covered in tests.

In this section we first describe briefly the general formulation and modeling used presently for investigating combustion instabilities in liquid rocket engines. We then summarize one form of approximate analysis that has proved effective in other applications and for deriving some useful theoretical results. It is unnecessary here to cover numerical simulations that are treated quite thoroughly in this volume in Chapter 18 by Habiballah and in Chapter 19 by Grenda et al.

A. Equations for Unsteady Motions in Combustion Chambers

We shall neither derive equations of motion here nor deal with the most general forms. Our chief purpose is to convey as thoroughly as possible the main ideas and to outline the basis for the more important theoretical results. Even though the instabilities observed are mainly oscillations of the gaseous phase, it is essential to account for the presence of the liquid phase as well. To simplify the discussion, we lump the liquid fuel and oxidizer together as a single liquid phase and represent the multicomponent gas mixture as a single average gas.[3,4] Thus, we treat a reacting two-phase mixture for which the equations of conservation are as follows.

Conservation of mass:

$$\frac{\partial}{\partial t}(\rho_g + \rho_\ell) + \nabla \cdot (\rho_g \boldsymbol{u}_g + \rho_\ell \boldsymbol{u}_\ell) = w_{eg} + w_{e\ell} \tag{8}$$

Conservation of momentum:

$$\frac{\partial}{\partial t}(\rho_g \boldsymbol{u}_g + \rho_\ell \boldsymbol{u}_\ell) + \nabla \cdot (\rho_g \boldsymbol{u}_g \boldsymbol{u}_g + \rho_\ell \boldsymbol{u}_\ell \boldsymbol{u}_\ell) = \nabla \cdot \tau_g + \boldsymbol{m}_{eg} + \boldsymbol{m}_{e\ell} \tag{9}$$

Conservation of energy:

$$\begin{aligned} &\frac{\partial}{\partial t}(\rho_g e_{g_0} + \rho_\ell e_{\ell_0}) + \nabla \cdot (\rho_g \boldsymbol{u}_g e_{g_0} + \rho_\ell \boldsymbol{u}_\ell e_{\ell_0}) \\ &\quad = \nabla \cdot (\tau_g \cdot \boldsymbol{u}_g) - \nabla \cdot \boldsymbol{q}_g + Q + Q_e \end{aligned} \tag{10}$$

Subscript g refers to the gas species and subscript ℓ to the liquid; subscript 0 denotes the stagnation state. External sources are denoted by subscript e. These terms do not normally appear in the literature but are included here because they may be used to represent the presence of active control, a subject to be addressed later. Stress tensor τ_g is assumed to depend on properties of the gas phase and may be written as the sum of the isotropic pressure and the viscous stress tensor τ_v,

$$\tau_g = -p\boldsymbol{I} + \tau_v \tag{11}$$

All forces exerted in the flow by external influences are represented by the momentum sources $\boldsymbol{m}_{eg}$ and $\boldsymbol{m}_{e\ell}$. Distributed forces (gravity, electromagnetic, etc.) are not an issue in the problems considered here; nonzero values of $\boldsymbol{m}_{eg}$ and $\boldsymbol{m}_{e\ell}$ will arise due to momentum transfer associated with flow of material through the boundary of the combustion chamber and interactions associated with material injected for actuating control. Internal heat flow is represented by $\boldsymbol{q}_g$, sufficiently well approximated by Fourier's law for the gas phase; Q represents heat addition in the gas phase associated with combustion processes not accompanying conversion of the condensed phase to gas. Hence, the symbol e_{g_0} stands for stagnation thermal energy only, containing no part related to chemical processes. For a mixture of perfect gases, each species having constant specific heat,

$$e_{g_0} = C_v T_g + \tfrac{1}{2} u_g^2 \tag{12}$$

where C_v and u_g strictly stand for values mass averaged over all gaseous species.

Let w_ℓ denote the rate at which the liquid phase is converted to gas, and Eq. (8) can be written as the sum of the following two equations.

Conservation of mass (gas phase):

$$\frac{\partial \rho_g}{\partial t} + \nabla \cdot (\rho_g \boldsymbol{u}_g) = w_\ell + w_{e_g} \tag{13}$$

Conservation of mass (liquid phase):

$$\frac{\partial \rho_\ell}{\partial t} + \nabla \cdot (\rho_\ell \boldsymbol{u}_\ell) = -w_\ell + w_{e\ell} \tag{14}$$

The following manipulations are directed to eventually writing a nonlinear wave equation governing the behavior of waves in the mixture. Elasticity (the spring constant) for wave propagation is provided by the compressibility of the gas only, whereas the mass of an elementary oscillator in the medium is the sum of the masses of gas and liquid in a unit volume.

Equation (8), the equation for the total density of the mixture, can be rewritten as

$$\frac{\partial \rho}{\partial t} + \nabla \cdot (\rho \boldsymbol{u}_g) = \mathcal{W} + \mathcal{W}_e \tag{15}$$

where $\rho = \rho_\ell + \rho_g$, the total external source of mass is

$$\mathcal{W}_e = w_{eg} + w_{e\ell} \tag{16}$$

and

$$\mathcal{W} = -\nabla \cdot (\rho_\ell \delta \boldsymbol{u}_\ell) \tag{17}$$

The slip velocity between the condensed phase and the gas is defined as $\delta \boldsymbol{u}_\ell = \boldsymbol{u}_\ell - \boldsymbol{u}_g$. Now expand the momentum equation (9) and substitute the definitions to give the following.

Conservation of momentum (gas phase):

$$\rho_g \frac{\partial \boldsymbol{u}_g}{\partial t} + \rho_g \boldsymbol{u}_g \cdot \nabla \boldsymbol{u}_g + \nabla p = \nabla \cdot \tau_v + \boldsymbol{F}_\ell - (\sigma + \sigma_e - \boldsymbol{m}_e) \tag{18}$$

where

$$\boldsymbol{F}_\ell = -\rho_\ell\left[\frac{\partial \boldsymbol{u}_\ell}{\partial t} + \boldsymbol{u}_\ell \cdot \nabla \boldsymbol{u}_\ell\right] \tag{19}$$

$$\sigma = (\boldsymbol{u}_g - \boldsymbol{u}_\ell)w_\ell = -\delta \boldsymbol{u}_\ell w_\ell \tag{20}$$

$$\sigma_e = \boldsymbol{u}_g w_{eg} + \boldsymbol{u}_\ell w_{e\ell} \tag{21}$$

$$\boldsymbol{m}_e = \boldsymbol{m}_{eg} + \boldsymbol{m}_{e\ell} \tag{22}$$

Equation (19) is the momentum equation for the condensed phase, and σ represents the rate at which momentum is supplied to the newly created gas phase by the gases already present. Hence $-\sigma$ in Eq. (18) represents the force exerted on the gas phase by the vaporizing condensed phase.

To Eq. (18) add $\rho_\ell[\partial \boldsymbol{u}_g/\partial t + \boldsymbol{u}_g \cdot \nabla \boldsymbol{u}_g]$ to find

$$\rho\left[\frac{\partial \boldsymbol{u}_g}{\partial t} + \boldsymbol{u}_g \cdot \nabla \boldsymbol{u}_g\right] + \nabla p = \nabla \cdot \tau_v + \delta \boldsymbol{F}_\ell - (\sigma + \sigma_e - \boldsymbol{m}_e) \tag{23}$$

where

$$\delta \boldsymbol{F}_\ell = -\rho_\ell\left[\frac{\partial \delta \boldsymbol{u}_\ell}{\partial t} + \delta \boldsymbol{u}_\ell \cdot \nabla \delta \boldsymbol{u}_\ell + \delta \boldsymbol{u}_\ell \cdot \nabla \boldsymbol{u}_g + \boldsymbol{u}_g \cdot \nabla \delta \boldsymbol{u}_\ell\right] \tag{24}$$

is the force of interaction between the condensed and gas phases.

More elaborate manipulations eventually lead to the energy equation for the temperature of the gas phase,

$$\begin{aligned}\rho \bar{C}_v\left[\frac{\partial T_g}{\partial t} + \boldsymbol{u}_g \cdot \nabla T_g\right] &+ p\nabla \cdot \boldsymbol{u}_g \\ = (\tau_v \cdot \nabla)\boldsymbol{u}_g &- \nabla \cdot \boldsymbol{q}_g + (Q + Q_e) + \boldsymbol{u}_g \cdot (\sigma + \sigma_e - \boldsymbol{m}_e) \\ &+ \delta Q_\ell + w_\ell \delta e_0 + \delta \boldsymbol{u}_\ell \cdot \boldsymbol{F}_\ell - (e_{g0} w_{eg} + e_{\ell o} w_{e\ell})\end{aligned} \tag{25}$$

where

$$Q_\ell = -\rho_\ell\left[\frac{\partial e_\ell}{\partial t} + \boldsymbol{u}_\ell \cdot \nabla e_\ell\right], \qquad \text{and} \qquad \delta e_0 = e_{\ell_0} - e_{g0} \tag{26}$$

Corresponding to Eq. (24), the heat exchange between the two phases is

$$\delta Q_\ell = -\rho_\ell C_\ell\left[\frac{\partial \delta T_\ell}{\partial t} + \delta \boldsymbol{u}_\ell \cdot \nabla \delta T_\ell + \delta \boldsymbol{u}_\ell \cdot \nabla T_g + \boldsymbol{u}_g \cdot \nabla \delta T_\ell\right] \tag{27}$$

where $\delta T_\ell = T_\ell - T_g$. The mass-averaged specific heat for the mixture is defined as

$$\bar{C}_v = \frac{1}{\rho}(\rho_g C_v + \rho_\ell C_\ell) = \frac{C_v + C_m C_\ell}{1 + C_m} \tag{28}$$

and $C_m = \rho_\ell/\rho$ is the liquid-phase loading, the fraction of mass in unit volume as liquid.

To simplify writing, we shall hereafter use the symbol $\boldsymbol{u}$ instead of $\boldsymbol{u}_g$ for the gas velocity, T instead of T_g for the gas temperature, and drop the overbars on mass-averaged thermodynamic properties. Thus, the three basic equations for unsteady motions in a two-phase mixture are as follows.

Conservation of mass:

$$\frac{\mathrm{D}\rho}{\mathrm{D}t} = -\rho\nabla\cdot\boldsymbol{u} + \mathcal{W} + \mathcal{W}_e \tag{29}$$

Conservation of momentum:

$$\rho\frac{\mathrm{D}\boldsymbol{u}}{\mathrm{D}t} = -\nabla p + \mathcal{F} + \mathcal{F}_e \tag{30}$$

Conservation of energy:

$$\rho C_v\frac{\mathrm{D}T}{\mathrm{D}t} = -p\nabla\cdot\boldsymbol{u} + \mathcal{Q} + \mathcal{Q}_e \tag{31}$$

where the sources are

$$\mathcal{W} + \mathcal{W}_e = -\nabla\cdot(\rho_\ell\delta\boldsymbol{u}_\ell) + w_{eg} + w_{e\ell} \tag{32}$$

$$\mathcal{F} + \mathcal{F}_e = \nabla\cdot\tau_v - \sigma + \delta\boldsymbol{F}_\ell - (\sigma_e - \boldsymbol{m}_e) \tag{33}$$

$$\begin{aligned}\mathcal{Q} + \mathcal{Q}_e = {} & (\tau_v\cdot\nabla)\boldsymbol{u} - \nabla\cdot\boldsymbol{q}_g + Q + \boldsymbol{u}\cdot\sigma_e + \delta Q_\ell + \delta e_0 w_\ell + \delta\boldsymbol{u}_\ell\cdot\boldsymbol{F}_\ell \\ & + \boldsymbol{u}\cdot(\sigma_e - \boldsymbol{m}_e) - \left(e_{g_0}w_{e_g} + e_{\ell_0}w_{e_\ell}\right) + Q_e\end{aligned} \tag{34}$$

The substantial derivative is defined in terms of the gas velocity,

$$\frac{\mathrm{D}}{\mathrm{D}t} = \frac{\partial}{\partial t} + \boldsymbol{u}\cdot\nabla \tag{35}$$

The equation of state appropriate to this formulation is

$$p = \rho R T_g \tag{36}$$

where R is the mass-averaged gas constant, equal to $\bar{C}_p - \bar{C}_v$. From the preceding equations, the equation for the pressure can be derived,

$$\frac{\mathrm{D}p}{\mathrm{D}t} = -\gamma p\nabla\cdot\boldsymbol{u} + \mathcal{P} + \mathcal{P}_e \tag{37}$$

with

$$\mathcal{P} + \mathcal{P}_e = \frac{R}{C_v}Q - RT\nabla\cdot(\rho_\ell\delta\boldsymbol{u}_\ell) + p\frac{\mathrm{D}\ell_n R}{\mathrm{D}t} + \left(\frac{R}{C_v}Q_e + RTw_e\right) \tag{38}$$

Equations (29–31) and the equations of state (36) describe the motions of the two-phase mixture. With the possibility of small modifications or extensions to accommodate many species, these are the basic equations for all analyses of combustion instabilities. To obtain solutions, in addition to setting boundary conditions, the condensed phase must be treated separately with Eqs. (14), (19), and (27), with suitable laws for the force and heat exchange between the two phases.

We shall not pursue the matter of properly posed problems, a subject addressed in this volume in those chapters dealing with numerical simulations. Moreover, although the equations derived here can be used as the basis for nonlinear analysis, we shall restrict ourselves here to a short outline of linear stability analysis that will serve sufficiently well for later discussion of some special topics. A comprehensive discussion of nonlinear oscillations in rocket engines is given in Chapter 13 of this volume.

B. Wave Equation and Approximate Solution

A wave equation may be constructed by analogy with a procedure followed in classical acoustics. Because nonlinear behavior will not be treated herein, we shall work from the beginning with the linearized forms of the governing equations. Write all variables as sums of mean and fluctuating parts, $p = \bar{p} + p'$, etc., and assume that the mean values are independent of time. Substitute in Eqs. (30) and (37) and ignore terms of second order and higher in the fluctuations to find

$$\bar{\rho}\frac{\partial \boldsymbol{u}'}{\partial t} = -\bar{\rho}(\bar{\boldsymbol{u}} \cdot \nabla \boldsymbol{u}' + \boldsymbol{u}' \cdot \nabla \bar{\boldsymbol{u}}) - \nabla p' + \mathcal{F}' + \mathcal{F}_e' \tag{39}$$

$$\frac{\partial p'}{\partial t} = -\gamma \bar{p} \nabla \cdot \boldsymbol{u}' - \gamma p' \nabla \cdot \bar{\boldsymbol{u}} - \bar{\boldsymbol{u}} \cdot \nabla p' + \mathcal{P}' + \mathcal{P}_e' \tag{40}$$

We have also assumed that the mean pressure and density are uniform. Now differentiate Eq. (40) with respect to time, substitute Eq. (39), and rearrange the result to find

$$\nabla^2 p' - \frac{1}{\bar{a}^2}\frac{\partial^2 p'}{\partial t^2} = h + h_e \tag{41}$$

where

$$h + h_e = -\bar{\rho}\nabla \cdot (\bar{\boldsymbol{u}} \cdot \nabla \boldsymbol{u}' + \boldsymbol{u}' \cdot \nabla \bar{\boldsymbol{u}}) + \frac{1}{\bar{a}^2}\bar{\boldsymbol{u}} \cdot \nabla \frac{\partial p'}{\partial t}$$
$$+ \frac{\gamma}{\bar{a}^2}\frac{\partial p'}{\partial t}\nabla \cdot \bar{\boldsymbol{u}} + \nabla \cdot \mathcal{F}' - \frac{1}{\bar{a}^2}\frac{\partial \mathcal{P}'}{\partial t} + \nabla \cdot \mathcal{F}_e' - \frac{1}{\bar{a}^2}\frac{\partial \mathcal{P}_e'}{\partial t} \tag{42}$$

Terms identified by subscript e always correspond.

The boundary condition is set on the normal gradient of the pressure by taking the scalar product of Eq. (39) with the outward normal vector $\boldsymbol{n}$, giving

$$\boldsymbol{n} \cdot \nabla p' = -f - f_e \tag{43}$$

where

$$f + f_e = \bar{\rho}\frac{\partial \boldsymbol{u}'}{\partial t} \cdot \boldsymbol{n} + \bar{\rho}(\bar{\boldsymbol{u}} \cdot \nabla \boldsymbol{u}' + \boldsymbol{u}' \cdot \nabla \bar{\boldsymbol{u}}) \cdot \boldsymbol{n} - \mathcal{F}' \cdot \boldsymbol{n} - \mathcal{F}_e' \cdot \boldsymbol{n} \tag{44}$$

The first term f is commonly rewritten by introducing a response or admittance function characterizing the unsteady behavior of the boundary in response to imposed fluctuations.

Only gasdynamical interactions between the mean and fluctuating velocity fields are shown explicitly in Eqs. (41–44). All other processes are hidden in $\mathcal{P}'$ and $\mathcal{F}'$, those directly associated with external (control) actions being contained in $\mathcal{F}_e'$ and $\mathcal{P}_e'$.

Traditional calculations of linear behavior are based on the differential equations (39) and (40), augmented by the linearized equation of state and possibly by linearized forms of other equations given in Sec. IV.A. In the remainder of this chapter we will use the integrated form based on spatial averaging described, for example, by Culick and Yang.[4] The fundamental idea is that the actual problem defined by Eq. (41) with its boundary condition, Eq. (43), differs in some sense by small amounts from the classical acoustics problem defined by setting $h = f = h_e = f_e = 0$. That is, the influences of all processes are small perturbations. This formulation, then, has the form naturally suited to an iteration procedure readily constructed by introducing a Green's function as first accomplished by Culick.[15] However, the identical first-order result (the only order legitimately considered with these equations) is obtained more directly in the following way.

The unperturbed problem defines the normal modes, with mode shapes ψ_n and natural frequencies $\omega_n = \bar{a}k_n$, where k_n is the wave number,

$$\nabla^2\psi_n + k_n^2\psi_n = 0 \tag{45}$$

$$\boldsymbol{n}\cdot\nabla\psi_n = 0 \tag{46}$$

The idea is then to compare the perturbed (actual) and unperturbed (classical) problems quite literally by comparing their difference in a spatially averaged sense. Multiply Eq. (41) by ψ_n, Eq. (45) by p', and subtract and integrate over the volume of the chamber to give

$$\int\left[\psi_n\nabla^2 p' - p'\nabla^2\psi_n\right]\mathrm{d}V - \frac{1}{\bar{a}^2}\int\left[\psi_n\frac{\partial^2 p'}{\partial t^2} - \omega_n^2 p'\psi_n\right]\mathrm{d}V = \int\psi_n(h + h_e)\,\mathrm{d}V$$

Upon application of Green's theorem and substitution of the boundary conditions, Eqs. (43) and (46), we find

$$\frac{1}{\bar{a}^2}\int\psi_n\left[\frac{\partial^2 p'}{\partial t^2} + \omega_n^2 p'\right]\mathrm{d}V = -\left[\int\psi_n h\,\mathrm{d}V + \oint\!\!\oint\psi_n f\,\mathrm{d}S\right] - \left[\int\psi_n h_e\,\mathrm{d}V + \oint\!\!\oint\psi_n f_e\,\mathrm{d}S\right] \tag{47}$$

The preceding calculations are formally correct and introduce no errors. Now we make the approximation that the pressure field in the actual chamber can be expanded in a series of normal modes for the classical unperturbed problem,

$$p' = \bar{p}\sum_{n=1}^{\infty}\eta_n(t)\psi_n(\boldsymbol{r}) \tag{48}$$

The error incurred in the results arises because $\psi_n(\boldsymbol{r})$, and therefore the right-hand side, do not satisfy the correct boundary condition (43) set on p' in the actual problem. When the series is used in the given integrals, the error is attenuated, and for other formal reasons not covered here (see, e.g., Morse and Feshback[16]) the use of Eq. (48) is entirely legitimate. The inaccuracy is not as serious an inaccuracy as one might think. We assume also, not a practical constraint, that the normal modes are orthogonal,

$$\int \psi_m \psi_n \, \mathrm{d}V = E_n^2 \delta_{mn} \tag{49}$$

Substitution in Eq. (47) leads finally to the second-order equations for the amplitudes $\eta_n(t)$,

$$\frac{\mathrm{d}^2 \eta_n}{\mathrm{d}t^2} + \omega_n^2 \eta_n = F_n + F_{ne} \tag{50}$$

and the forcing terms are

$$F_n = -\frac{\bar{a}^2}{\bar{p} E_n^2} \left\{ \int h \psi_n \, \mathrm{d}V + \oiint f \psi_n \, \mathrm{d}S \right\} \tag{51}$$

$$F_{ne} = -\frac{\bar{a}^2}{\bar{p} E_n^2} \left\{ \int h_e \psi_n \, \mathrm{d}V + \oiint f_e \psi_n \, \mathrm{d}S \right\} \tag{52}$$

With proper interpretation and suitable modeling of the processes represented in functions h, f, h_e, and f_e, the representation of combustion instabilities by Eqs. (50–52) has wide applications to all propulsion systems. Relatively little has been done with this formulation applied specifically to liquid rocket engines but, as we shall see shortly, it is quite easy to obtain general results equivalent to those obtained with methods more familiar in this field.

An important point to emphasize is that the approach taken here provides a useful viewpoint as well as a framework useful for both theoretical and practical computations. According to Eqs. (48) and (50), any unsteady motion may be regarded as a collection of oscillators, one associated with each normal mode of oscillation appearing in the synthesis, Eq. (48). The problem then comes down to determining the time evolution of the amplitudes $\eta_n(t)$ of the oscillators (modes). Although we have assumed linear motions here, the same formal representation applies to nonlinear motions, so long as the amplitudes do not become so large as to violate the approximations taken to give the governing differential equations.

V. Some Results of Analysis

The point has been emphasized that with the analysis involving spatial averaging, as constructed in the preceding section, we can accommodate practically any process active in real combustion. Of course, some approximations are involved, and it is necessary to work out explicit representations or models of the processes in question, but the situation is quite different from that which arises when one chooses to solve the partial differential equations governing the motion. In that case, numerical solutions accentuate errors in modeling because of the differentiation required. Moreover, it is easier with the framework available here to discuss quite general results in a relatively simple and clear fashion. With a few exceptions, we will not be concerned with detailed examples in the remainder of this chapter.

A. Linear Stability

As remarked earlier, there is really only one problem of linear stability, and it is readily solved within the present analysis. According to the principles of Fourier analysis, with essentially no practical restrictions, any function can be synthesized as a finite or infinite series of orthogonal functions. Here, we apply the idea to a pressure record represented as a sum of the normal modes $\psi_n(\boldsymbol{r})$ of a chamber, with time-dependent amplitudes $\eta_n(t)$, Eq. (48). Finding the normal modes requires solving the classical problem of the scalar wave equation for a closed volume that has the same shape as the combustion chamber in question, with the exception that the exhaust nozzle is closed at its entrance. Both the mode shapes $\psi_n(\boldsymbol{r})$ and the natural frequencies are found in this calculation, the solution to Eqs. (45) and (46).

For the problem of linear stability, the amplitudes are assumed to have the form

$$\eta_n(t) = \hat{\eta}_n e^{i\bar{a}K_n t} \tag{53}$$

where the actual complex wave number is

$$K_n = (\Omega_n - i\alpha_n)/\bar{a} \tag{54}$$

Now the perturbed mode corresponding to the nth classical mode may either grow ($\alpha_n > 0$) or decay ($\alpha_n < 0$) in time. The actual frequency is

$$\Omega_n = \omega_n + \delta\omega_n \tag{55}$$

Hence, the problem of linear stability comes down to determining the frequency shift $\delta\omega_n$ and growth constant α_n for each perturbed mode. The basis for the calculation is the linearized form of the equations already treated.

Substitute Eqs. (53–55) in Eq. (50), with $F_{ne} = 0$ to find*

$$(\Omega_n - i\alpha_n)^2 = \omega_n^2 - \hat{F}_n/\hat{\eta}_n$$

where, for linear behavior, we write $h = \hat{h}e^{i\bar{a}K_n t}$ and $f = \hat{f}e^{i\bar{a}K_n t}$, so that $F_n = \hat{F}_n e^{i\bar{a}K_n t}$. For the usual case, $\alpha_n \ll \Omega_n$ and $(\Omega_n - \omega_n)/\omega_n \ll 1$, which must be satisfied to be consistent with approximations already made. Then separating the real and imaginary parts of the last equation, with the definition (51), gives

$$\Omega_n = \omega_n + \frac{\bar{a}^2}{2\omega_n \bar{p} E_n^2}\left\{\int \frac{\hat{h}^{(r)}}{\hat{\eta}_n}\psi_n \,\mathrm{d}V + \oiint \frac{\hat{f}^{(r)}}{\hat{\eta}_n}\psi_n \,\mathrm{d}S\right\} \tag{56}$$

$$\alpha_n = \frac{-\bar{a}^2}{2\omega_n \bar{p} E_n^2}\left\{\int \frac{\hat{h}^{(i)}}{\hat{\eta}_n}\psi_n \,\mathrm{d}V + \oiint \frac{\hat{f}^{(i)}}{\hat{\eta}_n}\psi_n \,\mathrm{d}S\right\} \tag{57}$$

---

*The result and therefore Eqs. (56) and (57) are obtained only if no linear coupling of modes is contained in Eqs. (51) and (52) after the expansion (48) and the corresponding representation of the velocity fluctuation have been substituted. Linear coupling can be formally eliminated by constructing a new set of orthogonal modes. However, normally the corrections will contribute only higher order terms in Eqs. (56) and (57), and so no error is committed by simply ignoring linear coupling throughout the analysis.

in which superscripts (r) and (i) denote real and imaginary parts. These two formulas are the solution to the problem of linear stability. In particular, setting $\alpha_n = 0$ defines the stability boundary in terms of the various parameters arising in the functions h and f. That result is the basis for the Standard Stability Prediction Program[17] written for the U.S. Air Force and widely used by industrial and governmental organizations to analyze the stability of unsteady motions in solid propellant rockets.

B. Rayleigh's Criterion

Probably the most widely quoted general principle in the field of combustion instabilities is the criterion formulated by Rayleigh in 1878[18,19]:

> If heat be periodically communicated to, and abstracted from, a mass of air vibrating (for example) in a cylinder bounded by a piston, the effect produced will depend upon the phase of the vibration at which the transfer of heat takes place. If heat be given to the air at the moment of greatest condensation, or be taken from it at the moment of greatest rarefaction, the vibration is encouraged. On the other hand, if heat be given at the moment of greatest rarefaction, or abstracted at the moment of greatest condensation, the vibration is discouraged.

Roughly, the gist of the principle is that heat addition tends most strongly to drive acoustic waves if the energy is added in the region of space where the oscillating pressure reaches greatest amplitude and is in phase. Although the principle seems fairly obviously true, these remarks do not serve as proof. The formulation described can be used not only to establish the criterion but also to extend its application to all processes and nonlinear behavior as well.[3,20] Zinn[21] has recently discussed Rayleigh's criterion but only for linear heat addition.

We take advantage of the fact that each acoustic mode has an associated oscillator having unit mass and natural frequency ω_n, whose motion (i.e., time evolution of its amplitude η_n) is described by Eq. (50). The energy of a simple oscillator is the sum of its kinetic and potential energies,

$$\mathcal{E}_n = \tfrac{1}{2}\left(\dot{\eta}_n^2 + \omega_n^2 \eta_n^2\right) \tag{58}$$

We interpret Rayleigh's criterion to be a statement concerning the change $\Delta\mathcal{E}_n$ of the energy in one cycle. The product of the force $(F_n + F_{ne})$ times the velocity $\dot{\eta}_n$ is the power input, and so

$$\Delta\mathcal{E}_n(t) = \int_t^{t+\tau_n} (F_n + F_{ne})\dot{\eta}_n \, \mathrm{d}t' \tag{59}$$

In general, $\Delta\mathcal{E}_n(t)$ varies with time, increasing from cycle to cycle for an unstable mode.

Indeed, for linear motions, if all processes are taken into account, $\Delta\mathcal{E}_n$ is proportional to the growth constant α_n, and this extended form of Rayleigh's criterion is equivalent to the principle of linear stability,[3,4]

$$\Delta\mathcal{E}_n = 2\pi\omega_n\alpha_n \tag{60}$$

Consequently, the precise form of Rayleigh's stated criterion is found by taking only the part of F_n that represents heat addition and setting $F_{ne} = 0$. The result is[20]

$$\Delta\mathcal{E}_n(t) = (\gamma - 1)\frac{\omega_n^2}{\bar{p}E_n^2}\int \psi_n \, \mathrm{d}V \int_t^{t+\tau_n} \eta_n Q' \, \mathrm{d}t' \tag{61}$$

in which Q' must be expressed in terms of real quantities. Thus, instead of the complex form we write in the case Q' proportional to p',

$$Q' = Q_0\psi_n(\beta_1\eta_n + \beta_2\dot{\eta}_n) \tag{62}$$

where β_1 and β_2 are real. The formula (61) leads to

$$\Delta\mathcal{E}_n = \pi(\gamma - 1)\beta_1\omega_n Q_0 \tag{63}$$

Thus, $\Delta\mathcal{E}_n$ is positive if $\beta_1 Q_0 > 0$, requiring that some portion of the heat addition must be in phase with the pressure fluctuation, as stated by Rayleigh.

Combination of Eqs. (60) and (61) gives the convenient formula for the growth constant due to heat addition,

$$\alpha_n = \frac{(\gamma - 1)\omega_n}{2\pi\bar{p}E_n^2}\int \psi_n \, \mathrm{d}V \int_t^{t+\tau_n} \eta_n Q' \, \mathrm{d}t' \tag{64}$$

where we have used the fact that $\eta_n Q'$ has period $\tau_n = 2\pi/\omega_n$ to shift the limits of the integral. We should also note that whereas the time derivative $\partial Q'/\partial t$ appears as the source in the wave equation (41) and in the oscillator equations (50), Q' itself occurs in Eqs. (61) and (64), also as a consequence of periodicity as explained by Culick.[20]

C. Response Factors

Particularly during the time when a great deal of effort was being devoted to problems of combustion instabilities in engines intended for the Apollo vehicle, the notion of a response factor was frequently applied. Although the various definitions used may differ in detail, the essential idea is common to all: a response factor is a quantity whose numerical value is a measure of the destabilizing effect of a process whose behavior is approximated by a linear model. It appears that all definitions of response factors are equivalent to the definition of the growth (or decay) constant α_n given by formula (57) through the approximate analysis formulated in Sec. IV. The interpretation of α_n follows from its definition and the definition of the energy in an acoustic wave. For small amplitude motions, the energy per unit volume is the sum of kinetic and potential energies,

$$\tfrac{1}{2}\rho(u')^2 + \tfrac{1}{2}(p')^2/\rho \tag{65}$$

To the approximation required in this analysis, $\boldsymbol{u}'$ and p' can be replaced by their classical, unperturbed relations; for the nth mode,

$$p_n' = \bar{p}\hat{\eta}_n e^{(i\omega_n+\alpha_n)t}\psi_n; \qquad \boldsymbol{u}_n' = i\frac{\bar{p}}{\bar{\rho}}\frac{\hat{\eta}_n}{\omega_n}e^{(i\omega_n+\alpha_n)t}\nabla\psi_n \tag{66}$$

The total time-averaged energy in the chamber is

$$\langle \mathcal{E}_n \rangle = \frac{1}{\tau_n} \int_t^{t+\tau_n} \mathrm{d}t' \int \mathrm{d}V \left[\frac{1}{2}\bar{\rho}(\mathrm{Re}\{\boldsymbol{u}_n'\})^2 + \frac{1}{2}\frac{(\mathrm{Re}\{p_n'\})^2}{\bar{\rho}} \right] \tag{67}$$

Substitution of Eq. (66) and carrying out the integrals gives

$$\langle \mathcal{E}_n \rangle = \mathcal{E}_{n0} e^{2\alpha_n t} \tag{68}$$

where $\mathcal{E}_{n0}$ is a constant. Hence,

$$\alpha_n = \frac{1}{2}\frac{1}{\langle \mathcal{E}_n \rangle}\frac{\mathrm{d}\langle \mathcal{E}_n \rangle}{\mathrm{d}t} \tag{69}$$

is half of the fractional rate of change of the time-averaged energy in the mode.

Since it is a result for linear behavior, the formula (57) for α_n is a sum of terms on the right-hand side, with each term being associated with a single process. For example, consider only the first term in f on the right-hand side of Eq. (44) and ignore all others in $h + h_e$ and $f + f_e$; then Eq. (57) becomes

$$\alpha_n = \frac{-\gamma}{2\omega_n E_n^2} \oiint \frac{1}{\hat{\eta}_n}\left(\frac{\partial \hat{\boldsymbol{u}}'}{\partial t} \cdot \boldsymbol{n} \right)^{(i)} \psi_n \, \mathrm{d}S \tag{70}$$

where the superscript caret means taking the factor multiplying $e^{i\omega_n t}$. It is common practice in acoustics to represent the response of a surface with an admittance function A. Here, that model implies assuming that the velocity normal to the surface induced by an imposed pressure fluctuation is proportional to that fluctuation,

$$\boldsymbol{u}' \cdot \boldsymbol{n} = A(\bar{a}/\bar{p})p' \tag{71}$$

where the ratio $\bar{a}/\bar{p}$ has been introduced to make A dimensionless. [Note that $\boldsymbol{n}$ is positive outward. A negative sign should be added to the definition of the admittance function if $\boldsymbol{u}$ is positive inward. See Ref. 4 for a thorough discussion.] Because $A = A^{(r)} + iA^{(i)}$ is complex, it follows that

$$\begin{aligned} \frac{\partial \boldsymbol{u}'}{\partial t} \cdot \boldsymbol{n} &= (A^{(r)} + iA^{(i)})\frac{\bar{a}}{\bar{p}}\frac{\partial p'}{\partial t} \\ &\approx (A^{(r)} + iA^{(i)})(\bar{a}/\bar{p}) i\omega_n \hat{\eta}_n \psi_n e^{i\omega_n t} \end{aligned}$$

and

$$\left(\frac{\partial \hat{\boldsymbol{u}}'}{\partial t} \cdot \boldsymbol{n} \right)^{(i)} = (\bar{a}/\bar{p})\omega_n A^{(r)} \hat{\eta}_n \psi_n$$

The formula (70) becomes

$$\alpha_n = \frac{-\gamma \bar{a}}{2\bar{p}E_n^2} \oiint A^{(r)} \psi_n^2 \, \mathrm{d}S \tag{72}$$

This is a familiar result; with the definition of the growth constant used here, α_n is negative (stabilizing) if the real part of the admittance function is positive. Moreover, those parts of the surface contribute most where the mode shape has the greatest amplitude. These conclusions amount to a Rayleigh's criterion for the response of a surface.

Similar special formulas can be deduced for processes in volume. Each formula for a contribution to the net growth constant is proportional to a response factor whose form depends on the definitions chosen. A response method, then, consists generally in computing the various response factors and summing the results. In short, a response method is entirely equivalent to a computation of linear stability that, as we have shown in the preceding section, amounts to the most general form of Rayleigh's criterion.

D. Time Lag (n, τ) Representation

Shortly after oscillations were encountered in early tests of liquid-propellant rocket engines in the United States, von Kármán suggested representing the mechanism with a time lag.[22] The idea was also independently introduced by Gunder and Friant[23] in 1950 and was subsequently developed to the greatest degree possible by Crocco and his students at Princeton; see Ref. 1 for thorough coverage of the subject.

The essential idea is that a finite interval—the time lag—exists between the time at which an element of propellant enters the chamber and the time at which it burns and releases its chemical energy. Such a lag must always exist, and there is no unique value because combustion is distributed throughout the chamber. If the pressure has the time dependence $p' = \bar{p}\psi_n(\boldsymbol{r})\sin\omega_n t$, then the simplest way to introduce a time lag is to assume that the fluctuation of energy release has the form

$$Q' = \hat{Q}(\boldsymbol{r})\sin\omega_n(t-\tau) \tag{73}$$

where τ represents the time lag. Substitution in the form of Rayleigh's criterion, Eq. (61), gives

$$\begin{aligned} \Delta\mathcal{E}_n &= 2\pi\omega_n\alpha_n \\ &= (\gamma-1)\frac{\omega_n^2}{\bar{p}E_n^2}\int\psi_n(\boldsymbol{r})\hat{Q}(\boldsymbol{r})\,\mathrm{d}V\int_t^{t+\tau_n}\sin\omega_n t'\sin\omega_n(t'-\tau)\,\mathrm{d}t' \end{aligned} \tag{74}$$

and α_n is

$$\alpha_n = (\gamma-1)\frac{1}{2\bar{p}E_n^2}\int\psi_n(\boldsymbol{r})\hat{Q}(\boldsymbol{r})\,\mathrm{d}V\cos\omega_n\tau \tag{75}$$

If $\hat{Q}$ is itself proportional to the local pressure amplitude ψ_n, then the integral is positive, and α_n is positive if $0 < \omega_n\tau < \pi/2, 3\pi/2 < \omega_n\tau < 5\pi/2, \ldots$. The process is destabilizing if τ is in the ranges of

$$0 < \tau < \frac{\pi}{2\omega_n}; \qquad \frac{3\pi}{2\omega_n} < \tau < \frac{5\pi}{2\omega_n}; \qquad \cdots \tag{76}$$

Although the oscillating behavior of α_n given by Eq. (75) is not in any sense fundamentally unacceptable, many examples of attempts to infer time lags for real processes suggest that such a result is unrealistic.[24,25] Part of the difficulty comes from the assumption usually made (in the absence of other information) that the time lag is independent of frequency. Even on intuitive grounds, one should expect that behavior is not the case. In at least one case for the response of a burning solid propellant, both theory and experiment show that any time lag introduced to represent the behavior must depend significantly on frequency.

It appears that all applications of the time-lag model are based on the reasoning[1,3] given by Crocco. The principal result is the formula for the fluctuation of conversion of liquid to gas, the source term in the continuity equations (13) and (14),

$$w'_\ell = \bar{w}_\ell n\left[\frac{p'(t)}{\bar{p}} - \frac{p'(t-\tau)}{\bar{p}}\right] \tag{77}$$

where the interaction index n represents the sensitivity of propellant combustion to pressure oscillation. No details of physical processes are treated in the construction of this result, although of course they are implied in the definitions of n and τ. To analyze linear stability, assume that a single mode is present and that $p' = \bar{p}\hat{\eta}_n e^{i\omega_n t}\psi_n(\boldsymbol{r})$, $w'_\ell = \hat{w}_\ell e^{i\omega_n t}$ to give

$$\hat{w}_\ell = \bar{w}_\ell n(1 - e^{-i\omega_n\tau})\hat{\eta}_n\psi_n(\boldsymbol{r}) \tag{78}$$

It is perhaps best to interpret Eq. (78) as a particular two-parameter representation of the complex function $\hat{w}_\ell$. The real and imaginary parts of the formula are

$$\hat{w}_\ell^{(r)} = n\bar{w}_\ell(1 - \cos\omega_n\tau)\hat{\eta}_n\psi_n(\boldsymbol{r}), \qquad \hat{w}_\ell^{(i)} = n\bar{w}_\ell\sin(\omega_n\tau)\hat{\eta}_n\psi_n(\boldsymbol{r}) \tag{79}$$

Formally, the $(n-\tau)$ model simply replaces the two functions $\hat{w}_\ell^{(r)}$ and $\hat{w}_\ell^{(i)}$ with the two quantities n and τ. There is some advantage in having even a crude understanding of the physical basis for this form. Any advantage is lost if the model is taken too seriously without recognizing its serious limitations, the chief one being the fact that n and τ are not constant in reality and, thus, the serious uncertainties in w'_ℓ are not clarified by simply introducing n and τ.

There have been successful applications of the idea, notably in laboratory experiments at Princeton reported by Crocco et al.[26] and as a means of correlating extensive data taken in some development programs during the 1960s.[2] The way in which the results (78) or (79) are used in applications can be explained very simply with formulas we have derived with the approximate analysis. With the external contributions dropped, the general formulas (56) and (57) for linear stability are

$$\Omega_n = \omega_n - \frac{1}{2\omega_n}\left(\frac{\hat{F}_n}{\hat{\eta}_n}\right)^{(r)} \tag{80}$$

$$\alpha_n = \frac{1}{2\omega_n}\left(\frac{\hat{F}_n}{\hat{\eta}_n}\right)^{(i)} \tag{81}$$

It is a simple matter to trace the contribution of the mass source w_ℓ from Eq. (79) to Eqs. (51) and (52). Obviously, there will be a term linear in the fluctuation w'_ℓ; in fact, it is $\partial w'_\ell/\partial t$ that appears so that the term in $\hat{F}_n$ is proportional to $i\hat{w}_\ell$. The details are unimportant here; the essential point is that Eqs. (80) and (81) take the form

$$\Omega_n = \omega_n + C_3 \int \psi_n \hat{w}_\ell^{(i)}\, dV - C_4 \tag{82}$$

$$\alpha_n = C_1 \int \psi_n \hat{w}_\ell^{(r)}\, dV - C_2 \tag{83}$$

On the stability boundary, $\alpha_n = 0$, and after substitution of Eq. (79), we can write the two formulas for the two unknowns n and τ:

$$n \sin \omega_n \tau = \left[(\Omega_n - \omega_n + C_4) \Big/ \left(C_3 \int \psi_n^2 \bar{w}_\ell \, dV \right) \right] \tag{84}$$

$$n(1 - \cos \omega_n \tau) = \left(C_2 / C_1 \int \psi_n^2 \bar{w}_\ell \, dV \right) \tag{85}$$

The right-hand sides are functions of the parameters defining the system according to injector type, geometry, dominant processes, and so forth. In practice, the idea is to vary those parameters and find the loci of values in such a way that small changes in one direction produce unstable oscillations and that changes in the other direction (i.e., to the other side of the locus) cause small disturbances to be attenuated. Then the right-hand sides of Eqs. (84) and (85) are known on the stability boundary, and the two equations can be solved for n and τ. Further discussion and some results may be found in the early work by Crocco et al.[26] and in Ref. 2.

It may at times be useful to introduce the n–τ model providing its limitations are understood, but the time has passed when this should be a major part of any analysis of combustion instabilities. With the increasing growth of large numerical simulations, any serious need for the n–τ model is rapidly disappearing. Money, time, and effort are now much better spent on more fundamental investigations of the processes actually present in an engine.

VI. Stability Rating

As interesting and challenging as understanding combustion instabilities is, the eventual purpose in practice is to guarantee their nonexistence. Theory, analysis, and laboratory work should establish the basic laws governing unsteady behavior, that is, scaling laws—dependence on geometry and other parameters that define the system, generally representing as thoroughly as possible linear and nonlinear behavior. If the task could be carried out to completion, then the basis would exist for minimum development costs of stable engines, with test programs planned mainly to confirm expected performance and stability. The actual case is still far from the ideal, and important testing procedures devised to establish stability characteristics of engines remain an important—and expensive—part of any development program. Such procedures define the process called stability rating.

As is the case for much of this subject, Ref. 2 remains the most complete reference for stability. Little seems to have changed in the past 20 years except for improvements in instrumentation, data acquisition, and data processing. Although other methods have been used to investigate special aspects of stability (e.g., oscillations involving the propellant supply system), here we shall discuss briefly only the two basic methods used routinely to determine the stability of high-frequency wave modes in a chamber. Bombing a chamber by initiating small explosive charges, the older of the two, was introduced in the 1950s. The second method is called temperature ramping in which the temperature of one of the propellants, usually the fuel, is increased more or less linearly in time until an unstable motion develops spontaneously, or, less commonly, the transient following a small explosion becomes a sustained oscillation. Because ample guidance is available elsewhere for applying these methods, there is no need here for elaborating details. Rather, only a general context with examples is provided.

We have emphasized that there are two general kinds of stability, linear and nonlinear. Stability rating in the linear sense means determining the stability boundary in some space of parameters. For example, when the n–τ model of unsteady combustion is used, the essential parameters are, of course, n and τ. As explained in Sec. V.D, however, the values of n and τ can only be inferred as functions of parameters that can be controlled experimentally, e.g., oxidizer-fuel ratio or geometrical properties of the chamber. Determining n and τ from experimental work requires a model, normally that producing Eq. (77), and a linear stability theory that gives, for example, relations such as Eqs. (84) and (85). Thus, the procedure is truly semiempirical, based on a mixture of theoretical and observational results. As used in engine development, methods of stability rating do not appeal to any theoretical results. A limited number of parameters are changed (e.g., explosive charge size or propellant temperature), and the stability of motions is then assumed in some simple terms, such as rate of decay or growth of oscillations. Frequency shifts are normally small, or cannot be interpreted meaningfully, so that the best one can do in rating linear stability is to determine the rate of growth or decay.

Nonlinear stability is intrinsically a far more difficult matter than linear stability. There is no satisfactory, comprehensive theory for nonlinear combustion instabilities even though much is known in respect to the stability of nonlinear dynamical systems in general. Nonlinear instability means that a linearly stable system becomes unstable to a sufficiently large disturbance. We are concerned here with systems that contain sources and sinks of energy. The problem of nonlinear instability is, therefore, not simply an elaboration of the motion of a rotating pendulum in a gravitational field; in that case, small initial disturbances produce small-amplitude to-and-fro oscillations but sufficiently large disturbances cause the pendulum to execute continuing circular motions.

For rocket engines, the only practical means of assessing nonlinear stability is based on detonating small explosive charges. A typical strategy involves a sequence of charges, increasing in size to reach a condition at which the disturbance no longer decays but continues as more-or-less periodic large-amplitude motions. Empirical rules have been constructed to define what constitutes acceptable dynamic stability of a chamber.

VII. Passive and Active Control of Combustion Instabilities

The ultimate practical purpose of studying combustion instabilities is to develop sufficient understanding to be able to control the amplitudes at acceptable levels. Quite generally, there are only two strategies to follow: change the basic design of the system (geometry, injector type, etc.) or introduce some sort of control. Because constraints set for other reasons often prevent sufficient changes in design, much effort has necessarily been exerted to devise means of control, always by passive means in operational propulsion systems.

A. Passive Control

The term "passive control" applies to the installation of baffles, resonators, or acoustic liners that suppress acoustic waves in the chamber. The essential point is to force the resonance to occur in frequency ranges where the driving mechanisms are inadequate to sustain oscillations or to directly damp the mechanical energy of unsteady motions. A secondary, but potentially more significant reason for the effectiveness of baffles, remarked upon in connection with the F-1 engine in Sec.II.A, is the possibility for shadowing regions of sensitive processes from disturbances. Figure 4 illustrates the idea that radial baffles extending axially into the chamber cause certain modes to be suppressed. A single baffle placed along a diameter, for example, will discourage spinning modes. These symmetrically placed radial baffles act against tangential oscillations. The effectiveness of baffles is limited by the practical constraint that they cannot extend too far from the injector face because structural integrity may be sacrificed and flow losses may be unacceptable.

Whatever the device used, limitations always arise because of its frequency response. Resonators may be the most obvious example, typically showing a fairly narrow peaked response whose height is reduced and width is increased for larger amplitude motions. Although the data on the behavior of passive devices taken in scale models at room temperature is inexpensive to acquire and usually serves as a useful guide in design, care must be exercised when extrapolating these results to practical rocket engines due to the differences in temperature and associated speed of sound. The performance of these devices is most conveniently expressed directly as an attenuation constant or, if more detailed behavior is known, an impedance or admittance function may be defined and used in a formula such as Eq. (72).

B. Active Control

Active control of combustion instabilities has received considerable attention in the past 10 years even though it is not a new concept. Tsien[27] suggested that the chugging instability in a liquid rocket motor could be stabilized by controlling the supply of propellant. Figure 9 is a sketch explaining his proposal. The main item is a feedback control loop based on sensing the pressure and controlling the line capacitance. The system may be modeled with the simple analysis outlined in Sec. IV. With a time-lag model of combustion, Eq. (1) becomes

$$\frac{\mathrm{d}^2 p'}{\mathrm{d}t^2} + 2\alpha \frac{\mathrm{d}p'}{\mathrm{d}t} + \omega_0^2 p' = \beta p'(t-\tau) + u(t) \tag{86}$$

where $u(t)$ is the control force. In the chugging mode, the chamber pressure is practically uniform so that p' here, indeed, represents the pressure measured everywhere. Taking the Laplace transform of Eq. (86), we find the relation between the transform $P(s)$ of the pressure and that of the control force $U(s)$

$$P(s) = [G/(1 - GQ)]U(s) \tag{87}$$

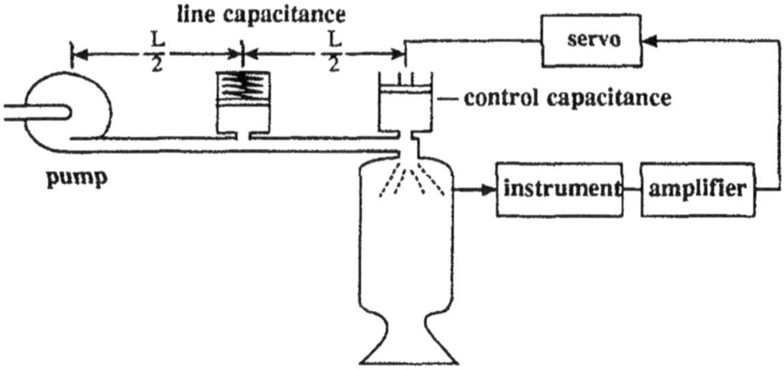

Fig. 9 Feedback control of chugging instability.

where G is the transfer function of the chamber (or the plant in the terminology of control theory),

$$G(s) = 1/\left(s^2 + 2\alpha s + \omega_0^2\right) \tag{88}$$

and

$$Q = \beta e^{-s\tau} \tag{89}$$

The result (87) has exactly the form interpreted with the block diagram drawn in Fig. 3. Well-known methods of classical control theory may be applied to investigate the performance and stability of this feedback system, although things are complicated somewhat by the presence of the time lag.

Tsien's method[27] of active control has never been successfully applied in practice. When it was first proposed, the available instrumentation and transducers were likely inadequate. Subsequently, the problem of chugging, most familiar in the form of the POGO instability, has been satisfactorily solved by passive control in operational systems.

Contemporary ideas of active control began about 10 years ago as an outgrowth of the field of noise control. A common and successful method of noise control is based on the idea of antisound[28] in which destructive interference is caused by injecting an appropriate acoustic field to cancel unwanted noise. Almost all demonstrated applications of active control have been based on that idea. Culick has provided a brief review of the work through 1988 in Ref. 3. Similar work has continued, largely associated with the development of air-breathing engines such as afterburners and gas-turbine combustors.

As emphasized earlier, the analysis described in Sec. IV has a form that is naturally adapted to the application of modern control theory. In the control field, the subject is called control of distributed systems. The source terms denoted by subscript e in the analysis accommodate any practical means of control, including actions in the chamber and at the boundary. The only systematical work on the control of combustion instabilities based on this approach appears in Fung[29] and in associated publications.[30–32]

In this context, any analysis of active control begins with the set of coupled oscillator equations (50). The theoretical work is then divided into two main parts: modeling the dominant physical processes and designing a controller. The most difficult part of the problem is the modeling, which by necessity also includes modeling the actuators used. Combustion systems offer especially challenging problems in this respect because they are intrinsically nonlinear and contain substantial noise, and the activity of control changes the characteristics of the plant being controlled. Another way of expressing the last point is that there are inevitably uncertainties in the value of the parameters characterizing the plant.

If linear behavior is assumed, and the system has been satisfactorily modeled, then the available methods of modern control theory can be applied. It is a much more difficult matter if nonlinear processes are taken into account. For practical applications, the most serious obstacle is probably actuation. There seems to be general agreement that if active control is to succeed with operational systems, actuation will likely involve control of the propellant supply. The few results available were obtained with gas-filled laboratory devices. Control with liquid propellants is evidently much more difficult, with no successes reported.

A fundamental reason for difficulties of controlling the propellant supply, and therefore combustion processes, is the lack of understanding of the internal processes: modeling is again a central problem. It is likely that significant time lags are unavoidable; indeed, time lags are probably variable and non-uniform in space. Whereas active control of combustion instabilities seems to be a promising approach, success with large-scale systems is by no means assumed. Much research remains to be accomplished.

VIII. Concluding Remarks

Most of the progress in the area of combustion instabilities in liquid rocket engines was achieved prior to the early 1970s. Although there have been some notable advances since then, there has been relatively little change in basic understanding. This is the case partly because there has been no pressing need for a large, vigorous, and continuing research program that involves both theory and experiment. A significant exception has been the ONERA work in France, begun in 1981 as a result of flight failures of the Ariane vehicle. The relatively modest size of the program has probably limited the rate of accomplishments, but the sustained effort is unique in the field.

As in virtually all fields of terrestrial fluid mechanics, any successful research effort must comprise three parts: experiment, theory, and numerical simulation (CFD). The rocket propulsion community has generally been slow to take advantage of the rapid advances in computing resources and computational fluid mechanics, but that situation has been changing in the past several years. Some of the chapters in this volume summarize recent progress and suggest what is possible in the near future.

It is essential in a thorough investigation of combustion instabilities that all relevant driving and damping processes be accounted for. Stability depends on the balance of energy gains and losses; no matter how attractive a particular mechanism may appear, a proper conclusion can be reached only in the context of a framework treating all aspects of the problem. Experimental results, of course, necessarily contain all physical and chemical processes. Computing resources have advanced to the stage at which most relevant processes can be included in numerical simulations.

More traditional analysis, such as that outlined in this chapter, remains a central part of the basis for interpreting experimental results, understanding observed behavior, and for theoretical work. There is a broad spectrum of problems to be treated, having many applications in combustion devices generally. It is fundamentally important that analytical work be integrated as thoroughly as possible with numerical simulations and experimental results. Purely theoretical work in this field is practically impossible. The problems are so complicated that analysis must be motivated by observations and, as we have emphasized in the introductory remarks, it is a fruitful strategy to incorporate from the beginning certain global aspects of observed behavior and then develop an analysis. Numerical simulations are in that respect closer to a level of proceeding directly from first principles. Therefore, an important role for numerical simulations is to provide a basis for assessing the validity of approximate results. Accurate confirmation leads to confidence in extending the analysis to situations for which numerical simulations are prohibitively expensive.

There is presently a large mismatch between available theories and experimental results. Theoretical and analytical results for linear behavior are the most advanced. Indeed, for all practical purposes, one might argue that linear theory is now complete. That is not to claim that highly accurate results can be obtained because many processes are still imperfectly understood; modeling is inaccurate, and necessary ancillary data are often uncertain as well. Even though the basic theory is well understood, there are relatively few experimental results available to check the theory in detail.

On the other hand, the greater part of observational data have to do with nonlinear behavior for which the theory is in a very primitive state. Particularly outstanding in this regard is the matter of nonlinear instability, the general characteristic that is investigated with the experimental technique called bombing. Nearly 30 years have passed since Crocco and his students first began treating nonlinear behavior, yet there is really no practical question that can be satisfactorily answered in any useful and general sense.

There are two basic approaches to the problem of nonlinear behavior. The first, entirely mathematical, concerns the behavior of solutions to nonlinear differential equations. Great progress has been made recently in nonlinear dynamical systems; many of the methods and formal results are probably applicable to combustion instabilities, but only limited use has been made of the theory. Two topics in particular are clearly relevant: the theory of subcritical bifurcations, which includes the phenomenon called triggering or nonlinear instability; and the very broad matter of nonlinear behavior with noise.

The second basic aspect of nonlinear behavior is the physical theory of the dominant nonlinear processes. Remarkably little attention has been paid to this subject, other than qualitative descriptive commentary. Yet nonlinear models of physical behavior must be constructed to be consistent with observations if the theoretical work is to produce meaningful results. Nonlinear gasdynamics is always present, is easily modeled accurately, and should therefore always be included in any nonlinear analysis. In liquid rocket engines, observational results suggest, we believe, that in addition to gasdynamics, the dominant nonlinear processes are most probably associated with the fluid mechanics of injected propellants and subsequent spray field. The sensitivity of these processes is enhanced further by supercritical vaporization and combustion processes. Cold-flow tests are a necessary first step, and many have been made, but the influences of large rates of heat addition associated with combustion processes are probably too severe to be ignored.

The general matter of control merits much continued attention. Although the results of passive control devices are fairly well understood, the possible application of active control should be pursued vigorously. This is not simply a matter of wrapping a feedback loop around the system. Intelligent—and therefore successful—use of control will be based on thorough understanding of the available control theory and realistic modeling of the system. Results obtained to date with laboratory devices do not indicate whether active control will be either effective or ineffective. The subject remains in its initial stages.

Acknowledgments

The authors are indebted to Josef M. Wicker for many helpful discussions of feedback loop of combustion instability.

References

[1]Crocco, L., and Cheng, S. I., *Theory of Combustion Instability in Liquid Propellant Rocket Motors,* AGARD monograph, No. 8, Butterworths, London, 1956.

[2]Harrje, D. T., and Reardon, F. H. (eds.), *Liquid Propellant Rocket Combustion Instability,* NASA SP-194, 1972.

[3]Culick, F. E. C., "Combustion Instabilities in Liquid-Fueled Propulsion System—An Overview," AGARD Conf. Proceedings No. 450, 1988.

[4]Culick, F. E. C., and Yang, V., "Prediction of the Stability of Unsteady Motions in Solid Propellant Rocket Motors," *Nonsteady Burning and Combustion Stability of Solid Propellants,* edited by L. Deluca and M. Summerfield, Vol. 143, Progress in Astronautics and Aeronautics, AIAA, Washington, DC, 1992, Chap. 18, pp. 719–780.

[5]Oefelein, J. C., and Yang, V., "Comprehensive Review of Liquid-Propellant Combustion Instabilities in F-1 Engines," *Journal of Propulsion and Power,* Vol. 9, No. 5, 1993, pp. 657–677.

[6]Wicker, J. M., Yoon, M. W., and Yang, V., "Linear and Nonlinear Pressure Oscillations in Baffled Combustion Chambers," *Journal of Sound and Vibration,* Vol. 184, 1995, pp. 141–172.

[7]Elverum, G., Jr., Staudhammer, P., Miller, J., Hoffman, A., and Rockow, R., "The Descent Engine for the Lunar Module," AIAA Paper 67-521, 1967.

[8]Yodzis, C. W., "Engines for Manned Spacecraft," AIAA Paper 68-567, 1968.

[9]Anon, *Liquid Rocket Engine Injectors,* NASA SP-8089, 1976.

[10]Ryan, H. M., Anderson, W. E., Pal, S., and Santoro, R. J., "Atomization Characteristics of Impinging Liquid Jets," *Journal of Propulsion and Power,* Vol. 11, No. 1, 1995, pp. 135–145.

[11]Hsieh, K. C., Shuen, J. S., and Yang, V., "Droplet Vaporization in High-Pressure Environments, I: Near Critical Conditions," *Combustion Science and Technology,* Vol. 76, 1991, pp. 111–132.

[12]Shuen, J. S., Yang, V., and Hsiao, C. C., "Combustion of Liquid-Fuel Droplets at Supercritical Conditions," *Combustion and Flame,* Vol. 89, 1992, pp. 299–319.

[13]Lafon, P., Yang, V., and Habiballah, M, "Pressure-Coupled Vaporization and Combustion Responses of Liquid Oxygen (LOX) Droplets in Supercritical Hydrogen Environments," AIAA Paper 95-2432, 1995.

[14]Yang, V., Hsiao, C. C., Shuen, J. S., and Hsieh, K. C., "Droplet Behavior at Supercritical Conditions," *Recent Advances in Spray Combustion,* edited by K. K. Kuo, Progress in Astronautics and Aeronautics, AIAA, Washington, DC, 1995.

[15]Culick, F. E. C., "High Frequency Oscillations in Liquid Rockets," *AIAA Journal,* Vol. 1, No. 5, 1963, pp. 1097–1104.

[16]Morse, P. M., and Feshback, H., *Methods of Theoretical Physics,* McGraw–Hill, New York, 1953, Chap. 11.

[17]Nickerson, G. R., Culick, F. E. C., and Dang, L. G., "Standard Stability Prediction Method for Solid Rocket Motors," Software and Engineering Associates, Inc., AFRPL TR-83-017, Carson City, NV, 1983.

[18]Rayleigh, Lord, "The Explanation of Certain Acoustical Phenomena," *Royal Institution Proceedings,* Vol. VIII, London, 1878, pp. 536–542.

[19]Rayleigh, Lord, *The Theory of Sound,* Vol. II, Dover, New York, 1945, p. 226.

[20]Culick, F. E. C., "A Note on Rayleigh's Criterion," *Combustion Science and Technology,* Vol. 56, 1987, pp. 159–166.

[21]Zinn, B. T., "Pulsating Combustion," *Advanced Combustion Methods,* edited by F. J. Weinberg, Academic, London, 1986, Chap. 2, pp. 113–181.

[22]Summerfield, M., "A Theory of Unstable Propulsion in Liquid Propellant Rocket Systems," *ARS Journal,* Vol. 21, 1951, pp. 108–114.

[23]Gunder, D. F., and Friant, D. R., "Stability of Flow in a Rocket Motor," *Journal of Applied Mechanics,* Vol. 17, 1950, pp. 327–333.

[24]Heidmann, M. F., and Wieber, P. R., "Analysis of n-Heptane Vaporization in Unstable Combustor with Traveling Transverse Oscillations," NASA TN D-3424, 1966.

[25]Heidmann, M. F., and Wieber, P. R., "Analysis of Frequency Response Characteristics of Propellant Vaporization," NASA TN D-3749, 1966.

[26]Crocco, L., Grey, J., and Harrje, D. T., "Theory of Liquid Propellant Rocket Combustion Instability and Its Experimental Verification," *ARS Journal,* Vol. 30, 1960, pp. 159–168.

[27]Tsien, H. S., "Servo-Stabilization of Combustion in Rocket Motors," *ARS Journal,* Vol. 22, 1952, pp. 256–263.

[28]Ffows-Williams, J. E., "Anti-Sound," *Proceedings of the Royal Society, London,* Vol. A395, 1984, pp. 63–88.

[29]Fung, Y. T., "Active Control of Linear and Nonlinear Pressure Oscillations in Combustion Chambers," Ph.D. Thesis, Dept. of Mechanical Engineering, Pennsylvania State Univ., University Park, PA, 1991.

[30]Fung, Y. T., Yang, V., and Sinha, A., "Active Control of Combustion Instabilities with Distributed Actuators," *Combustion Science and Technology,* Vol. 78, 1991, pp. 217–245.

[31]Yang, V., Sinha, A., and Fung, Y. T., "State Feedback Control of Pressure Oscillations in Combustion Chambers," *Journal of Propulsion and Power,* Vol. 8, No. 1, 1992, pp. 66–73.

[32]Fung, Y. T., and Yang, V., "Active Control of Nonlinear Pressure Oscillations in Combustion Chambers," *Journal of Propulsion and Power,* Vol. 8, No. 6, 1992, pp. 1282–1289.

Chapter 2

Instability Phenomena in Liquid Oxygen/Hydrogen Propellant Rocket Engines

James Hulka*
Aerojet, Sacramento, California 95813
and
John J. Hutt†
NASA Marshall Space Flight Center, Huntsville, Alabama 35812

I. Introduction

IN this chapter, high-frequency combustion instability phenomena of liquid oxygen/hydrogen propellant rocket engines from the United States are summarized. Experimental data are compiled and discussed from programs at Aerojet,[1-16] Pratt and Whitney Aircraft,[17-27] Rocketdyne,[28-45] NASA Lewis Research Center (LeRC),[46-72] NASA Marshall Space Flight Center (MSFC),[73-75] and elsewhere.[76,77] Included are data and results from the flight engines RL10, J-2, and the Space Shuttle main engine (SSME); engine development programs for J-2, J-2S, Aerospike, M-1, XLR-129, SSME, and the Space Transportation main engine (STME); and investigative programs in government and industry. Although omissions have no doubt occurred, the objective of this summary was to compile and include as much material as possible.

The combustion stability of many of these programs is represented by a particular injection element type—the concentric or coaxial orifice—and particular conditions for developing high-frequency acoustic instability–low-hydrogen injection temperature. The concentric-orifice injection element was used in cryogenic oxygen/hydrogen rocket testing as early as the 1940s and later became the element of choice on all flight engine injectors. Usually designed for high-hydrogen injection temperatures, this element (and others) consistently developed spontaneous high-frequency acoustic combustion instabilities when operating at sufficiently low-hydrogen injection temperatures.

The ability to reproduce these instabilities promoted a number of investigative research programs, notably at the NASA LeRC, in which the low-hydrogen

*Senior Engineer, Advanced Development Engineering.

†Aerospace Engineer, Propulsion Laboratory.

temperature aspects were studied. Various mechanisms describing the instabilities were theorized as engine parameters were correlated and empirical stability boundaries defined. Since lowering hydrogen injection temperature at constant flow rate increased the fluid density and reduced the hydrogen injection velocity, one of the earliest engine stability parameters of importance to be recognized was the hydrogen-to-oxygen injection velocity ratio. Operating at high-injection velocity ratios was found to be one of the most effective means to stabilize liquid oxygen/hydrogen combustion for concentric-orifice element injectors. An empirical evaluation of velocity ratio is provided in this chapter.

Another important engine stability parameter to be recognized was the hydrogen injection pressure drop, or flow resistance. At a sufficiently low pressure drop across the hydrogen injection orifice, the orifice would be sensitive to combustion oscillations which could induce oscillations in hydrogen flow rate, leading to the onset of injection-coupled instabilities. Operating at high-hydrogen pressure drop was not a general effective means for stabilization, however, since cases of instability were found with high-hydrogen pressure drop, indicating coupling with other instability mechanisms was possible.

The influence of these parameters and many others are discussed in this chapter. The technical discussion is divided into three sections. A historical overview of liquid oxygen/hydrogen propellant rocket engine testing and development in the United States is first presented. Instability phenomenology is discussed in the next section in four categories based on the injector element type: 1) plain concentric-orifice elements (shear coax), 2) swirling concentric-orifice elements (swirl coax), 3) oxidizer showerhead elements, and 4) oxidizer impinging-jet elements. Last, a brief review of instability phenomena with combustion chamber stabilization devices is presented.

II. Historical Overview

In the United States, cryogenic oxygen/hydrogen propulsion technology was pioneered in the late 1940s, covering areas of materials technology, propellant handling, test operations, combustor heat transfer, and injector design.[1,76,77] Multitube concentric-orifice element injectors were developed during these very early programs and operated without any indication of combustion instability. They were subsequently used in the development of liquid oxygen/hydrogen propellant rocket engines for flight, which began in the late 1950s with the RL10 and J-2 engines.

The 67-kN (15,000-lbf) thrust RL10 engine[18] was under preliminary development by Pratt and Whitney Aircraft for the upper stage engine of the Air Force Centaur when main development was transferred to NASA. A plain-jet concentric-orifice element injector, with a hydrogen transpiration-cooled faceplate and oxidizer injection tubes flush with the injector face, was designed soon after the NASA contract award in 1958. This version, RL10A-1, was certified for flight in 1961 and flown on the Atlas Centaur in 1963, and became the first production liquid oxygen/hydrogen engine in the United States. There were no combustion instabilities observed during development or preliminary flight rating tests,[17] a fact later attributed to the high-hydrogen injection temperature (161 K).[17] Later testing during investigative programs disclosed that an engine modified for liquid hydrogen operation could be made unstable.[22,23] The injector design was modified in later versions (RL10A-3) by swirling the oxidizer jets in the core elements to increase injector performance.[18]

Development of the J-2 engine under contract to NASA[29–31] began at Rocketdyne in 1960, following some predevelopment.[28] Initial testing was conducted with full size ringed manifold injectors derived from liquid oxygen/kerosene Atlas hardware.[29] Most injectors used showerhead elements with uncooled faces, so injector characteristic velocity performance was poor. More important, all of the ring manifold injector patterns tested initially at conditions simulating the near-liquid hydrogen temperatures expected during the engine start sequence developed spontaneous high-frequency combustion instabilities. This was considered a major technical problem.[29] Radial baffles were installed and appeared to eliminate the instability in most cases, so baffle cooling studies were initiated. Later work showed that a divergence ring from the injector face was also stabilizing at low-hydrogen temperatures, although cooling the ring with these injector patterns proved difficult. Combustion instabilities were also induced by explosive charges at higher hydrogen temperatures than could be induced by hydrogen temperature ramping.

The first concentric-orifice element J-2 injector with a transpiration-cooled faceplate was tested in late 1961.[29] Although still unstable at near-liquid hydrogen temperatures, this injector type was found to be more stable with colder hydrogen injection temperatures than any plain-orifice element injector yet tested without baffles, because vibration levels measured with chamber-mounted accelerometers were considerably reduced. Injector performance was found to be initially similar to the plain-orifice element patterns.

The central oxidizer tubes of these first concentric-orifice elements were flush with the injector face. Further exploration to improve combustion stability and injector performance led to confining the initial unlike propellant contact with various mixing cup designs extended from the injector face. Improvements in combustion stability but difficulty in cooling led to recessing the oxidizer tube back from the injector face. The first recessed concentric-orifice element, introduced in mid-1963, increased injector characteristic velocity performance over nonrecessed by 2% and reduced hydrogen injection temperature for initiating instability to 39 K at mainstage conditions and 32 K at simulated engine start transient conditions.[29] Operation above these threshold temperatures was resistant to the detonation of 200 grain bombs.[29]

This recessed version was eventually selected for the preliminary flight rating test (PFRT) design. The flight rating test (FRT) design added a metering orifice at the inlet of the oxidizer tube, which increased the overall oxidizer pressure drop and improved the low-frequency (chug) combustion stability margin.[29] This change had no noticeable effect on the high-frequency stability margin.[29] The first J-2 engine was flown in 1966.

Despite extensive component development stability testing, engine stability testing of the 1025-kN (230,000-lbf) thrust J-2 engine[31] found a dynamically unstable region between 30 and 50% of full thrust. Acoustic instability was readily induced by thermally detonated nondirectional bombs and did not damp until the engine was shut down.[31] The tested areas outside this region were stable. Between 75 and 100% of full thrust, significant disturbances were difficult to produce, and those produced always damped. At conditions below 20% of full thrust during the oxidizer manifold priming of the start transient, low-frequency (300–400 Hz) chugging occurred, and disturbances were produced which excited acoustic modes; however, oscillations always damped shortly after the manifolds were fully primed. The region between 50 and 75% of full thrust was untested.

The J-2S engine development began in 1966 as an experimental J-2 engine, adding a tap-off engine cycle and uprating thrust to 1180 kN (265,000 lbf), with an accompanying increase in chamber pressure from 5430 KPa (787 psia) to 8690 KPa (1260 psia).[32] A weak but persistent instability made a late appearance in the J-2S program after nearly 6000 s of engine firings had been accumulated.[32] The growth rates of the oscillations were very low and amplitudes were nondestructive. Extensive engine diagnostic testing isolated the cause of the oscillation to coupling between the main injector oxidizer tube flow passage, the injection element recess area just downstream of the oxidizer tube, and the main chamber combustion zone. The oxidizer tubes defined an oscillation frequency which the recessed area, as a precombustor, could amplify. Coupling with main chamber acoustics was unnecessary to create and maintain the instability, apparently only synchronizing the individual injection element responses. The instability was eliminated by adding inserts to the oxidizer tubes,[32] which increased the pressure drop and retuned the hydraulic resonances.

In addition to these flight engine development programs, many investigative combustion stability programs were conducted from the late 1950s through the early 1970s that focused on obtaining fundamental information and developing preventive measures for instability. One of the most informative was the testing conducted at the NASA Lewis Research Center,[46–72] much of it at 89-kN (20,000-lbf) thrust with concentric-orifice element injectors. This research program systematically varied a large variety of geometrical and operational parameters using liquid oxygen/hydrogen propellants and provided the basis for comparison of a significant variety of parameters. Combustion stability margin was rated mainly by hydrogen injection temperature ramping.

Concentric-orifice element injectors were initially found to be more stable with higher fuel-to-oxidizer injection velocity ratios and the use of recessed oxidizer tubes,[57] results comparable to previous J-2 engine injector development experience.[29,30] Later investigation of the stabilizing effects of many element design details that did not change the hydrogen or oxygen injection velocities, such as impingement angle, oxidizer tube base thickness, and, in part, the recess, suggested that the influence of the hydrogen on stability was due to the flow resistance of the hydrogen, especially through the injection pressure drop.[69,78] Identification of the effects on stability of injector pattern layout,[70] combustion chamber geometry,[64,67,70,71] and oxidizer response[55,56,79] led to further refinement of the influence of combustor parameters.

This work supported a cooperative effort with Aerojet Liquid Rocket Co. for the development of the M-1 engine,[8,9,61,66] the largest liquid oxygen/hydrogen propellant thrust chamber tested in the United States. The 6670-kN (1.5×10^6-lbf) thrust M-1 engine was an upper stage concept terminated in advanced component development. The injector used an injection element derived from the J-2 element,[66] although in a thrust chamber more than twice the diameter. Acoustic instability was initiated by hydrogen injection temperature ramping. The instability appeared abruptly when reducing hydrogen injection temperature and stabilized at a higher hydrogen injection temperature when increasing the temperature from unstable conditions.[9] Different baffle lengths did not change the hydrogen injection temperature at instability,[9,66] similar to experience with J-2 engine development injectors with already superior stability characteristics.[29]

Other liquid oxygen/hydrogen investigative research programs at Aerojet included cryogenic hydrogen technology demonstrations in the 1940s[1] and late

1950s,[2] feasibility of staged-combustion engine cycles,[3] combustion stability investigations at high-chamber pressure,[4–7] advanced injector concepts including platelet injectors,[10,11] development of platelet injectors for candidate SSME preburners,[14] and candidate auxiliary propulsion systems for the Space Shuttle.[15] Cryogenic oxygen/hydrogen propulsion technology was initiated at Aerojet in the late 1940s,[1] evolving from gaseous oxygen/hydrogen rocket research and vehicle performance analyses showing the importance of propellant density. Concentric-orifice element injectors were developed to preclude injector face burning but resulted in the highest injector characteristic velocity performance of all concepts tested.[1] Testing with hydrogen resumed in the late 1950s using Titan hardware for feasibility studies of large engines.[2]

The staged-combustion feasibility program[3] tested hardware that showed the general feasibility of the staged-combustion cycle concept that had been previously recommended for development of higher pressure, higher performance liquid rocket engines. The high-chamber pressure program[4–7] investigated combustion stability at high-chamber pressure (up to 18 MPa), systematically exploring combustion dynamics and providing an experimental evaluation of the sensitive timelag combustion instability theory, using both plain-orifice and concentric-orifice element injectors. The advanced injector concepts program[10,11] developed the transverse excitation chamber, a subscale research tool specially designed to measure the frequency sensitivity of injection elements. Candidate SSME preburners[14] were tested with platelet injector modules with typical orifice dimensions of 0.05 cm. Overpressures of full chamber pressure from 15 to 80 grain bombs damped within 10 ms in both single and combined module demonstrations. Injectors from the Titan hydrogen feasibility program[2] were tested in large "onion-shaped" combustion chambers to provide hot gas for component development on the nuclear propulsion programs NERVA[12] and PHOEBUS-2.[13] High-harmonic mode instabilities occurred in a number of tests.[12]

At Pratt and Whitney Aircraft, liquid oxygen/hydrogen investigative and development programs included feasibility of high-pressure thrust chambers and turbomachinery,[19,20] stability testing with liquid and cold gaseous hydrogen,[21,22] investigation of acoustic, liners,[23,24] and further characterization of the staged combustion rocket engine cycle.[25,26] The high-pressure feasibility program[19,20] investigated injector performance and chamber cooling techniques at high pressure (up to 26 MPa) with plain-orifice and slot element injectors. Most tests were conducted with ambient temperature hydrogen, all of which were stable.[19,20] A pump-fed test at 222-kN (50,000-lbf) thrust with 60-K hydrogen and a circular slot triplet injector developed an apparent instability at the first tangential frequency midway through the test, persisting until shutdown.[20] Warpage of the slot injector through overheating had reduced the hydrogen injector pressure drop to 1/3 of its original value and reduced thrust-based impulse performance 10% during the instability.[20]

The liquid and cold gaseous hydrogen program[21,22] studied the effects of injection momentum ratio, injection velocity ratio, and combustion chamber contraction ratio on stability as functions of the hydrogen injection temperature, using concentric-orifice elements in slab motors and a modified RL10 injector. The mechanical damping devices[23,24] program characterized the stabilizing aspects of combustion chamber acoustic liners. An RL-10 injector modified for less hydrogen injection area that was unstable with liquid hydrogen was stabilized with use of various acoustic liners.[23,24]

Early advanced development of the staged-combustion cycle was demonstrated at 1110-kN (250,000 lbf) thrust at Pratt and Whitney Aircraft.[25,26] Both preburner and main chamber injectors used dual-inlet oxidizer swirl concentric-orifice elements, with the preburner elements laid out in clusters and the main injector elements in radial spraybars.[25] Chamber pressure perturbations from nondirectional 40 and 80 grain charges were readily damped in both combustion chambers at 20 and 100% of full thrust, and with preburner hydrogen injector temperature of 67 K (Ref. 25). The preburner injector layout was changed during the XLR-129 engine development[26] to a hexagonal pattern. Static and pulse tests were stable with a hydrogen injection temperature of 67 K, but nonacoustic low-frequency combustion instability occurred below 25% of full thrust.[26]

At Rocketdyne, liquid oxygen/hydrogen investigative and development programs during this era included photographic combustion studies in large two-dimensional chambers,[33] advanced development of the annular Aerospike[34–37] and linear Aerospike engines,[38–40] development of toroidal thrust chamber concepts,[41] evaluation of high-pressure regeneratively cooled thrust chambers,[42] and investigation of combustor effects on thrust chamber performance.[43] Development of the annular injector of the 1110-kN (250,000-lbf) thrust plug nozzle annular Aerospike Engine,[34–37] nearly 250 cm in diameter, was conducted using segments. Segment tests with 100 to 320-K hydrogen injection temperature and concentric-orifice and oxidizer-fuel-oxidizer (OFO) triplet elements were stable.[34,35] Tests with multiple segments exhibited mixed stability characteristics, with some tests recovering from a bomb in 40 ms (Ref. 34) and some tests displaying moderate amplitude 1800–2000-Hz oxidizer feed system oscillations.[35] Full-scale Aerospike engine tests[35,36] displayed the 1800–2000-Hz oscillation and also high-amplitude (60% of chamber pressure) rotating oscillations in the annular combustion chamber. The amplitudes of these oscillations were later reduced by increasing oxidizer injector pressure drop, isolating oxidizer manifold compartments, and modifying compartment baffles.[36,37]

Segment tests on the linear Aerospike engine were all stable with ambient–500-K hydrogen injection temperature, using triplet and concentric-orifice element injectors.[38,39] Full-scale tests were conducted with no evidence of instability.[40]

Many of these investigations combined to provide the basis for the latest large liquid oxygen/hydrogen propellant flight engine developed in the United States, the 2090-kN (470,000-lbf) thrust Space Shuttle main engine, manufactured by Rocketdyne. The three injector designs, main injector, hydrogen turbopump preburner, and oxygen turbopump preburner, were derived from the J-2S, with added complexity in the main injector manifolding due to the high velocities from using the preburner hot gas as the main injector fuel. Preburners were evaluated for temperature uniformity[45] and dynamic stability; tests with bombs were all stable, with and without acoustic resonators but all with injector face baffles. A moderate vibration occurred with some engine tests at the hydrogen turbopump preburner, at a frequency corresponding to the second axial resonance of the oxidizer injection tubes; however, chamber pressure oscillation amplitudes were negligible. Combustion stability was a major consideration in the design of the main injector[44] so that injector face baffles and chamber perimeter Helmholtz acoustic resonators were included, along with variable oxidizer injection tube lengths to preclude J-2S style instabilities. Nondirectional bomb perturbations in tests with and without the injector face resonators were quickly damped at power levels from 30 to 107% of full power level (FPL).[44] These tests were recently repeated with an experimental injector with baffles and Helmholtz resonators removed[75] and were equally stable. Additional bomb tests conducted at previously untested regions of the start transient, down to 15% FPL, were also stable. These tests continue the trend that all liquid oxygen/hydrogen staged-combustion tests in the United States have been stable in high frequency.

Efforts in the 1980s to reduce production and launch costs of a heavy lift launch vehicle culminated in the proposal of the Space Transportation main engine (STME), a high-chamber pressure (16-MPa) gas generator cycle with main chamber hydrogen injection temperature of 110 K. Subscale injector component tests were conducted with oxidizer swirl concentric-orifice element injectors,[16,27,74] and all were stable, including some with 50-K hydrogen injection temperature and 13-grain nondirectional bombs.

III. High-Frequency Combustion Instability Phenomena

In the initial rocket test programs with liquid oxygen/hydrogen propellants, combustion was reported to be smooth and without discernable oscillations.[1,76,77] There were no instabilities encountered during the development of the RL10 engine,[17] which had a nominal hydrogen injection temperature of 161 K. In one early instability study, oscillations were detected but were sporadic and irregular.[47] In feasibility studies at Aerojet with Titan hardware,[2] however, sustained high-frequency combustion instability was generated and was found easier to excite at fuel-rich mixture ratios with colder hydrogen than at higher mixture ratios with warmer hydrogen. The investigation was not extensive enough to separate the effects of mixture ratio and hydrogen temperature.

In the early phases of the J-2 engine injector development,[29] injector component tests at low-mixture ratio with liquid hydrogen to simulate start transient conditions were always unstable. Tests near the nominal operating conditions with hydrogen injection temperature of 111 K, including bomb stability tests, however, were always stable, even with unbaffled patterns. Hence, stability rating at nominal conditions was abandoned on the J-2 program. At sufficiently high hydrogen temperatures, all injectors could be stabilized. The capability to operate stably over the entire expected hydrogen injection temperature range of 28–167 K was determined to be the most important stability criterion for the J-2 engine. To evaluate stability margin, tests were created which gradually reduced the hydrogen injection temperature until instability occurred, while maintaining constant chamber pressure and mixture ratio. This method, called hydrogen injection temperature ramping, is shown in Fig. 1.[57] A lower hydrogen injection temperature at which acoustic instability occurred implied increased combustion stability margin.

Many physical mechanisms have been hypothesized to explain how temperature ramping causes instability, but all appear to have a limited range of applicability. Engine development, rather, has focused on mapping the stability boundaries with engine parameters to ensure that engines are not designed in regions that are susceptible to combustion instability. The injection velocity ratio, for example, was one of the earliest recognized engine parameters.

The effects of velocity ratio and other rocket engine parameters on combustion instability are discussed in this section. Since the injector element types are strongly influenced by these engine parameters, the discussion is divided into sections based on element type.

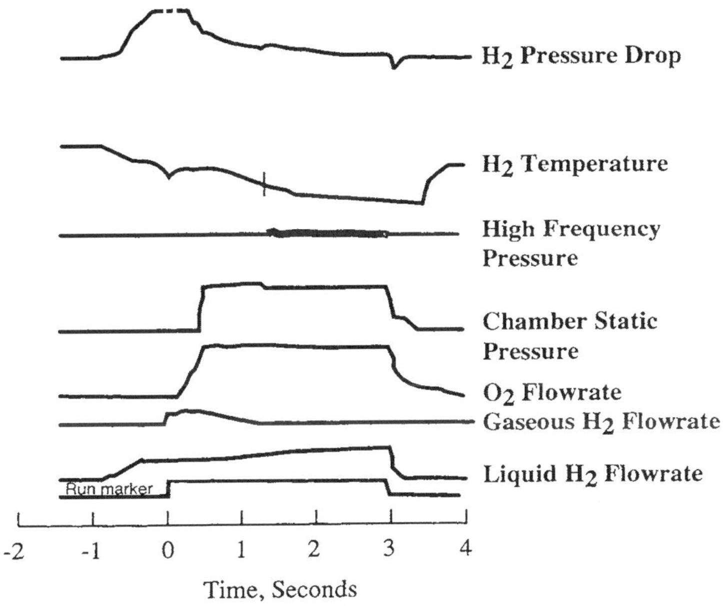

Fig. 1 Oscillograph traces from typical hydrogen injection temperature ramping test.[57]

A. Phenomena with Concentric-Orifice Element Injectors

The concentric-orifice injection element type is used on all flight liquid oxygen/hydrogen propellant rocket engines from the United States, along with a hydrogen transpiration-cooled porous metal face. Geometric and operating conditions of these flight thrust chambers are listed in Table 1. Because of their extensive test history, these designs may be considered to represent very inherently stable conditions.

1. Plain Liquid Oxidizer Jet (Shear Coax)

Parallel fuel injection. The J-2 and SSME flight production injectors listed in Table 1 use a concentric-orifice injection element where the liquid oxidizer jet flows full from the injection tube and is not swirled, and the fuel is injected from the concentric annulus parallel to the axis of the oxidizer flow. This element is known conventionally as a shear coax. The liquid oxygen injection plane is usually recessed from the injector face approximately one oxidizer tube exit orifice diameter. On the SSME, the oxidizer is diffused radially due to a slight internal taper at the oxidizer tube exit.

Although used on the very earliest injector testing,[1] the shear coaxial element evolved significantly on the J-2 engine injector development,[29] and current versions of these elements retain all of the basic features derived there. Early in the J-2 program, hydrogen injection temperature was found to influence injector performance and combustion stability of concentric-orifice elements through the momentum and injection velocity ratios,[29] although the nature of the development program limited the generation of investigative correlations.

Table 1 Flight production hardware parameters

Parameter	RL10A-3	J-2	SSME preburners Fuel pump	SSME preburners Oxidizer pump	SSME main chamber
Vaccum thrust, kN	67	1000	N/A	N/A	2090
Chamber pressure, KPa	2082	5309	34120	34090	20770
O/F chamber mixture ratio	5.01	5.71	0.893	0.653	6.01
Chamber diameter, cm	26.06	47.27	26.49	18.87	45.06
Contraction ratio	2.97	1.58	N/A	N/A	2.96
Chamber length, cm	33.0	45.7	15/33.0[a]	14.8/31.2[a]	35.6
Number of elements	216	614	264	120	600
Oxidizer jet type	Rbbn swr	Plain	Plain	Plain	Plain
Oxidizer jet exit diameter, cm	0.201	0.462	0.226	0.196	0.478
Oxidizer flow rate/element, kg/s	0.064	0.331	0.117	0.091	0.594[b]
Fuel injection temperature, K	161	108	154	154	841[c]
Fuel inlet pressure drop					
Oxidizer, % of chamber pressure	20.0	23.5	22.7	19.1	17.6
Fuel, % of chamber pressure	15.0	12.9	7.4	7.5	6.1
F/O injection velocity ratio	15.4	9.7	12.6	13.0	11.3
Combustor stabilization devices	none	none	3 radial baffles	3 radial baffles	5 radial/1 hub baffles + Helmholtz resonators

[a]Distance to top hat/first stage turbine nozzle.
[b]Oxidizer orifice flow rate (not total oxidizer flow rate).
[c]Hot gas mixture ratio = 0.756.

Investigative testing at Pratt and Whitney Aircraft using a rectangular thrust chamber[21,22] began to develop these relationships in more systematic experimentation. Stability rating was determined by directional pulse testing at fixed hydrogen injection temperatures. Generally, higher values of hydrogen temperature, momentum ratio, velocity ratio, and contraction ratio were found to be stabilizing. The effect of contraction ratio agreed with some later published results[10,62] but contradicted others.[67,70] The best combination of parameters to separate the data was suggested by Fig. 2.

These results agreed with stabilization of the full-scale M-1 thrust chamber, found to depend on maintaining a high-injection velocity ratio, as shown in Fig. 3 for a variety of combustion chamber conditions.[9] Stability rating was determined with hydrogen temperature ramping. The combustion oscillation at the onset of instability appeared more abruptly than usually found with temperature ramping tests.[9] Also, combustion was stabilized at a higher hydrogen injection temperature when increasing the temperature from unstable operation.[9]

Studies with parallel hydrogen injection were limited in the NASA LeRC testing with concentric-orifice elements,[58,72] but some appeared as geometrical variations. The effect of oxidizer tube exit wall thickness for one design was least stable

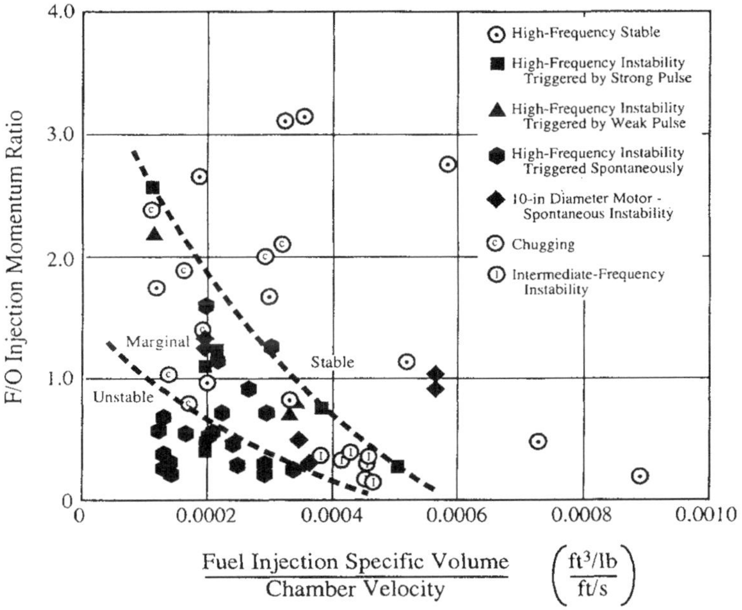

Fig. 2 Combined effects correlation for shear coaxial elements with parallel fuel.[22]

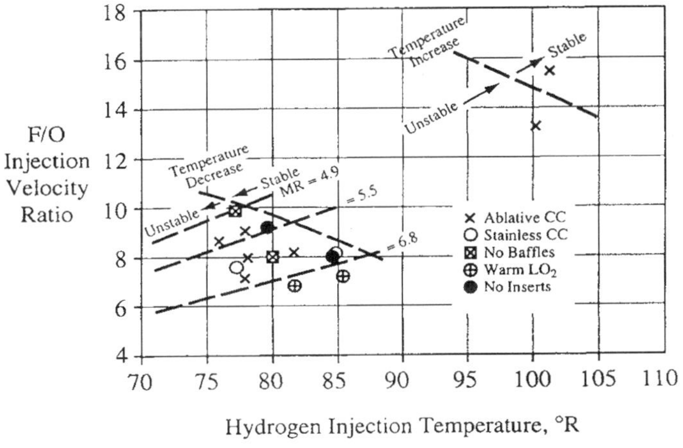

Fig. 3 Injection velocity ratio stability limits from M-1 main injector testing.[9]

at an intermediate thickness.[72] Also, continuous recessing of the oxidizer tube had continuous improvement in stability margin, even to depths of five times the orifice diameter.[72] On J-2 engine injector development, however, increasing the exit wall thickness degraded combustion stability, and continuous recessing did not continually improve stability.[29]

Injection velocity ratio was found to be a key parameter defining stability in many of these studies. To evaluate the effect of injection velocity ratio on stability of shear coaxial elements, data from the J-2 engine injector development,[29] the M-1 subscale[61] and full-scale[9] main injector development, NASA LeRC,[58,72] Pratt and Whitney,[22] and SSME[44] were compiled. Figure 4 shows a stability mapping of mixture ratio and fuel-to-oxidizer injection velocity ratio. Stable operation has occurred for injection velocity ratios as low as 1. The maximum unstable velocity ratio at main chamber mixture ratios, with a large variety of design and operating variations, however, has been about 10, which occurred on the M-1 full-scale injector testing. The M-1 instability data, above the diagonal line shown on Fig. 4, are separated from the other instability data, suggesting that chamber diameter is affecting these results. Additionally, not shown on Fig. 4, velocity ratios of about 15 were required to stabilize the M-1 main injector when increasing hydrogen injection temperature from unstable conditions, suggesting hysteresis exists in these stability boundaries.[9]

Higher velocity ratios were also unstable during J-2 engine injector development[29] for main chamber injectors operating off design at low-mixture ratio and low-hydrogen injection temperature and pressure, simulating start transient conditions. These data are shown on Fig. 4 at mixture ratios between 1 and 2. Figure 5 shows a stability mapping of hydrogen injection temperature and fuel-to-oxidizer injection velocity ratio. Instabilities with hydrogen temperatures as high as 83 K; and stable operation with liquid hydrogen, have occurred. The J-2 engine injector development instabilities with the highest velocity ratios occurred with hydrogen injection temperatures ranging from 28 to 42 K and chamber pressures of about 1.5 MPa. At these conditions, hydrogen density can vary two or three times smaller or larger with temperature differences of only ±5.6 K about a mean of 39 K. Hence, the potential uncertainty in hydrogen injection velocity for these conditions can be large. Two-phase flow was probably present in these tests. All higher temperature instabilities are found to have occurred at velocity ratios between 3 and 5.

Impinging fuel injection. The majority of liquid oxygen/hydrogen investigative stability testing used concentric-orifice injection elements where the liquid oxidizer jet flowed full from the injection tube and was not swirled, and the hydrogen was injected from the concentric annulus convergent to the axis of the oxidizer flow, impinging on the oxidizer shortly after injection.

A large variety of geometrical and operational parameters were evaluated at the NASA LeRC with impinging fuel concentric-orifice elements.[57–60, 62–65, 67–72] An investigation changing propellant injection areas at constant mixture ratio[57] found that reducing the fuel-to-oxidizer injection area ratio was stabilizing because the hydrogen injection temperature at instability was reduced. The effects of injection velocity parameters for these data are shown in Fig. 6.[57] A fuel-to-oxidizer injection velocity ratio of 6.5 and not absolute hydrogen injection velocity or fuel-to-oxidizer velocity differential was found to best define a combustion stability boundary regardless of injection temperature. This boundary held

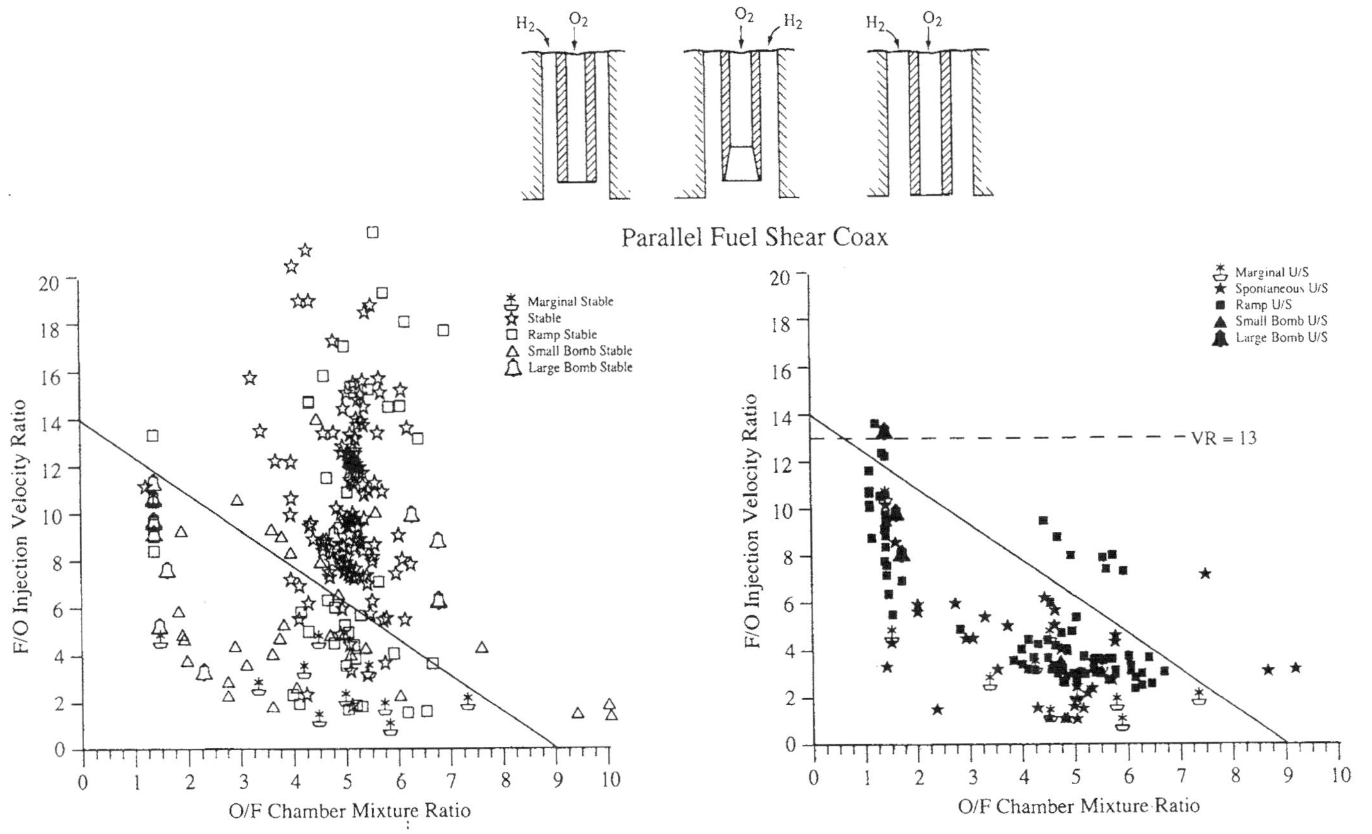

Fig. 4 Combustion stability mapping of injection velocity ratio and chamber mixture ratio for shear coaxial elements with parallel fuel (data from Refs. 9, 22, 29, 44, 58, 61, and 72).

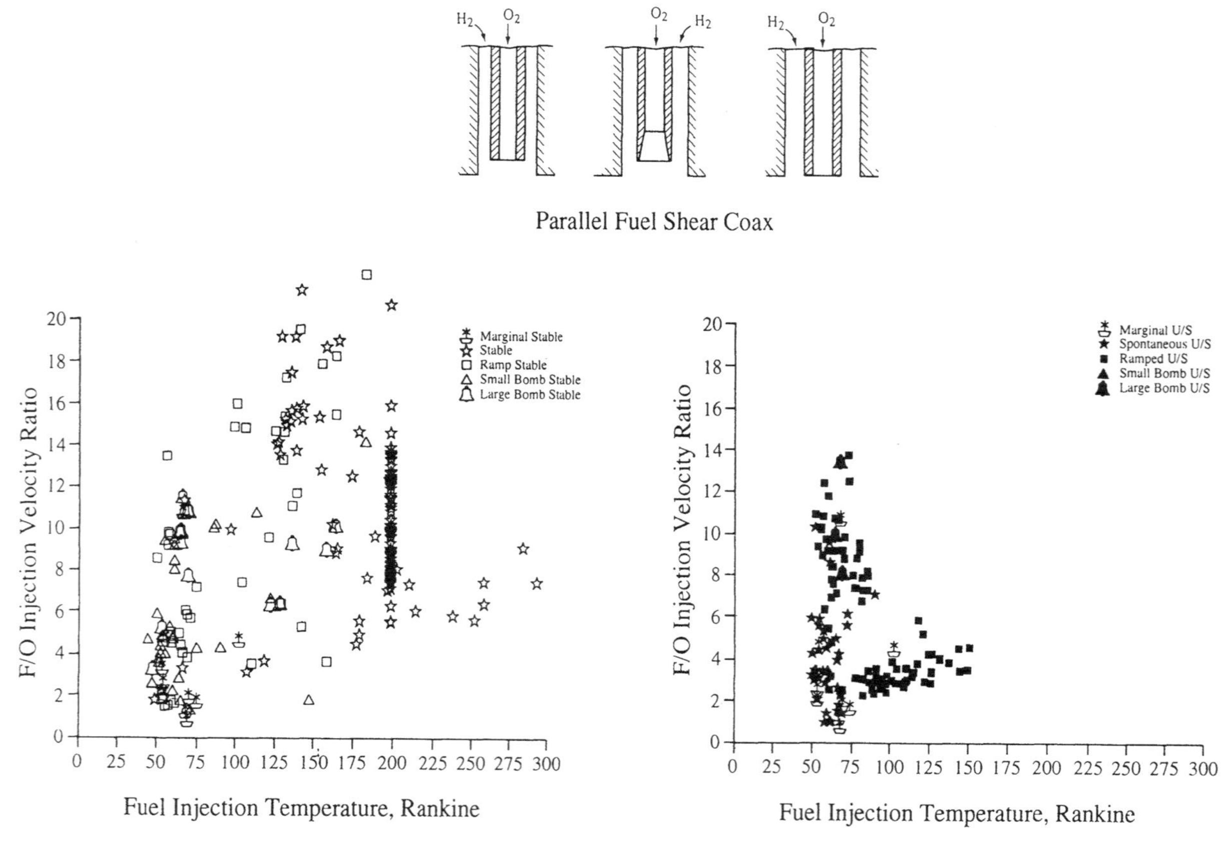

Fig. 5 Combustion stability mapping of injection velocity ratio and hydrogen injection temperature for shear coaxial elements with parallel fuel (data from Refs. 9, 22, 29, 44, 58, 61, and 72).

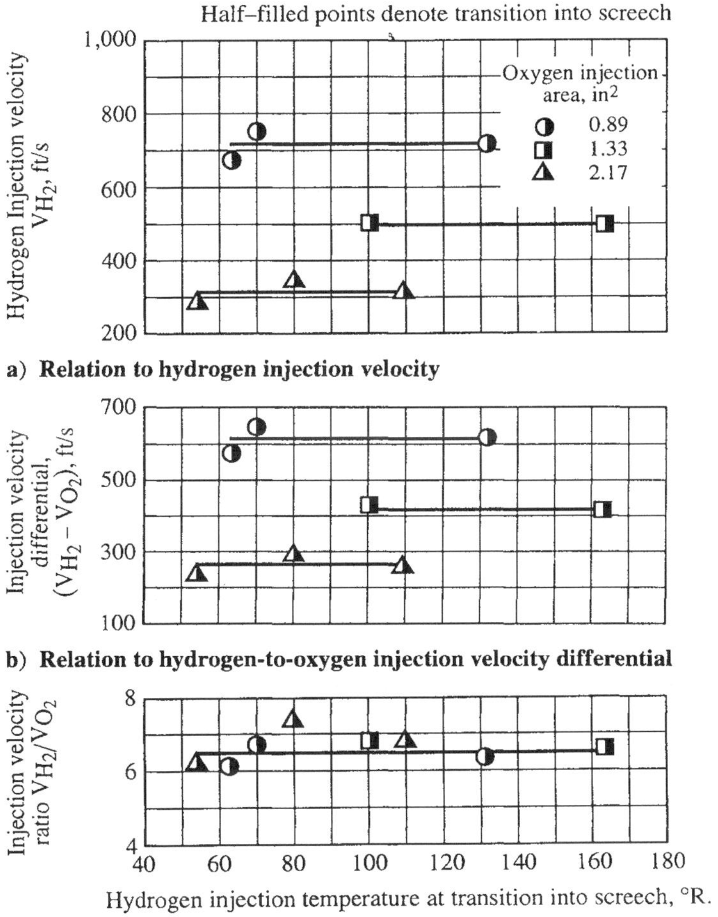

a) **Relation to hydrogen injection velocity**

b) **Relation to hydrogen-to-oxygen injection velocity differential**

c) **Relation to hydrogen-to-oxygen injection velocity ratio**

Fig. 6 Effects of injection velocity parameters on hydrogen injection temperature at instability for shear coaxial elements with impinging fuel, at chamber mixture ratio = 5 (Ref. 57).

for straight-bore, taper-reamed, and counterbored oxidizer tubes but was determined at constant chamber pressure, mixture ratio, flow-per-element, and chamber contraction ratio.[57]

In subsequent testing at the NASA LeRC, other factors were found that clearly influenced this velocity ratio stability boundary. Recessing the oxidizer tube was always stabilizing by lowering the hydrogen injection temperature at instability for fixed injection area injectors,[57,72] for example, the one shown in Fig. 7.[72] Characteristic velocity performance was also improved. The increase in hydrogen density associated with lower temperatures reduced the critical velocity ratio from 6.5 to about 4.0 at the recessed liquid oxygen tube exit plane for the same mixture ratio. These results were consistent with the J-2 engine injector development, where recessing was stabilizing for both mainstage and start transient conditions.[29]

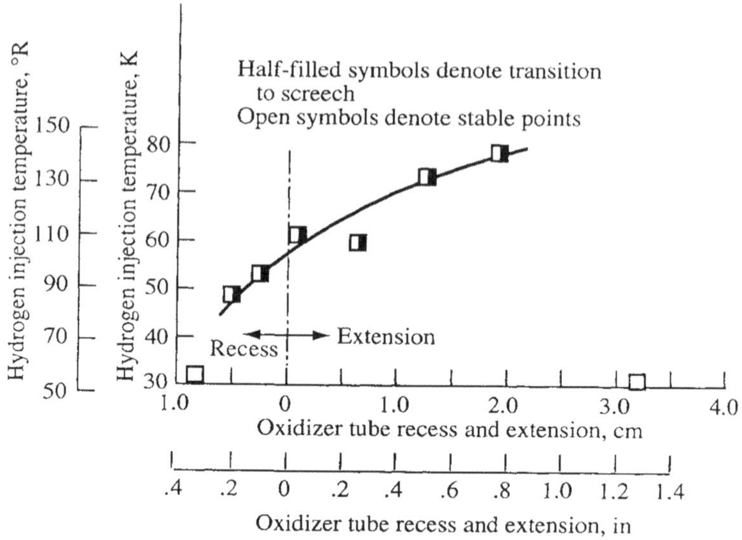

Fig. 7 Effect of oxidizer tube recess on hydrogen injection temperature at instability for shear coaxial elements with impinging fuel.[72]

Extended, rather than recessed, oxidizer tubes were stabilizing by lowering the hydrogen injection temperature at instability[62]; characteristic velocity performance in this case was reduced. This result was not consistent with J-2 engine injector development testing[29] nor with later testing at the NASA LeRC.[72]

Higher weight flow per element (or thrust per element) was found to be stabilizing in many studies.[58,67,68] In Ref. 58, all 11-kN thrust-per-element injectors (about 3.4-kg/s weight flow per element) were found to be stable, although low performing, even at velocity ratios as low as 1.0 and near-liquid hydrogen temperatures. In Ref. 68, lower thrust-per-element injectors were hydrogen temperature ramped unstable, as shown in Fig. 8,[68] but higher thrust-per-element injectors could not

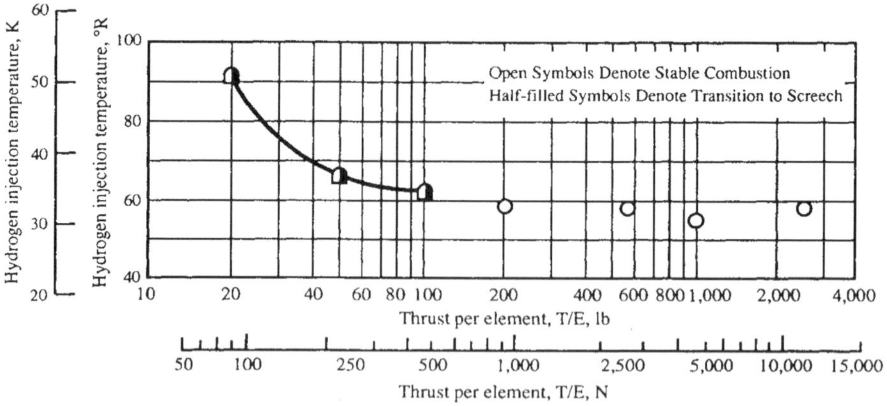

Fig. 8 Effect of thrust per element on hydrogen injection temperature at instability for shear coaxial elements with impinging fuel, at chamber pressure = 1380–2070 KPa (200–300 psia) and chamber contraction ratio = 1.89 (Ref. 68).

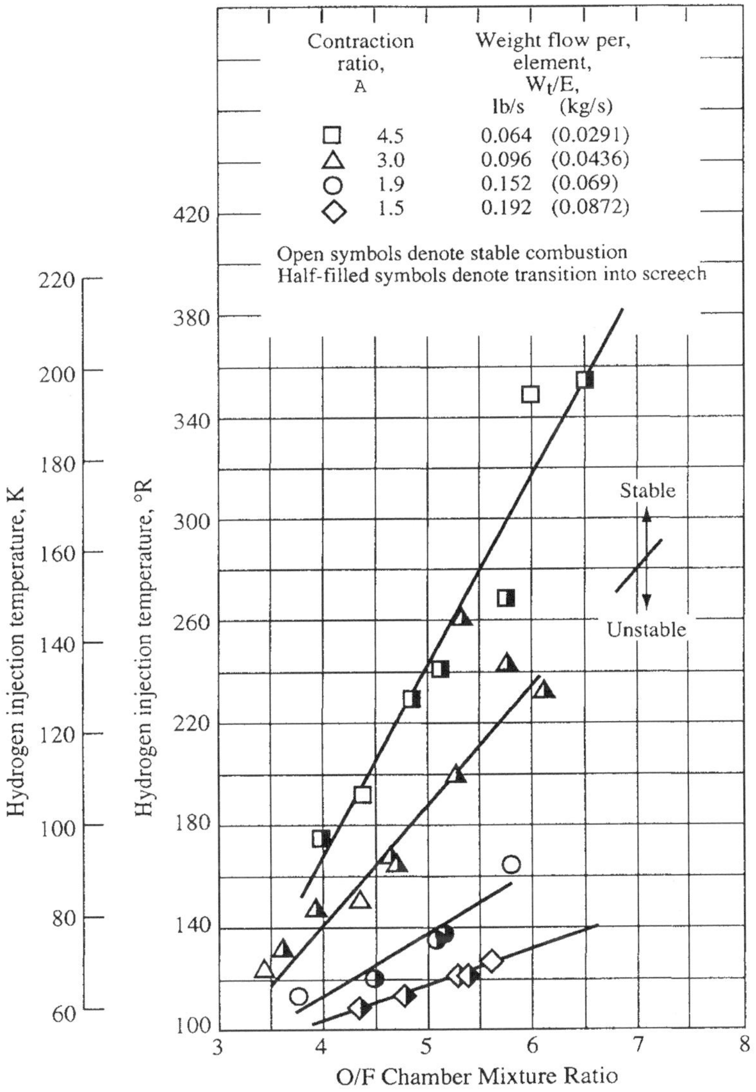

Fig. 9 Effects of weight flow per element and chamber contraction ratio on hydrogen injection temperature at instability for shear coaxial elements with impinging fuel, at chamber pressure = 2070 KPa (300 psia) (Ref. 67).

even be pulsed unstable. Weight flow per element combined with injection velocity ratio correlated reasonably well for this engine size.[68] In Ref. 67, higher weight flows per element were stabilizing, as shown in Fig. 9,[67] although weight flow per element was increased at constant chamber pressure with reduced chamber contraction ratios. Reducing chamber contraction ratio and chamber pressure at constant flow rate per element was found to be stabilizing, as shown in Fig. 10.[67] Throttling chamber pressure by lowering flow rate per element at constant chamber contraction ratio was found to have no effect on stability.[67] The combined effect

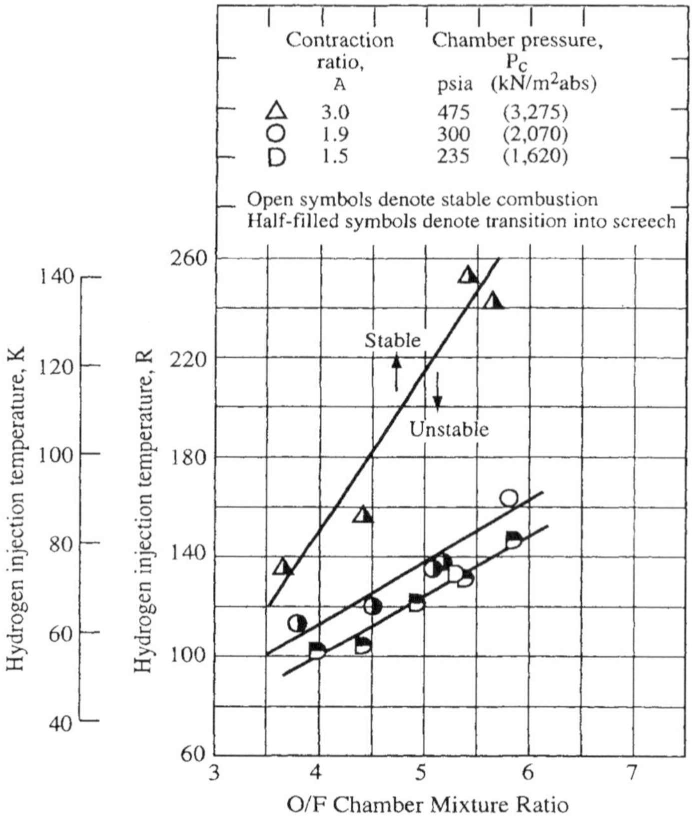

Fig. 10 Effects of chamber pressure and chamber contraction ratio on hydrogen injection temperature at instability for shear coaxial elements with impinging fuel, at weight flow per element = 0.069 kg/s (0.15 lbm/s) (Ref. 67).

of these results, along with the stabilizing aspect of weight flow per element, was that increasing contraction ratio was destabilizing.

The test series from Ref. 67 obtained some of the highest unstable hydrogen injection temperatures ever measured for concentric-orifice element injectors, with one test unstable at 195 K. The consistent conditions at instability were low-flow rate per element, high-contraction ratio, and high-mixture ratio.

In other testing at the NASA LeRC, reducing the oxidizer injection temperature was found to be stabilizing.[67] Increasing the impingement angle of the annular hydrogen on the oxidizer jet was found to be stabilizing with no effect on characteristic velocity performance.[72] There was no discernible effect on stability due to hydrogen annulus concentricity.[72] Use of a porous faceplate instead of a solid faceplate was found to be stabilizing.[62] The addition of combustion chamber periphery hydrogen film cooling was marginally destabilizing.[62] The reduction, however of the radial distribution of active elements, i.e., increasing the distance from the outer row of elements to the chamber wall, was also found to be destabilizing, independent of thrust chamber diameter and the number of injection elements.[70] A lower hydrogen injection temperature at instability for the smallest radial distribution

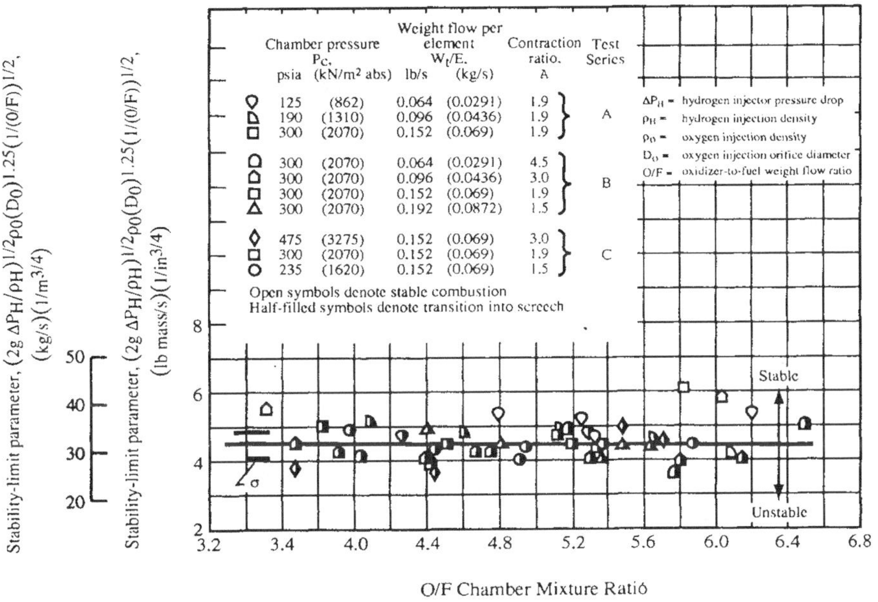

Fig. 11 Stability correlation of NASA LeRC 421 element injector with shear coaxial elements with impinging fuel.[67]

(largest gap between elements and the chamber wall) was considered the result of a change in transverse mode of instability from first tangential to first radial.[70] Decreasing hydrogen injection temperature was also found to be destabilizing to the longitudinal mode in a continuously variable length combustor.[71]

Various correlations have been formulated to explain the variations in these data. One correlation developed from a hydrogen response model[78] is shown in Fig. 11.[67] This correlation was initially generated to help explain the stabilizing effect of recessing the oxidizer tubes. Successful correlation of the data suggested that the contribution of the hydrogen system to stability was due to the hydrogen injection pressure drop,[69] which agreed with the hydrogen response model.[78] Variations in geometrical and operational parameters all affected stability by influencing hydrogen flow resistance[67] or pressure drop.[69] The contribution of the oxygen system to stability was suggested to be due to oxygen density and jet diameter,[67] as shown in Fig. 11. Measurements however, of the periodic behavior of the jet shape during instability[55,56] suggested instead the significant influence of the overall jet dynamics.[79]

An extensive stability investigation with impinging fuel concentric-orifice injection elements was conducted at Aerojet at high-chamber pressure (up to 18 MPa) and hydrogen injection temperatures from 33 K to 69 K.[4–7] Stability rating was conducted with tangential pulsing in the combustion chamber. The measured effects of injection velocity ratio and mixture ratio on combustion stability, suggested as a portrayal of combustion and acoustic processes, are shown in Fig. 12 for low-chamber pressures[7] (7 MPa). Similar to Fig. 2, regions of pulse sensitive operation were identified. At higher chamber pressures (17 MPa), the stable region was enlarged, which was attributed to reduced relative pulse overpressures and the

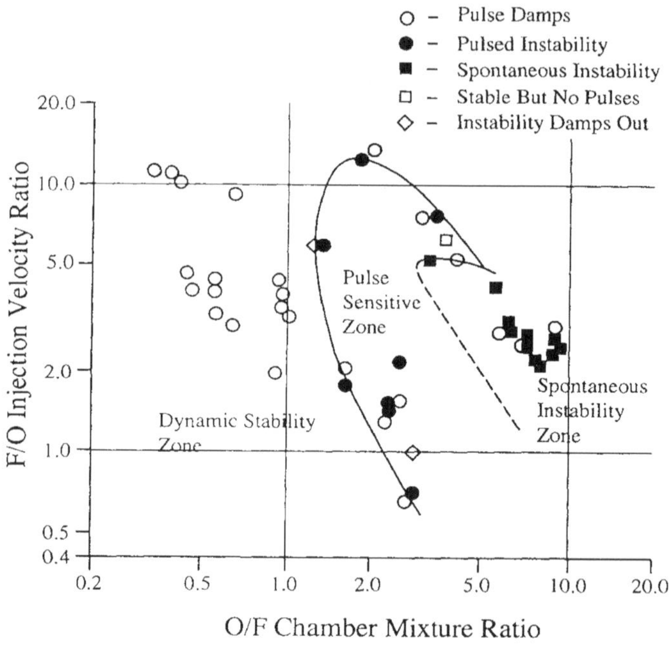

Fig. 12 Effects of injection velocity ratio and chamber mixture ratio on stability for shear coaxial elements with impinging fuel, at chamber pressure = 6200–8270 KPa (900–1200 psia).[7]

stabilizing effects at increased contraction ratio. This result contradicts Fig. 10. Since both higher chamber pressure and lower mixture ratio were obtained with contraction ratio increase, however, these results are not exactly comparative.

Impinging fuel concentric-orifice element stability data from the NASA LeRC[58, 62, 64, 67, 68, 70–72] and Aerojet[4–7] were compiled to compare with the parallel fuel shear coaxial data. The significantly increased geometrical and operational variations in the impinging fuel tests increased the maximum hydrogen injection temperature at instability from 83 K in Fig. 5 to 195 K.

The maximum unstable fuel-to-oxidizer velocity ratio at main chamber mixture ratios, however, has increased only from 10 to 13, as shown in Fig. 13, although most of the instabilities are still less than 8, similar to Fig. 4. The high-velocity ratio instabilities have occurred with much warmer hydrogen injection temperatures than in Fig. 5, mostly due to Ref. 67 data as discussed. The smaller number of stable tests in Fig. 13 than in Fig. 4 is due to the nature of investigative stability testing. For each unstable test due to temperature ramping, however, there are implied stable regions at immediately higher velocity ratios, which are not shown in Fig. 13, but would significantly increase the number of stable data.

2. *Swirled Liquid Oxidizer Jet (Swirl Coax)*

In many concentric-orifice element injectors tested in engine development and investigative programs, the liquid oxidizer was swirled before injection. Swirling was created either by free vortex flow from tangential injection at the entry of the oxidizer tube, in which the swirling flow could be hollow, or by forced vortex flow

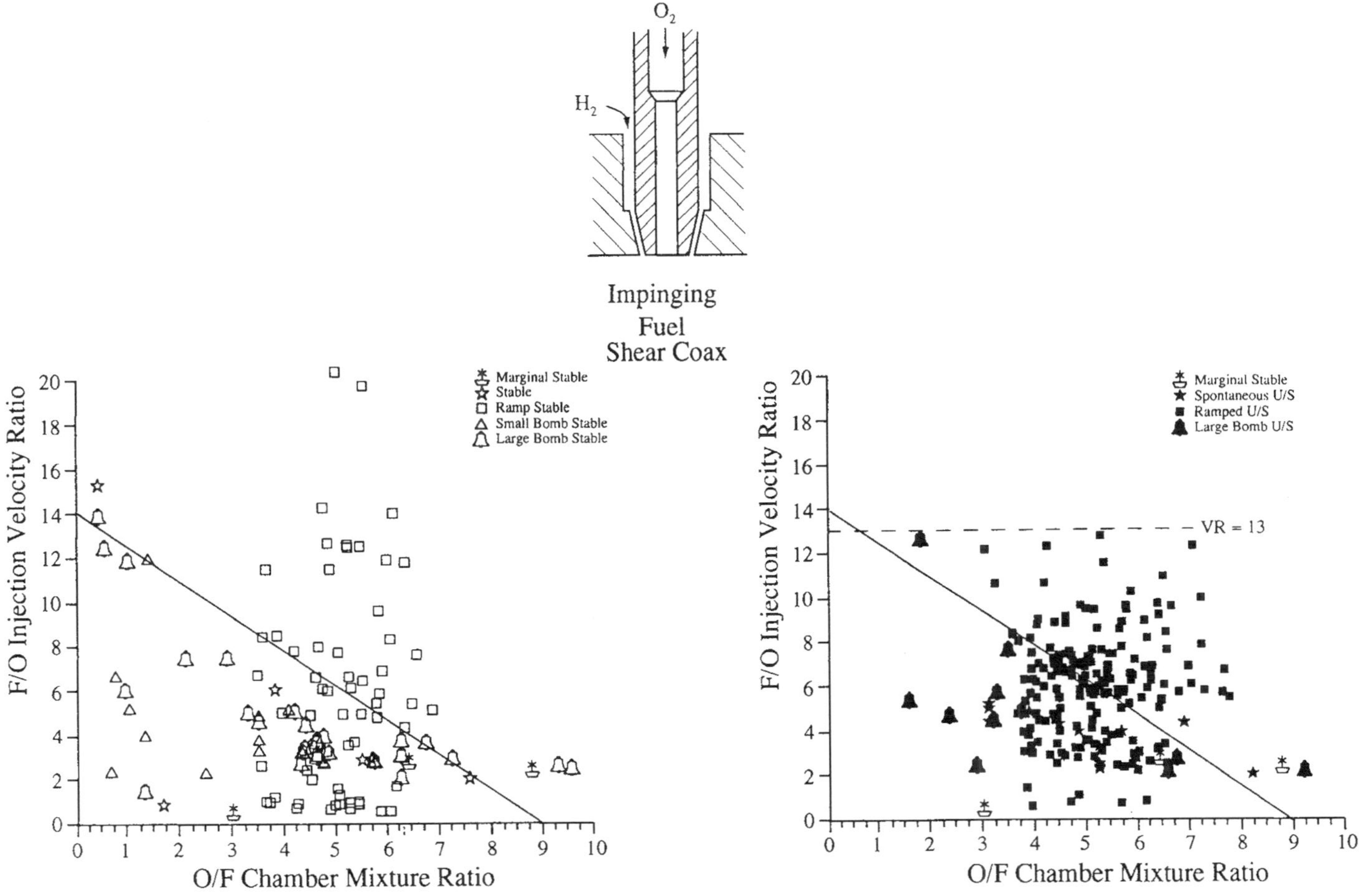

Fig. 13 Combustion stability mapping of injection velocity ratio and chamber mixture ratio for shear coaxial elements with impinging fuel (data from Refs. 4–7, 57, 58, 62, 64, 67, 68, and 70–72).

through a helical ribbon inside the oxidizer tube, in which the swirling flow could be either full or hollow, depending on geometry and the intensity of swirl. This design is known conventionally as a swirl coax. Injection discharge coefficients as low as 0.2 are possible with highly swirled injection. The RL10A-3 engine is the only flight engine with swirl coaxial elements, containing helical ribbon swirlers in all elements except those on the inner row and the outer row near the combustion chamber wall.[18]

Stability test data with swirled jet concentric-orifice element injectors were compiled from the J-2 engine injector development,[29] NASA LeRC,[58] Pratt and Whitney Aircraft,[21,25,26] and STME development subscale injector tests.[16,27,74] In all these hardware designs the hydrogen was injected parallel to the axis of the oxidizer flow.

Two high-temperature instabilities occurred with a helical swirler installed at the oxidizer tube exit.[21] All other instabilities occurred on the J-2 engine injector development,[29] where a variety of oxidizer swirled elements were tested but with the swirl never at the oxidizer tube exit.

A stability mapping of mixture ratio and fuel-to-oxidizer injection velocity ratio is shown in Fig. 14. This figure appears functionally similar to Fig. 4, with instabilities concentrated at low-velocity ratio and low-mixture ratio. Instabilities at main chamber mixture ratios have occurred at lower-velocity ratios than on Fig. 4, however, indicating a larger stable region, albeit one with different element operating characteristics and fewer data points.

B. Phenomena with Plain-Orifice Element Injectors

Although they have never appeared on flight production engines, plain orifice element injectors (also known as multiorifice element or ring manifold injectors) have been widely tested with liquid oxygen/hydrogen propellants. Many programs included this type of element because of manufacturing simplicity or prior experience from design and fabrication of liquid/liquid propellant injectors. Between 1958 and 1972, over 1000 plain-orifice element liquid oxygen/hydrogen tests were conducted in the United States. About one-quarter of these were from the J-2 engine injector development,[28,29] one-quarter from Aerospike development,[34–40] one-quarter from various programs at Aerojet,[2–8,10–15] and the others from various early programs at the NASA LeRC[46–49,51] and Pratt and Whitney Aircraft.[19,20]

About half of the tests were conducted with an oxidizer showerhead element, with either a hydrogen showerhead or a like-on-like doublet element. Of the remaining half, about two-thirds were conducted with OFO triplets, and the rest conducted with pentads (four-on-one OOFOO), like-on-like oxidizer doublets with various hydrogen injection methods, and slot elements.

There are sufficient data to compare the stability characteristics of these elements with concentric-orifice elements and, hence, compare the influences of the propellant state and the element types. The following discussion is divided into two categories based on whether the oxidizer is injected as a showerhead jet or self-impinges in some form.

1. *Showerhead Liquid Oxidizer Jets*

Showerhead oxidizer jet element test data were compiled from J-2 engine injector development[29] and various programs at Aerojet.[2,5–8] Hydrogen was injected

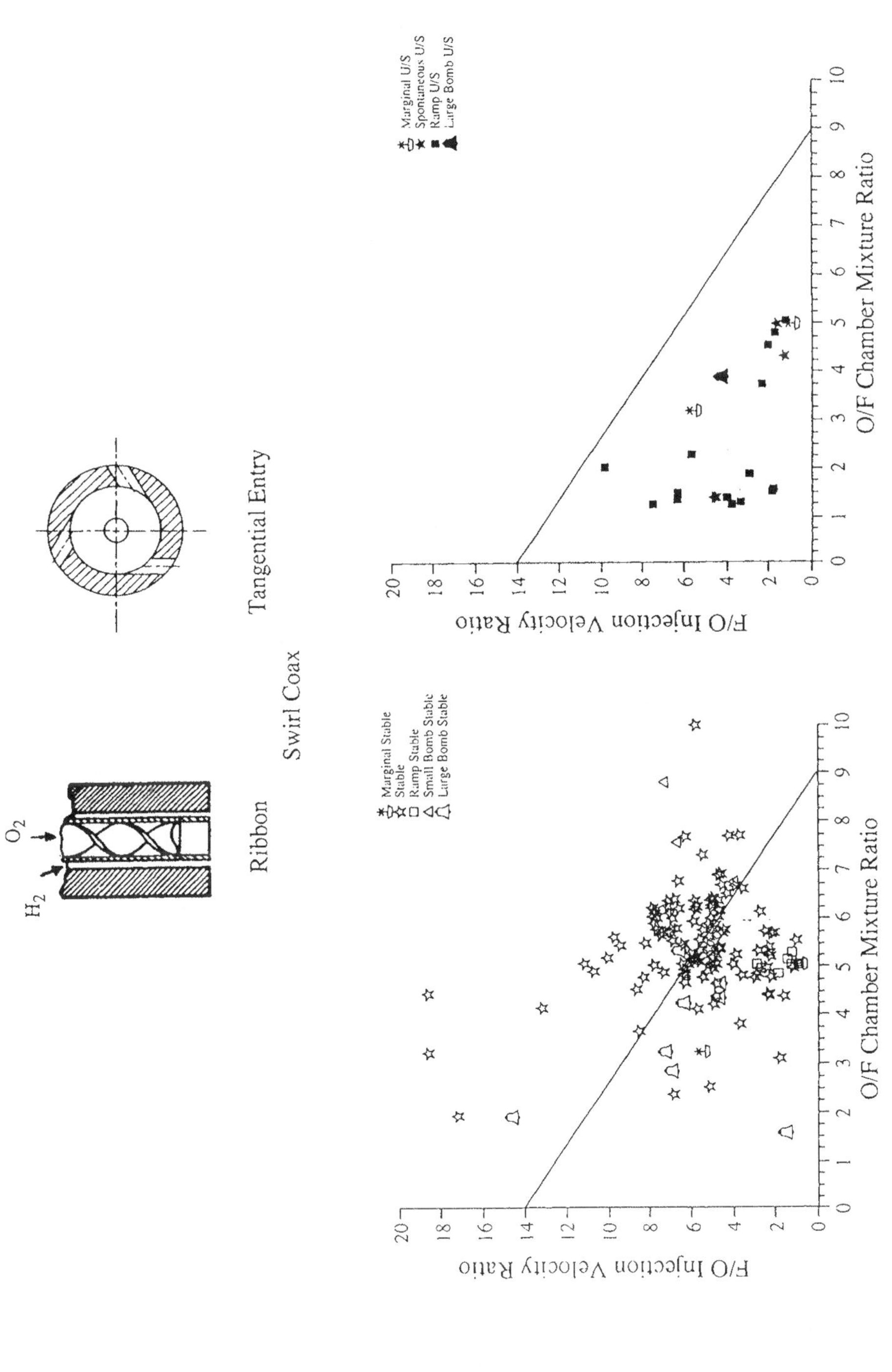

Fig. 14 Combustion stability mapping of injection velocity ratio and chamber mixture ratio for swirl coaxial elements with parallel fuel (data from Refs. 16, 21, 25–27, 29, 58, and 74).

mostly with showerheads and like-on-like doublets. Low-mixture ratio and low-hydrogen injection temperature were found to be less stable. A stability mapping with mixture ratio and injection velocity ratio is shown in Fig. 15. The test data are clearly separated into two regions, with instability occurring at low-mixture ratio and low-velocity ratio, exactly the same conditions found previously for concentric-orifice injector elements shown in Figs. 4, 13, and 14. This connection suggests the mechanisms for instability are similar between showerhead and concentric-orifice elements.

2. *Impinging Liquid Oxidizer Jets*

Test data for OFO triplet elements were compiled from the J-2 engine injector development,[29] the combustor effects on performance program,[43] and the stability characterization program,[10,11] which used the transverse excitation chamber to generate stability data. A stability mapping with mixture ratio and injection velocity ratio is shown in Fig. 16. Instabilities have occurred at high mixture ratios, high hydrogen injection temperatures, and high injection velocity ratios (up to 25). These conditions are significantly different from any of the unstable conditions presented previously, suggesting the instability mechanisms for this element are different from those for concentric-orifice elements or showerhead elements. This conclusion will be revisited with data from Aerospike development testing[34–40] and early NASA LeRC testing,[47,48] which were not included in Fig. 16.

Like-on-like oxidizer doublets were tested during the J-2 engine injector development,[29] and various Aerojet programs.[2,8,12,14,15] Insufficient data are available for meaningful presentation in the previous manner. The extensive use of like-on-like doublets with liquid/liquid propellants, however, suggests a comparison with these data. A comparison with combustion stability of storable propellant engines with like-on-like injection elements, using an empirical stability correlation developed by Hewitt (in Ref. 80), is shown in Fig. 17. The characteristic stability parameter d/V, represents the sensitive or less volatile propellant, which usually is the fuel, but in this case is the liquid oxygen. The slanted lines represent the empirical instability of the storable propellant engine stability data, adjusted for oxygen/hydrogen speed of sound. Correlation of the liquid oxygen/hydrogen stability is found to be good, suggesting the instability mechanism of the like-on-like element is consistent with different propellants and not dependent on the propellant state.

C. Combustion Chamber Stabilization Devices

Stabilization (or energy dissipation) devices have not had as extensive use or development with engines using liquid oxygen/hydrogen propellants as with other propellant types, because of the early recognition of stable configurations. As with other rocket engine systems, however, the addition of damping devices can improve combustion stability in liquid oxygen/hydrogen engines, especially for those with generally poor stability characteristics. Some stability devices, however, will not improve the stability margin of engines with already superior stability characteristics.

Radial baffles used in early injector configurations during J-2 engine injector development improved the combustion stability of injectors with poor stability characteristics at low-hydrogen injection temperatures, for both mainstage and

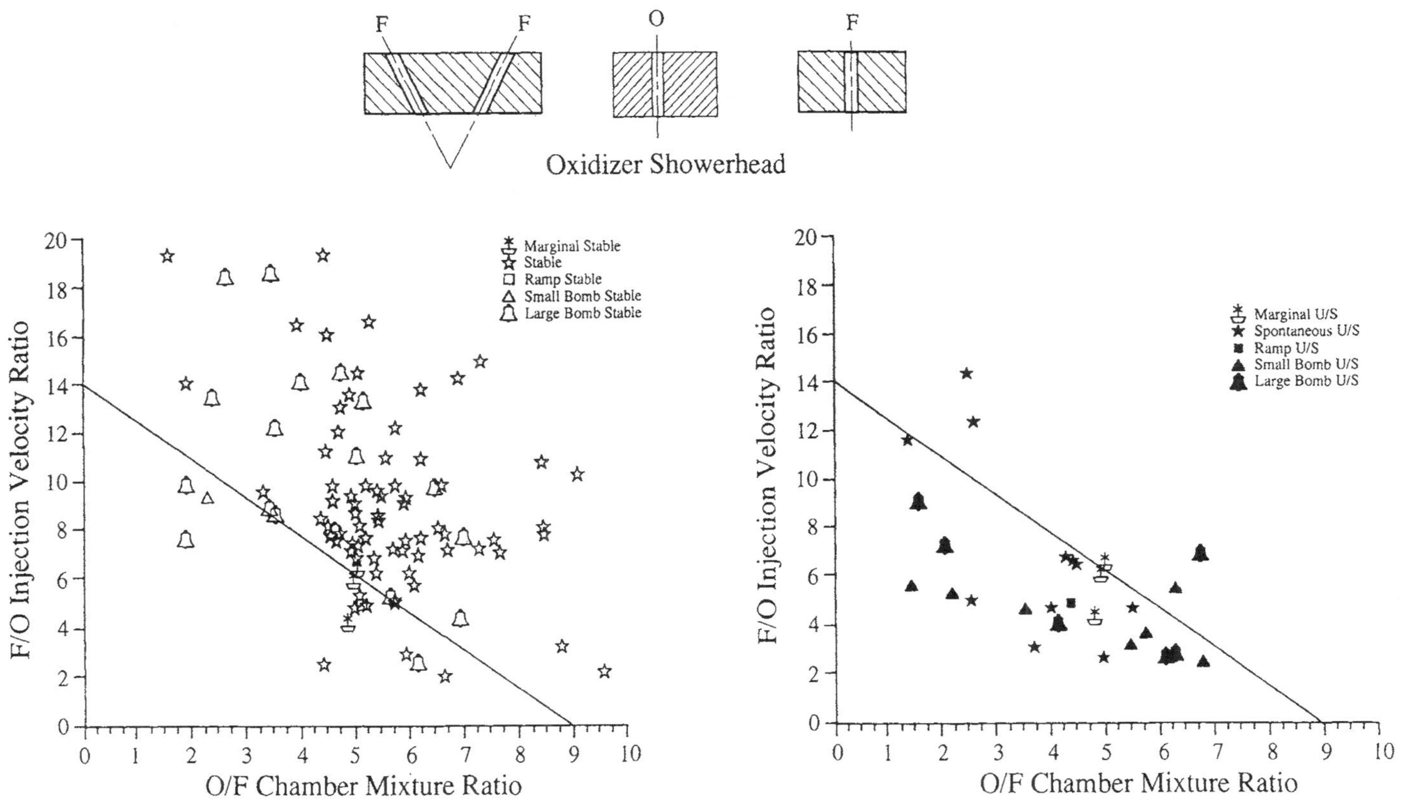

Fig. 15 Combustion stability mapping of injection velocity ratio and chamber mixture ratio for showerhead oxidizer jet elements (data from Refs. 2, 5–8, and 29).

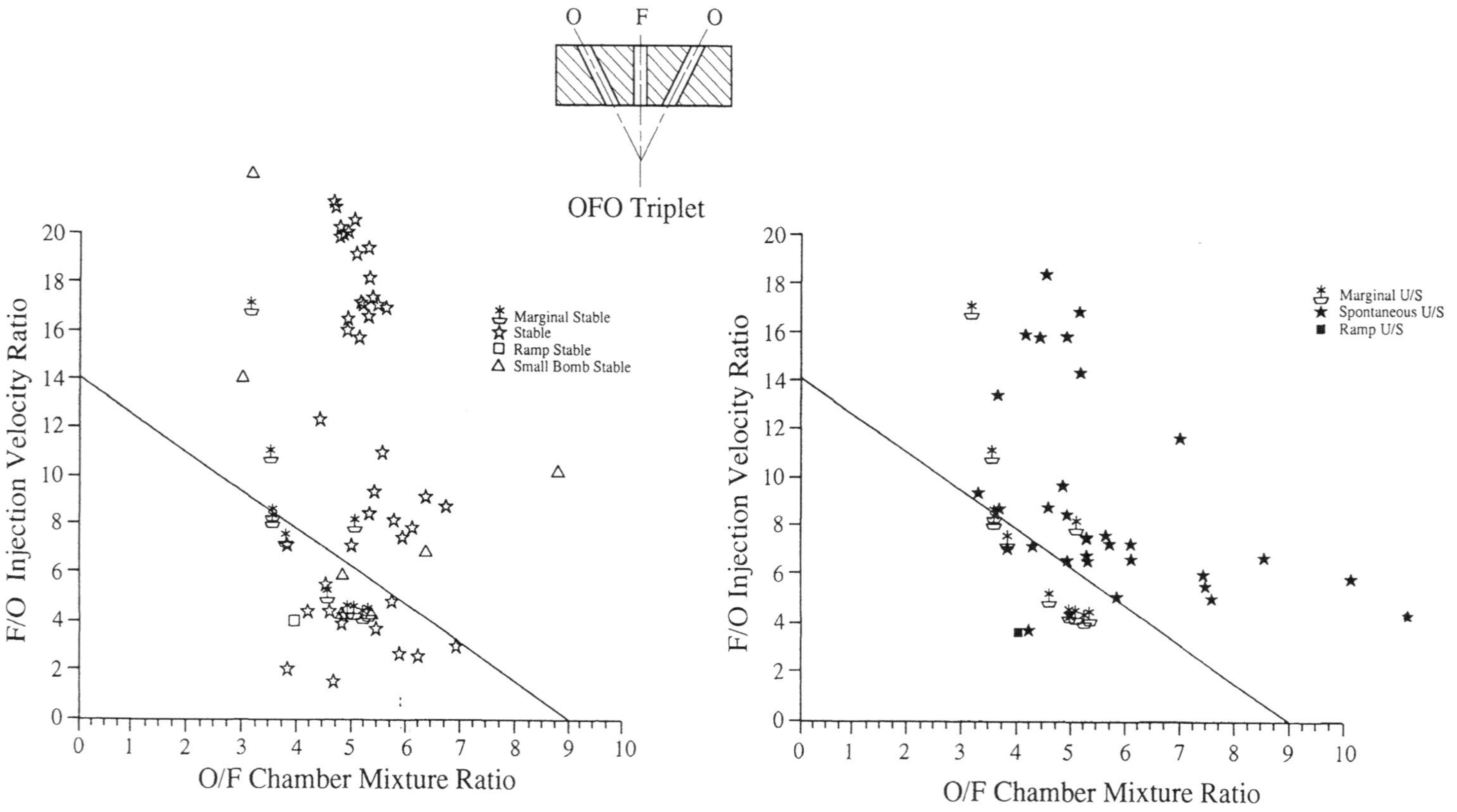

Fig. 16 Combustion stability mapping of injection velocity ratio and chamber mixture ratio for OFO triplet elements (data from Refs. 10, 11, 29, and 43).

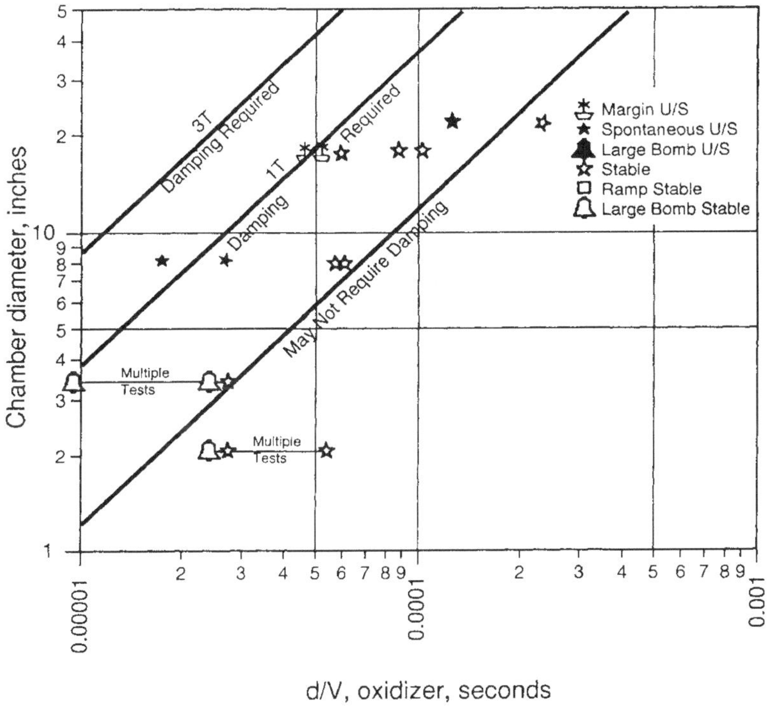

Fig. 17 Liquid oxygen/hydrogen propellant engine stability with like-on-like oxidizer doublet elements, using Hewitt correlation for storable propellant engine stability (data from Refs. 2, 8, 12, 14, 15, and 29).

simulated start transient conditions.[29] Baffles tested at the NASA LeRC[65] also reduced the hydrogen injection temperature at instability for a more unstable injector, for example, the one shown in Fig. 18. A variety of baffle patterns successfully stabilized this injector.

The presence of long or very short baffles, however, did not change the hydrogen injection temperature at instability for the M-1 main injector.[9] A temperature of 44 K was unstable from temperature ramping with both 10-cm-long baffles (10% of the chamber diameter) and 1.5-cm-long baffles (1.5% of the chamber diameter). The peak-to-peak amplitudes of the chamber pressure oscillations during instability, however, were reduced by 50–85% with the longer baffles. On the SSME main injector, the presence of baffles did not influence dynamic stability.[75] These results agreed with experience on J-2 engine injector development with injectors with already superior stability characteristics.[29]

Both Pratt and Whitney Aircraft[23,24] and NASA LeRC[59] showed stabilizing aspects of combustion chamber wall acoustic liners in liquid oxygen/hydrogen propellant engines. Liners typically included an axial array of Helmholtz resonators. At Pratt and Whitney Aircraft, an RL10 injector modified for liquid hydrogen injection that was unstable without a liner was stabilized with a liner.[23] At NASA LeRC, a liner reduced the hydrogen injection temperature at instability, as shown in Fig. 19.[59] The presence of the Helmholtz resonators on the SSME main injector, however, did not affect dynamic stability.[44,75]

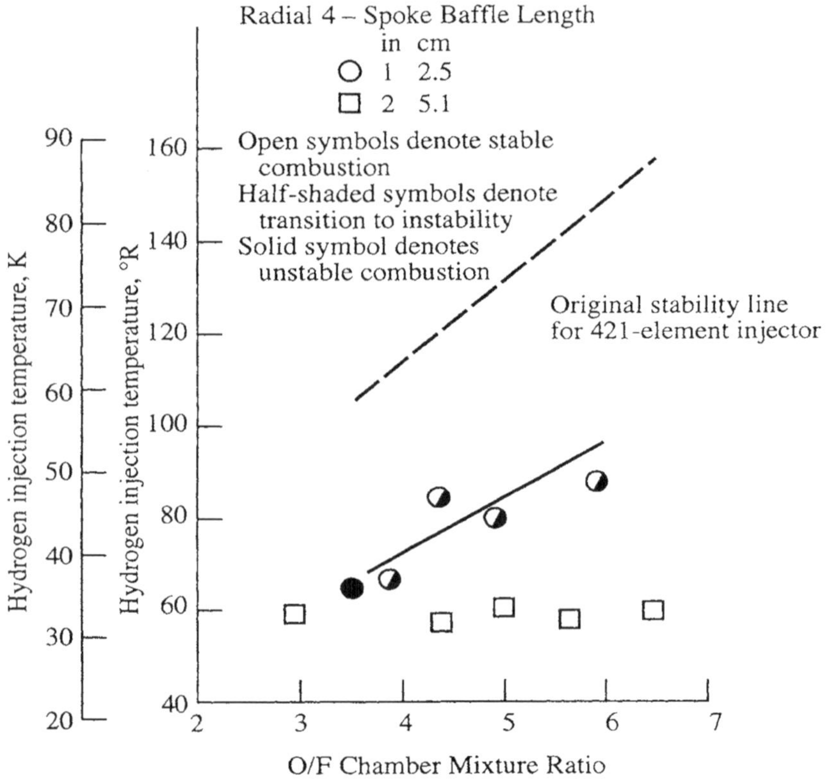

Fig. 18 Effect of baffles on hydrogen injection temperature at instability for shear coaxial elements with impinging fuel.[65]

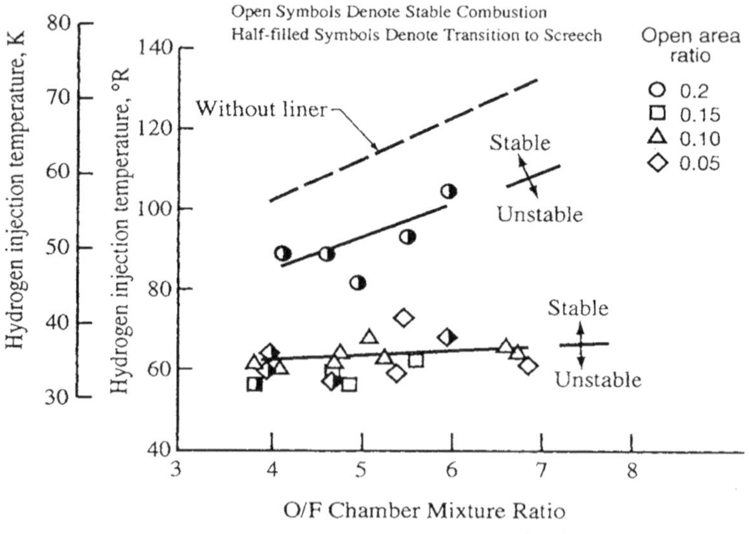

Fig. 19 Effect of various acoustic liners on hydrogen injection temperature at instability for shear coaxial elements with impinging fuel.[59]

Finally, a divergence ring from the injector face was found to be stabilizing on the J-2 engine injector development for injectors with poor stability characteristics during the start transient.[29] For more stable injectors, it made much less difference.

IV. Conclusions

The combustion stability history of liquid oxygen/hydrogen propellant rocket engines in the United States is largely represented by the use of concentric-orifice element injectors, which were found to be consistently unstable at off-design conditions with low-hydrogen injection temperatures. All flight engine injectors use concentric-orifice elements with hydrogen transpiration-cooled faces, and all operate with hydrogen injection temperatures in excess of 110 K.

Stability boundaries determined by concentric-orifice element injector testing were initially defined by hydrogen-to-oxygen injection velocity ratio. All flight engine injectors operate with hydrogen-to-oxygen injection velocity ratios in excess of about 10. In this chapter, a comparison of concentric-orifice element injector testing suggested the maximum hydrogen-to-oxygen velocity ratio to develop instability to be about 13. These occurred mostly in the J-2 engine injector development testing at low-chamber pressure and mixture ratio, with a possibility of two-phase hydrogen flow, and in NASA LeRC testing with low-flow rate per element, high-nozzle contraction ratio, and high-mixture ratio. These tests at NASA LeRC provided the most radical departure from the general characteristics, generating instabilities with hydrogen injection temperatures much higher than normal operation for the flight engines, and much higher than usual for investigative testing.

Stabilization at lower velocity ratios was determined in engine development and technology investigations to be possible with many means, including oxidizer tube recessing, higher flow rate per element, higher hydrogen pressure drop, impinging the hydrogen on the oxidizer, and swirling the oxidizer. These results led to the development of stability correlations at NASA LeRC defined by hydrogen injection response, influenced mostly by hydrogen pressure drop. Later, oxygen response and chamber responses were added for more complete formulations of combustion stability.

The general success of concentric-orifice elements in the early engine development precluded widespread later use of plain-orifice elements, except for investigative programs. An empirical examination of the stability characteristics of plain-orifice elements was included in this chapter to attempt to separate the effects of the injection elements from those of the propellant state. The instability phenomena with plain-orifice element injectors appeared to depend on the element type. Showerhead element instability characteristics appeared similar to concentric-orifice element characteristics, especially regarding injection velocity ratio. OFO triplet characteristics were different from concentric-orifice elements, with capability for instability at high injection velocity ratios. Like-on-like oxidizer doublet data compared well with an empirical stability correlation developed for storable propellant injectors.

Stability aids in the combustion chamber have improved the stability of more unstable concepts but generally have not improved the stability margin of more stable concepts for liquid oxygen/hydrogen propellant rocket engines.

References

[1]Osborn, G. H., and Stinnett, W. D., "Development of a 3000-Lb Thrust Liquid-Hydrogen–Liquid-Oxygen Rocket Motor," Aerojet Engineering Corp., Rept. RTM-54, Contract NOa(s) 8496, Azusa, CA, Aug. 29, 1949.

[2]Dodge, C. H., Kirincich, R. S., and Shurley, L. A., "Feasibility of a High Energy Fuel as a Rocket Propellant, Volume 3," Aerojet General Corp., Rept. 0215-07-F, Contract AF33(616)-5990, Sacramento, CA, Jan. 30, 1959.

[3]Miller, B. C., "Cryogenic Staged-Combustion Feasibility Study, Final Report," Aerojet General Corp., Rept. 2485, Contract NAS 8-2644, Azusa, CA, Feb. 28, 1963.

[4]Anon., "High Chamber Pressure Operation for Launch Vehicle Engines, Final Report," Aerojet General Corp., Rept. 4008-F-1, Contract NAS8-4008, Sacramento, CA, April 15, 1963.

[5]Anon., "High Chamber Pressure Operation for Launch Vehicle Engine Program, Final Report," Aerojet General Corp., Rept. 4008-SA4-F, Contract NAS8-4008, Supplemental Agreement No. 4, Sacramento, CA, June 1, 1964.

[6]Anon., "An Experimental Investigation of Combustion Stability Characteristics at High Chamber Pressure, Interim Report," Aerojet General Corp., Rept. 11741-I, Contract NAS 8-11741, Sacramento, CA, Sept. 5, 1965.

[7]Anon., "An Experimental Investigation of Combustion Stability Characteristics at High Chamber Pressure, Final Report, Phase II," Aerojet General Corp., Rept. 11741/SA6-F, Vol. 1, Contract NAS8-11741, Sacramento, CA, Aug. 25, 1966.

[8]Ito, J. I., "Development of LO2/LH2 Gas Generators for the M-1 Engine, Technology Report," Aerojet General Corp., NASA CR 5-4812, AGC 8800-59, Contract NAS 3-2555, Sacramento, CA, June 1, 1966.

[9]Barsotti, R. J., Datsko, S. C., Louison, R., Kovach, R. J., Miller, D. J., and Pulliam, W. P., "Development of Liquid Oxygen/Liquid Hydrogen Thrust Chamber for the M-1 Engine, Technology Report," Aerojet General Corp., NASA CR-54813, AGC 9400-5, Contract NAS3-2555, Sacramento, CA, May 15, 1968.

[10]Anon., "Stability Characterization of Advanced Injectors, Final Report on Phase I," Aerojet General Corp., Rept. 20672-P1F, Contract NAS 8-20672, Sacramento, CA, Oct. 25, 1968.

[11]Anon., "Design Guide for Stable H2/O2 Combustors, Volume 1: Design Application," Aerojet General Corp., Rept. 20672-P2D, Contract NAS 8-20672, Sacramento, CA, May 1970.

[12]Heinicke, D. R., "NERVA Injector Development For CY 63," Aerojet General Corp., Rept. TCR 9660-002, Contract SNP-1, Sacramento, CA, Jan. 1964.

[13]Anon., "Design Report of PHOEBUS-2 Multi-orifice Injector," Aerojet General Corp., Rept. RP-TM-0012 Space Nuclear Propulsion Office, Cleveland Extension, Contract SNPC-35, Sacramento, CA, Feb. 1967.

[14]Anon., "Space Shuttle Main Engine Phase C/D Program Definition Documentation, Engine Design Definition Report, Volume III, Combustion Devices," Aerojet Liquid Rocket Co., DR No. SE-275, Contract NAS 8-26188, Sacramento, CA, April 21, 1971.

[15]Blubaugh, A. L., and Schoenman, L., "Extended Temperature Range ACPS Thruster Investigation, Final Report," Aerojet Liquid Rocket Co. CR-134655, Contract NAS 3-16775, Sacramento, CA, Aug. 1, 1974.

[16]Hulka, J., Schneider, J. A., and Dexter, C. E., "Performance and Stability of a Booster Class LOX/H2 Swirl Coaxial Element Injector," AIAA Paper 91-1877, June 1991.

[17]Harrje, D. T., and Reardon, F. H. (eds.), "Liquid Propellant Rocket Combustion Instability," NASA SP-194, 1972, p. 359.

[18]Mayers, J. L., "Design Report for RL10A-3 Rocket Engine," Pratt and Whitney Aircraft PWA FR-324, Contract NAS 8-2690, West Palm Beach, FL, Jan. 31, 1962.

[19]Anon., "Final Report for the Period 26 June 1961 to 26 June 1962 on the High Pressure Rocket Engine Feasibility Program," Pratt and Whitney Aircraft PWA FR-467, Contract AF 04(611)-7435, West Palm Beach, FL, Sept. 27, 1962.

[20]Kah, C. L. C., and Lewis, G. D., "High Pressure Rocket Engine Feasibility Program, Final Report," Pratt and Whitney Aircraft, PWA FR-1171, Contract AF 04(611)-7435, West Palm Beach, FL, Dec. 10, 1964.

[21]Anon., "Summary Report on an Investigation of Combustion Instability for Liquid and Cold Gas Hydrogen Propellants," Pratt and Whitney Aircraft PWA FR-1005, Contract NAS8-11024, West Palm Beach, FL, June 19, 1964.

[22]Anon., "Investigation of Combustion Instability with Liquid Oxygen and Liquid or Cold Gaseous Hydrogen Propellants, Final Report, Phase II," Pratt and Whitney Aircraft PWA FR-1374, Contract NAS8-11024, West Palm Beach, FL, May 14, 1965.

[23]Anon., "A Summary of the Suppression of Combustion Oscillations with Mechanical Damping Devices, Summary Report of Tasks III through VI," Pratt and Whitney Aircraft PWA FR-1115, Contract NAS8-11038, West Palm Beach, FL, Nov. 28, 1964.

[24]Anon., "A Study of the Suppression of Combustion Oscillations with Mechanical Damping Devices, Phase II Summary Report," Pratt and Whitney Aircraft, PWA FR-1922, Contract NAS8-11038, West Palm Beach, FL, July 15, 1966.

[25]Atherton, R. R., "Advanced Cryogenic Rocket Engine Program Staged-Combustion Concept, Final Report, Volumes I through III," Pratt and Whitney Aircraft Tech. Rept. AFRPL-TR-67-298, Contract AF 04(611)-11401, West Palm Beach, FL, Dec. 1967.

[26]Atherton, R. R., "Air Force Reusable Rocket Engine Program, XLR129-P-1, Final Report, Volume I," Pratt and Whitney Aircraft, Tech. Rept. AFRPL-TR-71-1, Contract F04611-68-C-0002, West Palm Beach, FL, Jan. 1971.

[27]Peterson, E. L., Rozelle, R., and Borgel, P. J., "Characterization and Wall Compatibility Testing of a 40 K Pound Thrust Class Swirl Coaxial Injector and Calorimeter Combustion Chamber," AIAA Paper 91-1873, June 1991.

[28]Baker, D., and Grant, K., "Summary Report of Exploratory Liquid Oxygen and Liquid Hydrogen Thrust Chamber Tests," Rocketdyne Rept. R-1396, Rept. WADC-TR-59-6, Contract AF 33(616)-6102, Canoga Park, CA, Jan. 1959.

[29]Anon., "Program Quarterly Progress Reports," Rocketdyne Repts. R-2600-1–R-2600-19, Contract NAS8-19, Canoga Park, CA, Dec. 1960–May 1965.

[30]Anon., "Development History of the 200,000- and 225,000-Pound-Thrust J-2 Rocket Engines, Final Report," Rocketdyne Rept. R-6700, Contract NAS8-19, Canoga Park, CA, Dec. 13, 1966.

[31]Anon., "Development History of the 230,000-Pound-Thrust J-2 Rocket Engine, Final Report," Rocketdyne Rept. R-6800, Contract NAS8-19, Canoga Park, CA, Sept. 21, 1967.

[32]Anon., "4400-Hz Vibration Investigation, Final Report," Rocketdyne Rept. R-8742, Contract NAS8-25156, Canoga Park, CA, June 30, 1971.

[33]Combs, L. P., and Hoehn, F. W., "Steady-State Rocket Combustion of Gaseous Hydrogen and Liquid Oxygen, Part I: Experimental Investigation," Rocketdyne Research Rept. 64-24, Canoga Park, CA, June 1964.

[34]Anon., "Advanced Engineering Program System and Dynamics Investigation (Aerospike), Final Report, 1 January 1966 to 15 August 1967 (Volume II)," Rocketdyne Rept. R-7167, Contract NAS8-19, Canoga Park, CA, Aug. 15, 1967.

[35]Lary, F. B., "Advanced Cryogenic Rocket Engine Program, Aerospike Nozzle Concept, Final Report, Volume II," Rocketdyne Tech. Rept. AFRPL-TR-67-280, Contract AF 04(611)-11399, Canoga Park, CA, Jan. 1968.

[36]Anon., "Advanced Engine Aerospike Experimental Program, Interim Report (1 July 1967 to 31 May 1968), Volume II," Rocketdyne Rept. R-7475, Canoga Park, CA, Contract NAS8-20349, July 15, 1968.

[37]Anon., "Advanced Engine Aerospike Experimental Program, Thrust Chamber Performance (Volume 2: Final Report)," Rocketdyne Rept. R-7760, Canoga Park, CA, Contract NAS8-20349, July 8, 1969.

[38]Anon., "O_2/H_2 Advanced Maneuvering Propulsion Technology Program, Water-Cooled Segment Testing, Final Report," Rocketdyne Rept. R-8906, Rept. AFRPL-TR-72-25, Contract FO4611-67-C-0116, Canoga Park, CA, April 1972.

[39]Huang, D. H., "O_2/H_2 Advanced Maneuvering Propulsion Technology Program, Regeneratively Cooled Segment/Sector Evaluation," Rocketdyne Rept. R-9281, Rept. AFRPL-TR-73-84, Contract FO4611-67-C-0116, Canoga Park, CA, Dec. 1973.

[40]Sobin, A. J., and Martinez, A., "Linear Rocket Engines for Advanced Space Transportation Systems," AIAA Paper 75-1251, Oct. 1975.

[41]Lauffer, J. R., "Toroidal Combustion Chamber Evaluation Final Report," Rocketdyne Rept. R-7633, Contract NAS8-21038, Canoga Park, CA, Dec. 2, 1968.

[42]Goalwin, D. S., "A High-Pressure Regeneratively Cooled Thrust Chamber, Volume 1: Analysis, Test, and Evaluation," Rocketdyne Tech. Rept. AFRPL-TR-68-226, R-7646, Contract AF 04(611)-67C-0093, Canoga Park, CA, Aug. 1969.

[43]Nurick, W. H., and Hines, W. S., "Experimental Investigation of Combustor Effects on Rocket Thrust Chamber Performance, Final Report," Rocketdyne Rept. R-9315, Contract NAS W-2106, Canoga Park, CA, Sept. 1973.

[44]Pieron, R. A., "Verification Complete Package, SSME Thrust Chamber Assembly Combustion Stability Testing," Rocketdyne Rept. RSS-303-30, Contract NAS 8-27980, Canoga Park, CA, June 10, 1977.

[45]Anon., "Design Verification Complete Package, Hot Gas Temperature," Rocketdyne Rept. DVS-SSME-305E, Contract NAS 8-27980, Oct. 23, 1979.

[46]Auble, C. A., "A Study of Injection Processes for Liquid Oxygen and Gaseous Hydrogen in a 200-Pound-Thrust Rocket Engine," Lewis Flight Propulsion Lab., NACA RM E56I25a, Jan. 9, 1957.

[47]Baker, L., Jr., and Steffen, F. W., "Screaming Tendency of the Gaseous-Hydrogen–Liquid-Oxygen Propellant Combination," Lewis Flight Propulsion Lab., NACA RM E58E09, Sept. 30, 1958.

[48]Heidmann, M. F., and Baker, L. Jr., "Combustor Performance with Various Hydrogen-Oxygen Injection Methods in a 200-Pound-Thrust Rocket Engine," Lewis Flight Propulsion Lab., NACA RM E58E21, Sept. 30, 1958.

[49]Rothenberg, E. A., Kutina, F. J., and Kinney, G. R., "Experimental Performance of Gaseous Hydrogen and Liquid Oxygen in Uncooled 20,000-Pound-Thrust Rocket Engines," NASA Memorandum 4-8-59E, March 1959.

[50]Hersch, M., "Combined Effect of Contraction Ratio and Chamber Pressure on the Performance of a Gaseous-Hydrogen–Liquid-Oxygen Combustor for a Given Propellant Weight Flow and Oxidant-Fuel Ratio," NASA Lewis Research Center, NASA TN D-129, Feb. 1961.

[51]Drain, D. I., Schum, H. J., and Wasserbauer, C. A., "Relations of Combustion Dead Time to Engine Variables for a 20,000-Pound-Thrust Gaseous-Hydrogen Liquid Oxygen Rocket Engine," NASA Lewis Research Center, NASA TN D-851, June 1961.

[52]Hersch, M., "Effect of Interchanging Propellants on Rocket Combustor Performance with Coaxial Injection," NASA Lewis Research Center, NASA TN D-2169, Feb. 1964.

[53]Conrad, E. W., Hannum, N. P., and Bloomer, H. E., "Photographic Study of Liquid-Oxygen Boiling and Gas Injection in the Injector of a Chugging Rocket Engine," NASA Lewis Research Center, NASA TM X-948, Dec. 1964.

[54]Wanhainen, J. P., Antl, R. J., Hannum, N. P., and Mansour, A. H., "Throttling Characteristics of a Hydrogen-Oxygen, Regeneratively Cooled, Pump-Fed Rocket Engine," NASA Lewis Research Center, NASA TM X-1043, Dec. 1964.

[55]Heidmann, M. F., "Oxygen-Jet Behavior During Combustion Instability in a Two-Dimensional Combustor," NASA Lewis Research Center, NASA TN D-2725, March 1965.

[56]Heidmann, M. F., "Oscillatory Combustion of a Liquid-Oxygen Jet with Gaseous Hydrogen," NASA Lewis Research Center, NASA TN D-2753, March 1965.

[57]Wanhainen, J. P., Parish, H. C., and Conrad, E. W., "Effect of Propellant Injection Velocity on Screech in 20,000-Pound Hydrogen-Oxygen Rocket Engine," NASA Lewis Research Center, NASA TN D-3373, April 1966.

[58]Hannum, N. P., and Conrad, E. W., "Performance and Screech Characteristics of a Series of 2500-Pound-Thrust-per-Element Injectors for a Liquid-Oxygen–Hydrogen Rocket Engine," NASA Lewis Research Center, NASA TM X-1253, July 1966.

[59]Wanhainen, J. P., Bloomer, H. E., Vincent, D. W., and Curley, J. K., "Experimental Investigation of Acoustic Liners to Suppress Screech in Hydrogen-Oxygen Rockets," NASA Lewis Research Center, NASA TN D-3822, Feb. 1967.

[60]Heidmann, H. F., Sokolowski, D. E., and Diehl, L. A., "Study of Chugging Instability with Liquid-Oxygen and Gaseous-Hydrogen Combustors," NASA Lewis Research Center, NASA TN D-4005, June 1967.

[61]Scott, H. E., Bloomer, H. E., and Mansour, A. H., "M-1 Engine Subscale Injector Tests," NASA Lewis Research Center, TN D-4053, July 1967.

[62]Wanhainen, J. P., Hannum, N. P., and Russell, L. M., "Evaluation of Screech Suppression Concepts in a 20,000-Pound Thrust Hydrogen-Oxygen Rocket," NASA Lewis Research Center, NASA TM X-1435, Aug. 1967.

[63]Hersch, M., and Rice, E. J., "Gaseous-Hydrogen–Liquid-Oxygen Rocket Combustion at Supercritical Chamber Pressures," NASA Lewis Research Center, NASA TN D-4172, Sept. 1967.

[64]Bloomer, H. E., Wanhainen, J. P., and Vincent, D. W., "Chamber Shape Effects on Combustion Instability," NASA Lewis Research Center, NASA TM X-52361, Oct. 1967.

[65]Hannum, N. P., Bloomer, H. E., and Goelz, R. R., "Stabilizing Effects of Several Injector Face Baffle Configurations on Screech in a 20000-Pound-Thrust Hydrogen-Oxygen Rocket," NASA Lewis Research Center, NASA TN D-4515, April 1968.

[66]Dankhoff, W. F., Johnsen, I. A., Conrad, E. W., and Tomazic, W. A., "M-1 Injector Development—Philosophy and Implementation," NASA Lewis Research Center, NASA TN D-4730, Aug. 1968.

[67]Wanhainen, J. P., Feiler, C. E., and Morgan, C. J., "Effect of Chamber Pressure, Flow Per Element, and Contraction Ratio on Acoustic-Mode Instability in Hydrogen-Oxygen Rockets," NASA Lewis Research Center, NASA TN D-4733, Aug. 1968.

[68]Salmi, R. J., Wanhainen, J. P., and Hannum, N. P., "Effect of Thrust Per Element on Combustion Stability Characteristics of Hydrogen-Oxygen Rocket Engines," NASA Lewis Research Center, NASA TN D-4851, Oct. 1968.

[69]Conrad, E. W., Bloomer, H. E., Wanhainen, J. P., and Vincent, D. W., "Interim Summary of Liquid Rocket Acoustic-Mode-Instability Studies at a Nominal Thrust of 20,000 Pounds," NASA Lewis Research Center, NASA TN D-4968, Dec. 1968.

[70]Wanhainen, J. P., and Morgan, C. J., "Effect of Injection Element Radial Distribution and Chamber Geometry on Acoustic-Mode Instability in a Hydrogen Oxygen Rocket," NASA Lewis Research Center, NASA TN D-5375, Aug. 1969.

[71]Morgan, C. J., and Sokolowski, D. E., "Longitudinal Instability Limits with a Variable-Length Hydrogen-Oxygen Combustor," NASA Lewis Research Center, NASA TN D-6328, April 1971.

[72]Hannum, N. P., Russell, L. M., Vincent, D. W., and Conrad, E. W., "Some Injector Element Detail Effects on Screech in Hydrogen-Oxygen Rockets," NASA Lewis Research Center, NASA TM X-2982, Feb. 1974.

[73]Bailey, C. R., "A Preliminary Investigation of Oxidizer-Rich Oxygen-Hydrogen Combustion Characteristics," NASA MSFC, TN D-3729, Dec. 1966.

[74]Elam, S. K., "Subscale LOX/Hydrogen Testing with a Modular Chamber and a Swirl Coaxial Injector," AIAA Paper 91-1874, June 1991.

[75]Dennis, H. J., Jr., Hutt, J., and Nesman, T., "Stability Testing of a Modified Space Shuttle Main Engine," AIAA Paper 91-1983, June 1991.

[76]Baker, D. I., "Experimental Investigation of Gaseous Hydrogen and Liquid Oxygen in a Rocket Motor of 100-Pound Thrust," Jet Propulsion Lab., Rept. 4-48, Contract W-04-200-ORD-455, California Inst. of Technology, Pasadena, CA, Jan. 14, 1949.

[77]Johnston, H. L., and Doyle, W. L., "Development of the Liquid Hydrogen-Liquid Oxygen Propellant Combination for Rocket Motors," Ohio State Univ., Tech. Rept. 7, Research Foundation, Dec. 1951.

[78]Feiler, C. P., and Heidmann, M. F., "Dynamic Response of Gaseous-Hydrogen Flow System and Its Application to High-Frequency Combustion Instability," NASA Lewis Research Center, NASA TN D-4040, June 1967.

[79]Heidmann, M. F., and Groeneweg, J. F., "Analysis of the Dynamic Response of Liquid Jet Atomization to Acoustic Oscillations," NASA Lewis Research Center, NASA TN D-5339, July 1969.

[80]Anderson, W. E., Ryan, H. M., and Santoro, R. J., "Combustion Instability Phenomena of Importance to Liquid Bi-Propellant Rocket Engines," Pennsylvania State Univ., Propulsion Engineering Research Center, Rept. TR-91-009, Univ. Park, PA, Oct. 1991.

Chapter 3

Instability Phenomena in Liquid Oxygen/Hydrocarbon Rocket Engines

Jeffrey A. Muss*
University of California, Berkeley, Berkeley, California 94720

I. Introduction

THE earliest liquid rocket engines used liquid oxygen (LOX) as the oxidizer and various hydrocarbons (HC) as fuels. The propellants were readily available (liquid hydrogen was not), and the theoretical performance characteristics calculated by Olberth in 1917 indicated HC fuels were ideal for booster applications.[1] Goddard's first successful rocket flight in 1926 was powered using the combination of LOX and gasoline.[1] The German V-2 of World War II, the first production liquid rocket, used LOX and an aqueous ethanol solution (75% ethanol by volume) as propellants.[1,2] V-2s captured at the close of World War II served as building blocks for both American and Soviet liquid rocket development programs during the early Cold War period of the late 1940s and early 1950s.[1] The heavy-lift requirements of the early ballistic missile programs necessitated the replacement of the aqueous ethanol fuel with a kerosene petroleum distillate in both countries (known as RP-1 in the United States and T-1 in the former Soviet Union). By the early 1960s, however, storable propellant combinations, such as A-50/N_2O_4, had replaced LOX/RP-1 in many of the American missiles since the vehicle could be stored in a fueled configuration. The original LOX/RP-1 versions of the missiles did, however, serve as the building blocks for the launch vehicles that would open space to mankind. The American Saturn, Delta, and Atlas launch vehicles all use the LOX/RP-1 propellant combination in the high-thrust first-stage booster engines. In the Soviet Union, LOX/RP-1 has been used almost exclusively for launching large payloads. It is the propellant combination for both booster and upper stages of the Soyuz and Proton launch systems.[3,4] The development of LOX/RP-1 engines in the United States essentially ceased with the cancellation of the Apollo space project in the late 1960s and the subsequent development and activation of the LOX/H_2 powered Space Shuttle in the 1970s. The notable exception was Rocketdyne's adaptation of the H-1 engine for use as a booster for the Delta space launch vehicle.[5]

---

*Research Assistant, Department of Mechanical Engineering.

One of the first clear observations of combustion instability in liquid fueled rockets was noted in 1955 in conjunction with the development of the engines for the Thor and Atlas ballistic missiles.[5–7] Prior to this time, many rocket engines were damaged or destroyed during tests and launches,[2,8] but the cause of the failure is not clearly attributed to the feedback coupling between the fluid mechanics and the combustion process that is currently categorized as combustion instability. Detailed studies of the problems of combustion instability and solution techniques accompanied the development of the liquid rocket engines for Jupiter/Thor, Atlas, Titan II, and Saturn 1B and V.[5,9,10] Recent reports released from the former Soviet Union indicate that similar problems accompanied the development of many of their liquid rocket systems.[4]

Although these propulsion systems were successful, their development was a long, arduous task. The development process for LOX/RP-1 systems in particular has been characterized by a struggle to achieve stable combustion while maintaining high combustion performance. Often the stability and performance were acceptable in development versions of the engine only to become wildly unstable during engine qualification or flight operation. The American Atlas (MA-1)/Thor(MB-1) and F-1 and the Soviet RD-0110 engines were all plagued by long, costly re-engineering programs.[4,10] The final designs were as much the result of trial and error as an application of firm understanding of the underlying physics. In spite of this apparent drawback, LOX/RP-1 is still considered an attractive propellant combination for booster applications due to its high theoretical specific impulse, relative to storable propellants, and exceptional bulk density relative to other propellant combinations.

In this chapter, the recent advances in the understanding of combustion characteristic of liquid-fueled rocket engines using LOX/hydrocarbon (LOX/HC), particularly LOX/RP-1, propellants are reviewed. The state of knowledge as captured in NASA SP-194 is reviewed, with a special eye toward combustion characteristics and design recommendations for LOX/HC engine systems. The results from recent experimental programs focused on performance and combustion stability issues are chronicled. The development of critical analytical injector design models is discussed. These models facilitate the relationship between injector design features and performance and stability characteristics. As will be shown, recent results indicate that the stability characteristics of oxidizer-fuel-oxidizer(O-F-O) triplet elements are better than previously thought, especially in high thrust LOX/RP-1 applications. The triplet elements will be shown to provide high performance with manageable stability, in contrast to like-doublet injectors where performance goals are often compromised for stability.

II. Understanding of NASA SP-194

In 1967, Princeton University, under contract to NASA and the U.S. Armed Forces, began assembling a comprehensive reference book on the state-of-the-art knowledge on liquid rocket combustion, with a focus on combustion instability.[7] Published in 1972, NASA SP-194 is the most thorough work on combustion instability in liquid rocket engines to date. The present section summarizes the state of knowledge at the time of SP-194's publication with a special focus on aspects of LOX/HC combustion. Recent advances in the understanding of LOX/HC combustion stability characteristics, particularly LOX/RP-1 combustion characteristics, will then be discussed. Information on developments in the former Soviet

Union available in the open literature are included, although little information is available about the stability characteristics of the staged-combustion LOX/RP-1 systems currently in use. This discussion will not include information from recent LOX/CH_4 studies[11–13] since the engine configurations tend to more closely resemble LOX/H_2 systems.

SP-194 is a comprehensive compilation of combustion instability knowledge. It includes both a history of combustion instability problems in liquid rocket engine development and approaches to the solution of these problems. The chapters discuss the origins and categorization of combustion instabilities, theoretical models to describe stability behavior, guidelines to preclude unstable designs, and experimental techniques for evaluating the stability characteristics of engines.

A historical review of SP-194 indicates that there were two major impediments to a fuller understanding of the relationship between design features and combustion stability characteristics. Without this critical information, the ability to rapidly affect changes in engine designs in response to combustion instability problems was severely hindered. The limited computational power of the early 1970s severely limited the ability to apply all but the simplest models to practical engines. These limitations precluded running detailed parametric studies to understand the effect of input parameters on stability model results. The other significant shortfall was the absence of mechanistic models, or even good correlations, to describe the influence of injector element and thrust chamber design features on the response characteristics of the combustion process. The absence of such models limited the development of clear causal relationships between design and operating conditions and stability behavior. It should be clearly stated from the beginning that these problems affected the development of all rocket engine systems and were not confined to LOX/HC systems.

The following subsections will focus on the understanding contained in SP-194 as it relates to the causal relationship between design features and combustion stability behavior, with a special interest in the relation between injector element characteristics.

A. Causes of Combustion Instabilities

At the time of the writing of SP-194, it was clearly understood that combustion instability was the result of coupling between the fluid mechanics (acoustics) in the combustion chamber and the combustion process. The Rayleigh criterion was recognized to be the mechanism through which the combustion could supply the energy to sustain flow oscillations. It was known that a system was required to dissipate oscillatory energy more quickly than it is supplied at all frequencies to have a stable design.

Instabilities were categorized as nonacoustic or acoustic. Nonacoustic or chug instabilities are those where the characteristic wavelength is significantly longer than any characteristic dimension of the combustor. Chug instabilities had been successfully modeled using a lumped parameter chamber and total time lags, although time lags were usually inferred after the fact from engine test data. Acoustic instabilities, both intermediate and high frequency, were assumed to be supported predominantly through combustion mechanisms, although coupling with the feed system was acknowledged as a major driving mechanism for intermediate frequency instabilities. Some of the proposed mechanisms for coupling between combustion and chamber pressure oscillations included resurge, fluctuating drop vaporization and chemical kinetic rates, and oscillations in the primary and secondary atomization processes. A number of models were developed to describe these processes, but computational complexity usually constrained analyses to investigating the effect of only one phenomenon at a time. It is interesting to note that high-frequency coupling between pressure oscillations and the atomization process was attributed to the interaction between combustion gas velocity perturbations and the atomizing propellant,[14] and most analyses ignored propellant injection flow rate fluctuations resulting from chamber pressure oscillations.

It was felt that stabilization could be achieved by modifying either the combustion process or the chamber acoustics and flowfield. The objective of these changes was to decouple the combustion process from the chamber acoustics, at least enough so that the oscillatory energy input was not sustained. The acoustic field could be modified using a combination of thrust chamber baffles and acoustic cavities. The combustion process could be modified by changing the axial distribution of liquid-phase propellants, distributing the propellant away from pressure antinodes, or isolating liquid streams from the region of vigorous combustion. A common technique used to modify the combustion distribution was to increase the vapor fraction of the most volatile propellant by increasing its injection velocity while decreasing the velocity of the less volatile propellant. This provided both an increased length over which the least volatile propellant vaporized and a reasonably high-velocity gas phase that could presumably isolate the vaporizing drops from acoustic oscillations.

There were only limited models that could relate the combustion response to injector and chamber design and operating conditions. The most widely used of these models were based on the drop vaporization model of Priem and Heidmann,[15] and they were capable of predicting the vaporization-limited combustion response given an initial propellant drop size distribution and a mean axial combustion product distribution. It should be understood that the drop size distribution of sprays formed by practical injectors is disperse, and modeling the spray as being monodisperse is a gross simplification that could possibly lead to completely unreliable stability predictions. There were, furthermore, few models or correlations that related injector design features and operating conditions to the resulting drop size distribution. Rupe[16,17] and his coworkers at the Jet Propulsion Laboratory (JPL) studied many aspects of the mixing of sprays created with impinging elements and Priem and Heidmann[15] included some simple correlations between orifice diameter and drop size based on data originally collected by Ingebo.[18] With the exception of Abbe et al.,[19] who used Ingebo's drop size correlations as a basis with which to compare the stability characteristics of full-scale combustors that used different injector designs, few works attempted to analyze the causal relationship between element design features and combustion stability. Another major shortcoming was the absence of a model that could predict the effect of thrust chamber baffle geometry on combustion stability. This was especially significant because most large engines of the time required thrust chamber baffles for stable operation.

SP-194 offers a series of design guidelines to bridge the shortage of component models. Some of these are based on empirical correlations of stability data, whereas others are historic rules of thumb. The most notable of the empirical correlations are Reardon's for pressure sensitive interaction index n and sensitive time lag τ. These cover a range of propellant types, injector styles and operating conditions, but the sparseness of data for LOX/HC propellants led to the lumping of data for similar injectors, e.g., non-hypergolic impinging elements, and wide error bars. The data for n plotted against orifice diameter D_j decrease rapidly from nearly 3 at the smallest orifice diameters and asymptote toward 0.6 for orifices larger than 5 mm, but a simple correlation could not be determined. Time lag τ was found to correlate with the nozzle entrance Mach number M_e, reduced pressure parameter β_p, and D_j

$$\tau = 0.165 D_j^{1/2} M_e^{-1/3} \beta_p^{-1} \tag{1}$$

The strong effect of operating mixture ratio (or injection velocity) on time lag is not included in Eq. (1).

Although empirically correlated data were scarce, the design guidelines for LOX/HC engines were even more sweeping and general. Limited data from tests where RP-1 was partially substituted for the hypergolic fuel Hyballine indicated that hypergols may be more stable than LOX/RP-1, implying that combustion instability problems might be easier to solve in engines using hypergolic propellants than in LOX/RP-1 engines. This perception may have resulted, in part, from the fact that many hypergolic systems were designed for low-to-moderate thrust applications and, thus, they could be designed for operation with moderate diameter combustors at low chamber pressures. There is, however, substantial evidence to contradict this belief,[9] since the development of hypergolic engines seem to have a high incidence of chug and longitudinal stability problems, and most hydrazine-based fuels were known to have a strong propensity to produce "pops" that can trigger instabilities during engine operation.

B. Injection Element Design Guidelines

It was clearly understood that each type of injector element has its own characteristic response dependent on its design and operating parameters and that the response was independent of the chamber response. There are several subsections of SP-194 concerned with the relative suitability of various injector element configurations for LOX/HC injection. As noted before, the focus was on liquid HCs so there is no discussion of liquid–liquid coaxial injectors, even though they were widely used in the Soviet Union at the time.[4] Element specific design practices that could enhance stability characteristics are also noted.

Since most stability theories only differentiated between like and unlike impinging elements through their axial combustion distributions, it is not surprising that SP-194 concluded that "unlike-impinging jet injectors appear to be more unstable than like-impinging types." This behavior was attributed to more rapid mixing of the fuel and oxidizer for unlike impinging elements and the resultant reduction in the axial combustion distribution. It should be noted that most of the LOX/HC data supporting this premise was obtained on LOX/ethanol research combustors with relatively fine injector elements.

The like-on-like (LOL) impinging element, or like doublet, consisting of like propellant streams that impinge and atomize prior to mixing of the fuel and oxidizer sprays, is the preferred element type for LOX/HC engines. Most LOL data presented in SP-194 were from LOX/RP-1 booster type elements, including the LOL doublets used in the Atlas, Thor, and H-1 engines, and the fuel doublet + LOX triplet configuration used in the F-1 engine. The roughness of the combustion process was found to be significantly reduced if the element layout caused the atomized oxidizer and fuel streams to come into contact, and the "stability is maximized where unlike fans interact."

However, the unlike propellant mixing characteristics of many LOL element designs resulted in poor overall mixing and low combustion efficiencies (90–95%).[20] There are claims that larger orifices (>2.5 mm) have poorer azimuthal distributions of spray that could lead to clumping of spray around the impingement centerline. The clumping was thought to delay propellant mixing in LOL designs, effectively enhancing the stability by stretching axial combustion distribution. This is a notable difference between like and unlike elements, since clumping in the unlike case will not necessarily delay mixing or axially distribute combustion. The significance of this perceived difference in stability characteristics of like and unlike impinging element types is somewhat suspect since Reardon's correlations of n and τ indicate that all impinging non-hypergolic elements have roughly "the same stability characteristics."

There was no agreement on how to adjust combustion characteristics to improve stability. The goal was a balance of frequency response characteristics with combustion distribution and, possibly, a movement of the sensitive combustion zone. The size of drops in the propellant spray was known to have a large influence on the element's stability characteristics. The proper balance of injection velocity V_j was not clear, since increasing V_j produces smaller drops but also may spatially distribute combustion further. The key design guideline for enhanced stability characteristics in LOX/HC engines was that the LOX injection velocity should be at least three times larger than the less volatile fuel injection velocity. This would presumably cause the vaporizing fuel drop to be surrounded by an atmosphere of high-speed oxidizer vapor during the early drop lifetime, when drop vaporization rates are most sensitive to velocity perturbations. This observation may apply equally as well for hypergolic propellant combinations.

There are several discussions on the importance of design of the injection element manifolds. These observations apply to both like and unlike element types, and to all liquid-propellant combinations. Manifold hydraulics, orifice edge conditions, and relative angles between manifold flows and the injector element inlets have all been shown to have strong influences on the propellant stream characteristics, including flow rate through individual elements, stream impingement accuracy, and the spray size and velocity distributions. Contoured orifices were known to increase the orifice discharge coefficients potentially leading to accurate stream impingement and good spray quality as well as efficiently using the available injection pressure drop.

III. Advancements Since SP-194

Interest in LOX/HC rocket engines was revived in the late 1970s with the envisioned need for heavy-lift capabilities. Vehicle trade studies touted booster engines using a LOX/heavy hydrocarbon (HHC) propellant combinations rather than LOX/H_2 because the fuel is easier to store and handle than H_2 and the increased fuel density could result in smaller vehicles or larger payload capacity.[21] The new engines were expected to operate at high chamber pressures (15–20 MPa) in order to maximize thrust capabilities through optimized exit nozzle expansion ratios.

There were, however, many concerns about the risks associated with the development of such booster class LOX/HC engines. The development of the F-1 had experienced severe problems with combustion instability, performance, thrust chamber cooling, and engine durability.[10] Since the 1970s, advances in understanding in many key areas have been achieved through a combination of experimental technology development programs, advances in computational capabilities, and the development of design feature based spray characterization models. The next subsection chronicles the goals and experimental results of some of these technology programs. This is followed by a description of some of the advances in analysis tools in the area of injection characterization and, finally, a discussion of the net contributions to injection element design practices.

A. Experimental Combustor Studies

Beginning in the late 1970s a series of technology development programs was conducted in an effort to resolve the many of the technical concerns raised about the development of large, high-pressure LOX/HC engines. Most of the studies were conducted with classical impinging elements, although Aerojet built several platelet injectors[22] and TRW studied operation of pintle injectors with LOX/RP-1.[23] The bulk of the work focused on stability and performance concerns, but programs also addressed issues associated with gas generator design[24] and thrust chamber cooling.[25,26]

In the mid-1980s the Air Force and NASA organized several research programs to address performance and combustion stability risks identified from the development of the F-1 engine and Space Shuttle Main Engine (SSME). These programs were strongly focused toward devising methods for lowering the risks and costs of full-scale engine development. The following sections discuss the experimental findings of these programs.

1. NASA Lewis Research Center (1978–1979)

Pavli[27] and his coworkers at NASA Lewis Research Center (LeRC) conducted a test program to investigate the suitability of the Priem and Heidmann[15] generalized length vaporization model for the prediction of performance of heavy hydrocarbon fuels. Testing was conducted using LOX as the oxidizer and RP-1, JP-10, and liquefied natural gas (LNG) as fuels. The combustor consisted of a 0.14-m-diam chamber operating at 4.1 MPa chamber pressure. Three basic injector element types were evaluated: like-on-like doublet, O-F-O unlike triplet, and an F-O-O-F unlike split triplet.

The analysis indicated that the model was in good agreement with the performance data when the combustion was vaporization limited. The data indicated that unlike triplet elements were capable of providing excellent combustion efficiency ($\eta C^* > 99\%$) with acceptable high-frequency stability characteristics. Performance with the LOL injector was significantly lower with all propellant combinations. A portion of the performance loss with the like doublet was attributed to poor interpropellant mixing. In addition to the high-vaporization efficiency, the triplet injectors were found to produce good mixing with a relatively coarse pattern.

A radial-inlet 1/4-wave acoustic cavity sized to damp the first tangential (1T) mode was included at the injector plane in some of the tests. Since the focus of this program was performance modeling, no dynamic combustion stability testing (e.g., bombing) was performed. No spontaneous transverse instabilities were observed with the like doublet or the O-F-O triplet injector operating on LOX/RP-1, even without the cavities present. Spontaneous instabilities were observed without the acoustic cavity present for the O-F-O triplet burning LNG (1T) and JP-10 [first longitudinal (1L)], and the LOL burning JP-10 (1T). Nonlinear 1T instabilities were also noted with the split triplet elements using RP-1 fuel.

2. LOX/Hydrocarbon Injector Characterization Program (1985–1991)

The LOX/hydrocarbon injector characterization program (LHICP) was an Air Force-sponsored technology program focused on the development of analytical methods to reduce development risks, and the associated cost, of booster class LOX/HC engines. Of particular interest were techniques whereby full-sized injector elements could be tested in reduced-scale combustion chambers to significantly enhance the confidence in analytical model predictions of full-scale engine performance and stability characteristics. The program used LOX/RP-1 in large thrust per element O-F-O triplet (5.34 kN/element) and LOL (2.23 kN/element) injectors at high chamber pressures (13.8 MPa). Both subscale (0.18 MN) and full-scale (1 MN) combustors were tested.[28,29] The designs were intended to provide a wide range of potentially unstable operating conditions characteristic of high-performance booster engines, with the triplet injector expected to be higher performing but much more unstable than the LOL injector. The fabrication and testing portions of the program yielded several important observations.

In spite of the large difference in element thrust, the O-F-O triplet element resulted in a significantly higher nominal combustion efficiency than the LOL (99% vs 95%). The performance of the triplet injector showed little sensitivity to changes in mixture ratio, whereas the like-doublet injector exhibited a positive correlation between the oxidizer injection pressure and combustion efficiency. The performance of the triplet injector showed 1.5 and 4.5% decreases in performance as the combustion chamber length was reduced from 20 to 13 and 8 in. respectively.

The O-F-O triplet element showed better dynamic stability characteristics than the LOL, driving only a 1T instability in the subscale chamber, whereas the doublet drove both the 1T and second tangential mode (2T) instabilities. The O-F-O exhibited the only spontaneous instability. Dynamically stable operation was demonstrated for both injectors when 1T monotuned cavities, 1T/2T bituned cavities, or quarter-wave cavities with thrust chamber baffles were included.

The large orifice triplet elements ($D_j = 3.18$ mm) afforded simplicity in manufacturing over the compound angle of the finer, canted, like-doublet element. It is important to note that the O-F-O element was designed with identical diameter, contoured inlet fuel and oxidizer orifices to minimize misimpingement of propellant streams. This was predicted to result in high intraelement mixing and reduced drop size variation. The normalized oxidizer injection pressure drop ($\Delta P_{j,o}/P_c$) was much larger than what was historically used, about 41%. This may be responsible for the high intraelement mixing efficiency and the enhanced stability characteristics when compared to the triplet designs tested during the ROCCID and HHC programs, described next.

3. Rocket Engine Analytical Design Methodology Development Program (1988–1993)

The subscale (0.18 MN) thrust chamber from the LOX/HC injector characterization program was reworked in conjunction with validation testing for the

NASA Lewis Research Center sponsored LOX/hydrocarbon rocket engine analytical design methodology development program (ROCCID).[30] The objective of the injector redesign was to create a triplet design that could realize high performance (>99%) in a reduced chamber length (0.36- vs 0.51-m chamber length) and that had the capability to drive higher mode acoustic instabilities [2T and the first radial (1R)] in the absence of damping devices. The injector element orifice diameters were reduced to 2.29 mm whereas the number of elements was increased from 39 to 105. The unlike impingement height was slightly reduced so the reduced diameter contoured orifices would mate with the existing propellant manifold channels. The nominal $\Delta P_{j,o}/P_c$ was reduced to about 0.23.

Manufacturing difficulties precluded detailed evaluation of the stability characteristics of the new design, but instabilities as high as the 1R mode were observed. More notable, however, was the dramatic decrease in overall combustion efficiency. Predicted combustion efficiencies were in excess of 99%, whereas measured values ranged from 93 to 95% at nominal design conditions. These differences were attributed to poorer than anticipated propellant atomization and mixing. The reduction in propellant mixing was attributed to interaction of the propellant fans of adjacent elements before the atomization process was complete.[30]

4. *Heavy Hydrocarbon Main Injector Technology Program (1986–1991)*

In the late 1980s Rocketdyne evaluated several injector patterns in conjunction with a low-cost design concept for large, high-pressure, baffled combustors using LOX/RP-1 heavy hydrocarbon (HHC).[31] The overall concept was to develop small, stable, high performing combustion modules or cans, and then combine several cans in a single thrust chamber to create a high-thrust engine. Each can would act essentially as an isolated baffle pocket. Six injector patterns were evaluated for performance, stability, and wall heat transfer characteristics in a 0.09-m-diam chamber. The injector designs included a reduced element size derivative of the H-1 engine (fuel doublets and oxidizer triplets),[20] an O-F-O triplet with unlike orifice diameters, two LOL layouts (box and circumferential fans), and two LOX showerhead/fuel doublet configurations. The O-F-O injector was predicted to be more stable than the H-1 derivative, but the analysis did not consider the possibility of injection coupling, which did occur, causing 1L instabilities. The combustion chamber included 7 L-shaped Helmholtz resonators at the injector face tuned to the 1T mode (~8 kHz).

All injectors showed low-to-moderate performance (89–95% C^* efficiency) and demonstrated spontaneous chamber pressure oscillations (5–50% peak to peak) near the 1L frequency (about 1300 Hz). The largest oscillations occurred for the O-F-O triplet design, with significant oscillations existing prior to main-stage ignition. Comparisons of the phase of the oscillations for the O-F-O showed a strong correlation between pressure fluctuations in the main chamber and the oxidizer manifold, indicating the injection pressure drop may have been inadequate to preclude pressure wave coupling between the combustion chamber and the oxidizer manifold. The amplitude of the 1L oscillations was dramatically reduced for the O-F-O triplet injector when the combustor length was reduced from 0.49 to 0.34 m. The reduction in oscillation amplitude was accompanied by a significant increase in performance (98–99% vs 93% C^* efficiency), although predictions indicated that performance was mixing limited at 95%.

The dynamic stability of the H-1 derivative injector was evaluated using combustion bombs with and without the resonator ring present. The induced oscillations damped within 5 ms for all cases. Subsequent testing was performed on the other injectors with the Helmholtz resonator blocked closed. The circumferential fan LOL was shown to be statistically unstable in the 1T mode, whereas the box doublet LOL was bombed unstable in the 1T mode. The transition of a 1L instability into a depressed frequency 1T was also observed with the 1T monotuned cavity included. The O-F-O was shown to be dynamically stable in the tangential and radial modes when 1T cavities were used.

A canted fan LOL injector (1.47-mm oxidizer, 0.94-mm fuel, 14-deg included cant angle) was tested in a 0.145-m chamber. The design included a bituned (1T/2T) axial inlet quarter-wave acoustic cavity. The performance was significantly higher than the previous tests (97% C^* efficiency), but this injector was spontaneously unstable (30% peak to peak) in the 1T mode and bombed unstable (50% peak to peak) in the third tangential (3T) mode. Test data indicated the temperature in the acoustic cavities were much higher than the design value. This caused the characteristic frequencies of the resonators to differ significantly from the resonant frequencies of the chamber.

5. *Pressure Fed Booster (1988–1992)*

The largest LOX/HC engine tested in the United States during this period was the pressure fed booster,[32] designed and built by Aerojet for NASA Marshall Space Flight Center. The full-scale engine was designed to produce 3.4 MN of thrust at nominal operating conditions, and a wide range of throttling conditions were envisioned. The nominal chamber pressure, 5.0 MPa, was constrained by 7.6-MPa maximum vehicle tank pressure. A subscale combustor (0.49-m chamber diameter, 1-m chamber length, 0.9-MN thrust at 3:1 expansion) was designed and tested with integral radial inlet quarter-wave acoustic cavities. The injector consisted of 12 injection modules. Each module contained five O-F-O triplet injection elements with equal diameter 6.1-mm orifices. Like the triplet injector in the LHICP (see the foregoing), this injector was designed with a high relative injection pressure drop for the oxidizer circuit ($\Delta P_{j,o}/P_c \sim 0.45$). Adjustable flow rate fuel film cooling (FFC) showerhead elements were included at the periphery of the injector.

Performance testing indicated a core C^* combustion efficiency ηC^* of 97–99% at chamber pressures between 5.0 and 6.9 MPa, while overall engine performance was 95–97% ηC^* with 15% FFC. The performance was shown to increase nearly linearly as either of the propellant injection pressure drops was increased. The injector was statistically stable but dynamically unstable in the 1T mode without acoustic cavities. Tests with a monotuned and bituned cavity showed dynamic stability at the nominal operating point. Unstable operation also occurred in the 1T, 2T, and 3T modes at high chamber pressures (~6.9 MPa) or low mixture ratios. It is interesting to note that there were no spontaneous instabilities reported regardless of the presence of acoustic cavities.

IV. Injection Process Modeling for Combustion Stability Analysis

The most important of the component models developed are those that relate engine operating conditions and injector design features to propellant spray characteristics. These characteristics are the critical inputs to all performance and combustion response models. Foremost of there is the relationship between injector element design parameters and propellant spray characteristics. As with the early work in this field, the main goal of many of the recent modeling efforts was to determine the influence of injector operating and design conditions on propellant spray characteristics so that combustor performance, rather than combustion stability, could be reliably predicted.

The earliest work yielded simple correlations for the dependence of some characteristic drop size and spray distribution on the orifice diameter and element type for impinging injector type.[14-17] Some of the first comprehensive studies were conducted by Dickerson et al.[33,34] at Rocketdyne using hot wax sprays to characterize spray distributions and the influence of impinging element design features. This work was continued into the mid-1970s with increasing geometric sophistication, and both liquid/liquid impinging and gas/liquid coaxial element types were examined. Multidimensional spray distribution characteristics were inferred from flow distribution measurements obtained using a patternator. Ferrenberg et al.[35] compiled a comprehensive summary of the studies undertaken through 1982.

Over the same time span, Ito at Aerojet developed analytical and hybrid analytical/empirical models that related propellant atomization characteristics to element design features for many types of impinging elements.[36] These models permitted the prediction of both spray characteristics relevant to performance (median drop size and distribution) and several time lags, including the impingement and atomization components. The mechanism for analytical prediction of time lag components was of critical importance to the development of predictive combustion stability analysis tools. Fang[37] used the time lags predicted with Ito's models to analyze a range of chug and longitudinal mode combustion instabilities.

Combustion instability data correlation performed at Aerojet also indicated that the sensitive time lag could be correlated with the time it took for the less volatile propellant (usually fuel) to become 20% vaporized.[36,37] This value is much smaller than the 50% vaporization time proposed in SP-194. This correlation was developed from storable propellant engine data, mainly MMH/N_2O_4 using the Priem and Heidmann[15] generalized length vaporization correlation. The 20% vaporization concept was subsequently found to apply to a wide range of propellant combinations and operating conditions.

Subsequent work at Aerojet by Hewitt has shown a strong correlation between the observed combustion response frequency and the ratio of the injector orifice diameter to propellant injection velocity (D_j/V_j) of LOL injection elements. Whereas most of the data used to establish the correlation were from storable propellant engine systems, Anderson et al.[38] extended the experimental database on which the correlation is based to include the maximum combustion response frequency of LOL injectors using LOX/HC propellants (Fig. 1). The unstable points on Fig. 1 correspond to the highest frequency of combustion instability that was observed in the combustor, whereas the stable points on Fig. 1 correspond to the calculated frequency of the lowest chamber acoustic mode that was not excited by combustion. The correlation as shown in Fig. 1 demonstrates the importance of the injector operating conditions in determining the overall combustion stability characteristics of the engine.

The O-F-O element has a distinct advantage over the LOL element in that it can provide a very high interpropellant mixing efficiency. As demonstrated in the LHICP and LeRC programs already discussed, C^* efficiencies greater

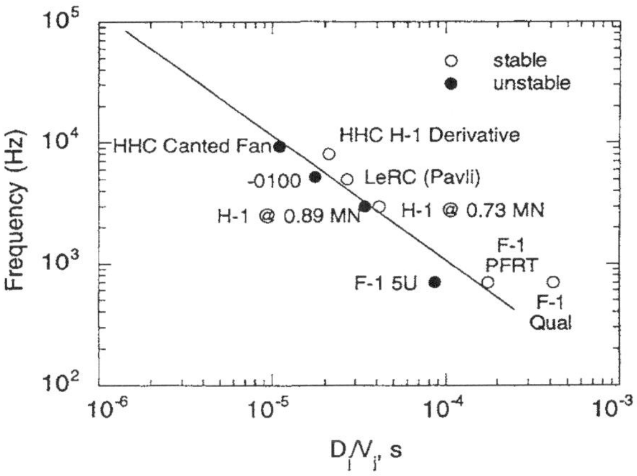

Fig. 1 Empirical correlation for the maximum response frequency of like-doublet injector elements burning LOX/HC propellants, adapted from Anderson et al.[38]; correlating parameter is the ratio D_j to V_j for the least volatile propellant (fuel). Solid line separating stable from unstable operating points estimates the maximum frequency that can be sustained through burning coupling mechanism.

than 99% can be achieved with well-designed O-F-O element-based injectors. Furthermore, O-F-O element designs that produce optimal levels of performance (equal LOX and LHC orifice diameters, equal LOX and LHC dynamic heads) are readily achieved with LOX/LHC propellants because of typical propellant mass ratios and typical propellant densities. However, most likely because of the increased energy release rate promoted by the enhanced mixing of the O-F-O element, the O-F-O element also appears to be more prone to spontaneous high-frequency instabilities than does the LOL element. This perceived tendency

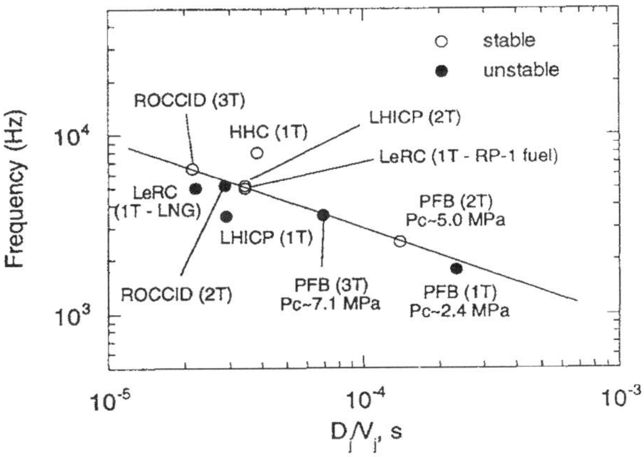

Fig. 2 Empirical correlation for the maximum response frequency of O-F-O injector elements burning LOX/RP-1 propellants, fashioned after Anderson et al.[38]; correlating parameter is the ratio of D_j to V_j for the least volatile propellant (fuel). Solid line separating stable from unstable operating points estimates the maximum frequency that can be sustained through burning coupling mechanism.

has discouraged rocket developers from using O-F-O elements, and there is a much greater experience level with LOL elements.

It should be noted that much of the perception about the relative stability of O-F-O and LOL elements comes from their comparison on an equivalent thrust-per-element basis, i.e, an O-F-O element can drive higher modes of instability than can a LOL element that injects the same amount of propellant. Since the O-F-O is higher performing than the LOL, however, O-F-O elements with higher thrust-per-element ratios can be used. In light of the ability of the correlation shown in Fig. 1 to match the stability characteristics of LOL elements, the correlation was extended to determine the maximum frequency response characteristics of O-F-O triplet elements with LOX/HC propellants in Fig. 2. Some reasonable trend is seen here as well, but the limited amount of available data makes a clear distinction less certain. It is important to note that these correlations identify the maximum frequency the element can support through a pure burning coupled mechanism, and it is not capable of predicting the stability characteristics if injection coupling mechanisms become important.

V. Conclusions

The importance of the development of analytical models capable of predicting the characteristics of the spray and combustion response characteristics can not be overemphasized. These models can be used in conjunction with relatively simple steady-state combustion and chamber response models to optimize engine performance and stability characteristics in a short period of time. They also provide tools with which analysts can prescribe hardware modifications to change the performance and stability characteristics of existing hardware designs.

There is strong evidence that large thrust elements can be designed to provide both high performance and good stability characteristics using LOX/RP-1 as propellants. However, it appears that LOL injector designs must trade performance for stability. The LOL injector designs usually provide moderate performance at best if manageable combustion stability characteristics are to be achieved (e.g., Atlas, F-1, HHC, LHICP). The results of the LHICP, LeRC, and pressure fed booster programs indicate, however, that large O-F-O triplet elements can provide very high performance while also possessing manageable stability characteristics. The element response characteristics presented in Figs. 1 and 2 indicate that larger orifice diameters or small injection velocities are required to keep the maximum response frequency low enough (e.g., below the chamber's 3T frequency) to be damped with a combination of acoustic cavities and simple thrust chamber baffle patterns in large-diameter, high-thrust engines. Low injection velocity and low injection pressure drop can lead to injection coupling in triplet designs and poor mixing in doublet designs. Increasing the orifice diameter results in reduced element density, which can result in large mixing losses in like-doublet designs. Since triplet elements result in intimate mixing at impingement, low element density does not seem to significantly affect performance. Low element density can, however, compromise injector face cooling.

The triplet test data indicates that the requirement of equal orifice diameter for the fuel and oxidizer may be desirable but that it is not sufficient condition for good performance. High injector impedance (large $\Delta P_{j,o}/P_c$) seems to be essential to preclude coupling through the feed system. The use of contoured inlet element designs provides for efficient use of the available injection pressure drop and may partially offset the need for large injection pressure drops. Sparse patterns do not seem to cause performance problems when triplet elements are used, and they may actually lead to better performance and stability characteristics. Triplet and LOL element layouts must carefully consider interaction between adjacent element spray fans since this close packing appears to cause atomization and mixing problems.

Finally, the D_j / V_j correlations presented in Figs. 1 and 2 can be used in conjunction with the relationship for the resonant frequency of modes in the chamber to guide the injector element sizing early in the development phase for large engines. This type of correlation permits trade studies between stability characteristics, damping device requirements, active cooling, and injector performance to be rapidly conducted.

References

[1]Winter, F. H., *Rockets into Space,* Harvard Univ. Press, Cambridge, MA, 1990.

[2]Dornberger, W., *V-2,* Viking Press, 1954.

[3] *Jane's Spaceflight Directory,* Jane's Pub., London, 1984 pp. 268 ff.

[4]Rubinsky, V. R., "Analysis of Problems of Operating Process High-Frequency Stability Improvement in the Chamber of LOX-Kerosene Liquid Rocket Engine with Gas Generator Power Cycle, First International Symposium on Combustion Instability in Liquid Rocket Engines," Pennsylvania State Univ., University Park, PA, Jan. 1993.

[5]Fuller, P. N., and Minami, H. M., "History of the Thor/Delta Booster Engine," *History of Liquid Rocket Engine Development in the United States 1955–1980,* edited by S. E. Doyle, Advances in Aeronautical Sciences History Series, Vol. 13, 1992, pp. 39–51.

[6]Smith, A. S., and Minami, H. M., "History of the Atlas Engines," *History of Liquid Rocket Engine Development in the United States 1955–1980,* edited by S. E. Doyle, Advances in Aeronautical Sciences History Series, Vol. 13, 1992, pp. 53–59.

[7]Harrje, D. T., and Reardon, F. H. (eds.), *Liquid Propellant Rocket Combustion Instability,* NASA SP-194, 1972.

[8]Rosen, M. W., *The Viking Rocket Story,* Harper and Brothers, New York, 1955.

[9]Anon., "Gemini Stability Improvement Program (GEMSIP)," USAF Contract AF 04(695)-517, Aug. 31, 1965.

[10]Anon., "History: Project First, F-1 Combustion Stability Program," Rocketdyne Rept. R-5615, Canoga Park, CA, 1964–1967.

[11]Bailey, R., "Test Evaluation of Oxygen-Methane Main Injectors," 1987 JANNAF Propulsion Meetings, 1987.

[12]Meagher, G. M., and Muss, J. A., "Swirl Coaxial Injector Element Characterization for Booster Engines," Advanced Earth-to-Orbit Technology Conf., Huntsville, AL, May 1988.

[13]Jensen, R. J., Dodson, H., and Trueblood, B., "Oxygen/Methane Combustion Stability Investigation," Advanced Earth-to-Orbit Technology Conf., Huntsville, AL, May 1988.

[14]Heidmann, M. F., and Groeneweg, J. F., "Analysis of the Dynamic Response of Liquid Jet Atomization to Acoustic Oscillations," NASA TN-D-5339, July 1969.

[15]Priem, R. J., and Heidmann, M. F., "Propellant Vaporization as a Design Criterion for Rocket-Engine Combustion Chambers," NASA TR-67, 1960.

[16]Rupe, J. H., "The Liquid Phase Mixing of a Pair of Impinging Streams," Progress Rept. 20-195, Jet Propulsion Lab., California Inst. of Technology, Pasadena, CA, Aug. 1953.

[17]Rupe, J. H., "A Correlation Between the Dynamic Properties of a Pair of Impinging Streams and the Uniformity of the Mixture Ratio Distribution in the Resulting Spray,"

Progress Rept. 20-209, Jet Propulsion Lab., California Inst. of Technology, Pasadena, CA, March 1956.

[18]Ingebo, R. D., "Drop-Size Distributions for Impinging-Jet Breakup in Airstreams Simulating the Velocity Conditions in Rocket Combustors," NACA TN-4222, March 1958.

[19]Abbe C. J., McLaughlin C. W., and Weiss R. R, "Influence of Storable Propellant Liquid Rocket Design Parameters on Combustion Instability," *Journal of Spacecraft and Rockets,* Vol. 5, No. 5, 1968, pp. 584–590.

[20]Anon., *Liquid Rocket Engine Injectors,* NASA SP-8089, March 1976.

[21]Frankenfeld, J. W., "High Performance Hydrocarbon Fuels," NASA CR-159480, Dec. 1978.

[22]Schoenman, L., and Gross, R. G., "Design, Fabrication Test and Delivery of a High-Pressure Oxygen-RP-1 Injector, Final Report," NASA Contract NAS8-33651, Aerojet Liquid Rocket Rept. 33651F, Sacramento, CA, Sept. 1981.

[23]Stoddard, F., "Inherent Stability of Central Element Coaxial Liquid-Liquid Injectors," Propulsion Engineering Research Center Fifth Annual Symposium on Space Propulsion, Pennsylvania State Univ., University Park, PA, Sept. 1993.

[24]Hernandez, R., and Mercer, S. D., "Carbon Deposition Characteristics of LO2/Hydrocarbon Propellants," AIAA/ASME/SAE/ASEE 23rd Joint Propulsion Conf., AIAA Paper 87-1885, Monterey, CA, 1987.

[25]Price, H. G., and Masters, P. A., "Liquid Oxygen Cooling of High Pressure LOX/Hydrocarbon Rocket Thrust Chambers," NASA TM-88805, Nov. 1986.

[26]Masters, P. A., Armstrong, E. S., and Price, H. G., "High-Pressure Calorimeter Chamber Tests for Liquid Oxygen/Kerosene (LOX/RP-1) Rocket Combustion," NASA TP-2862, 1988.

[27]Pavli, A. J., "Design and Evaluation of High Performance Rocket Engine Injectors for Use with Hydrocarbon Fuels," NASA TM-79319, Sept. 1979.

[28]Muss, J. A., and Pieper, J. L., "Performance and Stability Characterization of LOX/Hydrocarbon Injectors," AIAA/ASME/SAE/ASEE 24th Joint Propulsion Conf., AIAA Paper 88-3133, Boston MA, 1988.

[29]Pieper, J. L., "LOX/Hydrocarbon Rocket Engine Analytical Design Methodology Development and Validation, Task 1.0 Report," Contract NAS3-25556, Aerojet TechSystems, Sacramento, CA, April 1989.

[30]Pieper, J. L., and Walker, R. E., "LOX/Hydrocarbon Rocket Engine Analytical Design Methodology Development and Validation," NASA CR-191058, May 1993.

[31]Arbit, H. H., Tuegel, F. E., and Dodd, F. E., "Heavy Hydrocarbon Main Injector Technology Program, Final Report," NASA Contract NAS8-36369, Rocketdyne Div., Rockwell International Rept. RI/RD91-118, Canoga Park, CA, April 1991.

[32]Faulkner, C., and Dunn, G. M., "Pressure Fed Thrust Chamber Technology Program, Final Report," NASA Contract NAS8-37365, Aerojet Propulsion Div., Sacramento, CA, 1992.

[33]Dickerson, R., Tate, K., and Barsic, N., "Correlation of Spray Injector Parameters with Rocket Engine Performance," Final Technical Rept., Air Force Rocket Propulsion Lab., AFRPL-TR-68-147, North American Aviation/Rocketdyne, Canoga Park, CA, Sept. 1968.

[34]Dickerson, R., Tate, K., and Nurick, W., "Correlation of Spray Injector Parameters with Rocket Engine Performance," Special Technical Rept., Air Force Rocket Propulsion Lab., AFRPL-TR-68-11, North American Aviation/Rocketdyne, Canoga Park, CA, Jan. 1968.

[35]Ferrenberg, A., Hunt, K., and Duesberg, J., "Atomization and Mixing Study," Contract NAS8-34504, Rockwell International/Rocketdyne Div. Rept. RI/RD85-312, Canoga Park, CA, 1985.

[36]Muss, J. A., Nguyen, T. V., and Johnson, C. W., "User's Manual for Rocket Combustor Interactive Design (ROCCID) and Analysis Computer Program," NASA CR-187109, May 1991.

[37]Fang, J. J., "Application of Combustion Time-lag Theory to Combustion Stability Analysis of Liquid and Gaseous Propellant Rocket Engines," AIAA 22nd Aerospace Sciences Meeting, AIAA Paper 84-0510, Reno, NV, 1982.

[38]Anderson, W. E., Ryan, H. M., and Santoro, R. J., "Combustion Instability Phenomena of Importance to Liquid Bipropellant Rocket Engines," 28th JANNAF Combustion Meeting, San Antonio, TX, 1991.

Chapter 4

Combustion Instability in the RD-0110 Engine

Vitaly R. Rubinsky*
Chemiautomatics Design Bureau (CADB), Voronezh, Russia

I. Introduction

THE third stage of the Soyuz space vehicle is powered by the RD-0110, a LOX-kerosene liquid propellant rocket engine. During both development and acceptance testing, high-frequency combustion instability was observed. The four-chamber RD-0110 engine (Fig. 1) operates on a gas generator cycle. Propellants are fed to the combustion chamber by the turbopump, whose turbine operates on the combustion products of LOX and kerosene produced by a gas generator. Propellant ignition in both the chamber and in the gas generator is by a pyrotechnic device. Engine control and adjustment is provided by a regulator, a throttle, and a set of valves. The main technical data of this engine are presented in Table 1.

While designing and developing the chamber for RD-0110, a great deal of consideration was given to the analysis of the mixing system, whose scheme and parameters determine the stability and quality of operating process. During the initial injector design analysis stage, a great deal of consideration also was given to the problems of ensuring chamber operating process high-frequency stability using conventional techniques.[1,2] The well-known methods of stabilization, i.e., "extension" of the combustion zone (energy release) along the length or radius of the chamber, changing the mixing elements flow rate/atomizing characteristics, and mounting antipulsation partitions (baffles) at the injector, also were considered.

Proceeding from the idea of the propellants' interaction at the initial section of the chamber and its influence on the dynamics of propellant burnout near the injector, it was supposed that bipropellant centrifugal atomizers had real advantages in this point. It was expected that the propellant burnout dynamics could be influenced by changing the interaction between the oxidizer and fuel cones. At the same time, the necessity of providing the high quality of the propellant mixing system to obtain the requisite specific impulse was taken into account.

During thrust chamber development, occurrences of high-frequency combustion instability were observed to be dependent on the use of different atomizer designs. Studies of the problem confirmed the theory that stability was determined by the interaction between the atomizing cones of oxidizer and fuel, including the core

*Chief, Combustion Devices Department.

Table 1 RD-0110 engine main technical data

Parameter	Value
Thrust in vacuum R_v, kN	297.926
Secific impulse in vacuum I_s, m/s	3197
Engine mass m, kg	412.5
Engine specific mass $\bar{m}$, kg/kN	1.384
Mixture ratio $K^{\dot{m}}$	2.2
Chamber pressure P_k, MPa	6.816
Nozzle exit pressure P_a, MPa	0.00727
Gas generator pressure P_{gg}, MPa	5.78
Fuel pump outlet pressure P_{fpo}, MPa	13.24
Oxidizer pump outlet pressure P_{opo}, MPa	9.81
Gas generator temperature T_{gg}, K	1048
Turbine rotation speed n, rpm	18,400
Thrust range ΔR, %	(−9.5)–(+7)
Mixture ratio range $\Delta K^{\dot{m}}$, %	(−14)–(+11)
Propellant: oxidizer-LOX	
fuel-kerosene T-1	
Operation time τ, s	240

Fig. 1 Liquid rocket engine RD-0110.

and peripheral regions of the cone. Spray cone interaction is in turn determined by the design of the centrifugal atomizers. Eventually, an injector that comprised bipropellant centrifugal atomizers was developed and mastered that provided high-quality mixing and high-frequency stability of the chamber operating process over a wide range of pressure and mixture ratio.

High-frequency combustion instability was not observed during the final stages of the chamber off-line development, during the complex development of the whole RD-0110 engine, or during its operation in the third stage of the Soyuz rocket carrier. However, although very seldom (one case per 60–80 engine tests), high-frequency combustion instability began to be observed during engine start in the bench approval tests performed by the production plant. Combustion instability was not noticed during main operation. Studies of these problems were carried out with the goal of eliminating high-frequency instability during the start without changing the engine start scheme or the injector design.

Acoustic and model studies revealed the possibility of stabilizing the operating process by the arrangement of several short longitudinal ribs, made from combustible material, at the inner surface of the chamber; this measure was taken for further development to eliminate high-frequency instability during the start of the RD-0110. It was supposed that the ribs would provide high-frequency stability during the start and burnout for the moment the engine is to transition to the main operating mode. Off-line chamber tests and engine bench tests showed that placement of six longitudinal felt ribs completely eliminated the chamber operating process high-frequency instability during the start of the RD-0110 engine.

This chapter presents the analysis of the high-frequency combustion stability improvement undertaken at different stages of the RD-0110 development as well as the methodology of organization and carrying out of such an analysis. The results of these studies, which were carried out at the time of the chamber's creation, and the conclusions made on their basis were the foundation for further chamber development.

The chapter uses the materials of scientific and technical reports made at the Chemiautomatics Design Bureau and the Research Institute of Thermal Processes. After the completion of these studies the works in Refs. 3–5 appeared. They analyzed the stabilization methods for the chamber's operating process of the liquid rocket engines (LRE). These methods are to some extent similar to the mechanisms that were realized during the creation of the chamber for the RD-0110 engine.

II. High-Frequency Combustion Stability Studies During Off-Line Chamber Development

A. Description and Main Technical Data of the Chamber

The chamber is a welded undetachable construction consisting of two parts, the injector and the chamber itself (Figs. 2 and 3). The chamber injector (Figs. 4 and 5) consists of 91 atomizers (Fig. 6), firing bottom, deflector, and injector housing. The components of the injector are joined by vacuum brazing. The fuel goes through the chamber's cooling duct into the injector and then between the injector face plate and deflector into the atomizer's cavity. This scheme provides reliable cooling of the injector faceplate.

The chamber itself consists of the chamber unit, a throat insert, and a nozzle unit. The chamber unit consists of the outer and inner shells of the cylinder and

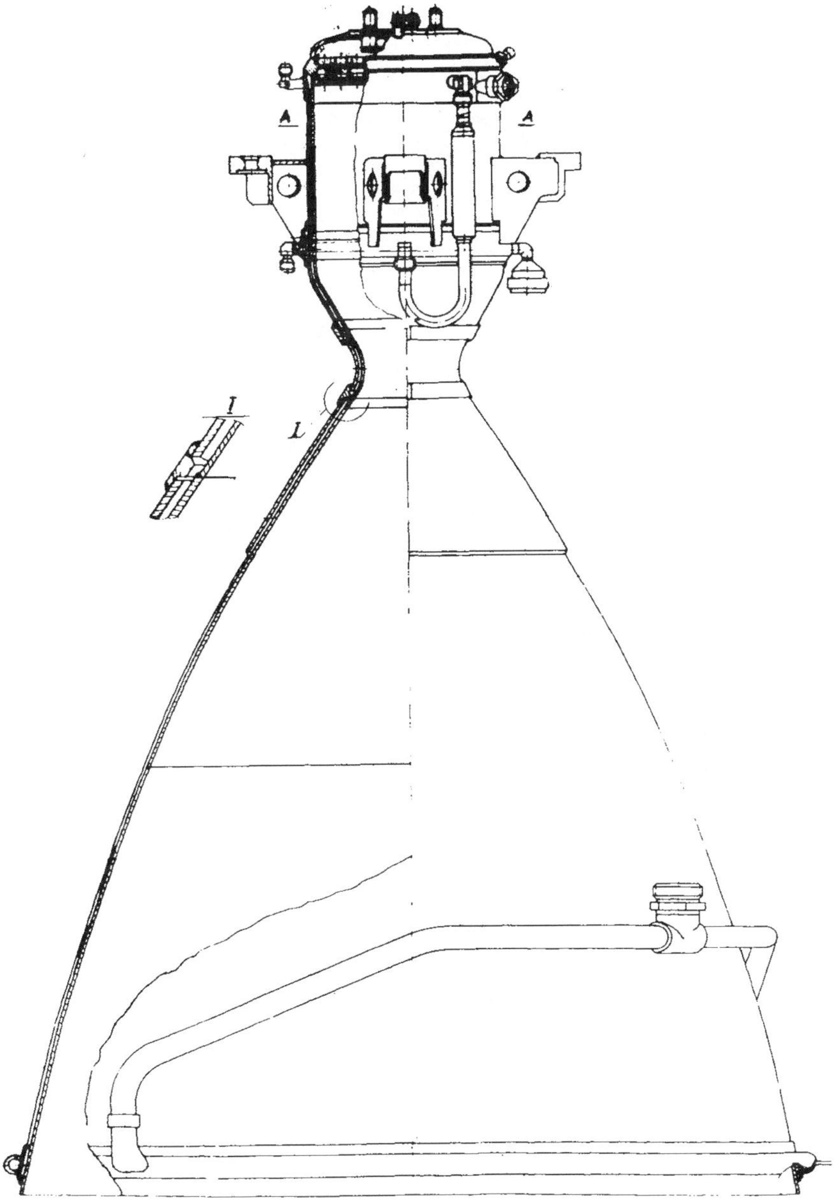

Fig. 2 Chamber of the RD-0110 engine.

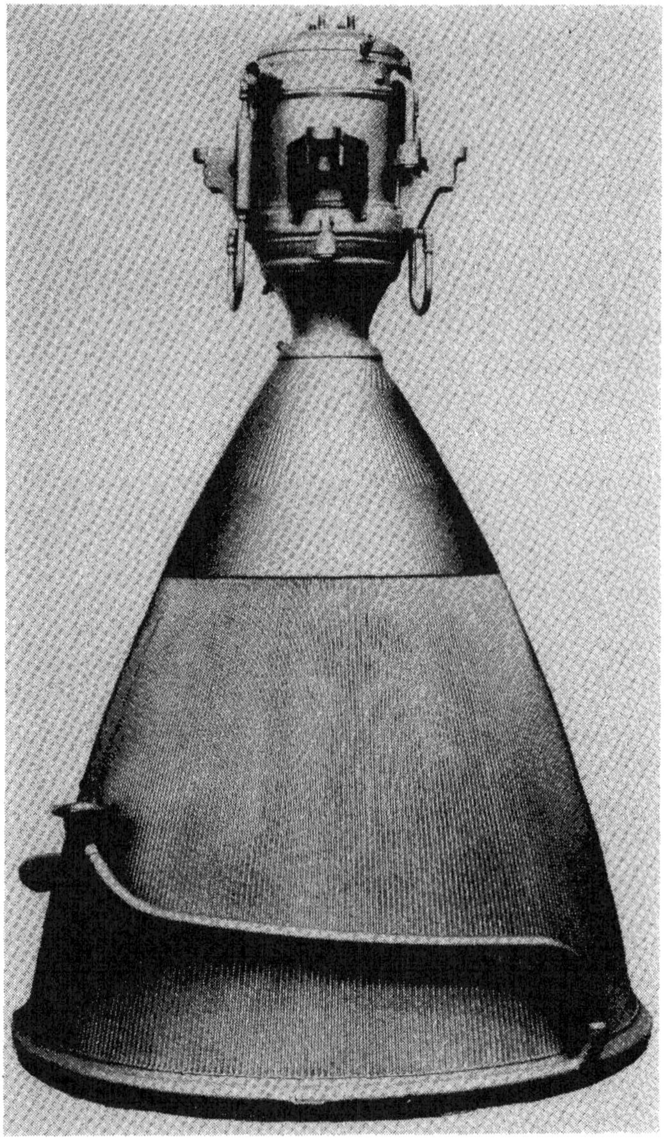

Fig. 3 Chamber of the RD-0110 engine.

nozzle inlet section; corrugated jackets, which form the cooling duct and are located between outer and inner shells of cylinder and nozzle inlet; and the curtain ring of inner cooling (Fig. 7). All of the shells are joined by vacuum brazing. The insert into the throat section (Figs. 8 and 9) is a brazed unit, consisting of the inner copper shell, with the milled cooling grooves, and the outer jacket. The supersonic nozzle (excluding the 0.256-m section from the copper insert) comprises two shells, an inner shell and a corrugated jacket which forms the nozzle cooling duct. The corrugated jacket has flanged edges for a two-sided connection with the outer jackets. At the nozzle exit section there is a ring manifold, made from a tube, to

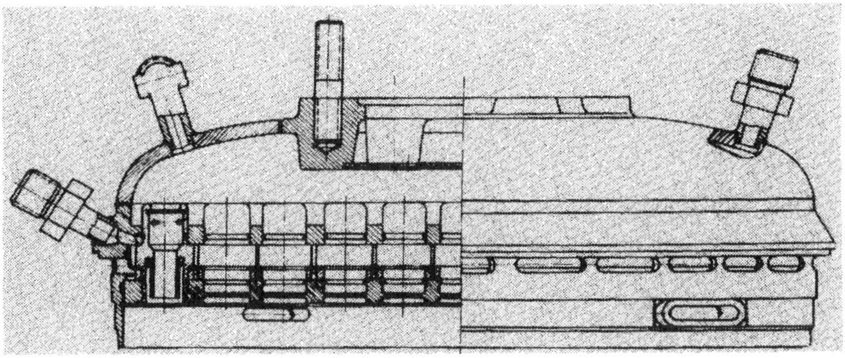

Fig. 4 Injector.

Fig. 5 Injector.

Fig. 6 Bipropellant centrifugal atomizers.

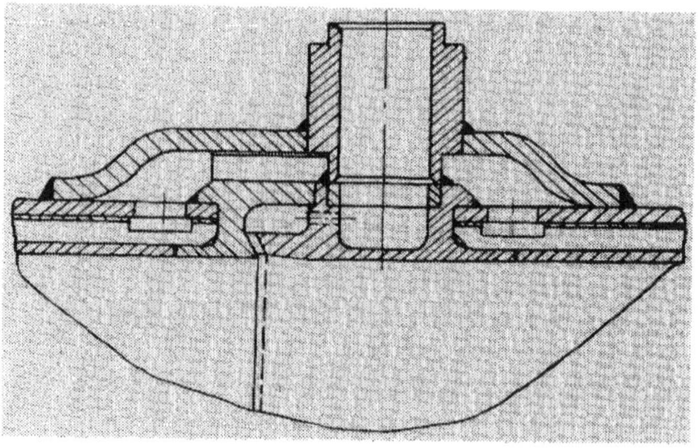

Fig. 7 Cooling curtain ring.

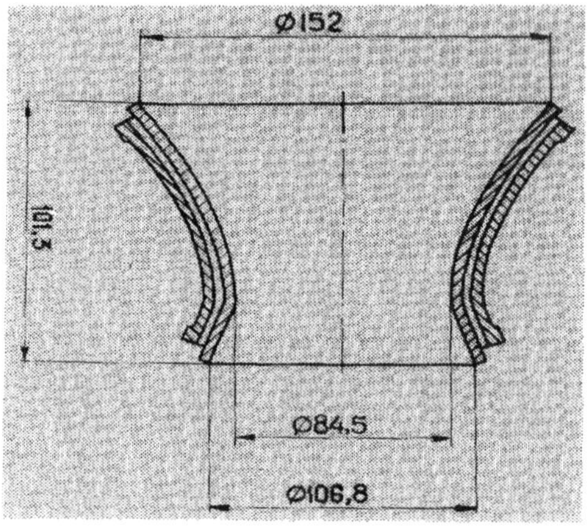

Fig. 8 Insert into throat section.

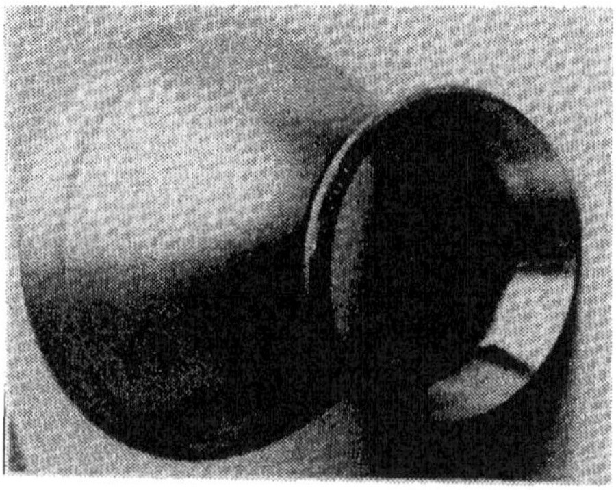

Fig. 9 Insert into throat section.

Table 2 RD-0110 engine main technical data

Parameter	Value
Vacuum thrust R_v, kN	73.142
Vacuum exhaust velocity I_s^v, m/s	3279
Engine mass m, kg	38.5
Chamber specific mass $\bar{m}$, kg/kN	0.526
Propellant flow rate, kg/s	
total $\dot{m}_\Sigma$	22.3
oxidizer $\dot{m}_o$	15.7
fuel $\dot{m}_f$	6.6
Fuel film cooling $\bar{\dot{m}}_c$, %	3
Chamber pressure P_k, MPa	6.816
Nozzle exit pressure P_a, kPa	7.27
Chamber diameter D_c, m	0.180
Chamber throat diameter D_{th}, mm	84.5
Nozzle exit diameter D_{ex}, m	0.764
Area ratio $\bar{F}_a$	81.64
ΔP oxidizer injection ΔP_{oa}, MPa	0.427
ΔP fuel injection ΔP_{fa}, MPa	0.696
Propellant mass flux q, g/(s-cm^2)	88.28
Gas residence time τ_r, ms	1.25
Lifetime τ, s	500

bring the fuel into the chamber's cooling duct. Jackets and corrugations are joined by welding and vacuum brazing. All of the shells and jackets of the chamber are made of chromium-nickel steel. The width of inner shells is 1.0 mm, the width of corrugated jackets is 0.5 mm, and the width of the outer jackets is 1.5 mm. The main technical data of the chamber are presented in Table 2.

B. Selection and Bases of the Scheme, Design, and Variants of Injector and Atomizers

Efficient combustion was required taking into account the specific impulse requirements of the RD-0110. Single-propellant and bipropellant atomizers were analyzed. To get the most uniform mixture ratio at the initial section of the chamber (in the scale of spacing between the atomizers), the bipropellant atomizers were chosen. The following points were taken into consideration. In principle, injectors with single-propellant atomizers can provide quite high (small-scaled) uniformity of propellant distribution. The experience of chamber mastering and development showed that, to reach such a goal, one should use a relatively great number of single-propellant atomizers in the injector to provide fine atomization of oxidizer and fuel. It is simpler, however, to get a uniform mixture ratio at the initial section by using fewer bipropellant atomizers, especially with internal mixing. It should be pointed out that the considerations leading to the choice of single-propellant atomizers were connected with the great advantage of such atomizers in case they needed modification and improvement. Such modifications are required in the process of chamber development if there is a necessity to solve the problems related to operating process high-frequency stability.

Thus, bipropellant atomizers are more convenient for the designer during the chamber development process because they do not need variants modifications, which change the design of atomizers and injector. A basic injector design incorporating the bipropellant atomizers was developed. This injector was subjected to the tests with three types of bipropellant atomizers: centrifugal atomizers (CS), centrifugal-spray atomizers (C), and spray atomizers (S). The location of atomizers is shown in Fig. 10.

For bipropellant centrifugal atomizers, the length of the oxidizer atomizer nozzle face recess relative to the fuel atomizer nozzle face, L, was taken as a main factor influencing the interaction of propellants at the initial section of the chamber. According to the results of hydraulic tests of centrifugal atomizers with size

Index of injectors	Atomizers lay-out diagram	Denotation	Type of atomizers	Number of atomizers
C1 C2 C3 C4 C5	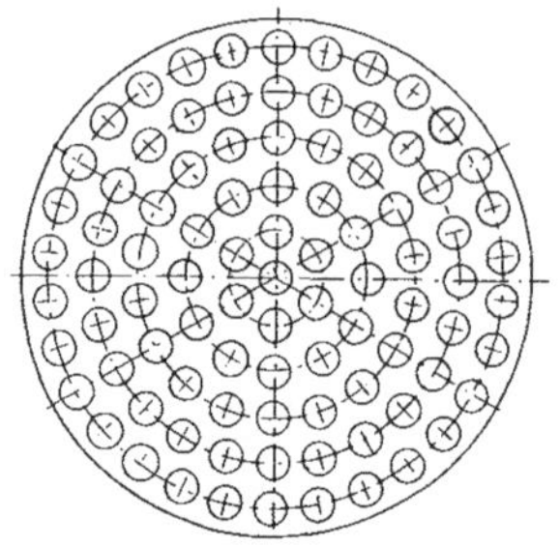		centrifugal	91
CS1	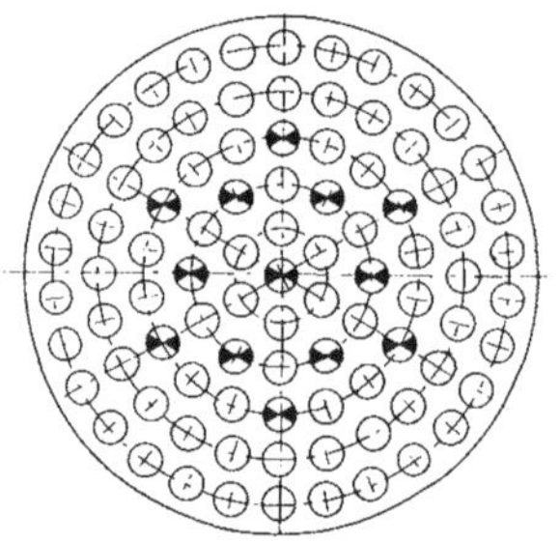		centrifugal	78
			spray	13
S1	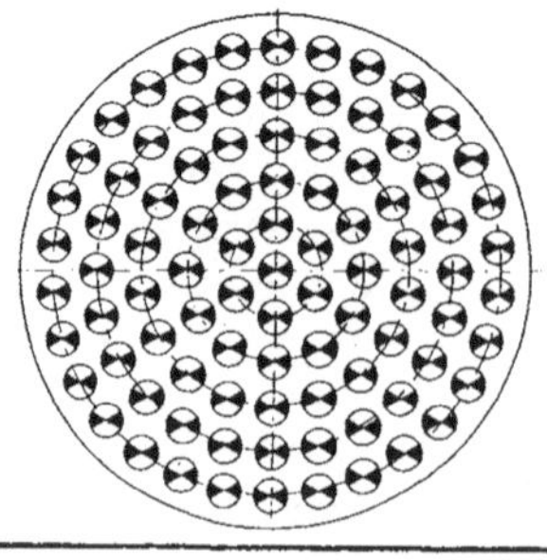		spray	91

Fig. 10 Scheme of atomizer layout at the injectors.

Table 3 Variants of injectors

Injector index	L periphery, mm	L center, mm
C1	5.0	5.0
C2	2.5	2.5
C3	1.5	1.5
C4	5.0	1.5
C5	2.5	1.5

$L = 5.0$ mm, mixing and interaction of propellants began in the fuel atomizer nozzle (emulsion atomizer). At the size $L = 1.5$ mm, mixing and interaction of propellants took place outside the atomizer (outer mixing atomizer). At the size $L = 2.5$ mm, oxidizer and fuel spraying cones had a contact near the outlet edge of fuel atomizer nozzle.

Thus, it was supposed that the dimension L can influence the completeness and high-frequency pressure oscillations in the chamber. Five variants of injectors C were developed (Table 3). Design and hydraulic parameters of all C atomizers were identical.

In case increases in the operating process high-frequency stability margins were required for the chamber with injectors C, the injector CS was developed. In its center there were 13 bipropellant spray atomizers with impinging jets of oxidizer and fuel (at a 30-deg angle) to extend the fuel component's initial interaction zone. The geometrical dimensions of the joint places for the spray atomizers were the same as for the joint places of the centrifugal atomizers, which made the process of injector production easier. A variant of injector S with bipropellant spray atomizers (which were used in injector CS) was developed for the case where the operating process high-frequency stability would not be realized on the base of injectors C and CS. The calculated mixture ratio for atomizers of injectors C, CS, and S were kept equal to the value of 2.85. Figure 11 presents the design and parameters of atomizers C and S.

C. Results

Off-line firing tests of the chambers were carried out in three stages. The aim of the first stage was the selection of a high-performance injector, providing high quality and stability of operating process in the chamber pressure range $P_c = 3.0$–8.0 MPa. The tests were done with the short chamber without supersonic section of the nozzle. Start of the chamber into the main operating mode ($P_c = 7.0$ MPa, $O/F = 2.44$) was done through the preliminary stage ($P_c = 3.5$ MPa, $O/F = 2.44$ or $O/F = 1.0$). Test duration was up to 20 s.

The second stage was carried out to develop reliable cooling. It was done in the chamber with a shortened supersonic section of the nozzle (Fig. 12), providing constant outflow of combustion products under Earth conditions.

The third stage was carried out with the full-sized chamber. The main aim of the tests was to confirm the serviceability of the chamber's final design during its life and the evaluation of specific impulse under the calculated conditions of combustion products outflow from the nozzle ($P_a = 7.27$ kPa). The test stand for the third stage was equipped with a thrust measuring device and an exhaust device to provide conditions of the combustion product outflow from the nozzle.

	Component	Geometrical characteristic	Spray-cone angle	Pressure drop	flow rate
		–	anple. deprees	MPa	g/sec.
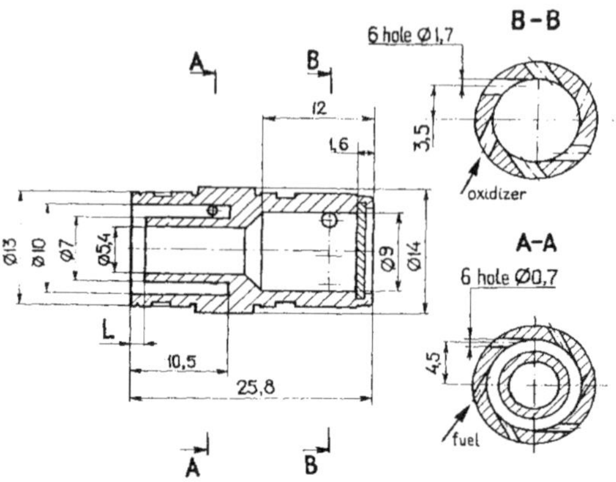	oxidizer	2	80	0,426	172,9
	fuel	24,5	135	0,696	64,8
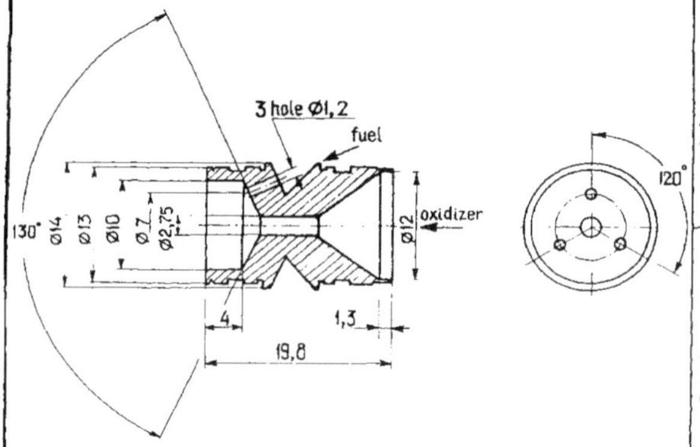	oxidizer	–	–	0,426	172,3
	fuel	–	–	0,696	64, 8

Fig. 11 Atomizer design and parameters.

The firing tests of the chambers with injectors C1 and C2 were ended by their failure (destruction) by oscillations that appeared in the starting mode with frequency $f = 4000$ Hz and amplitude more than 1.5 MPa. The duration of the chamber's operation was 0.04–0.1 s. The calculated value of the first tangential mode for the chamber was $f = 3900$ Hz. Thus, it was evident that tangential mode high-frequency oscillations arose during the tests. Tests of the chambers with reserve variants of injectors CS and S were also ended by their destruction during the start because of the tangential mode high-frequency oscillations. Thus, the approach of high-frequency instability elimination by extending the propellant interaction section in injectors CS and S gave no positive results in the operating process conditions of the oxygen-kerosene chamber of RD-0110. The pattern of chamber (with tested mixing heads) failure by high-frequency oscillations was

Fig. 12 Chamber with shortened nozzle.

practically the same: injector breaking away from the chamber, pipe connection breaking away, and the chamber's inner shell cracking and burning out. For injectors C1 and C2, the oxidizer atomizer nozzles were deformed, and there were no fusion (melting) marks; for injectors CS and S there was fusion (melting) of spray atomizers. Figure 13 demonstrates typical cases of chamber failure by high-frequency oscillations.

Tests of chambers with injector C3, however, were stable during the start and in the range of $P_c = 4.0–8.0$ MPa. The tests of chambers with the injectors C3 resulted in decreased values of specific impulse; therefore, tests of chambers with injectors C4 and C5, in which the dimension L was changed to activate the

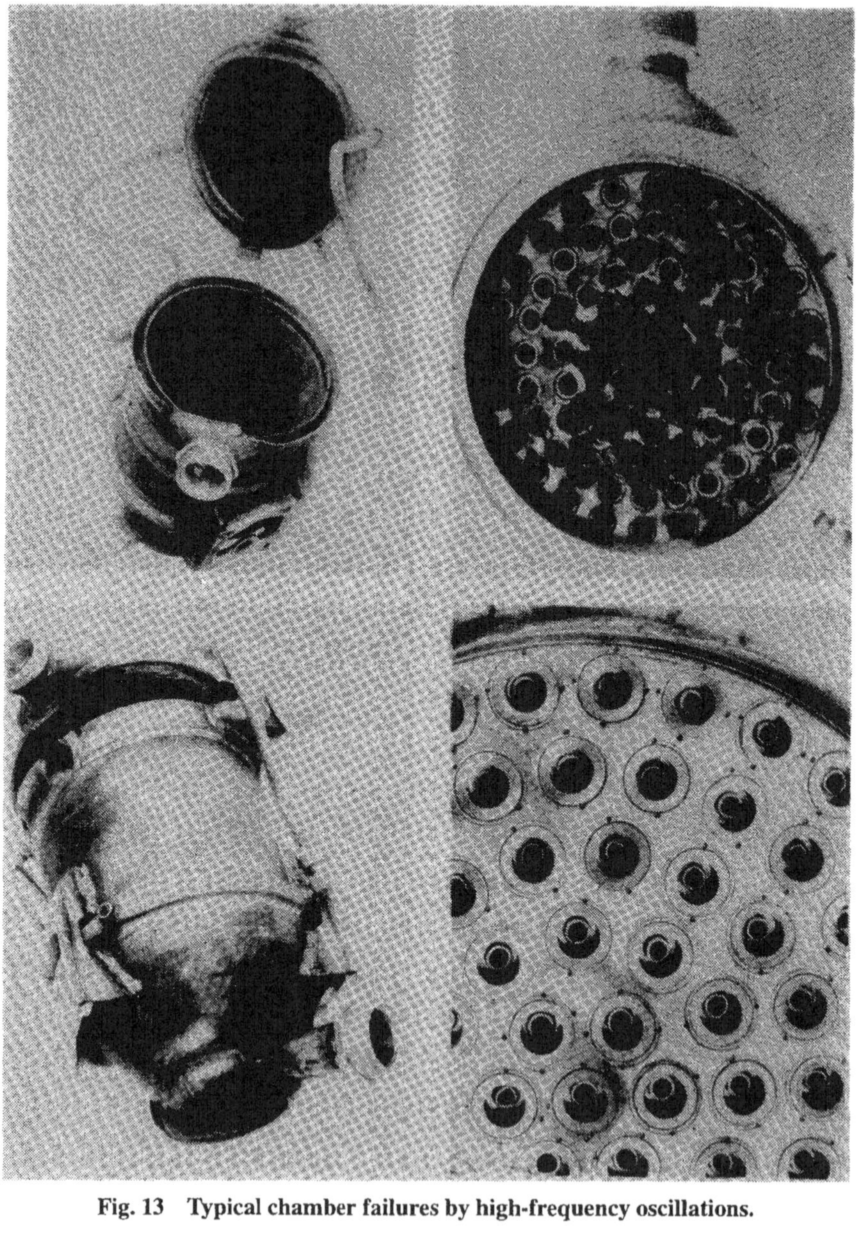

Fig. 13 Typical chamber failures by high-frequency oscillations.

propellants interaction in the initial zone, were undertaken. Tests of chambers with injector C4 were ended by the failure by high-frequency oscillations of the tangential mode during the start. The chamber with injector C5 successfully went through the tests, with a wide range of starting conditions and P_c and O/F at the main operating mode value. The specific impulse of the chamber with injector C5 was higher than the chamber with injector C3 by 10–15 m/s and met the technical requirements.

Results of the firing tests of the chambers with injectors C1–C5 confirmed the idea of the influence of propellant interaction in the initial zone (determined by the interlocation of oxidizer and fuel atomizing cones of two-component centrifugal atomizer L) on combustion efficiency and high-frequency stability. Thus, the dynamics of propellant burnout near the injector, combustion efficiency, and high-frequency combustion stability can be purposefully influenced by the change of L in bipropellant centrifugal atomizers located in the center or periphery of the injector.

The studies revealed that mixing between the oxidizer and fuel atomizing cones inside the atomizer over the whole injector field (variants C1 and C2) results in a short initial zone and the dynamics and volumetric energy release rate can create favorable conditions for replenishment of tangential mode oscillations. To change the energy release behavior in the initial zone and to eliminate the conditions for lateral oscillations initiation, it is sufficient to push the mixing zones of the spray cone–spray cone interactions out of the atomizers to the center of the injector. Possibly under these conditions, the mechanism of energy-release "extension" in the center of the injector begins to work, which leads to the stabilization of the combustion process. The analogous processes take place when the peripheral atomizers influence the propellant's atomizing cones in the same way (variants C3–C5).

Further development tests of RD-0110 confirmed the high-frequency stability and completeness of the operating process in the chamber with injector C5, which was accepted as a standard in the design documentation.

III. High-Frequency Stability Improvement During Engine Start

A. Analysis of High-Frequency Stability at RD-0110 Delivery

At the first stage of RD-0110 engine delivery to customers, during the firing approval tests performed by the production plant, one of the engine's chambers failed 0.04–0.1 s after start. The time from the beginning of the process to the moment the engine goes into the main operating mode was not higher than 0.18 s. The pattern of failure was similar to the pattern of the chamber's failure with injector C1 during off-line development tests and indicated that the cause was the appearance of the first tangential mode high-frequency oscillations in the chamber. Such cases were rare, one chamber out of four chambers per engine per 70–80 engines, i.e., one out of every 280–320 chambers. Chamber failure by high-frequency oscillations did not occur during the RD-0110 engine operation on the rocket-carrier Soyuz.

A careful analysis of production peculiarities and possible deviations in the chambers and engines that had high-frequency instabilities and start peculiarities was carried out. No deviations were revealed that were thought to influence high-frequency combustion stability. Prior experience with oxygen-kerosene engine development showed that high-frequency combustion stability is sensitive to the mixture ratio during start. Thus, special firing tests that provided the measurement of oxygen and fuel flow rate through each chamber to evaluate the mixture ratio deviation range for the four chambers during the start were carried out. The average mixture ratio in these tests was $0.9 \pm 15\%$. The spread in the four chambers of one engine also was 15% from the average O/F value of the engine. The total spread of O/F during start was 0.7–1.1 for all of the tests. Taking into account these results, off-line firing tests of the chambers were carried out with changes of O/F during start of 0.6–1.3 over the chamber pressure range 1.5–5 MPa. The aim of the tests was to reveal the stability margin during the start.

Chambers with extreme combinations of oxidizer and fuel atomizer pressure drop were also produced and were tested successfully. To simulate the high-frequency instability, tests with the changes of outer and inner factors at the start of operating mode were carried out; the changes were directed at decreasing the high-frequency instability margins. The engines went through the tests successfully. Thus, according to the results of great number of studies and analyses, the origins of the rare high-frequency combustion instabilities that occurred during the start of the RD-0110 engine were still undiscovered. In so far as the RD-0110 engine operated well on the rocket-carrier Soyuz, the cases of high-frequency instability during approval tests at the production plant were possibly connected with rare occurrences of a certain combination of production engine peculiarities and test stand peculiarities, which are difficult to detect.

Taking into account the fact that the chamber worked well in the main operating mode, it was necessary to find a method of high-frequency instability elimination which would not need a change in the engine start cyclogram (including the start on the rocket-carrier stage), or changes of injector design, which would lead to a great volume of work for engine finishing and mastering. It was also very desirable to find a solution that could be used in engines already produced after their nonsophisticated finishing.

A method that met these requirements and improved high-frequency stability of chamber operating process during the start consisted of the following. Several longitudinal "soft" ribs of middle height and made of combustible material (technical felt) were stuck to the inner cylindrical surface of the chamber beginning from the injector. It was supposed that the ribs would provide reliability at the start and would then burnout up to the moment the engine is to go onto the main operating mode.

B. Results of Model Studies on Effectiveness of Ribs

To analyze chamber acoustic characteristics (including the characteristics of the chamber with ribs), a special study was carried out. A chamber purged by air was periodically perturbed in time. Instantaneous measurements of pressure oscillations and injector flow rate oscillations and their phase shifts were measured. To actuate the chamber, an acoustic pulse generator (pulsator) was placed upstream of the injector. Acoustic characteristics of the chamber were determined by the amplitude-phase-frequency characteristic (normalized impedance), which is indicative of the amplitude of nondimensional injector flow rate pulsations,

$$K_A e^{i\phi} = \frac{p'/\bar{p}}{\dot{m}'/\bar{\dot{m}}}$$

where p' and $\dot{m}'$ are the pulsating components of pressure and flow rate and $\bar{p}$ and $\bar{\dot{m}}$ are the means of these values. K_A is the chamber amplitude-frequency characteristic and ϕ the phase characteristic.

The function K_A has the form of a resonance curve. Its maximum value ($K_A^{\max}$) corresponds to the chamber resonance frequency and signifies the acoustic gain factor. A reduction of $K_A^{\max}$ can be explained by additional dissipation of acoustic energy. The studies showed that mounting of the longitudinal ribs led to a considerable dissipation process. Thus, the felt ribs with the height of 26–28 mm decreased the resonance the acoustic gain factor by approximately three times. Figure 14 demonstrates the amplitude-frequency characteristics of the chamber with and without ribs for comparison. The nondimensional frequency of disturbance input to the chamber, $\omega R_k/C^*$, where R_k is the chamber radius, C^* the sound velocity in the nozzle throat section, and ω the oscillation frequency, is marked at the horizontal axis, and the ratio of the dimensionless amplitude of the pressure fluctation to the dimensionless injector flow rate fluctuation amplitude is marked at the vertical axis. The maximum value of this characteristic signifies a resonance acoustic gain factor of the first tangential mode oscillations. The acoustic studies showed that the main factor that influences the effect of ribs is the rib height. Neither the number of ribs nor the rib width have a considerable effect. There is a definite length of ribs beginning from which their effectiveness is practically the same. The acoustic study also showed that felt ribs are 50–100% more effective than metallic ribs of the same height.

The following is an explanation of the peculiarities of soft rib behavior in the chamber. Under the influence of oscillations, a submerged body will initiate periodic compression and rarefaction of the surrounding fluid and will be the source of acoustic waves. Energy, carried by these waves, will be obtained from the kinetic energy of the moving body. If the body is immovable and the acoustic wave is incident on it, there would be scattering of sound; together with the incident wave there would appear additional (scattered) waves. If the length of the acoustic wave is large in comparison with the dimensions of the body, the scattered wave can be considered to be a wave radiated by the body. The difference consists in the following: we are dealing not with the body movement in the fluid but with the movement of the fluid relative to the body.

Proceeding from such ideas, the mechanism of acoustic energy dissipation in the chamber with soft ribs can be described in the following way. The rib, on which the acoustic wave is incident (in this case, the wave has the first tangential mode of oscillations), scatters the wave. In so far as the scattered acoustic field will be determined by the geometric dimensions of the rib, the scattered wave pattern will have higher oscillation modes than the main movement. These oscillation modes in the resonance frequency should attenuate very quickly while moving along the chamber. The supposition concerning the higher performance of soft ribs made of felt can then be made: there is an additional dissipation mechanism, one that is possibly connected with the friction loss of felt on fibers under the rib oscillations or with the gas viscosity in the interporous space of the felt.

Special studies were carried out to check the effectiveness of felt ribs under firing test conditions. The special model chamber, working with the propellant combination nitrogen tetraoxide + tonka (50% triethylamine, 50% xylidine) was used. The results of the tests showed the presence of the chamber start mode, which always had high-frequency oscillations of the tangential mode. The chamber model was started both with and without ribs. Sensors registered the moment of onset of high-frequency oscillation. The chamber model was equipped with eight ribs,

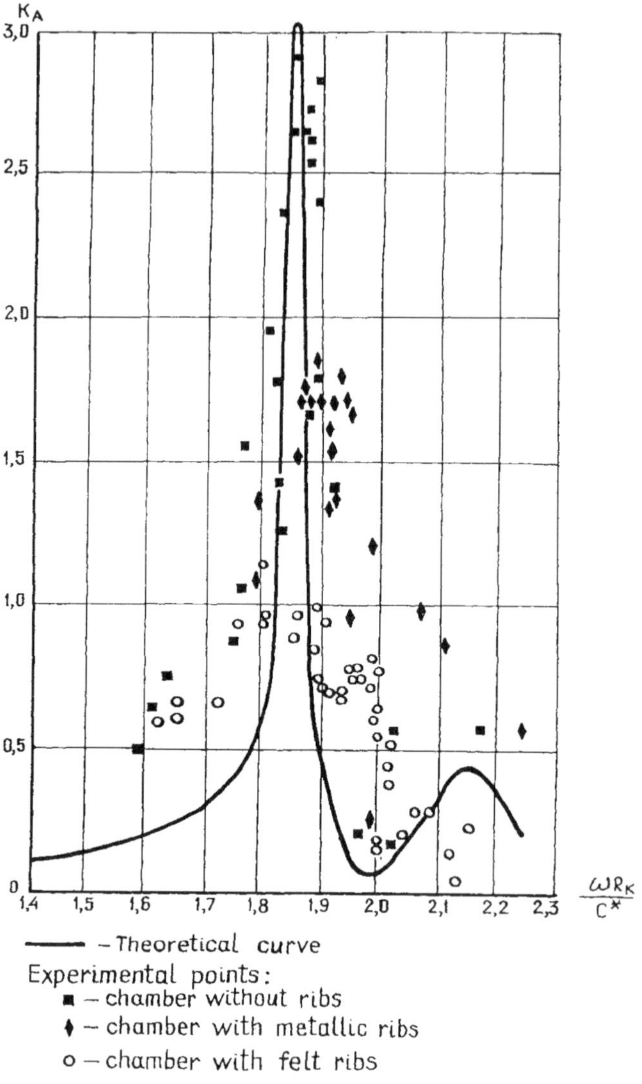

Fig. 14 Amplitude-frequency characteristic of the chamber.

0.25 m long, 20 mm wide, and 40 mm high. The ribs were stuck by a heat-resistant adhesive.

To get information about the moment of rib burnout or detachment during tests, thermocouples with open junctions were put into the ribs. The thermocouple fit 2–3 mm deep into the rib. If the ribs burned off or were torn off by the flow, the thermocouple became bare and burned down. During the tests of the chambers with the ribs, the high-frequency oscillations were noticed practically at the moment of rib burnout. In the chamber without ribs, the amplitude of the high-frequency oscillations increased quite quickly. In the chamber with ribs, oscillation amplitude increased smoothly while the process of rib burnout was going on. Figure 15 presents the results of the tests.

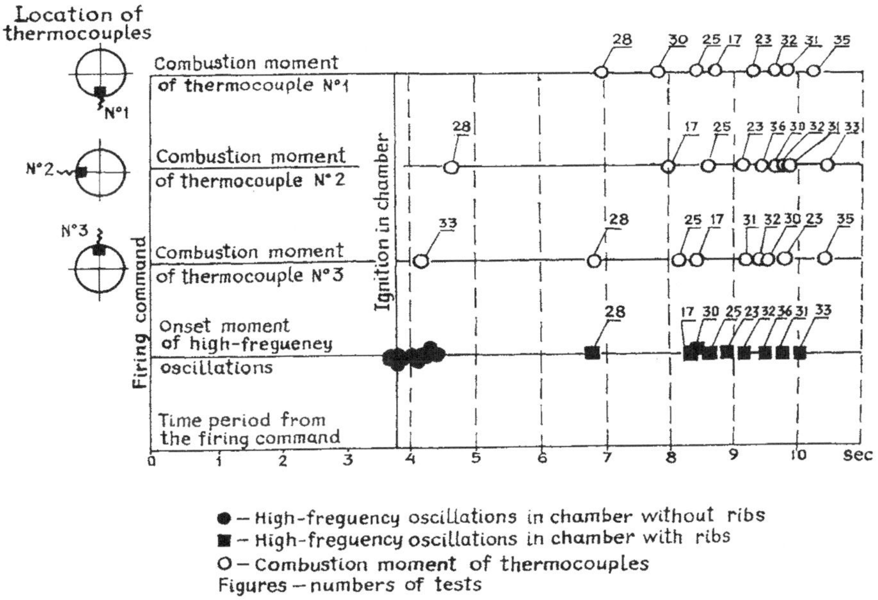

Fig. 15 Time when high-frequency oscillation occurs in the chamber with and without ribs.

As shown in Fig. 15, the high-frequency oscillations began 3–5 s later in the chamber equipped with felt ribs. The influence of rib number, height, and length on the suppression of high-frequency oscillations was checked by demonstrating that shortening rib length from 0.25 to 0.125 m and shortening rib height from 40 to 25 mm did not change the suppression effectiveness of high-frequency oscillations during the start. Decreasing the rib number from eight to four reduced the duration of stable operation by 1–1.5 s.

Tests were carried out to compare the effectiveness of soft ribs (made of felt) and stiff ribs (made of plastic and metal). Tests of a chamber with eight plastic ribs made of plastic ASTT and of the same size as the felt ribs were destroyed by high-frequency oscillations during the start of the chamber at 3.7 s and 3.8 s. The felt ribs stabilized the process up to 7–10 s.

IV. Test Results on the Use of Combustible Felt Ribs During Firing Tests in Chambers and Engines

Acoustic studies and firing tests confirmed the effectiveness of felt ribs for stabilization during chamber start up to the moment of rib burnout. Full-scale studies were also done in this direction during off-line bench tests of chambers and engines. During the bench tests of the chambers it was necessary to evaluate the effectiveness of combustible felt ribs for stability improvement for at least 0.18 s of the engine start and to allow a decision about rib integration into the standard design. Thus, it was necessary to master a reliable technology to bond the ribs, to define the influence of the rib-bonding process on the reliability of chamber cylindrical section inner wall cooling, to determine the explosion hazard of felt ribs at the initial stage of start under the conditions of advanced liquid oxygen flowing into the chamber when the cartridge igniters operate, and to confirm the biological shielding of ribs under the long-term storage.

The technology of rib production and adhesion was developed and mastered. It consisted of the selection of an adhesive that could perform satisfactorily in contact with stainless steel without special thermal treatment or great effort to press and stick the ribs, selection of a biological shield for long-term storage, and determination of the optimal felt type.

During the technological mastering and chamber tests, different types of adhesive were tried: adhesive VS-10T, epoxy underpaint hard putty E4020, hermetic Viksint with an underlayer of cold drying P11, and adhesive K300. Fitness of adhesive was determined by the time of stable chamber operation with the ribs, the adhesive and rib remainders at the chamber inner wall, and the state of the chamber inner wall after the tests. According to the analysis of preliminary tests, the best results were given by the VS-10T adhesive and hermetic Viksint with the underlayer of cold drying P11. The samples stuck by the VS-10T adhesive had high-mechanical strength. To bond the ribs, however, required a high drying temperature (180°C for 2 h) and great effort of pressing, which excluded the use of such an adhesive to bond the ribs in the ready-made engines. Also, the chamber life tests showed that the VS-10T adhesive sometimes did not completely burn off and formed carbonization at the chamber inner wall. Carbonizations disturbing the gas boundary layer lead to wall burnthrough behind ribs.

Hermetic Viksint with underlayer P11 does not need high bonding temperatures; it is polymerized at 20°C during 72 h. The joint was very strong and elastic. Satisfactory resistance and effectiveness of the ribs stuck by the hermetic Viksint with underlayer P11 was confirmed by the chamber firing tests. Reliability of the rib bond was controlled by the samples. Minimum breaking load values were obtained during shear tests of the steel-felt adhesive joint bonded by hermetic Viksint with underlayer P11. Felt samples $20 \times 20 \times 40$ mm were bonded to the plate from steel IX18H10T. The tearoff force for 25 samples was 65–90 kPa.

Chamber firing tests revealed that the rib resistance depends considerably on the type of felt and on the location of its fibers relative to the bond plane. Ribs made of 3–10-mm-wide layers with fibers parallel to the bond plane and bonded by VS-10T adhesive lost their effectiveness during the chamber firing tests. During the firing tests the ribs made of fibers parallel to the bond plane became larger and were torn off just after the start and consequently did not provide the necessary effect. A rib tear occurred in the felt. Ribs made of 2–15-mm-wide layers with fibers located perpendicular to the bond plane and bonded with hermetic Viksint with one elastic adhesive joint did not change their size, had full burnout, and suppressed the oscillations for more than 0.18 s after the chamber start. This design was considered to be an optimum one.

To provide the biological shield of felt ribs, different coverings were tested: water glass, aluminum enamel C0813, adhesive VS-10T, and hermetic Viksint. The best covering was a fine layer of hermetic Viksint; it was elastic and was not broken while ribs were bonded to the chamber, which allowed the opportunity to cover the ribs before their attachment to the chamber. All of the other coverings must take place after the ribs were attached to the chamber to exclude any damage when pressing the ribs by the rig during their bonding. This fact causes considerable technological difficulties and decreases the reliability of the rib biological shield.

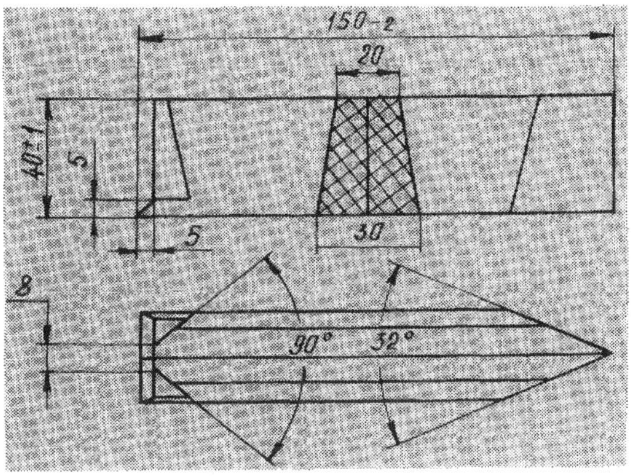

Fig. 16 Rib.

The firing tests showed that the width of covering influences the ribs' effectiveness: increasing the covering width up to 1.5–2.0 mm reduces their effectiveness. The optimum covering was one with 0.2–0.6 mm width.

After the technological mastering and preliminary firing tests (which determined the felt type for rib production, the type of adhesive for their sticking, the technological process, and the rib material) a design variant was chosen for the final stage of off-line chamber firing tests to confirm the effectiveness of ribs in enhancing stability during simulated engine start conditions. Six ribs were located in the chamber. The ribs had the following dimensions: length 0.150 m, height 40 mm of trapezoidal section, top base 20 mm, and lower base (bonding place) 30 mm, with a sharp trailing edge to smooth the gas flow (Figs. 16 and 17). To provide a shield from biological organisms, the ribs were covered by 0.2–0.6-mm-wide hermetic Viksint before their placement into the chamber. The ribs were stuck onto the chamber inner wall by the hermetic Viksint with underlayer P11. Figure 18 shows the chamber with the ribs.

Fig. 17 Set of ribs.

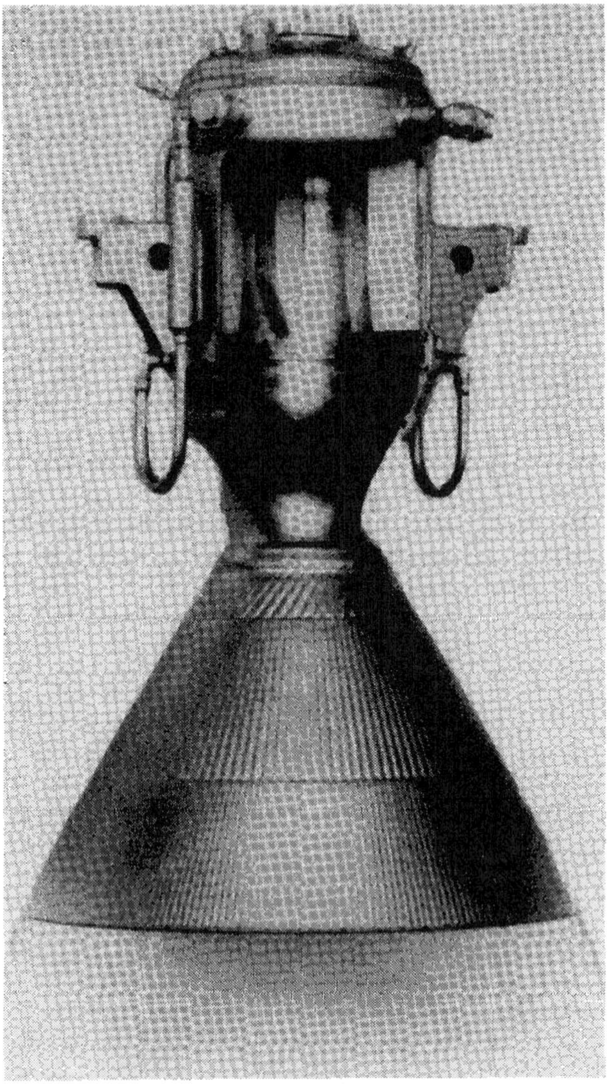

Fig. 18 Chamber with ribs.

Insofar as the high-frequency instability ocurrence rarely arose, it was necessary to develop a method to reliably produce high-frequency oscillations with large amplitude at the initial moment of chamber start up. A method was developed to produce high-frequency oscillations during chamber start up by the use of a cavitating tube (venturi) located at the oxygen inlet to the injector. The cavitating tube (venturi) had converging and diverging parts with a minimum section (throat) diameter of 7 mm. A pressure differential 10–12 MPa provided the cavitating mode oxygen outflow.

During the tests, asymmetric high-frequency oscillations with frequencies of 1100–1600 Hz and amplitudes more than 1.0 MPa arose. Such oscillations made the analysis of effective measures to improve high-frequency stability difficult. The moment of the onset of high-frequency oscillations also depended considerably on the cavitation intensity at the initial moment of chamber start. Such considerations led to the choice not to use this method of producing high-frequency oscillations.

During off-line mastering tests, injector C1 produced high-frequency tangential mode instability with amplitudes greater than 1.5 MPa at 0.04–0.11 s after the beginning of the chamber start-up process. Therefore, later studies of the effectiveness of ribs were performed in chambers with the injector C1. In so far as the effectiveness of ribs in stabilizing combustion during the start process could be confirmed only during the tests of one chamber without ribs (with high-frequency instability during the start) and with ribs (with high-frequency instability after the ribs tear off or burn out), it was necessary to master the chamber emergency switchoff before chamber failure due to high-frequency oscillations. For that, a signal was given to the emergency switchoff system from the pressure fluctation piezoelectric transducer, which was put in the combustion cavity in front of the injector atomizers. When the number of pressure fluctations with the frequency of the first tangential mode $f = 4000$ Hz and with an amplitude ≥ 0.5 MPa exceeded more than 15–20 cycles, the emergency switchoff system gave the signal to switchoff the chamber by closing a quick-acting oxidizer valve located just before the injector. Such a system provided a test shutdown within 0.04 s after the beginning of high-frequency oscillations and kept the chamber intact for repeat tests. The difference in time between the signal from the emergency switchoff system and the moment of ignition determined the time of stable chamber operation. A comparison of these differences during tests of the chambers with and without ribs led to a conclusion about the effectiveness of ribs for stabilization of the operating process. The results from the final stage of chamber off-line firing tests are presented in Table 4. Test results are given graphically in Fig. 19. As seen from the presented materials, during the tests of the chambers without ribs, high-frequency tangential-mode oscillations with frequency $f \sim 4000$ Hz developed 0.03–0.095 s after ignition. The presence of ribs increased the time of stable operation up to 0.237–0.304 s. Rapid increase in the amplitude of high-frequency oscillations began after the ribs burned out and tore off.

The moment the ribs tore off was measured at the beginning and at the end of the rib during the tests. For that, two magnets were placed into the felt ribs (at the

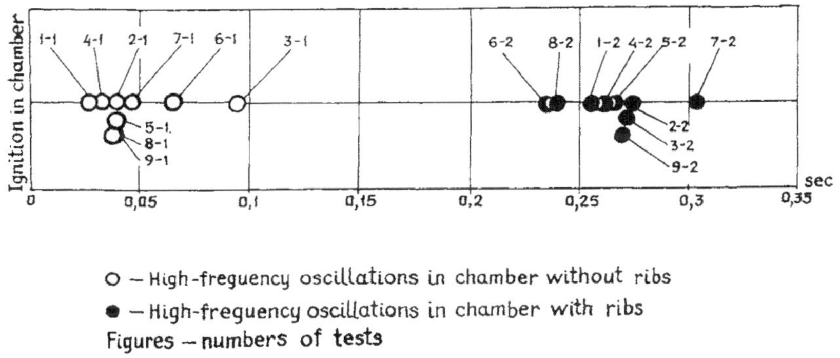

Fig. 19 Time when high-frequency oscillation occurs during the off-line tests of chambers with and without ribs.

Table 4 Effectiveness of ribs in damping oscillations

Test no.	Ribs	Ignition time, s	Oscillation inception time, s	Inception times, s	Time delay due to ribs, s
T1-1	absent	0.7	0.73	0.03	
T1-2	present	0.705	0.963	0.258	0.228
T2-1	absent	0.716	0.758	0.042	
T2-2	present	0.707	0.983	0.276	0.224
T3-1	absent	0.722	0.817	0.095	
T3-2	present	0.842	1.117	0.275	0.18
T4-1	absent	0.707	0.742	0.035	
T4-2	present	0.709	0.973	0.264	0.229
T5-1	absent	0.7	0.74	0.04	
T5-2	present	0.726	0.994	0.268	0.228
T6-1	absent	0.72	0.787	0.067	
T6-2	present	0.738	0.975	0.237	0.17
T7-1	absent	0.723	0.77	0.047	
T7-2	present	0.742	1.046	0.304	0.257
T8-1	absent	0.685	0.723	0.038	
T8-2	present	0.686	0.926	0.24	0.202
T9-1	absent	0.703	0.741	0.038	
T9-1	present	0.71	0.984	0.274	0.236

beginning and at the end of the rib). Blocks of hermetically sealed reed relays were stuck by C300 adhesive onto the chamber outer wall over the magnets that were placed in the ribs. The contacts of the hermetically sealed reed relays were kept in the closed position by the magnets. The measurement system registered the moment the ribs with magnets were torn off by recording the time the contacts broke.

The time of contact break, indicating the moment the ribs tear off, coincided with the time that the signal from the emergency switchoff system was given by sensing high-frequency oscillations with increasing amplitude. Thus, the felt ribs provided stability during the start up to the moment of rib burnout or tearoff. The delay of the appearance of the high-frequency oscillations during chamber start up is 0.17–0.257 s longer with ribs than in the chambers without ribs. Taking into account the fact that the time from propellant ignition to the moment the engine transitions into the main operating mode is less than 0.18 s and the ribs (according to the tests results) stay effective for between 0.237 and 0.304 s, the attachment of combustible felt ribs provided stability during the engine start.

When the engine goes on, the chambers start together with an oxygen advance and with working igniters within 0.4 s. To confirm that there was no explosion hazard and to ensure reliable operation of the chambers with ribs, off-line firing tests of the chambers were carried out with oxygen advance flowing into the chamber during the start up to 1.2 s with operating igniters. The chambers went through the tests successfully.

Vibrational and transport overload tests of the chambers with ribs and with preliminary thermostatting of the chambers were also carried out. The levels of vibrational and transport overload and thermostatting conditions corresponded to the engine requirements. Visual inspection after the tests indicated no problems. After the vibrational and transport overload tests, the chambers went through firing tests, during which the appearance of high-frequency oscillations was delayed by 0.24–0.28 s, corresponding to the results presented in Table 4. After the off-line chamber firing tests, qualification life tests of the three-engine RD-0110 were performed. The engines went through the tests successfully, and visual inspection of the chambers did not reveal any defects. The time range of rib tearoff during the engine start was 0.24–0.29 s, which also corresponded to the results of chamber off-line tests presented in Table 4.

The combustible felt ribs were included in the standard chamber documentation of the RD-0110 engine as a method of providing stability improvement during start. After the modification to include ribs, more than 2500 RD-0110 engines, or more than 10,000 chambers, went through the firing approval tests at the production plant. There was not one case of high-frequency combustion instability during the engine start. More than 1000 of the RD-0110 engines successfully took part in the start of the Sojuz rocket carrier, including manned spacecrafts.

References

[1]Crocco L., and Cheng, S.-I., *Theory of Combustion Process Instability in Liquid Rocket Engines,* IL, Moscow, 1958, pp. 59–69.

[2]Sheveljuk, M. I., *Theoretical Basis for Liquid Rocket Engines Design,* Oborongiz, Moscow, 1960, pp. 589–597.

[3]Harrje, D. T., and Reardon, F. G. (eds.), *Instability of Combustion Process in Liquid Rocket Engines,* Mir, Moscow, 1975, pp. 426–445.

[4]Alemasov, V. E., Dregalin, A. F., and Tishin, A. P., *Theory of Rocket Engines,* Mashinostroenie, Moscow, 1980, pp. 372–374.

[5]Natanzon, M. S., *Combustible Process Instability,* Mashinostroenie, Moscow, 1986, pp. 115–121.

Chapter 5

Instability Phenomena in Earth Storable Bipropellant Rocket Engines

Eric A. Hurlbert*
NASA Johnson Space Center, Houston, Texas 77058
and
Jing Liang Sun† and Baojiong Zhang‡
Shanghai Bureau of Astronautics, Shanghai, People's Republic of China

Nomenclature

K_i = gas solubility equilibrium constant
P_i = partial pressure of gas over the liquid
I_{sp} = specific impulse
i = gas
j = propellant
F = fuel
O = oxidizer

I. Introduction

EARTH storable bipropellant engines are used extensively in many areas of space flight. In the Apollo program, storable bipropellant engines landed humans on the moon and returned them safely to Earth. In current applications, the Space Shuttle uses Earth storable engines for maneuvering and deorbiting. Many launch vehicles, such as the Titan, Delta, Long March, and Ariane, use Earth storable propellants for boosters and upper stages. Earth storable propellants are also used in most unmanned spacecraft and satellites.

Dynamically stable combustion in Earth storable propellant engines has been obtained through extensive testing of injector element patterns and damping devices, such as baffles and acoustic cavities. Development of stable injector patterns

---

*Propulsion Engineer, Propulsion and Power Division.
†Professor and Launch Vehicle Chief Designer.
‡Professor.

Table 1 Thermodynamic and physical properties[1,2] at 298 K and maximum performance with N_2O_4

	Density, kg/m³/s	Freez. temp., K	Boil. temp., K	Crit. temp., K	Crit. press., MPa	Specific heat, J/gK	Viscos. centipoise, g/m-s	Therm. cond., cal/ cm-s-K	Surf. tens., dynes /cm²	I_{sp} max, s	Opt. mix. ratio
NTO	1454	262	294	431	10.1	1.57	0.396	3.13*e*-4	25		
N_2H_4	1014	275	387	653	14.7	3.09	0.910	7.86*e*-5	66	348	1.4
MMH	872	221	361	593	8.24	2.93	0.680	5.92*e*-4	34	344	2.3
UDMH	785	215	336	523	5.42	2.73	0.509	3.76*e*-4	24	342	2.7
A50	897	265	343	608	11.7	2.91	0.817	6.83*e*-4	29	345	2.1

and suitable damping devices occurred primarily in the 1960s, during the Apollo program. This chapter describes the development of storable propellant engines in the United States and China. Through these case studies, which include quantitative details of the combustion instabilities experienced, combustion instability phenomena in storable propellant engines can be examined.

Earth storable propellants comprise a family of propellant combinations, which use primarily nitrogen tetroxide (NTO) as the oxidizer, and hydrazine, monomethyl hydrazine (MMH), unsymmetrical dimethylhydrazine (UDMH), or their blends as the fuel. Earth storable refers to the characteristic of being liquid at standard temperature and pressure, and being relatively easy to store for periods of years. Table 1 lists the thermodynamic and physical properties of these propellants, with the maximum theoretical performance evaluated using NTO at an area ratio of 60:1 and optimum mixture ratio.[1,2]

II. History and Applications of Earth Storable Propellants

Nitrogen tetroxide was first experimentally used for rocket engine research in 1930 by Crocco in Italy and by Glushko in Russia.[3] The fuels were gasoline and toluene, respectively. Hypergolicity, the spontaneous ignition upon contact of oxidizer and fuel, was probably first discovered around 1937, by Walter in Germany, using hydrazine hydrate ($N_2H_4 \cdot H_2O$) and peroxide.[3] The word hypergolic was coined by Noeggerath during experimental work with Sanger and Mueller in Germany on nitric acid and fuels such as aniline.[3]

In the 1950s, hydrazine was considered for use as a propellant because of its high performance, high density, and hypergolicity with storable oxidizers. The only problem was the freezing point, which did not meet military specifications for cold weather applications. A U.S. Navy contract with Aerojet and Metallectro addressed this problem successfully.[3] This contract evaluated the hydrazine derivatives, such as monomethylhydrazine (MMH, CN_2H_6), unsymmetrical dimethylhydrazine (UDMH, $C_2N_2H_8$), and symmetrical dimethylhydrazine (SDMH).[3] MMH and UDMH were found to be suitable; however, SDMH proved to be a low performer. Aerojet later produced Aerozine 50 (A50), a 50/50 blend of hydrazine and UDMH.[3]

Storable propellants are used in applications that benefit from 1) long-term storability, 2) high-bulk density performance, and 3) hypergolicity. These attributes result in space vehicles that are simple, cost effective, and reliable. Because of the high density, it is practical to use tank pressure to feed the engine. A pressure-fed engine significantly reduces vehicle complexity, such that engine operation consists of opening only two engine valves. For cost and packaging reasons, many applications use a nonoptimum mixture ratio of 1.6 to make the oxidizer and fuel tanks identical. This results in a lower actual specific impulse, which typically ranges from 280 s to 334 s. Nevertheless, the high-bulk density still provides good overall vehicle performance, since the small tank volumes reduce vehicle structural mass and improve packaging in confined areas. The main drawback is the toxicity of Earth storable propellants, which requires measures to prevent human contact during testing and launch vehicle processing.

Titan 2 was the first major U.S. launch vehicle program to use Earth storable propellants (A50 and NTO).[4] Taking advantage of Earth storability, Titan 2 replaced the Titan 1 and Atlas, due to the possible launch delays inherent with fueling liquid oxygen and kerosene minutes before launch.[4] It went into service in 1962 and was later used in the Gemini program with engine stability improvements, which included baffles and injection distribution modifications.[5,6] Following the Gemini program, the Apollo program used NTO with A50 and MMH. In the 1950s, China began development of storable propellant engines, which led to the Long March Series I and II. In the 1970s, the Space Shuttle program switched entirely to MMH and NTO for on-orbit propulsion. Depending on the requirements of future vehicles, Earth storable bipropellant engines may again be used, if long-term storability, propulsion system simplicity, high reliability, and small packaging are desired.

III. Combustion Phenomena in Earth Storable Engines

Combustion in Earth storable bipropellant engines has some unique characteristics. One critical characteristic is hypergolicity, in that chemical reaction can start with contact of the liquids and can affect the atomization process. Many instabilities with Earth storable propellants can be related in some manner to 1) popping, 2) resurging, and 3) reactive stream separation. In addition, products that form prior to ignition can cause ignition spikes and disturb combustion. In pressure-fed engines, dissolved gas in the propellant can also affect combustion stability.

The injector type affects the manner in which the propellants come into contact, thereby also affecting combustion stability. Coherent stream injector elements include unlike doublets (UD), triplets [fuel-oxidizer-fuel (FOF) or oxidizer-fuel-oxidizer (OFO)], quadlets, and coaxial elements (pintle or multiple). Self-atomizing elements include the transverse like-on-like (TLOL) and splashplate elements. Examples of these elements can be found in the engine case studies later in this chapter. Additional discussions of storable propellant instabilities can be found in Hefner[6] and Abbe et al.[7]

A. Preignition Products

During engine start and prior to ignition, preignition products can form in the chamber. After ignition, these preignition products can detonate and produce a large disturbance to the combustion process that can affect combustion stability. During the Apollo program, Christos et al.[12] examined the explosivity of the preignition products at the U.S. Bureau of Mines, using ballistic mortar and other tests. Christos found that the preignition products had an explosive equivalence of 139% of TNT.

Mayer et al.[8] experimentally studied the preignition of NTO and MMH at low pressures from 10-mm Hg (0.2 psia) to 110-mm Hg (2.2 psia). Using mass spectrometry, Mayer et al.[8] found that unstable nitroamines are created during the preignition phase. Prior to ignition, a flameless reaction occurs between the fuel and NO_2, which produces a yellow fog in the chamber. Mayer et al.[8] reported that this fog consists of monomethyl nitroamide ($CH_3N_2HO_2$). Analysis of the residues of this fog found MMH nitrate. The mechanism proposed for the formation of MMH nitrate is

$$CH_3N_2H_3 + 2NO_2 \rightarrow 3H_2O + 2N_2 + CO$$

$$H_2O + 3NO_2 \rightarrow 2HNO_3 + NO$$

$$HNO_3 + CH_3N_2H_3 \rightarrow CH_3N_2H_3 \bullet HNO_3$$

Saad et al.[9] examined the mass spectra and infrared spectra of reaction product residues and found, in addition to MMH nitrate, methylnitrosamine (CH_3NHNO), which is unstable and decomposes to water, methanol, and nitrogen. Takimoto and Denault[10] also examined residues from thrusters and found MMH nitrate and ammonium nitrate. Interestingly, ammonium nitrate formed only during the steady-state firings. They proposed that the mixture containing MMH nitrate also contains heat-sensitive compounds, such as methyl azides and diazomethane, which can trigger sudden MMH nitrate decomposition. MMH nitrate has a low-vapor pressure which allows it to accumulate over successive pulses, especially when the thruster temperature is cold. Takimoto and Denault[10] found that three engine-operation factors increase the formation of MMH nitrate: 1) multiple pulses of short duration, 2) low propellant and engine temperatures, and 3) oxidizer leading the fuel in entering the chamber.

Hydrazinium nitrate and UDMH nitrate is obtained for the fuels, hydrazine and UDMH, respectively. Mayer et al.[8] reported that UDMH nitrate is unstable and decomposes much more readily, suggesting that UDMH may not form deposits which remain on the walls. In general for all Earth storable propellants, many different intermediates can be formed. Mills et al.[11] reported that low-activation energies are involved, which leads to the formation of many complex intermediates. Careful consideration of engine operating conditions and valve timing can limit the formation of preignition products.

B. Hypergolic Ignition and Overpressures

During ignition, large overpressures can occur due to the explosion of any preignition products. This is significant primarily because of the potential for damage to the engine or to the stability of the combustion process. Juran and Stechman[15] reported that large ignition overpressures are related to the accumulation of intermediate reaction products on cold chamber walls and that as thruster initial temperature increases, the pressure spike occurs less frequently.

The ignition of hypergolic propellants can be described as a thermal mechanism. The thermal ignition of an NTO and MMH mixture can be computed using the Semenov method or the Frank–Kamenetskii method as described by Glassman[13] and Gray.[14] These methods consider the balance between the rate of heat production by chemical reaction and the rate of heat conduction to the walls. When the rate of heat production exceeds the rate of heat conduction, a rapid reaction occurs, which is ignition.

Kerkam and Kahl[16] performed engine tests on the 10,000-lbf thrust A50/NTO Lunar Module descent engine. High-speed photography showed brown viscous low-vapor-pressure deposits on the chamber wall just prior to ignition and the pressure spike. The occurrence of pressure spiking can be controlled by 1) correct oxidizer lead/lag time, 2) engine and propellant temperature, and 3) minimization of manifold (dribble) volume.

C. Hypergolic Droplet and Spray Combustion

Combustion of hypergolic propellants can occur in many forms of homogeneous or heterogeneous reactions. Most of the chemical reaction probably occurs as a homogeneous reaction between fuel vapor and oxidizer vapor, typically in which liquid fuel droplets are vaporizing in a gaseous oxidizer-rich environment, since NTO has the lower boiling point. At cooler conditions, heterogeneous reactions between liquid fuel and oxidizer vapor or even homogeneous reactions between liquid fuel and liquid oxidizer can also be expected. Experimental studies have sought to characterize this process, since it can have large effect on performance and stability.

The contact of two hypergolic liquids was experimentally studied by Daimon et al.[18] to measure the time lag of ignition or explosion. The tests were performed by impacting droplets of fuel, at various velocities and drop size, onto a pool of NTO. Upon contact, a layer of fuel and NTO vapor forms due to the gasification of NTO. In this layer a turbulent combustion reaction occurs. The results also indicated that the time lag of ignition decreased as droplet diameter increased.

Allison and Faeth[19] studied the burning of suspended MMH, N_2H_4, and UDMH droplets at atmospheric pressure in gaseous oxygen. The d^2-evaporation law plots, where d is the droplet diameter, were determined for the burning rates. They found that the droplet burns in two distinct regions: the inner region as a monopropellant, and the outer region as a bipropellant. This is especially true for hydrazine and less so for MMH and UDMH.

Eberstein and Glassman[20] experimentally determined the decomposition rates of monopropellant gas phases of MMH, N_2H_4, and UDMH in an adiabatic flow reactor. They proposed that the two methyl radicals in UDMH increase the chain-branching reactions, which explains why UDMH decomposition has the lowest activation energy. In MMH, however, this chain branching does not occur; hence, MMH has the highest activation energy.

For actual injectors under various operating conditions, Lawver[17] photographed an MMH/NTO unielement unlike doublet. Lawver explained that a yellow flame region occurs near the impingement point due to NH_2 radical emission. Farther down from the impingement point, a blue region indicates methyl radical emission. MMH/NTO thrusters are observed to exhaust a blue-white flame from the nozzle. Lawver[17] also noted that hydrazine burns with a yellow flame over the entire combustion region, due to NH_2 radical emission.

This is a short list of some of the analytical models of the hypergolic spray combustion process that have been made. For an MMH/NTO pintle engine, Krishnan et al.[22] considered both propellants entering as droplets and vaporizing in a Lagrangian frame, which was then mapped into an Eulerian frame and reacted in a simplified single step. Chiu et al.[23] treated the entire process in an Eulerian frame using the codes GEMCHIP and BICOMB. Larosiliere et al.[21] used a Lagrangian–Eulerian approach and a two-step reaction for hydrazine and NTO and also considered the effects of reactive stream separation.

D. Reactive Stream Separation

Impinging streams of hypergolic propellants can be classified by the degree of 1) penetration, 2) mixing, and 3) separation. Penetration occurs when one propellant stream "shoots through" the other, which results in poor mixing. Mixing is the production of an uniformly distributed field of droplets that provides predictable smooth combustion. Blowapart or reactive stream separation (RSS) occurs when a gas layer forms between the impinging streams and prevents proper mixing.

Zung and White[24] and Lawver[17] used high-speed photography to view unielement impinging type injectors under actual engine operating conditions. Lawver studied blowapart for unlike doublets, triplets, and splashplate elements used in the Space Shuttle engines, which are discussed later in this chapter. Mixing is displayed in Fig. 1, and reactive stream separation is shown in Fig. 2, both for the Space Shuttle reaction control engine unlike doublet at actual operating conditions.[17] Reactive stream separation in coaxial injectors has recently been studied by Habiballah et al.[25]

Kushida and Houseman[26] identified criteria for predicting the separation of reactive streams that was based on pressure and the contact time as related to jet diameter and velocity. Lawver[17] correlated the regime of mixing and separation to the fuel orifice Reynolds number, because fuel vaporization was identified as the combustion rate limiting process. Lawver found that increasing chamber pressure and fuel velocity promoted separation. This study also concluded that the transition from mixed to separated modes should be avoided within the engine operational envelope.

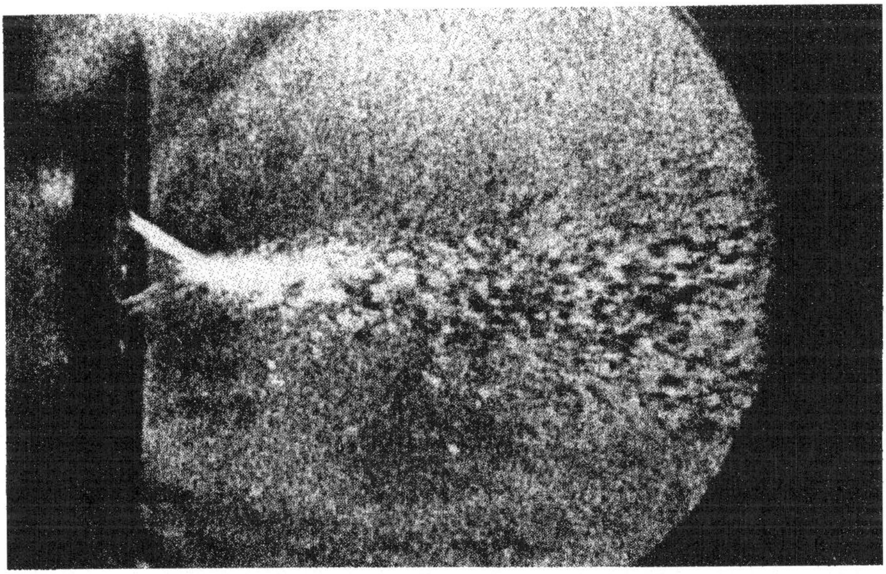

Fig. 1 Mixing of impinging MMH and NTO streams.

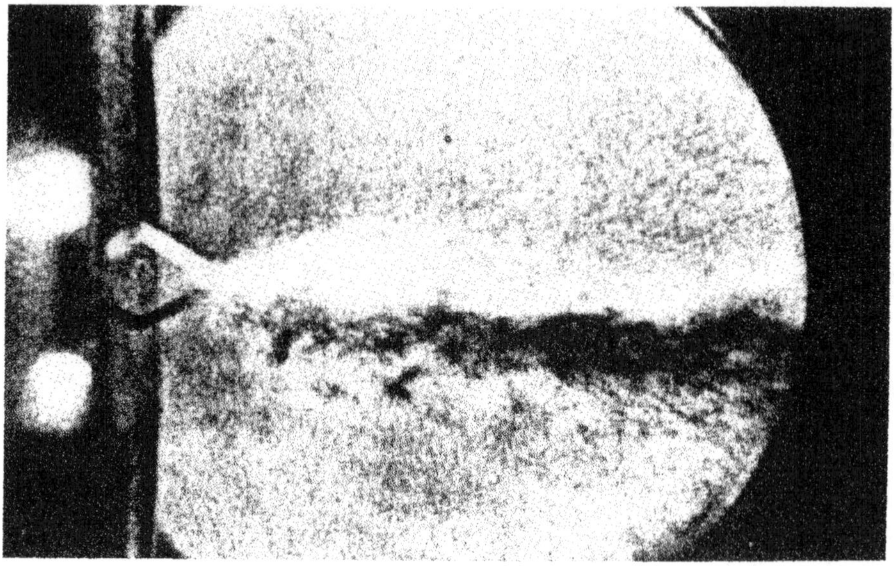

Fig. 2 Separation of NTO and MMH streams.

E. Popping

Perhaps one of the most unique characteristics of hypergolic propellants may be the tendency to generate "pops." These pops can trigger instabilities in the engine, much the same as "bomb" tests used in the stability rating of engines. Although other types of propellants may display this phenomena, popping occurs more frequently with storable propellants. Many of the engines discussed later in this chapter, such as the Apollo Service Module engine, have experienced varying degrees of popping.

Pops are the result of accumulation and subsequent explosion of intermediate products, which are probably nitrates of the fuel, similar to preignition products. The mechanism for the generation of a pop is of considerable interest, since pops have been observed to initiate instabilities. For impinging elements, popping occurs at equal fluid stream dynamic pressures, which causes the streams to simultaneously stagnate. In addition, the temperature range of 297–300 K displayed the highest popping rate, outside of this range the popping rate falls off.[27]

Mills et al.[11] postulated that two conditions must be present for popping to occur: 1) the propellant temperatures must allow the propellants to mix as liquids and 2) the time for full atomization must be long enough to allow ignition to occur. Visual observations of unielement tests indicated that pops occur in the ligaments. These ligaments were observed to explode before fully atomizing. It was also observed in injector tests that two types of pops can occur: 1) smaller explosions within the element and 2) larger two-phase detonations of the entire field of elements.

Based on streak photography observations, Houseman and Lee[28] proposed that popping and reactive stream separation are related. Lawver[17] reported that pops are caused by reaction intermediates formed at the impingement point by surface reactions that occur at low-fuel-vapor pressures or under conditions of high heat

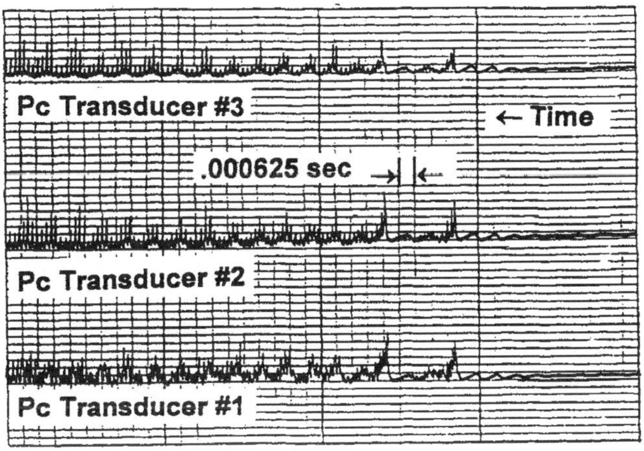

Fig. 3 Chamber pressure resulting from resurging.

loss, such as with large streams (>0.060 in. diameter) or cold propellants. The reaction intermediates form when the vapor separating the flow streams collapses and allows the liquids to mix. These reaction intermediates then move into a hot region where they detonate or pop.

F. Resurging

Resurging causes chamber pressure excursions that can eventually result in high-frequency instability. An example of resurging is shown in the pressure trace[29] in Fig. 3. This particular case is probably best described as a klystron effect,[27] in which sharp pulsing flow stream variations, possibly caused by engine chugging, cause the propellants to clump.[27] These clumps periodically form and then ignite, resulting in resurging that excites an acoustic resonance of the chamber. The example in Fig. 3 shows resurging continuing until it becomes a sustained high-frequency instability.

G. Dissolved Gases

The presence of dissolved gas in propellants is a potential problem for Earth storable bipropellant engines, which are pressure fed from screen acquisition tanks, rather than pump fed. In the propellant tank, pressurant gas dissolves in the propellant over time, until the propellant becomes saturated. During engine operation, these gases evolve from the propellants due to 1) the pressure drop in steady flow in the feed lines and the engine injector and 2) the feedline pressure transients at engine startup and shutdown. A significant quantity of helium can be produced which can effect the stability of the engine, as evidenced in the Space Shuttle primary reaction control thruster and the Apollo Lunar Module Ascent engine.

Chang et al.[30,31] provide experimentally determined relationships for the solubility of various gases in storable propellants. Solubility of gas in a liquid is determined from the application of Henry's law, which states that the solubility in terms of the mole fraction concentration x_i of a dissolved gas i is proportional to the partial pressure.

$$(\mathrm{ppm}_i) = K_i \cdot P_i$$

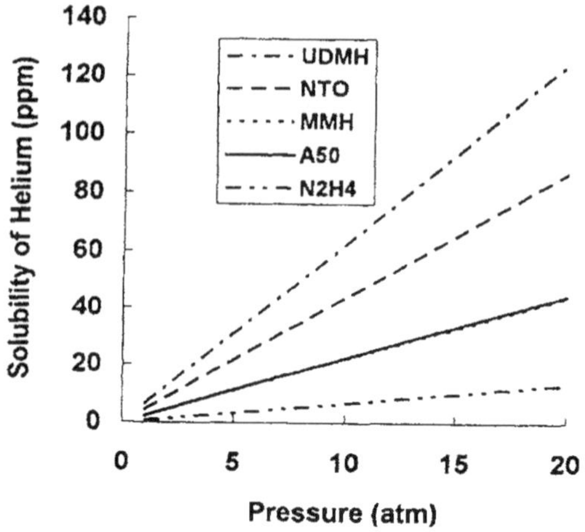

Fig. 4 Helium solubility in storable propellants.

The amount of dissolved gas in solution can be found using the preceding equation, assuming enough time has been allowed for complete saturation, which could be hours for a typical tank. Figure 4 shows that helium dissolves in the highest concentration in UDMH and NTO.

Visual observations suggest that gas is released rapidly from solution when the pressure is dropped. However, few data are available to determine quantitatively the time rate of evolution of gas from the propellant. Assuming the release is instantaneous as the pressure is dropped, however, a conservative estimate of the volume of gas created can be easily determined.

Perturbation of the combustion process can occur by 1) creating an imbalance resulting in improper mixing of the fuel and oxidizer streams, 2) reducing fuel film cooling, thereby creating a hot chamber, injector, and/or acoustic cavities (depending on configuration of the engine), 3) changing feed system acoustic properties, or 4) creating an oxidizer-rich environment in the chamber leading to an overpressure. In terms of changing the feed system acoustics, Juran and Zwick[32] explored the effects of dissolved gas on small engine injectors. The other possible effects of helium are discussed later in the chapter for the Space Shuttle primary reaction control thruster.

Most storable engine development and certification test programs include some type of entrained bubble and saturated propellant test. Smith[33] provides a method for saturating MMH and NTO propellants and for measuring the degree of saturation of the propellants.

IV. Storable Engine Case Studies

The engine damage, which was experienced in Apollo development programs, resulted in appreciation of combustion instability. Subsequent testing and experience provided some understanding, which has reduced the number of trial and error injector tests required for recent engines [e.g., the pump-fed Space Shuttle orbital maneuvering engine (OME)]. The following engine case studies for Apollo, Space Shuttle, and Long March programs describe this experience. Table 2 contains a summary of engine performance and injector design details for the case

Table 2 Engine data

Engine	Prop. ox fuel	Thrust chamber, N[lbf]	*Pc*, MPa [psia]	Mixture ratio	I_{sp}, s	Nozzle ratio	Char. length, cm	Chamb. diam, cm	Throat diam, cm
Gemini 1st Stage	NTO A50	956,320[b] [21,500]	5.41 [785]	2.00	287[b]	8.00	107.0	55.37	38.74
Gemini 2nd Stage	NTO A50	444,800[a] [100,000]	5.54 [804]	1.80	315[a]	49.20	97.0	36.83	23.19
Apollo SPS	NTO A50	31,184[a] [20,500]	0.68 [99]	1.60	311[a]	62.50	86.0	44.70	31.50
Apollo LEMDE	NTO A50	46,704[a] [10,500]	1.03 [150]	1.60	305[a]	53.00	67.0	28.83	18.24
Apollo LEMDE	NTO A50	44,480[a] [10,000]	0.77 [110]	1.60	304[a]	47.50	132.0	33.53	21.13
Apollo LEM Ascent	NTO A50	15,568[a] [3,500]	0.83 [120]	1.60	306[a]	45.60	77.0	19.79	11.63
Apollo LEM Ascent	NTO A50	15,568[a] [3,500]	0.85 [123]	1.60	309[a]	45.60	77.0	19.79	11.63
Space Shuttle PRCS	NTO MMH	3,870[a] [870]	1.05 [152]	1.60	281[a]	22.00	57.4	9.91	9.18
Space Shuttle OME	NTO MMH	26,688[a] [6,000]	0.86 [125]	1.65	316[a]	55.00	76.0	20.60	14.76
Space Shuttle Pump OME	NTO MMH	26,688[a] [6,000]	2.41 [350]	1.65	334[a]	150.00	90.3	13.97	8.53
Long March 3 FY-20	NTO UDMH	696,500[b] [156,587]	7.83 [1070]	2.21	263[b]	12.80	112.0	40.10	27.70
Long March 1 YF-1	IRFNA-27 UDMH	253,000[b] [56,880]	6.14 [890]	2.61	245[b]	10.10	190.0	42.00	18.50

[a]Vacuum. [b]Sea level. [c]Not available. [d]See text for details.

studies examined. Case studies of these engines provide some understanding of the complex phenomena and the approach taken to eliminate instabilities.

Unfortunately, other significant more recent engines, such as the XLR-132 and the Ariane L7, are not included in the following case studies. A discussion of the Ariane L7 engine is provided by Langel et al.[34] Until recently, small engines (less than 100 lbf) were believed to be immune to high-frequency instability, since the chamber acoustic modes are very high. A recent experience with a 5-lbf-thrust engine is discussed by Crone et al.,[35] which suggests that this is not necessarily always true.

A. Case Study 1: Apollo Service Propulsion System Engine

The Apollo service propulsion system (SPS) propelled the service module into lunar orbit and returned it to Earth using a single 91,180-N (20,500-lbf) thrust engine. The development of the Apollo SPS engine required tests on many different injector designs. Much of this was due to high-frequency instabilities that were experienced late in the program as a result of specification changes: 1) the mixture ratio was changed from 2.0 to 1.6, which resulted in increased

Table 2 Engine data (cont.)

Comb. eff.	Injector Delta-P, *% Pc*	Element type	Number element	Thrust/ element, N [lbf]	Primary ox orif., mm	Primary fu orif., mm	Film cooling % fuel	Baffle number H, mm	Cavity % Inj face area
0.972	—[c] —[c]	LD	568 ox 516 fu	890 [200]	3.05	2.03	14.5	none	none
0.974	17 ox 17 fu	Quad	1319 ox 118 fu	445 [100]	1.27	1.02	16.0	6 + hub 101	none
0.975	45	UD	575	178 [40]	1.04–1.085	1.04–1.95	none	5 + hub 101	none
0.972	26	FOF	165	283 [63.6]	1.96	1.24	8.7	3 Y 44	none
0.970	90 ox 22 fu	Pintle	1 ann. fu 36 ox	1232 [277]	—[d]	—[d]	none	none	none
0.962	25	FOF	84	185 [41.7]	1.40	0.84	none	3 Y 32	—[c]
0.970	26	FO	177	88 [19.8]	1.27–0.914	1.01–812	—[c]	3 Y 44	5.2%
—[c]	23 ox 30 fu	FO	84	46 [10.35]	0.13	0.09	25.0	none	8.0%
0.975	44 ox 48 fu	LOL	272	98 [22]	0.81	0.71	none	none	—[d]
0.990	50	Splash-plate	450	59 [13]	—[c]	—[c]	—[c]	none	29.0%
0.965	16 ox 14 fu	LD	607 ox 605 fu	575 [129]	2.70	2.30	6.4	6 + hub 100	none
0.940	21 ox 17 fu	Coaxial	247	1024 [230]	—[d]	—[d]	10.4	6 + hub 100	none

popping and 2) a new requirement to demonstrate dynamic stability using bomb tests was added.[36] The problem with high-frequency instabilities was resolved after additional injector modifications.

The BLOCK I engine development program, which specified a mixture ratio of 2.0, tested triplet, unlike-doublet, and quadlet elements in unbaffled injectors. Spontaneous instabilities occurred at the first tangential (1T) of 1650 Hz and the second tangential (2T) of 2700 Hz. When injector face baffles were installed, the high-frequency instability problem appeared to be solved. During a prequalification program, however, a pop caused an instability. This new problem was resolved by extending the baffle tips 2 in., as shown in the testing of the BLOCK I, injectors POUL 41-26 to POUL 41-39, in Table 3.

The BLOCK II engine development program added requirements for dynamic stability and a mixture ratio of 1.6. The first BLOCK II injector, labeled POUL 41-26 modification (Mod.) 0, utilized the 5-4-2 (blade number-hub height in inches-blade peripheral in inches) baffle. The rate of popping for this injector increased from 1.2 to 3.6 pops/min when the mixture ratio was changed from 2.0 to 1.6. These pops resulted in sustained high-frequency instabilities. Similarly, bomb tests also resulted in sustained instabilities.

Table 3 Aerojet Apollo SPS engine development summary

Injector (POUL no.)	BLOCK I 41-26	BLOCK I 41-22	BLOCK I 41-38	BLOCK I 41-39	BLOCK II 41-26 0	BLOCK II 41-26 IV
Element type	U-D	U-D	U-D	U-D	U-D	U-D
No. of elements	575	430	575	570	575	575
Thrust/elem., N	169	226.8	169	169	169	169
Ox orf D, mm	1.04–1.85	1.29–2.06	1.04–1.85	1.09–1.85	1.04–1.85	1.04–1.85
Fu orf D, mm	0.838–1.50	0.991–1.60	0.787–1.52	0.889–1.52	1.04–1.96	1.04–1.96
FRC, %	6.9	7.15	6.45	5.43	5.08	6.59
Imp. angle	60	30–60	60	60	60	60
Imp. dist., mm	3.81–12.2	4.57–13.0	3.81–12.2	3.81–12.2	3.81–12.2	3.81–12.2
C^* eff., %	96.4	96.3	96.8	97.3	97	97
Spontaneous	Unstable[a]	Stable[c,d]	Stable[c,d]	Unstable,[a,b] Stable[c,d]	Unstable[a]	Stable[b]
Bomb Instb.	Unstable[a], Stable[b]	Stable[a], Stable[b]	Unstable[a], Stable[b]	Unstable[a], Stable[b]	Unstable[a]	Stable[b]

[a]Stainless-steel chamber, 2-in. baffle. [b]Stainless-steel chamber, 4-in. baffle.
[c]Ablative chamber, 2-in. baffle. [d]Ablative chamber, 2-in. baffle.

The high-frequency instability problem was solved by increasing the baffle height. The POUL 41-26 Mod. II injector was tested with a 5-4-2 baffle and a 5-4-4 baffle. The 5-4-4 baffle increased the baffle height at the periphery of the chamber to 101.6 mm. Although the results showed an increase in pops/min from 17.8 to 29.5, the extra baffle height eliminated sustained instabilities.[37]

The popping problem was significantly improved with the POUL 41-26 Mod. IV injector, shown in Fig. 5. Testing found that the injector element was bistable in the operating regime.[38] The discharge coefficient would change dynamically and cause disturbances in the flow. The Mod. IV injector orifices were counterbored to eliminate this problem.

Acoustic tests, using a speaker and microphone, measured the modes of the baffled injector 5-4-4. Corrected for hot-fire conditions, the modes are 1T at 1400 Hz, 2T at 1940 Hz, first radial (1R) at 2050 Hz, 2T + baffle mode at 3000 Hz, 3T at 3200 Hz, 1R + baffle at 3600 Hz, and 1T + 1R + baffle at 5050 Hz. The actual observed modes of instabilities appeared to be about 3600 Hz and 5000 Hz.

In summary, increasing the baffle height by 50.8 mm (2 in.) at the periphery of the chamber appeared to be the most significant stabilizing factor. Overall, the 5-4-2 baffle accumulated 46 instabilities out of 4040 bomb tests, and the 5-4-4 baffle accumulated zero instabilities out of 1898 bomb tests. During the test program, it was also noted that the ablative flight engine chambers damped instabilities much better than the "work horse" steel engine chambers used in development tests.[39] This may have been due to the acoustic absorbency of the ablative material. Another significant observation was that popping increased with decreasing mixture ratio and increasing fuel film cooling, and that the orifice L/D affected the popping rate.

B. Case Study: Lunar Excursion Module Descent Engine

The Apollo lunar excursion module (LEM) descent engine landed the two astronauts on the moon. The LEM was propelled by a single engine, which was

Fig. 5 Apollo SPS engine injector.

supplied propellant by a redundant feed system and which had a requirement for 10:1 throttling. Two engine design programs, conducted in parallel, competed for this engine program. The final design selection was based on combustion stability.

1. Helium Injection Throttled Lunar Excursion Module Descent Engine

This engine had a unique throttling mechanism, which injected helium into the propellants to reduce the propellant flow rates while maintaining the pressure drop across the injector. It also had an advantage over other designs by having a fully redundant throttling system of propellant and helium valves. This engine design experienced significant low-frequency stability problems, however, possibly caused by the helium in the manifolds.[40]

The specified thrust of this engine was 46704 N, with a throttle ratio of 10:1. The engine chamber used an ablative liner made from a silica tape wrap. The injector consisted of 165 FOF triplet elements. In 1964, tests with helium injection revealed a tendency of the injector to buzz and pop below 70% thrust. The engine experienced a significant chug instability problem from 35–70% throttle at 30–500 cps. This chamber and feed system coupling was probably caused by the lower injector pressure drop at low-throttle settings and by the helium in the feed system, which changed the acoustic characteristics of the system.

In addition to the low-frequency instability, two types of popping occurred.[40] The first type was isolated pops of 0.698-MPa peak, followed by acoustic oscillations which damped in 20 ms. The second type, probably resurging, occurred in conjunction with the chugging. When the chamber pressure built up in amplitude, a pop occurred. The pop was then followed by a burst of acoustic oscillations which damped with time. Because of these problems, this engine was not selected.

2. *Variable Area Pintle Injector Throttleable Engine*

The selected LEM descent engine used a pintle injector. This engine development program was relatively free of combustion stability problems. The only problem detected was occasional rough combustion or random pressure spikes characterized by no discernible frequency. A nonredundant throttle mechanism varied the injector flow orifice size of the pintle injector, providing thrust from 4,448 N to 44,480 N at 304 s I_{sp} (Ref. 41). The feed system valves were also modulated to control propellant flow in addition to the pintle sleeve.

With the pintle injector, propellants are injected radially (orthogonal to axis) into the chamber at the center. A cross section is shown in Fig. 6. The oxidizer flows through the center of the pintle and out through 36 radial slots. The fuel

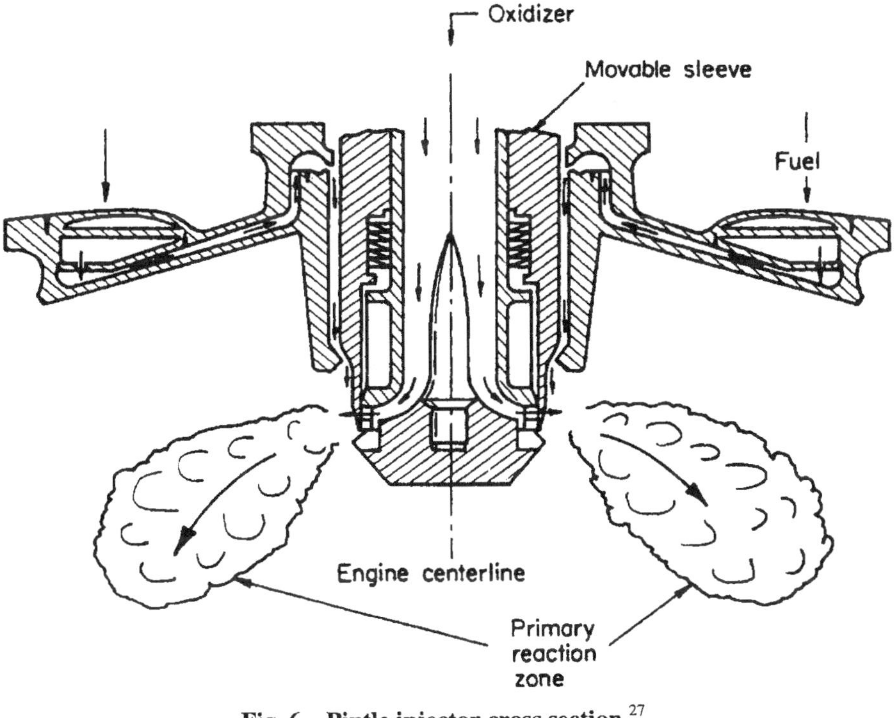

Fig. 6 Pintle injector cross section.[27]

flows in an axial annular sheet in an outer sleeve of the pintle. Two different pintle configurations were tested: 1) 36 primary and 36 secondary oxidizer slots and 2) 32 primary and 32 secondary oxidizer slots. Although the coarser 32 slot injector had a higher efficiency due to the ability of the coarser oxidizer stream to mix with the fuel sheet, this also produced a rougher combustion.

This engine was certified to be dynamically stable by bomb tests with peak pressures up to 1.03-MPa (150-psi) peak over *Pc*. No high-frequency instabilities were encountered. Pressure spikes or combustion disturbances were experienced at full thrust and at 30–50% thrust, depending on specific injection conditions. The rough combustion was associated with very high-fuel-injection velocities and low-oxidizer-injection velocities.[42]

C. Apollo Lunar Module Ascent Engine

This engine boosted the LEM from the lunar surface to rendezvous with the orbiting Command and Service Module. Although the single ascent engine was fed by a redundant feed system, an engine malfunction due to chamber or injector structural failure was a major concern, especially to the astronauts. This dictated that the engine be dynamically stable. The engine specifications were for 15,568-N thrust at a mixture ratio of 1.6 NTO/A-50.

The development injector consisted of 84 FOF triplets with 90 fuel-rich unlike doublets at the wall for cooling. The early design did not include baffles. Although the engine initially performed well according to specifications, later on in the program, it developed spontaneous instabilities. When bomb testing was added to the specification, the engine experienced significant stability problems.

The injector, without baffles, had three self-induced instabilities and numerous instabilities during bomb testing. One test run exhibited 12 s of high-frequency instability that resulted in a roughing and bubbling of the chamber wall. The characteristic high-frequency instability occurred at frequencies of 3500 Hz and 7000 Hz at amplitudes of 0.303–1.289 MPa (44–187 psi) peak to peak. Table 4 is a summary of the stability of various engine configurations.

The injector was redesigned in 1965 to include baffles.[43] Numerous tests were conducted using 3- and 4-bladed baffles of different lengths. The baffle design selected used three equally spaced blades. The test results of steel and ablative chambers showed that significant damping was available in ablative chambers. In

Table 4 LEM ascent engine development summary[a]

Injector	B3L	B8L1A	B3L4A	B8L1A	B8L1A	E2B
Element	F-O-F	F-O-F	F-O-F	F-O-F	F-O-F	F-O-F
No. of elem.	92	92	92	92	92	92
No. FFC elem.	96	64	96	64	64	96
C^* eff., %	97.3	98.1	97.3	98.1	98.1	97
Baffle no.	None	None	3	4	3	3
Len., mm			31.8	25.4	31.8	31.8
Spon. stb.	U(SS)	U(SS)	U(SS)	U(SS)	U(SS)	U(SS)
Bomb stb.	U(SS)	U(SS)	U(SS)	U(SS)	U(SS)	U(SS) S(ABL)

[a]U, unstable; S, stable; SS, stainless steel; and ABL, ablative.

fact, high-frequency dynamic stability was achieved only in the ablative chamber.

Although there was success with eliminating high-frequency instability, a problem with low-frequency instability was observed at an amplitude of 0.552–0.621 MPa (80–90 psi) and 350–400 Hz in both the steel and ablative chambers. There is some evidence that gas in solution caused or contributed to the low-frequency problems.[43] This problem was not resolved.

In 1967, due to the stability problems with the initial injector, a backup engine injector for the LEM ascent engine was developed in parallel. This development program began in August of 1967, and culminated in the first lunar landing in July 1969.[44] This engine produced 15,568-N (3,500-lbf) thrust with an I_{sp}. of 3030 m/s (309 s) at a mixture ratio of 1.6 NTO/A50 and used the same chamber developed in the first injector program.

The injector consisted of 177 unlike doublets with a 3-blade baffle and slot acoustic cavities. The 44.5-mm-in-high baffle damped the first tangential mode of instability. The acoustic cavities, located at the outer edge of the injector, were designed to damp the first radial, third tangential, and higher modes.[45,46]

Chug stability was improved by adding a volume to the fuel inlet duct. This injector was successfully bomb tested a total of 302 times, with one exception. On the 99th bomb test, an instability occurred. This was attributed to a damaged injector with plugged and misimpinging orifices caused by the shrapnel from previous bomb testing.[44]

D. Case Study: Space Shuttle Primary RCS Thruster

The Space Shuttle primary reaction control system (PRCS) thruster provides control of the Space Shuttle during ascent, on-orbit maneuvers, and re-entry. The PRCS thruster can also de-orbit the Shuttle, if the two orbital maneuvering engines fail. It is a 3870-N (870-lbf) vacuum thrust engine operating at a mixture ratio of 1.6:1 NTO to MMH. Long-life requirements were prime considerations in the design. The duty cycle requirements for this engine are extreme: 1) continuous burn duration from 80 ms to 800 s, 2) total life of 12,800 s, and 3) and cycle life of 20,000 pulses. The engine has proven quite capable of performing this duty cycle.

An extensive ground test program revealed that when helium is ingested in a certain manner, instabilities may sometimes occur. This ground test data can help to understand the effects of helium ingestion on the combustion process. Fortunately, additional ground tests showed that the actual propellant conditions in flight may preclude the occurrence of damaging instabilities. The engine also proved very robust in that sustained instabilities were observed for 40–80 s before causing chamber damage. In addition, an instability protection system was installed to prevent any damage to the Shuttle from a chamber burn through.

The injector consists of a single ring of 84 unlike doublets and an outer ring of 42 fuel film cooling orifices at the chamber wall as shown in Fig. 7. The fuel film cooling accounts for 23–25% of the fuel flow. There are 42 quarter-wave acoustic cavities tuned for the first tangential (5600 Hz). Half of the cavities, 25.4 mm (1.0 in.) in depth, are tuned for startup conditions, and the other half, 27.9 mm (1.1 in.) in depth, for steady state.

Early development testing of this thruster showed that either without the acoustic cavities or with cavities of 4% total injector face area, the engine was spontaneously

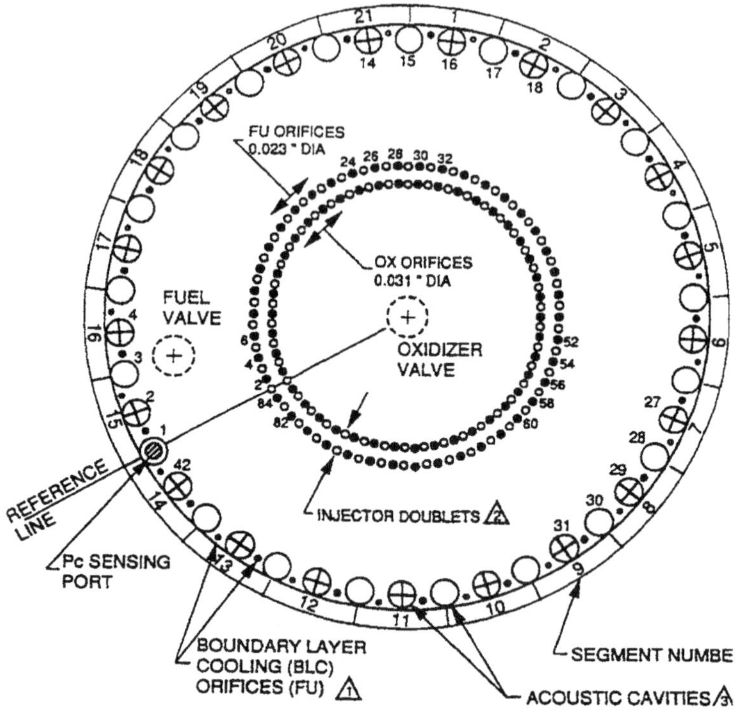

Fig. 7 Space Shuttle PRCS injector face.

unstable in the first tangential mode. Stability was obtained by increasing the area to 8%. Table 5 provides a summary of the development and qualification testing of the engine. During qualification in 1977, the engine was certified dynamically stable using bomb tests.

Later in the program, during the acceptance test of a flight thruster, an instability occurred. Helium bubbles trapped in the feed lines as the result of a poor purge and prime of the system are the suspected cause. This instability resulted in the program taking three basic actions to 1) protect the vehicle from instability, 2) determine the cause, and 3) screen production thrusters for stability problems.

Table 5 Space Shuttle RCS engine development summary

Engine configuration	Stability	Mode/ freq., Hz
No. damping devices	Unstable	1T/6000
4% Area acoustic Cavitles	Unstable	1T/6000
8% Area acoustic cavities	Stable	
Helium injected during start transient	Unstable	1T/6000
Helium saturated fuel and oxidizer	Stable	
Saturated fuel only	Unstable	1T/6000
Helium injected during steady, state firing	Stable	

In 1988, the thruster instability protection system (TIPS) was installed to protect the vehicle from chamber burn through. This system consisted of wrapping the valve lead wires around the chamber, in such a way that chamber burn through due to an instability would sever the wires and close the thruster valves. This was demonstrated to be successful by plugging the acoustic cavities on two engines and firing until burn through.

The second action was to determine the cause of the instability. After extensive testing, it was found that helium bubbles injected into the fuel at approximately 8 cc/s could trigger the instabilities. This was of considerable concern, since it is not unusual in a pressure-fed system to have gas bubbles. In order to prevent trapped helium bubbles in the flight system, procedures at the launch processing facility call for a vacuum to be pulled in the system and then to backfill with propellant.

The third action was to screen for manufacturing variation in production thrusters during acceptance tests by injecting helium at 8.19 cc/s steady state in a test consisting of 300 pulses. The acceptance criteria required that no instabilities occur. The criteria defined an instability as an engine acceleration exceeding 20 g at the thruster head end for a duration greater than 12 ms at the 1T of approximately 6000 Hz. In this screening test, the observed rate of instabilities with helium injection was less than 1%, typically around 0.1%. One thruster, which has been set aside for testing, had an observed rate of instability of up to 10%, so that it is statistically different.

In 1990, a test program was conducted at White Sands Test Facility (WSTF) to further examine the effects of helium on stability.[47] The test program found that the instability has several characteristics: 1) the instability always initiates during the start transient, 2) sometimes chugging and an 8000 Hz first longitudinal are seen during this start transient before transitioning to a 6000 Hz 1T, 3) the thrust drops 356–400 N, (80–90 lbf) during the instability, and 4) the injector flange temperature rise rate increases from approximately 5°C/s to 44°C/s.

Two engines were then subjected to sustained instabilities by continuing the pulse after initiation of the instability. The engines were found to be robust, in that sustained instabilities took 40–80 s to damage the chambers. This is significant in that most pulses on the Shuttle are less than 10 s.

Helium was found to be the most significant factor in causing instability. The ingestion of helium caused higher heating rates in the flange and rougher combustion, even when an instability did not occur. It was also observed that the rate of instabilities increased as the thruster flange temperature increased during repeated pulses with helium injected into the fuel. Based on this result, thruster heating could be postulated to affect the tuning of the acoustic cavities or the incidence of popping, thus increasing the number of instabilities.

Another series of tests at WSTF determined that how, when, and where the helium is injected is important. In the first test, when helium was injected into the fuel during the start transient, instabilities occurred. It should be noted that due to the helium injection apparatus, a significant quantity of helium accumulates at the thruster, between pulses, such that the peak flow rate of helium at startup is large (about 245–328 cc/s). In the second test, when helium was injected after the start transient using a slightly different apparatus, there were no instabilities.

In a third test, using fuel saturated with helium rather than helium injection, some albeit fewer instabilities occurred. In a fourth test, when both oxidizer and fuel propellants were saturated, there were no instabilities. The fuel side is obviously sensitive to helium ingestion, and helium ingestion in both the oxidizer and fuel seems to negate the effect. This reduces the probability of an instability on the Shuttle because both propellants are likely to be saturated, since helium dissolves in NTO more than in MMH.

In summary, the physical mechanism for the instability may involve disturbances in the fuel and oxidizer flow caused by helium ingestion. Unielement tests by Lawver[17] found that the PRCS injector element is sensitive to operating conditions from a reactive stream separation standpoint. The higher heating rates experienced during helium ingestion in the engine may also detune the acoustic cavities and increase susceptibility to disturbances. Recently, a limited number of tests suggested that an oxidizer valve lead can cause instabilities. A similar experience is reported by Crone et al.[35] for a 5-lbf engine in which an oxidizer lead caused overpressures that resulted in instabilities. Similarly in the PRCS, helium ingested in the fuel during the start transient would cause an oxidizer lead to some extent. However, direct measurement of an overpressure has not been made with the current low-response instrumentation for chamber pressure.

This case study demonstrates that the solution to a stability problem can require an extensive understanding and consideration of the conditions which cause the instability and a method to deal with the results. The PRCS engine has proven to be quite acceptable in both ground and flight testing based on the fact that 1) the probability of instability is low even under the worst-case conditions, 2) the actual flight conditions are less severe, 3) the engine is robust to damage from instabilities, 4) the time required to damage a chamber is longer than the typical pulse time, and 5) an instability protection system is installed.

E. Case Study: Space Shuttle Orbital Maneuvering Engine

The orbital maneuvering engine performs orbit circularization and de-orbit burns for the Space Shuttle. It was selected over the originally proposed LO_2/LH_2 RL10-based system, when the orbiter configuration required a small propulsion system package.[48] The relatively smooth development of the OME was a result of the extensive predevelopment effort, which included the Apollo experience. The OMS engine is rated at 26,688-N (6,000-lbf) thrust, and operates at a mixture ratio of 1.65:1 NTO/MMH with 316-s I_{sp}.

After trade studies showed that Earth storable propellants best met Space Shuttle requirements, the early OMS engine development program focused on which storable propellant, cooling method, and acoustic damping device should be used. The optimum design was determined to include an engine chamber regeneratively cooled by MMH, acoustic cavities rather than baffles for easier cooling and longer cycle life, and like doublets rather than unlike doublets or triplets for improved performance and stability.[48]

The early development tests focused on different injection element types, element arrangements, injector manufacturing, and depths of the acoustic cavities, as shown in Table 6. The transverse like-on-like (TLOL), 97.9 N (22 lbf) per element, was found to be stable and, hence, was selected for the design. The injector face is formed by platelets, as shown in Fig. 8. The elements are like-on-like doublets, arranged in 16 alternating rows.

There are 12 acoustic cavities: 8 each 1.5 in. deep for the 1T mode and 4 each 0.5 in. deep for the 2T, 1R, and 3T modes. The significant chamber acoustic modes are 1L, 1100 Hz; 1T, 3200 Hz, 1R, 6400 Hz, 2T, 5200 Hz, and 3T, 7200 Hz.

Table 6 Aerojet OMS engine development summary

Engine configuration	Stability	Mode/Freq., Hz
57.8 N (13 lbf/elem.), unlike doublet	Unstable	
97.9 N (22 lbf/elem.), like doublet	Stable	
867 Element X doublet	Resurge	
22 lbf/elem. V doublet	Unstable	1L
1000+ mixed elements	Unstable	
Acoustic cavities tuned for 1T, 3T	Unstable	2T
22 lbf/elem., TLOL	Stable	

During development of the acoustic cavities, some 2T instabilities occurred when the acoustic cavities were tuned only for the 1T and 3T chamber modes. The chug *Pc* limit for this engine was found to be 0.552 MPa (80 psia). The engine is certified dynamically stable per CPIA 247.

F. Case Study: Pump-Fed Orbital Maneuvering Engine Development Engine

This development engine, one of the most recent storable engines developed for NASA, provides improved performance, 334 s, as compared to the pressure-fed OME, and potentially provides an engine for orbital transfer vehicles or upper

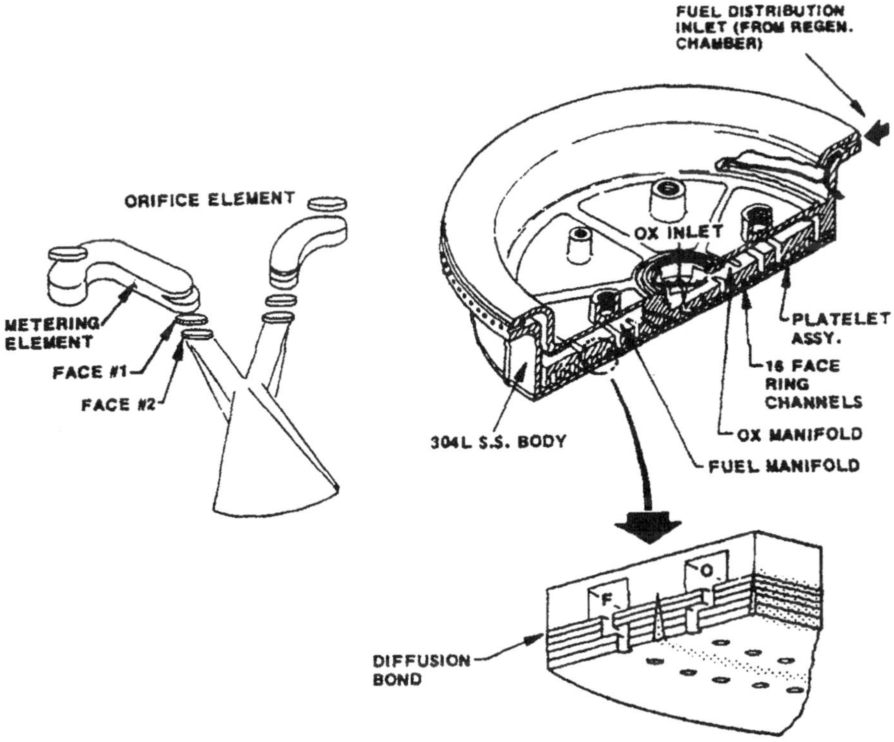

Fig. 8 Space Shuttle OMS engine injector.

stages. Development and test of the injector, which included bomb testing, was successfully completed in 12 months.

Performance and stability design analysis code[49] was used to examine different injectors and size the engine. The chamber pressure was chosen to be 2.413 MPa (350 psia). For the injector selection process, the OMS engine TLOL platelet injector, a quadlet, and a splashplate injector were evaluated analytically. The analysis showed that the splashplate injector was the best injector from a performance standpoint. A stability analysis showed that a chamber diameter of 13.97 cm (5.5 in.) was acceptable, and better than larger diameters. Acoustic cavity depth 22.86 mm (0.9 in.) and injector delta-P were also determined. To achieve the desired goal of 99% energy release efficiency (ERE) without encountering a first longitudinal instability, chamber lengths of 30.48–35.56 cm (12–14 in.) were selected for testing.

The test program was conducted on three different splashplate injectors and three chamber lengths. The 30.48-cm (12-in.) chamber did not achieve 99% ERE. The program goal of 99% ERE was achieved after only 18 total test firings, when the 14-in. chamber was tested. The injectors were all found to be stable above chamber pressures of 1.38 MPa (200 psia). The selected injector was bomb tested and displayed the required damping. In off-nominal tests below 1.38 MPa (200 psia), the first longitudinal instability occurred as predicted by analysis. This successful engine development program demonstrated an improved understanding of Earth storable propellant combustion stability.

G. Case Study: Long March FY-20 Engine

All launch vehicle engines of the Long March (LM) series, except the upper stage engine of LM-3, use storable propellant combinations. The most typical is the first-stage engine, FY-20 of LM-3. The FY-20 is a gas generator driven pump-fed NTO/UDMH engine. Because the FY-20 has never failed in its flight missions, its observed reliability is 100%. Early in the development phase, however, the engine often encountered destructive high-frequency combustion instabilities, which severely affected the development program. The basic performance characteristics of the FY-20 engine and its major components are listed in Table 2.

1. G Concept Injector for the FY-20

The original injector design of the FY-20 is called the G concept injector. This injector consists of like-on-like doublets distributed on 15 concentric rings (two rows on each ring). Oxidizer and fuel rings are arranged alternately. The injector face includes a 100-mm-high baffle with six blades and a hub used to eliminate transverse high-frequency combustion instabilities. During early tests on baffle configurations, sometimes one mode of instability would be eliminated, but another would become dominant. Decreasing the chamber contraction ratio and increasing the L/D tended to excite the longitudinal instability.

Later during off-nominal tests on the G concept injector, intense structural vibrations frequently occurred, with the composite vibration level on the injector head sometimes exceeding 600 g. The intense vibration often caused mechanical failures, such as duct breaks and component weld cracks. The vibration once caused an early shut down of the engine by suddenly closing a valve. Engine exhaust flame flashing could also be seen when the intensive vibration took place.

For storable hypergolic propellants, the propellant inlet temperature is an important factor that can affect combustion stability.[27] During mid-summer, with NTO temperatures higher than 16°C (61°F), statistical data indicated that the incidence of intense oscillations was approximately 66%. At higher engine thrust and mixture ratios, the vibration level increases slightly. A fast engine start with high pump inlet pressures also caused intense vibrations. The severe vibrations often caused a decrease in thrust and performance. The specific impulse loss was about 2%. In general, performance degradation occurs only with low-frequency combustion instability and lower order longitudinal acoustic mode instability. The intense vibration mainly caused structural damage to hardware. Severe erosion of the chamber wall did not occur in most of the chambers tested.

To efficiently monitor combustion instability, accelerometers to measure vibration were placed on the thrust chamber, turbopump, gas generator, and heat exchanger of FY-20 engine. According to the axial vibration recorded for the G concept injector, a predominant vibration of 886–900 Hz existed. At this frequency, the acceleration reached 100–300 g, sometimes even much higher. Pressure measurements upstream of the injector indicated that the frequency of chamber pressure oscillation was identical to the predominant frequency of mechanical vibration. This frequency is consistent with the first longitudinal acoustic mode calculated according to the combustion chamber geometrical sizes and parameters of combustion gas products. The first longitudinal oscillation frequency is not too high (900 Hz), because the combustion chamber size is large. This instability may also have coupled with the fluid pressure oscillation in the propellant feed system.

2. *S Concept Injector for the FY-20*

Two methods were attempted to suppress the first longitudinal acoustic mode combustion instability in the FY-20 engine: 1) modifying the feed system impedance and 2) modifying the injector axial energy release distributions along the combustion chamber. At first, a cavitating venturi, a throttling orifice, and a hydraulic capacitor were used in the oxidizer ducts to attempt to suppress instability, but test results showed that these were ineffective. The approach to modify the energy release distribution was then attempted. The energy release of combustion was evenly distributed along the chamber axis by modifying the injector element pattern, orifice diameter, flow intensity distribution, and injector pressure drop. This increased the impedance against the pressure perturbance, thereby changing the combustion response. This modified injector is called the S concept injector.

In comparison with the G concept injector, the S concept injector features several changes to the injector element pattern, the orifice diameters, and the flow intensity distribution. The flow intensity is distributed as a steep slope in the core section with a special high-flow intensity within the fourth ring, as shown in Fig. 9. The mass flow rate of the high flow-intensity area is 17% of the total propellant flowrate. The flow intensity is then increased at the number 11 and 12 rings to 1.7 times the average flow intensity of the area, which gives the distribution of flow intensity a hump shape.

The FY-20 engine utilizes a regeneratively cooled baffle, hence, injection at the baffle tips can not be used for axial energy release distribution. The flow intensity adjacent to the radial and circumferential baffles, however, is increased to 1.47 and 1.29 times the fan-shape area average flow intensity, respectively. Thus, the

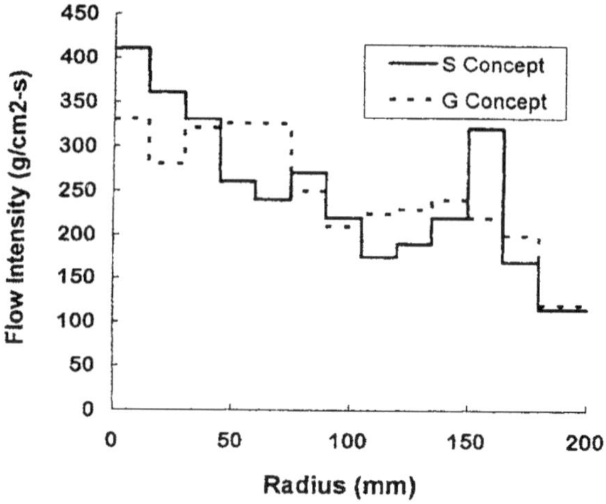

Fig. 9 FY-20 engine injector flow intensity distribution.

mass flow rate adjacent to the baffles is 31.9% of the total propellant mass flow rate.

The impingement angles of the basic injecting elements for the G concept injector are 50 deg for oxidizer and 60 deg for fuel. The impingement angles for the S concept injector were reduced to 30 deg for oxidizer and 50 deg for fuel. Adjacent to the baffles, the impingement angle was reduced from 60 deg for the G concept injector to 30 deg for the S concept injector. This moves the impingement points farther from the injector face and distributes the combustion energy release along the chamber axis, thus forming an uneven flame front so as to stabilize the dynamic combustion process.

The number of orifices in the S concept injector is almost equal to that of the G concept injector. Modification of the S concept injector is mainly to the orifice diameter of the injector elements in the high-flow-intensity areas. The orifice diameters were enlarged from 2.1 to 2.3 mm for the fuel and from 2.3 mm to 2.7 mm for the oxidizer. The pressure drop across the injector was reduced to maintain the same total flow rate. At the nominal mass flow rate, the fuel and oxidizer injecting pressure drops were reduced from 1.1466 MPa to 0.8869 MPa and 1.4994 MPa to 1.2397 MPa, respectively. Both enlarging the orifice diameter and reducing the injector pressure drop resulted in increased droplet diameter and facilitated the even distribution of the energy release along the chamber axis. The mixture ratio distribution for the S concept injector is somewhat more uniform than that of the G concept injector in order to compensate for the performance loss caused by modifying the flow intensity.

The axial energy release profile[51] calculated for the FY-20 injector was based on a flow tube combustion model and a droplet vaporization model, and considered UDMH exothermic decomposition and NTO endothermic dissociation mechanisms. In comparison with the G concept injector, the S concept injector has a more uneven combustion flame front and a more even energy release distribution. About one-half of the total propellant energy is released in the region farthest from the injector face. Based on the axial distribution of energy release, the sensitive lag theory of Crocco and Cheng[52] can be used to calculate the stability boundary.

The stability of the S concept injector is enhanced because the energy release is more distributed, and the sensitive lag time is increased.

3. *Testing of the S Concept Injector*

During tests of the S concept injector, no combustion instabilities were encountered at ordinary environmental temperatures. During test firings conducted in mid-summer ($T > 20°C$), however, four occurrences of intense vibrations were encountered for the S concept injector. As indicated in Ref. 27, when the temperature of the propellant combination, NTO/hydrazine and the like is higher than 16°C, popping is easily excited and can trigger combustion instability. Combustion instability may also be induced by unsteady flow of NTO at a temperature close to its boiling point.

Additional modifications to the S concept injector were made to 1) further modify the mass flow-intensity distribution along the circumferential and radial baffles, and 2) make the pressure drop of the fuel injecting orifice greater than that of the oxidizer injecting orifice. These modifications improved the injection momentum ratio of the two propellants and more evenly distributed the energy release distribution along the chamber axis.

It should be pointed out that the performance of the engine with the S concept injector is identical to that with the G concept injector because the characteristic length L^* of the combustion chamber is fairly large ($L^* = 1.12$ m) and the mixture ratio distribution is more uniform. Test firings of the modified S concept injector were successfully repeated in the high-temperature season and at high-thrust levels without occurrence of instability.

H. Case Study: Long March YF-1 Engine

The first-stage engine of the LM-1 launch vehicle YF-1 is a pump-fed IRFNA-27*/UDMH rocket engine with a gas-generator power cycle and with coaxial injection elements. (IRFNA-27 consists of 27% NTO, 71.56% HNO_3, and 1.8% H_2O.) Table 2 lists the basic performance characteristics of the latest YF-1 engine and its major components. Early in its development, however, the YF-1 engine utilized the IRFNA-20/MA-50 propellant combination (IRFNA-20 consists of 20% NTO, 77.5% HNO_3 and 2.5% H_2O; MA-50 consists of 50% triethylamine and 50% dimethyl l-benzeamine).

In the initial development, a series of instabilities, causing serious erosion of the injector face, resulted in engine failures. The mechanical vibration acceleration measured on the combustion chamber reached about 300 g. Initially, methods to improve cooling and reinforce components were used, but no obvious effects were observed. Through repeated tests and analysis, it was concluded that the failures were caused by high-frequency combustion instability.

In general, damage to the combustion chamber took place during transitions, e.g., at engine start, operating regime transition, or shutdown. The abrupt change in combustion chamber operating parameters was accompanied by large random disturbances that resulted in instability. These phenomena indicated that the margin of combustion stability was low and could not withstand disturbances.

Instantaneous burnout of the injector face and film cooling ring along the periphery of the combustion chamber often occurred. Erosion traces on the chamber surface were smooth, not rough and uneven as that caused by insufficient cooling and so it was obvious that a spinning gas stream existed simultaneously with an increasing heat flux. The bulging and dropping off of the injector element covers indicated that high-shock levels also existed in the combustion chamber.

A high-amplitude-pressure oscillation occurred in the combustion chamber and injector manifold at a frequency of 1500 Hz. This frequency is identical to that of the first tangential acoustic mode. The combustion chamber has a strong tendency to trigger transverse combustion instabilities for three reasons: 1) the cylindrical combustion chamber has a large contraction ratio of 5.15:1, 2) the injector face has no baffle, and 3) the flow intensity distribution is uniform.

1. Liquid-Phase Division Modification of the YF-1 Injector

The liquid-phase division method was used to reduce the occurrence of transverse high-frequency instabilities.[53] In liquid-phase division, the entire injector face is divided into several zones by a number of injecting elements which have different atomization characteristics from the basic elements. In this way, energy release is dispersed in the combustion chamber space, and pressure wave propagation is damped because of an insufficient supply of energy.

The liquid-phase division injector pattern is arranged in a honeycomb or concentric circle pattern, as shown in Fig. 10. The principle of division arrangement requires that the energy release points be arranged apart from the oscillatory wave crest and that the atomization characteristics of division injecting elements and basic coaxial elements differ significantly, as shown in Fig. 11. The division elements utilize a double component impinging/coaxial combination. About 30% of the fuel passes through a swirler to form a centrifugal atomization cone which protects the injector face. Division elements are arranged in six rows radially and in two circles peripherally.

To enhance the effects of liquid phase division, the mass flow rate of the division elements is about 1.5 times that of the basic elements. At the same time, the large mass flow rate double-component centrifugal elements, of which the flow rate is 1.5 times that of the basic element, are placed in the core area of the injector face. Thus, the hump-shape distribution of flow intensity is formed radially and peripherally.

After applying liquid-phase division, the results of test firing improved greatly. Within a large range of operating parameters, such as thrust ranging from 78.4 kN

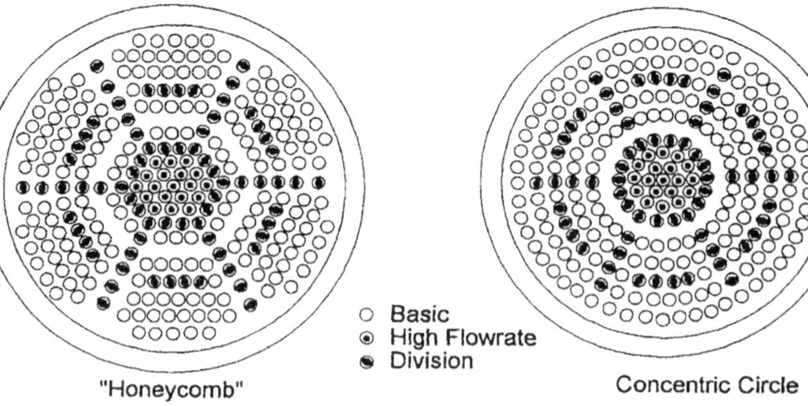

Fig. 10 YF-1 injector pattern.

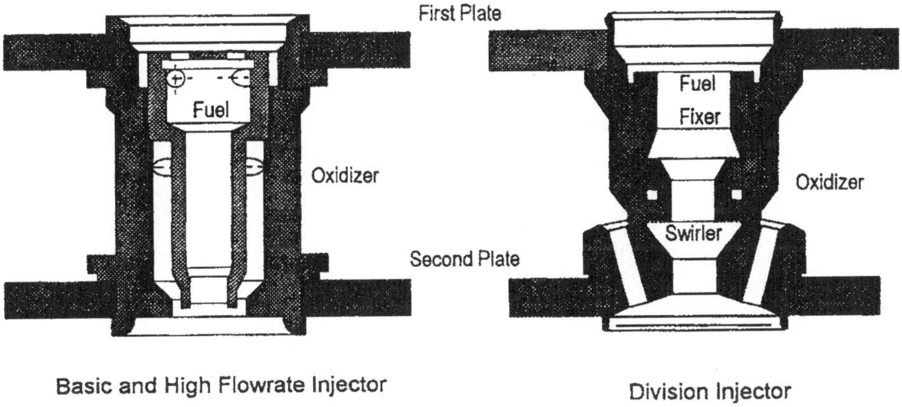

Fig. 11 YF-1 basic, high flow rate, and division injection elements.

(17,625 lbf) to 267.5 kN (60,140 lbf), and mixture ratio ranging from 3.67 to 5.92, the engine was able to operate in a stable manner. The probability of successful test firings reached 94%. No erosion traces were found in the combustion chamber tested. The combustion stability of the YF-1 injector was improved by liquid-phase division.

2. *IRFNA-27/UDMH Propellants with YF-1 Injector*

When the propellant combination of the YF-1 engine was changed to IRFNA-27/UDMH, the liquid-phase division method continued to be used because of its significant effects in reducing combustion instability. At the same time, other modifications were made, such as changing the structure of the injector elements, increasing the rows of radial division elements, and improving the flow intensity distribution. But in all five test firings, a high-frequency combustion instability at a frequency of 1600 Hz occurred during the transients (three during start and two during shutdown). The difference in propellant properties changed the sensitive combustion time lag and weakened the ability of liquid-phase division to stabilize combustion.

Although the physical concepts of ignition time delay, combustion time lag, and sensitive time lag are different, they have some inherent relation to the propellant activity. The ignition delay time of the two propellant combinations is listed in Table 7.The ignition delay time of IRFNA-27/UDMH is several times less than that of IRFNA-20/MA-50. Also, UDMH decomposes at a different temperature, which can produce popping and excite combustion instability.

Table 7 Ignition delay time of the two propellant combinations

Propellant	Pressure, kg/cm^2	Temp., °C	Ignition delay time, ms
IRFNA-20/MA-50	1	4–14	16
	0.1	4–14	17
IRFNA-27/UDMH	1	4–14	3
	0.1	4–14	4

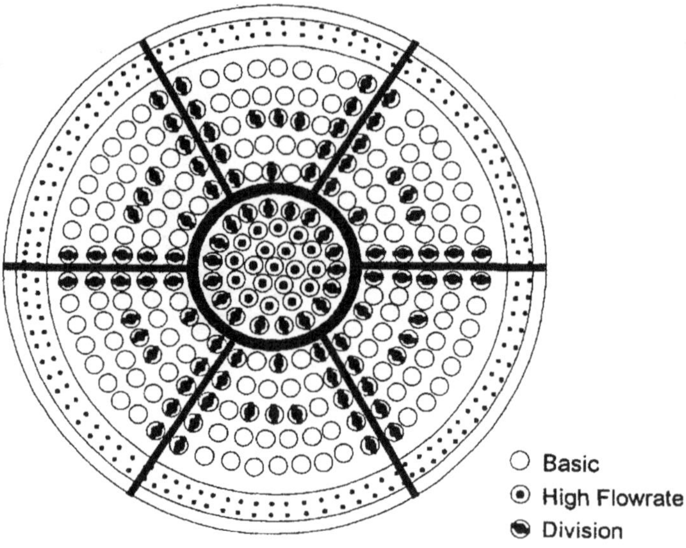

Fig. 12 YF-1 injector with baffles.

3. Stabilizing the YF-1 Injector with Baffles

To eliminate high-frequency combustion instability, a regeneratively cooled baffle was combined with liquid-phase division. As shown in Fig. 12, a 100-mm-high regeneratively cooled baffle is mounted on the injector face. In a number of test firings, with the thrust increasing from 186.2 to 274.4 kN, no combustion instabilities occurred. Very stable combustion was obtained. The original liquid-phase division and flow-intensity distribution remained unchanged. The baffle concepts were also investigated in another model engine, where the baffle proved very effective in eliminating the transverse mode oscillation.[54]

It should be pointed out that the number of the liquid-phase division elements has an impact on engine performance, as depicted in Table 8. It was once considered that decreasing the number of the liquid-phase division elements, or even removing them might improve the specific impulse while maintaining the stability margin. Three test firings were conducted to demonstrate the feasibility of the concept, but all were unstable. This indicated that the liquid-phase division does have a significant influence on combustion stability even when baffles are used.

Artificial disturbances, using a pulse gun, were induced to rate the YF-1 engine combustion stability. The powder pulse generators, each 3 g of P-45, are mounted

Table 8 Effect of the number of the liquid phase elements on engine performance

No. of elements	I_{sp}, s	ηI_{sp}s	ηc^*	No. tests
60	250.8	0.943	0.959	1
66	248.9	0.938	0.954	3
78	247.3	0.934	0.950	5
90	244.9	0.920	0.939	1
102	244.3	0.918	0.944	27

on two sides of the baffle cavities, within the baffle cavities and outside the cavities. The chamber pressure is measured by two water-cooled piezoresistive high-frequency pressure transducers. The sensing membrane of one of the transducers is in contacted with the combustion gas in order to calibrate the influence of the mechanical vibration on the performance of the transducer. The test firings were conducted at a main-stage thrust of 245 kN (55,000 lbf). The pulses were initiated during the phases of engine start, main stage, operating regime transition, and shutdown. During the test, the maximum amplitude of pressure was 1.3121 MPa (190 psia), but after rapid attenuation the normal operating state returned in 11 ms. No destructive high-frequency transverse combustion instabilities were excited. The bomb tests demonstrated that the YF-1 engine has a definite stability.[55]

Combining the baffle with the liquid-phase division on the FY-20 and YF-1 engines has proved to be a very effective method to obtain stability, since the axial and transverse energy release distributions are directly related to combustion instabilities. If these methods are properly used, high-engine performance and stability can be best achieved.

References

[1]Schmidt, E. W., *Hydrazine and Its Derivatives: Preparation, Properties, and Applications,* Wiley, New York, 1984.

[2]Wright, A. C., "USAF Propellant Handbooks Nitric Acid/Nitrogen Tetroxide Oxidizers Vol II," Martin Marietta Corp., AFRPL-TR-76-76, 1977.

[3]Clark J., *Ignition !,* Rutgers Univ. Press, New Brunswick, NJ, 1972.

[4]Von Braun, W., and Ordway, F. I. III, *History of Rocketry and Space Travel,* Thomas Y. Crowell, New York, 1969.

[5]Reardon, F. H., Mcbride, J. M., and Smith, A. J., "Effect of Injection Distribution on Combustion Stability," *AIAA Journal,* Vol. 4, No. 3, 1966, pp. 506–510.

[6]Hefner, R. J., "Review of Combustion Stability Development with Storable Propellants," *Journal of Spacecraft and Rockets,* Vol. 3, No. 7, 1966, pp. 1046–1051.

[7]Abbe, C. J., McLaughlin, C. W., and Weiss, R. R., "Influence of Storable Propellant Liquid Rocket Design Parameters on Combustion Instability," *Journal of Spacecraft and Rockets,* 1968, pp. 584–590.

[8]Mayer, S. W., Taylor, D., and Schieler, L., "Preignition Products from Storable Propellants at Simulated High Altitude Conditions," Aerospace Corp., TR-0158(9210-02)-1, El Segundo, CA, Nov. 1967.

[9]Saad, M. A., Detweiler, M., and Sweeney, M., "Analysis of Reaction Products of Nitrogen Tetroxide with Hydrazine under Nonignition Conditions," *AIAA Journal,* Vol. 10, No. 8, 1972, pp. 1073–1078.

[10]Takimoto, H., and Denault, G. C., "Combustion Residues from NTO–MMH Motors," Aerospace Corp., TR-0066(5210-10)-1, Sept. 1969.

[11]Mills, T. R., Breen, B. P., Tkachenko, E. A., and Lawver, B. R., "Transients Influencing Rocket Engine Ignition and Popping," NAS 7-467, Interim 14 Month Report, April 1969.

[12]Christos, T., Miron, Y., James, H., and Perlee, H., "Combustion Characteristics of Condensed Phase Hydrazine Type-Fuels with Nitrogen Tetroxide," *Journal of Spacecraft and Rockets,* Vol. 4, No. 9, 1967, pp. 1224–1229.

[13]Glassman, I., *Combustion,* Academic, New York, 1987

[14]Gray, P., and Sherrington, M. E., "Self-Heating and Spontaneous Ignition in the Combustion of Gaseous Methylhydrazine," *Journal of Chemical Society, Faraday,* Vol. 70, No. 4, 1974, pp. 740–751.

[15]Juran, W., and Stechman, R. C., "Ignition Transients in Small Hypergolic Rockets," *Journal of Spacecraft and Rockets,* Vol. 5, No. 3, 1968, pp. 288–292.

[16]Kerkam, B. J., and Kahl, R. C., "Hypergol Engine Restart Characteristics," Boeing Co., NASA Manned Spacecraft Center.

[17]Lawver, B. R., "High Performance NTO/Amine Elements Blowapart," NASA Johnson Space Center, Aerojet Liquid Rocket Co., Contract NAS 9-14186, March 1979.

[18]Daimon, W., Gotoh, Y., and Kimura, I., "Mechanism of Explosion Induced by Contact of Hypergolic Liquids," *Journal of Propulsion and Power,* Vol. 7, No. 6, 1991, pp. 946–952.

[19]Allison, C. B., and Faeth, G. M., "Decomposition and Hybrid Combustion of Hydrazine, MMH, and UDMH as Droplets in a Combustion Gas Environment," *Combustion and Flame,* Vol. 19, 1972, pp. 213–226.

[20]Eberstein, I. J., and Glassman, I., "The Gas Phase Decomposition of Hydrazine and its Methyl Derivatives," *Tenth Symposium (International) on Combustion,* (Princeton Univ.), Combustion Inst. Pittsburgh, PA, 1965, pp. 365–374.

[21]Larosiliere, L., Litchford, R., and Jeng, S.-M., "Hypergolic Bipropellant Spray Combustion and Flow Modeling in Rocket Engines," 26th AIAA/ASME/SAE/ASEE Joint Propulsion Conf., AIAA Paper 90-2238, July 1990.

[22]Krishnan, A., Przekwas, A. J., and Gross, K., "Orbital Maneuvering Vehicle (OMV) Thrust Chamber Performance," CFD Research Corp. CFRDC Rep. 4075/1, NAS8-37196, July 1990.

[23]Chiu, H. H., Jiang, T. L., Gross, K., and Krebsbach, A. N., "Numerical Analysis of Bipropellant Combustion in Orbit Maneuvering Vehicle Thrust Chamber," NAG-106, March 1989.

[24]Zung, L. B., and White, J. R., "Combustion Process of Impinging Hypergolic Propellants," NASA CR-1704, Lewis Research Center, May 1971.

[25]Habiballah, M., Dubois, I., Gicquel, P., and Foucaud, R., "Experimental Study of Combustion Processes Involved in Hypergolic Propellant Coaxial Injector Operation," 28th AIAA/ASME/SAE/ASEE Joint Propulsion Conf., AIAA Paper 92-3388, July 1992.

[26]Kushida, R., and Houseman, J., "Criteria for Separation of Impinging Streams of Hypergolic Propellants," Jet Propulsion Lab., Western States Section, Combustion Inst. WSCI-67-38, Oct. 1967.

[27]Harrje, D. T., and Reardon, F. H., "Liquid Propellant Rocket Combustion Instability," NASA SP194, 1972.

[28]Houseman, J., and Lee, A., "Popping Phenomena with the Hydrazine Nitrogen Tetroxide Propellant System," *Journal of Spacecraft and Rockets,* Vol. 9, No. 9, 1972, pp. 678–681.

[29]Phillips, D. G., "Analysis of Test Data from RCS Primary Thruster Instability Testing at WSTF," Rockwell International, TMOS-A109, Downey, CA, Dec. 1984.

[30]Chang, E. T., Gokcen, N. A., and Poston, T. M., "Solubilities of Gases in Simple and Complex Propellants," *Journal of Physical Chemistry,* Vol. 70, 1966.

[31]Chang, E. T., Gokcen, N. A., and Poston, T. M., "Thermodynamic Properties of Gases in Propellants. II. Solubilities of Helium, Nitrogen, and Argon Gas in Hydrazine, Methylhydrazine, and Unsymmetrical Dimethyl Hydrazine," *Journal of Physical Chemistry,* Vol. 72, No. 2, Feb. 1968.

[32]Juran, W., and Zwick, E. B., "Dissolved Gas and Helmholtz Resonance in Rocket Injectors," *Journal of Spacecraft and Rockets,* Vol. 6., No. 3, 1969, pp. 347–349.

[33]Smith, I. D., "The Determination of the Helium Saturation Level of Monomethylhydrazine and Mon-3 Nitrogen Tetroxide," NASA White Sands Test Facility, TR-244-001, April 1979.

[34]Langel, G., Laudien, E., Pongratz, R., Habiballah, M., Grand-Perret, S., and Vinbgert, L., "Combustion Characteristics of the Ariane 5 L7 Engine," 42nd Congress of the International Astronautical Federation, IAF-91-250, Oct. 1991.

[35]Crone, C., Stechman, C., and Woll, P., "Combustion Instability in Small Earth Storable Bipropellant Engines," 29th AIAA/ASME/SAE/ASEE Joint Propulsion Conf., AIAA Paper 93-1960, June 1993.

[36]Anon., "Service Propulsion System Flight Support System Handbook—Apollo & Skylab," NASA Johnson Space Center, Propulsion and Power Div., 1967.

[37]McBride, J. M., and Hefner, R. J., "Combustion Stability Characteristics of the Apollo SPS and Transtage Liquid Rocket Engines," Chemical Propulsion Information Agency, 4th ICPRG Combustion Conf., Oct. 1967.

[38]Valentine, R. S., Rossi, F. S., and Kromrey, R. V., "Fluid Dynamic Effects of Apollo Engine Pressure Spikes," *Journal of Spacecraft and Rockets,* Vol. 5, No. 1, 1968, pp. 31–35.

[39]Anon., "Apollo Engine Contractors Combustion Stability Meeting," NASA Johnson Space Center, Houston, TX, Aug. 1965.

[40]Anon., "LEM Descent Engine Characteristic Report," Rocketdyne, Contract NAS9-1100, Canoga Park, CA, Nov. 1964.

[41]Anon., "Lunar Module Descent Engine (LMDE)," TRW Systems, Power Systems Div., Feb. 1968.

[42]Anon., "LEM Descent Engine, Apollo Contractor Injector Review Meeting," NASA Johnson Space Center, Horston, TX, Aug. 1965.

[43]Senneff, J. M., and Morgante, P. J., "Combustion Stability Investigation of the LEM Ascent Engine," Bell Aerosystems Chemical Propulsion Information Agency, Laurel, MD, Nov. 1966.

[44]Harmon, T. J., "Apollo Lunar Module Ascent Propulsion—No Second Chance," 28th AIAA/ASME/SAE/ASEE Joint Propulsion Conf., AIAA Paper 92-3125, July 1992.

[45]Oberg, C. L., Wong, T. L., and Ford, W. M., "Evaluation of Acoustic Cavities for Combustion Stabilization," Rocketdyne, NASA CR-115087, July 1971.

[46]Anon., "Lunar Module Ascent Engine Program Review," NASA Johnson Space Center, Houston, TX, July 1968.

[47]Ross, B. G., "Primary RCS Thruster Stability Analysis Program—Final Report," Rockwell International Corp., NAS9-14000, Downey, CA, Oct. 1992.

[48]Gibson, C., and Humphries, C., *Orbital Maneuvering System Design Evolution,* Space Shuttle Technical Conf., NASA Johnson Space Center, Houston, TX, 1985, pp. 639–655.

[49]Bertolino, J. D., "Uprated OME Component Demonstration Program Final Report," Gencorp Aerojet, Propulsion Systems Plant, NAS 9-17215, June 1993.

[50]Sun, J. L., and Zhang, B. J., "Combustion Stability Analysis for the SD60 Engine," Shanghai Bureau of Astronautics, PRC, Oct. 1963.

[51]Zhuang, F. C., "Combustion Chamber Calculation for Model for the FY-20 Engine," Shanghai Bureau of Astronautics, PRC, Sept. 1987.

[52]Crocco, L., and Cheng, S. I., "Theory of Combustion Instability in Liquid Propellant Rocket Motors," AGARD Monograph, No. 8, Butterworths, London, 1956.

[53]Sun, J. L., "Combustion Instability for Liquid Propellant Rocket Engine," Research and Study No. 3, Shanghai Bureau of Astronautics, PRC, 1963.

[54]Sun, J. L., and Zhang, B. J., "Combustion Instability Study of the 5D61A Engine," Shanghai Bureau of Astronautics, PRC, 1964.

[55]Zhu, L. C., "High Frequency Combustion Oscillation Research and Summary for the YF-1 Engine," Shanghai Bureau of Astronautics, PRC, Dec. 1971.

Chapter 6

Coaxial Injector Atomization

Lucien Vingert,* Pierre Gicquel,* Daniel Lourme†
ONERA, Châtillon, 92320 France
and
Luc Ménoret‡
Société Européenne de Propulsion, Vernon, France

Nomenclature

A = cross section
D = diameter
$\bar{D}$ = mean diameter
D_M = mass median diameter
D_{32} = Sauter mean diameter
K = model constant
$\dot{m}$ = mass flow rate
Q = volume flow rate
V = velocity
λ = wavelength
μ = dynamic viscosity
ρ = density
σ = surface tension

Subscripts
g = gas
j = jet
l = liquid
r = relative

---

*Research Engineer, Energetics Department.
†Coordinator of ONERA Space Activities.
‡Development Engineer, Combustion Components Department.

I. Introduction

COAXIAL twin fluid injectors are widely used in rocket engines, especially for cryogenic engines like the Space Shuttle main engine (SSME) or the European Ariane 5 Vulcain engine, using liquid oxygen and gaseous hydrogen as propellants. The basic mechanisms that atomize the liquid oxygen (LOX) jet into a spray of polydispersed droplets are not yet well modeled nor even completely identified. For this reason the majority of previous work on this subject was purely experimental. Unfortunately, the operating conditions in which the spray characterizations are performed are often very far from actual engine conditions (e.g., cold nonreactive flows instead of hot-burning environment, simulants that are not very representative of LOX, subcritical instead of supercritical pressure). The empirical laws are, therefore, often limited to the operating range where they were derived and do not fit the data obtained with different geometries, or even with the same geometry in different experimental conditions, and consequently often do not apply to actual engine conditions. Despite those limitations, the experiments are useful to compare different designs of a given injector in similar operating conditions, to compare atomization efficiency, and to rate injector performance and combustion stability. A given engine can be stable or unstable by a single injector modification that affects the droplet size. The Ariane Viking engine, for example, has become stable by increasing the mean droplet diameter.

Another important objective of atomization and spray characterizing is the development of computer codes to study the behavior of complete combustion chambers. Because of the prohibitive computation times, it is not realistic to describe all of the elementary processes of atomization or the individual droplets. It is much easier to take spray characteristics near the injectors as the boundary condition at the inlet of the computational domain. Therefore, a purely experimental characterization of atomization may be sufficient. This does not mean that numerical simulation and theoretical studies of atomization processes are of no practical interest. To the contrary, droplet size and velocity as inputs to a chamber model are difficult measurements, and theoretical values, if available, would avoid the problems of nonrepresentative experimental conditions.

The current work includes a literature survey on theoretical aspects and experimental work, part of which is taken from previous reviews performed by Ferrenberg et al.[1] in 1985 and by Ghafourian et al.[2] in 1991. The experimental work on coaxial rocket engine injectors was conducted mainly in the 1960s and early 1970s with a frozen wax technique and in the 1980s with optical particle sizing, first with particle sizers based on the Fraunhofer diffraction theory and more recently with phase Doppler particle analysers (PDPA). Recently a few theoretical approaches have been developed, most often based on linearized equations, but also original ideas like a maximal entropy formalism that seems to be very promising. Numerical simulations of atomization process using the two basic techniques of volume of fluid (VOF) and jet embedding (JE) may also become more important with the increasing performance of modern computers.

The second part of this chapter presents the approach retained at ONERA to characterize the coaxial injectors of both the gas generator and the thrust chamber of the Vulcain engine. This philosophy is, in fact, a progressive empirical method consisting of three steps. First, a series of cold-flow tests were conducted in a pressurized vessel fitted with windows at various chamber pressures up to 30 bar, using the Malvern particle sizer. Water was used to simulate the LOX and nitrogen or helium to simulate the hydrogen. Second, cold-flow tests were conducted with LOX, at chamber pressures between 1 and 10 bar. The third step used alternative simulants with intermediate physical properties between water and LOX to characterize the change from water to LOX. This third step also aimed at increasing the experimental range of Reynolds and Weber numbers to reach the high values corresponding to engine conditions. Nitrogen or helium were used in all tests to simulate hydrogen.

The last section of the chapter presents the industrial manufacturer's point of view. There the work still to be done to characterize injectors, and to help the manufacturer in designing a new injection element when starting the development of a new engine, will be discussed.

II. Literature Review

A. Theoretical Investigations

1. Early Work

The first theoretical work on the rupture of a round liquid jet into droplets was by Lord Rayleigh[3] as early as 1878. He studied a low-speed liquid jet flowing into vacuum. In this case, the surface tension force is dominating the process, and drops are pinched off from the end of the jet with a diameter comparable to that of the jet. From potential energy considerations, Rayleigh showed that the jet is unstable to symmetrical disturbances with wavelengths larger than the jet circumference. He derived an expression for wavelength and growth rate of the disturbance. Rayleigh's theory was extended by Weber[4] in 1931, who included the liquid viscosity effect and derived an expression for the most favorable wavelength for drop formation in the case of a viscous jet. The analysis was improved by Tyler[5] who added an expression for the size of the droplets issued from this type of low-speed jet breakup.

Ohnesorge[6] in 1936 first identified different regimes of jet breakup by means of a dimensional analysis. The two most important dimensionless numbers in atomization process are the Reynolds number *Re*, which represents the ratio of inertial to viscous forces, and the Weber number *We*, which represents the ratio of inertial to surface tension forces. The Ohnesorge number is defined as $Z = We^{1/2}/Re$,

$$Re = \frac{\rho_l V_l D_j}{\mu_l} \qquad We = \frac{\rho_l V_l^2 D_j}{\sigma} \qquad Z = \frac{\mu_l}{\sqrt{\sigma \rho_l D_j}}$$

By plotting Z vs *Re*, the low-velocity and the high-velocity regions can be distinguished. In the low-velocity region, the jet breakup is due to the action of surface tension forces (Rayleigh regime). In the second region, the influence of aerodynamical (inertial) forces increases exponentially with *Re* at constant Z, so that it rapidly dominates the process. In a practical case (injector geometry and propellant given), the orifice diameter as well as the fluid properties are fixed. The Ohnesorge number is consequently defined, so that the initial jet velocity (the injected mass flow rate) becomes the most important variable.

Nukiyama and Tanasawa[7] studied coaxial injectors in 1938 and 1939. They collected drops on a glass slide coated with a lubricant and took photomicrographs to measure the drop sizes. They developed a correlation relating the drop sizes to operating conditions and also gave an expression of the droplet size distribution function,

$$N(D) = \frac{\beta D^{\alpha}}{\Gamma[(\alpha+1)/\beta]\bar{D}^{\alpha+1}} \exp\left[\left(-\frac{D}{\bar{D}}\right)^{\beta}\right] \tag{1}$$

where $N(D)$ is the number of droplets having diameter D, and α and β are two model constants. In Nukiyama and Tanasawa's experiment α was found equal to 2, but this is not a general case. This expression is rather rarely used in droplet sizing applications, despite the fact that it often fits the data. Most authors prefer the expression of Rosin and Rammler,[8] which can be written in the same terms with $\alpha = \beta - 4$. The latter expression leads to easier mathematics and simple calculations of mean diameters, but one must notice that the Nukiyama–Tanasawa expression can be derived from maximal entropy formalism, as we will see in a following section. Concerning Nukiyama and Tanasawa's empirical correlation for the Sauter mean diameter (SMD), which can be written

$$SMD = 0.585 D_j^{0.5} We^{0.5} + 53.Z^{0.45} D_j^{0.225} (Q_l/Q_g)^{1.5} \tag{2}$$

one can notice that it was expressed in the shape $SMD = SMD_1 + SMD_2$ with the first term depending on the Weber number and the second on the Ohnesorge number and on the air-to-liquid flow rate ratio. This type of expression, with various exponents, was used by many subsequent investigators of air-blast atomizers.

A coaxial injector is different from a round liquid jet because of the presence of the high-velocity coflowing gaseous jet, which induces disturbances on the surface of the liquid jet in the immediate vicinity of the injection plane. As a first approximation, the jet surface, compared to those small-scale disturbances, can be considered as an infinite flat interface. Consequently, to describe the mechanisms that take place in the atomization process, the instability of surface waves induced by wind must be modeled. This phenomenon is considered to be dominating whereas others, like cavitation and turbulence, are neglected. Kelvin[9] analyzed in 1871 the influence of wind on waves in water under the effects of surface tension and gravity. The Kelvin–Helmholtz instability was defined where waves are generated due to the static pressure perturbation of the air. The pressure and the velocity of the air vary in antiphase at the top and the bottom of the waves, according to Bernoulli's theorem. Therefore, waves are generated on the water surface when the aerodynamical forces overcome the surface tension and the gravity forces that tend to reduce the disturbance. Kelvin derived that the smallest velocity of frictionless air which can raise a ripple on rigorously quiescent frictionless water is 6.6 m/s. Taylor[10] considered the planar wave growth at the surface of a viscous liquid fluid of infinite depth. He showed that if the liquid density is much higher than the surrounding gas density, the growth rate increases and shorter wavelengths become dominant as $(\rho_l/\rho_g)(Re_l/We_l)^2$ increases. Taylor's analysis constitutes the starting point for the main subsequent wave growth models.

After the waves have broken from the main jet surface, they undergo additional disruption, called secondary atomization. As long as a high relative velocity exists between the surrounding gas flow and some large drops, formed during the jet or ligament breakup, those drops further break up during their flight, into smaller and more stable droplets. Hinze[11] presented a theoretical analysis of droplet breakup. He described the deformation of a liquid sphere caused by normal pressure distribution resulting from a flow of gas over its surface. Hinze postulated the existence of a critical Weber number as a criterion for drop breakup, $We = \rho_g D_l V_r^2/\sigma$, where D_l is the original drop diameter. This critical Weber number is approximately 10 for low-viscosity and 20 for high-viscosity liquids. Hinze[11] also gave an expression for the breakup time. This breakup time, together with the initial droplet velocities, is important in combustion applications because it gives the distance to where the secondary atomization takes place compared to the flame position. If droplet breakup occurs before the droplet encounters the flame front, the result of the secondary atomization has to be considered as the correct input for combustion computational fluid dynamics (CFD) codes, where droplets are injected in the computational domain upstream boundary. If the breakup time is too long, the secondary atomization should be modeled in the code, and the result of the primary atomization constitutes the right input.

2. *Identification of Phenomena, Classification of Breakup Regimes, and Dimensional Analysis*

As indicated by Rice[12] the object of atomization is to increase liquid surface area to accelerate burning. This surface area is increased by various means, until the process becomes unstable, leading to disintegration of the liquid surface into droplets. Such a breakup of round or planar liquid jets is a result of the complex interaction between inertial, viscous, and surface tension forces that can be summarized as follows: liquid viscosity has a damping effect on the growth of disturbances on the surface, aerodynamic forces tend to promote them, and surface tension tends to pull the liquid together. Pressure oscillations and the turbulent nature of the injected fluids and their environment also influence the dynamics of atomization.

Various means may be used to increase the liquid surface area, including the use of a high-speed gas stream in twin fluid gas/liquid atomizers, like the coaxial elements used in cryogenic rocket engines. Similar atomizers are widely used in gas turbine technology. They were reviewed by Lefebvre[13] who classified them in the air-blast and the air-assist types. In the air-blast atomizer, the liquid is atomized by the air that is naturally present in the system and that participates in the combustion, whereas in the air-assist atomizer, an auxiliary air supply, most often at high pressure, is used. A coaxial rocket injector is a combination of the two. Air-blast or air-assist atomizers can also be classified as plain-jet and prefilming types. In the plain-jet type, the liquid is first attenuated into one or several fine jets, before it encounters the gas stream, whereas in the prefilming type, it is attenuated into a thin sheet inside the atomizer. In both cases, the growth of infinitesimal disturbances, generated within the atomizer and/or due to aerodynamic interaction with the surrounding gas, are assumed responsible for the deformation of the original surface of the liquid. The relative motion with respect to the atomizing gas then causes an instability of the liquid stream and a tendency to form waves that rapidly grow in amplitude to critical values at which the liquid stream undergoes a breakup into ligaments and drops.

The hypothesis, however, that the atomization occurs only through aerodynamic interaction between the liquid and the surrounding gas has not found universal acceptance. It is argued that a fully turbulent jet can break up without the application of any external force. Reitz and Bracco[14] considering the case of a high-speed single jet, indicated that several mechanisms, acting simultaneously, are responsible for the jet breakup in the atomization regime. The jet breakup process may occur within the nozzle itself, due to liquid turbulence, so that the radial component of velocity in turbulent pipe flow could cause the immediate disruption of the jet at the nozzle exit. Liquid cavitation phenomena inside the nozzle could create large amplitude pressure disturbances in the flow, leading to atomization, so that cavitation bubbles may enhance the jet breakup process. The jet breakup may be due to the velocity profile relaxation: high-velocity laminar liquid jets could be more unstable than fully developed turbulent jets because of the rearrangement in the cross-sectional velocity profile of the laminar jet once the constraint of the nozzle wall is removed at the nozzle exit; this redistribution of energy within the jet gives rise to radial velocity components which disrupt the jet. this effect is minimal with fully developed turbulent jets having flat exit velocity profiles. The atomization may also be due to the acceleration experienced by the liquid in the boundary layer on the nozzle wall exit caused by the abrupt change in the boundary condition.

All of these mechanisms refer to the atomization regime of liquid jets, which first has to be defined. In his classification of breakup regimes of a single round liquid jet, in terms of breakup length (or intact liquid core length), Reitz[15] summarized the conclusions of several previous investigators and distinguished four breakup regimes with increasing Reynolds number at constant Ohnesorge number. (For given fluids, liquid and ambient gas, and a fixed nozzle diameter, that is equivalent to increasing jet velocity.) These regimes are (see Fig. 1) as follows.

1) In the Rayleigh jet breakup regime, the surface tension force, which induces axisymmetric oscillations of the jet surface, is mainly responsible for breakup; drops are pinched off from the end of the jet with diameter exceeding that of the jet.

2) In the first wind-induced regime, the aerodynamic forces start playing a role; unstable waves at long wavelengths grow along the jet; drops with diameter of the order of the jet diameter are pinched off from the end of the jet.

3) In the second wind-induced regime, drops, produced by the unstable growth of short wavelength surface waves, are torn off from the jet surface and are smaller than the jet diameter.

4) In the atomization regime, the jet surface appears to break up immediately at the nozzle exit into very small drops.

Surface tension, which enhances the two first regimes, opposes the third one. Liquid viscosity seems to have no effect in the second wind-induced regime.

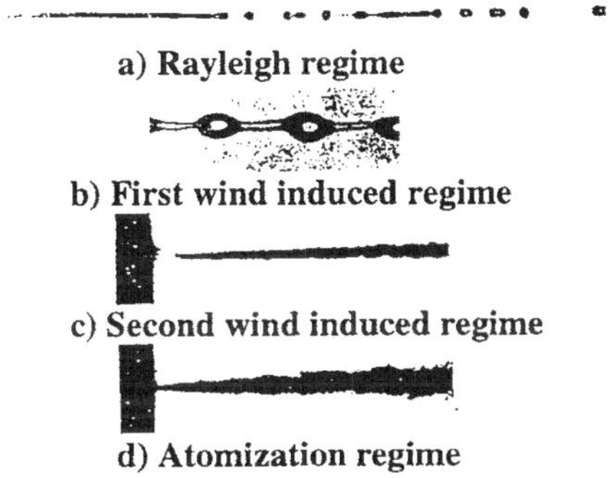

Fig. 1 Breakup regimes of a single liquid jet in quiescent air (Ref. 30).

The mechanisms involved in the atomization regime are still unknown. Defining the Weber number as $We = \rho_g d V_l^2/\sigma$, where d is the nozzle diameter and V_l the initial jet velocity, the second wind-induced regime occurs for $0.4 \leq We \leq 13$ and the atomization regime for $We \geq 13$. Some authors situate the limit for atomization at $We \geq 40$, which is not a very important difference when one knows that, for example, in the practical case of the Vulcain engine coaxial injectors, this number is of the order of magnitude of 10^3.

In the wave growth concept, it is generally assumed that the sign of the relative velocity between the gas and the liquid has no effect; i.e., the phenomena are assumed to be essentially the same in the case of a high-speed liquid jet injected into quiescent air and in the case of a low-speed liquid jet exposed to a high-speed coflowing gas stream. As indicated by Lefebvre[16] this might not be exactly true, because in pressure atomizers the various flow phenomena, such as turbulence and cavitation that promote the self-destruction of the liquid jet or sheet and greatly enhance atomization, are highly velocity dependent. In air-blast atomizers, where the liquid speed is as low as possible to increase the relative gas/liquid velocity, liquid turbulence and cavitation are limited, so that atomization consequently becomes almost solely dependent on the air-to-liquid mass flow rate ratio and on the momentum of the atomizing gas. This suggests that a model where only aerodynamic effects are taken into account should be sufficient for coaxial rocket injectors.

Gomi,[17] reviewing previous experimental work on coaxial injectors and adding his own results (Fig. 2), classified the breakup regimes of such atomizers in three categories, depending on the mixture ratio (Fig. 3): $\dot{m}_l/\dot{m}_g \leq 0.2$, where the relative velocity determines the drop size; $0.2 < \dot{m}_l/\dot{m}_g \leq 1$, where the relative velocity and the mixture ratio determine the drop size; and $1 < \dot{m}_l/\dot{m}_g$, where many parameters affect drop sizes. He[17] explained this dependence of drop size by the

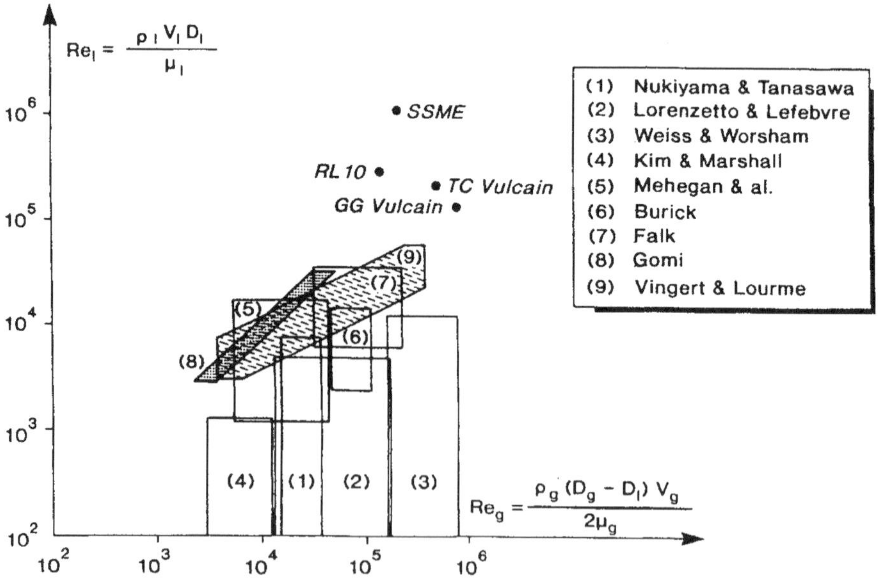

Fig. 2 Experimental range of coaxial injectors investigations.[17,70]

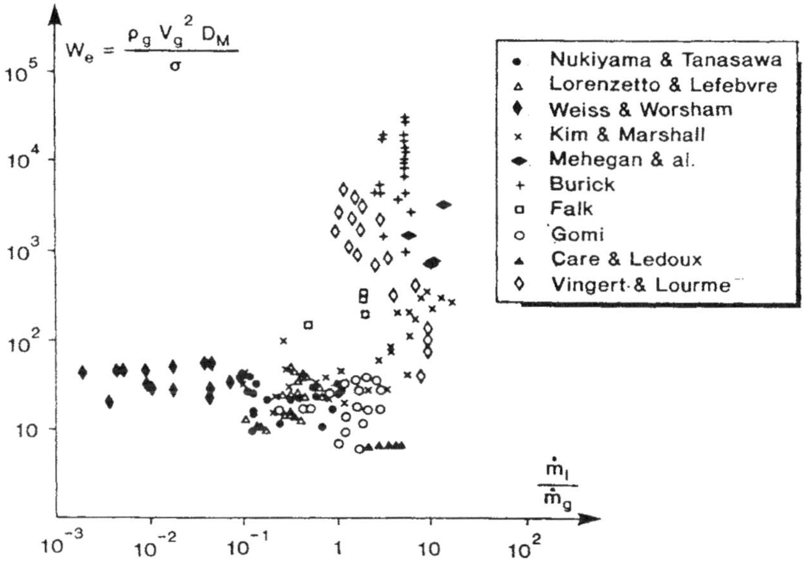

Fig. 3 Correlation of experimental data.[17,70]

potential core, length of the annular gas jet relative to the length of the atomizing region. At a small mixture ratio, the atomization of liquid is completed within the potential core and then drop sizes are determined by the relative velocity. At a large mixture ratio, the atomization terminates downstream because of the rapid decrease of the effective gas velocity which acts on the liquid surface.

Buschulte[18] collected the results of several investigators and plotted the ratio of maximum droplet diameter to jet diameter ($D_{\max}/D_j$) vs an aerodynamical Weber number defined as

$$We_{ae} = \frac{\sigma}{\rho_g D_j (V_g - V_l \cos\alpha)^2 \sin^2\alpha} \tag{3}$$

where α is the angle between the jet core velocity and the gas stream velocity. Notice that this Weber number is the inverse of the classical definition. He found that all data whose aerodynamic Weber numbers were less than 0.1 could be correlated by a mean value expression obtained by a least square method,

$$D_{\max}/D_j = 2.2 We_{ae}^{0.5} \tag{4}$$

However, the scatter of the data was as large as an order of magnitude around this mean value.

A new classification of breakup regimes for coaxial airblast injectors was introduced by Farago and Chigier,[19] who carried out an intensive experimental program, using a spark photograph technique, to visualize the rupture of round liquid jets, with or without coflowing gas stream. The Weber number was varied over a range of six orders of magnitude, whereas the Reynolds number was varied over a range of two orders of magnitude. Thus,

$$10^{-3} \le We = \frac{\rho_g V_r^2 D_j}{\sigma_l} \le 10^3 \qquad 2.10^2 \le Re_l = \frac{\rho_l V_l D_j}{\mu_l} \le 2.10^4$$

Detailed analysis of the photographs showed that jet disintegration can be divided into the following three categories (Fig. 4).

1) Rayleigh breakup has the same definition as given earlier. This category can be divided in two subgroups: axisymmetric breakup ($We < 15$) and nonaxisymmetric ($15 < We < 25$).

2) The jet disintegration is produced via membrane-type ligaments ($25 < We < 70$). The round jet develops into a thin sheet (membrane), which forms Kelvin–Helmholtz waves and breaks up into drops, with diameter smaller than the jet diameter. The jet disintegration is completed in one or two wavelengths.

3) The jet disintegration is produced via fiber type ligaments ($100 < We < 500$). Fibers are formed and then peel off the jet. The fibers break into drops via the Rayleigh mechanism. The liquid core is accelerated and develops waves. Breakup occurs via the nonaxisymmetric Rayleigh mechanism or by breaking into ligaments. The ligaments may then break up into smaller drops or ligaments by one of the three disintegration modes. The completion of jet disintegration needs many wavelengths. The produced drops are an order of magnitude smaller than the jet diameter. This category can be divided into two submodes: pulsating jet disruption, which is the normal submode, and superpulsating jet disruption, which is connected to an extremely high periodical change between low- and high-density regions in the spray ($150 < We < 500$).

Care was taken during these experiments to avoid any vibration or pulsation in the feedlines, and the atomizers were designed to uncouple the jet rupture process from the upstream conditions. Despite those precautions, the atomization process was always pulsating, which seems to be a fundamental aspect of atomization. The third (fiber type) process is the most important in practical applications. In this mode, the largest proportion of the liquid is atomized in the secondary atomization region, which is not taken into account in the theories dealing with the onset of the instabilities and the primary atomization region based on wave instabilities at the axisymmetric intact liquid surface. One should also remember that due to the high density of atomizing hydrogen gas and very low surface tension of liquid oxygen, there is still at least one order of magnitude between the highest investigated Weber numbers and the nominal conditions of rocket engine injection.

Lefebvre[20] distinguished two atomization mechanisms in air-blast atomizers, the classical atomization, where the droplet formation is due to wave instability on the liquid surface, and the prompt atomization, which occurs if the gas stream impacts the liquid jet with high angles of incidence or if the gas velocity is unusually high. In this disruption mode, the time necessary for the liquid jet instabilities to develop and form growing waves is not available. Instead, the liquid jet is rapidly torn into small fragments, so that the final drop size does not depend on the initial jet diameter nor on the gas density. The following correlation for the mean drop sizes produced under prompt atomization was proposed:

$$\frac{D_{32}}{D_j} = \frac{1.5}{1 + CWe/(1 + 1/ALR)} \tag{5}$$

where C is an empirical constant, ALR is the air-to-liquid mass flow rate ratio ($ALR = \dot{m}_g/\dot{m}_l$) and the Weber number is defined as $We = \rho_l V_g^2 D_j/\sigma_l$.

Farago[21] tried to find the maximal possible number of common non-dimensional groups for six different types of aerodynamical atomizers based on two categories.

a) Axisymmetric Rayleigh disintegration; $u_w = 1.02$ m/s, $u_a = 22.9$ m/s

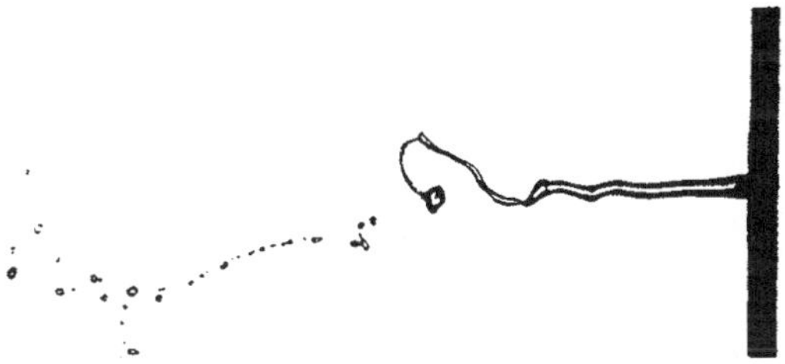

b) Nonaxisymmetric Rayleigh breakup; $u_w = 0.643$ m/s, $u_a = 32.4$ m/s

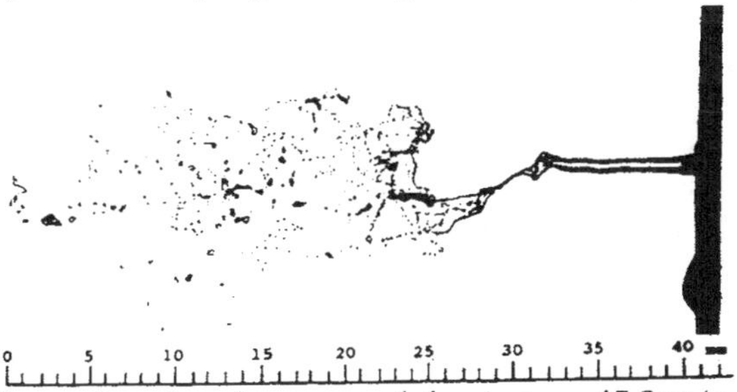

c) Membrane-type breakup; $u_w = 1.4$ m/s, $u_a = 45.8$ m/s

d) Fiber-type breakup (pulsating); $u_w = 12.6$ m/s, $u_a = 183$ m/s

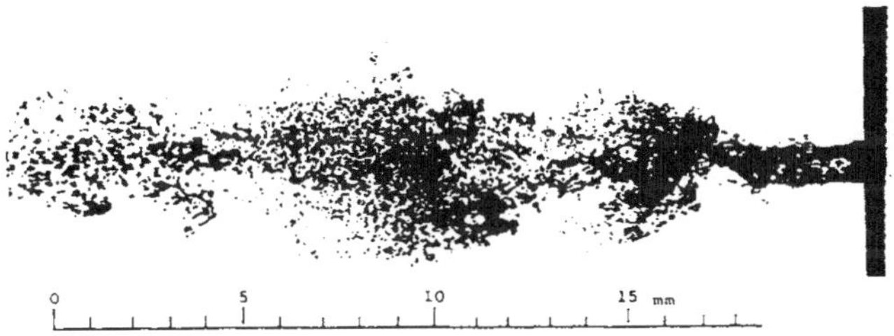

e) Fiber-type breakup (superpulsating submode); $u_w = 6.4$ m/s, $u_a = 183$ m/s

Fig. 4 Breakup regimes of a water jet in a coaxial air stream; emerging jet diameter = 1 mm (Ref. 19).

Coaxial twin fluid atomizers, two-dimensional atomizers (thin-liquid sheet between two parallel gas streams) and thin annular liquid sheet between two gas streams constitute the category A, where nine variables are assumed necessary for a complete description. Disruption of a single liquid drop in a gas stream, atomization of a round liquid jet in quiescent air, and thin liquid sheet ejected into quiescent air constitute the category B, where only seven variables are involved. The identified independent variables for the liquid are the hydraulic diameter, the density, the mean velocity, the dynamic viscosity, and the surface tension. For the gas, they are the mass flow (or an appropriated diameter), the density, the mean velocity, and the dynamic viscosity. The most convenient equation for category A where only common groups are used, is, according to Farago,

$$\frac{\bar{D}}{D_j} = K\left(\frac{\rho_l}{\rho_g}\right)^{b+d+j+f}\left(\frac{V_l}{V_g}\right)^{c+d+j+f} Re_l^{-d-j}\, We_{\mathrm{ae}}^{-e}\left(\frac{\mu_g}{\mu_l}\right)^{j}\left(\frac{\dot{m}_g}{\dot{m}_l}\right)^{f} \tag{6}$$

It is obvious from this review that the mechanisms that are involved in the atomization processes of coaxial injectors are not well identified. Several processes seem to be involved simultaneously, their relative importance depending on the specific design or on the operating range of each atomizer. A complete theoretical description valid in all cases may, therefore, be difficult to achieve. Nevertheless, different more or less sophisticated models are available, which are worth being examined, even if their applicability is limited.

3. *Linear Models*

The successive linear models that can be found in the literature are all improvements at different stages of Taylor's analysis.[10] They examine the stability of a liquid jet surface due to an infinitesimal axisymmetric disturbance with a Fourier form: $X = X_0 e^{ikz+\alpha t}$, where k is the wave number and α the wave amplitude exponential growth rate, which is governed by the dispersion equation described in detail by Levich[22] or Sterling and Sleicher,[23] for instance. This dispersion equation is obtained from the linearized Navier–Stokes equation by considering the aerodynamic effects with some simplifying assumptions and using the linear stability theory. The result is that α depends on the local relative velocity, on the gas and liquid densities, on the kinematic viscosity, and on the wave number.

Two important references are the works of Mayer[24] and Dickerson and Schuman.[25] Mayer analyzed the wind-induced capillary waves on a flat, deep liquid surface and obtained an expression for average drop size by assuming that when a wave grows to an amplitude comparable to its wavelength, the crest severs and forms droplets of size comparable to the wavelength,

$$\bar{D} = B_A\left(\frac{\mu_l\sqrt{\sigma_l/\rho_l}}{\rho_g V_g^2}\right)^{2/3} \tag{7}$$

Adelberg[26,27] extended this analysis by including the effect of fluid acceleration on the disturbance wave speed and found that acceleration waves become dominant for large fluid acceleration. This type of breakup should have negligible influence with subsonic flows. Dickerson and Schuman[25] added an expression for the atomization rates of jets and drops. The mass flow rate torn off from the jet surface was given by

$$\frac{\dot{m}}{A} = C_A \left(\frac{\rho_g^2 V_r^4 \mu_l \rho_l}{\sigma} \right)^{1/3} \tag{8}$$

where the constants C_A, ranging from 0.003 to 3 and B_A, had to be experimentally determined.

Dombrowski and Johns[28] considered the stability of a viscous liquid sheet subject to asymmetric waves, complementing Squire's results[29] on an inviscid moving sheet. They[28] gave equations for the sheet thickness, for the wave growth rate of the most unstable frequency, as well as for the corresponding wave number. The breakup of the liquid sheet due to the wave growth concept occurs when the total growth of the wave f, which is given by the natural logarithm of the ratio of wave amplitude to initial disturbance amplitude, approaches the constant value of $12 (f/f_0 = 12)$. The balance of forces caused by gas pressure, surface tension, liquid inertia, and viscosity, which controls the breakup mechanism of the sheet, can be written as follows:

$$2\rho_g n V_r^2 - 2\sigma n^2 - \rho_l \tau \left(\frac{\partial f}{\partial t} \right)^2 - \mu_l \tau n^2 \left(\frac{\partial f}{\partial t} \right) = 0 \tag{9}$$

where V_r is the relative velocity, f the total growth of wave, t the time, τ the sheet thickness, and n the wave number of the disturbance imposed on the liquid stream,

$$n = \frac{F^2 \rho_g V_r^2}{2\sigma} \tag{10}$$

where F is the ratio of wave growth of liquid to that of inviscid liquid. The maximum value of F is given by

$$1 - F^2 - V F^3 = 0 \tag{11}$$

where

$$V = \frac{\mu_l \rho_g V_r^2}{2} \sqrt{\frac{\tau}{2\rho_l \sigma^3}} \tag{12}$$

Reitz and Bracco[30] presented a unified first-order linear theory of the breakup of liquid jets, summarizing the work of Levich[22] and Sterling and Sleicher.[23] They assumed that atomization was primarily due to aerodynamic interaction between the liquid and the gas leading to unstable wave growth on the liquid surface. For the special case of slip at the gas-liquid interface and imposed infinitesimal axisymmetric surface displacement, a nondimensional dispersion equation was derived, which also included the physical and dynamical parameters of the liquid jet and the surrounding gas. Mechanisms of breakup in different regimes could be related to limiting cases in this analysis. At least three regimes were identified, depending on the gas/liquid relative velocity. An equivalent work was presented by Lin and Lian,[31] who showed that the instability of the liquid jet is characterized by three independent parameters: the Reynolds number, the Weber number, and the gas-to-liquid density ratio.

In Taylor's analysis,[10] the gas-to-liquid density ratio was assumed to be very small, but in actual rocket engines conditions, the injected hydrogen is at low temperature and its density is, therefore, not negligible compared to the liquid density. Solving the dispersion equation with ρ_g/ρ_l varying from 10^{-3} to 10^{-1}, Krülle et al.[32] recently showed that this ratio can have significant influence on the surface wave growth. Increasing gas density, at constant liquid density, had an obvious destabilizing effect on the liquid surface. Also, most probable surface wavelengths were shifted to lower values. The existence of gas velocity profiles, however, leads to an increase of the most probable wavelengths. This latter effect might be substantial and overcome the former, but the result remains that the computed wavelengths and related drop sizes are small compared to the most often reported experimental values which are of the order of 10^{-4} m (100 μm).

Caré and Ledoux[33] developed a complete linear model of the instability of a liquid jet in a coflowing air current. They considered the case of swirling and non-swirling flows for both fluids and also discussed the influence of a confinement for the gas flow. But they did not include the effects of boundary-layer velocity profiles and used the equations of inviscid fluids. Because of the high relative liquid-gas velocities this model is, in fact, asymptotic to the classical Kelvin–Helmholtz theory for capillary waves at a simple interface. The dominant wavelength Λ of maximum growth Ω is then found,

$$\Lambda = \frac{3\pi\sigma}{\rho_g V_r^2} \qquad \Omega = \frac{2\rho_g V_r^3}{3\sqrt{3}\sigma}\left(\frac{\rho_g}{\rho_l}\right)^{1/2} \tag{13}$$

and the propagation velocity of the wave c is

$$c = (\rho_g V_g + \rho_l V_l)/\rho_l \tag{14}$$

which reduces to V_l when the gas-to-liquid density ratio is small.

Caré and Ledoux[33] continued the analysis by deriving equations for the mean drop diameter produced by the breakup process. For this they assumed that at breakup the wave had formed a cylindrical ligament of diameter $\Lambda/2$ and amplitude $A\Lambda$, where A is an unknown constant. By writing the conservation of mass and of surface tension for a ligament and the definition of *SMD*

$$\sum_i N_i D_i^3 = 3A\Lambda^3/8 \tag{15}$$

$$\sum_i N_i D_i^2 = A\Lambda^2 \tag{16}$$

$$D_{32} = \sum_i N_i D_i^3 \Big/ \sum_i N_i D_i^2 \tag{17}$$

and eliminating A they obtained

$$D_{32} = 3/4\Lambda \tag{18}$$

This model was compared to measurements performed near the liquid core (at 5 mm) where it was assumed that no secondary atomization had occurred and good agreement was found.

The number of drops formed dN_d on a length dx of liquid jet during the time dt can be written

$$d^2N_d = C_1 \frac{64\pi\, A a}{9 t_b \Lambda^2} \mathrm{d}x\, \mathrm{d}t \tag{19}$$

where C_1 is another unknown constant, a is the local jet diameter, and t_b the lifetime of a ligament that can be written as $t_b = C_2/\Omega$. This is equivalent to $X/X_0 = \exp C_2$, where X is the wave amplitude at rupture and X_0 the initial amplitude and C_2 a third constant. Assuming that the jet velocity is constant, equal to V_l, and writing the mass balance in the liquid core, the jet diameter evolution becomes then

$$\frac{\mathrm{d}a}{\mathrm{d}x} = -0.78\, A\, C_1/C_2\, \Omega \Lambda / V_l \tag{20}$$

In case of total atomization, i.e., all of the liquid is torn off from the jet as droplets, a cone length could be calculated (the average jet surface is assumed to have a conical shape)

$$L = \frac{2\sqrt{3} a_0 V_l}{9\pi\, C\, V_r} \left(\frac{\rho_l}{\rho_g}\right)^{1/2} \tag{21}$$

Finally, Caré and Ledoux[33] gave an expression for the initial spray angle that is comparable to similar expressions given by Reitz and Bracco[14] and also by several other authors,

$$tg\frac{\theta}{2} = \frac{2\pi\, A\, V_r}{\sqrt{3}\, V_l} \left(\frac{\rho_g}{\rho_l}\right)^{1/2} = A\frac{\Lambda\Omega}{V_l} \tag{22}$$

These linear models have proved capable to describe accurately low-speed jet breakup regimes but remain limited in the more practical case of high-speed regimes, especially the atomization regime, because they do not account for the multifarious mechanisms that are involved. Only the initial disturbance may be related to them. Moreover, it is argued that the wave can not be considered small when its amplitude becomes equal to the wavelength and when the rupture occurs. In fact, the formation of a ligament is a highly non-linear phenomenon.

4. *Derivation of Droplet Size Distribution Functions*

The linear models predict a mean droplet size related to operating conditions but do not predict any kind of polydispersion of droplet sizes. In combustion applications, however, it is critical to know not only the mean diameters but also the size distribution function, because this provides information on the combustion zone extent. Of particular concern for combustion stability, a combustor will have different behavior if the combustion zone is concentrated near the injection plate than if it is distributed over a large volume in the chamber.

Few authors have discussed the problem of polydispersion in sprays. Simmons[34] made an attempt to derive the shape of the distributions when the Sauter mean diameter (*SMD*) or the mass median diameter (*MMD*) were known. He used the results of several experimental works (more than 2000 tests on about 100 different nozzles) on gas turbine injectors, where the droplet sizes were measured either with a frozen wax technique or by optical sizing (video imaging). Simplex pressure atomizers as well as different types of air-assist and air-blast atomizers were used, and it was noted that in all cases, the ratio of *MMD* to *SMD* was $1.2 \pm 5\%$.

More recently, a new approach has been used, based on a maximum entropy formalism (also called information theory) to derive the shape of the droplets sizes distributions from the first principles: mass, momentum, and energy conservation. Two cases have been studied: the rupture of a thin flat liquid sheet due to the growth of surface waves by Sellens and Brzustowski[35] and the case of an atomizing jet by Li and Tankin.[36–38] The maximum entropy formalism deals with the probabilities of discrete states whose behavior is constrained by the expected values of several quantities $g_{i,j}$, which are functions of the set of states j, and which constitute the physical input to the problem. The probability is obtained by maximizing the entropy, which is the measure of uncertainty arising from information theory, subject to these constraints. The result is $p_j = \exp(-\lambda_0 - \Sigma \lambda_i g_{i,j})$ with the normalization constraint $\Sigma_{pj} = 1$. The λ_i are arbitrary Lagrange multipliers that have to be evaluated in each particular application. This can be extended to the case of a continuous formulation by assuming a uniform discretization of the solution space. The variable f is then a probability density function (PDF), such that $p_j = f \mathrm{d}\psi_j$, where $\mathrm{d}\psi$ is the small element of solution space ψ corresponding to state j. In the case of the breakup of a thin liquid sheet, the constraints are developed from conservation of mass, momentum, surface, and kinetic energy.

Sellens and Brzustowski[35] separated the surface and the kinetic energy in order to take into account some element of irreversibility. Source terms account for transfer between the various modes of energy. A first-order Taylor series expansion transforms the four constraints together with the normalization constraint in a set of five equations that can be solved by means of a Newton–Raphson root solving technique extended to five dimensions, giving the solution for f as a PDF in the size-velocity space. Integrating f from zero to a given maximum permitted velocity v_m over the velocity dimension of the space yields the size PDF. This drop size distribution, called maximum entropy (ME) distribution, was compared to Rosin–Rammler and to upper limit (from Mugele and Evans[39]) distributions for different values of the sheet thickness. All three distributions showed similar features, the Rosin–Rammler and the ME being very close in shape.

Sellens[40] later extended this analysis to an axisymetrical case by introducing a second component of velocity and applied it to a practical case of hollow cone swirl injector. This type of injector produces a conical liquid sheet which self-atomizes. The predicted distribution was compared to local measurements made with a phase Doppler particle analyser. Very good agreement was found in the initial sections of the spray, where the drag forces had not yet influenced the distributions. As the predicted distribution is in non-dimensional form, some parameters have to be adjusted to get the distribution over a mean diameter related to operating conditions. Consequently, the model has to be combined with another model able to predict the liquid sheet (or the jet, if applied in this case) behavior and the corresponding parameters (thickness, velocity, wave number, etc.) before the final breakup and atomization processes. Sellens[40] obtained these parameters by empirical means in the case of his swirl injector.

Li and Tankin[36] first showed that, with some natural assumptions, the droplets' size distribution produced by a single atomizing jet is a form of the Nukiyama–Tanasawa[7] distribution where the distribution parameter is equal to 3. The derived distribution was compared to some experimental work, and a good agreement was found for the mean diameters as well as for the shape of the distribution. In a second step, Li and Tankin[37,38] extended their analysis by including the velocity distribution of the droplets. Their mathematical procedure was very similar to Sellens' except that Sellens[40] wrote the conservation of kinetic and surface energy separately during the breakup process of the liquid sheet. The obtained equations showed the dependence of droplet size distribution on the Weber number, but actual calculations revealed a very small dependence. This small dependence indicates the predominance of the inertial forces over the surface tension effects in the practical spray formation at high injection velocities and confirms the validity of the first step leading to the Nukiyama–Tanasawa distribution, which corresponds to an infinite Weber number. Nevertheless, care has to be taken that although the nondimensional Sauter mean diameter variation with the Weber number is small, this variation will be increased by the order of magnitude of the mass mean diameter, if the dimensional form is used.

Recent work by Durst and Bhatia[41] showed that a one-dimensional hyperbolic distribution function not only provides an excellent fit to measured size data but also sheds some light on the physical processes that cause the drop size variations downstream of the nozzle. The very satisfactory results obtained by this approach encouraged them to apply a two-dimensional hyperbolic distribution to the coupled droplet size and velocity distributions.

The distribution derived from the information theory, for liquid sheets rupture by Sellens[40] and for single jet disintegration by Li and Tankin,[36–38] could apply to coaxial rocket engines injectors, but the source terms which appear in the conservation equations for momentum and energy may have a significant importance, although they are equal to zero in the aforementioned fully developed works.

5. *Numerical Simulations of Atomization Process*

The direct simulation of the surface dynamics and jet breakup was recently attempted by several authors, for instance, Fromm,[42] Childs and Mansour,[43] and Shokoohi and Elrod.[44] These analyses show encouraging results, but they were limited to low-speed, long-wave surface disturbances. Because of the horrendous computer time requirements, it may not be feasible to perform today the direct numerical nonlinear simulation of high-speed, short-wavelength atomization process. The alternative is to use approximate modeling techniques, consisting of solving the mean liquid jet flow as well as the gas flow from nonlinear transport equations and computing the breakup rate from semiempirical formulas.

Such a simulation of atomization process in a gas turbine prefilming type airblast injector was presented by Rizk and Mongia.[45] The computer code includes the simulation of the turbulent airflow through and exiting from the atomizer, as well as the successive processes of fuel filming and breakup, secondary atomization, and turbulent fuel dispersion. For the airflow calculation, the turbulent Navier–Stokes equations with source terms for phase exchanges were solved in a boundary-fitted, curvilinear grid. The droplet trajectories were computed by means of a Lagrangian calculation coupled with the Eulerian computation of the gas flow.

The fundamental processes of the fuel atomization include the liquid formation into a sheet and the subsequent disintegration of this sheet into ligaments due to the liquid surface instability. The process is governed by the balance of forces caused by gas pressure, liquid inertia, viscosity, and surface tension and is modeled according to the equations derived by Dombrowski and Johns,[28] which give the sheet thickness, the total growth of wave, the wave number, the time taken for breakup, and the average ligament length. The average initial drop size diameter at ligament breakup is estimated from the conservation of mass of the disintegrated part corresponding to a full wavelength.

The procedure adopted for the calculation of the wave growth on the ligament uses the same outlines as for the sheet rupture. The initial droplet velocity at their formation is assumed to equal the velocity of the ligament at the breakup point. Once the droplets are formed, they are allowed to deform under the effect of relative velocity according to the Weber number level experienced by the drop. The deformation of the drops is modeled considering a change of sphericity with time toward circular disk shapes. The deformed drops may undergo secondary atomization, which is modeled using the stripping breakup concept of Jenkins and Booker.[46] This concept of breakup of a liquid film of uniform thickness into ligaments and drops leads to the formation of a single-size drop spray. To overcome this limitation, the liquid film thickness at breakup, as well as the ligament exit angle and disintegrated length, are considered to vary around the mean value, with a Gaussian probability distribution. The spray characteristics (mean velocities and Sauter mean diameters) predicted by the model were successfully compared to phase Doppler measurements at different downstream axial and radial positions.

For coaxial rocket engine injectors, the first step should be replaced by the calculation of the round jet diameter evolution, but the wave instability concept may remain valid, at least for certain operating conditions, as was shown by Caré and Ledoux.[33] Even if the liquid core is actually very short, and if the ligament formation can not be evaluated by the theory of Dombrowski and Johns,[28] the second step of atomization, namely, the rupture of the ligaments into droplets, should be essentially the same, especially if the assumption of Farago and Chigier,[19] that the final step of atomization is always the same, is true.

Reitz[47] presented an application of the multidimensional numerical model KIVA to study the atomization and vaporization of a liquid jet injected from a round hole into a compressed gas. The liquid was injected in the computational domain as "blobs" with sizes equal to the nozzle exit diameter. The breakup of the blobs and of the resulting drops was modeled using a stability analysis for liquid jets. The product drops were distinguished from the parent drop by having different drop sizes. This had a significant effect on the fuel vapor distribution in the spray. Drop collision and coalescence were also accounted for, as well as the effect of drops on the gas turbulence. Since these phenomena were handled as subgrid scale processes, specific submodels were used. The drop size resulting from the primary atomization of the blobs was determined from the wavelength of unstable waves on the surface of the blobs using the first-order linear theory of Reitz and Bracco.[30] The dispersion relation, however, was not solved in the KIVA code. Instead, the maximum growth rate Ω and the corresponding wavelength Λ, valid for Ohnesorge numbers less than 1 and for gas-to-liquid density ratios less than 0.1, were curve fit from numerical solutions. The atomization process was modeled by postulating that new drops were formed with drop radius r from a parent drop or blob with radius a with

$$r = B_0\Lambda \qquad \text{for } B_0\Lambda \leq a$$
$$r = \min[(3\pi a^2 W/2\Omega)^{0.33}, \quad (3a^2\Lambda/4)^{0.33}] \qquad \text{for } B_0\Lambda \geq a \tag{23}$$

where B_0 was a constant equal to 0.61 and W the local relative velocity. The second breakup procedure was only allowed once for each injected parcel, to prevent product drops from undergoing further unrealistic size changes. The rate of change of drop radius due to drop breakup was given by

$$\frac{da}{dt} = -\frac{(a-r)\Lambda\Omega}{3.726 B_1 a} \tag{24}$$

where B_1 was chosen equal to 10, but it must be regarded as an adjustable model constant, possibly with a different value in each breakup regime.

These models were developed for gas turbine or diesel engine injectors, which have different design features from rocket engine elements. A model specially adapted to the latter purpose was developed by Liang et al.[48] and Liang and Schuman.[49] The advanced rocket injector combustor code (ARICC) is based on the volume of fluid technique, which accounts for the mutual volumetric displacement of the gas, the liquid, and the droplets coexisting in the computational domain by means of the variable $F = v_g/(v_g + v_l)$, representing the ratio of the cell volume occupied by the gas over the total cell volume, which appears in most of the balance and transfer equations. It is also possible to deduce the location and orientation of a gas-liquid interface cutting arbitrarily across the computational grid, and then to apply the atomization model on a local basis. The atomizing jet is thus explicitly modeled, which is in contrast to the CFD codes in which the atomization model, or its results, is used as an upper boundary condition. The advantage of this more sophisticated model is that a more realistic description is obtained in situations where the atomization process takes place over an extended region of the combustor. However, due to computer costs in practical cases, this model, which needs a very fine grid in the vicinity of the liquid jet core, can only be used in a small section of the combustion chamber, near the injection faceplate. In a first version of ARICC presented in 1986, the atomization model was the same as in the earlier coaxial injection combustion model (CICM) code of Sutton.[50] This model was based on Mayer's analysis[24] and can be expressed as follows: the mean diameter of droplets issued from the jet stripping mechanism and the local mass stripping rate are

$$\bar{D} = B_A\left(\frac{\mu_l\sqrt{\sigma/\rho_l}}{\rho_g V_r^2}\right)^{2/3} \tag{25}$$

$$\dot{M}_A = C_A\left[\frac{\mu_l(\rho_g V_r^2)^2}{\sigma/\rho_l}\right]^{1/3}\pi D_j \tag{26}$$

where D_j is the local jet diameter.

In 1990[49] this model was replaced by the wave growth model of Reitz,[47] which is more general, but reduces to Mayer's with very large Weber numbers and viscous liquids. Using Reitz's equations, the complex growth rate ω and the wavelength Λ having the maximum growth rate Ω were computed from the liquid and gas properties and from the local relative gas-liquid velocity W, which has to be evaluated properly first. The atomization model was then postulated in the same terms used by Reitz: if $B_A \Lambda$ is smaller than or equal to the jet radius, then

$$r = B_A \Lambda$$

if it is bigger, then

$$r = \min[(3\pi a^2 W / 2\Omega)^{0.33}, (3a^2 \Lambda / 4)^{0.33}]$$

where r is the radius of the formed droplet and a the local jet radius given by

$$\frac{da}{dt} = \frac{-\Lambda \Omega}{3.788 C_A} \tag{27}$$

The two empirical constants that appear in the preceding equations have to be optimized through numerical experimentation. In fact, two sets of constants may be used, one for the primary atomization of the jet, and one for the secondary atomization of the droplets. Suggested values for the 1990 model[49] are, for the jet atomization,

$$C_A = 0.5, \qquad B_A = 5.0$$

and for the droplets breakup,

$$C_A = 10.0, \qquad B_A = 0.61$$

The initial droplet velocity is equal to the jet velocity V_l in the jet axis direction plus a radially outward component of $V_l \cdot \tan(\theta/2)$ where θ is randomly distributed between 0 and Θ where $\tan(\theta/2) = 0.188 \Lambda \Omega / \nu_l$.

The jet-embedding technique proposed by Przekwas[51] uses the multigrid concept by employing the regular grid for the gas phase, and the embedded, surface-conforming, adaptative grid for the injected liquid jet core. In work presented in 1989,[52] the conservation equations describing the liquid jet-core flow were numerically solved in a one-dimensional adaptative, surface-conforming grid based on the JE technique. Gravity and internal viscous effects were neglected, but the viscous friction at the gas-liquid interface was accounted for. The solution of the JE method was coupled with the wave breakup model to compute the wave growth and droplet separation process along the predicted jet-core trajectory. Two formulations (linear and approximate nonlinear theory of Yuen[53]) of the wave stability models were used, but the second one was used only for low-speed jet simulation. For the liquid jet atomization, the ligaments or drops were assumed to separate from the liquid jet surface due to the growth of surface wave above critical levels.

As indicated in a preceding section, the linear analysis examines the stability of a liquid jet surface due to an infinitesimal axisymmetric disturbance with the Fourier form: $X = X_0 e^{ikz + \alpha t}$, which is governed by the dispersion equation. By imposing an infinitesimal axisymmetric disturbance with a spectrum of wavelengths, the dispersion equation was solved to obtain the growth rates for the amplitude of all waves. The criterion for ligaments stripping from the growing waves is based on the assumption that when a given wave has grown to an amplitude larger than its wavelength, the crest breaks off as a ligament. The drop size does not necessarily correspond to the fastest growing wavelength since it is possible for some shorter wavelength λ, with smaller growth rates, to take less time to reach their own critical amplitude. This time and the corresponding stripping mass of a ligament were expressed as

$$t_n = \ell_n(\lambda_n / X_0)/\alpha_n \tag{28}$$

$$m_n = \rho_l a(\pi \lambda_n)^2/8 \tag{29}$$

The calculation was applied successfully to a low-speed jet (Rayleigh regime) and also to a high-speed single round jet with encouraging results.[52] When applied to a coaxial LOX/H_2 rocket injector jet (without combustion) it gave results comparable to the ARICC code. The predicted intact-core length was about 10 injector diameters.

6. *Analytical and Semiempirical Models*

Numerical models combining multidimensional gaseous-phase flow simulation with particle tracking may predict detailed features of a spray, but such a complete numerical simulation often needs several hours of computer time and is not suitable for intensive use at the beginning of development of a new atomizer. Therefore, there is a great need for analytical and simplified tools giving a first reasonable approximation in the preliminary phases of a new injector design. Later designs can be improved and optimized with a sophisticated numerical model.

Rizk and Mongia[54] developed a semianalytical model based on the same assumptions as their earlier described numerical model. The same steps of the atomization process are followed, namely, the formation of a thin sheet or fine jets, the appearance of liquid instability, the formation of rapidly growing waves, breakup into ligaments that in turn break up into droplets, secondary atomization, and finally droplet trajectory calculations. Modeling of the atomization process remains exactly the same as in the complete numerical model, which applies Dombrowski and Johns' theory[28] for the sheet or jets and ligaments breakup. The secondary atomization and the droplet trajectories equations that are ordinary differential equations are also the same as in the complete model. The only difference is that the air flowfield is not computed but estimated with analytical formulas based on the concept of velocity components decay along the spray axis of Beer and Chigier,[55] and on detailed measurements in the flowfield of an air swirler. For a coaxial rocket injector, similar expressions can be found in the model of Yatsuyanagi[56] described later. The assumptions that the sheet thickness and the gas velocity vary according to a Gaussian distribution to produce a polydisperse spray are replaced by the use of a modified Rosin–Rammler distribution function for the droplets sizes. The two parameters $\bar{D}$ and N of this distribution are deduced from the final Sauter mean diameter,

$$\bar{D} = (0.91\ SMD)^{1.18} \tag{30}$$

$$N = \bar{D}/(0.126\bar{D} + 1.2) \tag{31}$$

The modified Rosin–Rammler equation can be written as

$$1\text{-}G = \exp\left[-\left(\frac{\ell_n D}{\ell_n \bar{D}}\right)^N\right] \tag{32}$$

whereas the original equation is

$$1 - G = \exp[-(D/\bar{D})^N] \tag{33}$$

where G is the cumulative volume fraction, i.e., the fraction of the total volume contained in drops of diameter less than D. Assuming that the liquid density is constant, the volume and the mass fraction are equal.

Yatsuyanagi[56] developed an analytical model for coaxial liquid/gas jets, where the atomization process was modeled with Mayer's[24] and Dickerson and Schuman's laws[25] for the mean diameter and the mass stripping rate but with some corrective factors. For the gas velocity, three regions were distinguished in the jets. In the downstream region, far from the injector outlet, the gas velocity profile was given by an expression proposed by Abramovitch[57]

$$u_g(z,r) = u_c(z)\left[1 - \left(\frac{r}{R_g(z)}\right)^{1.5}\right]^2 \tag{34}$$

where $R_g(z)$ is the outer boundary of the gaseous jet and $u_c(z)$ is the maximum velocity on the jet axis. In the first region, near the injector outlet and along the liquid core, the gaseous velocity profile had a minimum on the jet axis and two maxima located at a radius r_m given by

$$r_m = 0.93(D_g - D_l) \tag{35}$$

The velocity profile for $r \leq r_m$ was then expressed as

$$\frac{u_{\max} - u}{u_{\max} - u_c} = \left[1 - \left(\frac{r}{r_m}\right)^{1.5}\right]^2 \tag{36}$$

and for $r \geq r_m$

$$\frac{u}{u_{\max}} = \left[1 - \left(\frac{r - r_m}{R_g - r_m}\right)^{1.5}\right]^2 \tag{37}$$

In the intermediate region, the same shape was used as in the downstream region. The fitting of experimental data in this zone was not as accurate but remained acceptable. The dispersed liquid (droplets) mass flow rate at each z location was given by a similar expression

$$\dot{m}_l(z,r) = \dot{m}_{lc}(z)\left[1 - \left(\frac{r}{2.874\bar{r}_l(z)}\right)^{1.5}\right]^3 \tag{38}$$

where $\bar{r}_l$ was defined by

$$\dot{m}_l(z,\bar{r}_l)/\dot{m}_{lc}(z) = 0.5 \tag{39}$$

This model was successfully compared to velocity and mass flow rate measurements in water/nitrogen coaxial jets. A very similar model was used at ONERA (for instance, see Barrère[58]) to calculate the upstream droplet size distribution from downstream measurements.

Yeung[59] described a similarity analysis of the far region of a spray, i.e., outside the region where the liquid core exists and where droplets are formed. Following Schlichting's approach[60] for turbulent jet analysis, he[59] obtained a similarity solution in the case of a flat jet with the assumptions that the droplets are rigid spheres, that the gravity is negligible, and that the droplet equilibration time is very large, which means that the drag force does not affect the droplet motion over a sufficiently large distance. In this case, the droplet motion is entirely inertia dominated, their velocity remains constant, and their trajectories are straight lines. This may be considered a simplistic model, but it gives a first approximation for the velocity field, which may be sufficient in cold-flow experiments.

Several other authors have been interested in the downstream variations of spray characteristics, sometimes taking into account vaporization,[61] sometimes neglecting it.[62,63] Their models were derived for atomizers quite different from coaxial rocket engine injectors, but the main ideas can be easily adapted. Such models may be very useful when one is interested in estimating the initial droplet size distribution near their formation zone from measurements performed rather far downstream, which is most often the case due to limitations of the sizing technique.

B. Experimental Investigations

Experimental characterizations of coaxial rocket injector elements have been reviewed by Ferrenberg et al.,[1] who presented in Appendix A of their report a complete state of the art in 1983. In their Appendix B,[1] they extended this survey by including the conclusions of Lefebvre,[13] who reviewed the experimental work on air-blast atomizers. They[1] stated that a suitable correlation for such atomizers should consist of the sum of two terms: $SMD = SMD_1 + SMD_2$, where the Weber number appears in SMD_1 and the Ohnesorge number in SMD_2. This correlation provides some theoretical basis for the *SMD* equations for air-blast injectors, because the major effects of aerodynamic, surface tension and viscosity forces are taken into account. In the general correlation equation suggested by Lefebvre, however, the factor $1 + \dot{m}_l/\dot{m}g$ also appears with an exponent between 0.5 and 1 for which there is no clear reason.

In experiments performed after Lefebvre's review, Jasuja[64] showed that the second term can be negligible for low-viscosity liquids. A suggested explanation is that at least two different mechanisms are involved in air-blast atomization. For low-viscosity liquids injected into a coflowing gas stream, capillary waves are produced on the liquid surface. These waves grow, become unstable, and break off as ligaments that disintegrate into droplets. For high-viscosity liquids, the capillary mechanism may not be valid. Instead, the liquid is drawn out in the form of long ligaments that break up relatively slowly, in regions of lower gas velocity, into larger size droplets. Water and liquid oxygen are considered low-viscosity fluids.

Ferrenberg et al.[1] also performed tests using droplet sizing interferometry (a technique adapted from laser Doppler velocimetry). Because of severe measurement difficulties (this technique can only be used in low-density sprays), they were unable to obtain measurements in a pressurized vessel. Therefore, they limited their initially ambitious program to investigations of the spray at atmospheric pressure, in unconfined environments. Nevertheless, the precautionary measures they indicated to avoid droplets recirculations in the pressurized chamber will serve subsequent investigators. They did not obtain any new correlation for representative mean diameters with this test program, but they indicated that among the available expressions of previous authors the correlation of Kim and Marshal[65] should be used for a first rough approximation of droplets sizes, even though when applied to their measurements, the predicted sizes were around 30% too small.

$$MMD = 5.3610^{-3}\frac{\sigma^{0.41}\mu_l^{0.32}}{\left(\rho_g V_r^2\right)^{0.57} A_g^{0.36}\rho_l^{0.16}}$$

$$+ 3.4410^{-3}\left(\frac{\mu_l^2}{\sigma\rho_l}\right)^{0.17}\frac{\dot{m}_l}{\dot{m}_g}V_r^{-0.54} \tag{40}$$

We can notice that when applied to the results presented in Sec. II of this chapter Kim and Marshall's expression effectively approaches the experimental values by 30–40% at atmospheric pressure but does not fit all of the values obtained at higher pressures.

To extrapolate the results of water (or other simulants) to actual propellants, Ferrenberg et al.[1] suggested to use the corrective factor proposed either by Ingebo[66]

$$\frac{\bar{D}_{\text{actual propellant}}}{\bar{D}_{\text{simulant}}} = \frac{(\mu_l\sigma/\rho_l)^{1/4}_{\text{actual propellant}}}{(\mu_l\sigma/\rho_l)^{1/4}_{\text{simulant}}} \tag{41}$$

or by Wolfe and Anderson[67]

$$\frac{\bar{D}_{\text{actual propellant}}}{\bar{D}_{\text{simulant}}} = \frac{\left(\mu_l^{1/3}\sigma^{1/2}\rho_l^{-1/6}\right)_{\text{actual propellant}}}{\left(\mu_l^{1/3}\sigma^{1/2}\rho_l^{-1/6}\right)_{\text{simulant}}} \tag{42}$$

Gomi[17] investigated the sprays of Japanese injectors, also at atmospheric pressure, using a diffracted light technique (Malvern particle sizer). He did not propose any correlation for his results but indicated qualitative trends that are consistent with other investigations.

The experiments of Farago and Chigier[68] and Eroglu and Chigier[69] attempted to distinguish the different breakup regimes by means of spark photographs of the initial region of the jet. Their conclusions are summarized in an earlier section of this chapter. Some phase Doppler measurements were performed and a maximum of *SMD* near the spray axis was observed in the downstream region. No correlation equation was given.

Vingert[70] presented the results of measurements performed in a pressurized vessel (up to 30 bar), with a Malvern particle sizer, and gave a correlation expression applying to European Vulcain engine injectors (see Sec. II). This expression is currently valid only for water tests, but work is continuing, using alternative fluids to check the influences of liquid density, viscosity, and surface tension. First tests with actual liquid oxygen were performed in 1991 and presented by Gicquel et al.[71]

The same injectors were tested at atmospheric pressure by Gaic et al.[72] The main objective of their work was to compare and to improve several measurement techniques, but they showed that good atomization can only be obtained at atmospheric pressure when the gas-to-liquid momentum ratio approaches its nominal value. This statement was used by Vingert[70] to increase the experimental range where the proposed correlation can be applied.

A similar injector was also investigated at atmospheric pressure by Caré and Ledoux,[33] who used these results to validate the linear model of Caré.[73] As it is often the case in such atmospheric pressure tests, the liquid-to-gas mixture and momentum ratios were far from actual rocket engine values. The agreement between the predicted and the measured mean diameters, however, was reasonable near the injector exit, typically 10 mm downstream and 5 mm from the spray axis, which is outside of the liquid core but very close to its initial breakup.

Recent experimental work, part of which is currently ongoing, was also presented by Cox,[74] Krülle et al.,[32] Zallen,[75] Zaller and Klem,[76] and Sankar et al.[77] Huynh et al.[78] are interested in a point that is often neglected but which is very important, especially in the field of combustion instability, namely, the influence of large-amplitude pressure waves interacting with the spray.

A major difficulty when comparing data of several investigators is the use of different measurement techniques, which often do not actually measure the same quantities. Ferrenberg and Varma[79] described the difference between temporal distributions obtained, for instance with the wax freezing method or by all the techniques derived from LDV, and spatial distributions obtained mainly from imaging techniques. Dodge[80] indicated the most important error sources when comparing the results of devices based on the diffracted light method and sizers based on the phase Doppler method, that are currently the most widely used techniques. He also indicated[81] a way to compute overall representative average diameters from local measurements. Inversion techniques have also been developed[82–84] to transform line-averaged data obtained with diffraction-based devices into local values given, for instance, by the phase Doppler technique.

III. Experimental Characterization of the Ariane 5 Vulcain Engine Coaxial Injector Sprays

A. General Strategy

ONERA has undertaken the characterization of the spray of a coaxial gas-liquid injector to provide input data for the numerical model PHEDRE, which is a computational fluid dynamics code specially intended to study the high-frequency (HF) combustion instabilities in liquid rocket engines.[85] Coaxial twin fluid atomizers have been extensively studied for several years, but the major part of the available literature deals with gas turbine injectors, which have rather different geometrical designs, different mixture ratio ranges, and lower flow capacities. Moreover, a theoretical model able to describe completely the mechanisms of droplet formation of this atomizer type is not yet available, although progress has been promising. Therefore, it was decided to investigate the spray of the Vulcain engine injectors (Fig. 5) by fully empirical means, because the atomization data were needed early in the HF study program.

At the onset, no cryotechnic research test facility was available to work directly on a liquid oxygen spray (nor even liquid nitrogen). Use of alternative fluids was selected to determine correlations suitable for extrapolation to actual propellants. The first step was to quantify the influence of geometry (LOX post diameter, recess length, annular gap thickness) by testing several elements in similar conditions, using water to simulate LOX and nitrogen or pressurized helium to simulate the cold dense gaseous hydrogen. These tests were also intended to determine the typical mean diameter and the spread parameter as analytical functions of the main parameters (injection velocities, chamber pressure, densities, etc.) In addition, since the boundary condition required by the chamber simulation model

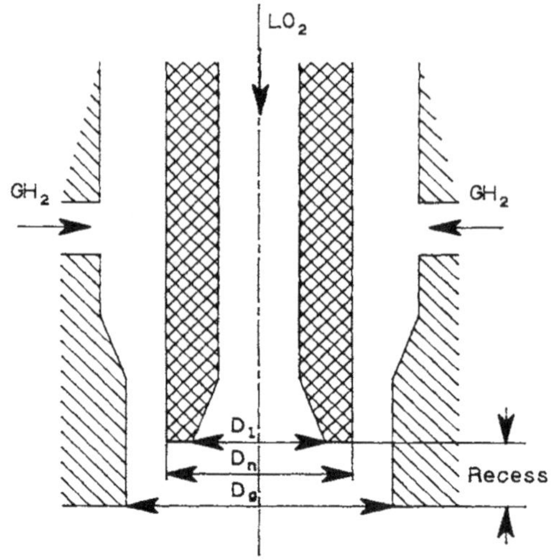

Fig. 5 Typical coaxial injector.

consists of atomization data as near as possible from the injector outlet, a data reduction technique has been developed that permits calculation of the upper mass flow rate drop size distribution (UMD) from the downstream measurements. The technique is specially adapted to coaxial injectors, but the model is very similar to the one described by Brault and Lourme[86] for impinging jet elements. Most of the assumptions on the spray structure are the same as those used by Yatsuyanagi.[56] The next step of this study was to investigate liquid properties effects by using various liquids with values of surface tension, viscosity, and density intermediate between those of water and LOX. Testing with LOX would complete the study.

During a previous characterization of the Viking engine impinging elements, the same strategy was used successfully. Some additional tests were performed with unsymmetrical dimethyl hydrazine (UDMH), which confirmed the correlations established with simulants.[87]

B. Water Tests

The water cold-flow test facility is located in the Fauga–Mauzac ONERA Centre (near Toulouse). Water is flowed vertically into a pressurized cylindrical vessel fitted with windows for the optical particle sizing devices. It is able to operate with constant internal pressures up to 30 bar. The injector can be moved horizontally and vertically, which permits measurements at different downstream distances as far as 200 mm and radial locations between 0 and 50 mm from the spray axis. To protect the windows against water splash, a weak airflow was blown from the windows toward the spray. In addition, deflectors, gutters, and capillarity traps were used to avoid the presence of a parasitic spray of small droplets carried back by the gas recirculating at the chamber bottom.

In this study, the Malvern 2600 HSD instrument based on Fraunhofer diffraction was used. For the Société Européene de Propulsion (SEP) injectors, most of the measurements were made with a 300-mm focal length for the receiving Fourier transform lens. This choice leads to a measurable droplet size range of 5.6–564 μm. For the Messerschmitt-Bölkow-Blohm (MBB) elements, a focal length of 600 mm was used, which sets the droplet size range between 11.2 and 1128 μm. Specific precautions were taken for the very dense spray measurements. The Malvern instrument is capable of uncorrected dense spray measurement up to obscuration values Q (defined as the ratio of extinguished light through the spray to the transmitted light when no spray is present) of 0.5. This range can be extended to higher values provided the bias effect due to multiple scattering is taken into account by appropriate corrective functions. The current work used equations derived for the Rosin–Rammler processing during the Viking study,[87] which are

$$D/D_0 = 0.3\cos[\pi/4(Q^5 + Q^{8.5})] + 0.7 \tag{43}$$

$$N = N_0 \quad \text{for} \quad Q < 0.8 - 0.1\,N \tag{44}$$

$$N = 0.8\frac{0.8 + N_0 - Q}{1.68 - 0.1N_0 - Q} \quad \text{for} \quad Q > 0.8 - 0.1N \tag{45}$$

where D_0 and N_0 are the measured Rosin–Rammler parameters provided by the Malvern software, and D and N are the corrected values. These functions have been validated for Q as high as 0.97, but they are not valid if local over concentrations higher than 10^{-3} exist somewhere on the optical path of the laser beam. Two different tomographic techniques[82,84] were used to determine the local concentration profiles in the sprays for representative working conditions (Fig. 6).The local values of volume concentrations did not exceed 5.10^{-4} for this testing. An additional method used to reduce the obscuration was to split the spray by moving sleeves inward, in order to reduce the thickness passed through by the laser beam. The edges of sleeves were empirically designed to minimize the perturbation of the flowfield inside the spray and to avoid the presence in the laser beam of the additional spray caused by impingement with the splitter. By dividing the vessel in two parts, the splitter also contributed to eliminate the recirculated droplets, created at the bottom of the vessel, from the probe volume. The vaporization of

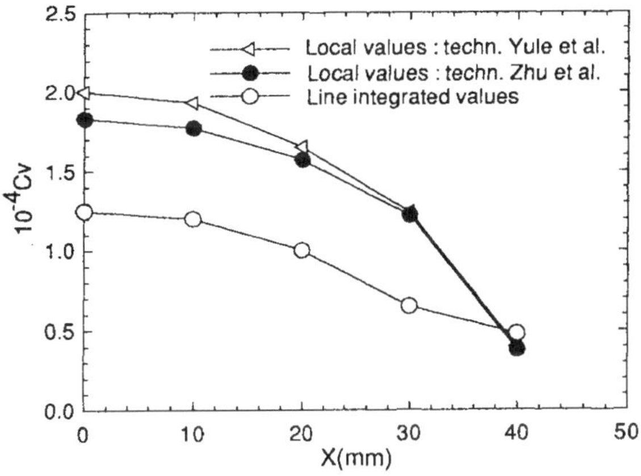

Fig. 6 Tomographical transformation of Malvern measurements; volume concentration profiles.

droplets (when volatile fluids are used) or the presence of a gas motion around the spray (intended to simulate the combustion gases) create refractive index gradients that can lead to beam deviations. The resulting perturbation of the signals received by the central photodiode shows an increase of obscuration, which is actually not related to an increase of spray density. When the variations of obscuration vs time are recorded during the "steady state" of an injector element operation, one can observe fluctuations that can reach 50% of the mean signal. The droplet sizes do not follow these fluctuations. To obtain stable measurements for the mean obscuration and diameters, a minimum of 200 MALVERN sweeps was chosen.

In actual conditions, the SEP and MBB injectors operate in a high-pressure (80–100 bar) environment with liquid oxygen and cold and dense gaseous hydrogen. Simulating the LOX with water can be satisfactory provided the water injection velocity is sufficient to ensure a fully turbulent flow in the LOX post before the first contact with the gas. A good similarity for the gas injection should consist in keeping the density ratio $n = \rho_g/\rho_l$ and the order of magnitude of the two velocities as well as their ratio $m = V_g/V_l$. The mass flow rate or mixture ratio r and the momentum ratio p would then be nominal also. These conditions can be realized easily for the MBB thrust chamber injectors, but some difficulty appears for the SEP elements due to the high density of the hydrogen, which is nearly liquid when injected into the gas generator (70 kg/m^3). Nitrogen pressure of 50 bar would be required to obtain this density, exceeding the test vessel capacity of 30 bar. For a given back pressure, the three ratios m, p, and r mentioned previously, can not be nominal simultaneously, but three different working conditions can be defined, each keeping one of these ratios. The test plan was then established with following rules: subsonic gaseous flow in the injector, complete atomization of the liquid near the outlet, and total mass flow rate in the nominal ±30% range for each element. It was not possible to achieve these three stipulations simultaneously when the back pressure was less than 15 bar (30 bar when helium was used instead of nitrogen) because the gaseous mass flow rate decreased drastically due to the density decrease, and its energy became insufficient to atomize the liquid jet. To increase the mass flow rate by increasing the velocity was not possible because the flow would have become supersonic. Visualizations at atmospheric pressure showed that the atomization was complete if the momentum ratio at injector outlet was nearly equal to its nominal value (Fig. 7). Keeping this ratio p constant reduced the mass flow rates at lower back pressures and, consequently, allowed a large range of injection velocities. When helium was used, very high injection velocities could be reached with still subsonic flow in the injector. Constant momentum ratio is intermediate between constant velocity ratio and constant mixture ratio. The similarity proposed by Gill,[88] keeping the ϕ parameter defined by Abramovitch, $\phi = (nm^2 - 1)(n + 1)/2(nm + 1)^2$ is intermediate between constant velocity ratio and constant momentum ratio.

About 200 tests have been performed at back pressures from 1 to 30 bar, with seven different elements for which the nominal operating conditions are shown in Table 1.

None of the droplet diameter correlations from the literature satisfactorily fit the experimental results, so the data were correlated with classical linear multiregression: $D_{32} = D_{32}^* \Pi(X_i/X_i^*)^{\alpha_i}$ with reference value D_{32}^* corresponding to specified reference data X_i^* for the design and operation variables X_i. For a single injector the most important parameter, which gave the maximum rate of variation τ explained by the correlation equation, was the relative injection velocity

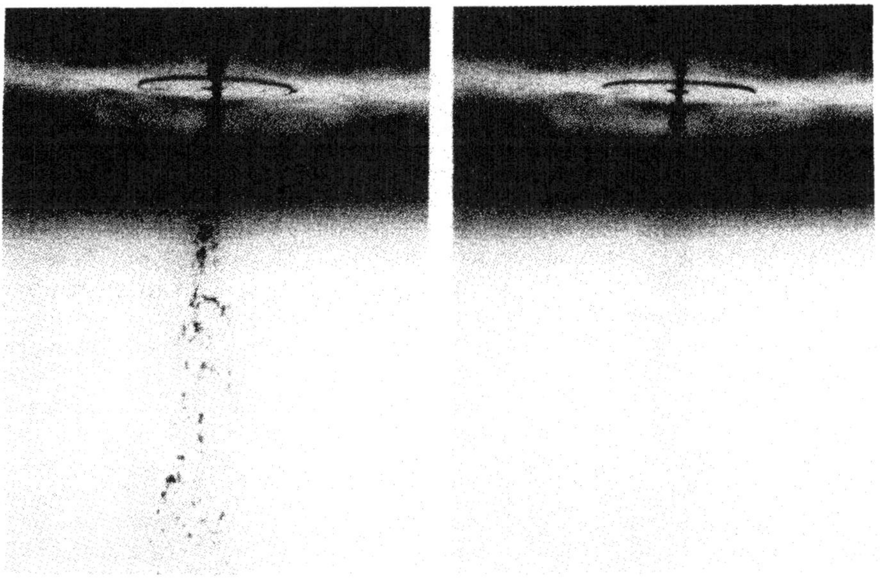

Fig. 7 Comparison of poor and good atomization.

$V_r = V_g - V_l$ (or V_g because $V_l \ll V_g$). This conclusion is consistent with results from literature. The influences of the injected gas density ρ_g and the liquid velocity V_l are much less significant and not well determined. To correlate the diameters measured for the different elements, three geometrical parameters, the LOX post inner diameter D_l, the annular gap outer diameter D_g, and the recess length R, had to be added. Results are summarized in Table 2, where for each test series are included the involved element, the number of tests, the α_i exponent values, and their standard deviation σ_i, the rate of variation explained by the correlation equation τ (for a confidence level of 95%), and the overall uncertainty on the equation σ. The final droplet diameter correlation expression with water and nitrogen or water and helium simulants, having a rate of variation explained of 86% and an overall uncertainty of 12%, is

$$D_{32} = 127\left(\frac{V_r}{160}\right)^{-0.93}\left(\frac{D_l}{2}\right)^{2.25}\left(\frac{D_g}{3}\right)^{-2.65}\left(\frac{R}{D_l}\right)^{-0.26} \tag{46}$$

Table 1 Nominal working conditions of tested elements

Injector type	V_g, m/s	V_l, m/s	ρ_g, kg/m^3	ρ_l, kg/m^3	$\dot{m}_g/\dot{m}_l$
SEP2	148	12	70	1140	1.11
SEP4	93	12	70	1140	1.11
MBB 1 and 2	228	10	24	1140	0.17
MBB4	153	18	24	1140	0.17

The PHEDRE code can handle several droplet classes at injection. Consequently, it is useful to include the whole distribution in reference conditions, instead of a single mean diameter. These correlations have been presented for the Sauter mean diameter, but similar expressions can be found for the Rosin–Rammler diameter $\bar{D}$ and for the mass median diameter D_M. $\bar{D}$ and D_M are related by $D_M = \bar{D}(\ln 2)^{1/N}$. The Rosin–Rammler exponent N can, therefore, be correlated in the form:

$$1/N = 1/N^* + \sum \alpha_i \ln(X_i/X_i^*)$$

Different equations of this type can be found, which fit the experimental values of N. To choose between them, the Sauter mean diameter and the mass median diameter calculated with the nominal values predicted for both Rosin–Rammler parameters were compared to the values obtained through the correlations directly established on D_{32} and D_M. The best agreement was obtained with

$$\frac{1}{N} = \frac{1}{2.15} - 0.09 \ln\left(\frac{\dot{m}_t}{60.5}\right) + 0.17 \ln\left(\frac{D_l}{2}\right) \tag{47}$$

where $\dot{m}_t$ is the total mass flow rate in grams per second. The rate of variation explained by this equation and the overall uncertainty are 60% and 7.6%.

The correlation expressions retained are plotted vs measured data in Fig. 8 for both mean diameters and Rosin–Rammler exponent.

C. Liquid Oxygen Tests

The initial results of size and velocity measurements under cryogenic conditions are presented in this section of the chapter. The objective of this test campaign was to approach the dimensionless numbers (Re, We, Z, etc.) that characterize the liquid and gas flows at the nominal operating point of the thrust chamber or the gas generator of the Vulcain rocket engine (see Table 3). All tests were conducted at the cryogenic facility of the Industrial Chemistry Laboratory of the Liège University (Belgium), which was built by the team of Pr. Germain at the request of SEP for the consolidation program H10. The objectives of this program were to further understand the HM7B cryogenic liquid rocket engine used to propel the third stage of Ariane 4 and, consequently, to increase its performance and reliability. Flow rate capabilities of the facility are 200 g/s of liquid oxygen or nitrogen, 50 g/s of gaseous nitrogen, and 2–3 g/s of gaseous helium. The injection head was designed and fabricated at ONERA to receive a coaxial injection element. It was set up on a pressurized spray chamber with three optical ports, fitted with three fused silica windows for optical applications. Liquid mass flow rate was controlled by a diaphragm orifice. Gas mass flow rate was controlled by a sonic throat. One thermocouple and one pressure probe were mounted on the chamber side to monitor the test conditions. The same one-half scale Vulcain gas generator coaxial injector element previously used on the water/nitogen testing at ONERA Fauga Mauzac with a pressurized spray chamber was used for the cryogenic testing. Use of the same injector allowed more direct comparisons between water/nitrogen tests and the LOX/nitrogen or LOX/helium tests. The spray characterization was performed with an Aerometrics one-component PDPA system, described by Bachalo and Houser,[89] used in a classical configuration. The laser emission was a helium-neon of 15-mW power. The transmitter was fitted

Table 2 Sauter mean diameter correlations

Injector	SEP 2		SEP 4		SEP 2 + 4		SEP 2		MBB 1		MBB 2	
Tests	36		51		87		21[a]		37		15	
τ, %	83		76		78		83		69		80	
σ, %	6.5		9.9		9.0		8.6		8.3		8.1	
Variables	α_i	σ_i	α_i	σ_i	α_i	σ_i	α_i	σ_i	α_i	σ_i	α_i	σ_i
V_r	−1.1	7.8	−0.78	8.0	−0.84	6.0	−0.77	10	−1.1	11	−1.2	14
D_l												
D_g					−2.6	8.2						
R												
Injector	MBB 1 + 2		MBB 1		ALL		SEP 2		ALL		ALL	
Tests	52		8[a]		139		57[b]		168		215[c]	
τ, %	77		51		89		93		89		86	
σ, %	8.2		3.7		9.3		7.7		9.9		12	
Variables	α_i	σ_i	α_i	σ_i	α_i	σ_i	α_i	σ_i	α_i	σ_i	α_i	σ_i
V_r	−1.1	9.0	−0.84	6.0	−0.90	5.2	−0.86	3.8	−0.91	3.3	−0.93	3.2
D_l					2.3	6.0			2.2	7.3	2.2	5.8
D_g					−0.27	7.8			−0.27	9.4	−2.7	5.9
R	−0.28	18			−0.29	18			−0.25	26	−0.30	21

[a]Water–helium tests.
[b]Water–N_2 and water–He tests.
[c]Including additional tests at lower pressures, tests with MBB 4 and tests with SEP 2 and SEP 4 at full scale.

Table 3 Dimensionless groups, nominal values, and experimental range

	$We = \frac{\rho_g V_r^2 Y}{\sigma_l}$	$\frac{\mu_l V_r}{\sigma_l}$	$Re_g = \frac{\rho_g V_g Y}{\mu_g}$	$Re_l = \frac{\rho_l V_l D_l}{\mu_l}$	$\rho g/\rho l$
Exp. range	1 → 19 10^3	2 → 12	8 → 200 10^3	9 → 40 10^3	2 → 40 10^{-3}
Nominal SEP	92×10^3	2.6	800×10^3	190×10^3	60×10^3
Nominal MBB	45×10^3	3.4	480×10^3	390×10^3	20×10^{-3}
	V_g/V_l	$\dot{m}_g/\dot{m}_l$	Mach	ϕ	$Z = \frac{\mu_l}{\sqrt{\rho_l \sigma_l D_l}}$
Exp. range	10 → 90	1 → 16	0.2 → 0.95	0.05 → 4	2 → 3 10^{-3}
Nominal SEP	15	1.1	0.17	1.7	1×10^{-3}
Nominal MBB	22	0.2	0.25	2.3	0.6×10^{-3}

with a standard focus lens of 500 mm. A standard receiver fitted with a 500-mm lens was placed in 30-deg forward scatter angle for collecting the scattered light.

Four pairs of liquid/gas simulants were tested, which were liquid nitrogen and liquid oxygen with room temperature nitrogen, liquid oxygen with room temperature helium, and liquid oxygen with cold (around 150 K) helium. Tests were conducted with three chamber pressures (10, 5, and 1 bar), different mass flow rates of LOX and gaseous nitrogen and a variable position of the PDPA probe area from 15 to 80 mm downstream in front of the LOX post (on the geometrical centerline). The feasibility of size and velocity measurements in severe conditions (pressure and cryogenic conditions) was demonstrated, but some of the initial tests were invalidated due to problems controlling the LOX injection temperature during the test. Sufficient accuracy on LOX temperature in the injection head was difficult at the considered pressures, and for some tests where the injected oxygen was too hot, cavitation occurred. Preliminary results of the most valid tests showed

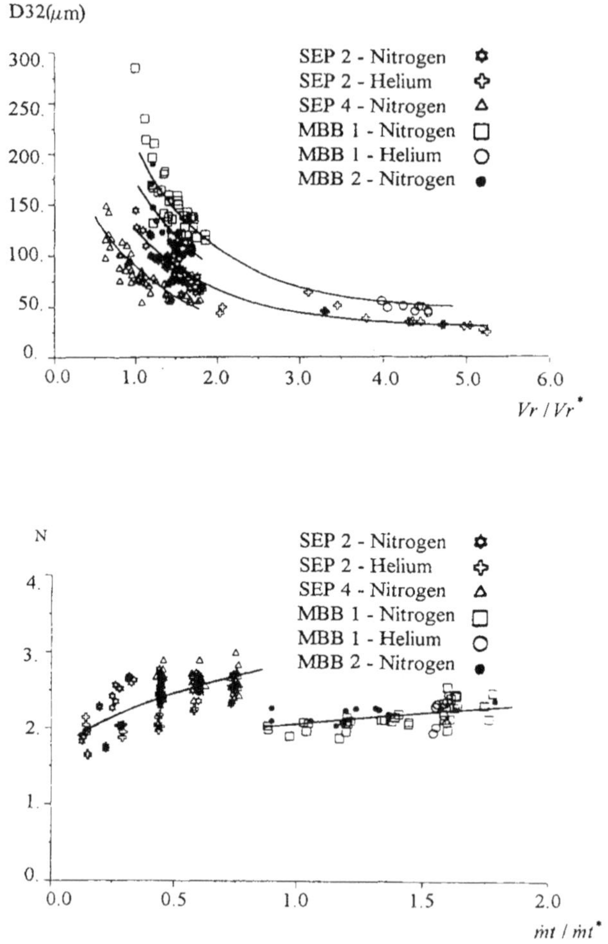

Fig. 8 Final correlation expressions for water tests.

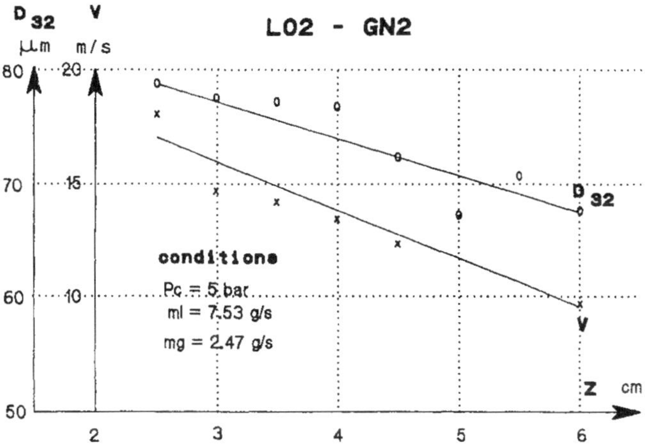

Fig. 9 **Evolution of *SMD* and $\bar{V}$ along the spray axis; $P = 5$ bar.**

a *SMD* between 50 and 80 μm. These are relatively large drops in spite of gaseous flow at room temperature and a spray chamber filled with pressurized gaseous nitrogen also at room temperature. These droplet sizes are not in good agreement with predictions from empirical relations taking into account the surface tension, density, and viscosity of the liquid oxygen. Ferrenberg and Varma[79] have suggested that the empirical relations found with simulants (more often water and nitrogen) be corrected with the ratio

$$K = \frac{\left(\mu_l^{1/3}\sigma^{1/2}\rho_l^{-1/6}\right)_{\text{actual propellant}}}{\left(\mu_l^{1/3}\sigma^{1/2}\rho_l^{-1/6}\right)_{\text{simulant}}} \tag{48}$$

Applying this criterion in our case yields $D_{32_{\text{LOX}}} \approx 0.3 D_{32_{\text{water}}}$; the water droplets are expected to be much larger than the LOX ones, which was not found to be the case. Figures 9 and 10 show the evolution of *SMD* and velocity along the spray axis. The LOX drop sizes here are of the same order as water droplets during

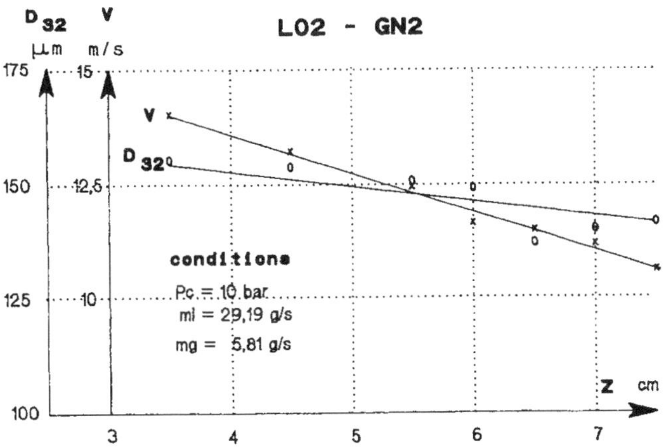

Fig. 10 **Evolution of *SMD* and $\bar{V}$ along the spray axis; $P = 10$ bar.**

Table 4 Comparison between predicted and measured *SMD*

	D_{32}, μm		
	Correlation		Measurement
$V_r = V_g - V_l$, m/s	H_2O	LO_2	LO_2
281	70	21	68.9
271	72	21	65
238	82	24.6	160
220	88	26.8	68.4
137	136	40	154.8

water/nitrogen tests and even larger if we consider the test at the maximum mass flow rate. Figure 11 shows data acquisition of a water/nitrogen test performed at 16 bar, with the probe area located at 200 mm downstream of the injector exit and a LOX/nitrogen test performed at 10 bar with the probe area at 80 mm, respectively. The liquid mass flow rates were the same in the two cases. The *SMD* is 91 μm with water and 205 μm with LOX.

The cryogenic data and the correlation previously obtained by fitting water/nitrogen or water/helium tests are compared in Table 4. The LOX/nitrogen and water/nitrogen tests were conducted with different back pressure and mass flow rates so some differences may be expected. At least two reasons can be found to explain this difference: first, the uncertainty of the position of the probe area in the spray, and whether it was always on the spray axis and, second, the vaporization effect. The small droplets vaporize very fast so that only the bigger ones are detected by the PDPA and taken into account to compute the Sauter mean diameter, which, consequently, tends to increase.

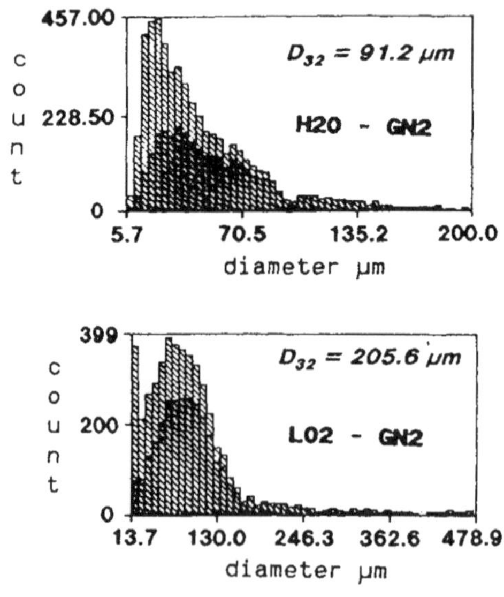

Fig. 11 Comparison between water and LOX test.

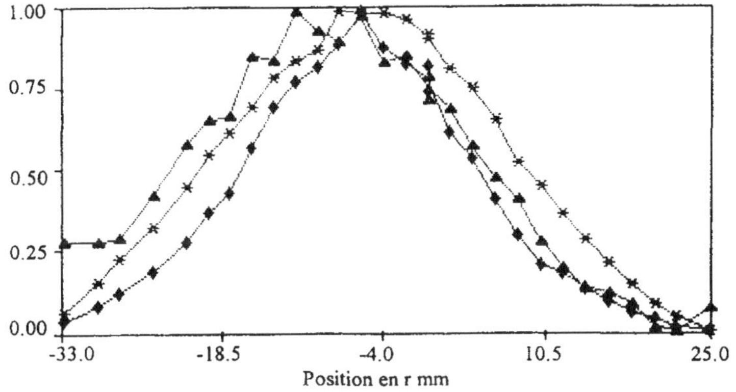

* V1 Mean	3.022 - 17.994 m/sec
▲ D32	61.720 - 92.713 μm
♦ Volume Flow Rate	3.0E-05 -1.1E-03 cc/sec

Fig. 12 Radial evolution of *SMD* on water test.

Another explanation may be the differences between the local PDPA measurements and the line-of-sight integrated MALVERN measurements. Recent tests were performed using the PDPA with water/nitrogen and the same injector at atmospheric pressure to check this point. The observed radial evolutions of mean axial velocity, of volume flux, and of *SMD* are plotted in Fig. 12. The *SMD* decreased from the spray axis to the periphery. A line-averaged *SMD* can be calculated that gives, according to Dodge,[81] a representative mean diameter for an axisymmetric particle field. With this the PDPA results can be compared to the line-of-sight measurements of the Malvern. The following equation is used to calculate line-averaged *SMD* with local point measurements:

$$D_{32_m} = \sum_{i=+R}^{i=-R} W_i \left(d_{30_s}\right)^3 \Big/ \sum_{i=+R}^{i=-R} W_i \left(d_{20_s}\right)^2 \tag{49}$$

where d_{30_s} is the spatial volume mean diameter, d_{20_s} the spatial surface mean diameter, and W_i is given by

$$W_i = \Delta R_i n_i / t_i A_i \tag{50}$$

where ΔR_i is the radial displacement step, n_i the number of droplets, t_i the acquisition time, A_i the probe area of radial position number i.

Figure 13 shows two evolutions of *SMD* along the axis. There is a good correlation between $V_r = V_g - V_l$ with the drop size, which decreases when V_r increases. This result is in agreement with the correlation given for water tests.

The feasibility of drop size and velocity measurements in a cryogenic spray of LOX was demonstrated. The results are self-consistent, but the PDPA point measurement of drop sizes are not in good agreement with the expected values deduced from previous MALVERN measurements in water tests. This difference is likely due to vaporization effects and to line of sight vs local measurements. The PDPA can provide the local velocity of the droplets, and thereby a possibility

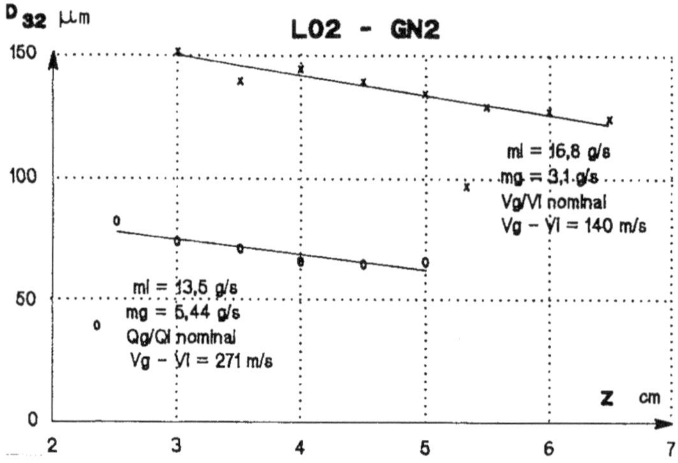

Fig. 13 Evolutions of *SMD* along the spray axis at two injection velocities.

of computing, with an appropriate spray model, the spray features in downstream sections where no PDPA measurements were made in the LOX spray but where all of the MALVERN results were obtained.

IV. Further Improvements, Industry's Point of View

A. Overall Guidelines

The complexity of characterizing the reactive spray from coaxial LOX/GH2 injectors for the Ariane 5 gas generator led SEP and CNES to finance ONERA's experimental work on the simpler problem of characterizing the non-reactive spray of coaxial injectors. The question remains, however, about what kind of knowledge can be gained from cold sprays that can be useful for hot-reactive sprays.

The following assumptions and related options are derived from analysis of the reactive problem, illustrated in Fig. 14.

Assumption 1: the flame does not occur in the recess region.

Assumption 2: the spray maps obtained in the nonreactive initial region have a major influence on the flame location.

Assumption 3:the features of atomization, downstream of the flame detachment position, are lying between two extreme cases. In case 1 the flame surface is positioned radially far from the LOX core, i.e., in the mixing layer between the outer gases and the outer part of a nonreactive annular two-phase jet; then, there will be a region of the LOX core where atomization is produced by the inner part of a nonreactive annular jet. In case 2 the flame reaches the LOX core, which means it is anchored either in the mixing layer between the LOX core and the inner part of a nonreactive annular two-phase jet, or (if the former case 1 allows such a downstream radial evolution) before the end of the LOX core.

These assumptions lead to the following options: cold-flow tests can provide representative spray maps in the nonreactive region and, consequently, be useful

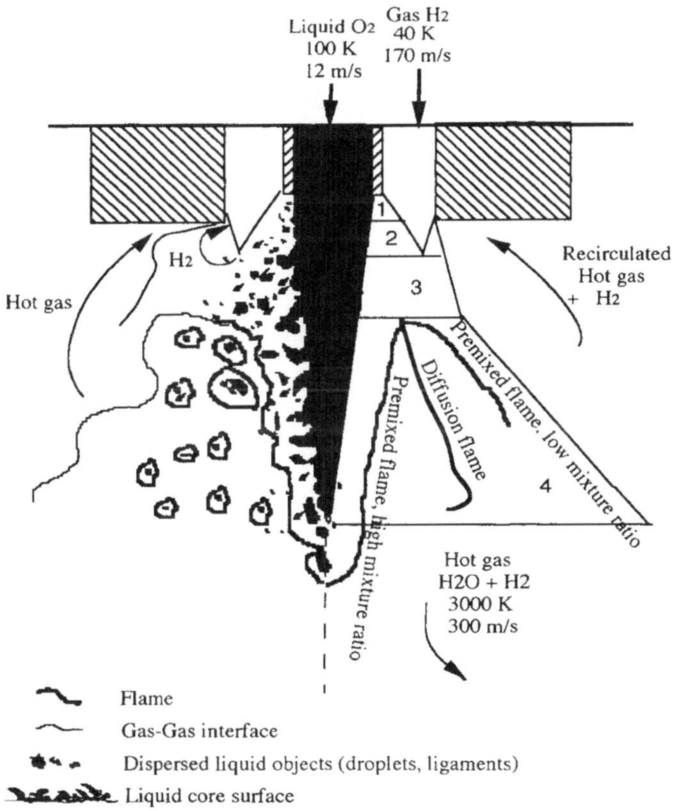

1 : liquid + cold gas mixing zone, non reactive, confined
2 : liquid + cold gas mixing zone, non reactive, non confined
3 : Spray + cold + hot gas mixing without burning
4 : Burning spray zone

Fig. 14 Coaxial injection flow phenomena.

for determining the flame position; cold-flow tests can be useful for reactive sprays where regions of case 1 type exist.

To fulfill the objectives, the cold-flow tests have to produce representative spray maps in the nonreactive regions, that is, in the recess zone and in the near outer region (regions 1 and 2 of Fig. 14). As the tests are often performed with fluids rather different from actual ones (ambient temperature instead of cryogenic, for instance), similarity rules are required to design representative tests. The overall strategy is illustrated in Fig. 15.

B. First Choice of Similarity Criteria

1. *Real Configuration Input Parameters*

The recess region spray field is characterized by inner interaction laws and by boundary conditions, mainly geometry and flow at the inlet section (see details in Tables 5 and 6).

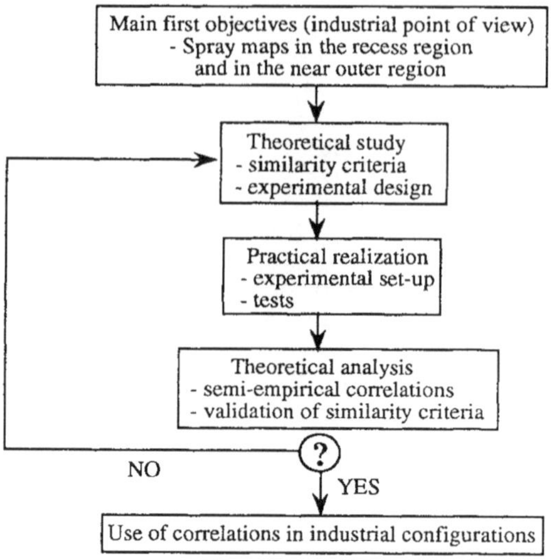

Fig. 15 Overall strategy for cold spray tests.

In the case of the Ariane 5 Vulcain gas generator, the spray is produced by the atomization of a central liquid oxygen jet (100 K, 12 m/s, $D_l = 2.5$ mm) by an annular gaseous hydrogen jet [40 K, 170 m/s, $Y = 0.5(D_g - D_l) = 0.5$ mm] and confined by a recess tube ($R = 3$ mm, $D_g = 4$ mm).

2. *Similarity Parameters*

The spray in the recess zone results from the mixing of two flows and, consequently, depends on the initial profiles of these two flows, mainly the velocity profiles. Hence, the similarity is obtained by keeping the Reynolds numbers Re_g and Re_l and the similarity of the upstream injector geometries (including roughness of the walls). This last condition, which is not always fulfilled practically, must be replaced by the similarity of the boundary-layer thicknesses, especially two momentum thicknesses that feed the two-phase wake of the LOX post, θ_g and θ_l, from where the criteria $\theta_g^* = \theta_g/b_g$ and $\theta_l^* = \theta_l/b_l$ came. Moreover, the Mach number condition ($M = 0.15$) can be relaxed to $M \leq 0.5$ and even more if we ensure that at least θ_g^* is kept.

The two-phase mixing layer is developing from a wake of initial equivalent width: $\bar{b}_s = \theta_g + b_s + \theta_l$.

The production of dispersed liquid phase (ligaments, droplets, etc.) from the continuous liquid phase (liquid core with turbulent surface) results from the interaction of two kinds of forces: 1) fragmentation forces (mainly gas acting on liquid), i.e., inertial, $\rho_g V_r^2 b^2$, and viscosity, $\mu_g V_r b$, depending on the scale b considered and 2) cohesion forces (mainly liquid acting on liquid), i.e., inertial, $\rho_l V_r^2 b^2$, and viscosity, $\mu_l V_r b$, and surface tension: σb.

Owing to the six degrees of freedom that are introduced here, and the units involved, three additional criteria must be considered. At this step of estimation, the density ratio $n = \rho_g/\rho_l$, the velocity ratio $m = V_g/V_l$, and the Weber number $We = \rho_g V_r^2 \bar{b}_s/\sigma_l$ were chosen for reason of a priori relevance and simplicity.

Table 5 Input parameters in actual industrial configuration, case of the Ariane 5 Vulcain gas generator

Geometry:		
$b_l \approx 1.25$ mm	$\alpha \approx 7$ deg	
$b_s \approx 0.1$ mm	$h_l \approx 0.1$ mm	
$b_g \approx 0.6$ mm	$h_r \approx 3$ mm	
$b_{11} \approx 0.8$ mm	roughness	
Operating parameters (approximate values):		
Fluids	Gas H_2	Liquid O_2
Generating conditions		
Pressure P, bar	100	100
Temperature T, K	40	100
Recess entry conditions		
Mass flow rates Q, kg/s	4	4
Density ρ, kg/m^3	60	1100
Dynamic viscosity μ, kg/m/s	8.7×10^{-6}	1.8×10^{-4}
Sound velocity a, m/s	1100	900
Surface tension σ, N/m		0.01
Velocity V, m/s	175	12
Chamber conditions		
Pressure P, bar	85	
Mixture ratio $Rm = Q_l/Q_g$	1	
Temperature T, K	900	
Molar mass M, g	1.4	
Sound velocity a, m/s	1600	
Chamber section A, cm^2	200	
Sonic throat section A_0 cm^2	20	
Characteristic velocity C^*, m/s	2100	

When the flow enters the chamber, it enters a new environment, and a new set of nondimensional numbers has to be introduced, relating the new to the previous environment parameters, i.e., the chamber conditions to the injection conditions; i.e.,

S^* = chamber section/injector total section
ρ^* = chamber outer gas density/injection gas density
μ^* = chamber gas dynamic viscosity/injection gas viscosity
M^* = injection gas velocity/chamber gas speed of sound.

Practically, for the recess and the near recess regions, the condition $M^* = 0.1$ may be relaxed to $M^* \leq 0.5$.

For the mean spray characteristics and turbulent main features of the spray, the control of the acoustic boundary conditions appears to be necessary. On the chamber side, it is preferable to have an anechoic acoustic condition. On the injector side, the problem is more complicated, as a consequence the following

Table 6 Similarity parameters, case of the Ariane 5 Vulcain gas generator coaxial injectors nonreactive spray field in the recess and near-recess regions

Solid geometry:		
Nondimensional form	b_g/b_l, b_s/b_l, h_r/b_l, etc.	
Operating parameters:		
Recess entry, $z = 0$		
Incoming flows	Gas	Liquid
Mach, $M = V/a$	0.15	
Reynolds, $Re = \rho V b/\mu$	8.10^5	1.10^5
Geometrical upstream similarity otherwise momentum thickness		
$\theta^* = \theta/b$	0.07	0.24
Mixing		
Equivalent wake reduced thickness, $\bar{b}_s^* = b_s/b$	4.5	
Weber, $We = \rho_g(V_g - V_l)^2\bar{b}_s^*/\sigma$	7×10^4	
Velocity ratio, $m = V_g/V_l$	15	
Density ratio, $n = \rho_g/\rho_l$	0.06	
Chamber		
$S^* = S_{\mathrm{ch}}/S_{\mathrm{inj}}$	25	
$\rho^* = \rho_{\mathrm{ch}}/\rho_{\mathrm{inj}}$	0.06	
$\mu^* = \mu_{\mathrm{ch}}/\mu_{\mathrm{inj}}$	2.8	
$M^* = M_{\mathrm{ch}}/M_{\mathrm{inj}}$	0.1	

options are possible: the most simple consists in utilizing anechoic injectors; the most representative tries to reproduce, at any relevant frequency, the admittances of the gaseous and liquid injectors, seen from the recess inlet section. The liquid quality must be known also (dissolved gases and microbubbles spectra), especially for the relatively low-pressure tests.

3. *Practical Use of Similarity Criteria*

This choice of similarity criteria can be used to build a test program for further tests using various liquids with lower surface tension and viscosity. They will permit extending the experimental ranges of Weber and Reynolds numbers in order to reach the real configuration values.

V. Conclusion

In the past, twin fluid coaxial injectors have been used extensively in cryogenic rocket engines, most generally providing high-propulsive efficiency and good combustion stability behavior. The basic atomization phenomena, however, which convert the primary liquid jets into droplets, are not yet fully understood and not yet well modeled, which is required if a numerical simulation of the complete engine is desired.

This chapter is an attempt to get a deeper insight into the state of the art of coaxial injection element characterization in cryogenic rocket engines. Relevant literature on theoretical aspects and experimental work was reviewed and discussed.

A number of theoretical investigations have been carried out on liquid jet atomization, but they have been mainly restricted to low-speed regimes and do not actually apply to rocket engine injectors for drop size prediction. Linear theories, in spite of recent improvements, remain limited in practical applications but can provide useful information on some parametric effects on the mean diameters and, with the help of the information theory, on the drop size distributions. Direct numerical simulations also are not currently suitable for the high-speed regimes of concern, but approximate modeling techniques such as volume-of-fluid and jet-embedding techniques are very promising.

Experimental investigations currently remain the best way to characterize a given injection element. The specific features of the coaxial design make the measurements quite impossible in engine operation conditions and difficult in cold flow ones, with simulation then questionable. Recent progress in diagnostic techniques, however, e.g., Malvern and PDPA, significantly improved the capability of test techniques.

The progressive approach followed by ONERA consisted in cold-flow tests using simulant fluids on both liquid and gas sides, then with LOX as the liquid. The parametric experiments were found to be self-consistent, but the transposition from one to the other did not agree with the findings of former investigators. Discrepancies could arise from different test situations and instrumentation as well as from undesired secondary mechanisms. Further tests with other simulant fluids in fixed experimental conditions are now planned to answer the question.

The industrial manufacturer's point of view has focused on the selection of similarity criteria especially with regard to the incoming flows, the two-phase wake mixing, and the chamber environment. Even if absolute values cannot yet be obtained for hot reactive spray characteristics, a good assessment of the relative influence of the major parameters on these characteristics seems to be reachable, which would considerably help to improve existing hardware or to design new injection elements.

References

[1]Ferrenberg, A., Hunt, K., and Duesberg, J., "Atomization and Mixing Study," NASA-CR-178751, Dec. 1985.

[2]Ghafourian, A., Mahalingam, S., Dindi, H., and Daily, J. W., "A Review of Atomization in Liquid Rocket Engines," AIAA 29th Aerospace Sciences Meeting, AIAA Paper 91-0283, Reno, NV, Jan. 1991.

[3]Rayleigh, Lord, "On the Instability of Jets," *Proceedings of London Mathematics Society,* reprint, Vol. 10, 1878, pp. 4–13.

[4]Weber, C., "Disintegration of Liquid Jets," *Zeitschrift fürAngewandte Mathematik und Mechanik,* Vol. 11 No. 2, 1931, pp. 136–159.

[5]Tyler, F., "Instability of Liquid Jets," *Philosophical Magazine of London,* Vol. 15, 1933, pp. 504–518.

[6]Ohnesorge, W., "Formation of Drops of Nozzles and the Breakup of Liquid Jets," *Zeitschrift für Angewandte Mathematik und Mechnik,* Vol. 16, 1936, pp. 355–358.

[7]Nukiyama, S., and Tanasawa, Y., "Experiments on the Atomization of Liquids in an Air Stream," *Transactions of the Japan Society of Mechanical Engineers,* Vol. 4, Pts. 1, 2, 1938, pp. 86–97, 138–143 and Vol. 5, Pts. 3, 4, 1939, pp. 63–67, 68–75.

[8]Rosin, P., and Rammler, E., "Laws Governing the Fineness of Powdered Coal," *Journal of the Institute of Fuel,* Vol. 7, 1933, p. 29.

[9]Kelvin, Lord (Thomson), "Hydrokinetic Solutions and Observations," *Philosphical Magazine of London,* Vol. 42, 1871, pp. 362–377.

[10]Taylor, G. I., "Generation of Ripples by Wind Blowing over a Viscous Liquid," *Collected Works of G. I. Taylor,* Vol. 3, edited by G. K. Batchelor, Cambridge, Univ. Press, Cambridge, England, UK, 1963, pp. 244–254.

[11]Hinze, J. O., "Critical Speeds and Sizes of Liquid Globules," *Applied Science Research,* Series A, Vol. 1 No. 9, 1948, pp. 273–288.

[12]Rice, E. J., "Mechanisms of Atomization," *Liquid Propellant Rocket Instability,* edited by D. T. Harrje and F. H. Reardon, NASA SP 194, 1972, pp. 49–55.

[13]Lefebvre, A. H., "Airblast Atomization," *Progressive Energy and Combustion Science,* Vol. 6, 1980, pp. 233–261.

[14]Reitz, R. D., and Bracco, F. V., "Mechanisms of Atomization of a Liquid Jet," *Physics of Fluids,* Vol. 25 No. 10, 1982, pp. 1730–1742.

[15]Reitz, R. D., "Breakup Regimes of a Single Liquid Jet," American Mathematical Society, AMS Rept. 1262, Feb. 1976.

[16]Lefebvre, A. H., "Properties of Sprays," *Particle and Particle Systems Characterization,* Vol. 6, No. 4, 1989, pp. 176–186.

[17]Gomi, H., "Pneumatic Atomisation with Coaxial Injectors: Measurements of Drop Sizes by the Diffraction Method and Liquid Phase Fraction by the Attenuation of Light Method," Ph.D. Thesis, Dept. of Aeronautical Engineering and Fuel Technology, Univ. of Sheffield, Sheffield, England, UK, Dec. 1984.

[18]Buschulte, W., "Liquid Atomization by Injector Nozzles," *International Symposium on Air Breathing Engine,* AIAA, Washington DC, 1989, pp. 977–986 (ISABE 89-7104).

[19]Farago, Z., and Chigier, N., "Morphological Classification of Disintegration of Round Liquid Jets," *Atomization and Sprays,* Vol. 2 No. 2, 1992, pp. 137–153.

[20]Lefebvre, A. H., "Twin-Fluid Atomization: Factors Influencing Mean Drop Size," *Atomization and Sprays,* Vol. 2, No. 2, 1992, pp. 101–119.

[21]Farago, Z., "Dimensional Analysis of Aerodynamical Atomization," *Proceedings of ILASS—Europe 7th Annual Conf.,* Inst. of Liquid Atomization and Spray Systems, 1991.

[22]Levich, V. G., *Physiochemical Hydrodynamics,* Prentice-Hall, Englewood Cliffs, NJ, 1962.

[23]Sterling, A. M., and Sleicher, C. A., "The Instability of Capillary Jets," *Journal of Fluid Mechanics,* Vol. 68, 1975, pp. 477–495.

[24]Mayer, E., "Capillary Mechanism of Liquid Atomization in High-Velocity Gas Streams," *Proceedings of 12th International Astronautical Congress,* Washington, DC, 1961, pp. 731–740.

[25]Dickerson, R. A., and Schuman, M. D., "Atomization Rates of Droplets and Jets," Heterogeneous Combustion Conf., AIAA Paper 63-498, Palm Beach, FL, Dec. 11–13, 1963.

[26]Adelberg, M., "Breakup Rate Penetration of a Liquid Jet in a Gas Stream," *AIAA Journal,* Vol. 5, No. 8, 1967, p. 1408.

[27]Adelberg, M., "Mean Drop Size Resulting from the Injection of a Liquid Jet into a High-Speed Gas Stream," *AIAA Journal,* Vol. 6 No. 6, 1968, p. 1143.

[28]Dombrowski, N., and Johns, W. R., "The Aerodynamic Instability and Disintegration of Viscous Liquid Sheets," *Chemical Engineering Science,* Vol. 18, 1963, pp. 203–214.

[29]Squire, H. B., "Investigation of the Instability of a Moving Liquid Film," *Journal of Applied Physics,* Vol. 4, 1953, pp. 167–169.

[30]Reitz, R. D., and Bracco, F. V., "Mechanisms of Breakup of Round Liquid Jets," *Encyclopedia of Fluid Mechanics,* Gulf, Houston, TX, 1986, Chap. 10, pp. 233–249.

[31]Lin, S. P., and Lian, Z. W., "A Unified Theory of the Breakup of Liquid Jets," *Proceedings of the 4th International Conference on Liquid Atomization and Spray Systems,* (Sendai, Japan), Fuel Society of Tokyo, Tokyo, 1988, p. 349.

[32]Krülle, G., Mayer, W., and Schley, C. A., "Recent Advances \spreadlong{-1}in H_2/O_2 High Pressure Coaxial Injector Performance Analysis," AIAA/ASME/SAE/ASEE 26th Joint Propulsion Conf., AIAA Paper 90-1959, Orlando, FL, July 16–18, 1990.

[33]Caré, I., and Ledoux, M., "Study of an Airblast Coaxial Atomizer. Experiments and Modelisation," *Proceedings of the 5th International Conference on Liquid Atomization and Spray Systems* (Gaithersburg, MD), Fuel Society of Tokyo, Tokyo, 1991, pp. 763–768.

[34]Simmons, H. C, "The Correlation of Drop Size Distributions in Fuel Nozzle Sprays. Part 1: The Drop Size/Volume Fraction Distribution," *Journal of Engineering for Power,* July 1977, pp. 309–314.

[35]Sellens, R. W., and Brzustowski, T. A., "A Prediction of the Drop Size Distribution in a Spray from First Principles," *Atomization and Spray Technology,* Vol. 1, No. 2, 1985, pp. 89–102.

[36]Li, X., and Tankin, R. S., "Droplet Size Distribution: A Derivation of a Nukiyama–Tanasawa Type Distribution Function," *Combustion Science and Technology,* Vol. 56, 1987, pp. 65–76.

[37]Li, X., and Tankin, R. S., "Derivation of Droplet Size Distribution in Sprays by using Information Theory," *Combustion Science and Technology,* Vol. 60, 1988, pp. 345–357.

[38]Li, X., and Tankin, R. S., "Prediction of Droplet Size and Velocity Distributions in Sprays using Maximum Entropy Principle," *Combustion Science and Technology,* Vol. 68, 1989, pp. 147–155.

[39]Mugele, R. A., and Evans, H. D., "Drop Size Distributions in Sprays," *Industrial and Engineering Chemistry,* Vol. 43, pp. 1317–1324.

[40]Sellens, R. W., "Prediction of the Drop Size and Velocity Distribution in a Spray, Based on the Maximum Entropy Formalism," *Particle and Particle Systems Characterization,* Vol. 6, No. 1, 1989, pp. 17–27.

[41]Durst, F., and Bhatia, J. C., "Description of Sprays Using Joint Hyperbolic Distribution in Particle Size and Velocity," Lehrstuhl für Strömungsmechanik, LSTM Rept. 241/T/88, Univ. of Erlangen, Erlangen, Germany, 1988.

[42]Fromm, J. E., "Numerical Calculation of the Fluid Dynamics of Drop-on-Demand Jets," *IBM Journal of Resource Development,* Vol. 28, No. 3, 1984, p. 322.

[43]Childs, R. E., and Mansour, N. N., "Simulation of Fundamental Atomization Mechanism in Fuel Sprays," AIAA 26th Aerospace Sciences Meeting, AIAA Paper 88-0238, Reno, NV, Jan. 1988.

[44]Shokoohi, F., and Elrod, H. G., "Numerical Investigation of the Disintegration of Liquid Jets," *Journal of Computational Physics,* Vol. 71, No. 2, 1987, pp. 324–342.

[45]Rizk, N. K., and Mongia, H. C., "Model for Airblast Atomization," *Journal of Propulsion and Power,* Vol. 7, No. 3, 1991, pp. 305–311.

[46]Jenkins, D. C., and Booker, J. D., "The Time Required for High Speed Air Streams to Disintegrate Water Drops," Ministry of Aviation, Aeronautical Research Council, CP 827, London, 1965.

[47]Reitz, R. D., "Modeling Atomization Processes in High-Pressure Vaporizing Sprays," *Atomization and Spray Technology,* Vol. 3, No. 4, 1987, p. 309.

[48]Liang, P. Y., Fischer, S., and Chang, Y. M., "Comprehensive Modeling of a Liquid Rocket Combustion Chamber," *Journal of Propulsion and Power,* Vol. 2, No. 2, 1986, pp. 97–104.

[49]Liang, P. Y., and Schuman, M. D., "Atomization Modeling in a Multiphase Flow Environment and Comparison with Experiments," AIAA 21st Fluid Dynamics, Plasma Dynamics and Lasers Conf., AIAA Paper 90-1617, Seattle, WA, June 1990.

[50]Sutton, R. D., Schuman, M. D., and Chadwick, W. D., "Operating Manual for Coaxial Injection Combustion Model," NASA-CR-129031, April 1974.

[51]Przekwas, A. J., "Spray Formation in Liquid Propellant Atomization Systems," *JANNAF Workshop on Atomization,* NASA Marshall Space Flight Center, Huntsville, AL, May 1987.

[52]Przekwas, A. J., Chuech, S., and Singhal, A., "Numerical Modeling for Primary Atomization of Liquid Jets," AIAA 27th Aerospace Sciences Meeting, AIAA Paper 89-0163, Reno, NV, Jan. 1989.

[53]Yuen, M. C., "Non-Linear Capillary Instability of a Liquid Jet," *Journal of Fluid Mechanics,* Vol. 33, Pt. 1, 1968, pp. 151–163.

[54]Rizk, N. K., and Mongia, H. C., "A Semi-Analytical Approach Validation for Airblast Atomizers," AIAA 28th Aerospace Sciences Meeting, AIAA Paper 90-0361, Reno, NV, Jan. 1990.

[55]Beer, J. M., and Chigier, N. A., *Combustion Aerodynamics,* Halsted Press Div., Wiley, New York, 1972.

[56]Yatsuyanagi, N., "Experimental and Analytical Study on Spray Field by Liquid/Gas Coaxial Injector Elements," Tech. Rept. TR 692, National Aerospace Lab., Sendai, Japan, 1978; in Japanese, Abstract in English.

[57]Abramovich, G. N., *Theory of Turbulent Jets,* Fizmatgiz, 1960; also MIT Press, Cambridge, MA, 1963.

[58]Barrère, M., "Simulation d'un Injecteur Coaxial (H_2-O_2). Expériences et Théorie Simplifiées (Modéle de Yatsuyanagi)," ONERA Rept. 15/6116 EY, Châtillon, France, Oct. 1986.

[59]Yeung, W. S., "Similarity Analysis of Gas-Liquid Spray Systems," *Journal of Applied Mechanics,* Vol. 49, Dec. 1982, pp. 687–690.

[60]Schlichting, H., *Boundary Layer Theory,* 7th ed., McGraw-Hill, New York, 1979.

[61]Chin, J. S., Durett, R., and Lefebvre, A. H., "The Interdependance of Spray Characteristics and Evaporation History of Fuel Sprays," *Journal of Engineering for Gas Turbines and Power,* Vol. 106, July 1984, pp. 639–644.

[62]Chin, J. S., Nickolaus, D., and Lefebvre, A. H., "Influence of Downstream Distance on the Spray Characteristics of Pressure-Swirl Atomizers," *Journal of Engineering for Gas Turbines and Power,* Vol. 108, Jan. 1986, pp. 219–224.

[63]Zhao, Y. H., Li, W. M., and Chin, J. S., "Experimental and Analytical Investigation on the Variation of Spray Characteristics Along Radial Distance Downstream of a Pressure Swirl Atomizer," American Society of Mechanical Engineers, ASME Paper 86-GT-51, Jan. 1986.

[64]Jasuja, A. K., "Plain-Jet Airblast Atomization of Alternative Liquid Petroleum Fuels Under High Ambient Air Pressure Conditions," American Society of Mechanical Engineers, ASME Paper 82-GT-32, Jan. 1982.

[65]Kim, K. Y., and Marshall, W. R., "Drop-Size Distributions from Pneumatic Atomizers," *AIChE Journal,* Vol. 17, No. 3, 1971, pp. 575–584.

[66]Ingebo, R., and Foster, H., "Drop Size Distributions for Crosscurrent Breakup of Liquid Jets in Airstreams," NACA TN 4087, Oct. 1957.

[67]Wolfe, H. E., and Anderson, W. H., "Kinetics, Mechanism and Resultant Droplet Sizes of the Aerodynamic Breakup of Liquid Drops," Aerojet-General Rept. 0395-04(18) SP, AD-437340, 1964.

[68]Farago, Z., and Chigier, N., "Parametric Experiments on Coaxial Airblast Jet Atomization," American Society of Mechanical Engineers, ASME Paper 90-GT-81, June 1990.

[69]Eroglu, H., and Chigier, N., "Initial Drop Size and Velocity Distributions for Airblast Coaxial Atomizers," *Journal of Fluids Engineering,* Vol. 113, Sept. 1991, pp. 453–459.

[70]Vingert, L., "Coaxial Injector Spray Characterization for the Ariane 5 Vulcain Engine," *Proceedings of ILASS—Europe 6th Annual Conf.,* (Pisa, Italy), Inst. of Liquid Atomization and Spray Systems, ONERA, Châtillon, France, 1990.

[71]Gicquel P., Vingert, L., and Millan, P., "Faisabilité de Mesures Granulométriques dans des Sprays Cryogéniques d'Injecteurs Coaxiaux," 6ème Congrès Francophone de Vélocimétrie Laser, Toulouse, France, Sept. 21–24, 1992.

[72]Gaic, P., Burnage, H., Yoon, S. J., and Lourme, D., "Distribution of Mean Concentration Drop Size and Velocity and Size-Velocity Correlations in the Spray of an Air-Assist Nozzle," *International Journal of Turbo and Jet-engines,* Vol. 3, No. 2, 1986, pp. 105–119.

[73]Caré, I., "Study of an Air-Assist Coaxial Injector," Ph.D. Thesis, Univ. of Rouen, Dept. of Sciences, Rouen, France, Dec. 1990 (in French).

[74]Cox, G. B., "Rocket Engine Injection Element Characterization," AIAA/ASME/SAE/ASEE 24th Joint Propulsion Conf., AIAA Paper 88-3135, Boston, MA, July 11–13, 1988.

[75]Zaller, M., "LOX/Hydrogen Coaxial Injector Atomization Test Program," NASA CR 187037, Oct. 1990.

[76]Zaller, M., and Klem, M., "Coaxial Injector Spray Characterization Using Water/Air as Simulant," NASA TM 105322, Nov. 1991.

[77]Sankar, S. V., Wang, G., Brena de la Rosa, A., Rudoff, R. C., Isakovic, A., and Bachalo, W. D., "Characterization of Coaxial Rocket Injector Sprays Under High Pressure Environments," AIAA 30th Aerospace Sciences Meeting, AIAA Paper 92-0228, Reno, NV, Jan. 1992.

[78]Huynh, C., Ghafourian, A., Mahalingam, S., and Daily, J. W., "Combustor Design for Atomization Study in Liquid Rocket Engines," AIAA 30th Aerospace Sciences Meeting, AIAA Paper 92-0465, Reno, NV, Jan. 1992.

[79]Ferrenberg, A. J., and Varma, M. S., "Atomization Data for Spray Combustion Modeling," AIAA/ASME/SAE/ASEE 21st Joint Propulsion Conf., Monterey, CA, July 1985.

[80]Dodge, L. G., "Comparison of Performance of Drop-Sizing Instruments," *Applied Optics,* Vol. 26, No. 11, June 1987.

[81]Dodge, L. G., "Representation of Average Drop Sizes in Sprays," *Journal of Propulsion and Power,* Vol. 4, No. 6, 1988, pp. 490–496.

[82]Yule, A. J., Seng, A. H., Felton, P. G., Ungut, A., and Chigier, N. A., "A Laser Tomographic Investigation of Liquid Fuel Sprays," *Proceedings of the 18th (International) Symposium on Combustion,* Combustion Inst., Pittsburgh, PA, 1981, pp. 1501–1510.

[83]Hammond, D. C., Jr., "Deconvolution Technique for Line-of-Sight Optical Scattering Measurements in Axisymmetric Sprays," *Applied Optics,* Vol. 20, No. 3, 1981, pp. 493–499.

[84]Zhu, H. M., Sun, T. Y., and Chigier, N., "Tomographical Transformation of Malvern Spray Measurements," *Atomization and Spray Technology,* Vol. 3, No. 2, 1987, pp. 89–105.

[85]Habiballah, M., Lourme, D., and Pit, F., "A Comprehensive Model for Combustion Stability Studies Applied to the Ariane Viking Engine," AIAA Paper 88-0086, 1988.

[86]Brault, F., and Lourme, D., "Experimental Characterization of the Spray Formed by two Impinging Jets in a Liquid Rocket Injector," *International Conference on Liquid Atomization and Spray Systems* (London), ONERA TP No. 1985-68, July 1985.

[87]Lourme, D., "Like-on-like Injector Spray Characterization for the Ariane Viking Engine," ONERA TP No. 1986-62, AIAA/SAE/ASME, Huntsville, AL, June 1986.

[88]Gill, G. S., "A Qualitative Technique for Concentric Tube Element Optimization, Utilizing the Factor (Dynamic Head Ratio—1)," AIAA Paper 88-0076, 1988.

[89]Bachalo, W. D., and Houser, M. J., "Phase Doppler Spray Analyser for Simultaneous Measurements of Drop Size and Velocity," *Optical Engineering,* Vol. 23 No. 5, 1984, pp. 583–590.

Chapter 7

Shear Coaxial Injector Spray Characterization

Michelle M. Zaller*
Sverdrup Technology, Inc., Brookpark, Ohio 44142
and
Mark D. Klem†
NASA Lewis Research Center, Cleveland, Ohio 44135

I. Introduction

IN combustors fed by shear coaxial injectors, injection of gas and liquid propellants and the formation of liquid droplets provide the initial conditions for calculating the propellant vaporization, mixing, and reaction rates. This makes propellant atomization a critical input to models that predict combustor performance and stability. Coaxial injector spray formation, however, is a complex process that is difficult to model. Analytical atomization models do not yet have the capability to accurately predict injector spray drop sizes. Drop size and velocity measurements, as well as local gas velocity measurements, are needed to validate atomization models, improve modeling capabilities, establish a coaxial injector spray database, and develop empirical correlations. Obtaining measurements of drop sizes produced by liquid propellant rocket injectors has been identified[1,2] as critical to improving the accuracy of computational models that predict the performance and stability margin of liquid propellant rocket engines. Drop size predictions made by available atomization models vary over several orders of magnitude,[3] and none of the models account for the recess of the liquid oxygen post, which separates the central liquid oxygen flow from the annular hydrogen flow and which has been shown to influence engine performance and stability.[4] Significant gains can be made by focusing on improving atomization models, because the drop size and velocity distributions produced at the completion of atomization are the initial conditions for vaporization, mixing, and reaction calculations. Without detailed spray drop size and velocity measurements, accurate liquid rocket engine performance and stability predictions cannot be made. Making accurate performance and stability predictions would contribute to reducing

*Test Engineer, High Thrust Technology Section, Lewis Research Center Group.

†Aerospace Engineer, Space Propulsion Technology Division.

the expensive testing performed in engine development programs, avoiding combustion instabilities, and optimizing the efficiency of new engines.

Unfortunately, the high-temperature and high-pressure environment of rocket engine combustors has made it difficult to obtain detailed measurements of drop size and velocity in hot firing engines. Because of the lack of hot-firing rocket combustor atomization data, other methods have been relied on to estimate the drop sizes occurring in hot firing engines. One such method has been to use theoretically derived atomization correlations with adjustable parameters in combustor models. These correlations were calibrated by forcing the overall model performance predictions to agree with actual rocket engine performance measurements. This adjustable parameter method assumes that all other processes in the hot-firing rocket engine, such as multiphase mixing and supercritical droplet vaporization, are well modeled. In fact, very little data exist to support models of many significant reacting spray processes. Therefore, this method of inferring drop size distributions based solely on overall performance data does not improve the ability of these models to predict injector spray drop size distributions.

Another technique used to estimate the drop sizes occurring in hot-firing rocket engines is to rely on cold-flow atomization data. Drop size correlations derived from nonreacting test results that have been adjusted to take into account fluid property differences are used as inputs to combustor models. The disadvantage of cold-flow data is that it is obtained at radically different conditions than those existing in a hot-firing combustor. The surface tension and viscosity of the simulants, which influence the drop size distributions, often do not match those of the reactive, hot-fire propellants. It is anticipated that qualitative trends predicted with cold-flow data have some relationship to hot-fire performance. Until extensive drop size measurements in hot-firing engines are performed, however, cold-flow data must be relied on to validate and improve current analytical models.

Because of the high cost and difficulty of obtaining drop size measurements in hot-firing rocket engines, it is desirable to develop cold-flow correlations relating drop size distributions to injection parameters. Optimally, cold-flow databases could be compared to correlations of drop sizes in hot-firing engines, giving a means of estimating performance and stability parameters in future development programs without resorting to hot-fire testing.

A. Previous Investigations

Liquid atomization in coaxial injector sprays is primarily due to the aerodynamic shear of the high-velocity gas stream on the surface of the liquid jet.[5,6] Waves of varying frequencies and growth rates form on the liquid jet surface. The dominant frequencies of these waves are functions of the gas and liquid stream relative velocities[7] and liquid jet turbulence levels.[8] Coupling between acoustic disturbances and the natural atomization frequencies is postulated to be a possible combustion instability mechanism.[9] Small mean drop size sprays have rapid vaporization rates that could couple with acoustic waves to induce combustion instability. Efforts have been made by numerous investigators to determine the influence of injection parameters on the drop sizes produced by coaxial injectors. These efforts have confirmed theoretical predictions that the relative velocity between the liquid jet and the surrounding gas stream has a major influence on the resulting drop sizes.

The droplet sizes resulting from the breakup of a liquid jet by a coaxial high-velocity gas stream have been determined by various researchers to depend on the liquid and gas properties, as well as the liquid orifice diameter, and liquid and gas flow rates and injection velocities. The relative influence of each of these parameters on the drop size distribution is not known. Extensive literature surveys are given by Ferrenberg et al.[10] on liquid propellant rocket injector characterization, and Lefebvre[11] on air-blast atomization studies. Some air-blast atomizers break up the liquid stream in a manner similar to coaxial injectors; however, the ratio of liquid to gas flow rate is typically an order of magnitude smaller in air-blast atomizers than in rocket injectors.

One technique used to characterize liquid spray drop size distributions was the wax droplet freezing technique. Hot-liquid wax was injected into ambient pressure air, and the resulting atomized wax droplets froze after being formed. The wax droplets were collected, sieved, and weighed, making it possible to measure large numbers of droplets simultaneously. Empirical correlations were used to account for property differences between the wax and the liquid being simulated. Wax freezing tests could not be used to measure the particle velocities. One difficulty associated with the use of hot wax was the selection of an injection temperature that permitted completion of atomization before the wax particles solidified. Weiss and Worsham[12] used the wax freezing technique to determine the drop sizes produced by the breakup of a liquid jet in a high-velocity coaxial airstream. The tests conducted by Weiss and Worsham[12] examined the influence of varying ambient pressure (air density) and liquid viscosity on the liquid breakup. Kim and Marshall[13] studied the drop size distributions resulting from plain jet air-blast atomizers while varying the wax viscosity. Falk[14] studied the atomization of wax by heated nitrogen gas in coaxial injectors. Although these experiments were conducted at ambient pressure, a gas acceleration ramps was used to simulate the effect of gas acceleration due to chemical reaction in real combustors. Unlike Weiss and Worsham,[12] Falk[14] did not use a sampling probe, and so the data reported by Falk[14] could not include information about the variation in drop size distributions as a function of axial and radial distance from the injector. Use of the wax droplet freezing technique had an advantage over optical spray diagnostics; since the sprays did not need to be kept optically thin, liquid flow rates similar to full-scale launch vehicle injector element flow rates were possible.

Lorenzetto and Lefebvre[15] used diffractively scattered light from a mercury arc lamp to estimate drop sizes in plain-jet air-blast atomizer sprays. Lorenzetto and Lefebvre estimated that their light scattering instrument could measure Sauter mean diameters as low as 20 μm. Gomi[16] applied a laser diffraction technique to low flow rate coaxial injector sprays. Velocities could not be measured with the laser diffraction technique, and only line-of-sight average drop sizes at one radial location were obtained. Like photographic techniques, laser diffraction instruments obtain spatially averaged measurements of drop size, and unless measurements are made for a number of radial locations, the variation of drop sizes throughout the spray cannot be computed. Vingert[17] applied laser diffraction to coaxial injector sprays in a pressurized chamber. Nitrogen and helium were used to atomize water jets. Measurements of sprays with higher flow rates were achieved by using spray splitters and by correcting data with high-obscuration values. Deconvolutions of the line-of-sight data were performed to estimate spray volume concentration profiles. Several different injector geometries and the effect of liquid post recess were studied, in addition to varying gas properties and gas and liquid flow rates and velocities.

Ferrenberg et al.[10] conducted an extensive study of coaxial injectors using a visibility/intensity droplet sizing interferometry (V/I DSI) technique.[18] The V/I DSI instrument was capable of simultaneously measuring droplet sizes and velocities. This study was limited, however, by the inability of the V/I DSI to reliably measure drop sizes below 18μm, handle a wide droplet velocity range, or make measurements in dense sprays. Use of the V/I DSI represented one of the first attempts to use a flux sensitive instrument to make rocket injector spray drop size and velocity measurements.

Sankar et al.[19,20] used a phase-Doppler particle analyzer (PDPA) to make detailed coaxial injector spray measurements. The phase-Doppler technique, a laser-based interferometric technique, was similar to the V/I DSI; measurements of drop size and velocity with high spatial resolution could be made. The PDPA represented an improvement over the V/I DSI technique, since the PDPA was insensitive to the relative and absolute intensities of the laser beams, could measure smaller droplets, and could handle a wide droplet velocity range. In the studies conducted by Sankar et al.,[19] the effects of varying gas and liquid flow rates, injection velocities, gas density, and post recess on the spray drop sizes, velocities, and mass fluxes were examined. Measurements were also made of vaporizing liquid nitrogen sprays at atmospheric pressure. Mean drop sizes measured in the liquid nitrogen sprays were much smaller than the water/nitrogen spray drop sizes. Sankar et al.[19] observed that the liquid nitrogen jets atomized without any gas flow. This probably indicated that the liquid nitrogen was boiling in the injector, which could have caused the observed improvement in atomization. PDPA measurements were obtained near the liquid jet breakup point for coaxial injector sprays by Eroglu and Chigier.[21] Goix et al.[22] performed PDPA measurements and obtained OH fluorescence images in reacting methanol/air sprays. Data were obtained for two flow conditions and three different axial locations. One of the tests shows Sauter mean diameter (SMD) radial profiles with a minimum at the spray center. This differs from cold-flow results obtained by other investigators,[19–21,23] which generally have flat profiles or maxima at the centerline.

B. Objectives

To obtain the coaxial injector spray data required to improve current modeling capability, a four-phase test program was initiated at the NASA Lewis Research Center. The ultimate goal of this program is to bridge the gap between cold-flow and hot-firing tests by attempting to correlate empirical drop size data acquired under cold-flow and hot-firing conditions. The first phase of the test program was intended to provide an estimate of the capabilities of the droplet sizing instrumentation and provide detailed mappings of the drop sizes produced by nonreacting water/air coaxial injector sprays as a function of liquid and gas flow rates and injector geometry. In the second phase of the program, drop size and velocity measurements were obtained in confined water/nitrogen coaxial injector sprays in a 50-mm-diam cylindrical chamber. Confining the spray permitted the effect of varying chamber pressure to be examined. The third phase of the proposed test program is to obtain drop size and velocity measurements in confined liquid nitrogen/helium sprays. Liquid nitrogen/helium will be used to simulate the fluid properties of liquid oxygen/hydrogen. The fourth phase of the proposed test program is to obtain drop size data in a LOX/hydrogen rocket engine using the same single-element coaxial injector. The results of the first and second phases of the test program are presented here. These results are expected to be relevant for engines operating at chamber pressures below the liquid propellant critical pressure.

II. Apparatus

A. Facility

The coaxial injector nonreacting spray tests were conducted at the NASA Lewis Research Center. The test facility is described elsewhere.[23] Domestic water at 520 kPa (75 psia) and shop air at 790 kPa (115 psia) were available for the first phase of testing. A 25-μm filter was used on the water and air supplies. For these nonreacting atomization tests, the shear coaxial injector was mounted to spray straight down. A catch basin was located 1.2 m (4 ft) downstream of the injector face. Both air and water flow rates and injection velocities were varied. For the confined spray tests, water at 790 kPa (115 psia) and air at 790 kPa (115 psia) were available. The cylindrical chamber was 50 mm in diameter $\times$ 0.3 m long. No coflowing gas was provided, and no attempt was made to prevent droplet recirculation. Three 25 $\times$ 62 mm windows were provided to allow optical access. Two of the windows were located at 0 and 150 deg so that the PDPA could be positioned to collect light scattered by the drops in a near forward direction. The chamber pressure for the confined spray tests was controlled using a back-pressure regulator.

B. Injector Hardware

The single-element, shear coaxial injector consisted of four parts: a liquid inlet manifold, liquid post, gas manifold, and faceplate (Fig. 1). Each liquid post had four fins to center it within the gas manifold. These fins were located 50 mm (2 in.) upstream of the injector face to minimize their effect on the gas velocity profiles at the injector exit. By changing out the liquid post and faceplate, several injectors could be evaluated using a single gas manifold. The injector elements, designed

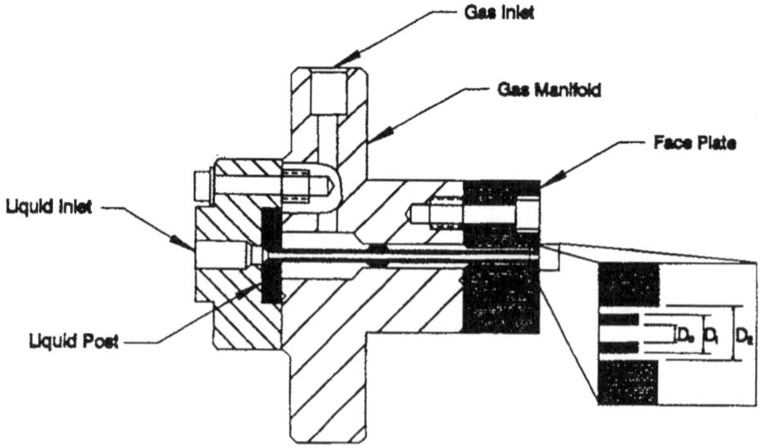

Fig. 1 Shear coaxial injector geometry.

Table 1 Coaxial injector configurations

Injector no.	D_2, mm (in.)	D_1, mm (in.)	D_0, mm (in.)
1	5.56 (0.219)	3.18 (0.125)	2.54 (0.100)
2	5.16 (0.203)	3.18 (0.125)	2.54 (0.100)
3	5.56 (0.219)	3.18 (0.125)	1.98 (0.078)
4	5.16 (0.203)	3.18 (0.125)	1.98 (0.078)
5	5.56 (0.219)	3.18 (0.125)	1.32 (0.052)
6	5.16 (0.203)	3.18 (0.125)	1.32 (0.052)

for a nominal 330-N (75-lbf) thrust at 5.5-MPa (800-psia) chamber pressure using liquid oxygen and gaseous hydrogen, were smaller than the space shuttle main engine (SSME) main chamber injector elements, but comparable in size to RL-10 injector elements. This injector size was selected to reduce the spray number density and to permit application of optical drop sizing diagnostics. The coaxial injector was operated conventionally, that is, liquid was flowed through the center post, and gas through the injector annulus. The water-to-air mass flow rate ratio ranged from 0.6 to 4.0.

Six injector geometries, with different liquid and gas injection areas, were selected for testing. These injector geometries are listed in Table 1, with illustration of the various diameters provided in Fig. 1. Different injector geometries were tested to examine the effect of varying the gas velocity independently of the gas flow rate on the atomization process. The data were taken using sufficiently low gas flow rates to avoid sonic flow at the injector exit. Conducting tests with the gas gap unchoked allowed the atomizing gas velocity to be varied and the unchoked gas flow in a hot-firing engine to be more closely simulated. The liquid flow rates were reduced from the injector design flow rates to correspond to the reduced (unchoked) gas flow rates.

C. Diagnostic Instrumentation

A phase-Doppler particle analyzer was used to obtain drop size measurements in the shear coaxial injector sprays. This instrument is similar to a laser Doppler velocimeter in its use of scattered laser light to obtain flowfield information. The theory of operation of the PDPA is described in the literature.[24] It measures the size and velocity of individual droplets as they pass through a small measurement volume. The measurement volume diameter must be larger than the droplet diameters, yet small enough to detect one droplet at a time at the measurement location. Various moments of the drop size distribution (such as volume mean diameter $\bar{D}_{30}$, and Sauter mean diameter $\bar{D}_{32}$) are computed from the acquired histograms, instead of fitting curves to the distribution shapes.

For these experiments, the PDPA was configured to provide an overall size measurement range from 4 to 350 μm. A 50:1 maximum diameter ratio setting was available for any given run within the 4–350-μm total size range. A monodisperse droplet generator was used to verify the performance of the phase-Doppler particle analyzer. Refractively scattered light was collected at 30 deg (with 0 deg defined as on-axis forward scatter). Two measurements of droplet size are made for each droplet. If these two measurements do not agree to within a specified tolerance, the droplet is determined to be aspherical, and the measurement is rejected. Signals with low signal-to-noise ratios, such as signals resulting from multiple particles simultaneously present in the probe volume, are also rejected.

A mass flux patternator was used in open spray tests to estimate the mass flux distribution in the shear coaxial injector sprays. The patternator consisted of 32 square tubes assembled in a row. Each tube was 6.4 × 6.4 mm (0.25 × 0.25 in.) square. Flexible tubes carried the liquid to collection containers. Measurements were made with the open tube ends 50 mm (2 in.) downstream of the injector face.

III. Procedure

Drop sizes and velocities were measured using the PDPA in a plane 50 mm (2 in.) downstream of the injector face. Drop sizes and velocities were measured at seven radial locations in the spray until 6000 valid drop sizes were obtained. The smallest drops had mean velocities that were about 20–40% higher than the large drop mean velocities. Data rates (number of drops measured per total measurement time) were 100 times higher in the spray center than at 16-mm (0.62-in.) radius for almost every flow rate and geometry. The radial location of the measurement volume was increased until the valid data rate fell below 60 samples /s. Most of the data presented here were acquired within a radius of 16 mm. For the unconfined sprays, validation rates varied from about 70–100%. Validation rates fell as low as 40% for the pressurized spray tests, largely due to spray accumulating on the windows. To reduce measurement uncertainty, multiple measurements were taken at each location for the confined spray tests. Although the 40% validation rate is low, the drop size and velocity mean values were unaffected. This was checked by acquiring atmospheric pressure data in the chamber, with and without the windows installed, and getting reasonable agreement.

The mean drop size measurement repeatability was determined to be affected mostly by the photomultiplier tube (PMT) gain. The effect of the PMT gain on the polydisperse spray measurements was examined by taking data at one spray location using different PMT high-voltage settings. Mean drop sizes decreased almost linearly with increasing PMT sensitivity, giving no indication of an appropriate PMT gain selection. It was decided to select the PMT high voltage that produced the maximum valid data rate. This criterion for the PMT setting worked well in dense portions of the spray, since the valid data rate reached a maximum value as the PMT gain was varied. While using this procedure, the repeatability of the mean drop sizes were estimated to be ±10% and velocities ±4%.

The effect of sample size on the mean drop sizes was determined by taking eight series of measurements at one spray location while varying the number of valid samples collected. The mean drop sizes varied widely for small sample sizes. As the sample size was increased, the standard deviation of the eight measurements of mean drop size reached an acceptable value at a sample size of 6000 (Fig. 2). Larger sample sizes did not lower the standard deviation significantly, so a minimum sample size of 6000 valid drops was selected for all of the tests.

The radial symmetry of the coaxial injector sprays was examined. By traversing across a diameter of the spray[12] tests were conducted since these tests exhibited a high degree of symmetry, the remaining tests were conducted by traversing along only half the spray diameter. A sample measurement illustrating the spray mean drop size and velocity profile symmetry for the entire spray width is shown in Figs. 3 and 4.

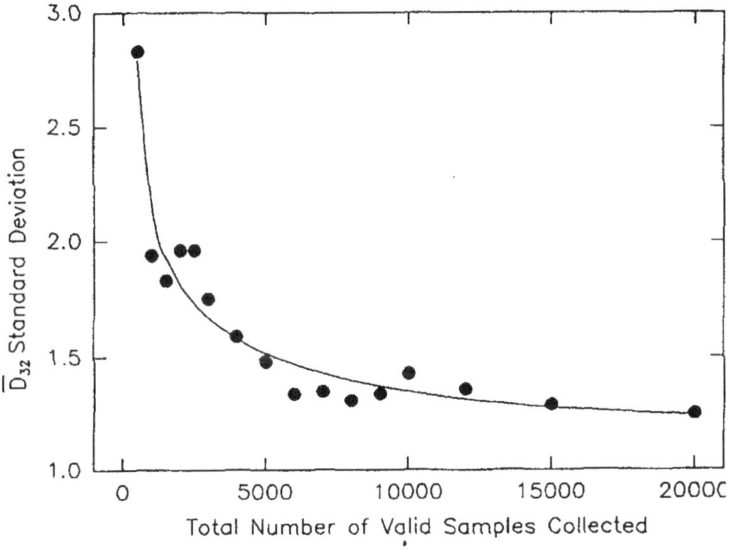

Fig. 2 Effect of sample size on $\overline{D}_{32}$ standard deviation for repeated measurements.

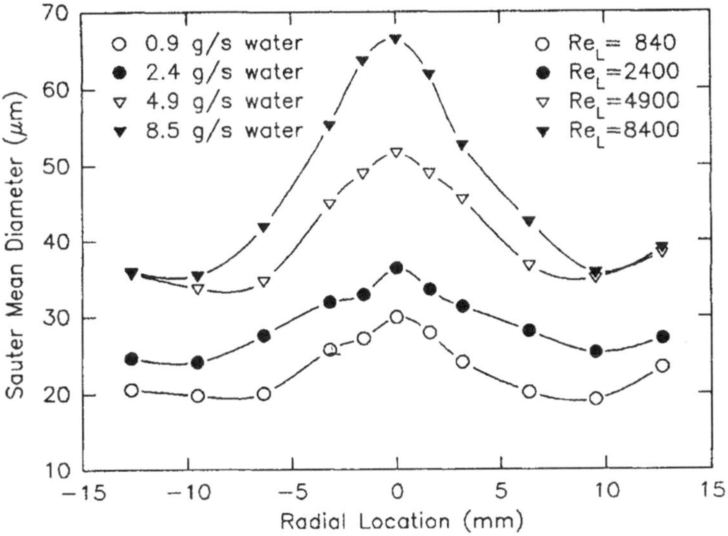

Fig. 3 Typical Sauter mean diameter profile for shear coaxial injector spray, injector 5, airflow rate 4 g/s, V_g = 200 m/s.

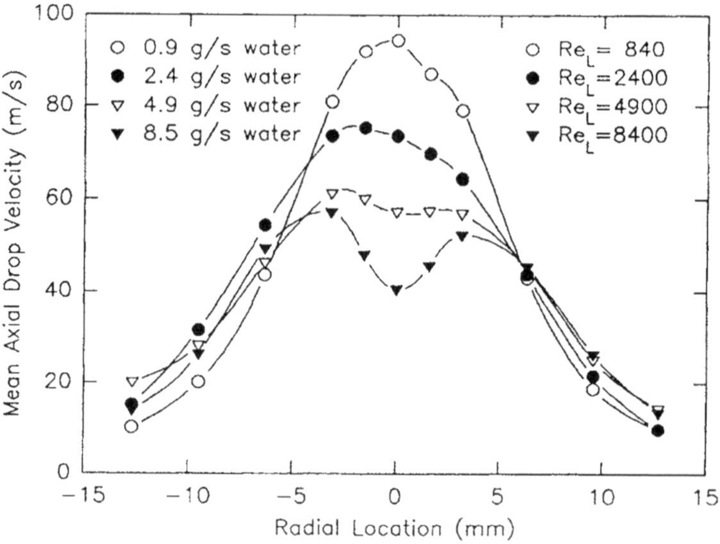

Fig. 4 Mean axial drop velocity for shear coaxial injector spray, injector 5, airflow rate 4 g/s, V_g = 200 m/s.

A tube patternator was used to measure the spray mass flux using injector 6 with several different water flow rates. The patternator tubes were covered before the spray was turned on. After the flow rates were set, the cover was removed. The water was carried in flexible tubes to the measurement containers, which were weighed before and after each run. Liquid was collected for between 1 and 5 min, with the shorter run times used with the higher water flow rates. Several tests were also conducted using one set of flow rates. The repeatability of the data was exceptionally high, with all of the runs with the same flow rates agreeing to within ±10%. Although the spray patternator clearly interfered with the spray, the collection efficiencies were between 80% and 100% of the flow meter value.

IV. Results and Discussion

A. Injection Parameter Effect on Spray

1. Gas Velocity Effect on Unconfined Spray

Drop size and drop velocity profiles were measured to determine the effect of varying gas and liquid flow rates on the injector spray. Sauter mean diameters and drop mean axial velocities are presented for injector 1 (see Table 1) as a function of radial location for a constant water flow rate of 0.9 g/s and three air flow rates of 3, 4, and 5 g/s (solid circles in Figs. 5 and 6). Mean drop sizes decreased with increasing air flow rate, in agreement with observations made by numerous investigators.[10–17, 19–23] SMD values at the edge of the spray were not observed to be minimum values. For the 3 g/s airflow rate, the SMD measured at the spray edge was actually 20% larger than the SMD measured in the center (Fig. 5). In tests performed by Sankar et al.,[19] slight increases in Sauter mean diameter were also observed at the spray edge. For all three airflow rates, the SMD profiles were flat, with little variation in SMD as a function

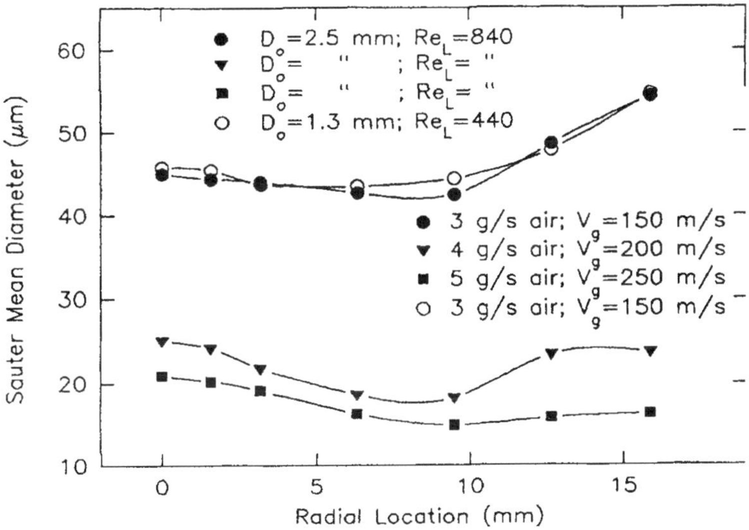

Fig. 5 Sauter mean diameter profiles for various airflow rates, injectors 1 and 5, water flow rate 0.9 g/s.

of radius. Mean axial drop velocities for the same cases are shown in Fig. 6. The mean axial drop velocities reached maximum values at the spray center, then decreased to approximately 7 m/s for all three airflow rates at 16-mm radius. Spray centerline velocities increased as the air injection velocity was increased. Similar trends in drop size and velocity were observed with the other 5 coaxial injector geometries.

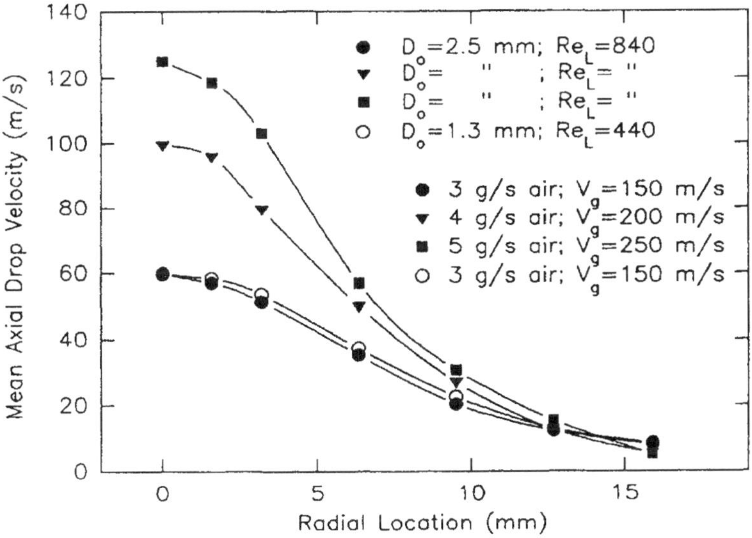

Fig. 6 Mean axial drop velocities for various airflow rates, injectors 1 and 5, water flow rate 0.9 g/s.

2. *Liquid Orifice Diameter Effect on Unconfined Spray*

Using the same water flow rate of 0.9 g/s and an airflow rate of 3 g/s, mean drop size and velocity data were measured using a smaller liquid orifice diameter (injector 5). These drop size and velocity curves (indicated by open circles in Figs. 5 and 6) are nearly identical to the data obtained for the same flow rates with injector 1, despite the factor of two decrease in the liquid orifice diameter. There was a corresponding factor of two decrease in the liquid jet Reynolds number ($Re_L = \rho_L V_L D_0/\mu_L$), which has often been used as a correlating parameter for shear coaxial injector sprays.[25,26] This indicates either a weak dependence of the spray drop sizes on liquid jet Reynolds number or the liquid jet Reynolds number might only serve to distinguish among liquid jet breakup or flow regimes. The influence of the liquid jet Reynolds number may be greater for turbulent liquid jet flows.

3. *Liquid Flow Rate Effect on Unconfined Spray*

A series of tests was conducted with injector 1 using an airflow rate of 3 g/s air and four different water flow rates (Fig. 7). The Sauter mean diameter curves in Fig. 7 are similarly shaped, despite the change in liquid flow rate. The SMD profile shapes are not strongly affected by the change in liquid jet Reynolds numbers, despite the occurrence of both laminar and turbulent liquid flows. For this airflow rate, the largest drop sizes occurred at the spray edge. Little variation in drop sizes is evident near the centerline, as compared with the data presented in Fig. 3, which shows a larger SMD variation along the centerline for the same water flow rates.

To measure spray mass flux, a tube patternator with 6.4-mm-square tubes was used with injector 6. The airflow rate was kept constant at 4 g/s, while the water flow rate was varied from 0.8 to 15 g/s. Collection efficiencies varied from 80

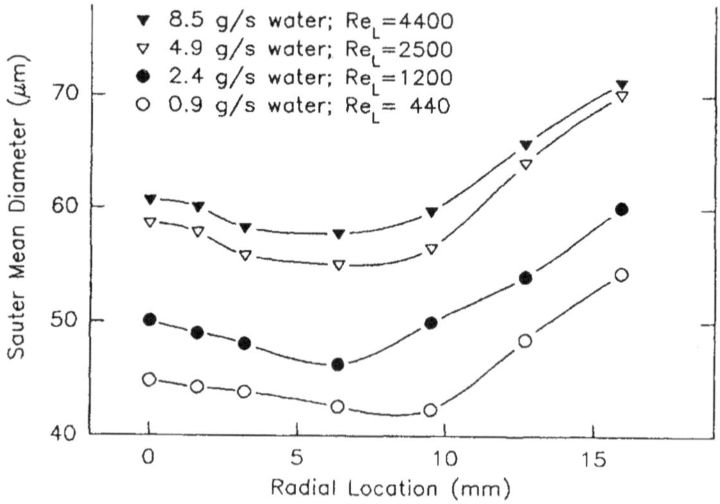

Fig. 7 Sauter mean diameter profiles for various water flow rates, injector 1, airflow rate 3 g/s, V_g = 150 m/s.

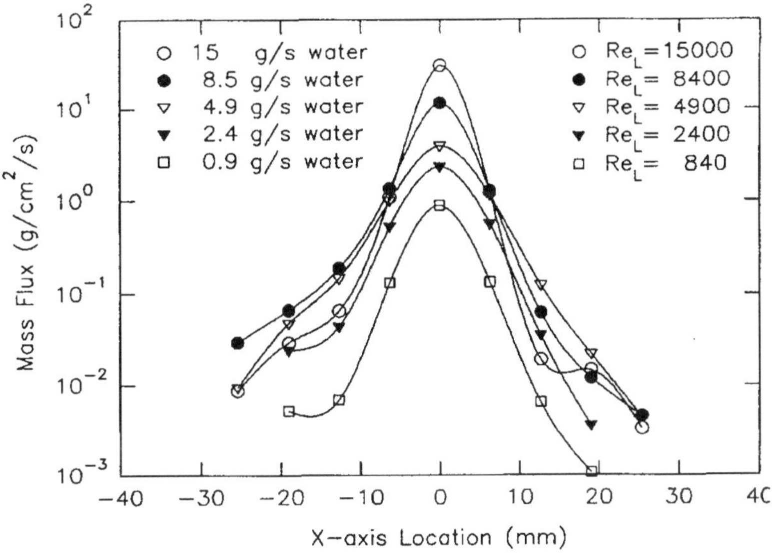

Fig. 8 Liquid mass flux for varying liquid flow rates using 6.4-mm-square tube patternator in open air spray, injector 6, airflow rate 4 g/s, V_g = 250 m/s.

to 100% of the flow meter value. Use of the patternator provided much better mass flux measurements than the phase Doppler for these sprays, especially at the higher flow rate cases, in which large, nonspherical liquid particles were probably present at the measurement plane. The mass fluxes are plotted against radius in Fig. 8. Between 60 and 90% of the spray mass for each case was collected in the center tube, indicating that the sprays are quite narrow, with spray axial velocity components much larger than the radial components.

4. *Post Recess Effect on Unconfined Spray*

The post was recessed in increments of 1.6 mm from 0.0 to 6.4 mm. The results presented in Figs. 9 and 10 were obtained using a water flow rate of 2.4 g/s and an airflow rate of 4 g/s. As the post recess was increased, the centerline SMD increased. At the same time, SMD measured at 12.7-mm radius decreased with increasing recess, with the exception of the 6.4-mm recess case. Therefore, the general effect of increasing post recess was to increase the slope of the SMD profile as a function of radius. For all cases with post recess, the drop velocity maximum did not occur at the centerline (Fig. 10).

5. *Confined and Unconfined Spray*

The effect of enclosing the spray within a chamber was examined. Spray mean drop size and velocity data are presented in Figs. 11 and 12 for injector 6. One set of tests was conducted without the chamber present. Another set of tests was performed using the 50-mm-diam × 0.3-m-long cylindrical chamber, with the chamber pressure maintained at atmospheric. The sprays were quite similar between 0- and 6.4-mm radius. Beyond 6.4-mm radius, the presence of the chamber walls affects the measured spray sizes and velocities. Drop sizes for the enclosed spray are larger than for the unenclosed spray, presumably due to overcounting of large drops recirculating within the chamber. Mean axial velocities were negative at the edge of the enclosed spray, confirming the presence of a recirculation zone.

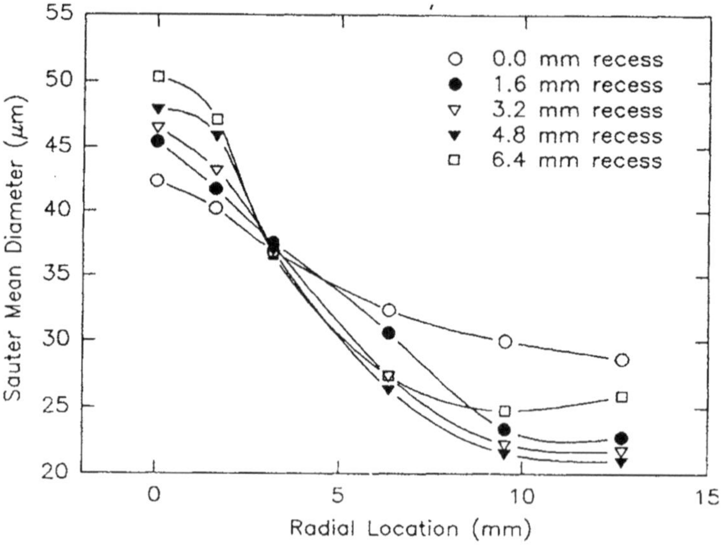

Fig. 9 Sauter mean diameter profiles for various liquid post recess, injector 6, airflow rate 4 g/s, water flow rate 2.4 g/s, Re_L = 2400.

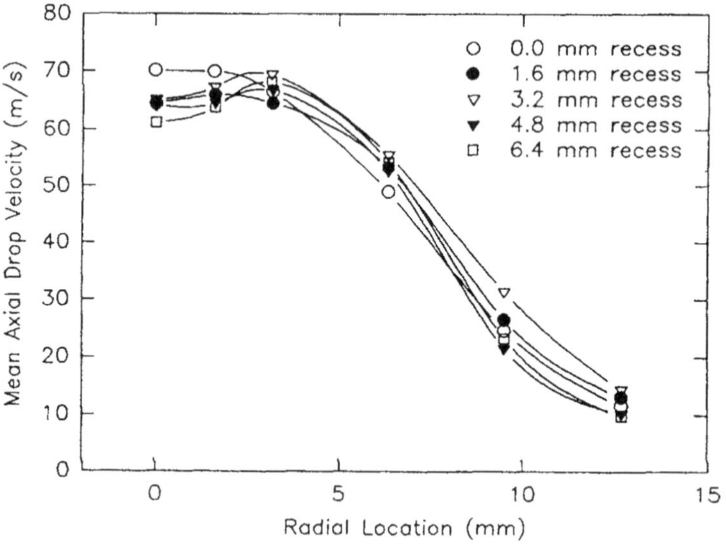

Fig. 10 Mean axial drop velocities for various liquid post recess, injector 6, airflow rate 4 g/s, water flow rate 2.4 g/s, Re_L = 2400.

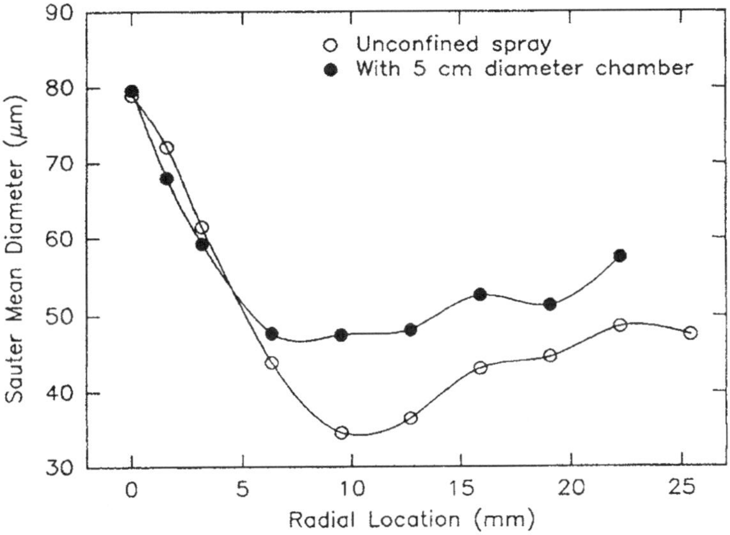

Fig. 11 Sauter mean diameter profiles for confined and unconfined spray, injector 6, airflow rate 4 g/s, V_g = 260 m/s, water flow rate 8.6 g/s, Re_L = 8500.

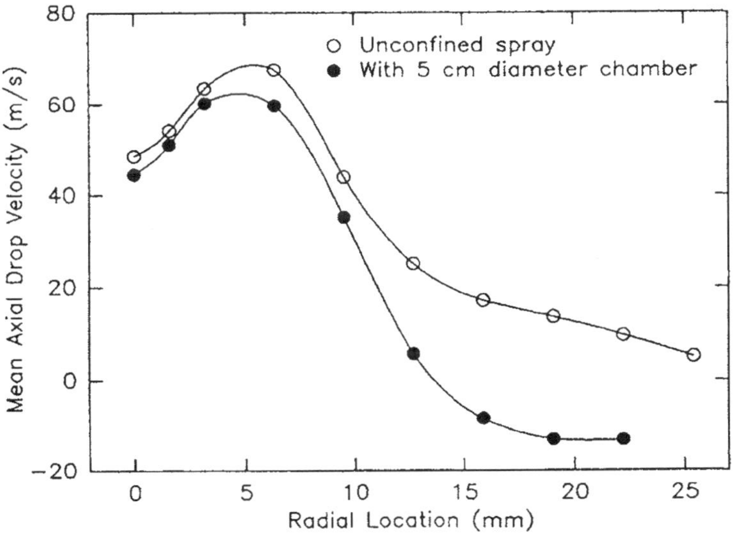

Fig. 12 Mean axial drop velocities for confined and unconfined spray, injector 6, airflow rate 4 g/s, V_g = 260 m/s, water flow rate 8.6 g/s, Re_L = 8500.

6. Constant $\rho_g V_g$ for Varying Chamber Pressures

Injector 6 was used with flow rates of 2.3 g/s water and 6.8 g/s air at both 300-kPa and 500-kPa chamber pressures (Figs. 13 and 14). Increasing the chamber pressure while keeping the gas and liquid flow rates constant resulted in larger drop sizes. Since the gas flow rate was the same for both cases, the gas injection velocity for the 500-kPa case was lower than the 300-kPa case, causing the drop sizes to increase with increasing chamber pressure. The drop velocities decreased with increasing chamber pressure near the spray center. Obtaining measurements was difficult due to spray collecting on the chamber windows. Multiple measurements were made at each radial location to minimize the error caused by the presence of the spray on the windows.

7. Constant V_g for Varying Chamber Pressures

Four tests were conducted within the pressurized chamber with 2.3 g/s water and varying airflow rates. The airflow rates were selected to keep the gas injection velocity constant while varying the chamber pressure. Injector 6 was used with chamber pressures from 100 to 500 kPa abs (Figs. 15 and 16). The mean drop sizes at the edge of the spray were similar for all four chamber pressures. Near the spray center, the drop sizes decreased with increasing chamber pressure, probably due to the increased gas stream momentum. The drop velocities were similar for all four cases.

8. Constant $\rho_g V_g^2$ for Varying Chamber Pressures

Additional tests were conducted for the same four chamber pressures using the same injector (injector 6) and water flow rate (2.3 g/s). The airflow rates were chosen to keep the gas stream dynamic head (gas density times gas injection velocity squared) constant for all four chamber pressures (Figs. 17 and 18). The

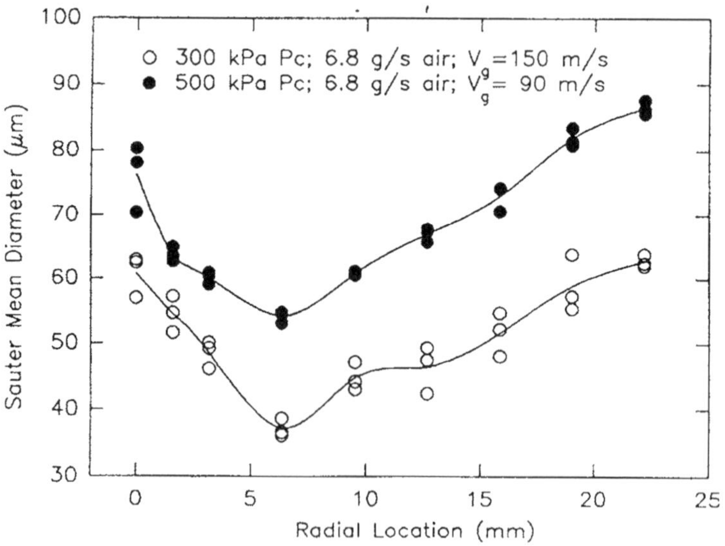

Fig. 13 Effect of chamber pressure on Sauter mean diameter profiles for constant gas flow rate, injector 6, water flow rate 2.3 g/s, Re_L = 2300.

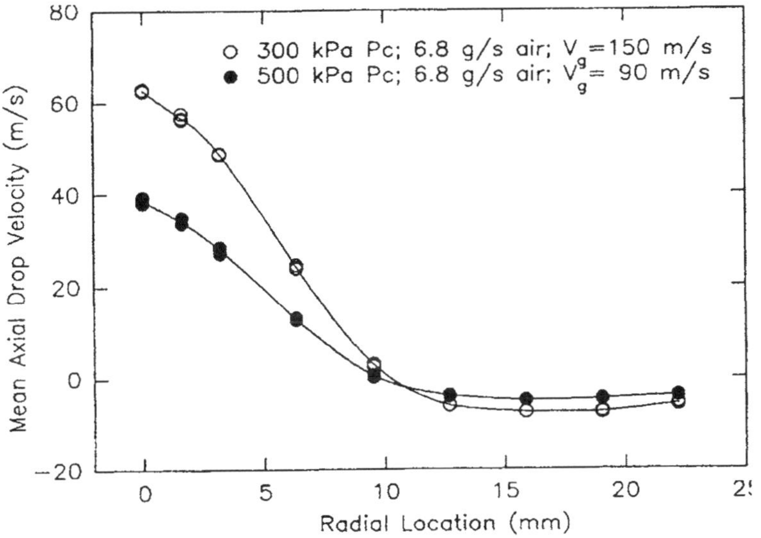

Fig. 14 Effect of chamber pressure on mean axial drop velocities for constant gas flow rate, injector 6, water flow rate 2.3 g/s, Re_L = 2300.

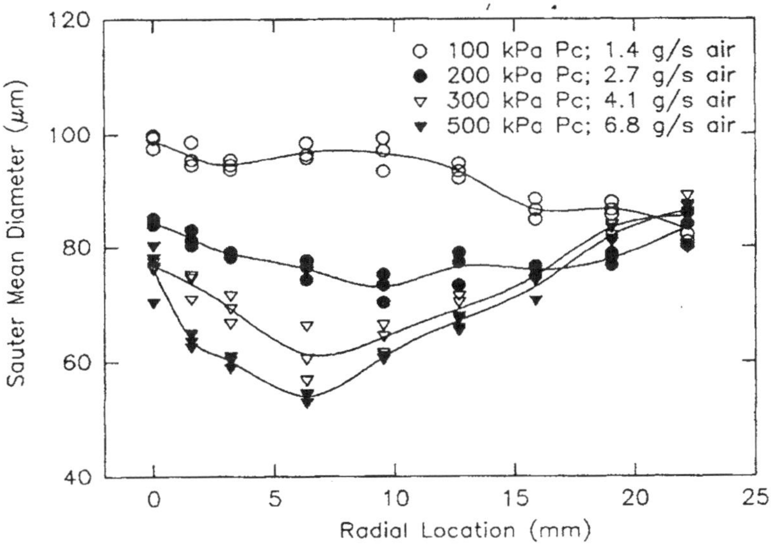

Fig. 15 Effect of chamber pressure on Sauter mean diameter profiles for constant gas injection velocity, injector 6, water flow rate 2.3 g/s, Re_L = 2300.

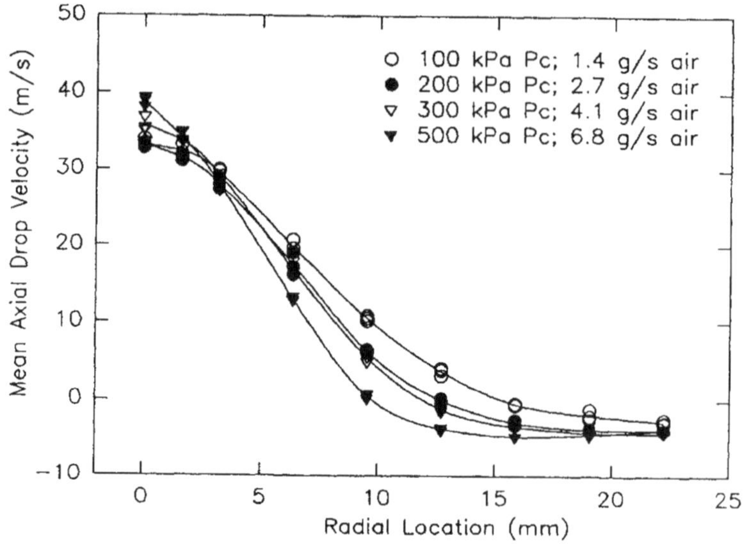

Fig. 16 Effect of chamber pressure on mean axial drop velocities for constant gas injection velocity, injector 6, water flow rate 2.3 g/s, Re_L = 2300.

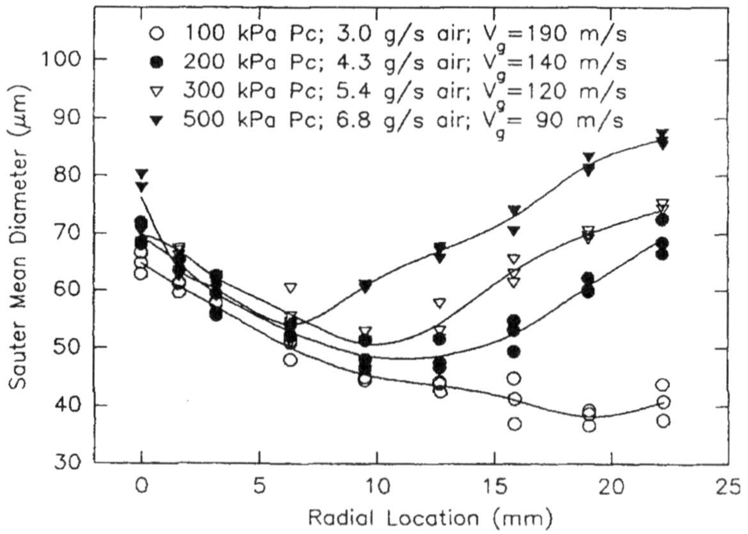

Fig. 17 Effect of chamber pressure on Sauter mean diameter profiles for constant gas $\rho_g V_g{}^2$, injector 6, water flow rate 2.3 g/s, Re_L = 2300.

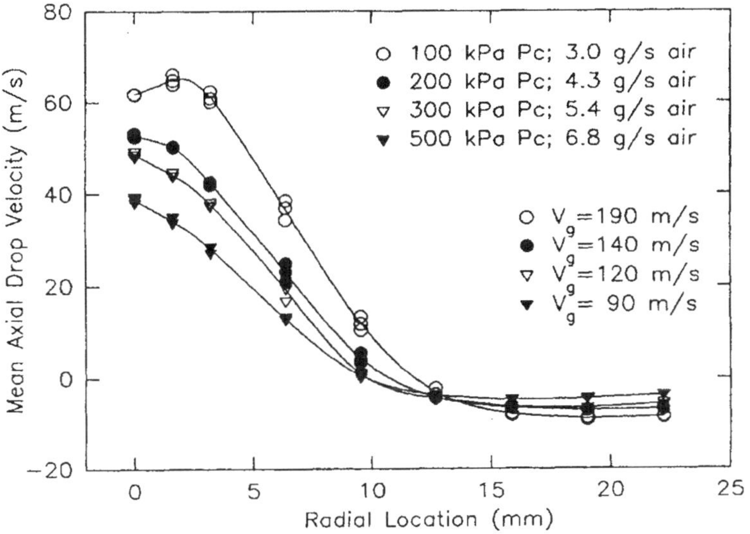

Fig. 18 Effect of chamber pressure on mean axial drop velocities for constant $\rho_g V_g{}^2$ injector 6, water flow rate 2.3 g/s, Re_L = 2300.

drop sizes stayed nearly the same at the spray center, despite the changes in airflow rate and density. This indicates that the gas stream dynamic head is more significant than the gas velocity or flow rate alone in determining the resulting drop sizes formed. A slight increase in drop sizes near the center with increasing chamber pressure was observed.

B. Composite Mean Drop Sizes

To make the drop size data obtained for this particular injector more universally applicable and to determine the relative influence of various injection parameters, a correlation between the injection parameters and a composite mean spray drop size was found. First, composite mean drop sizes were computed for each test case. These were based on $\bar{D}_{30}$ (volume mean diameter) instead of $\bar{D}_{32}$. $\bar{D}_{30}$ is more appropriate because it represents a drop with the average volume (or mass), instead of a drop with the average volume-to-surface area. The composite mean drop sizes were weighted averages of the volume mean drop sizes obtained at the seven radial locations sampled in these tests. The drop sizes at each radial location were weighted by the amount of time required to collect N samples, and the area of a ring representing an integrated area for each radial location: where A is the area of a representative ring at that radial location, subscript i represents each radial location, j is the number of radial locations, subscript C designates the composite value,

$$(\bar{D}_{30}^3)_C = \left(\sum_{i=1}^{j} \frac{A_i \bar{D}_{30_i}^3}{t_i} \bigg/ \sum_{i=1}^{j} \frac{A_i}{t_i} \right) \tag{1}$$

and t is the time taken to acquire N samples. For all of the tests presented here, N is 6000 samples. Equation (1) assumes the validation rate to be high, which was true for the unconfined spray data acquired in this study, and neglects the variation of the measurement cross-sectional area with droplet size.

Composite mean diameters from over 80 different tests were correlated with respect to the injection parameters using linear regression. A good fit to the data was obtained with the equation

$$\bar{D}_{30c} = 5.92 \times 10^{-5}\left(\frac{\sigma}{\mu_L}\right)^{0.75}\left(\frac{\rho_L}{\rho_g}\right)^{0.05}\left(\frac{\dot{m}_L}{\dot{m}_g}\right)^{0.25}\frac{V_L^{0.05}}{V_r^{0.8}} \tag{2}$$

where σ, μ, and ρ indicate surface tension, viscosity, and density. V is the injection velocity, $\dot{m}$ is the flow rate, and subscripts g and L refer to the gas and liquid phases, respectively. Gas and liquid velocities, flow rates, injection areas, and the gas density were varied. These variable ranges were the following: liquid orifice diameter D_0 1.3–2.5 mm, relative velocity V_r 80–290 m/s, liquid flow rate $\dot{m}_L$ 0.9–15 g/s, gas flow rate $\dot{m}_g$ 1.4–6.8 g/s, and liquid-to-gas flow rate ratio 0.3–4.2. No attempt was made to determine the influence of liquid properties, such as surface tension, density, and viscosity, on the atomization process. These fluid properties were included in the composite mean drop size correlation for nondimensionalization purposes. Also note that the relative velocity appears in the drop size correlation, even though the gas velocity was much larger than the liquid velocity, and the influence of relative velocity was not demonstrated conclusively. The fit of the correlation to the data was checked using linear regression. The square of the correlation coefficient, r^2, was 0.96. The large value of r^2 that was obtained indicates that the data and the equation are highly correlated. All of the data points are plotted against the correlation in Fig. 19.

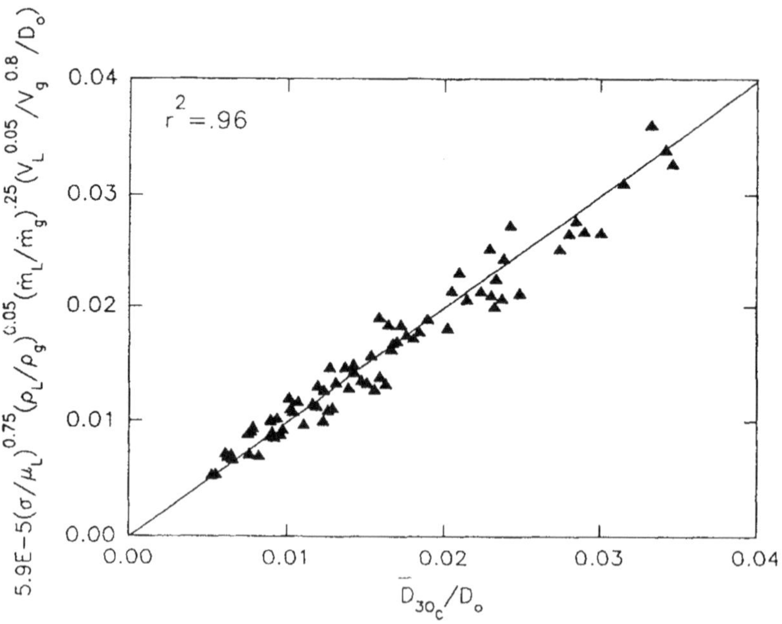

Fig. 19 Composite mean drop sizes plotted vs correlation.

The recess data were correlated separately. The influence of post recess was expressed as a fraction of the composite mean diameter value with no post recess,

$$\bar{D}_{30_{C,\mathrm{recess}}}(\bar{D}_{30_C})(1 + 0.18R - 0.278R^{0.5}) \tag{3}$$

where R is the dimensionless ratio of recess depth to liquid post outer diameter. This empirical correlation was based on data for $0 \leq R \leq 2$ and is plotted in Fig. 20.

Previous experiments have shown that the impact of liquid post recess on the drop sizes produced by coaxial injector sprays to be unclear or negligible. Burick[27] reported that recessing the LOX post typically increased overall mean drop sizes for post recesses of less than twice the post diameter. This is not in agreement with the data presented here. Falk[14] found that recessing the liquid post had very little effect on mean drop sizes; however, only one test with a recessed post was conducted in that study. Gomi[17] observed very slight differences in drop sizes between cases with zero recess and cases with a post recess equal to three post diameters. Sankar et al.[19] showed that increasing post recess had a variable effect on drop sizes. Other evidence has indicated that recessing the liquid post might produce smaller mean drop sizes. Hot-firing tests have shown that increased post recess leads to improved combustion efficiency.[4] It is possible that improved atomization was a factor in the observed increase in performance. The data reported here confirms this observation, although it does not take into account the differences between cold-flow and hot-firing testing. Other effects, such as burning in the cup formed by the liquid post recess, may also affect the drop size distributions in hot-firing rocket engines.

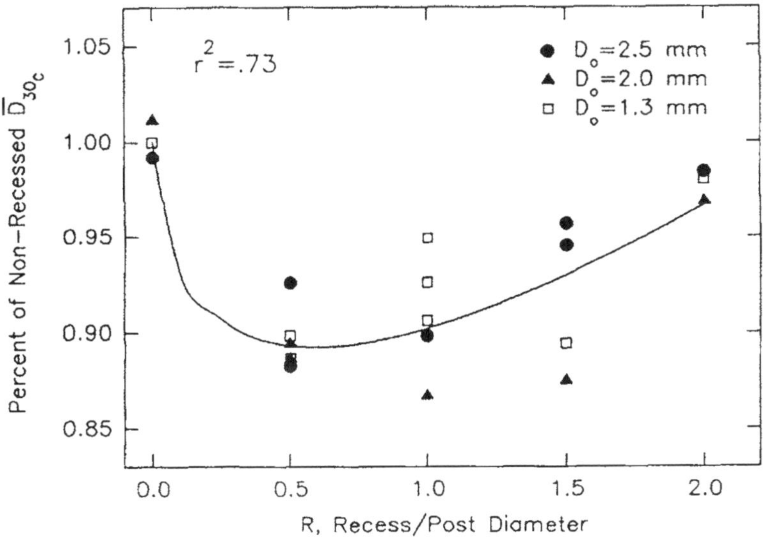

Fig. 20 Composite mean drop sizes for cases with liquid post recess normalized to cases without recess, injectors 2, 4, and 6, airflow rate 4 g/s, water flow rate 2.4 g/s, $Re_L = 2400$.

V. Concluding Remarks

Open air sprays produced by a single-element, shear coaxial injector flowing water and air were characterized using an Aerometrics phase-Doppler particle analyzer and a mass flux patternator. Sauter mean diameter and mean axial drop velocity profiles were examined in detail with the PDPA. The shapes of these profiles were demonstrated to be a function of the water and airflow rates and injection velocities. In all cases, increasing the gas stream velocity resulted in smaller drop sizes. The velocity profile shapes had maxima at the spray center for the lower gas and liquid flow rates; for higher water flow rates, the drops in the spray center did not have the maximum velocities. The liquid orifice diameter was varied by a factor of two in these tests, and was not observed to influence the drop sizes significantly. Sauter mean diameter distributions were flat for low water and airflow rates; the size distributions had maxima at the spray center for higher water flow rates. Recessing the injector post up to twice the injector post outside diameter caused the drop sizes measured at the centerline to increase.

The PDPA was used to make measurements of single-element, shear coaxial injector sprays within a pressurized chamber. Drop sizes measured within the chamber were larger than for open air sprays, especially at the spray edge. Velocities beyond 12-mm radius were negative for almost every test, indicating the presence of a recirculation zone. At higher chamber pressures, the airflow rate had to be increased in order to create a spray without the intact liquid core of the jet reaching the measurement plane. It was demonstrated that keeping the gas dynamic head constant (or Weber number) while varying the gas density created sprays with similar SMD near the spray center. The similarity, however, among SMD or mean velocity profiles obtained at different chamber pressures was not preserved by maintaining a constant gas dynamic head. Therefore, although maintaining the gas dynamic head was a reasonable correlating factor, it was not sufficient to guarantee complete similarity among sprays obtained at different chamber pressures.

A one-dimensional correlation was developed between a composite $\bar{D}_{30}$ and the injection flow rates and velocities for water/air shear coaxial injector sprays. Fluid properties were included in the correlation in order to nondimensionalize it. The relative velocity appears in the drop size correlation, despite the fact that the influence of liquid velocity alone was not demonstrated conclusively. This correlation should be able to predict shear coaxial injector spray drop sizes only for nonvaporizing, subcritical pressure sprays. It does predict the influence of many injection parameters, however, which should be relevant to other flow regimes. Once appropriate similarity parameters are established, it is expected to be useful for specifying initial drop sizes in reacting spray calculations. Additional testing will be required using fluids with a wide range of properties and a test matrix that includes lower gas/liquid velocity ratios in order to extend the correlation applicability.

References

[1]Jensen, R. J., "A Summary of the JANNAF Workshop on Liquid Rocket Engine Combustion Driven Instability Mechanisms," *26th JANNAF Combustion Meeting,* edited by D. L. Becker, Vol. 2, CPIA-PUBL-529-VOL-2, Chemical Propulsion Information Agency, Laurel, MD, 1989, pp. 273–288.

[2]Gross, K. W., "Liquid Engine Jet Atomization Workshop Report," *24th JANNAF Combustion Meeting,* edited by D. L. Becker, Vol. 2, CPIA-PUBL-476-VOL-2, Chemical Propulsion Information Agency, Laurel, MD, 1987, pp. 351–353.

[3]Zaller, M., "LOX/Hydrogen Coaxial Injector Test Program," NASA CR-187037, Oct. 1990.

[4]Wanhainen, J. P., Parish, H. C., and Conrad, E. W., "Effect of Propellant Injection Velocity on Screech in 20,000-Pound Hydrogen-Oxygen Rocket Engine," NASA TN D-3373, April 1966.

[5]Lin, S. P., and Lian, Z. W., "Mechanisms of the Breakup of Liquid Jets," *AIAA Journal,* Vol. 28, No. 1, 1990, pp. 120–126.

[6]Reitz, R. D., and Bracco, F. V., "Mechanism of Atomization of a Liquid Jet," *Physics of Fluids,* Vol. 25, No. 10, 1982, pp. 1730–1742.

[7]Eroglu, H., and Chigier, N., "Liquid Jet Instability in Coaxial Air Flow," *ICLASS-91 Proceedings of the Fifth International Conference on Liquid Atomization and Spray Systems,* Paper 78, July 1991, Gaithersburg, MD, pp. 703–710.

[8]Mayer, W., and Krülle, G., "Rocket Engine Coaxial Injector Liquid/Gas Interface Phenomena," AIAA Paper 92-3389, July 1992.

[9]Ghafourian, A., McGuffin, R., Mahalingam, S., and Daily, J. W., "Dynamic Response to Acoustic Perturbation of an Atomizing Coaxial Jet in a Liquid Rocket Engine," AIAA Paper 93-0232, Jan. 1993.

[10]Ferrenberg, A., Hunt, K., and Duesberg, J., "Atomization and Mixing Study," Rockwell International, RI/RD85-312, NASA Contract NAS8-34504, NASA CR-178751, Canoga Park, CA, Dec. 1985.

[11]Lefebvre, A. H., "Airblast Atomization," *Progress in Energy and Combustion Science,* Vol. 6, 1980, pp. 233–261.

[12]Weiss, M. A., and Worsham, C. H., "Atomization in High Velocity Airstreams," *ARS Journal,* Vol. 29, No. 4, 1959, pp. 252–259.

[13]Kim, K. Y., and Marshall, W. R., Jr., "Drop-Size Distributions from Pneumatic Atomizers," *AIChE Journal,* Vol. 17, No. 3, 1971, pp. 575–584.

[14]Falk, A. Y., "Coaxial Spray Atomization in Accelerating Gas Stream," Final Rept. NASA CR-134825, June 1975.

[15]Lorenzetto, G. E., and Lefebvre, A. H., "Measurements of Drop Size on a Plain-Jet Airblast Atomizer," *AIAA Journal,* Vol. 15, No. 7, 1977, pp. 1006–1010.

[16]Gomi, H., "Pneumatic Atomisation with Coaxial Injectors: Measurements of Drop Sizes by the Diffraction Method and Liquid Phase Fraction by the Attenuation of Light," National Aerospace Lab., NAL-TR-888T, Tech. Rept., Tokyo, Nov. 1985.

[17]Vingert, L., "Coaxial Injector Spray Characterization for the Ariane 5 Vulcain Engine," ONERA TP 1990-89, 1990.

[18]Hess, C. F., "A Technique Combining the Visibility of a Doppler Signal with the Peak Intensity of the Pedestal to Measure the Size and Velocity of Droplets in a Spray," AIAA Paper 84-0203, Jan. 1984.

[19]Sankar, S. V., Brena de la Rosa, A., Isakovic, A., and Bachalo, W. D., "Liquid Atomization by Coaxial Rocket Injectors," AIAA Paper 91-0691, Jan. 1991.

[20]Sankar, S. V., Wang, G., Brena de la Rosa, A., Rudoff, R. C., Isakovic, A., and Bachalo, W. D., "Characterization of Coaxial Rocket Injector Sprays Under High Pressure Environments," AIAA Paper 92-0228, Jan. 1992.

[21]Eroglu, H., and Chigier, N., "Initial Drop Size and Velocity Distributions for Airblast Coaxial Atomizers," *Journal of Fluids Engineering,* Vol. 113, No. 3, 1991, pp. 453–459.

[22]Goix, P. J., Cessou, A., Stepowski, D., and Edwards, C. F., "Structure of a Methanol/Air Coaxial Reacting Spray Near the Stabilization Region by OH Fluorescence Imaging and

Two-Component Phase-Doppler Interferometry," Sandia National Lab., DE92-005328, SAND-91-8714, Livermore, CA, 1992.

[23]Zaller, M., and Klem, M. D., "Coaxial Injector Spray Characterization Using Water/Air as Simulants," NASA TM-105322, Nov. 1991.

[24]Bachalo, W. D., and Houser, M. J., "Phase/Doppler Spray Analyzer for Simultaneous Measurements of Drop Size and Velocity Distributions," *Journal of Optical Engineering,* Vol. 23, No. 5, 1984, pp. 583–590.

[25]Huynh, C., Ghafourian, A., Mahalingam, S., and Daily, J. W., "Combustor Design for Atomization Study in Liquid Rocket Engines," AIAA Paper 92-0465, Jan. 1992.

[26]Faragó, Z., and Chigier, N., "Morphological Classification of Disintegration of Round Liquid Jets in a Coaxial Air Stream," *Atomization and Sprays,* Vol. 2, No. 2, 1992, pp. 137–153.

[27]Burick, R. J., "Atomization and Mixing of Gas/Liquid Coaxial Injectors," *Journal of Spacecraft and Rockets,* Vol. 9, No. 5, 1972, pp. 326–331.

Chapter 8

Impinging Jet Injector Atomization

William E. Anderson,* Harry M. Ryan,* and Robert J. Santoro†
Pennsylvania State University, University Park, Pennsylvania 16802

I. Introduction

HIGH-FREQUENCY combustion instability in rocket engines occurs when the energy released during combustion couples with an acoustic mode of the thrust chamber and causes amplification of the oscillating pressure field.[1] There is no exact or conclusive proof regarding the specific combustion-based mechanisms that control initiation and sustenance of resonant high-frequency pressure oscillations in liquid rocket engines and, thus, no substantiated predictive model capable of accounting for thrust chamber design and operation on combustion instability. Most work on the coupling mechanism has been based on the underlying assumption that it is traceable to a single one of the principle component processes: injection, atomization, vaporization, mixing, or chemical kinetics.

All of the component processes, with the exception of injection, are determined to a large degree by the event of primary atomization, i.e., the initial fragmentation of an intact liquid core into drops and ligaments. Propellant spray characteristics, which are a direct product of the atomization process, also determine the thrust chamber's combustion efficiency and hot-side heat transfer characteristics. Therefore, it is necessary to understand primary atomization on the basis of its importance in the practical realization of optimal rocket engine operation.

In chemical thrusters powered by liquid propellants, atomization is made to occur efficiently by the use of highly inertial conditions. In rocket engines, where propellants are injected at a high-dynamic head (anywhere from 0.1 and 3 MPa), the greatest source of inertia available for destabilization of an injected propellant stream is another propellant stream. Liquid jet impingement on another liquid jet is a common and effective method to atomize and mix propellants prior to combustion. A number of combinations have been used with up to five impinging jets,[2] but the prevalent element type is the like doublet (Fig. 1), where two geometrically and dynamically similar jets of the same propellant impinge

---

*Research Assistant, Department of Mechanical Engineering, Propulsion Engineering Research Center.

†Professor of Mechanical Engineering, Propulsion Engineering Research Center.

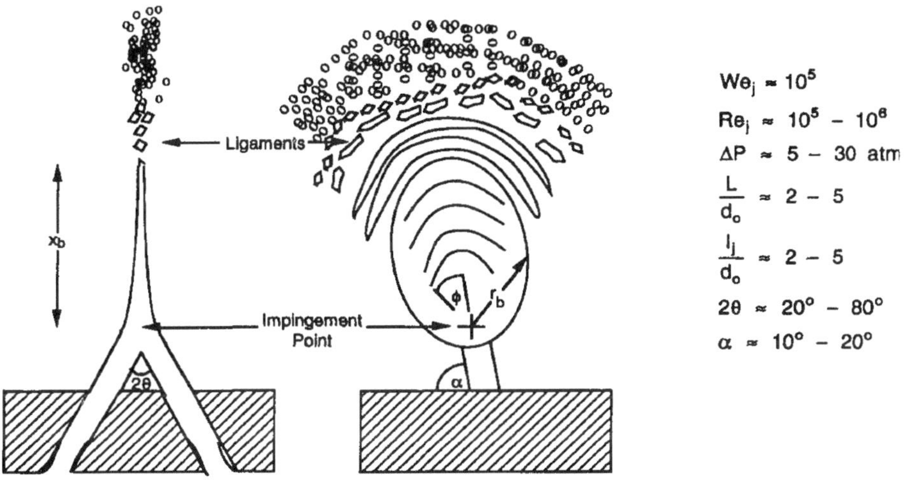

Fig. 1 Schematic of typical like doublet impinging jet injector element.

on each other at an oblique angle. Geometric and operating characteristics of a variety of like doublet elements used in production thrusters can be found in Ref. 2.

Impinging jet injector elements are the preferred injector geometry for rocket engines that use storable propellants or liquid hydrocarbons. These injectors are also a desirable substitute for coaxial injectors, typically used with gas/liquid propellant combinations, due to better atomization and mixing characteristics and inherently lower fabrication costs. Present design analysis for impinging jet injectors is not nearly as well developed as it is for coaxial injectors, however, primarily due to the lack of basic knowledge regarding the fundamental atomization mechanisms of impinging jets and their subsequent mixing and combustion processes. This situation is exacerbated by the fact that design and operational effects on combustion stability and performance for impinging jet injectors typically act in opposite directions, i.e., improved performance is achieved usually through sacrificing high-frequency stability margin, and vice versa.

The ability to control combustion instability is clearly dependent on the ability to understand primary atomization and the role it plays in the initiation and growth mechanisms of combustion instabilities in rocket engines. To date, there is not a clear, basic, and quantitative understanding of the processes that control atomization of impinging jets. The purpose of this chapter is to provide a concise report on the current state of knowledge on the primary atomization mechanisms of impinging liquid jets and to explore the roles that primary atomization may play in combustion stability. Results from previous experimental and theoretical studies on impinging liquid jets are summarized. Results from models using linear stability theory to predict certain measurable variables are presented. The results from more recent experimental work on impinging liquid jets are presented and compared with these theoretical results. Explanations for impinging jet behavior in different operating regimes are discussed. The role that primary atomization may play in combustion instability is discussed. In the end, some conjecture is provided regarding the operative mechanisms of impinging jet atomization, and the subject area is summarized.

II. Background

A. General Characteristics of Impinging Liquid Jets

The impingement of liquid jets is a very efficient method of atomization and mixing whereby the dynamic head of the propellant is used to destabilize an opposing liquid propellant stream, thereby fragmenting the liquid into ligaments and droplets. In the typical case of dynamically similar impinging jets, a very high-pressure region exists at the point where the axis streamlines of both jets converge and stagnate. After deceleration at the stagnation point where the static pressure is ideally equal to the total pressure in the propellant manifold, the fluid is accelerated nearly back to the jet's initial velocity. The high pressure in the stagnation region also causes the propellant to spread laterally, creating a fan. Shear occurs in the region between the stagnation streamline of the jet axis and the constant pressure streamline that defines the outer boundary of the impinging jet system. Drop velocity[3] and sheet velocity measurements[4] at atmospheric pressure for a variety of geometric and flow conditions indicate that the liquid downstream of the stagnation point moves at a speed nearly equal to the initial speed of the jets.

A schematic of the like doublet impinging jet injector is shown in Fig. 1. Equal diameter liquid jets impinge and form a fanlike elliptical liquid sheet. For the case of directly opposed jets, a circular disk would be formed after impingement. The angle between the jets, 2θ, is generally about 60 deg. The impingement point is typically about five length-to-diameter ratios (l_j/d_o) downstream of the injector face. Injection orifices are typically fabricated by twist drilling, electrical discharge machining, or laser drilling. In practice, the fan may be inclined at a small angle α toward an adjacent fan of the opposite propellant to induce large-scale mixing. The angular coordinate in the plane of the sheet is defined as ϕ. At typical rocket operating conditions, Reynolds numbers based on orifice diameter and liquid properties are on the order of 10^5–10^6; liquid Weber numbers $(We_l = \rho_l U_j^2 d_o/\sigma)$ indicating the balance between liquid inertial forces that tend to fragment the propellant and surface tension forces that tend to maintain the integrity of the liquid mass are on the order of 10^5; and injection pressure drops range from about 0.5 to 3.0 MPa (5 to 30 atm). A vena contracta usually forms downstream of the entrance region of the injector element and subsequently reattaches to the wall. Orifice discharge coefficients typically range between the thin orifice limit of 0.61 and 0.9. The entrance to the orifice is usually sharp, and fabrication cost considerations typically preclude rigorous specification of surface roughness or inlet contour.

The atomization process may be considered to take place in two general steps. Gross fragmentation of the intact liquid stream occurs during primary atomization under the influence of aerodynamic, viscous, inertial, and tensile forces. The shape and size of the liquid fragments are dependent on propellant injector geometry, propellant flow condition, and ambient conditions. During secondary atomization, the liquid fragments further disintegrate into a drop field that is dependent on local relative flow conditions and on the size and shape of the fragments formed during primary atomization. The general trend is for the drop Weber number $(We_D = \rho_g U_{\text{rel}}^2 d_D/\sigma)$ to move toward unity in a drop field in a quasi-equilibrium state.

Understanding the relative order of effect of stresses that can disrupt the intact liquid provides insight into the physics of atomization. Following Ranz,[5] order of magnitude estimates of stresses active in the atomization process of a like doublet fuel element in the F-1 engine[2] are presented in Table 1. For this injector, the

Table 1 Forces on a liquid element using F-1 qual injector as case study

Viscous forces		
Ambient fluid, N/m^2	$\mu_g V/L$	0.04
Liquid boundry layer, N/m^2	$\mu_l V/L$	3.6
Initial forces		
Ambient fluid, N/m^2	$\rho_g V^2$	2.0
Liquid, N/m^2	$\rho_l V^2$	62
Cohesive forces, N/m^2	σ/d	0.0002

fuel was RP-1, orifice diameter d_o was 7.1 mm, the injection velocity U_j was 17 m/s, and the total impingement angle 2θ was 30 deg. Relevant length scales L to estimate interfacial shear effects are based on the boundary layer depth δ; the calculation of L was based on a turbulent boundary layer with a nondimensional thickness δ/d_o that grows as $0.16Re_x^{-1/7}$ where x scales on the orifice diameter d_o. This leads to a boundary-layer depth estimate of $\delta \sim 0.06d_o$, or $L \sim 0.4$ mm for both liquid and gas. To estimate ambient fluid inertia, a maximum relative velocity is used with the assumption that a quiescent atmosphere surrounds the jets, thus, $V \sim 17$ m/s. The velocity related to liquid inertial forces is the relative velocity of the two jets; thus, here the relative velocity is 8.8 m/s. The liquid's cohesive forces were estimated with d based on an injection diameter of 7.1 mm and σ being equal to the surface tension of RP-1 in air, 1.5 mg/s^2. Only when length scales become extremely small do surface tension effects become important. Based on a local relative velocity of 17 m/s, drop Weber numbers of order one do not occur until drops become submicron size. A measured drop distribution in the dilute spray region of this type of injector would be centered somewhere greater than 100 μ. It is seen from Table 1 that liquid inertial forces are an order of magnitude higher than the next higher force, the aerodynamic stress. Therefore, it can be expected that the condition of the jet before impingement is very important in determining atomization characteristics and that jet dynamics will play the major role in the breakup of the liquid sheet.

B. Previous Experimental Studies

Most previous studies of impinging jet atomizers have emphasized cold-flow patternator measurements of the mixing characteristics[6–10] using water as a simulant, and drop size measurements using optical techniques either using water as a simulant again[9,10] or by using hot-wax flow and measurement techniques.[11] The results of these studies should be used carefully, since cavitation in the orifice[12] may have occurred in many tests, especially where the jet velocity was high or where a high back pressure was not established. A good summary of the atomization correlations published up until 1984 can be found in a report by Ferrenberg et al.[13]

George[14] used holography to measure monomethyl hydrazine (MMH) drops in a combusting nitrogen tetroxide/MMH environment at a chamber pressure of 156 psia. These measurements were compared with atmospheric cold-flow measurements using the same injectors and using water as a simulant. George[14] found that the data compared very poorly with Ingebo's correlation[15] for drop size from impinging jet atomizers. George[14] developed correlations for cold-flow and combusting condition drop size as well as a correlation for the change in drop size as a function of distance from the impingement point for the combusting condition. A companion study was conducted at the Air Force Aeronautics Laboratory where injector elements with the same geometry as George's were used in a full-scale rocket test firing.[16] It was likely, however, that in George's cold-flow tests the water jets were cavitating because of the high jet velocities and atmospheric back pressure used in the tests. Other visual studies of impinging jets under combustion flow conditions include holographic studies[17] and high-speed videographic studies.[18–20]

Heidmann et al.[21] undertook an extensive study of impinging jets to characterize the effects of orifice diameter d_o, jet velocity U_j, impingement angle 2θ, preimpingement length l_j, and liquid properties on the structure of the resulting spray. Liquid velocities between 4 and 30 m/s were studied for orifice diameters of 0.64, 1.02, and 1.45 mm (0.025, 0.040, and 0.057 in.) with L/d_o ratios of 80, 50, and 35, respectively. Impingement angles in the range $30 < 2\theta < 100$ deg were studied while preimpingement length was varied between 6 and 60 orifice diameters. Viscosity and surface tension effects were studied by using various glycerol solutions, varsol, and water.

From these studies, Heidmann et al.[21] determined that the liquid jet velocity and impingement angle were critical parameters in determining spray structure and also identified four spray regimes. For low-velocity jets ($U_j < 4$ m/s), a closed rim regime was observed in which a sheet is initially formed that subsequently contracts into a cylindrical stream. At higher velocities ($4 < U_j < 6$ m/s), a periodic drop regime was identified in which waves appeared on the surface of the sheet and regularly circumferentially spaced drops were shed from the periphery of the main sheet. At higher velocities ($6 < U_j < 9$ m/s), in the open rim regime, waves were seen on the surface of the sheet and the leading edge of the sheet fragmented into ligaments and drops. Both open rim and periodic drop patterns were seen at jet velocities between 5 and 10 m/s with periodic phenomena more pronounced for higher viscosity liquids. Above 10 m/s and to the highest velocities tested (~30 m/s), spray patterns were characterized as being in the fully developed regime, where periodic waves of drops projecting from the point of impingement were apparent. Heidmann et al. reported that sheet breakup length x_b increased with jet velocity over the periodic drop and closed rim regimes, and then approached a constant value in the open rim regime; they were not able to measure breakup length in the fully developed regime with any certainty.

Heidmann et al.[21] recognized that the observed periodicity would be an obvious mechanism of combustion instability if it occurred with the appropriate time scale and if some coherence could be observed between the temporal characteristics of droplet formation and the unsteady gas motions in the chamber. In their studies of the spray formed by two impinging liquid jets, Heidmann et al.[21] noted that the periodic shedding was dependent on injection velocity, impingement angle, and injection orifice diameter. Their study emphasized measuring the "wave frequency" of drops and ligaments and used a pulse meter to measure the rate at which a light beam directed normal to the plane of the sheet was interrupted. Heidmann et al.[21] reported that this wave frequency was linearly proportional to resultant sheet velocity (which was incorrectly defined as $U_j \cos\theta$) and that orifice diameter and preimpingement length had minor effects on the observed wave frequency. It is important to note that the occurrence of a well-defined single characteristic frequency reported by Heidmann et al.[21] could be determined only over a limited distance with respect to the impingement point and varied in location as operating conditions were changed.

Dombrowski and Hooper[4] studied the disintegration of the sheet formed by turbulent and laminar impinging water jets in air at velocities from about 7 to 20 m/s. Based on the results of Heidmann et al.[21] with respect to the effects of jet diameter and preimpingement length, they chose to study a single orifice diameter (0.5 mm) and preimpingement length (2 mm). A tube length of 200 mm ($L/d_o = 400$) was used to guarantee fully developed velocity profiles at the exit of the tube. The impingement angle was varied from 50 to 140 deg. They concluded that one of three mechanisms was operative in the disintegration of the sheet, primarily depending on the relative jet inertia and whether the jets were turbulent or laminar. For laminar jets at low velocities ($U_j < 10$ m/s) and low-impingement angles, where large drops were shed off a stationary rim, the leading edge of the sheet was in a state of equilibrium between inertial and surface tension forces. Above 10 m/s for both laminar and turbulent jets, waves formed locally at the point of liquid jet impact. For laminar jets, Dombrowski and Hooper[4] argued that these "impact waves" were associated with the high-velocity jet core, and that for laminar jets with velocities between 10 and 20 m/s, the impact waves were damped by the lower velocity fluid from the laminar jet periphery, which also comprised the peripheral regions of the sheet. Measured reductions in drop size were reasoned to be due to small disturbances introduced by the impact waves that reduced the thickness of the sheet's rim. Furthermore, they reported that at laminar jet velocities approaching 20 m/s, impact waves caused early breakdown of the sheet downstream of the impingement point, whereas waves amplified by aerodynamic stresses controlled the breakdown of the rest of the sheet. At high velocities and large impingement angles, impact waves predominated over the entire sheet. Dombrowski and Hooper[4] reported that in the equilibrium state, x_b increased with increasing jet velocity at a constant impingement angle, but for sheet breakdown controlled by either impact waves or aerodynamic forces, x_b decreased with increases in jet velocity at any given impingement angle.

Dombrowski and Hooper[4] reported significantly different behavior for turbulent jets. They attributed these differences to the difference in velocity profile between laminar jets and turbulent jets. For turbulent jets, impact wave effects were seen at velocities as low as 7 m/s for large impingement angles. Dombrowski and Hooper[4] attributed this specifically to the absence of a low-velocity region in peripheral regions of the sheet to damp the impact waves. Drop size decreased with increasing jet velocity and increasing impingement angle. The transition to impact wave-controlled breakup for the turbulent jet case was determined to occur for relative jet Weber numbers, $\rho_l U_j^2 d_j \sin^2\theta/\sigma$, greater than 100.

Huang[22] examined the breakup of axisymmetric liquid sheets formed by the impingement of two directly opposed water jets that covered jet Weber number regimes from 100 to 30,000. Huang[22] used a thin orifice that apparently produced slug flow jets and, thus, were laminar. Two dissimilar breakup regimes connected by a transition regime were documented. The first breakup regime, $100 < We_j < 500$, was characterized by stable liquid sheets, increasing breakup length with increasing jet velocity, and a semicylindrical liquid edge. Huang[22] measured the diameter of drops shed off the outer edge of the sheet from photographs and showed that they were comparable to drop diameters calculated according to Rayleigh jet instability theory using the thickness of the semicylinder at the edge for jet diameter. Huang[22] analyzed this regime by balancing outwardly directed inertial force and inwardly directed surface tension force, and derived a semiempirical relation for breakup radius r_b,

$$r_b = 0.167 d_j We_j / 2 \tag{1}$$

The transition regime, $500 < We_j < 2{,}000$, included the critical zone where r_b was a maximum. Past this point, r_b decreased with increasing jet velocity. Sinuous waves on the liquid sheet were first apparent in this regime. The second breakup regime, $We_j > 2{,}000$, was characterized by a predominance of antisymmetric waves and steadily decreasing r_b with increasing jet Weber number. Huang[22] analyzed breakup behavior in this regime using a model of a vibrating membrane and derived a semiempirical relation for r_b,

$$r_b = 14.2 d_j s^{-2/3} We_j^{-1/3} \tag{2}$$

C. Theoretical Models of Impinging Jet Atomization

Taylor[23] developed a model for low Weber number impinging jets that accurately predicts the overall geometry of the formed sheet and made detailed measurements of sheet thickness and lateral spread of the fan formed by jets colliding at impingement angles 2θ of 60, 90, and 120 deg, and at jet velocities up to 5.6 m/s to verify his model. His measurements indicated that the sheet thickness h at any radial location r on the sheet was of the form $rh = F(\theta, \phi)$, which is independent of U_s. Measured values of F indicated a symmetric distribution about its peak value at $\phi = 0$ deg. Taylor[23] noted that the geometry of the sheet formed by the impinging jets was related to the presence of stationary antisymmetric waves within a limiting radius. The limiting radius was defined as the radius where the Weber number based on local sheet thickness is unity. Earlier analysis by Taylor[24] showed that antisymmetric waves move at a speed of $(2/We_s)^{1/2} U_s$. Taylor[23] correctly noted that the developed sheet speed and the jet speed are the same. The position of antisymmetric waves on the sheet is defined by

$$\sin^2 \Psi = \frac{2\sigma}{\rho_l U_s^2 h} = \frac{2\rho r}{\rho_l U_s^2 F(\theta, \phi)} \tag{3}$$

where Ψ is the angle between the wave front and the radial vector from the impingement point. For directly opposed jets, $2\theta = 180$ deg, the formed sheet is circular and the curves satisfying Eq. (3) are cardioid waves.[23]

Taylor[23] reasoned that the lateral spread of the sheet is due to the high pressure generated around an effective stagnation point and is related to the destruction of the momentum of the impinging jets directed toward each other. Taylor calculated the force required to induce the lateral spread,

$$\rho_l U_j^2 \int_0^{\pi} F(\theta, \phi) \sin\phi \, d\phi \tag{4}$$

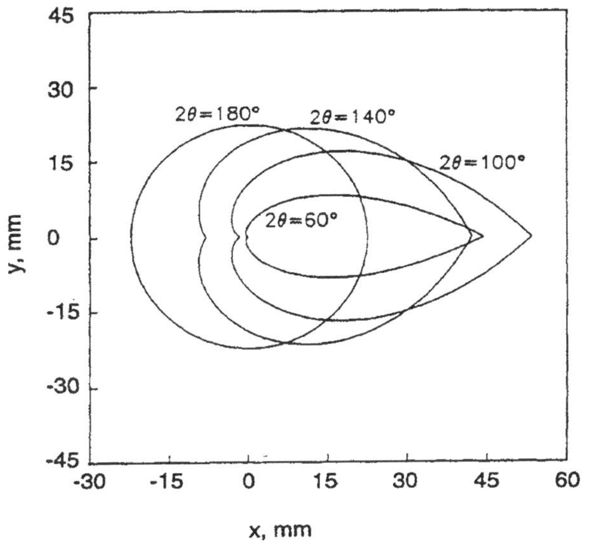

Fig. 2 Sheet shape predictions for water in air from the Taylor[23] model plotted as a function of impingement angle, 2θ with orifice diameter 0.51 mm and jet velocity 7.1 m/s.

and compared it with the force of impingement

$$\rho_l U_j^2 \sin\theta \int_0^{\pi} F(\theta, \phi)\, \mathrm{d}\phi \tag{5}$$

and found that the ratio of the force required for lateral spread to the impingement force was a constant independent of impingement angle equal to about 0.65. For the case of directly opposed jets, the ratio is $2/\pi$, or 0.64. Results from Taylor's model for sheet geometry are shown in Fig. 2. Taylor's results were used by Hasson and Peck[25] to verify an analytically derived solution of the sheet thickness distribution:

$$h = \frac{d_j^2 \sin^3\theta}{4r(1 - \cos\phi\cos\theta)^2} \tag{6}$$

Although it has been shown that Taylor's model for the formation of fan sheets by impinging jets and the resulting sheet shape is quite accurate for low Weber number conditions,[26] at higher Weber number conditions ($We_j > 2{,}000$), where the flow condition can be either laminar or turbulent, there has been much less success in developing an accurate and comprehensive model. This is particularly true for the case where the impinging liquid fans are turbulent. Because the atomization process starts with the primary breakup of the fan sheet formed by the impinging liquid jets, attention has focused mostly on single liquid–gas interfaces, constant thickness liquid sheets, or two-dimensional liquid sheets. Theoretical studies of liquid sheet atomization include numerical simulations,[27] vortex sheet discretization analyses of Kelvin–Helmholtz instabilities,[28,29] and linear stability theory-based analyses.[30–36]

The observed periodic nature of the breakup of both jets and sheets and the well-developed basis of linear stability theory has led to its extensive use in modeling primary atomization for both cases. Linear stability theory is based on the linear growth of small disturbances that can grow due to a variety of sources.[37] For conditions typical of a practical atomizer, the driving force behind the growth of the disturbances on the liquid surface is usually taken to be the aerodynamic stress, which tends to be greater in trough regions and less in crest regions.[38]

In the two-dimensional analysis of liquid sheets, the displacement on the surface of the liquid sheet, η, is modeled as a periodic function of position and time:

$$\eta = A\exp\{i(kx - \beta t)\} \tag{7}$$

In the temporal theory, waves on the liquid surface with initial small amplitude η_0 will grow exponentially according to

$$\eta/\eta_0 = e^{\beta_i t} \tag{8}$$

where β_i is the imaginary part of the complex growth rate factor β. Generally, the gas–liquid interface is regarded as inviscid, and η is calculated by solving the potential flow problem. In practice, β_i is calculated for a spectrum of wave numbers, and the sheet is assumed to take the shape of the fastest growing wave. Wave growth can occur through amplification of either sinuous (antisymmetric) or dilatational (symmetric) waves. Analyses have indicated that for high-speed sheets, sinuous waves always grow faster than dilatational waves.[33]

Squire[33] used this approach to analyze a constant thickness sheet. For high-sheet Weber numbers We_s and for disturbance wavelengths larger than the thickness of the sheet, he derived an equation for the fastest growing disturbance wavelength, λ_{max},

$$\lambda_{max} = \frac{2\pi}{k_{max}} = \frac{4\pi\sigma}{\rho_g U_s^2} \tag{9}$$

and λ_{max} had a displacement growth rate coefficient of

$$\beta_i = \frac{\rho_g U_s^2}{\sqrt{2\rho_l \sigma h}} \tag{10}$$

Thus, from Squire's analysis, the wavelength of the fastest growing disturbance is dependent only on the relative velocity between the gas and liquid for a given fluid combination, whereas β_i is dependent on relative velocity and sheet thickness. Squire[33] compared his analytical predictions of the fastest growing wavelength with apparent wavelengths measured from photographs of thin liquid films and found generally good agreement.

Taylor[23] also compared Squire's analysis to experimental results he obtained on the behavior of liquid sheets formed by impinging jets. He[23] approximated the three-dimensional sheet by considering only the center portion, $\phi = 0$, and calculated β_i from his measurements of the thickness distribution along that axis. He compared the predicted fastest growing wavelength to the wavelengths he observed through flash photography of the sheet and reported agreement to within 20%. This comparison was made for a sheet velocity of 5 m/s and was not validated over a range of velocities.

Dombrowski and Johns[34] used linear temporal stability theory to analyze the aerodynamic stress-induced amplification of surface perturbations on a thinning, two-dimensional, viscous liquid sheet moving with a velocity U_s relative to the ambient medium. The general dispersion equation they derived in quadratic form is

$$\beta_{i,nd}^2 + \frac{(kh)^2}{Re_s}\beta_{i,nd} + 2s(kh)^2\left(\frac{1}{sWe_s} - \frac{1}{kh}\right) = 0 \qquad (11)$$

where

$$\beta_{i,nd} = \frac{\beta_i h}{U_s}, \qquad Re_s = \frac{\rho_l U_s h}{\mu_l}, \qquad We_s = \frac{\rho_l U_s^2 h}{\sigma}$$

The first term in Eq. (11) is associated with momentum in the displacement direction, the second term rises from viscous forces, and the first and second parts of the third term account for surface tension forces and aerodynamic forces, respectively.

It may be reasoned that spatial stability theory, i.e., where disturbances grow in space, should be used for these cases rather than the temporal theory, for which it is generally simpler to develop solutions. Chuech[32] compared results from temporal and spatial analyses of disturbances growing on a two-dimensional liquid sheet by using an inertial frame of reference and determined that, for large-sheet Weber numbers, $We_s > 100$, the temporal and spatial theories give essentially the same results. At lower sheet Weber numbers, discrepancies start to appear between the two models.

The linear stability theory in itself is not able to predict when the sheet breaks up. Empirical observations of sheet breakup have led to the use of a generally accepted breakup constant to use in conjunction with linear stability theory to predict when breakup occurs[39]:

$$\int_0^{t_b} \beta_{i,m}\, dt = \int_0^{x_b} \frac{\beta_{i,m}}{U_s}\, dx = 12 \qquad (12)$$

Using this approach, Dombrowski and Johns[34] envisioned the sheet to fragment into ligaments at crest and trough points. The ligaments then contract into cylinders with diameter d_L given as

$$d_L = \sqrt{4h/k} \qquad (13)$$

A second-order relation for the breakup of sheets formed by fan and swirl nozzles was derived by Clark and Dombrowski[40] where the combined growth of sinuous and dilatational modes were considered. Their analysis indicated that the sheet would break at points corresponding to three-eighths and seven-eighths of the fundamental sinuous wavelength.

Further atomization into spherical drops is modeled as a surface tension-controlled mechanism,[34] where dilatational waves grow on the surface of the cylindrical ligament and one drop is formed per wavelength. For water in air,

$$d_D = \left(\frac{3\pi}{\sqrt{2}}\right)^{1/3} d_L \left\{1 + \frac{3\mu}{(\rho_l \sigma d_L)^{1/2}}\right\}^{1/6} \approx 1.89 d_L \qquad (14)$$

Anderson et al.[3] presented numerically computed solutions to Eq. (11) for the case of orifice diameter equal to 0.64 mm, impingement angle varying from 20 to 100 deg, and jet velocities ranging from 10 to 25 m/s. Following Taylor,[23] Anderson et al.[3] computed growth rates for wave numbers from 0 to 10 mm^{-1} at the centerline of the liquid sheet, $\phi = 0$ deg, where the sheet is thickest. Sheet thickness on the jet axis was calculated using Eq. (6). Temporal growth rates were integrated, and sheet breakup was considered to occur at the first point where the value of the integral in Eq. (12) exceeded 12. The fastest growing wavelength λ_{max} was also computed in this way. Drop size was calculated using Eq. (14).

Ryan et al.[26] compared the analysis for viscous, thinning sheets developed by Dombrowski and Johns[34] with Squire's analysis[33] of invisicid constant thickness sheets for water jets injected into atmospheric pressure air, which comprises the experimental conditions most commonly employed. Ryan et al.[26] showed that under these conditions Squire's simpler expression for the most unstable wave number, Eq. (9), results in less than 1% deviation from the most unstable wave number calculated using the dispersion equation of Dombrowski and Johns,[34] Eq. (11). Under these conditions of large sheet Weber numbers and low-pressure ratio s, Ryan et al.[26] showed that the growth rate factor of the most unstable wave number $\beta_{i,\max}$, computed using Squire's expression Eq. (10), is essentially the same as that obtained from using Eq. (11). Using the inviscid expressions of Squire[33] for the fastest growing disturbance and growth rate factor [Eqs. (9) and (10)], and Eq. (6) for sheet shape, Ryan et al.[26] identified a nondimensional scaling parameter based on the jet Weber number, $We_j = \rho_j U_j^2 d_o/\sigma$, and the half-impingement angle θ; $We_j(1-\cos\theta)^2/\sin^3\theta$. Use of this parameter collapses the theoretical dependence of x_b, λ_{max}, and d_D on orifice diameter, impingement angle, and velocity into a single curve only dependent on the scaling parameter. Results from the model are shown in Fig. 3, where nondimensional lengths, normalized by the orifice diameter d_o, are plotted against the geometrically scaled jet Weber number.

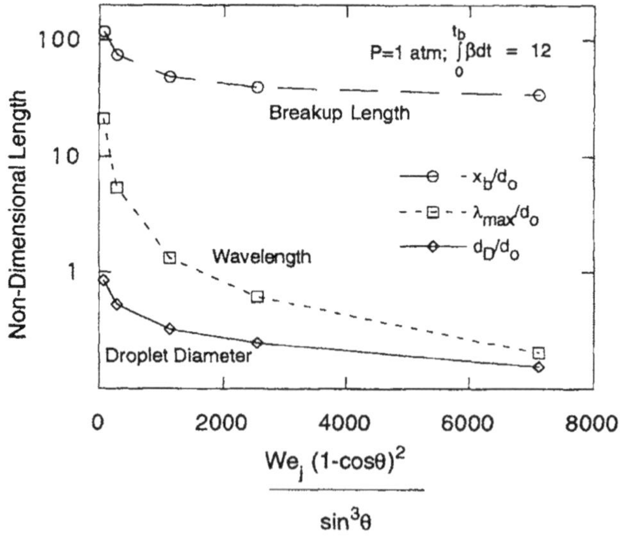

Fig. 3 Nondimensional lengths as a function of geometrically scaled jet Weber number as predicted by aerodynamic instability model.

The persistent use of the aerodynamic instability theory in modeling the atomization of impinging liquid jets may be related to the theory's ability to provide predictions of drop size that fairly match experimental measurements of drop size. Ibrahim and Przekwas[36] compared predicted drop-size using an aerodynamic instability theory with empirical correlations for drop size generated by impinging liquid jets, in which the mass median drop size was approximately proportional to the square root of the orifice diameter and approximately inversely proportional to the inverse of the sheet velocity, which can be shown to be similar to the analytical development for drop size shown earlier in Eqs. (9), (11), and (13). Direct comparisons between results of the linear theory and measured drop size have also been made that show reasonable agreement.[3,26]

The reduction in breakup length x_b with increasing sheet velocity also agrees with empirical data on liquid sheets formed by swirl or fan nozzles, which tend to show a $U_s^{-2/3}$ dependency[41] to U_s^{-2} dependency.[42] The velocity dependence of the breakup length at $We_j > 500$ predicted by linear stability-based theory, however, does not agree with the results reported by Heidmann et al.,[21] Huang,[22] or more current results obtained by Ryan et al.[26] The inability of the theory to represent the observed trends with respect to breakup length is consistent with the results of Dombrowski and Hooper[4] who suggested that impact wave effects and not aerodynamic force effects controlled sheet breakup for high velocity, turbulent impinging jet conditions. Dombrowski and Hooper[4] did argue, however, that breakup length should decrease with increasing jet velocity for both aerodynamic instability and impact wave mechanisms, although their measurements presented did not support this contention.

III. Experimental Measurements

A. Description of Experiments

A series of experiments have recently been undertaken to clarify the mechanism of impinging jet atomization.[26] The impinging jet injector apparatus was similar to that used by Heidmann et al.[21] and Dombrowski and Hooper.[4] Precision bore glass tubes were used as injectors because they allow visual inspection of the liquid within the tube and ensure that surface roughness is not a factor. Water flow through the glass tubes was driven by a nitrogen pressurized water tank. Individual rotameters were used to monitor the flow rate, and pressure gauges were used to measure the pressure drop across each glass tube. The impinging jet apparatus allowed for substitution of different diameter and length glass tubes as well as variation of impingement angle and preimpingement length and was based on optical mounts so that the tubes could be positioned such that the jets impinged precisely.

The impinging jets studied were formed with either 0.64-mm or 1.45-mm inner diameter glass tubes. In either case, the length of the glass tubes was 51 mm, which corresponds to L/d_o ratios of 80 and 35, respectively. A shorter L/d_o of 5 was also tested for the smaller inner diameter (0.64-mm) tube. The L/d_o ratios tested here, with the exception of the $L/d_o = 5$ case, are an order of magnitude greater than those used in actual rockets. The glass tubes used in the current set of experiments were attached to 6-mm outer diameter stainless-steel tubes with Swagelok fittings. The ratio of the inlet diameter to the glass tube diameter was approximately five, and the glass tube inlet orifice had either a sharp edge or a contour to allow the flow to remain laminar into a Reynolds number regime that typically produced turbulent flow.

The impinging jets were studied by image analysis and drop size/velocity measurements over a wide range of parameters. The 0.64-mm inner diameter ($L/d_o = 80$) tube at an impingement angle of 60 deg and a preimpingement length of 25 mm was used as the base case. The impingement angles (2θ) tested ranged from 40 to 100 deg in intervals of 20 deg. The preimpingement lengths were 2.5, 25, and 38 mm. At each of these conditions, the impinging jet sprays were studied at discrete velocities ranging from roughly 5 to 25 m/s.

Studies of high-velocity jets issuing from short L/d_o injectors must consider the potential for cavitation within the injector. Although no visual effects of cavitation were observed for the majority of the flow conditions tested here, the cavitation number for each flow condition was calculated and compared with Nurick's[12] critical cavitation number. Nurick[12] studied cavitation phenomena in short L/d_o (2–20) sharp-edged orifices with orifice diameters ranging from 1.52 mm to 3.18 mm. The cavitation number is defined as the ratio of the upstream pressure minus the vapor pressure of the fluid to the pressure drop across the orifice. Nurick[12] observed that for an L/d_o of 20 ($d_o = 3.18$ mm), cavitation occurred for cavitation numbers below 1.45. For the present case of an L/d_o of 80 ($D = 0.64$ mm), the cavitation number was between 1.1 and 1.4, and no gross cavitation effects were observed. Note that Nurick did not present data for L/d_o ratios longer than 20 but did observe that the critical cavitation number for the onset of cavitation does decrease for increasing L/d_o ratios.[12] Cavitation was observed for the $L/d_o = 5, d_o = 0.64$-mm glass tube at high velocities ($U_j > 25$ m/s). At this case, the cavitation number was 1.3 and was lower than the critical cavitation number of 1.8 observed by Nurick[12] for an L/d_o, $d_o = 3.18$ mm.

Drop size and velocity measurements were made using an argon-ion laser ($\lambda = 514.5$ nm) based fast Fourier transform (FFT) version of a phase-Doppler particle analyzer (PDPA) manufactured by Aerometrics, Inc. The theory behind the operation of the PDPA has been discussed extensively in the literature,[43] thus, only a few details concerning the system will be given here. For the present experiments, the collection optics module was oriented 30 deg off axis from the forward propagation direction of the laser beam, which is the best angle for measuring transparent drops.[43] The transmitting and collection optics defined a drop diameter measurement range between 40 and 1400 μm, the maximum range capable with the present system.

Drop size/drop velocity measurements at all parametric conditions were made at two locations downstream of the impingement point, 16 and 41 mm. At each downstream measurement location, drop size/drop velocity measurements were made at 6.4-mm intervals in the plane of the sheet along the line normal to the central axis of the sheet. Lefebvre[44] suggests that at least 5500 data points are necessary for $\pm 5\%$ accuracy in mean diameter measurements; consequently, 8000 drops were typically taken for the calculation of the various mean diameters.

A charge injection device (CID) solid-state camera with a resolution of 512×512 pixels was used to obtain digitized images of the spray field resulting from the two impinging jets. Instantaneous images of the flowfield were obtained using a strobe light with a time duration of 10 μs oriented to provide diffuse back lighting and high contrast. At each flow condition 17 images were recorded, and the length and width of the intact liquid sheets were measured from the digitized spray images. Standard deviations were calculated for each set of length and width measurements and are reported with the results. Following Heidmann et al.,[21] the length of the sheet was taken from the impingement point to the point where the sheet broke into drops, ligaments, or sheet vestiges. The width was defined as the maximum width of the complete sheet. It is noted that some photographs had cavity-like structures on the periphery of the sheet and that the width was measured upstream of this cavity. It is not clear how Heidmann et al.[21] defined the width of the sheet and, therefore, some ambiguity in definition probably exists between their sheet width results and the current results.

The three limiting cases of fully developed laminar jet flow, fully developed turbulent jet flow, and "plug" jet flow were studied. These cases were chosen because of their well-characterized velocity and turbulence intensity profiles. Absolute plug flow conditions were not obtained but were approached by using orifices with the short length-to-diameter ratio ($L/d_o = 5$) The short L/d_o orifices did not produce nonturbulent flow as a thin orifice would, and the short orifices were operated either in the laminar or turbulent flow regime. In addition to changing flow condition and L/d_o, other test parameters included impingement angle 2θ, orifice diameter d_o, and freejet length prior to impingement l_j. Measurements of breakup length, drop size and drop velocity distribution, and periodic wave structures are discussed subsequently.

B. Effect of Flow Condition

The flow condition of the jet before impingement, i.e., whether it was laminar or turbulent, had the major effect on atomization. To illustrate the importance of the initial conditions of the liquid jet, consider the instantaneous images of the sheets formed by laminar and turbulent jets impinging at full impingement angle 2θ of 60 deg shown in Fig. 4 taken under quiescent conditions and at atmospheric pressure. The resultant sheets have distinct characteristics. In the laminar case, Fig. 4a, small but prominent ripples on the surface of the sheet are seen just downstream of the impingement point, and after some distance the sheet suddenly disintegrates into droplets. Small and regular waves can be seen in the outer periphery of the liquid sheet. In many of the images of laminar impinging jets, incipient breakup occurred at midspan of the sheet, downstream of the impact waves. Drops are also seen shedding off the leading edge of the sheet. All of these observations are consistent with those made by Heidmann et al.[21] and Dombrowski and Hooper.[4]

In the turbulent case, Fig 4b, the amplitude of the waves that appear at impact seem much larger than in the laminar case. (Note that the scale in Fig 4b is about three times greater than that of Fig 4a.) Downstream, the sheet disintegrates into ligaments, with incipient breakup consistently occurring at the sheet periphery, where the sheet thickness is least. Periodicity is indicated by the regular spacing between disturbances on the liquid sheet and by the regular spacing between the detached ligaments. Examination of the detached liquid structures reveals irregular liquid shapes that contract into roughly cylindrical ligaments. The cylindrical ligaments appear to contract further into irregularly shaped drops that will eventually take a nearly spherical shape. From these images, it is reasonable to model the ligament-to-droplet formation process as being controlled by surface tension-driven instabilities, as given by Eq. (14).

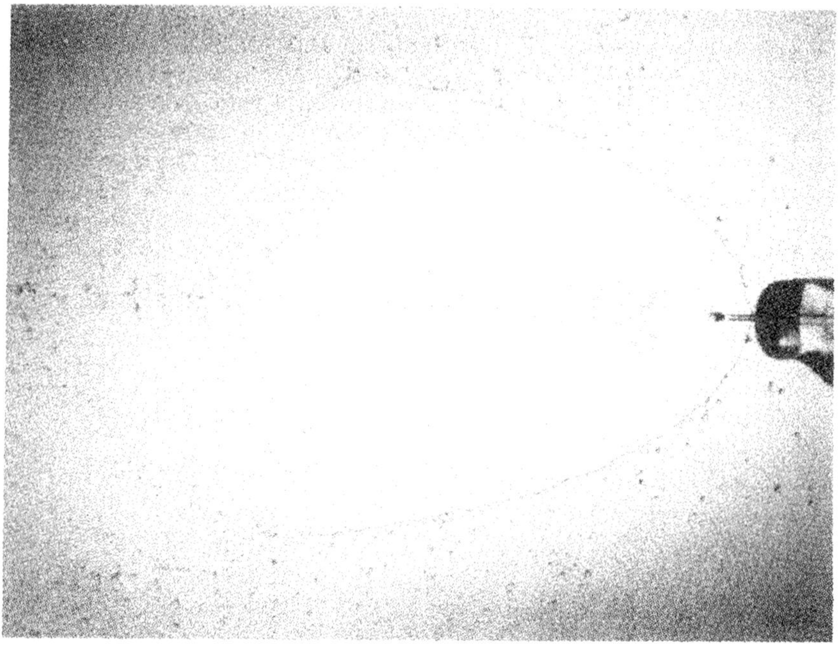

a)

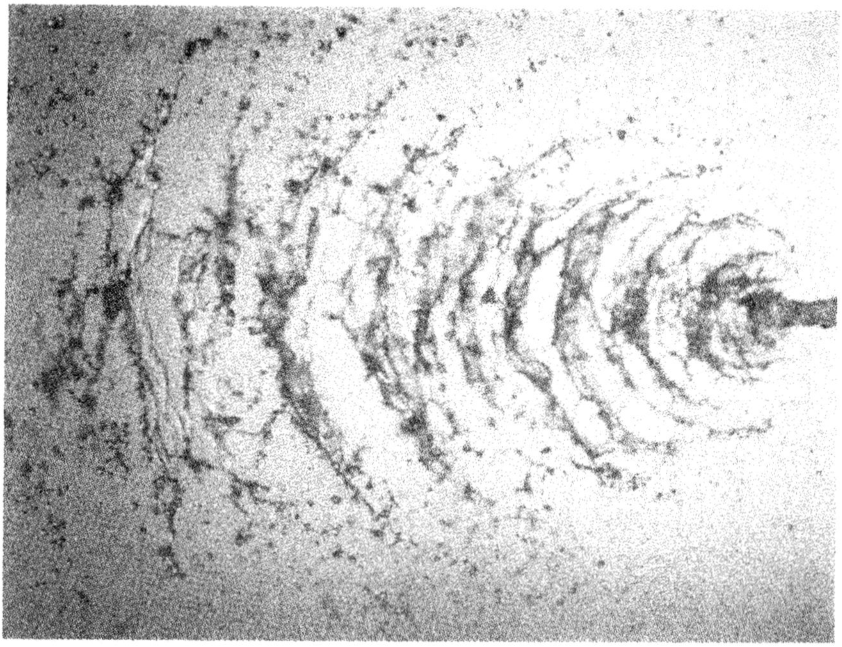

b)

Fig. 4 Instantaneous images of sprays formed by two water jets: a) resultant sheet formed by laminar jets from 0.51-mm inner diameter, L/d_o = 375 precision bore glass tubes, jet velocity 14.8 m/s and jet Reynolds number 7500 and b) resultant sheet formed by turbulent jets from 1.45-mm inner diameter, L/d_o = 80 precision bore glass tubes, jet velocity 12.2 m/s and jet Reynolds number 12,400.

C. Breakup Length

Length measurements of the intact sheet and of periodic structures were made from images such as those shown in Fig. 4. Nondimensional breakup length of sheet formed by impinging water jets from precision bore glass tubes, x_b/d_o, is plotted against jet Weber number in Fig. 5 for laminar and turbulent impinging jet cases. Clear differences between the two cases can be seen. The nondimensional breakup length of the turbulent impinging jets appears to have some dependence on impingement angle but appears to be relatively independent of jet Weber number. There is a strong dependence on jet Weber number for the laminar impinging jet case, and the general dependence of x_b on jet Weber number is consistent with that reported by Huang.[22] It is also interesting that the breakup length dependence on impingement angle for turbulent impinging jets is opposite that for laminar impinging jets.

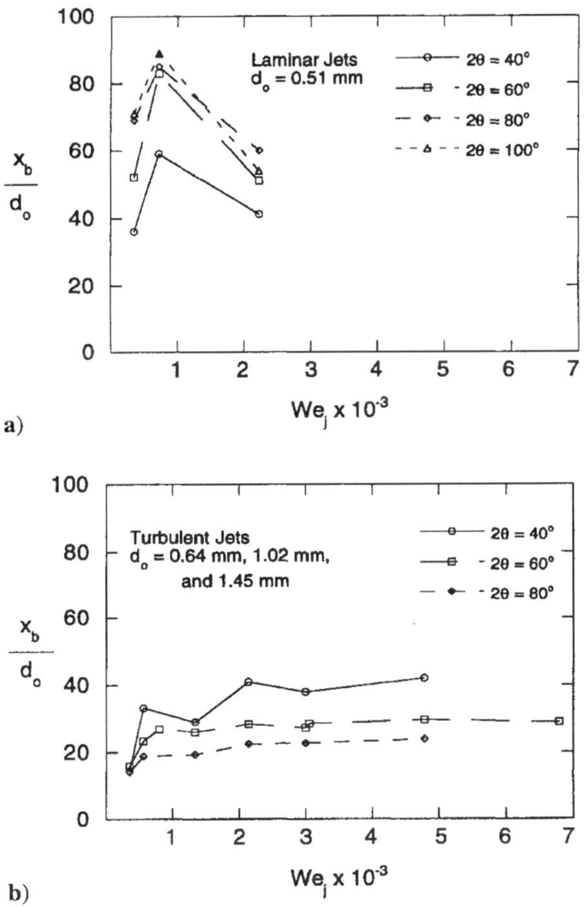

Fig. 5 Nondimensional breakup length of sheet plotted against jet Weber number We_j with impingement angle 2θ as a parameter: a) jets prior to impingement were laminar and b) jets prior to impingement were turbulent.

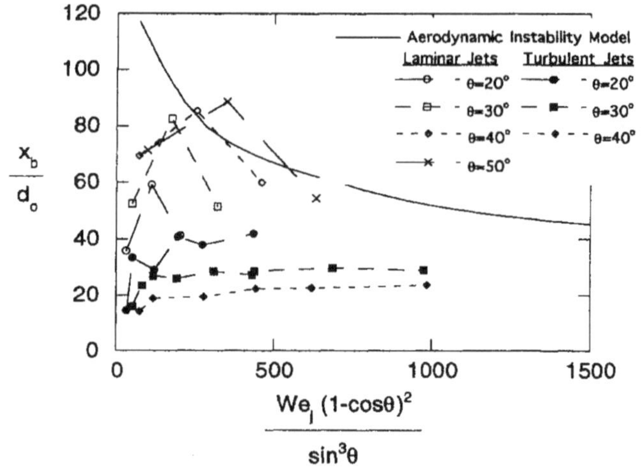

Fig. 6 Nondimensional breakup length of sheet formed by impinging water jets as a function of geometrically scaled jet Weber number; measurements are compared with results from aerodynamic instability model.

Comparisons of the measured nondimensional breakup length with predictions from the aerodynamic instability model used by Ryan et al.[26] are shown in Fig. 6 Again the differences between the turbulent impinging jet case and the laminar impinging jet case are evident. Laminar impinging jet breakup may be modeled relatively well by a combination of Taylor's model[23] at low jet Weber number and by the aerodynamic instability model at higher jet Weber number, but again the turbulent impinging jet breakup shows little dependence on the geometrically scaled jet Weber number, and the conclusion may be made that the aerodynamic instability model is not accounting for an important physical process.

D. Drop Size Distribution

Measurements of nondimensional arithmetic mean drop diameters are presented in Fig. 7 for the turbulent impinging jet case. The arithmetic mean drop diameter is used instead of the Sauter mean diameter because the few large drops present in the distribution drastically skews the Sauter mean diameter distribution toward large sizes. A comparison of measured mean drop size with the monodisperse drop size predicted by the aerodynamic instability model is also shown. It should be noted that the measurements using the PDPA instrument yield a mean value for the drop size distribution, whereas the theory effectively assumes that a monodispersed spray is formed. The model predictions match the experimental trend quite well and, as the model predicts, the measurements nearly collapse to a single curve when plotted against the geometrically scaled jet Weber number. An empirical correlation for normalized drop size as a function of the geometrically scaled jet Weber number $We_j f(\theta)$ was obtained from the experimental data shown in Fig. 7 with a correlation coefficient of 0.964,

$$\frac{d_D}{d_o} = 2.217 \cdot \{We_j f(\theta)\}^{-0.354} \tag{15}$$

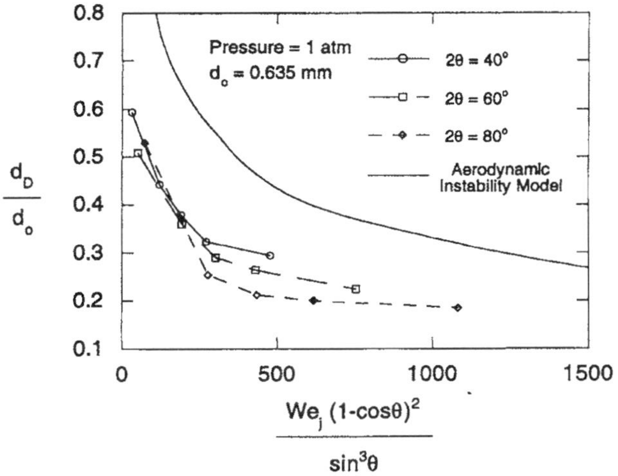

Fig. 7 Nondimensional drop diameter in spray formed by impinging water jets from precision bore glass tubes as a function of geometrically scaled jet Weber number; measurements are compared with results from aerodynamic instability model.

Drop size measurements were also obtained away from the centerline of the spray to assess the effect of the variation of the sheet thickness on drop size. The variation in the sheet thickness h with ϕ is represented in Eq. (6) through the cos ϕ dependence. The results for drop size measured by PDPA shown in Fig. 8 for impinging water jets injected into air have, therefore, been plotted as a function of ϕ for two orifice diameters, the 18.5 m/s jet velocity condition, various impingement angles, and two different downstream locations from the impingement point, $x = 16$ mm and $x = 41$ mm. As these results illustrate, for the two lower impingement angles, the measured arithmetic mean drop size decreases slightly as the absolute value of ϕ increases. This trend qualitatively agrees with results of Couto and Bastos–Netto,[45] who applied Eq. (6) to derive the drop size distribution dependence on ϕ. At the two highest impingement angles, however, no discernable decrease in D_{10} is observed with changes in ϕ, although analysis of Couto and Bastos–Netto[45] indicated that, for $\phi = 45$ deg, approximately a 30% reduction in drop size should be observed at an impingement angle of $2\theta = 90$ deg. These results suggest either that the expression for sheet thickness given by Eq. (6) is incorrect, particularly for higher impingement angles, or that the assumption that the sheet breaks up at a constant radial distance from the impingement point is incorrect.

PDPA measured mean drop velocities along the sheet centerline taken at $x =$ 16 mm and $x = 41$ mm downstream of the impingement point are shown in Fig. 9 for the $d_o = 1.45$-mm and $d_o = 0.64$-mm diam jets from precision bore glass tubes and for a variety of impingement angles. From the drop velocity measurements, it is seen that the sheet regains nearly all of the dynamic head of the jet before impingement, although a slight reduction in drop velocity with increasing impingement angle can be seen, which may be due to more deceleration resulting from drag on the smaller droplets that are formed at higher impingement angles and/or to less efficient dynamic pressure recovery at higher impingement angles.

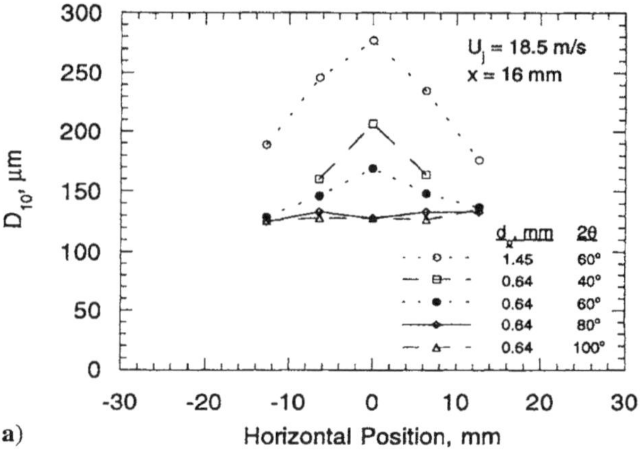

a)

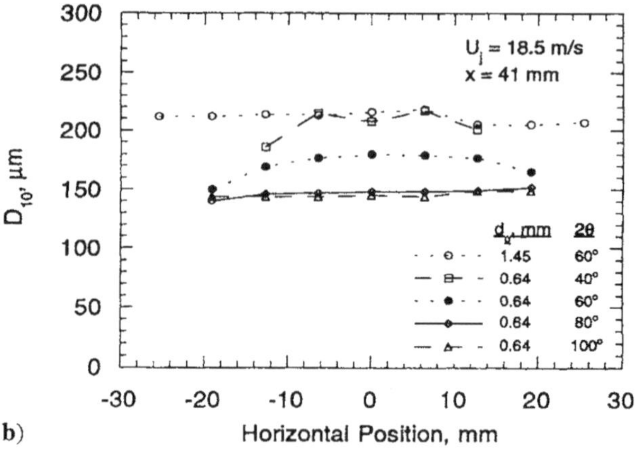

b)

Fig. 8 Arithmetic mean geometric drop size D_{10}, plotted against horizontal position away from jet axis plane (ϕ = 0) with impingement angle 2θ as a parameter: a) axial measurement location 16 mm downstream of impingement point and b) axial measurement location 41 mm downstream of impingement point.

E. Atomization Frequency

The measured distance normalized by orifice diameter of the separation between periodic structures for turbulent impinging jets is shown in Fig. 10 as a function of jet velocity. The wavelength appears to be primarily dependent on orifice diameter and independent of jet velocity. The observation of independence from jet velocity is contrary to the aerodynamic instability model predictions. The distance between detached ligaments is approximately constant at about four orifice diameters, and the distance between surface wave structures is approximately constant at about two orifice diameters. There was a rather large spread in the data, with a standard deviation of about ±35%.

The wavelength data was converted to atomization frequency simply by dividing the drop velocity (which was approximated as the initial jet velocity) by the measured wavelength. Atomization frequency data is presented in Fig. 11 based on ligament spacing. Also shown are wave frequency data from experiments of Heidmann

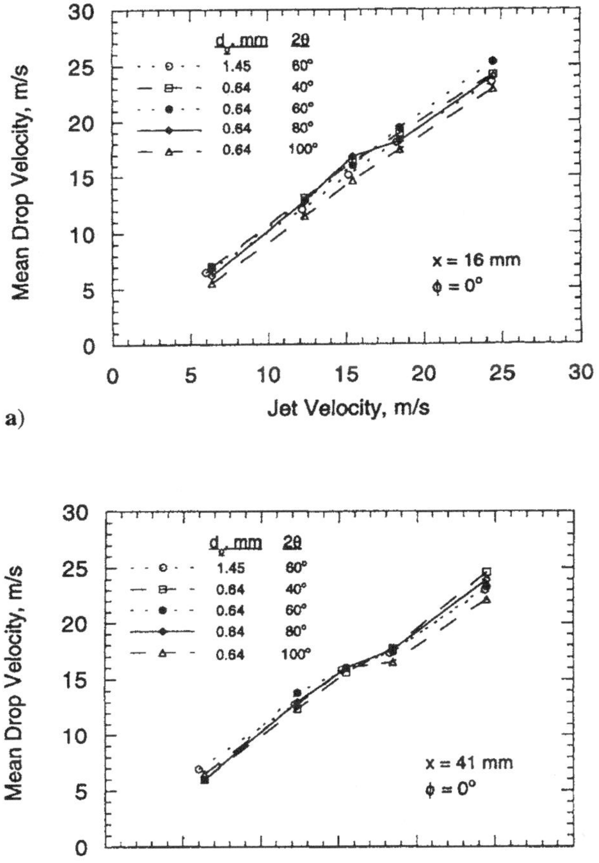

a)

b)

Fig. 9 Mean water drop velocity plotted as a function of impinging jet velocity U_j, along sheet axial plane ($\phi = 0$) with impingement angle 2θ as a parameter: a) axial measurement location 16 mm downstream of impingement point and b) axial measurement location 41 mm downstream of impingement point.

et al.[21] These atomization frequencies are quite similar to the maximum possible combustion instability frequency given by the Hewitt stability correlation for impinging jet atomizers[46] also shown in Fig. 11. The similar dependencies of maximum possible instability frequency and atomization frequency on d_o/U_j is significant in terms of primary atomization being a key process in combustion instability.

F. Orifice Length-to-Diameter Ratio and Impingement Length Effects

The effects of varying L/d_o on PDPA measured mean drop size and sheet shape for impinging water jets injected into air are shown in Figs. 12 and 13, respectively.

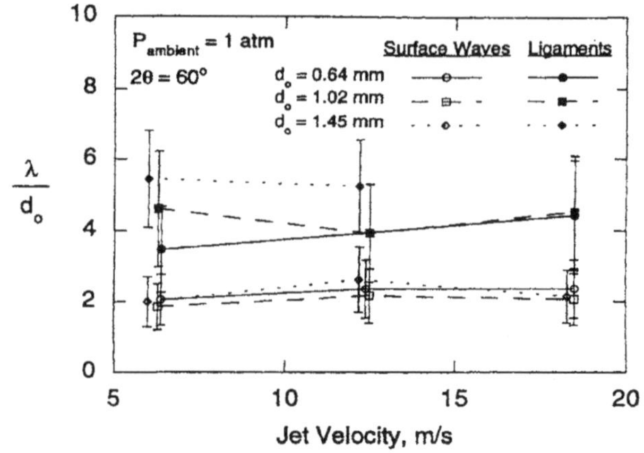

Fig. 10 Nondimensional averaged measurements of distance between adjacent disturbances on the sheet formed by impinging liquid jets and distance between consecutive ligaments detached from the liquid sheet with d_o as a parameter, impingement angle 2θ was 60 deg.

Figure 12 shows some effect of orifice L/d_o on mean measured drop size on the sheet centerline at a measurement point 16 mm downstream of the impingement point which seems to disappear at a measurement point 41 mm downstream of the impingement point. Orifice diameter of precision bore glass tubes was 0.64 mm and impingement angle 2θ was 60 deg. Figure 13 shows the effect of L/d_o on breakup length and sheet width of jets impinging from long L/d_o precision-bored glass tubes with orifice diameter 0.64 mm and short L/d_o drilled orifices at various injection velocities. Impingment length was 25 mm and impingement angle 2θ was 60 deg. No significant effects of orifice L/d_o on sheet shape are noticeable, which is in general agreement with other studies[47] on the effects of orifice L/d_o on mixing efficiency of two impinging jets.

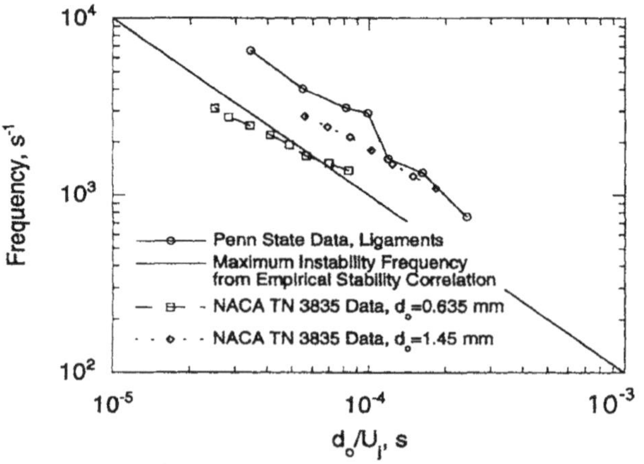

Fig. 11 Comparison of averaged measured atomization frequency with maximum combustion instability frequency as predicted by Hewitt stability correlation.[46]

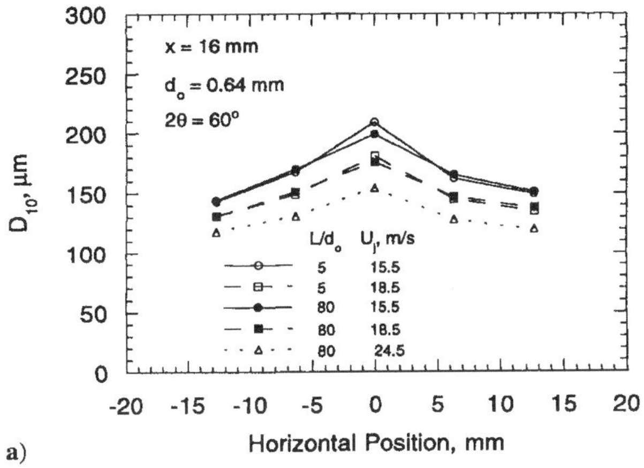

a)

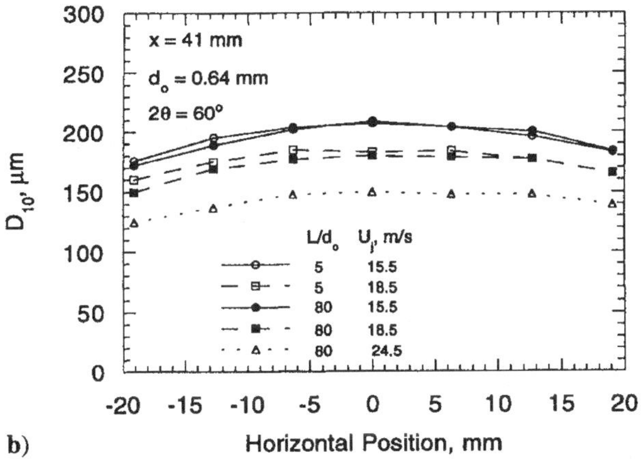

b)

Fig. 12 Arithmetic mean geometric drop size, D_{10}, plotted against horizontal position away from jet axis plane ($\phi = 0$) with orifice L/d_o and jet velocity as parameters: a) axial measurement location 16 mm downstream of impingement point and b) axial measurement location 41 mm downstream of impingement point.

Variations in the preimpingement length of turbulent jets were also studied in terms of its effects on sheet shape at breakup and drop size. As the impingement length of the jet increases, the effects of velocity profile relaxation and of any misalignment will become greater, as will the effects of any growing disturbances on the jets. Preimpingement length was varied from 4 to 60 orifice diameters; the highest values of preimpingement length should have allowed the jet velocity profile to become nearly flat. The visual appearance of the spray structure was sensitive to the preimpingement length, particularly in terms of the breakup length. Breakup length and sheet width measurements for the various preimpingement lengths tested are shown in Fig. 14. Again, orifice diameter of precision bore glass tubes was 0.64 mm and impingement angle 2θ was 60 deg. The results show that the effect of the preimpingement length decreases with increasing jet velocity. Increased preimpingement length favors longer breakup lengths and slightly narrower liquid sheets. Figure 15 shows drop size PDPA

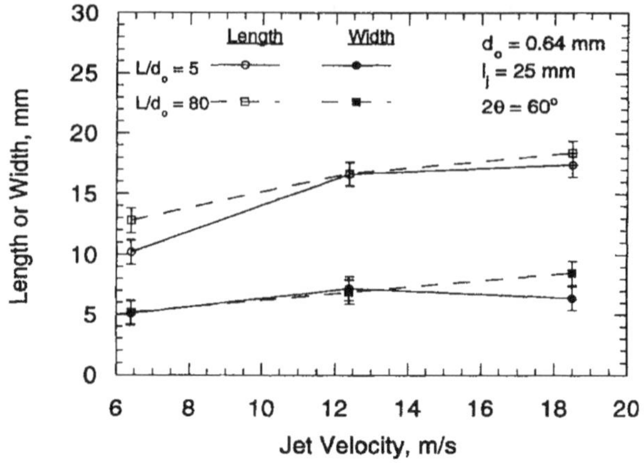

Fig. 13 Average measured liquid sheet length and measured sheet width at breakup plotted against jet velocity with orifice L/d_o as a parameter.

measurements for jet velocity $U_j = 18.5$ m/s and varying impingement lengths l_j as a function of position which indicate a tendency toward slightly smaller drops for longer preimpingement lengths, particularly away from the centerline of the spray. Orifice diameter, orifice L/d_o, and impingement angle are the same as in Fig. 14.

IV. Discussion

A. Practical Injector Systems

Some statements should first be made regarding the operation of practical injector systems in order to assess theoretical and experimental results just presented. As the propellant travels from the tank to its injection point, it travels through

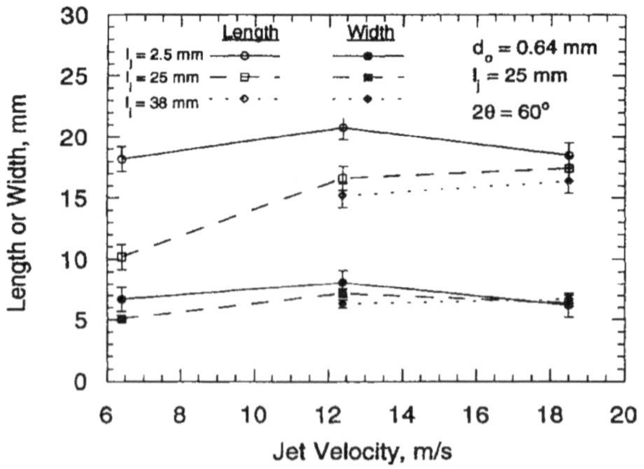

Fig. 14 Average measured liquid sheet length and measured sheet width at breakup plotted against jet velocity with impingement length as a parameter, orifice L/d_o was 80.

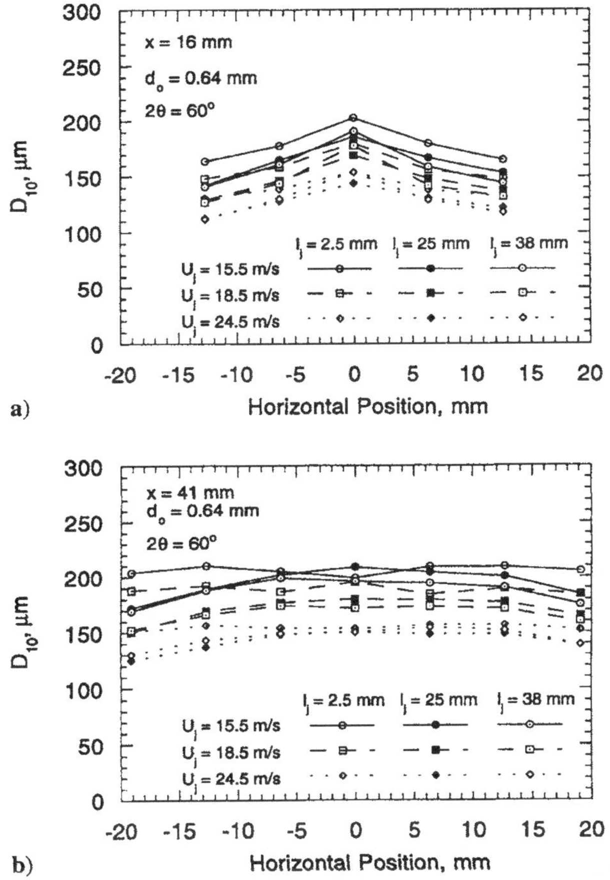

Fig. 15 Arithmetic mean geometric drop size D_{10}, plotted against horizontal position away from jet axis plane ($\phi = 0$) with impingement length and jet velocity as parameter: a) axial measurement location 16 mm downstream of impingement point and b) axial measurement location 41 mm downstream of impingement point.

filters, manifolds, metering orifices, and other flow channels which produce pressure drop and fluid particle rotation. Typically, the jets are analyzed as though they are irrotational; clearly, however, there is a significant amount of vorticity distributed throughout the fluid. Another important observation is the relatively short distances the fluid travels within the orifice and after it exits the orifice on its way to the impingement point. These distances, less than 10 orifice diameters, are insufficient to produce a developed flowfield. Requisite entry length/diameter ratios for developed flows can be estimated for $Re = 10^5$ anywhere from 12 (Ref. 48) to 2800 (Ref. 49). Engineering rule of thumb estimates place entry lengths at $L/d_o = 75$. At any rate, the jet flow at the orifice exit is not developed, and this must be taken into account in theoretical developments as well as in experimental design. The final important point regarding real injector designs is that the entrance to the orifice is usually sharp and that cost considerations related to manufacturing typically preclude rigorous specification of surface roughness or inlet contour. Since the effects of contour and roughness cannot be predicted with any confidence, the design engineer is hesitant to stipulate specific tolerances even though there is circumstantial evidence in real engine combustors that roughness and contour can effect performance and stability. Studies of orifice geometry effects on jet atomization[50,51] indicated that surface roughness, inlet contour, and orifice L/d_o all had a major effect on the atomization characteristics. These results do not appear to be consistent with results from impinging jet injectors just presented. Even though the results presented cover a relatively small range of geometric parameters, and only approached the very short L/d_o and l_j/d_o geometries used in practical injectors, they provide some evidence regarding the fundamental differences between atomization processes of plain jets and impinging jets.

B. Primary Atomization Mechanisms

It is clear from the results of Taylor[23] and Huang,[22] for example, that the breakup mechanism of impinging liquid jets at ambient atmospheric pressure and low jet Weber ($We_j < 500$) conditions is related to a balance between the liquid's inertial force pushing liquid outward and the liquid's cohesive forces that try to maintain the integrity of the sheet. For sheets formed by laminar jets at higher Weber number conditions, atomization may possibly be described by aerodynamic instability theory. For turbulent flow conditions representative of practical injectors, however, the situation is much more complicated. Comparison between experimental results at these conditions and predictions from the theory clearly indicate that there are important physical processes that are not being correctly treated.

The experimental results presented are all consistent with the conclusion by Dombrowski and Hooper[4] that impact waves control the breakup process of impinging liquid jets, although their contention regarding the effects of jet velocity profile on impact wave damping must still be regarded as conjecture. To understand the breakup mechanism, it is necessary to define the source of the impact waves that are reasoned here to be due to fluctuations of pressure and/or momentum at or near the impingement point. It would appear that these fluctuations are either due to disturbances existing within the jet prior to impingement, e.g., turbulence or jet instabilities, or are generated at the impingement point by jet–jet interactions, such as periodic movement about a mean stagnation point, which is a known feature of stagnation point flow.[52]

Although the immediate area of interest is breakup of turbulent jets because practical injector systems produce turbulent jets, a comparison between laminar and turbulent impinging jets may again serve some purpose in clarifying the source of impact waves, or at least help in identifying the appropriate steps that must be taken to resolve the issue. Recall that Dombrowski and Hooper[4] made no distinction between the source of impact waves that appeared on sheets formed either by high-velocity laminar impinging jets or by turbulent impinging jets but concluded that their appearance was due to velocity profile differences, i.e., the slow moving fluid near the outer radius of the laminar jet damped impact waves, whereas the turbulent jets with their "slug flow" profile provided less damping. More evidence regarding these effects could have been obtained by testing impinging laminar jets with flat distribution profiles acquired by using either thin orifices or long preimpingement lengths.

Results shown in Fig. 5 show the very significant differences in breakup length between laminar impinging jets and turbulent impinging jets. To more directly address the issue of impact wave formation, measurements of the distance between disturbances on the intact sheet and between adjacent detached ligaments were made for the laminar impinging jet case and compared to the data shown in Fig. 10 for turbulent impinging jets. Consistent measurements of wavelength were attainable only for laminar impinging jets with injection velocities greater than about 15 m/s. For laminar impinging jets, where $U_j = 17.9$ m/s, $d_o = 0.51$ mm, and $2\theta = 60$ deg, the measured mean normalized wavelengths λ/d_o on the intact sheet and between adjacent ligaments were 4.0 ± 0.3 and 5.5 ± 0.5, respectively. These numbers compare well with the detached ligament normalized mean wavelength measured for turbulent impinging jets, which ranged between 3.5 and 5.5. Normalized mean surface wavelengths were on the order of two for turbulent impinging jets.

To evaluate the possibility that disturbances existing on the preimpingement jets are sources of impact waves, it is useful to consider the notion that the disturbances leading to the generation of impact waves and subsequent breakup of impinging jets are esentially the same disturbances that lead to breakup in single jets. An excellent review in this area is provided by McCarthy and Molloy,[53] with some more recent work specifically oriented to the highly dynamic operating conditions of atomization provided by Reitz and Bracco[51] and Wu et al.[54] Reitz and Bracco[51] considered the effects of aerodynamically induced liquid surface disturbance growth, instabilities due to velocity profile relaxation, cavitation, and turbulence as potential mechanisms that control the breakup process of single jets. Reitz and Bracco[51] concluded that the aerodynamic instability model was capable of predicting many of the trends they observed, and they developed a physical model of the breakup process whereby disturbances on the surface of the liquid jet, which are present when the jet emerges from the orifice due to events occurring upstream of the orifice exit, are amplified selectively by aerodynamic forces. The view of single jet atomization that Reitz and Bracco[51] provided was that as the amplitude of these periodic disturbances grew to the same order of magnitude as their wavelength, a drop would be ejected off the liquid surface. The view of Wu et al.[54] of primary atomization was based on turbulence in the liquid, i.e., that large eddies with sufficient energy to penetrate through the liquid surface were the source of drops. In light of the observed similarity between the disturbances on sheets formed by the impinging high-speed laminar jets and impinging turbulent jets, there should be some doubt that turbulence plays the dominant role in the generation of impact waves. The comparison, however, was made only at one condition and should be made over a significant range of conditions before any definite statement related to the role turbulence plays in impact wave generation can be made.

To evaluate the possibility of aerodynamically induced jet instability as a contributing factor, the characteristics of growing disturbances on the freejet surface were calculated. Both long waves ($kd_j < 2$) and short waves ($kd_j > 2$) can grow on the jet.[38] The wavelength of the fastest growing short wave is dependent on relative velocity,

$$\lambda_{\max} = 3\pi\sigma/\rho_g U_j^2 \tag{16}$$

with a growth rate coefficient of

$$\beta_i = 0.4\rho_g U_j^3 \sqrt{s}/\sigma \tag{17}$$

The equation relating the growth rate of long waves on the jet was also derived by Levich[38]

$$\beta_i \sqrt{\frac{\sigma k^2}{2\rho_l r_j}\left(1 - s^2 - k^2 r_j^2\right) - \frac{\rho_g k^4 r_j^2 U_j^2 [\ell n(k r_j/2)]}{2\rho_l}} \tag{18}$$

In Eq. (18) s indicates the jet instability mode, where $s = 0$ for symmetrical long waves and $s = 1$ for helical long waves. For a 2-mm diam water jet in air traveling at 10 m/s, a plot of Eq. (18) reveals that the fastest growing symmetrical long wave would have a wavelength 3 times the jet diameter, and the fastest growing helical long wave would have a wavelength 4.2 times the jet diameter.

For sufficiently high velocities where the term containing U_j is much greater than the surface tension term, it can be shown[38] that the length of the fastest growing long wave (regardless of the instability mode) is dependent only on jet diameter,

$$\lambda_{\max} \approx 2d_j \tag{19}$$

and the wave has a growth rate coefficient of

$$\beta_i = U_j \sqrt{s}/d_j \tag{20}$$

The ratio of Eq. (17) to Eq. (20), which is the relative growth rate of short waves to long waves for high velocity jets, is essentially the gas Weber number, $We_g = \rho_g U^2 d_j/\sigma$. For the case of a 0.5-mm-diam water jet injected into air, $We_g = 1$ occurs at a velocity of about 4 m/s. Assuming a phase speed equal to the jet velocity and an impingement length of five orifice diameters, a disturbance on 0.5-mm-diam jet with an injection velocity of 20 m/s would grow by about 20%. It should be noted that since a long wave contains much more mass than does a short wave, a long wave disturbance produces a larger impact wave. It is also interesting to note that the measured detached ligament wavelengths shown in Fig. 10 for turbulent impinging jets and the surface disturbance and detached ligament wavelengths already described for the laminar impinging jet case are both similar to the long wavelength predicted by the Levich[38] theory, i.e., disturbance wavelengths around 2–4 times the jet diameter. Although the linear stability analysis for disturbance growth on jets is strictly applicable for laminar flows, stability theory has also been successfully applied to the growth of wavelike structures in turbulent flows.[55,56]

C. Relation of Primary Atomization to Combustion Instability

Given that the breakup periodicity of impinging liquid jets is quite similar to high-frequency combustion instability as shown in Fig. 11, the detailed mechanism of how the process couples with pressure oscillations to modify atomization frequency and generate the requisite interelement coherence across the chamber is still missing. In Levine's remarkable photographic study of LOX/ethanol combustion using impinging jet atomizers, he[18] reported large-scale liquid fan and liquid jet shattering during an actual combustion instability event. Most published investigations of this phenomena used high-amplitude acoustic sources to provide an acoustic field. Longitudinal waves in chambers have been shown to induce periodic jet shortening and droplet formation with frequencies coupled to the sound source.[57,58] Experiments by Miesse[57] with localized high-intensity sound fields oriented transverse to laminar liquid jets ($10 < L/d_o < 40$) indicated the acoustic field caused the droplets to disperse in a diverging sinusoidal fashion. Miesse[57] reported that the maximum dispersion of the jet occurred when the frequency of the sound was the natural frequency of the jet, f_j, which he determined empirically to have the approximate form $f_j \sim 0.03U_j/d_o$ over a frequency range of 10–200 Hz. The periodic effects were observed to decrease with increasing acoustic frequency and jet velocity. This frequency dependence on jet velocity and orifice diameter is the same as that shown in Fig. 10, albeit with frequencies about 3–6 times lower than those shown in Fig. 10. Experiments with three parallel liquid jets in the same cylindrical enclosure used for the longitudinal mode oscillations indicated that the jets were relatively unaffected by radial and tangential mode oscillations; this was considered to be at least in part due to the frequency of the transverse modes, which were not reported. Oscillatory jet displacement from the mean impingement point could also lead to an oscillatory inertial force at the impingement point to generate impact waves.

Buffum and Williams[59] studied turbulent jets in an acoustic field and reported the same sort of sinuous behavior observed by Miesse.[57] An analysis of the jet displacement revealed that the drag coefficient on the liquid jet was three times greater than expected based on solid body data. Tests in which the injector unit was mechanically vibrated demonstrated[60] that periodic intermittencies in a liquid stream and periodic droplet formation could be induced at frequencies up to 4500 Hz using a water jet from a 1.2-mm-diam orifice at velocities up to about 30 m/s. This frequency is very much in line with the atomization frequencies shown in Fig. 10. Thus, there is a significant amount of experimental evidence that supports the theory that acoustically or mechanically forced periodicity of the atomization process may be a driving mechanism to sustain combustion pressure oscillations in a rocket engine. It is natural to expect that acoustic waves can also interact with liquid sheet breakup. In the attempts of Heidmann et al.[21] to show an effect, however, no coupling was evident.

V. Summary and Conclusions

In summary, experiments contrasting laminar and turbulent impinging jet conditions clearly demonstrate that the jet conditions have a dramatic effect on the atomization process of impinging liquid jets. Specifically, the measured breakup lengths for laminar impinging jets are longer and displayed different trends as a function of jet velocity and impingement angle than those for turbulent impinging jets. In addition, the measured spray width for the turbulent impinging jet case remained approximately constant as a function of jet velocity and impingement angle for a fixed orifice diameter, unlike the laminar impinging jet case. The present results are in agreement with results of earlier studies at atmospheric pressure by Dombrowski and Hooper,[4] who speculated that these differences arose mainly from differences in velocity profiles between laminar jets and turbulent jets and the way that the impact waves were damped. The important question remains, however, of how the impact waves, which determine the breakup characteristics of turbulent impinging jets, are formed in the first place. It appears likely that the source of the impact waves are large pressure and momentum fluctuations that are either accentuated near the impingement point or are a direct result of the impingement. A quantitative assessment of the specific mechanisms controlling atomization remains to be established.

Comparisons of experimental results at atmospheric pressure with results from a linear stability-based theory indicates the use of the aerodynamic instability model for describing the breakup of a sheet formed by turbulent impinging jets appears to be incorrect, at least for atmospheric conditions. The model predicted a monotonically decreasing sheet breakup length with increasing Weber number and a strong dependence of disturbance wavelength on sheet velocity, neither of which were apparent from experimental measurements. It is interesting, however, that the aerodynamic instability model does produce results that match experimentally measured mean drop size quite well.

Variations in the preimpingement length have a measurable influence on the breakup length and drop size. Although these effects are small, they can be associated with the state of the liquid jet prior to impingement. Application of linear stability theory to single liquid jets indicated that long wavelength disturbances are similar to measured disturbances on sheets formed by impinging jets in terms of their magnitude and linear dependence on jet diameter. However, the magnitude of these variations appear to be insufficient to explain the observed difference in breakup length. Measurement of surface wave and periodic ligament formation phenomena for turbulent impinging jets indicates that the observed wavelengths are directly proportional to the orifice diameter and independent of jet velocity and impingement angle. The relationship between the periodic wave and ligament structures observed in these studies needs to be considered in light of their potential to initiate and sustain combustion instability phenomena associated with impinging jet rocket engine injectors. At any rate, combustion instability models should include the effects of the time-dependent nature of atomization, as well as the effects of a realistic distribution of drop sizes.

Based on the results of the studies presented in this chapter, approaches to modeling impinging jet atomization should focus on preimpingement jet conditions and the physics near the jet impingement point, specifically the source of the observed impact waves that appear to control turbulent impinging jet breakup. No current models address these phenomena. Experimentation at higher ambient pressures must be performed to better evaluate aerodynamic effects under conditions comparable to rocket engine environments. More quantitative theories based on the impact wave mechanism are needed for comparison with experimental results.

References

[1]Harrje, D. T., and Reardon, F. H. (eds.), *Liquid Propellant Rocket Combustion Instability*, NASA SP-194, Washington, DC, 1972.

[2]Anon., "Liquid Rocket Engine Injectors," NASA SP-8089, 1976.

[3]Anderson, W. E., Ryan, H. M., Pal. S, and Santoro, R. J., "Fundamental Studies of Impinging Liquid Jets," AIAA 30th Aerospace Sciences Meeting, AIAA Paper 92-0458, Reno, NV, Jan. 6–9, 1992.

[4]Dombrowski, N., and Hooper, P. C., "A Study of the Sprays Formed by Impinging Jets in Laminar and Turbulent Flow," *Journal of Fluid Mechanics*, Vol. 18, Pt. 3, 1964, pp. 392–400.

[5]Ranz, W. E., "Some Experiments on Orifice Sprays," *Canadian Journal of Chemical Engineering*, Aug. 1958, pp. 175–181.

[6]Riebling, R. W., "Effect of Orifice Length-to-Diameter Ratio on Mixing in the Spray from a Pair of Unlike Impinging Liquid Jets," *Journal of Spacecraft and Rockets*, Vol. 7, No. 7, 1970, pp. 894–896.

[7]Rupe, J. H., "The Liquid-Phase Mixing of a Pair of Impinging Streams," Jet Propulsion Lab., Progress Rept. 20-195, California Inst. of Technology, Pasadena, CA, Aug. 1953.

[8]Hoehn, F. W., Rupe, J. H., and Sotter, J. G., "Liquid-Phase Mixing of Bipropellant Doublets," Jet Propulsion Lab., NASA Tech. Rept. 32-1546, California Inst. of Technology, Pasadena, CA, Feb. 1972.

[9]Vassallo, P., Ashgriz, N., and Boorady, F. A., "Effect of Flow Rate on the Spray Characteristics of Impinging Water Jets," *Journal of Propulsion and Power,* Vol. 8, No. 5, 1992, pp. 980–986.

[10]Hautman, D. J., "Spray Characterization of Impinging Rocket Injectors," 26th JANNAF Propulsion Meeting, Cleveland, OH, May 1989.

[11]Knight, R. M., and Nurick, W. H., "Correlation of Spray Dropsize Distribution and Injector Variables," Rocketdyne Rept. R-7995, Contract NAS7-726, Jet Propulsion Lab., California Inst. of Technology, Pasadena, CA, Sept. 1969.

[12]Nurick, W. H., "Orifice Cavitation and Its Effects on Spray Mixing," *Journal of Fluids Engineering,* Dec. 1976, pp. 681–687.

[13]Ferrenberg, A., Hunt, K., and Duesberg, J., "Atomization and Mixing Study," NASA Marshall Space Flight Center, Contract NAS8-34504, 1985.

[14]George, D. J., "Rocket Injector Hot Firing and Cold Flow Spray Fields," AIAA/SAE 9th Propulsion Conf., AIAA Paper 73-1192, Las Vegas, NV, Nov. 5–7, 1973.

[15]Ingebo, R. D., "Drop Size Distributions for Impinging Jet Breakup in Airstreams Simulating the Velocity Conditions in Rocket Combusors," NASA TN 4222, 1958.

[16]Abbe, C. J., McLaughlin, C. W., and Weiss, R. R., "Influence of Storable Propellant Liquid Rocket Design Parameters on Combustion Instability," *Journal of Spacecraft and Rockets,* Vol. 5, No. 5, 1968, pp. 584–590.

[17]Wuerker, R. F., and Matthews, B. J., "Producing Holograms of Reacting Sprays in Liquid Propellant Rocket Engines," TRW Rept. 68.4712. 2-017, 1968.

[18]Levine, R. S., "Experimental Status of High Frequency Liquid Rocket Combustion Instability," Tenth Symposium (International) on Symposium, Combustion Inst., Pittsburgh, PA, 1965, pp. 1083–1099.

[19]Berman, K., and Cheney, S. H., "Rocket Motor Instability Studies," *Jet Propulsion,* Vol. 25, No. 10, 1955, pp. 513–518.

[20]Rossman, T. G., "A High-Speed and High-Resolution Photographic Technique for the Observation of Propellants Injected into a Firing Combustion Chamber," Air Force Office of Scientific Research, AFOSR Doc. No. TN59-8, 1959.

[21]Heidmann, M. F., Priem, R. J., and Humphrey, J. C., "A Study of Sprays Formed by Two Impinging Jets," NACA TN 3835, March 1957.

[22]Huang, J. C. P., "The Break-Up of Axisymmetric Liquid Sheets," *Journal of Fluid Mechanics,* Vol. 43, Pt. 2, 1970, pp. 305–319.

[23]Taylor, G. I., "Formation of Thin Flat Sheets of Water," *Proceedings of the Royal Society of London, A,* Vol. 259, 1960, pp. 1–17.

[24]Taylor, G. I., "The Dynamics of Thin Sheets of Fluid II. Waves on Fluid Sheets," *Proceedings of the Royal Society of London, A,* Vol. 253, 1959, pp. 296–312.

[25]Hasson, D., and Peck, R. E., "Thickness Distribution in a Sheet Formed by Impinging Jets," *AIChE Journal,* Sept. 1964, pp. 752–754.

[26]Ryan, H. M., Anderson, W. E., Pal, S., and Santoro, R. J., "Atomization Characteristics of Impinging Liquid Jets," *Journal of Propulsion and Power,* Vol. 11, No. 1, 1995, pp. 135–145.

[27]Childs, R. E., and Mansour, N. N., "Simulation of Fundamental Atomization Mechanisms in Fuel Sprays," *Journal of Propulsion and Power,* Vol. 5, No. 6, 1989, pp. 641–649.

[28]Rangel, R., and Hess, C., "Nonlinear Spatial Instability of a Fluid Sheet," AIAA 28th Aerospace Sciences Meeting, AIAA Paper 90-0118, Reno, NV, Jan. 8–11, 1990.

[29]Rangel, R. H., and Sirignano, W. A., "Nonlinear Growth of Kelvin–Helmholtz Instability: Effect of Surface Tension and Density Ratio," *Physics of Fluids,* Vol. 31, No. 7, 1988, pp. 1845–1855.

[30]Lin, S. P., Lian, Z. W., and Creighton, B. J., "Absolute and Convective Instability of a Liquid Sheet," *Journal of Fluid Mechanics,* Vol. 220, 1990, pp. 673–689.

[31]Li, X., and Tankin, R. S., "On the Temporal Instability of a Two-Dimensional Viscous Liquid Sheet," *Journal of Fluid Mechanics,* Vol. 226, 1991, pp. 425–443.

[32]Chuech, S. G., "A Comparative Study of Temporal and Spatial Instabilities of a Two-Dimensional Liquid Sheet," *Heat and Mass Transfer in Spray Systems,* HTD-Vol. 187, American Society of Mechanical Engineers, New York, 1991, pp. 19–25.

[33]Squire, H. B., "Investigation of the Instability of a Moving Liquid Film," *British Journal of Applied Physics,* Vol. 4, 1953, pp. 167–169.

[34]Dombrowski, N., and Johns, W. R., "The Aerodynamic Instability and Disintegration of Viscous Liquid Sheets," *Chemical Engineering Science,* Vol. 18, 1963, pp. 203–214.

[35]Meyer, J., and Weihs, D., "Capillary Instability of an Annular Liquid Jet," *Journal of Fluid Mechanics,* Vol. 179, 1987, pp. 531–545.

[36]Ibrahim, E. A., and Przekwas, A. J., "Impinging Jets Atomization," *Physics of Fluids A,* Vol. 3, No. 12, 1991, pp. 2981–2987.

[37]Drazin, P. G., and Reid, W. H., *Hydrodynamic Stability,* Cambridge Univ. Press, New York, 1981.

[38]Levich, V. G., *Physicochemical Hydrodynamics,* Prentice-Hall, Englewood Cliffs, NJ, 1962, Chap. 11.

[39]Dombrowski, N., and Hooper, P. C., "The Effect of Ambient Density on Drop Formation in Sprays," *Chemical Engineering Science,* Vol. 17, 1962, pp. 291–305.

[40]Clark, C. J., and Dombrowski, N., "Aerodynamic Instability and Disintegration of Inviscid Liquid Sheets," *Proceedings of the Royal Society of London, A,* Vol. 329, 1972, pp. 467–478.

[41]Briffa, F. E. J., and Dombroski, N., "Entrainment of Air Into a Liquid Spray," *AIChE Journal,* Vol. 12, No. 4, 1966, pp. 708–717.

[42]Ostrach, S., and Koestel, A., "Film Instabilities in Two-Phase Flows," *AIChE Journal,* Vol. 11, No. 2, 1965, pp. 294–303.

[43]Ibrahim, K. M., Werthimer, G. D., and Bachalo, W. D., "Signal Processing Considerations for Laser Doppler and Phase Doppler Considerations," Fifth International Symposium on the Application of Laser Techniques of Fluid Mechanics, Lisbon, Portugal, July 9–12, 1990.

[44]Lefebvre, A. H., *Atomization and Sprays,* Hemisphere, Philadelphia, PA, 1989.

[45]Couto, H. S., and Bastos-Netto, D., "Modeling Drop Size Distribution from Impinging Jets," *Journal of Propulsion,* July–Aug. 1991, pp. 654–656.

[46]Anderson, W. E., Ryan, H. M., and Santoro, R. J., "Combustion Instability Phenomena of Importance to Liquid Bipropellant Rocket Engines," 28th JANNAF Combustion Meeting, San Antonio, TX, Oct. 1991.

[47]Riebling, R. W., "Effect of Orifice Length-to-Diameter Ratio on Mixing in the Spray from a Pair of Unlike Impinging Liquid Jets," *Journal of Spacecraft and Rockets,* Vol. 7, No. 7, 1970, pp. 894–896.

[48]Latzko, H., *Zeitschrift für Angewandte Mathematik und Mechanik,* Vol. 1, 1921, p. 277.

[49]Coulson, J. M., and Richardson, J. F., *Chemical Engineering,* Vol. 1, 1956.

[50]Ohrn, T. R., Senser, D. W., and Lefebvre, A. H., "Geometrical Effects on the Breakup of High Velocity Liquid Jets," *Heat and Mass Transfer in Fires and Combustion Systems*, HTD-Vol. 148, edited by W. L. Grosshandler and H. G. Semerjian, American Society of Mechanical Engineers, New York, 1990, pp. 3–10.

[51]Reitz, R. D., and Bracco, F. V., "Mechanisms of Breakup of Round Liquid Jets," *Encyclopedia of Fluid Mechanics*, Vol. 3, edited by N. P. Cheremisinoff, 1984, pp. 233–249.

[52]Batchelor, G. K., *An Introduction to Fluid Dynamics*, Cambridge Univ. Press, Cambridge, England, UK, 1967.

[53]McCarthy, M. J., and Molloy, N. A., "Review of Stability of Liquid Jets and the Influence of Nozzle Design," *Chemical Engineering Journal*, Vol. 7, 1974, pp. 1–20.

[54]Wu, P.-K., Tseng, L.-K., and Faeth, G. M., "Primary Breakup in Gas/Liquid Mixing Layers for Turbulent Liquids," *Atomization and Sprays*, Vol. 2, 1992, pp. 295–317.

[55]Tam, C. K. W., and Chen, K. C., "A Statistical Model of Turbulence in Two-Dimensional Mixing Layers," *Journal of Fluid Mechanics*, Vol. 92, 1979, pp. 303–326.

[56]Chan, Y. Y., "Wavelike Eddies in a Turbulent Jet," *AIAA Journal*, Vol. 15, No. 7, 1977, pp. 992–1001.

[57]Miesse, C. C., "The Effect of Ambient Pressure Oscillations on the Disintegration and Dispersion of a Liquid Jet," *Jet Propulsion*, Oct. 1955.

[58]Reba, I., and Brosilow, C., "Combustion Instability: Liquid Stream and Droplet Behavior, Part III, The Response of Liquid Jets to Large Amplitude Sonic Oscillations," Wright Air Development Center, Tech. Rept. 59-720, Wright–Patterson AFB, OH, Sept. 1960.

[59]Buffum, F. G., and Williams, F. A., "Response of Turbulent Liquid Jets to Transverse Acoustic Fields," *Proceedings of the 1967 Heat Transfer and Fluid Mechanics Institute*, edited by P. A. Libby, D. B. Olfe, and C. W. Van Atta, 1967, pp. 247–276.

[60]MacCormack, P. D., "A Driving Mechanism for High Frequency Combustion Instability in Liquid Fuel Rocket Engines," *Journal of the Royal Aeronautical Society*, 1964, pp. 633–637.

Chapter 9

Aerodynamic Effects on Primary and Secondary Spray Breakup

P.-K. Wu,* L.-P. Hsiang,† and G. M. Faeth‡
University of Michigan, Ann Arbor, Michigan 48109-2118

Nomenclature

a	= drop acceleration
C_D	= drop drag coefficient
$\bar{C}_D$	= mean drop drag coefficient over breakup period
C_i	= empirical coefficient
d	= drop or injector diameter
d_c	= cross stream drop diameter
$d_{c\max}$	= maximum cross stream drop diameter
$e_{p\ell}$	= ellipticity of drop
Eo	= Eötvös number, $a\rho_f d^2/\sigma$ or $a\vert\rho_f - \rho_g\vert d^2/\sigma$
g	= acceleration of gravity
ℓ	= characteristic eddy size
ℓ_K	= Kolmogorov length scale of turbulence
MMD	= mass median drop diameter
Oh	= Ohnesorge number of a drop, $\mu_f/(\rho_f d_0\sigma)^{1/2}$
Oh_{cr}	= Ohnesorge number of core drop, $\mu_f/(\rho_f d_b\sigma)^{1/2}$
Oh_d	= jet Ohnesorge number, $\mu_f/(\rho_f d\sigma)^{1/2}$
p	= pressure
Re	= Reynolds number, $\rho_g du/\mu_g$
Re_{fd}	= jet Reynolds number, $\bar{u}_0 d/\nu_f$
$Re_{f\Lambda}$	= Reynolds number, $\bar{u}_0\Lambda/\nu_f$
r	= radial distance
SMD	= Sauter mean diameter
SMD_p	= Sauter mean diameter after primary breakup for a merged primary breakup

*Research Fellow, Department of Aerospace Engineering.
†Graduate Student Research Assistant, Department of Aerospace Engineering.
‡Professor, Department of Aerospace Engineering.

t = time
t^* = characteristic breakup time, $d_0(\rho_f/\rho_g)^{1/2}/u_0$
u = streamwise relative velocity
u' = streamwise absolute velocity
u_p = streamwise drop velocity
v = radial velocity
v_ℓ = radial velocity associated with eddy of size ℓ
v_p = radial drop velocity
We = initial Weber number of drop, $\rho_g d_0 u_0^2/\sigma$
We_{cr} = Weber number of core drop, $\rho_g d_b(u_0' - u_b)^2/\sigma$
We_{fd} = jet Weber number, $\rho_f d\bar{u}_0^2/\sigma$
We_{gd} = jet Weber number, $\rho_g d\bar{u}_0^2/\sigma$
x = streamwise distance
δ = liquid boundary-layer thickness
Λ = radial integral length scale
μ = molecular viscosity
ν = kinematic viscosity
ρ = density
σ = surface tension
τ_b = characteristic secondary breakup time, $(\rho_f/\rho_g)^{1/2}SMD/u_0$
τ_R = characteristic Rayleigh breakup time, $(\rho_f SMD^3/\sigma)^{1/2}$

Subscripts and Superscripts

b = property at end of breakup
cr = critical conditions for onset and end of breakup
f = liquid-phase property
g = gas-phase property
i = at point of breakup initiation
o = initial condition
$(\bar{\ })$ = time-averaged mean property
$(\tilde{\ })$ = mass-averaged mean property
$(\bar{\ })'$ = time-averaged rms fluctuating property

I. Introduction

PRIMARY and secondary breakup are important classical multiphase flow problems that have motivated numerous past studies. Primary breakup involves the initial formation of drops and other liquid fragments at the surface of a liquid. Primary breakup is important because it controls the initial dispersion of liquid into the gas phase and, through the strong effect of drop sizes on interphase transport rates, the subsequent mixing properties of sprays. Secondary breakup involves any subsequent breakup of drops or liquid fragments present as dispersed liquids. Secondary breakup is important because drops after primary breakup are intrinsically unstable to secondary breakup, which affects subsequent mixing rates by influencing drop sizes as well. In addition, effects of liquid heating or acceleration of the gas phase can lead to conditions where drops undergo secondary breakup. In spite of their importance, however, current understanding of primary and secondary breakup is limited due to problems of making measurements within dense dispersions of drops. In particular, measurements of primary breakup involve observations near an undulating liquid surface having protruding ligaments, and frequently generating irregular liquid fragments, which prevents the use of methods convenient for spherical drop fields such as Fraunhofer diffraction and phase Doppler anemometry.[1,2] Additionally, secondary breakup involves the individual drop generating a relatively localized drop field that is not convenient to characterize using line-of-sight or single-point measurements. Recent development of pulsed holography techniques for measurements within dense sprays, however, has provided an alternative experimental approach that has removed some of the experimental limitations.[1,2] Thus, the objective of this chapter is to briefly describe what has been learned about primary and secondary breakup using pulsed holography thus far and to highlight aspects of these phenomena that merit more attention in the future.

Harrje and Reardon[3] review early work on primary and secondary breakup as related to the problem of liquid rocket engine combustion and combustion instability. Other more general reviews of aspects of liquid atomization processes are available as well.[4–9] This chapter will focus specifically on the properties of primary and secondary breakup disclosed by pulsed holography measurements that are not treated in these earlier reviews. The studies of primary breakup focused on breakup along the surfaces of round liquid jets in still gases, whereas the studies of secondary breakup were limited to simple shock-wave disturbances. Thus, the findings do not directly address the complexities of most practical atomization processes. In particular, many instances of primary breakup involve thin liquid sheets, interacting with gas flows having a range of relative velocities, which will not be considered here. Furthermore, practical processes of secondary breakup in sprays often involve gradual variations of disturbances imposed on drops, which will not be addressed as well. Nevertheless, primary breakup of relatively deep liquids, along the surface of round liquid jets, and secondary breakup following shock-wave disturbances, are classical liquid breakup problems that provide well-defined experimental conditions which help to simplify the interpretation of measurements. References 3–9 should be consulted for an introduction to the more extensive literature concerning general processes of primary and secondary breakup.

Some aspects of secondary breakup affect primary breakup; in particular, conditions can be encountered where processes of primary and secondary breakup overlap or merge. Thus, secondary breakup is considered before primary breakup in the following. The discussion is brief, see Refs. 8–15 for additional details.

II. Secondary Breakup

A. Introduction

Because of numerous applications, secondary breakup has received significant attention in the past. Early work in the field is reviewed by Harrje and Reardon,[3] Giffen and Muraszew,[4] Clift et al.,[5] and Hinze[16]; therefore, the following discussion will be limited to more recent studies. The definition of the onset of secondary breakup, secondary breakup dynamics, and the outcome of secondary breakup will be considered in turn.

Most earlier work on secondary breakup has at least touched on the definition and conditions for the onset of various breakup regimes. The breakup regime observed at the onset of secondary breakup has been termed bag breakup; it involves deflection of the drop into a thin disk normal to the flow direction, followed by deformation of the center of the disk into a thin balloonlike structure extending in the downstream direction, both of which subsequently divide into drops (see Refs. 16–23 for photographs of all the breakup regimes discussed here). The shear breakup regime is observed at higher relative velocities; it involves deflection of the periphery of the disk in the downstream direction, rather than the center, and stripping of drops from the periphery of the disk. The transition between the bag and shear breakup regimes involves complex breakup processes, with portions of this regime termed parachute breakup, chaotic breakup, bag-jet breakup, transition breakup,[19,24–26] etc.; this regime will be denoted the multimode breakup regime in the following. A complex breakup mechanism also has been observed at very large relative velocities, which is called the catastrophic breakup regime[21,22]; however, this regime is associated with very strong disturbances, such as detonation waves, rather than spray processes and will not be considered here.

Existing observations of secondary breakup have generally involved liquid/gas density ratios greater than 500 and Reynolds numbers greater than 100. For these conditions, Hinze[16] has shown that transitions between breakup regimes are largely functions of the Weber ($We = \rho_g d_0 u_0^2/\sigma$) and Ohnesorge ($Oh = \mu_f/(\rho_g d_0 \sigma)^{1/2}$) numbers, which are measures of the ratios of drop drag and liquid viscous forces to surface tension forces, respectively. He found that progressively larger disturbances, i.e., larger Weber numbers, were required for the onset of breakup as Ohnesorge number increased because viscous forces in the liquid tend to inhibit drop deformation at large Ohnesorge number, which is the first step in the breakup process. In fact, viscous forces essentially suppressed secondary breakup for the available range of Weber number ($We < 40$) for $Oh > 2$ (Ref. 16). Among others, Loparev[25] showed that the properties of the disturbances also affected the onset of breakup, with more slowly applied disturbances requiring higher values of Weber number for breakup at a particular value of Ohnesorge number. Subsequent considerations will be limited to shock-wave disturbances to avoid this complication. Borisov et al.[24] proposed an alternative breakup regime map in terms of Weber and Reynolds numbers ($Re = \rho_g d u/\mu_g$), considering both the bag and shear breakup regimes, which is best suited to conditions where $Oh \ll 1$. Krzeczkowski[19] extended the breakup regime map of Hinze[16] to locate transitions to the bag, bag jet, multimode (also called transition breakup) and shear breakup regimes as a function of Weber and Ohnesorge numbers.

Another aspect of secondary breakup that has been studied is the time required to complete breakup. Liang et al.[27] summarize past measurements of breakup times, including the findings of Simpkins and Bales[28] and Ranger and Nicholls[20] for shear breakup, Reinecke and Mckay[21] and Reinecke and Waldman,[22] for catastrophic breakup—all for shock-wave disturbances at large liquid/gas density ratios and low Ohnesorge number. For these conditions, breakup times could be normalized by a characteristic breakup time, $t^* = d_0(\rho_f/\rho_g)^{1/2}/u_0$, finding that the normalized breakup time was constant within experimental uncertainties over a large range of Weber number ($10 < We < 10^6$) that includes both the shear and catastrophic breakup regimes.

Finally, due to the problems of observing drops after secondary breakup, there is very little information available about the outcome of secondary breakup even though this information is vital for understanding the structure of dense sprays.[9]

An exception is some limited results reported by Gel'fand et al.[18] for the bag breakup regime. A bimodal drop size distribution was observed with small drops resulting from breakup of the bag and large drops associated with breakup of the liquid ring at the base of the bag. However, this information is too limited to provide general guidance about the drop sizes produced by secondary breakup.

The preceding review indicates that there are several gaps in the literature concerning secondary breakup. In particular, conditions for the onset of various breakup regimes have been defined reasonably well by Krzeczkowski,[19] but analogous deformation regimes have not been defined. Results at large Ohnesorge numbers where liquid viscosity effects are important are also limited. Breakup times and drop deformation have been studied as well, however, available information is limited for the bag and transition breakup regimes that are important for drop breakup in dense sprays. Finally, measurement problems have limited study of the outcome of secondary breakup so that virtually no information is available for this critical breakup property. Thus, the objectives of the present investigation were to extend the earlier work to provide measurements of the onset of various deformation and breakup regimes, the evolution of breakup processes, and the resulting drop sizes and velocities after secondary breakup. Phenomenological descriptions of these processes were used to help interpret the data. The measurements involved pulsed shadowgraph photography and holography limited to conditions which are representative of sprays near atmospheric pressure: $\rho_f/\rho_g > 500$ and $Re > 100$. The following discussion is brief, more details can be found in Refs. 10–12.

B. Experimental Methods

1. Apparatus

The measurements of secondary breakup properties were carried out using a shock tube apparatus with the driven section open to the atmosphere.[10–12] The driven section had a rectangular cross section (38 mm wide $\times$ 64 mm high) and a length of 6.7 m with the test location 4 m from the downstream end. This provided test times of 17–21 ms in the uniform flow region behind the incident shock wave. The test location had quartz windows mounted flush with the interior side walls of the shock tube in order to allow observation of drop breakup. Drop breakup was observed in air initially at 98 kPa and 297 $\pm$ 2 K in the driven section of the shock tube with shock Mach numbers in the range 1.08–1.31.

Two different drop generator systems were used for the shock tube experiments. Operation at low and moderate Ohnesorge numbers involved use of a vibrating capillary tube drop generator which generated a stream of drops. Electrostatic deflection of a fraction of the drops out of the stream yielded a drop stream with sufficient separation so that interactions between drops during secondary breakup were eliminated.

Drop generation at large Ohnesorge numbers required a different approach because such drops are very difficult to form. Instead, a low Ohnesorge number drop was formed from a liquid solution consisting of a volatile solvent having a low viscosity and a nonvolatile viscous liquid. This drop then was levitated at the test location until the volatile solvent evaporated away. This yielded a small drop of a highly viscous liquid to provide the desired large Ohnesorge number condition. For work reported here, Dow Corning 200 Fluids were used for the viscous liquids with n-heptane used as the volatile solvent.

2. *Instrumentation*

Drops were observed using pulsed shadowgraph photographs and motion pictures to study the overall dynamics of breakup and single- and double-pulsed holography to study the outcome of breakup. The pulsed shadowgraph photographs were obtained using a xenon lamp having a 1-μs pulse duration. The image was recorded using a Graphlex camera at a 6:1 magnification. These photographs were obtained in a darkened room with a variable time delay between shock passage and operation of the flash lamp so that various portions of the breakup process could be observed.

Motion picture shadowgraphs were obtained using a copper vapor laser having a 30-nm-pulse duration, in conjunction with a 35-mm rotating drum camera. Laser triggering was controlled to yield 5–14 motion picture shadowgraphs with 120–160 μs separation times between images.

The holocamera and reconstruction systems were similar to those described by Ruff et al.[1,2] An off-axis holocamera arrangement was used with a 2–3:1 magnification of the hologram image and a 20-ns-laser pulse time. This arrangement was coupled with reconstruction optics that allowed drop diameters as small as 25 μm to be measured with 5% accuracy and drops as small as 10–15 μm to be observed. Reconstruction of double-pulsed holograms yielded two images of the spray with separation times as short as 1 μs so that drop velocities could be measured.

The diameters of mildly irregular drops were found by measuring their maximum and minimum diameters. It then was assumed that the drop had an ellipsoidal shape; taking the effective diameter to be the diameter of a sphere having the same volume as the ellipsoid. More irregular liquid fragments were sized by finding the cross-sectional area and perimeter of the image and proceeding as before for an ellipsoid having the same properties. The velocity of each drop was found by measuring the distance between the centroid of its two images and dividing by the known time between laser pulses. Results at each test condition were summed over at least three realizations, in order to provide drop size and velocity correlations.

3. *Test Conditions*

Test drops of water, *n*-heptane, mercury, two Dow Corning 200 Fluids, and various glycerol mixtures were studied. Initial drop diameters were in the range 150–1550 μm. Ranges of other variables are as follows: $\rho_f/\rho_g = 580$–12,000, $Oh = 0.006$–560, $We = 0.5$–680, and $Re = 340$–15,760. The Reynolds number range of these measurements is higher than conditions where gas viscosity plays a strong role on drop drag properties, e.g., within this Reynolds number range, the drag coefficient of solid spheres only varies in the range 0.4–0.6 (Refs. 10–12).

C. Presentation of Results on Secondary Breakup

1. *Deformation and Breakup Regimes*

The presentation of secondary breakup results will begin with definition of deformation and breakup regime transitions in order to help organize the remainder of the findings. The deformation and breakup regime map, showing transitions as functions of Weber and Ohnesorge numbers similar to Hinze[16] and Krzeczkowski[19] is illustrated in Fig. 1. Present evaluation of the onset of breakup (the transition to

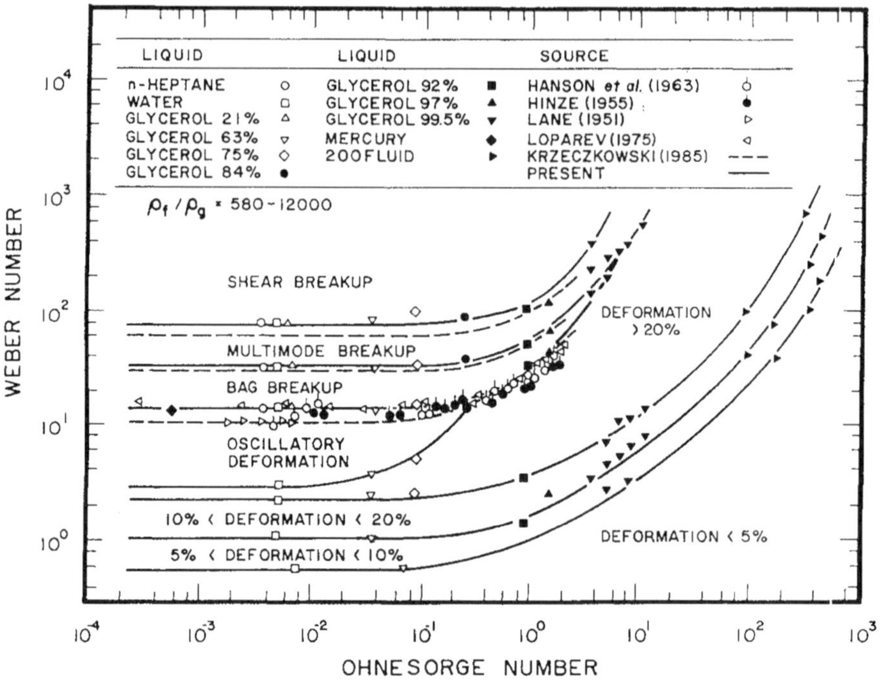

Fig. 1 Drop deformation and breakup regime map for liquid/gas density ratios greater than 500.

the bag breakup regime) is identical to the findings of Hinze[16] and Krzeczkowski[19] within experimental uncertainties. Present results also agree quite well with transitions found by Krzeczkowski[19] to the shear breakup regime and the multimode breakup regime (which Krzeczkowski[19] called the transition breakup regime). In view of the somewhat subjective identifications of breakup regimes and their transitions, this level of agreement is quite satisfying.

Observations of transitions to the various deformation and oscillatory deformation regimes illustrated in Fig. 1 have not been reported before. The definition of transition to the various deformation regimes was taken to be the condition where the drop deformed so that the ratio of its maximum (cross stream) dimension to its initial diameter was within the given range. Following these transitions, there was a range of Weber numbers at each Ohnesorge number where the drop decayed back to a spherical shape much like an overdamped oscillation. For $Oh > 0.4$, this regime ended by the onset of bag breakup, however, for $Oh < 0.4$, there was a range of Weber numbers where the drop oscillated with progressively decaying ratios of maximum to initial diameters before the bag breakup regime was reached; this regime is denoted the oscillatory deformation regime in Fig. 1.

The most striking feature of the flow regime map of Fig. 1 is that progressively higher Weber numbers are needed for the various transitions as Ohnesorge number increases. In fact, for $We < 1000$, breakup is no longer possible for $Oh > 10$ whereas 5% deformation is no longer possible for $Oh > 1000$. Recalling that Ohnesorge number characterizes the ratio between liquid viscous forces and surface tension forces, the inhibition of deformation and breakup at large Ohnesorge numbers clearly is due to increased damping by liquid viscous forces. This increased viscous damping slows the deformation process so that drag forces can reduce relative velocities and the potential for breakup. This high-Ohnesorge number regime is encountered during spray combustion processes at high pressures, where values of surface tension become small but viscosity remains finite as the drop surface nears its thermodynamic critical point. Thus, the findings illustrated in Fig. 1 suggest that drops at these conditions would not necessarily shatter due to small surface tension near the thermodynamic critical point as often thought[9]; instead, they might only deform or even remain spherical. However, additional study of drop breakup processes at high pressure is needed before definitive conclusions about this behavior can be obtained. In particular, specific drop trajectories across the flow regime map depend on atomization and mixing properties of the spray whereas near-critical drop processes involve much lower values of ρ_f/ρ_g than those considered in Fig. 1.

2. *Breakup Times*

The discussion of deformation and breakup regime transitions highlights the importance of breakup time, which is defined as the time between initiation of the velocity disturbances and the last fragmentation of drop liquid. In particular, as drop velocity relaxation times and breakup times approach one another, the propensity for drop breakup decreases due to reduction of relative velocities between the drop and the gas. Present measurements of breakup times, along with earlier measurements for shock-wave disturbances from Refs. 20–22, 28, and 29, are plotted as a function of Weber number in Fig. 2. The breakup times t_b in the figure are normalized by the characteristic breakup time for shear breakup defined by Ranger and Nicholls[20] as follows:

$$t^* = d_0(\rho_f/\rho_g)^{1/2}/u_0 \tag{1}$$

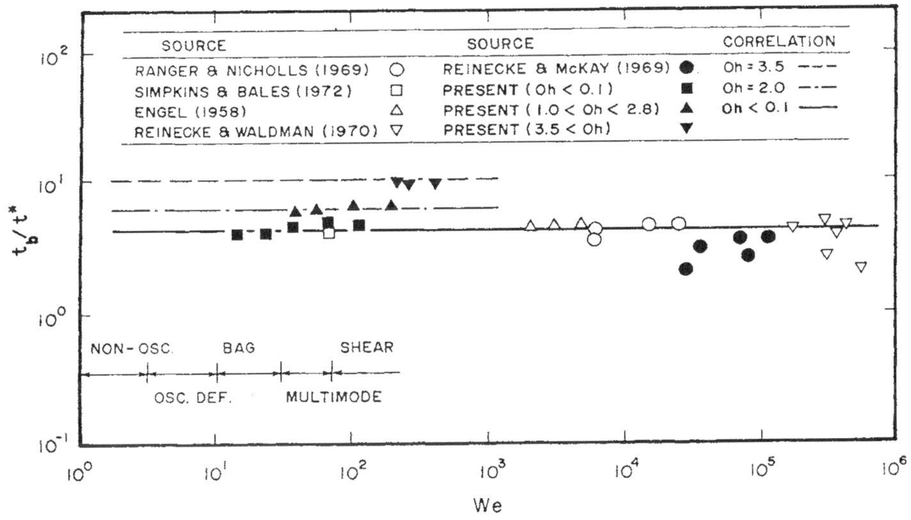

Fig. 2 Drop breakup times as a function of Weber and Ohnesorge numbers.

Except for the present results, which are grouped according to Ohnesorge number, the measurements are for $Oh < 0.1$ and effects of liquid viscosity are small. Thus, the deformation and breakup regimes at small Ohnesorge numbers are illustrated in the figure for reference purposes.

A remarkable feature of the breakup time results of Fig. 2 at $Oh < 0.1$ is that t_b/t^* varies very little even though Weber number varies over a large range (roughly 10–10^6) and a variety of breakup regimes are involved. In fact, the breakup time correlation of Ranger and Nicholls[20] developed for the shear breakup regime

$$t_b/t^* = 5.0 \tag{2}$$

provides a reasonably good correlation of all of the measurements for $Oh < 0.1$. When the present results for $Oh > 0.1$ are considered, however, it is seen that t_b/t^* progressively increases with increasing Ohnesorge number. This reflects the importance of liquid viscosity on breakup evident from the breakup regime map of Fig. 1.

3. *Drop Deformation*

The first stage of drop deformation, in the period where the drop flattens and first reaches a maximum cross stream dimension, was studied due to its influence on drop velocity relaxation and breakup. In particular, the distortion of the drop should affect its drag properties, and thus relative velocities during the breakup process, which undoubtedly plays a role in the onset of breakup.

Drop distortion correlates reasonably well as a linear function of time with the maximum distortion reached at roughly $t/t^* = 1.6$ when $Oh < 0.1$ (Ref. 10). Notably, measurements plotted by Wierzba and Takayama[23] exhibit very similar behavior. Measurements of drop distortion at $Oh > 0.1$, however, show progressive delay in the time required for the drop to reach maximum distortion.

The next parameter of interest is the maximum cross stream diameter of the drop $d_{c\,\max}$. An approximate expression for the variation of $d_{c\,\max}$ with flow conditions can be obtained for conditions where effects of liquid viscosity are small, $Oh < 0.1$, by considering the interaction between surface tension and pressure forces when the drop is drawn into a flattened shape. This yields the following expression for the maximum deformation[10]:

$$d_{c\,\max}/d_0 = 1 + 0.19 We^{1/2}, \qquad Oh < 0.1, \qquad We < 10^2 \tag{3}$$

Figure 3 is an illustration of present measurements of $d_{c\,\max}/d_0$ as a function of Weber number, with Ohnesorge number as a parameter. The correlating expression of Eq. (3) for $Oh \leq 0.1$ also is plotted in the figure; it is evident that Eq. (3) provides a reasonable fit of the data. Effects of increasing Ohnesorge number can be seen, with $d_{c\,\max}/d_0$ tending to decrease at a particular Weber number as Ohnesorge number is increased. Because the deformation motions of the drop cease at the point where $d_{c\,\max}$ is reached, this behavior is not thought to be a direct effect of viscous forces on the force balance fixing $d_{c\,\max}$. Instead, the increased time of deformation due to effects of liquid viscosity is a more probable mechanism. This allows drag forces to act for a longer time before the maximum deformation condition is reached, which tends to reduce the relative velocity, and correspondingly $d_{c\,\max}$.

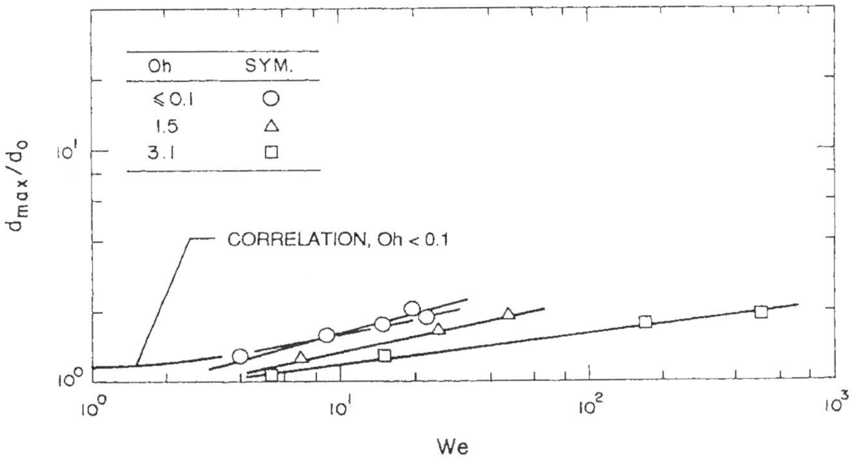

Fig. 3 Maximum cross stream drop diameters prior to breakup as a function of Weber and Ohnesorge numbers.

4. Drop Drag

Drop drag properties were found by measuring the motion of the centroid of the drop in the uniform flowfield behind the shock wave. This approach is only approximate because it neglects the forces involved as the mass of the drop is redistributed during drop deformation, however, this effect is not expected to be large for present test conditions because characteristic velocities in the liquid phase are small as discussed subsequently.[10]

The experiments to find C_D involved the initial deformation of the drops up to the time $d_{c\,\max}$ was reached for $Oh < 0.1$ and for a moderate range of Reynolds number (1000–2500) where effects of Reynolds number on the drag of the drops are expected to be small.[8] Thus, it was found that C_D largely was a function of the degree of deformation of the drop for present test conditions. To highlight this behavior, the results are plotted in terms of d_c/d_0 in Fig. 4. Measurements of C_D for solid spheres and thin disks, drawn from White[30] for the same range of Reynolds number as the present tests, also are illustrated on the plot. In spite of the relatively large uncertainties of the measurements, the trend of the data is quite clear: for d_c/d_0 near unity, C_D approximates results for solid spheres and then increases to approach results for thin disks at $d_c/d_0 = 2$. Thus, behavior in the period observed appears to be dominated by distortion of the drop, rather than internal circulation which would cause reductions of C_D from values appropriate for solid spheres. This seems reasonable because liquid-phase velocities are relatively small for present test conditions. In particular, conservation of momentum considerations suggest that characteristic liquid-phase velocities are on the order of $(\rho_g/\rho_f)^{1/2}u_0$, see Ranger and Nicholls.[20] Thus, for $(\rho_f/\rho_g) > 500$, liquid velocities are roughly 5% of the initial relative velocity.

5. Drop Sizes

Measurements of drop sizes after breakup were limited to conditions where $Oh < 0.1$. This was necessary in order to capture the entire drop field after breakup on a single hologram, because larger values of the Ohnesorge number yielded

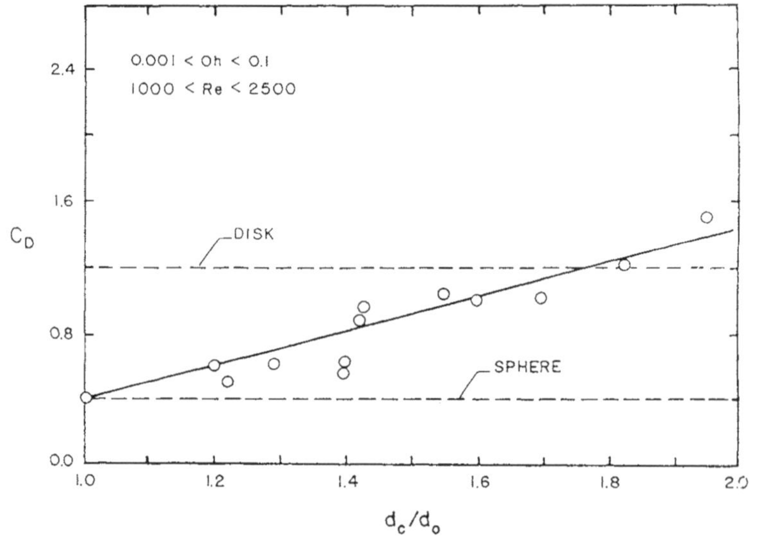

Fig. 4 Drop drag coefficient prior to breakup as a function of the normalized cross stream diameter.

regions containing drops that were too large for the present optical arrangement. The measurements included $We < 10^3$, which corresponds to the bag, transition, and shear breakup regimes.

Past work on the structure of dense sprays and processes of primary breakup of nonturbulent and turbulent liquids[2, 13, 14] indicated that local drop size distributions generally satisfied the universal root normal distribution function of Simmons,[31] with *MMD/SMD* = 1.2 (see Belz[32] for a discussion of the properties of the root normal distribution function). This vastly simplifies the presentation of data because the root normal distribution only has two moments, and with *MMD/SMD* equal to a constant, is specified entirely by the Sauter mean diameter alone. Thus, initial measurements of drop sizes after breakup focused on evaluating the root normal distribution function.

Typical measurements of drop size distributions are illustrated in Fig. 5 for bag breakup. Results for the other breakup regimes were similar, although for shear breakup trends illustrated in Fig. 5 they were achieved only after removing the core or drop-forming drop from the distribution as discussed in the following (the core drop will be handled separately). The results are plotted in terms of the root normal distribution function, with the function itself illustrated for values of *MMD/SMD* = 1.1, 1.2, and 1.5. The data are somewhat scattered at large drop sizes because of the limited number of large drops. The results are represented reasonably well by the universal root normal distribution function with *MMD/SMD* = 1.2, similar to the earlier findings.[2, 13, 14] In contrast to the present findings, Gel'fand et al.[18] observed a bimodal distribution of drop sizes after bag breakup for the two conditions they considered. The reason for this discrepancy is unknown and clearly merits additional study.

A correlating expression for the Sauter mean diameter after secondary breakup is obtained by noting the similarity between primary breakup of nonturbulent liquids[13] and secondary breakup due to stripping originating from the windward

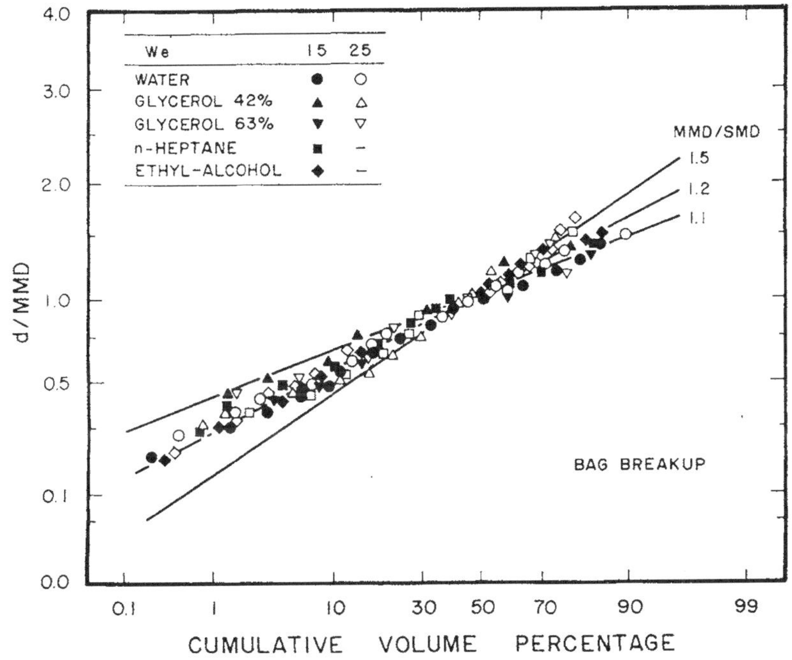

Fig. 5 Distribution of drop diameters after bag breakup.

side of the drop in the shear breakup regime. The configuration for secondary breakup in the shear breakup regime is shown in Fig. 6, where the core (drop-forming) drop is illustrated. It is assumed that the relative velocity at the time of breakup can be represented by the initial relative velocity, that drop sizes after breakup are comparable to the thickness of the boundary layer as it reaches the periphery of the drop, that the liquid boundary layer is laminar, and that the Sauter mean diameter is dominated by the largest drop sizes in the distribution. The length of development of the liquid boundary layer is taken to be proportional to d_0, which should be the condition tending to yield the largest drop sizes.[10]

Present measurements of the Sauter mean diameter after secondary breakup are plotted in terms of the variables from the secondary breakup model in Fig. 7. These results are for $Oh < 0.1$ and $We < 10^3$, including the bag, transition, and shear breakup regimes. A correlation of the data based on the approximate theory also is shown on the plot. The best empirical fit of the data to the theoretical variables is as follows:

$$\rho_g SMD\, u_0^2/\sigma = 6.2(\rho_f/\rho_g)^{1/4}[\mu_f/(\rho_f d_0 u_0)]^{1/2} We \tag{4}$$

with the correlation coefficient of the fit being 0.91. It should be noted, however, that ρ_f/ρ_g does not vary greatly over the present test range, and additional measurements are needed to explore density ratio effects.

It is probably fortuitous, and certainly surprising, that a single correlation can express the Sauter mean diameter after bag, transition, and shear breakup since the three breakup mechanisms appear to be rather different. On the other hand, similar behavior for the three breakup regimes is consistent with the observation

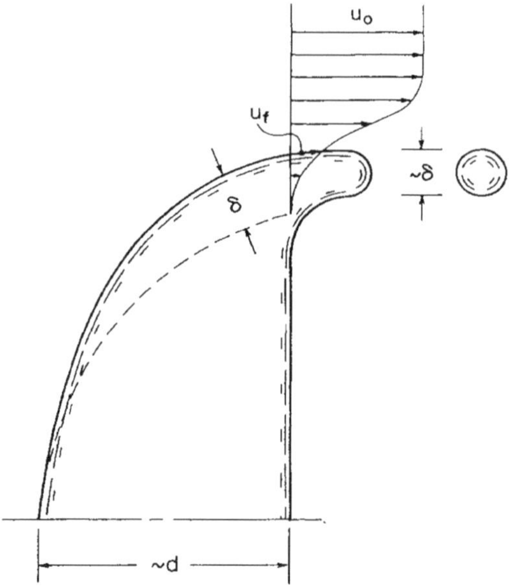

Fig. 6 Shear breakup process of drops.

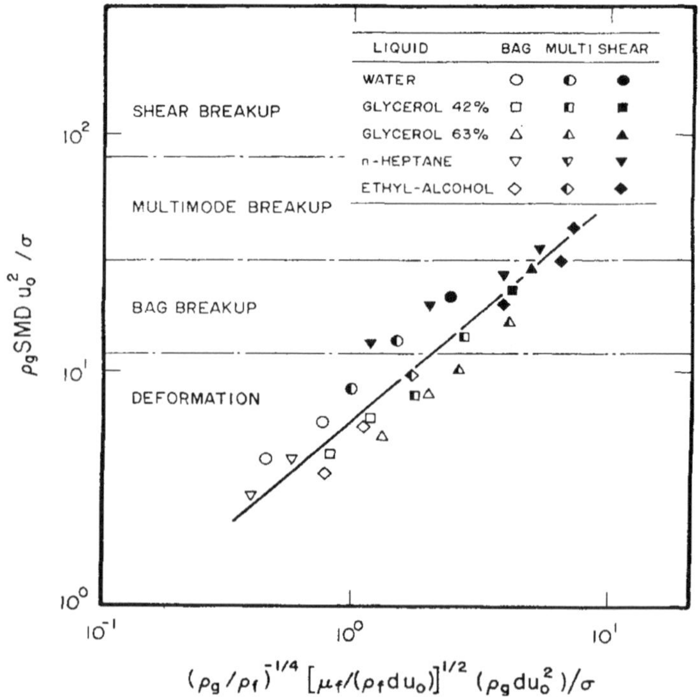

Fig. 7 Correlation of the *SMD* after secondary breakup.

that their breakup times correlated in the same way, as discussed in connection with Fig. 2

6. *Core Drop Properties*

The velocity and size of the core drop at the end of breakup must be known in order to treat it separately from the rest of the drop population after shear breakup. Since drops stripped from the core drop were not observed to undergo subsequent breakup, the end of shear breakup coincides with the end of drop stripping from the core drop. In the following, the velocity of the core drop at the end of breakup will be considered first. Given this information about core drop velocities, subsequent stability considerations yield a method for correlating measurements of core drop sizes. These considerations are limited to the streamwise velocity of the core drop because these drops did not exhibit appreciable cross stream velocities during the present experiments (aside from their cross stream motion due to operation of the drop generator system).

To assist data correlation, a simplified analysis was used to estimate core drop velocities at the end of breakup. The major assumptions of the analysis[11,12] are as follows: virtual mass, Bassett history, and gravitational forces were ignored; gas velocities were assumed to be constant; mass stripping from the core drop was ignored; and a constant average drag coefficient was used over the period of breakup.

Based on the preceding assumptions, the following expression for the absolute velocity of the core drop at the end of breakup, $u'_b = u_0 - u_b$, can be found[11,12]:

$$(u'_b/u_0)(\rho_f/\rho_g)^{1/2}(1+3C) = 3\bar{C}_D(t_b/t^*)/4 \tag{5}$$

where

$$C = (3\bar{C}_D t_b/4t^*)(\rho_g/\rho_f)^{1/2} \tag{6}$$

Earlier work has shown that $t_b/t^* = 5$ for $10 < We < 10^6$ and $Oh < 0.1$ (Refs. 10 and 27). Thus, the right-hand side of Eq. (5) should be a constant if a constant average value of C_D for the shear breakup process can be found. A reasonable correlation of the present measurements of u_b/u_0 was obtained by taking $\bar{C}_D = 5$, which is comparable to values observed near the maximum deformation condition.[10] This yields $3\bar{C}_D(t_b/t^*)/4 = 19$ on the right-hand side of Eq. (5).

The measurements of core drop velocities at the end of breakup, normalized as suggested by Eq. (5), are plotted as a function of Weber number in Fig. 8. Measurements are shown for all of the drop liquids over the test range of the shear breakup regime, $100 < We < 600$, along with the fitted predictions of Eq. (5). The correlation for u'_b/u_0 is relatively independent of Weber number over this range and is in fair agreement with Eq. (5), based on the estimates of t_b/t^* and $\bar{C}_D$ discussed earlier.

The velocity measurements indicated that the relative velocity of the core drop at the end of breakup was 30–40% lower than the initial relative velocity. This implied that the local Weber numbers of the core drop when breakup ended generally were greater than the critical Weber number for the onset of drop breakup due to shock-wave disturbances ($We = 13$). Thus, the criterion for the end of drop stripping from the core drop differs from the criterion for the onset of breakup.

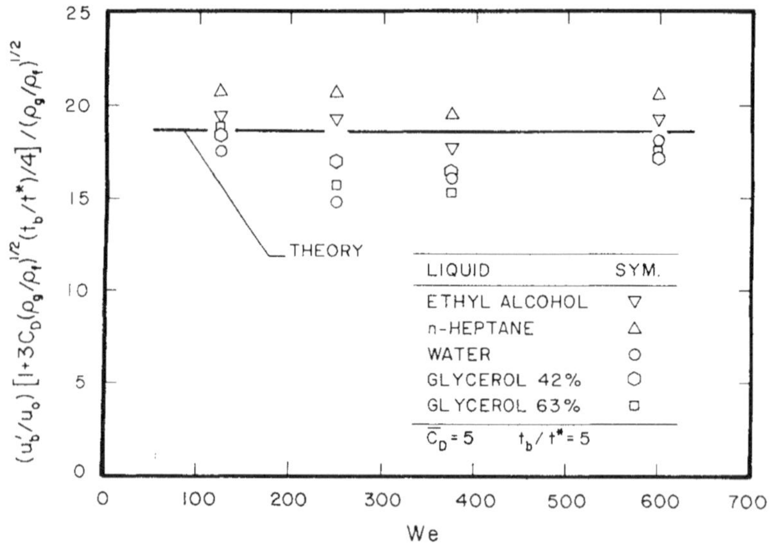

Fig. 8 Velocities of the core drop at the end of shear breakup.

The dynamic state of a drop at the start of secondary breakup, where the drop is round and the drop liquid is motionless, clearly differs from the state of the core drop when shear breakup ends, where the drop is deformed and liquid motion associated with drop stripping is present. Thus, it is not surprising that local Weber numbers of the core drop at the end of shear breakup were different from (and generally exceeded) the critical Weber number associated with the onset of secondary breakup. Instead, conditions defining the end of drop stripping for shear breakup appear to be related more closely to the onset of breakup for more gradual drop motions, like the breakup of freely falling drops. This correspondence is exploited in the following to find a criterion for the end of core drop stripping and a method for estimating the size of the core drop at this condition.

The deformation and size of freely falling drops generally are correlated in terms of the Eötövs number Eo. The appropriate expression when the drop acceleration a is due to gas motion relative to the drop is as follows[5]:

$$Eo = a|\rho_f - \rho_g|d^2/\sigma = a\rho_f d^2/\sigma \tag{7}$$

where the latter approximation follows because $\rho_f/\rho_g \gg 1$ for present test conditions. It is anticipated that drop stripping ends when a critical value of Eötvös number Eo_{cr} is reached, based on the behavior of freely falling drops. The values of Eo_{cr} were found for all shear breakup conditions, as defined in Refs. 11 and 12, and are plotted as a function of Weber number in Fig. 9. The critical Eötvös number of the core drop at the end of shear breakup is seen to be relatively independent of Weber number and liquid type over the range of the measurements, yielding a mean value, $Eo_{cr} = 16$. This behavior also is similar to the breakup requirements of freely falling drops, as discussed later.

The previous considerations suggest that Weber number, Eötvös number, and time all are factors in drop breakup events at low-Ohnesorge number. These interactions are highlighted by the local values of critical Weber number, critical Eötvös number, and t_{cr}/t^* for various breakup events summarized in Table 1.

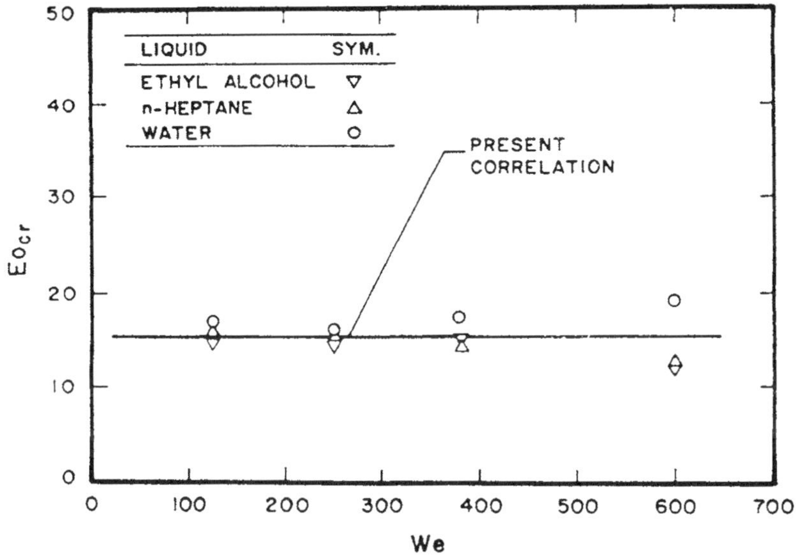

Fig. 9 Critical Eötvös number of the core drop at the end of shear breakup.

For abrupt disturbances, like the onset of secondary breakup due to shock-wave disturbances, local values of critical Weber numbers and critical Eötvös numbers were estimated at the time of breakup using measured values of drop drag in the deformation period.[11,12] Thus, these values are lower than criteria normally given for breakup regime transitions due to drop acceleration prior to the onset of breakup. Similar to the normalized breakup time, the normalized time at the onset of breakup for shock-wave disturbances is a constant over a wide range of Weber number, $t_{cr}/t^* = 1.6$. For this process, drops in the deformation period have local values of Weber and Eötvös numbers that exceed the limits for the onset of breakup, however, breakup does not begin until the drop has had time to deform and achieve a dynamical condition in the liquid that allows drops to separate from the parent drop. The characteristics of critical Weber number, critical Eötvös number and t_{cr}/t^* are somewhat different for gradual disturbances.[33–36] In the case

Table 1 Criteria for secondary breakup processes[a]

Process	We_{cr}	Eo_{cr}	t_{cr}/t^*
Abrupt (shock-wave) disturbances:			
Start of bag breakup, in gases	8–23	24–70	1.6
Start of multimode breakup, in gases	23–53	70–160	1.6
Start of shear breakup, in gases	53 <	160 <	1.6
Gradual disturbances:			
Start of bag breakup, in gases[b]	11–13	11	large
Start of bag breakup, in liquids[b]	7–9	15	large
End of shear breakup, in gases	8–29[c]	16	5.5

[a]$Oh < 0.05$, We_{cr} and Eo_{cr} subsequently increase with increasing Ohnesorge number.
[b]Freely falling drops in a motionless environment, from Refs. 33–36.
[c]Present test range with wider range probable.

of bag breakup for a freely falling drop, Eötvös number, based on the actual drop acceleration, is a maximum at the start of free fall, whereas t_{cr}/t^* is large due to the relatively slow acceleration of the drop. Thus, liquid properties are roughly quasisteady at each relative velocity condition and breakup only occurs when forces on the drop surface due to drag are too large to be stabilized by surface tension, i.e., when the Weber number of the drop reaches a critical value. Finally, the end of drop stripping for shear breakup also involves near quasisteady liquid behavior with the dynamical state of the drop being stabilized by surface tension once the forces on the surface, represented by the drop acceleration, became lower than a critical value represented by the critical Eötvös number. A range of critical Weber number is associated with this condition due to the large variation of the drag coefficient with the degree of deformation of the drop. Thus, various breakup events are associated with required minimum values of Weber number, Eötvös number, and time, with one of these parameters frequently serving as the controlling parameter for a particular process.

7. *Drop Velocities*

The last aspect of secondary breakup considered was the drop-size/velocity correlation at the completion of secondary breakup. These results will include all of the drops in the bag and multimode breakup regimes. The core drop will be excluded in the shear breakup regime, however, because its properties have already been established. A simplified analysis, similar to the approach used for core drop velocities, was used to assist the correlation of drop velocity data.[11,12] Although the analysis was rather crude, it did yield results that were useful for correlating the velocity measurements. Similar to consideration of core drop velocities, attention was limited to streamwise drop velocities.

The drop-size/velocity correlation based on the variables from the simplified analysis is plotted in Fig. 10. The test results involve a variety of drop liquids over the bag, multimode, and shear breakup regimes. Results for the core drops have been shown as dark symbols, for comparison purposes, but they are not included in the following drop-size/velocity correlation. The measured results clearly are independent of the breakup regime and can be correlated reasonably well using the following empirical fit based on the simplified analysis[11,12]:

$$u_0/u_b - 1 = 2.7\left[(\rho_g/\rho_f)^{1/2} d_0/d\right]^{2/3} \tag{8}$$

Equation (8) appears to be rather robust over the three breakup regimes, the range of liquid types, and the flow conditions considered during the present experiments. Even the results for core drops are in fair agreement with Eq. (8), however, the specific relationships for core drops discussed earlier provide a better estimate of their velocities. Additionally, the results indicate that relative velocities are reduced 30–70% over the period of breakup, with the smallest drops experiencing the largest reduction of relative velocity due to their smaller relaxation times. Finally, this variation of drop velocities with size implies that the secondary breakup process extends over a considerable region of space; for example, core drops move 30–40 initial droplet diameters during the period of breakup.[11] Thus, secondary breakup corresponds to a rate process, somewhat like drop vaporization, rather than to an instantaneous process that can be characterized by jump conditions.

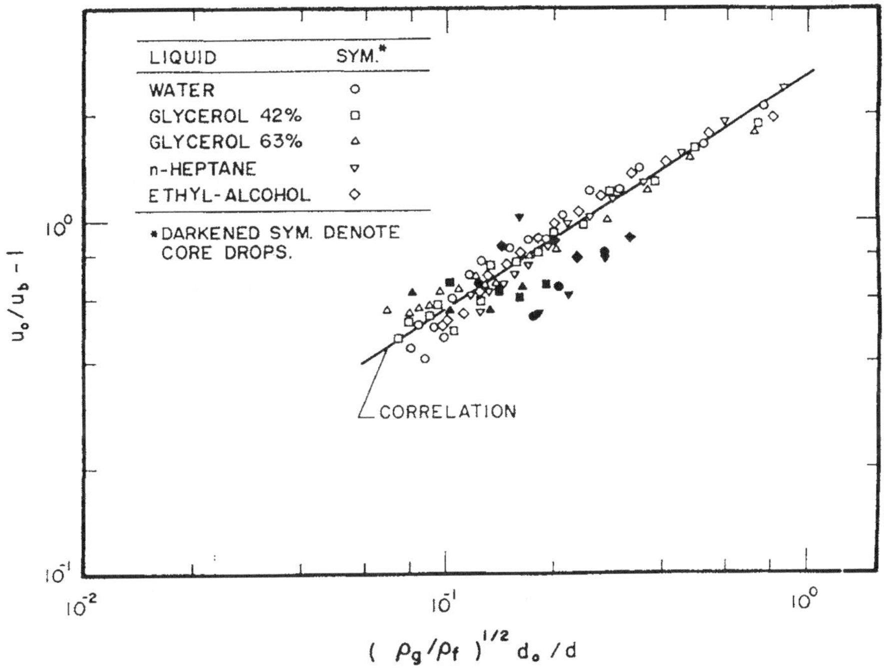

Fig. 10 Correlation of drop velocities with drop size at the end of secondary breakup.

D. Conclusions

Drop deformation and secondary breakup after shock-wave initiated disturbances were studied, considering a variety of liquids in air at normal temperature and pressure ($We = 0.5$–600, $Oh = 0.006$–600, $\rho_f/\rho_g = 580$–12000, and $Re = 340$–15760). The major conclusions of the study are as follows.

1) Drop deformation and breakup begin at $We \sim 1$ and 10, respectively, at low Ohnesorge number. Effects of liquid viscosity, however, inhibit deformation and breakup at high Ohnesorge numbers, e.g., for $We < 1000$, breakup and deformation were no longer possible for $Oh > 10$ and 1000, respectively. This inhibition of deformation and breakup should be important for high-pressure combustion processes where drops reach large Ohnesorge numbers as the liquid surfaces approach the thermodynamic critical point. see Fig. 1.

2) Drop size distributions after secondary breakup satisfy the universal root normal distribution of Simmons[31] with $MMD/SMD = 1.2$, similar to other observations in dense sprays and after primary breakup.[1,2,13] An exception was shear breakup where the core drop must be treated separately. Thus, the size distribution is completely defined by the Sauter mean diameter alone, see Fig. 5.

3) The Sauter mean diameter after secondary breakup could be correlated in terms of a characteristic liquid boundary-layer thickness for all three breakup regimes, see Fig. 7 and Eq. (4).

4) The streamwise velocity and size of the core drop after shear breakup were correlated successfully based on simplified considerations of drop motion during breakup, and the observation that the Eötvös number at the end of drop stripping was a constant, i.e., $Eo_{cr} = 16$, see Figs. 8 and 9 and Eq. (5).

5) The relative streamwise velocities of the drop liquid are significantly reduced during secondary breakup (30–70%, depending on drop size) due to the large drag coefficients during the drop deformation stage and the reduced relaxation times of smaller drops. These effects were correlated successfully based on variables found from simplified analysis of drop motion, see Fig. 10 and Eq. (8).

6) At low Ohnesorge numbers, criteria for various drop breakup processes can be represented by critical values of Weber number, Eötvös number, and t/t^*. Although certain minimum values are required for all three parameters, reaching a critical local value of one of the parameters tends to control the onset of particular breakup events: t_{cr}/t^* for the onset of breakup after a shock-wave disturbance, critical Weber number for the onset of breakup of a freely falling drop, and critical Eötvös number for the end of drop stripping from the core drop during shear breakup.

Aside from the deformation and breakup regime map, the present findings generally were limited to $Oh < 0.039$, with results concerning core drop properties limited to $Oh < 0.011$. Increasing the Ohnesorge number tends to impede drop deformation and breakup processes and should modify secondary breakup behavior considerably from the results observed during the present study. The liquid/gas density ratios of present work also were relatively limited and are most appropriate for sprays at atmospheric pressure. Effects of both Ohnesorge number and ρ_f/ρ_g clearly merit additional study in order to better understand secondary breakup properties for practical combusting sprays, particularly at the conditions of liquid rocket engines. Finally, the rate aspects of secondary breakup, e.g., the evolution of dispersed drop liquid as a function of time, needs to be explored because secondary breakup extends over a significant region in space and time and is not completely characterized by jump conditions.

III. Turbulent Primary Breakup

A. Introduction

Primary breakup to form drops near liquid surfaces is an important fundamental process of sprays. Unfortunately, current understanding of primary breakup is limited due to problems of observing the process in dense spray environments, effects of secondary breakup that modify drop size distributions prior to the drop field reaching conditions where drop sizes can be measured readily, and effects of liquid-phase disturbances that can vastly alter primary breakup properties. Recently, pulsed holography techniques have provided optical access to primary breakup properties; thus, findings using these methods will be discussed in the following.[13–15] This work has been limited to primary breakup along the surfaces of round liquid jets in still gases. This region corresponds to the near-injector dense-spray region of pressure-atomized sprays for atomization breakup conditions, in the absence of significant gas velocities relative to the injector passage.

Past studies of pressure-atomized sprays have established that spray properties are influenced by turbulence at the jet exit. First of all, the early studies of pressure atomization by De Juhasz et al.[37] and Lee and Spencer[38] showed that both atomization quality and mixing rates differed for laminar and turbulent flow at the jet exit. Next, Grant and Middleman[39] and Phinney[40] observed that jet stability and the onset of breakup were affected by turbulence at the jet exit as well. Additionally, Hiroyasu et al.[41] studied the length of the all-liquid core near the jet exit, which resembles the potential core region of single-phase jets. They conclude that turbulence generated in the flow passage has a significant effect on jet breakup properties, with even relatively short injectors exhibiting effects of turbulence due to disturbances generated at the reattachment points of separated flow regions. Chehroudi et al.[42] criticized the measurements of all-liquid core lengths of Hiroyasu et al.,[41] typically finding values roughly half as long. Nevertheless, several other studies with reasonably good characterization of turbulence properties at the jet exit and alternative methods for measuring all-liquid core lengths and jet properties also have reported effects of jet exit turbulence on liquid jet breakup properties.[43–49]

The recent studies of Ruff et al.[1,2,43] and Tseng et al.[44,45] also have helped to quantify effects of turbulence at the jet exit on the mixing rates and structure of the dispersed phase region of pressure-atomized sprays. These studies involved liquid jets in still air at various pressures with both nonturbulent slug flow and fully developed turbulent pipe flow at the jet exit. Measurements of liquid volume fraction distributions showed much faster mixing rates for fully developed turbulent pipe flow than for nonturbulent slug flow. Additionally, drop sizes after primary breakup were larger for turbulent than nonturbulent liquids even though the other properties of these flows were nearly identical. Finally, drop sizes were relatively independent of liquid/gas density ratios for values greater than 500. However, drop sizes progressively decreased as liquid/gas density ratios were reduced below 500, so that aerodynamic effects appear to be limited to this low-density ratio regime.

Other studies also found that disturbances within the liquid phase, i.e., the presence of liquid turbulence, has dominated observations of breakup in pressure-atomized sprays and that aerodynamic effects are not very important at liquid/gas density ratios typical of observations at normal pressure and temperature. First of all, Hoyt and Taylor[46–48] found that the breakup of liquid jets in air at atmospheric pressure was associated with the presence of turbulent boundary layers along the walls near the jet exit. Additionally, large changes in the aerodynamic environment, including both coflowing the counterflowing air, had little effect on breakup properties. Finally, Wu et al.[49] observed the complete suppression of primary breakup along the liquid surface when boundary layers that develop along the injector passage walls for slug flow conditions are removed; instead, the entire liquid column breaks up as a whole at some distance from the jet exit, due to growth of large-scale disturbances. Thus, even the so-called slug flows considered in the past,[1,2,13,44,45] probably involved effects of breakup enhanced by turbulent boundary layers near the liquid surface.

The preceding review of the literature shows that liquid properties affect primary breakup on the surface of the round jet and that aerodynamic effects, or at least the onset of conditions where they are important, are controlled by the liquid/gas density ratio and are not very significant for surface breakup at liquid/gas density ratios greater than 500. Thus, the objective of the present investigation was to observe primary breakup along liquid surfaces with well-defined liquid disturbance levels at various liquid/gas density ratios. Measurements involved round liquid jets in still gases with fully developed turbulent pipe flow at the jet exit. Effects of flow dynamics and liquid physical properties were studied by considering various gases as the jet environment. The measurements included pulsed shadowgraph photography and holography to find the location of the onset of turbulent primary breakup as well as drop sizes and velocities after primary breakup as a function of distance from the jet exit. Additionally, phenomenological theories were used to help interpret and correlate the results. The following discussion is brief; more details can be found in Refs. 14, 15, and 49.

B. Experimental Methods

1. Apparatus

The measurements of primary breakup properties along the surface of round liquid jets in gases were carried out using an injector within a windowed test chamber. The injector consisted of a pneumatically driven piston/cylinder arrangement containing a 600-ml sample of the test liquid. The outlet of the cylinder had a rounded contraction to prevent cavitation, followed by a constant-diameter passage having a length-to-diameter ratio ≥ 41, to yield nearly fully developed turbulent pipe flow at the jet exit. Injection was vertically downward with the liquid collected at the bottom of the test chamber and then discarded.

The windowed test chamber was cylindrical with a diameter of 300 mm and a length of 1370 mm. The chamber could be evacuated and refilled with various gases at pressures of 1 and 2 atm in order to change ρ_f/ρ_g while avoiding problems of cavitation and flashing at low-chamber pressures. Total test times were 200–10,000 ms, which was sufficient due to short flow development and data acquisition times. Jet exit velocities were calibrated using an impact plate.

2. Instrumentation

Instrumentation consisted of pulsed shadowgraph photography and single- and double-pulsed holography to measure drop sizes and velocities. These observations were confined to the region near the liquid surface in order to characterize the outcome of primary breakup. The holocamera was used for pulsed shadowgraph photography, operating in the single-pulse mode with the reference beam blocked to yield a shadowgraph rather than a hologram; these measurements were used to measure primary breakup properties near the onset of breakup. The arrangement of the holocamera as well as measurement techniques were identical to the secondary breakup measurements discussed in Sec. II. Results at each test condition were summed over 20–400 liquid fragments, to provide drop size and velocity correlations.

3. Test Conditions

The experiments involved relatively large, 3.6–9.5-mm-diam, liquid jets in still gases. Test liquids included water, *n*-heptane, and various glycerol mixtures. Test gases included helium, air, and freon 12 at pressures of 1–2 atm. Jet exit velocities were in the range 16–109 m/s, yielding the following ranges of jet and primary breakup dynamic parameters: $\rho_f/\rho_g = 104$–6230, $Re_{fd} = 90{,}000$–$780{,}000$, $We_{gd} = 12$–3790, $We_{fd} = 60{,}000$–$1{,}090{,}000$, and $Oh_d = 0.0011$–0.0052. Most test conditions involved $We > 8$ and $We_{gd} > 40$, which places them in the atomization breakup regime defined by Miesse[52] and Ranz,[53] where primary breakup should begin right at the jet exit. Actually, the onset of primary breakup always occurred at some distance from the jet exit, as will be discussed later in more detail.

C. Presentation of Results on Primary Breakup

1. Drop Size Distributions

Drop size distributions after primary breakup were measured for all test conditions. It was found that they satisfied the universal root normal distribution function with the ratio $MMD/SMD = 1.2$, similar to results after secondary breakup, discussed earlier. These results also are consistent with numerous other observations in dense sprays.[1,2,31,43–45] Thus, drop size distributions will be specified by the SMD diameter alone in the following.

2. Drop Velocities

The measurements of mass- (volume-) averaged streamwise and cross stream drop velocities after primary breakup are plotted as a function of distance from the jet exit in Fig. 11. To highlight potential aerodynamic effects, results for $\rho_f/\rho_g > 500$ are shown as open symbols whereas results for $\rho_f/\rho_g < 500$ are shown as darkened and half-darkened symbols.

The normalized drop velocities illustrated in Fig. 11 provide crude correlations in terms of x/d but other factors probably are involved as well. Near the jet exit, $\bar{u}_p/\bar{u}_0$, has a value of near 0.6, which increases to roughly 0.9 for $x/d > 20$. This behavior is consistent with drop velocities after turbulent primary breakup being roughly the same as streamwise liquid velocities near the liquid surface. Thus, the lower streamwise velocities are expected near the jet exit due to the retarding effect of the passage wall, evolving to higher values farther downstream where liquid velocities become more uniform. Superimposed on this behavior,

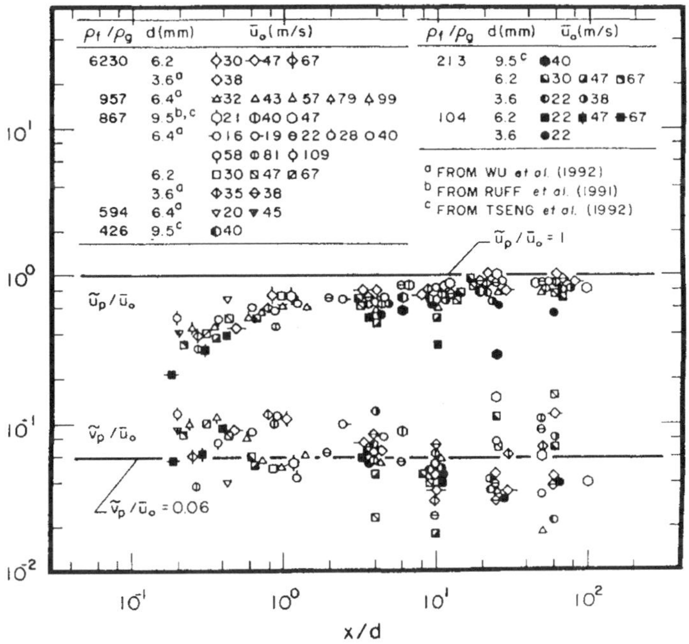

Fig. 11 Mass-averaged drop velocities after primary breakup as a function of distance from the injector exit.

however, is a clear trend that measurements for $\rho_f/\rho_g < 500$ have significantly lower mean streamwise velocities than the higher density ratio conditions—on the average roughly 20–40% lower. This can be attributed to aerodynamic drag if it is recalled that gas velocities near the liquid surface tend to be relatively low because the dispersed-phase region is very dilute (liquid volume fractions are less than 0.1%) so that momentum transfer to the gas phase is not very effective and relative velocities between the phases are large.[1,2]

Similar trends with respect to density variations are much less clear for the cross stream velocity component $\bar{v}_p/\bar{u}_0$ in Fig. 11, which are comparable to cross stream velocity fluctuations in the liquid. In particular, maximum values of $\bar{v}'_0/\bar{u}_0 = 0.058$ for fully developed turbulent pipe flow, even in the region near the wall[50,51] which is comparable to the cross stream drop velocity results seen in Fig. 11. An explanation of why the cross stream velocity component does not exhibit a reduction due to drag at low values of ρ_f/ρ_g, similar to the streamwise velocity component, is provided by an aerodynamic effect other than drag. In particular, the classical aerodynamic theories of primary drop breakup are based on the idea that flow velocities in the radial direction are increased for protuberances from the liquid surface due to acceleration of the gas over the tip of the protuberance,[52–56] which can be seen from the sketch appearing in Fig. 12. This yields a radial pressure drop across the protuberance that should increase the radial velocities of drops near the surface when aerodynamic effects are significant. This mechanism would tend to compensate for drag effects in the radial direction. Thus, it is plausible that radial drop velocities are not significantly changed by the presence of aerodynamic effects within the rather large scatter of the measurements.

3. *Onset of Breakup*

Phenomenological analysis to find drop properties at the onset of breakup was based on the configuration illustrated in Fig. 12. The onset mechanism was

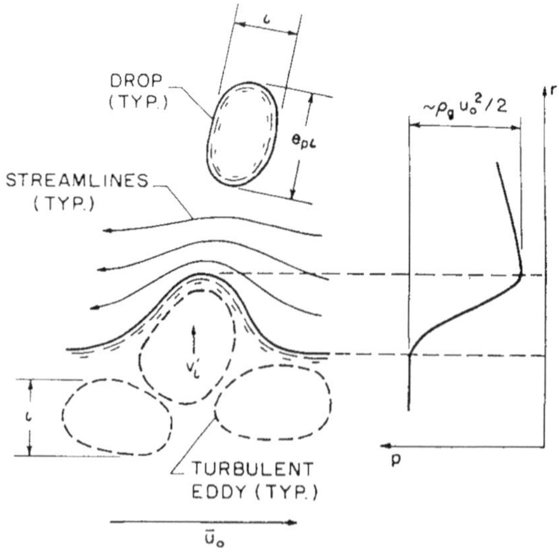

Fig. 12 Aerodynamically enhanced turbulent primary breakup at a liquid surface.

assumed to involve the formation of a drop from a turbulent eddy having a characteristic size ℓ, and a characteristic cross stream velocity relative to the surrounding liquid v_ℓ. The eddy is shown with an elongated shape because length scales in the streamwise direction are larger than in the cross stream direction for turbulent pipe flow.[50,51] The eddy is assumed to be convected in the streamwise direction at the local mean velocity, which is taken to be $\bar{u}_0$ based on the results discussed in connection with Fig. 11. The drop formed by the eddy also is assumed to have a diameter comparable to ℓ.

Drops formed at the onset of turbulent primary breakup are the smallest drops that can be formed by this mechanism.[14] The smallest drops that can be formed are either comparable to the smallest or Kolmogorov scales of turbulence, ℓ_K, or the smallest eddy that has sufficient mechanical energy to provide the surface energy needed to form a drop, whichever is larger. For present test conditions, ℓ_K was in the range 1–10 μm, which is much smaller than the smallest observed drop sizes; therefore, only the second criterion will be considered here, even though the first criterion may be relevant for some applications. The second criterion for the smallest drop that can be formed can be approximately quantified based on energy considerations. The mechanical energy available to form a drop includes the kinetic energy of an eddy of characteristic size, ℓ_i, relative to its surroundings, plus the added mechanical energy due to the pressure drop caused by acceleration of the surrounding gas over the tip of the protuberance, as illustrated in Fig. 12. It is reasonable to assume that ℓ_i is in the inertial range of the turbulence spectrum, which implies[53]

$$v\ell_i \sim \bar{v}_0'(\bar{v}_0'\ell_i/\Lambda)^{1/3} \tag{9}$$

where variations of turbulence properties in the liquid have been ignored, similar to Ref. 14. Formulation of these ideas, setting $SMD_i \sim \ell_i$, and assuming that turbulence properties in the liquid can be approximated by jet exit turbulence properties, yields the following implicit equation for SMD_i (Ref. 15):

$$\frac{SMD_i}{\Lambda}\left[1 + C_{sa}\left(\frac{\rho_g}{\rho_f}\right)\left(\frac{\bar{u}_0}{\bar{v}_0'}\right)^2\left(\frac{\Lambda}{SMD_i}\right)^{2/3}\right]^{3/5} = C_{si}\left(\frac{\bar{u}_0}{\bar{v}_0'}\right)^{6/5} We_{f\Lambda}^{-3/5} \tag{10}$$

where C_{sa} and C_{si} are empirical proportionality constants. With fully developed turbulent pipe flow at the jet exit, $\bar{v}_0'/\bar{u}_0$ essentially is a constant[50,51]; therefore, the effect of the aerodynamic enhancement term in Eq. (10) largely is controlled by the liquid/gas density ratio. This helps support past observations that the onset of aerodynamic effects depends on the liquid/gas density ratio rather than dynamic properties related to the liquid velocity.[14,45]

Present measurements of SMD_i are plotted in terms of the variables of Eq. (10) in Fig. 13. In doing this, C_{sa} was optimized to a value of 0.04, based on taking $\bar{v}_0'/\bar{u}_0 = 0.058$ for fully developed turbulent pipe flow.[51] In addition to present results, measurements from Ref. 14 also are shown on the plot, which involve a variety of liquids for conditions where aerodynamic effects are not important. As before, test conditions for $\rho_f/\rho_g > 500$, where aerodynamic effects could not be identified, are shown as open symbols. The correlation of the data for all conditions generally is within the scatter anticipated from experimental uncertainties. The power of $We_{f\Lambda}$ from the correlation of the data is not $-3/5$ as suggested by

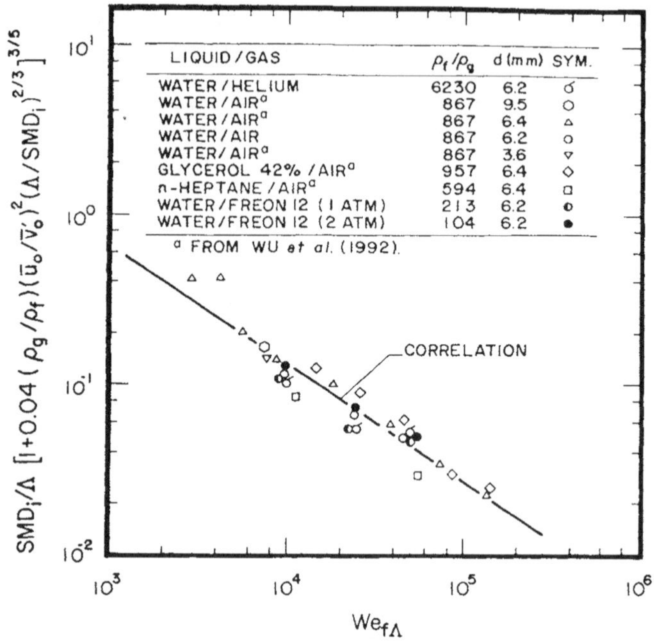

Fig. 13 ***SMD*** **at the initiation of turbulent primary breakup as a function of** $We_{f\lambda}$**.**

Eq. (9), however, and can be represented by the following empirical fit that is shown on the plot:

$$\frac{SMD_i}{\Lambda}\left[1+0.04\left(\frac{\rho_g}{\rho_f}\right)\left(\frac{\bar{u}_0}{\bar{v}_0'}\right)^2\left(\frac{\Lambda}{SMD_i}\right)^{2/3}\right]^{3/5} = 76We_{f\Lambda}^{-0.69} \tag{11}$$

Given SMD_i, the approach to find the location of the onset of turbulent primary breakup, x_i, was similar to Ref. 14. To find x_i, it is assumed that the drop-forming eddy convects along the liquid surface with a streamwise velocity $\bar{u}_0$ based on the results of Fig. 11. In addition, it was assumed that the onset of breakup was given by the time for growth of ligaments until their breakup into drops, taken to be the Rayleigh breakup time. The Rayleigh breakup time for a ligament of length ℓ_i, assuming that effects of liquid viscosity are small, is proportional to $(\rho_f \ell_i^3/\sigma)^{1/2}$, as discussed in Refs. 13 and 14. This yields the following expression for x_i (Ref. 15):

$$\frac{x_i}{\Lambda}\left[1+C_{sa}\left(\frac{\rho_g}{\rho_f}\right)\left(\frac{\bar{u}_0}{\bar{v}_0'}\right)^2\left(\frac{\Lambda}{SMD_i}\right)^{2/3}\right]^{9/10} = C_{xi}\left(\frac{\bar{u}_0}{\bar{v}_0'}\right)^{9/5} We_{f\Lambda}^{-4/10} \tag{12}$$

where C_{xi} is a constant of proportionality. Similar to Eq. (9), aerodynamic effects on x_i from Eq. (12) are controlled by the liquid/gas density ratio alone.

Both the present measurements of x_i and those of Ref. 14 are plotted in terms of the variables of Eq. (12) in Fig. 14, adopting $C_{sa} = 0.04$ as before. The correlation of the data over the full range of aerodynamic effects is reasonably good in view of

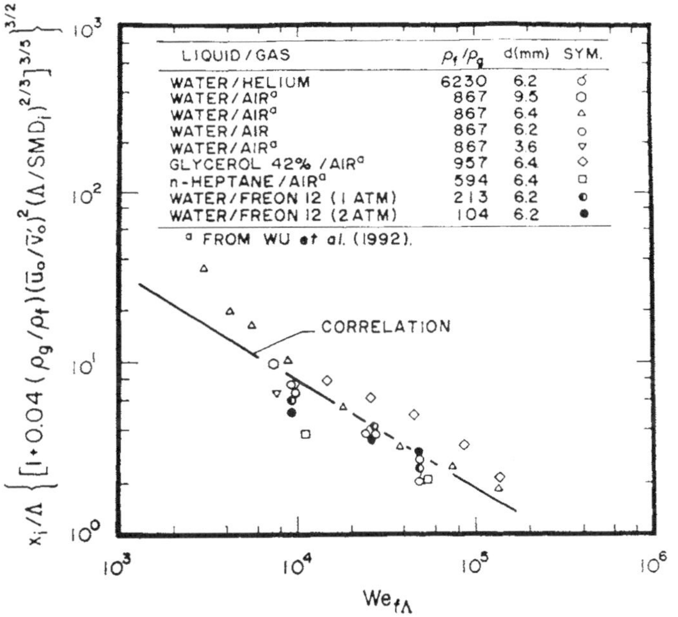

Fig. 14 Length to initiate turbulent primary breakup as a function $We_{f\Lambda}$.

the relatively large experimental uncertainties of x_i. As before, however, the power of $We_{f\Lambda}$ for the correlation of the data is not -0.4 as suggested by Eq. (12) but can be represented better by the following empirical expression, which is shown on the plot:

$$\frac{x_i}{\Lambda}\left[1+0.04\left(\frac{\rho_g}{\rho_f}\right)\left(\frac{\bar{u}_0}{\bar{v}_0'}\right)^2\left(\frac{\Lambda}{SMD_i}\right)^{2/3}\right]^{9/10} = 2570 We_{f\Lambda}^{-0.63} \qquad (13)$$

4. *Drop Sizes*

Aerodynamic secondary breakup times scale proportional to $\ell_i(\rho_f/\rho_g)^{1/2}/\bar{u}_0$ for an object of size ℓ_i, if the velocity of the gas near the liquid surface is assumed to be small.[10] Thus, for conditions where aerodynamic effects are important, Rayleigh breakup times increase more rapidly than do secondary breakup times as ℓ_i, increases. This implies a tendency for secondary and primary breakup to occur simultaneously, or to merge, as distance from the jet exit increases—a mechanism that dominated drop sizes near the liquid surface when aerodynamic effects are important, except near the onset of turbulent primary breakup.

Within the merged primary and secondary breakup regime, turbulent primary breakup properties will be found using the results of Ref. 14 whereas aerodynamic secondary breakup properties will be found from Eq. (4). The variation of *SMD* with distance from the jet exit for turbulent primary breakup was found assuming that the *SMD* was proportional to the largest drop that could be formed at a particular position with the time of breakup determined from the Rayleigh breakup time. Thus, the combined formulation yields the following expression for the *SMD* after merged primary and secondary breakup[15]:

$$\rho_g SMD\bar{u}_0^2/\sigma = C_s C_{sx}^{1/2}(x/\Lambda)^{1/3}(\rho_g/\rho_f)^{3/4} We_{f\Lambda}^{5/6} Re_{f\Lambda}^{-1/2} \tag{14}$$

where C_s and C_{sx} are empirical constants.

Present measurements of aerodynamic turbulent primary breakup, along with some results from Ref. 45, are plotted according to the variables of Eq. (14) in Fig. 15. The following best fit correlation of present measurements also is shown in the plot:

$$\rho_g SMD\bar{u}_0^2/\sigma = 12.0\left[(x/\Lambda)^{1/3}(\rho_g/\rho_f)^{3/4} We_{f\Lambda}^{5/6} Re_{f\Lambda}^{-1/2}\right]^{1.17} \tag{15}$$

Additionally, adopting the best fit values of C_s and C_{sx} in Eq. (14), the following theoretical prediction for *SMD* as a function of distance is obtained:

$$\rho_g SMD\bar{u}_0^2/\sigma = 12.9(x/\Lambda)^{1/3}(\rho_g/\rho_f)^{3/2} We_{f\Lambda}^{5/6} Re_{f\Lambda}^{-1/2} \tag{16}$$

with the standard deviation of the constant of 21%. Thus, the differences between Eqs. (15) and (16) are not statistically significant, within experimental uncertainties. Present measurements also are in good agreement with the predictions. The measurements of Tseng et al.[45] however, definitely yield smaller drop sizes after merged primary and secondary breakup. An explanation of this behavior is that these flows involved the highest concentrations of drops after turbulent primary breakup within the database, so that drop properties near the liquid surface could be influenced by small drops migrating from other parts of the multiphase mixing layer due to turbulent dispersion.

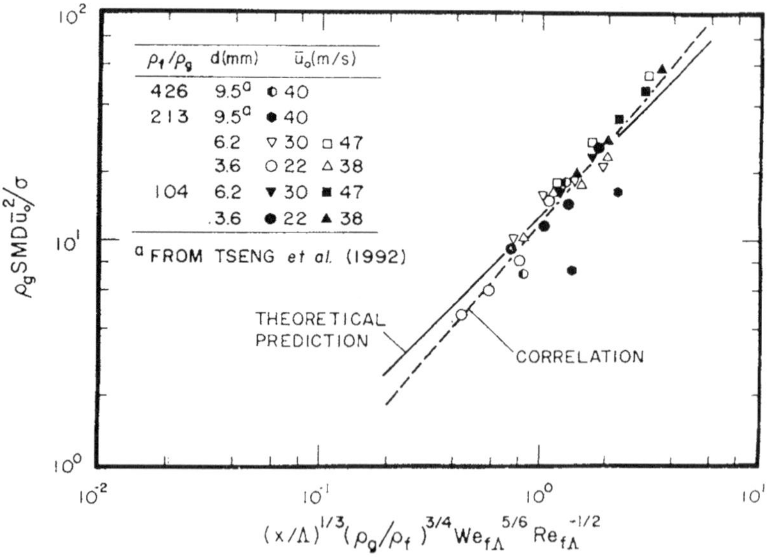

Fig. 15 *SMD* after merged turbulent primary and secondary breakup as a function of distance from the injector exit.

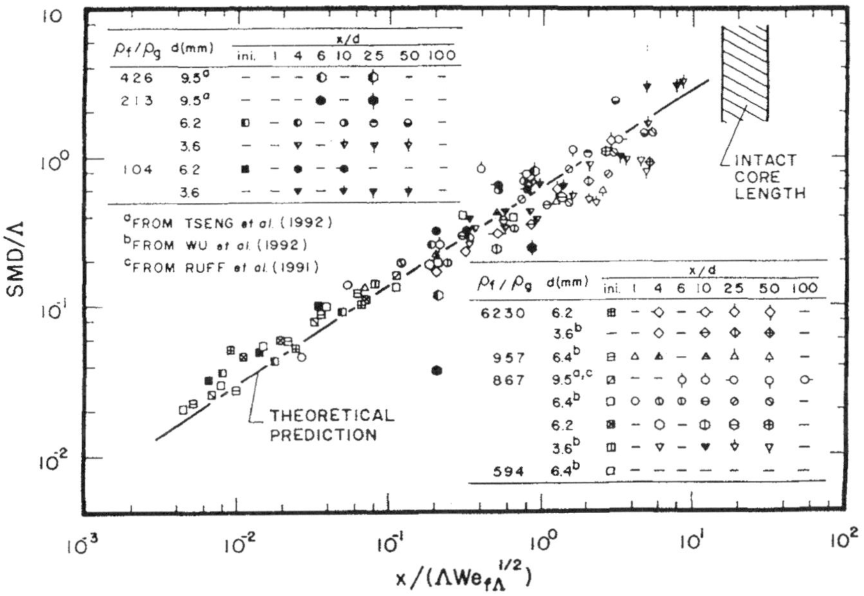

Fig. 16 ***SMD*** **after turbulent primary breakup as a function of distance from the injector exit, including inversion of merged secondary breakup effects.**

The consistency of the merged primary and secondary breakup process also can be examined by inverting the process; namely, by computing the drop sizes that should have been observed after primary breakup in the absence of aerodynamic effects, SMD_p, and comparing these results with other measurements at similar conditions. To do this, the nonaerodynamic turbulent primary breakup correlation was refitted to yield[14]

$$SMD_p/\Lambda = 0.65\left[x/\left(\Lambda We_{f\Lambda}^{1/2}\right)\right]^{2/3} \tag{17}$$

This implies reasonably consistent measurements of turbulent primary breakup with respect to Ref. 14, where the constant of Eq. (17) was 0.64. Then from the fit of the measurements illustrated in Fig. 15, the relationship between *SMD* and SMD_p is given by

$$SMD_p/\Lambda = 3.91 \times 10^{-3}(\rho_g/\rho_f)^{1/2} Re_{f\Lambda}(SMD/\Lambda)^2 \tag{18}$$

The measurements illustrated in Fig. 15, where primary and secondary breakup occurred, were inverted to get SMD_p, using Eq. (18). These results are plotted in Fig. 16, along with previous results for turbulent primary breakup where aerodynamic effects are not important from Ref. 14. Aside from the few results of Ref. 45 mentioned earlier in connection with Fig. 15, the inverted aerodynamic turbulent primary breakup measurements are seen to be in good agreement with the other results, supporting the idea of merging of primary and secondary breakup.

An interesting feature of the results of Fig. 16 is the progressive increase of *SMD* with distance from the injector. The largest *SMD* values observed in Fig. 16 approach the order of magnitude of the radial integral scale of the turbulence.

Whereas larger drops are feasible for liquid surfaces in general, this is not the case for round liquid jets due to the finite length of the liquid core. This is indicated in the figure using the correlation for liquid core length from Grant and Middleman.[39] The fact that the *SMD* approaches Λ near the end of the liquid core is consistent with the liquid column breaking up as a whole in this region.

5. Breakup Regimes

Present measurements suggested three regimes of turbulent primary breakup: 1) nonaerodynamic turbulent primary breakup, 2) aerodynamically enhanced turbulent primary breakup, observed at onset conditions, and 3) aerodynamic turbulent primary breakup, which involves merging of turbulent primary and secondary breakup. The results also indicted that the boundaries of these regimes are fixed by the liquid/gas density ratio and the relative magnitudes of characteristic Rayleigh breakup times of ligaments and the secondary breakup times of liquid fragments. The breakup times used to define these regimes were based on the *SMD* after primary breakup, or after the primary breakup stage of merged primary and secondary breakup, for conditions beyond the onset of breakup for present data. Thus, the characteristic Rayleigh breakup time was taken to be $\tau_R \sim (\rho_f SMD^3/\sigma)^{1/2}$, whereas the characteristic secondary breakup time was taken to be $\tau_b \sim (\rho_f/\rho_g)^{1/2}\ SMD/\bar{u}_0$. Then using Eq. (18) to eliminate *SMD* from the ratio, the characteristic time ratio was taken to be

$$\tau_R/\tau_b = (\rho_g/\rho_f)^{1/2}(x We_{f\Lambda}/\Lambda)^{1/3} \tag{19}$$

The resulting turbulent primary breakup regimes based on present measurements, as well as those from Refs. 1, 14, and 45, are illustrated in terms of ρ_f/ρ_g and τ_R/τ_b in Fig. 17. The total set of measurements yields $\rho_f/\rho_g = 500$ as the aerodynamic/nonaerodynamic transition, and $\tau_R/\tau_b = 4$ as the enhanced-aerodynamic/merged transition.

D. Conclusions

Primary breakup along the surface turbulent liquids was studied, considering liquid jets in still gases with fully developed turbulent pipe flow at the jet exit ($\rho_f/\rho_g = 104$–6230, $Re_{fd} = 90{,}000$–$780{,}000$, $We_{gd} = 12$–3790, $We_{fd} = 60{,}000$–$1{,}090{,}000$, and $Oh_d = 0.0011$–0.0052). The major conclusions of the study are as follows:

1) The presence of aerodynamic phenomena for turbulent primary breakup largely is controlled by the liquid/gas density ratio. When this ratio is less than 500, aerodynamic phenomena affect both conditions at the onset of breakup and drop sizes and velocities (to a lesser extent) after breakup.

2) Aerodynamic enhancement of the onset of turbulent primary breakup was due to the aerodynamic pressure reduction over the tips of protruding liquid elements. This effect assists the kinetic energy of a corresponding liquid eddy relative to its surroundings to provide the surface tension energy needed to form a drop, thus allowing smaller drops to form. Phenomenological analysis based on these ideas yielded reasonable correlations of onset properties, Eqs. (11) and (13), for the enhanced aerodynamic breakup regime (see Figs. 13 and 14).

3) For conditions where secondary breakup times become small in comparison to Rayleigh breakup of turbulence-induced ligaments protruding from the surface,

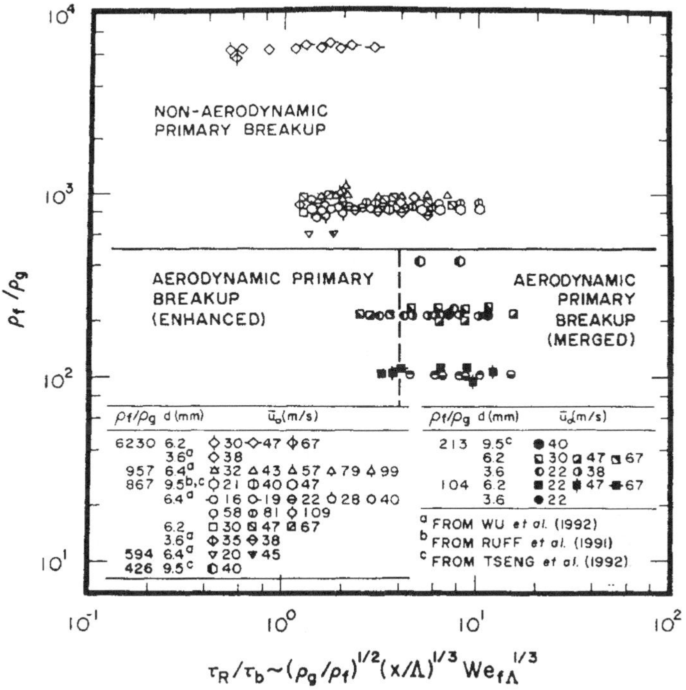

Fig. 17 Turbulent primary breakup regime map.

processes of primary and secondary breakup merge yielding smaller drops than when aerodynamic effects are absent. The reduction of drop sizes at these conditions correlated reasonably well with results for the secondary breakup of drops due to shock disturbances, yielding the correlation of Eq. (16), see Fig. 15.

4) Drop size distributions after aerodynamic turbulent primary breakup approximated Simmons'[31] universal root normal distribution with *MMD/SMD* = 1.2, similar to observations of other drop breakup processes as well as drops in the multiphase mixing layers of pressure-atomized sprays. Additionally, mass-averaged drop velocities after aerodynamic turbulent primary breakup approximate mean and rms velocity fluctuations of the liquid in the streamwise and cross stream directions, respectively, although there was a tendency for streamwise velocities to be somewhat reduced by aerodynamic effects.

A major issue still open involves primary breakup of nonturbulent liquids and the relevance of the classical primary breakup theories of Taylor[54] and Levich.[55] Current findings, and the observations of others,[46–49] all indicate that it is difficult to observe the nonturbulent primary breakup mechanism. The main problems are effects of liquid disturbances, the intrusion of secondary breakup, and weak aerodynamic effects for most liquids at normal pressure and temperature where measurements are most convenient. Until these experimental difficulties are resolved, understanding of this important primary breakup mechanism will remain limited. This is problematical because the classical nonturbulent aerodynamic primary breakup mechanism is likely to dominate breakup processes within combusting sprays at the pressures typically found in practical transportation and propulsion systems.[3–9] Finally, the rate properties of primary breakup are not well known, which limits capabilities for rational models of the near-injector region of sprays.

Acknowledgments

This research was sponsored by the Air Force Office of Scientific Research, Grant 89-0516 and F49620-92-J-0399, under the technical management of J. M. Tishkoff; and by the Office of Naval Research, Grant N00014-89-J-1199, under the technical management of G.D. Roy. The U.S. Government is authorized to reproduce and distribute copies for governmental purposes notwithstanding any copyright notation thereon.

References

[1]Ruff, G. A., Bernal, L. P., and Faeth, G. M., "Structure of the Near-Injector Region of Non-Evaporating Pressure-Atomized Sprays," *Journal of Propulsion and Power,* Vol. 7, No. 2, 1991, pp. 221–230.

[2]Ruff, G. A., Wu, P.-K., Bernal, L. P., and Faeth, G.M., "Continuous- and Dispersed-Phase Structure of Dense Non-Evaporating Pressure-Atomized Sprays," *Journal of Propulsion and Power,* Vol. 8, No. 2, 1992, pp. 280–289.

[3]Harrje, D. T., and Reardon, F. H., "Liquid Rocket Combustion Instability," NASA SP-194, 1972, pp. 49–55.

[4]Giffen, E., and Muraszew, A., *The Atomization of Liquid Fuels,* Wiley, New York, 1953.

[5]Clift, R., Grace, J. R., and Weber, M. E., *Bubbles, Drops and Particles,* Academic, New York, 1978, pp. 26, 339–347.

[6]Lefebvre, A. H., *Atomization and Sprays,* Hemisphere, New York, 1989, pp. 27–78, 201–272.

[7]Chigier, N. A., "The Physics at Atomization," *Proceedings of the Fifth International Conference on Liquid Atomization and Spray Systems,* NIST SP-813, National Inst. of Standards and Technology, Washington, DC, 1991, pp. 1–15.

[8]Faeth, G. M., "Mixing, Transport and Combustion in Sprays," *Progress in Energy and Combustion Science,* Vol. 13, No. 4, 1987, pp. 293–345.

[9]Faeth, G. M., "Structure and Atomization Properties of Dense Turbulent Sprays," *Twenty-Third Symposium (International) on Combustion,* Combustion Inst., Pittsburgh, PA, 1990, pp. 1345–1352.

[10]Hsiang, L.-P., and Faeth, G. M., "Near-Limit Drop Deformation and Secondary Breakup," *International Journal of Multiphase Flow,* Vol. 18, No. 5, 1992, pp. 635–652.

[11]Hsiang, L.-P., and Faeth, G. M., "Drop Properties After Secondary Breakup," *International Journal of Multiphase Flow,* Vol. 19, No. 5, 1993, pp. 721–735.

[12]Hsiang, L.-P., and Faeth, G. M., "Drop Deformation and Breakup Due to Shock Wave and Steady Disturbances," *International Journal of Multiphase Flow* (to be published).

[13]Wu, P.-K., Ruff, G. A., and Faeth, G. M., "Primary Breakup in Liquid/Gas Mixing Layers," *Atomization and Sprays,* Vol. 1, No. 4, 1991, pp. 421–440.

[14]Wu, P.-K., Tseng, L.-K., and Faeth, G. M., "Primary Breakup in Gas/Liquid Mixing Layers for Turbulent Liquids," *Atomization and Sprays,* Vol. 2, No. 3, 1992, pp. 295–317.

[15]Wu, P.-K., and Faeth, G. M, "Aerodynamic Effects on Primary Breakup of Turbulent Liquids," *Atomization and Sprays,* Vol. 3, No. 3, 1993, pp. 265–289.

[16]Hinze, J. O., "Fundamentals of the Hydrodynamic Mechanism of Splitting in Dispersion Processes," *AIChE Journal,* Vol. 1, No. 3, 1955, pp. 289–295.

[17]Hanson, A. R., Domich, E. G., and Adams, H. S., "Shock-Tube Investigation of the Breakup of Drops by Air Blasts," *Physics of Fluids,* Vol. 6, No. 8, 1963, pp. 1070–1080.

[18]Gel'fand, B. E., Gubin, S. A., and Kogarko, S. M., "Various Forms of Drop Fractionation in Shock Waves and Their Special Characteristics," *Inzhenerno-Fizicheskii Zhurnal,* Vol. 27, No. 1, 1974, pp. 119–126.

[19]Krzeczkowski, S. A., "Measurement of Liquid Droplet Disintegration Mechanisms," *International Journal of Multiphase Flow,* Vol. 6, No. 2, 1980, pp. 227–239.

[20]Ranger, A. A., and Nicholls, J. A., "The Aerodynamic Shattering of Liquid Drops," *AIAA Journal,* Vol. 7, No. 2, 1969, pp. 285–290.

[21]Reinecke, W. G., and McKay, W. L., "Experiments on Waterdrop Breakup Behind Mach 3 to 12 Shocks," Sandia Corp. Rept. SC-CR-70-6063, 1969.

[22]Reinecke, W. G., and Waldman, G. D., "A Study of Drop Breakup Behind Strong Shocks with Applications to Flight," Avco Rept. AVSD-0110-70-77, 1970.

[23]Wierzba, A., and Takayama, K., "Experimental Investigation of the Aerodynamic Breakup of Liquid Drops," *AIAA Journal,* Vol. 26, No. 11, 1988, pp. 1329–1335.

[24]Borisov, A. A., Gel'fand, B. E., Natanzon, M. S., and Kossov, O. M., "Droplet Breakup Regimes and Criteria for Their Existence," Inzhenerno-Fizicheskii Zhurna Vol. 40, No. 1, 1981, pp. 64–70.

[25]Loparev, V. P., "Experimental Investigation of the Atomization of Drops of Liquid under Conditions of a Gradual Rise of the External Forces," *Izvestiya Akademii Nauk SSSR, Mekhanika Zhidkosti i Gaza,* No. 3, 1975, pp. 174–178.

[26]Lane, W. R., "Shatter of Drops in Streams of Air," *Industrial Engineering Chemistry,* Vol. 43, 1951, pp. 1312–1317.

[27]Liang, P. Y., Eastes, T. W., and Gharakhari, A., "Computer Simulations of Drop Deformation and Drop Breakup," AIAA Paper 88-3142, 1988.

[28]Simpkins, P. G., and Bales, E. J., "Water-Drop Response to Sudden Accelerations," *Journal of Fluid Mechanics,* Vol. 55, No. 4, 1972, pp. 629–639.

[29]Engel, O. G., "Fragmentation of Waterdrops in the Zone Behind an Air Shock," *Journal of Research of the National Bureau of Standards,* Vol. 6, No. 3, 1958, pp. 245–280.

[30]White, F. M., *Viscous Fluid Flow,* McGraw–Hill, New York, 1974.

[31]Simmons, H. C., "The Correlation of Drop-Size Distributions in Fuel Nozzle Sprays," *Journal of Engineering for Power,* Vol. 99, No. 3, 1977, pp. 309–319.

[32]Belz, M. H., *Statistical Methods in the Process Industries,* Wiley, New York, 1973, pp. 103–104.

[33]Merrington, A. C., and Richardson, E. G., "The Break-Up of Liquid Jets," *Proceedings of Physical Society of London,* Vol. 59, 1947, pp. 1–13.

[34]Finlay, B. A., Ph.D. Thesis, Univ. of Birmingham, England, UK, 1957.

[35]Ryan, R. T., "The Behavior of Large Low-Surface-Tension Water Drops Falling at Terminal Velocity in Air," *Journal of Applied Meterology,* Vol. 15, 1976, pp. 157–165.

[36]Hu, S., and Kintner, R. C., "The Fall of Single Drops Through Water," *AIChE Journal,* Vol. 1, 1955, pp. 42–48.

[37]DeJuhasz, K. J., Zahm, O. F., Jr., and Schweitzer, P. H., "On the Formation and Dispersion of Oil Sprays," Engineering Experiment Station, Bulletin No. 40, Pennsylvania State Univ., University Park, PA, 1932, pp. 63–68.

[38a]Lee, D. W., and Spencer, R. C., "Preliminary Photomicrographic Studies of Fuel Sprays," NACA Tech. Note 424, 1932.

[38b]Lee, D. W., and Spencer, R. C., "Photomicrographic Studies of Fuel Sprays," NACA Rept. 454, 1933.

[39]Grant, R. P., and Middleman, S., "Newtonian Jet Stability," *AIChE Journal,* Vol. 12, No. 4, 1966, pp. 669–678.

[40]Phinney, R. E., "The Breakup of a Turbulent Jet in a Gaseous Atmosphere," *Journal of Fluid Mechanics,* Vol. 60, Pt. 4, 1973, pp. 689–701.

[41a]Hiroyasu, H., Shimizu, M., and Arai, M., "The Breakup of a High Speed Jet in a High Pressure Gaseous Environment," Univ. of Wisconsin, Madison, ICLASS-82.

[41b]Arai, M., Shimizu, M., and Hiroyasu, H., "Break-Up Length and Spray Angle of High Speed Jet," Univ. of Wisconsin, Madison, WI, ICLASS-85, 1985, p. IB/4/1.

[42]Chehroudi, B., Onuma, Y., Chen, S.-H., and Bracco, F. V., "On the Intact Core of Full Cone Sprays," Society of Automotive Engineers, SAE Paper 850126, 1985.

[43]Ruff, G. A., Sagar, A. D., and Faeth, G. M., "Structure of the Near Injector Region of Non-Evaporating Pressure-Atomized Sprays," *AIAA Journal,* Vol. 27, No. 7, 1989, pp. 901–908.

[44]Tseng, L.-K., Ruff, G. A., and Faeth, G. M., "Effects of Ambient Gas Density on the Structure of Liquid Jets in Still Gases," *AIAA Journal,* Vol. 30, No. 6, 1992, pp. 1537–1544.

[45]Tseng, L.-K., Wu, P.-K., and Faeth, G. M., "Dispersed-Phase Structure of Pressure-Atomized Sprays at Various Gas Densities," *Journal of Propulsion and Power,* Vol. 8, No. 6, 1992, pp. 1157–1166.

[46]Hoyt, J. W., and Taylor, J. J., "Waves on Water Jets," *Journal of Fluid Mechanics,* Vol. 88, Pt. 1, 1977, pp. 119–123.

[47]Hoyt, J. W., and Taylor, J. J., "Turbulence Structure in a Water Jet Discharging in Air," *Physics of Fluids,* Vol. 20, Pt. 2, No. 10, 1977, pp. S253–S257.

[48]Hoyt, J. W., and Taylor, J. J., "Effect of Nozzle Boundary Layer on Water Jets Discharging in Air," *Jets and Cavities,* edited by J. H. Kim, O. Furuya and B. R. Parkin, ASME-FED, Vol. 31, American Society of Mechanical Engineers, New York, 1985, pp. 93–100.

[49]Wu, P.-K., Miranda, R. F., and Faeth, G. M., "Effects of Initial Flow Conditions on Primary Breakup of Nonturbulent and Turbulent Liquid Jets," AIAA Paper 94-0561, Jan. 1994.

[50]Schlichting, H., *Boundary Layer Theory,* 7th ed., McGraw–Hill, New York, 1979, p. 599.

[51]Hinze, J. O., *Turbulence,* 2nd ed., McGraw–Hill, New York, 1975, pp. 427 and 724–742.

[52]Miesse, C. C., "Correlation of Experimental Data on the Disintegration of Liquid Jets," *Industrial Engineering Chemistry,* Vol. 47, No. 9, 1955, pp. 1690–1697.

[53]Ranz, W. E., "Some Experiments on Orifice Sprays," *Canadian Journal of Chemical Engineering,* Vol. 36, No. 8, 1958, pp. 175–181.

[54]Taylor, G. I., "Generation of Ripples by Wind Blowing Over a Viscous Liquid," *The Scientific Papers of Sir Geoffrey Ingram Taylor,* edited by G.K. Batchelor, Vol. 3, Cambridge Univ. Press, Cambridge, England, UK, 1963, pp. 244–254.

[55]Levich, V. G., *Physicochemical Hydrodynamics,* Prentice-Hall, Englewood Cliffs, NJ, 1962, pp. 639–646.

[56]Reitz, R. D., and Bracco, F. V., "Mechanism of Atomization of a Liquid Jet," *Physics of Fluids,* Vol. 25, No. 10, 1982, pp. 1730–1742.

[57]Tennekes, H., and Lumley, J. L., *A First Course in Turbulence,* MIT Press, Cambridge, MA, 1972, pp. 113–124.

[58]Weber, C., "Zum Zerfall eines Flussigkeitsstrahles," *Zeitschrift für Z Angewesen Mathematik und Mechanik,* Vol. 2, 1931, pp. 136–141.

Chapter 10

Injector Flame Stabilization Effects on Combustion Instability

A. F. Agarkov,* K. P. Denisov,† M. L. Dranovsky,‡ I. A. Zavorohin,§ V. N. Ivanov,* V. P. Pikalov,¶ and A. A. Shibanov§

Scientific Research Institute of Chemical Machine Building, NIICHIMMASH, Moscow, Russia

Nomenclature

A = oscillation amplitude
A_0 = pressure initial peak of pulse disturbance
A_{cr} = critical oscillation amplitude at which oscillation becomes sustained
A_{col} = first peak in disturbance device cold tests
A_n = root mean square amplitude of natural oscillation (noise), i.e., of pressure oscillation of the acoustic mode under study
$A_{n,max}$ = maximum amplitude of natural oscillation
$B(\tau)$ = correlation function
$B(0)$ = correlation function at $\tau = 0$
D_c = chamber diameter
E_1, E_2 = fluxes of oscillatory energy entering the system and dissipated by the system, respectively
ΔE = net energy flux, $E_2 - E_1$
E_Σ = total oscillatory energy of the system
f = oscillation frequency
f_0 = resonant frequency, $\omega_0/2\pi$
$K_{\dot{m}}$ = mixture ratio
l_{ch}, d_{ch} = length and diameter of transmission channel connecting disturbance device with combustion chamber

*Senior Engineer, Research Division.
†Vice-General Director.
‡Senior Research Scientist, Research Division.
§Senior Researcher, Research Division.
¶Head, Research Division.

$\dot{m}_\Sigma$ = propellant mass flow rate
N = empirical constant used to set a value of disturbance required for stability evaluation
n = disturbance parameter, $A_0/2A_{n,\max}$
P_c = combustion chamber pressure
P' = pressure pulsations at a certain location in combustion chamber
$R(\tau)$ = correlation coefficient, $B(\tau)/B(0)$
$S(\omega)$ = signal power spectrum
T = oscillation period, $2\pi/\omega$
T' = fluctuation value for instantaneous period
t = time
$\bar{x}$ = mathematical expectation, average value of random value
Y = operating parameter, e.g., propellant flow rate
Δ_f = filter width at a level of 0.5 times the maximum amplitude-frequency characteristics
δ = damping factor
δT = oscillation decrement
δT_e = effective, amplitude-averaged, decrement value
$\varepsilon_{\delta T}$ = relative error of decrement estimation
τ = time shift
τ_r = relaxation time of oscillations caused by artificial disturbances, i.e., the time of their amplitude decrease by a factor of e
ϕ = oscillation phase
ω = oscillation angular frequency

I. Introduction

THERE are various procedures to estimate liquid rocket engine (LRE) combustor stability to high-frequency (acoustic) oscillations.[1–4] The procedure devised at the Scientific Research Institute of Chemical Machine Building (NIICHIMMASH) and some other organizations is based on the combined processing and analysis of records of natural pressure oscillation (noise) and artificial pulse disturbance introduced into the chamber. Such an approach raises the validity of the estimate of stability and provides additional information about LRE combustor processes. The stability estimation procedure is a part of LRE development technology devised at NIICHIMMASH and is aimed at providing engine stability to high-frequency oscillations in a specified range of operation modes.

The technology includes the following stages: 1) laboratory tests of hydraulic, acoustic, and combusting models of combustion chamber and gas generator units; 2) combustion chamber and gas generator component stand tests in actual regimes of the engine assembly; and 3) checking of stability characteristics constancy. Corresponding characteristics defining the stability of the combustion process are devised for all of the development stages.

At the first stage some tentative information should be promptly obtained about combustion process stability with various geometries of actual injection elements (or injectors). This information can be obtained at benign simulated conditions: low pressures and low mass flow rates with simulators instead of actual propellants, if necessary. The tentative information on stability is used for selecting the most promising injector versions and for devising and implementing changes to the initial injector configuration with the object of stability improvement. The model test results are processed and extended to actual conditions using parametric criteria obtained on the basis of theoretical analysis and on the characterization of energy conditions of exciting self-oscillations.

The qualification of combustion process stability in model combusting tests is carried out by estimating the extent to which the regimes being simulated depart from the boundaries of the self-oscillating excitation region. A comparative estimation of various injector geometries is based on the relative position of the boundaries of the instability region, the hysteresis region, and on other characteristics taken for estimating combustion process stability under actual conditions.

After the transition to stand tests under actual conditions, simplified combusting models can be used to obtain predictive information related to a proposed change to some selected, promising injection element or injector head that is already being tested in actual regimes. Besides defining self-excited oscillation region boundaries for various injector geometries, studies of stability characteristics, gas passage length effects for gas/gas and gas/liquid injection elements, and the influence of harmonic oscillations transmitted from the feed system on the combustion zone properties (the combustion zone may enhance or reduce the oscillations) are performed with combusting models. In injector hydraulic tests at atmospheric and high-pressure discharge conditions, injector dynamic characteristics and injector inlet pressure oscillation influences on atomization parameters may be estimated. Acoustic models may be used for obtaining optimum characteristics of various passive damping devices (acoustic absorbers, dampers) and of active devices preventing the excitation of self-oscillations for which the kinetic energy of flows over straighteners, baffles, and so on may be a source.

At the second stage, an optimum (from the point of view of providing sufficient stability margins under actual conditions of disturbance effects caused by the operation of actual engine components) version of the injector is selected from some of the most promising ones chosen at the first stage. Stability characteristics in actual regimes are defined from parameters of amplitude [frequency spectra at the level of natural chamber pressure disturbances ("noise")] and from the chamber process response to external artificial pulse disturbances generated by a special reusable device. Stability characteristics are derived from measurements of inherent pressure oscillations and from the response to artifically imposed pressure disturbances. The pressure oscillation decrement δT, corresponding to the oscillations in the combustion chamber reaction zone at natural acoustic frequencies, is used as a combustion process stability characteristic. Relaxation time τ_r at a certain relative maximum value of pressure pulse disturbance peak is also used as a stability characteristic defined by the process response to pulse disturbances. The qualification of chamber process stability is usually performed by statistically comparing these experimental values of stability characteristics with corresponding limit values, based on the practical experience of applying this procedure obtained from prior research and development programs.

The third stage, stability characteristics checking, is carried out in engine manufacturing and operation. Amplitude characteristics of pressure oscillations in cavities upstream of the injector and vibrational accelerations of chamber and gas generator members are used. The limit values of these characteristics are specified for a certain engine on the basis of statistical data accumulated.

Problems of choosing stability characteristics for LRE actual tests and the methods and tools for determining these characteristics are discussed in the present chapter.

II. Stability Characteristics

A complete theoretical description of dynamic combustion processes in a rocket engine has not been developed. Therefore, an empirically based estimate of combustion chamber and gas generator process stability must be used. The stability degree of a combustion process can be determined from the response of the process to disturbances introduced into the chamber or from natural disturbances present in the chamber.

To obtain a quantitative estimate of stability, the following research work should be performed.

1) Optimum characteristics describing the combustion chamber and gas generator processes should be chosen.

2) Methods of using these characteristics for determining chamber process stability margins for different types of engines should be developed and tested.

3) Relevant instruments and devices for determining stability characteristic numerical values should be developed.

4) The connection of stability characteristics with engine design and operating parameters should be studied.

For choosing optimum stability characteristics, the linearized solution of the combustion chamber's response to both pulsed and stationary random disturbances will be examined. The response of chamber pressure P to a pulse action (for example, combustion chamber or gas generator pressure oscillations) can be presented as the sum of expressions of the form

$$P' = A_i \exp\left(-\int_0^t \delta_i \, dt \right) \sin\left(\int_0^t \omega_i \, dt + \varphi_i \right) \tag{1}$$

where A_i and ϕ_i are the initial amplitude and phase of the ith mode of the natural oscillation, and δ_i and ω_i are the oscillation damping factor and frequency governed by the system parameters.

Quantities δ and ω, in general, depend on time, i.e., $\delta = \delta(t)$ and $\omega = \omega(t)$. These quantities characterize pressure oscillations in combustion chambers at any given time, which is why certain definite values of the quantities correspond to definite characteristic states of the oscillatory system being considered. Negative and positive values of δ indicate unstable and stable states, respectively. The stability boundary is indicated by $\delta = 0$. This allows one to employ the value of δ as a characteristic of the oscillatory system state that can be found in experiments and, in particular, as the system stability characteristic. In LRE development the problem of finding design and operating parameter ranges that correspond to certain combustion process states, i.e., certain states of the oscillatory system being investigated, is of great interest. If these parameters influence the oscillation characteristics, some relationship between them and the values $\delta(t)$ and $\omega(t)$ should exist.

On the basis of available experimental data for stationary modes, it is found that the frequency of pressure oscillations in the combustion chamber and gas generator varies only slightly; this allows one to take the oscillation frequency to be constant or to introduce a certain correction for frequency variation and to use only one characteristic of oscillation, i.e., the damping factor δ.

The damping factor has a definite physical meaning; its value equals half of the ratio of the difference between oscillatory energy flux E_1 entering the system and oscillatory energy flux E_2 dissipated by the system to the total energy E_Σ present in the system,

$$\delta = (E_2 - E_1)/2E_\Sigma \tag{2}$$

Thus, the stability characteristic analysis can be associated with the analysis of the energy present in the oscillatory system and the oscillatory energy fluxes entering the system and dissipated by the system.

LRE combustion chambers or gas generators, being nonlinear oscillatory systems, can be represented as two members: an active member and a passive one with direct and feedback coupling. The active member is a source of oscillatory energy, and the passive one is its absorber. Nonlinear properties of these members appear in the dependence of the incoming flux of oscillatory energy $E_1(A)$ and the energy flux dissipated by the system $E_2(A)$ on the oscillation amplitude A. The relationship of these fluxes of oscillatory energy in a certain system defines its stability characteristics. The combustion chamber, being a nonlinear oscillatory system with volume distributed parameters, has some peculiarities that make it different from nonlinear oscillatory systems with lumped parameters. Some of these peculiarities are 1) the existence of a large number of oscillation forms with distinct amplitudes, many of which are realized simultaneously; 2) a possibility of interaction between various modes of oscillation; and 3) parameter fluctuations in the chamber. These peculiarities require defining the oscillation modes for which characteristics should be considered and for which the stability should be estimated.

The parameter fluctuations cause some scatter of oscillatory energy flux and stability characteristic values. To make the concepts and methodological approaches to stability rating clearer, however, in this section oscillations in combustion chambers will be treated from the deterministic standpoint, that is, only mean values of energy fluxes and stability characteristic will be used. Let us consider some peculiarities of the oscillation amplitude dependence of energy fluxes $E_1(A)$ and $E_2(A)$.

The change pattern of the energy flux entering the oscillatory system $E_1(A)$ is influenced by the propellant transformation conditions and, hence, should be different for combustion chambers with different propellant-phase states (gas–gas, gas–liquid, or liquid–liquid). Besides, the maximum level of the generated oscillatory energy flux depends on the oscillation period, on the amount of energy convected by propellants entering the system during this period, on the propellant transformation efficiency downstream of the combustion front, etc. As a consequence, the curve for energy flux $E_1(A)$ may have one or several more or less distinct bends.

For the qualitative analysis required here, let us plot energy fluxes entering the system and being dissipated by the system against the amplitude of a certain oscillation mode (Fig. 1). Oscillatory energy losses in a combustion chamber (gas generator) include energy losses through the injector, nozzle, and chamber side surfaces, as well as losses due to the chamber medium heterogeneity, viscous effects, etc. Some processes connected with the interaction of disturbances and the oscillatory system active member (atomization, droplet vaporization, and so on) also cause oscillation energy dissipation.

Each of the curves shown in Fig. 1 corresponds to certain design and operating parameters of the system. Three possible cases of oscillation amplitude dependence of oscillatory energy fluxes are shown. The net energy characteristic, which is the value of the oscillatory excess energy flux $\Delta E(A) = E_2(A) - E_1(A)$, defines the amplitude dependence of the damping factor. As E_Σ is always positive, the sign of the damping factor in Eq. (1) is defined by the sign of the difference $\Delta E(A)$.

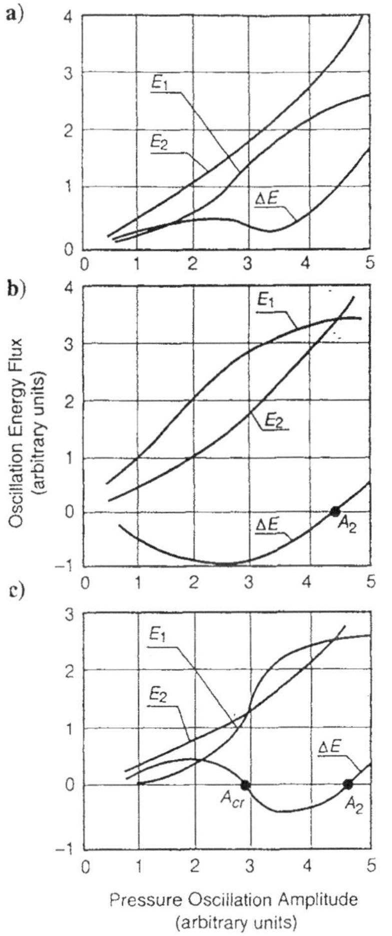

Fig. 1 Energy flux entering the oscillatory system E_1, energy flux dissipated by the system E_2, and their difference, $\Delta E = E_2 - E_1$, as a function of the oscillation amplitude: a) absolute stability, b) self-exciting system, and c) initial disturbance above A_{cr}.

The curve shown in Fig. 1a corresponds to an absolutely stable oscillatory system, and the curve in Fig. 1b corresponds to a self-exciting system with transition to the mode of stable self-oscillations. The curve in Fig. 1c represents the most common case; self-oscillations are excited in such a system when an initial disturbance of an amplitude above the critical value A_{cr} occurs.

From Fig. 1c, it can be seen that besides the damping factor there is one more characteristic that defines the system stability: the relation between the amplitude of the initial disturbance affecting the oscillatory system and the critical amplitude A_{cr}.

Variations in the operating mode of the combustion chamber cause changes in the relation between the energy flux entering the system and the oscillatory energy flux being dissipated in the system. These changes then result in a change of the relation $\delta(A)$.

Some possible versions of the curve $\delta = F(A)$ position change are shown in Fig. 2. The case of Fig. 2a corresponds to an absolutely stable system. In the case of Fig. 2b, the damping factor δ becomes zero at the oscillation amplitudes that are above the noise level A_n in the combustion chamber. With the mode changing further, an increase of the amplitude range in which δ is below or equal to zero is possible (Figs. 2c and 2d).

It can be seen from Figs. 2 that oscillation excitation is possible in two cases.

1) The oscillation amplitude of random disturbances exceeds the A_{cr} value (Fig. 2c). This case will be further referred to, as in most papers, as hard excitation.

2) The damping factor δ becomes zero at an amplitude corresponding to the level of combustion noise (Fig. 2d). This case of oscillation excitation will be referred to as soft excitation.

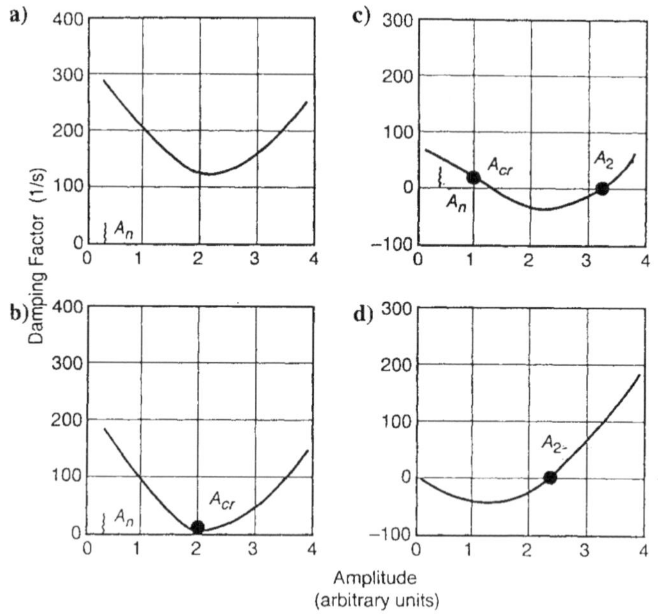

Fig. 2 Position variation of damping factor as a function of oscillation amplitude with approaching instability boundaries.

In the case shown in Figs. 2c (if the disturbance exceeds A_{cr}) and 2d, the oscillation amplitudes will change in such a way as to finally reach the A_2 level that corresponds to stable oscillations.

In practice, it is convenient to replace the dimensional value of the damping factor with a dimensionless quantity in the form of the product of this dimensional value and the oscillation period T; the dimensionless quantity is the oscillation decrement δT.

Let us determine how the dependence of pressure oscillation amplitude on decrement changes with time. Assuming the random processes in a oscillatory system to be stationary, we may write for the rms value of the amplitude change with time

$$\overline{[A(t+\tau)-A(t)]^2} = \overline{A^2(t+\tau)} + \overline{A^2(t)} - 2A(t+\tau)A(t)$$
$$= 2\overline{A^2(t)} - 2B_A(t) = 2\overline{A^2}(t)[1 - R_A(\tau)] \tag{3}$$

where $A^2(t)$ is the rms amplitude, $B_A(\tau)$ is the correlation function of the envelope $A(t)$, $R_A = B_A(\tau)/B_A(0)$ is the correlation coefficient of the envelope $A(t)$, and $B_A(0)$ is the correlation function $B_A(\tau)$ at $\tau = 0$.

The correlation coefficient $R_A(\tau)$ of the envelope of a random process at a simple oscillatory system outlet is known[5] to have the following form:

$$R_A(\tau) = 4/5[1 + (1/4)e^{-2\delta\tau}] \tag{4}$$

Substituting Eq. (4) into expression (3) gives

$$\overline{[A(t+\tau)-A(t)]^2}/\overline{A^2(t)} = 2/5(1 - e^{-2\delta\tau}) \tag{5}$$

From Eq. (5) it is evident that the rms value of the amplitude change for a fixed time interval decreases with a decrease in the damping factor δ, i.e., the time dependence of oscillation amplitude becomes smoother.

Let us consider the connection between the oscillation amplitude and the decrement. Let a wide-band random signal $x(t)$ (which for simplicity is assumed to be stationary white noise) be applied to the inlet of an oscillatory system with the following pulse transient function:

$$h(t) = \begin{Bmatrix} e^{-\delta t}\sin\omega t & t > 0 \\ 0 & t < 0 \end{Bmatrix} \tag{6}$$

Then at the system outlet we have for the signal $y(t)$

$$y(t) = \int_{-\infty}^{\infty} h(t-\tau)x(\tau)\,\mathrm{d}\tau$$

$$y^2(t) = \int_{-\infty}^{\infty}\int_{-\infty}^{\infty} h(t-\tau)h(t-\tau')x(\tau)x(\tau')\,\mathrm{d}\tau\,\mathrm{d}\tau'$$

or keeping in mind that the white noise correlation function is

$$\overline{x(\tau)x(\tau')} = \delta(\tau - \tau')\overline{x^2}(t)$$

where $\delta(\tau - \tau')$ is the Dirac delta function, we get

$$\overline{y^2(t)} = \overline{x^2(t)} \int_0^\infty h^2(\tau)\, d\tau \tag{7}$$

Substituting Eq. (6) into Eq. (7) gives

$$\begin{aligned}\overline{y^2(t)} &= \overline{x^2(t)} \int_0^\infty e^{-2\delta\tau} \sin^2 \omega\tau\, d\tau \\ &= \overline{x^2(t)} \frac{1}{2} \int_0^\infty e^{-2\delta\tau}(1 - \cos 2\omega\tau)\, d\tau = \frac{\overline{x^2(t)}}{4\delta}\end{aligned} \tag{8}$$

The integration low limit is taken to be zero as the system starts at the time moment $\tau = 0$. Presenting the random oscillatory process $y(t)$ as $y(t) = A(t)\sin[\omega t + \varphi(t)]$ results in

$$\overline{y^2(t)} = \overline{A^2(t)\sin^2[\omega t + \varphi(t)]} = \overline{A^2(t)/2} \tag{9}$$

Substituting Eq. (8) into Eq. (9) gives

$$\overline{A^2(t)} = \overline{x^2(t)}/2\delta$$

or

$$\sqrt{\overline{A^2(t)}} = (1/\sqrt{\delta})\left(\sqrt{\overline{x^2(t)}}/\sqrt{2}\right) \tag{10}$$

Thus, the rms value of the oscillation amplitude at the outlet of an oscillatory system to the inlet where a random wide-band signal is applied is inversely proportional to the square root of the damping factor δ of this system.

As indicated by Eq. (10), as the chamber process approaches the soft excitation mode (instability boundary), the decrement (damping factor) decrease is accompanied by a rise in the amplitude of acoustic pressure oscillations. As the system approaches the instability boundary, the energy spectrum parameters vary as well. The oscillations become more regular, i.e, the energy of the oscillations at their natural acoustic frequencies, corresponding to a higher quality factor of the system, gets relatively higher. The change in the spectrum of oscillation energy may be characterized by a number of parameters, all of which can be used in certain cases to indicate that the system is approaching the instability boundary. These parameters provide more information as they become more closely connected with δT.

A hypothetical dependence of pressure oscillation amplitude on operating parameter Y is presented in Fig. 3. In the parameter range $Y \in Y_{cr1}, \dots, Y_{cr2}$, the given dependence has a hysteresis character. The ratio A_{cr}/A_n defines how far the noise level $A_n(Y)$ is from the amplitude $A_{cr}(Y)$, resulting in excitation of oscillations of the operating parameter (Y). Artificial disturbances should be applied to the combustion chamber for determining A_{cr}. In practice, some problems arise while determining A_{cr}. Either the critical value of the amplitude could not be attained even at high-disturbance amplitudes or the combustion chamber was damaged as a result of attaining A_{cr}. Thus, it is taken that the margin to hard excitation of oscillations is sufficient if 1) after applying a pulse disturbance, pressure oscillations decay and 2) the disturbance parameter $n = A_0/2A_{n,\max} \geq N$. Here,

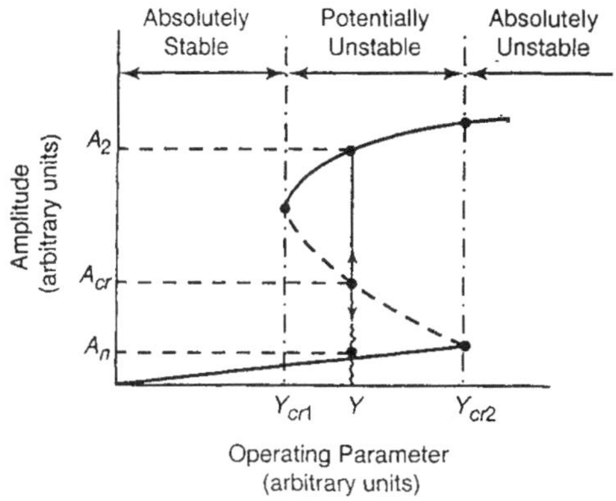

Fig. 3 Pressure oscillation amplitude in combustion chamber against operating parameter Y defining the combustor process stability (bifurcational diagram); Y_{cr1} and Y_{cr2} denote the critical values of Y.

the initial pressure peak under pulse disturbance A_0, the maximum value of the total high-frequency signal within a certain time interval before application of the pressure pulse disturbance $A_{n,\max}$, and a constant N are found in experiments. The experience of evaluating LRE combustion chamber process stability has shown that this characteristic is informative enough and represents the objective reality adequately.

If A_{cr} cannot be obtained, which is the most frequent case in practice, it is necessary to show that combustion chamber pressure pulse disturbances that exceed the noise level by a certain factor (N) decay sufficiently fast.

For characterizing the decay velocity of pressure oscillations caused by pulse disturbances, the relaxation time τ_r (i.e., the time of oscillation amplitude decrease by a factor of e behind an initial transient region corresponding to a constant time of τ_1) may be used. It should be noted that in evaluating chamber process stability with relaxation time the maximum pressure peak due to a pulse disturbance should distinctly differ from combustion chamber natural noise pressure peaks but not so much so as to distort the essential characteristics of the chamber pressure. Applying disturbances that greatly distort the chamber process may give incorrect information as high disturbances affect the propellant flow to the combustion chamber.

A quantity inversely proportional to the relaxation time is directly proportional to the effective (amplitude-averaged) decrement δT_e. Thus, the stability characteristics that are defined from natural noises δT and from pulse disturbance τ_r have the same physical essence based on the damping velocity of pressure disturbances of a different nature (natural or artificial). Sufficiency of the selected stability characteristics and their actual values are confirmed by multiple data obtained with experimental and real LREs.

As an example to demonstrate the pattern of the effective decrement variation (Figs. 1 and 2) against amplitude, some experimental results are given in Fig. 4. The study was carried out in a gas/liquid combustion chamber with nitrogen tetroxide as an oxidizer and UDMH as a fuel. The chamber diameter was 0.26 m. The injector had 91 pneumatic injection elements of 18 mm diam with eight 1.2-mm holes in the walls for feeding the fuel. The oscillation effective decrement was determined by introducing disturbances of a prescribed amplitude by a shock tube

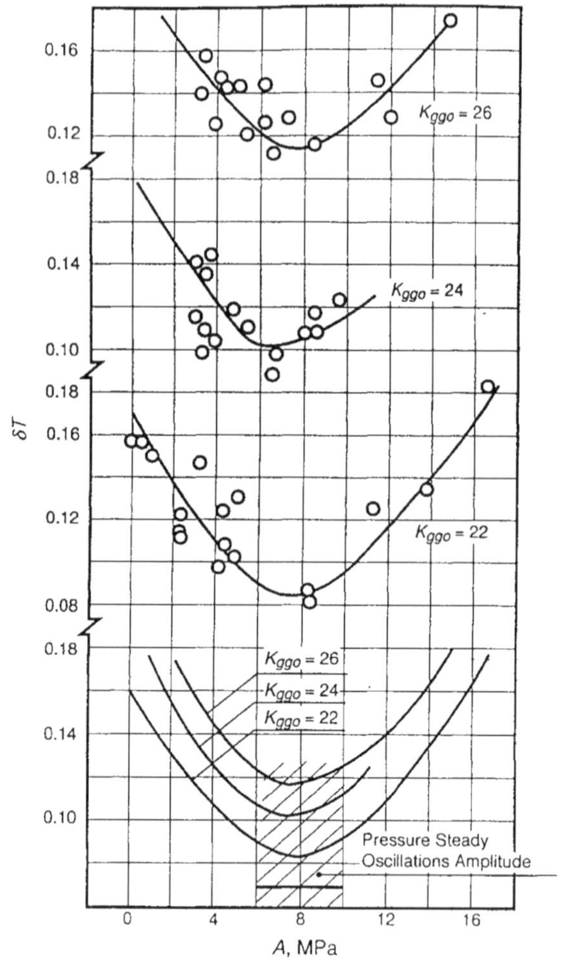

Fig. 4 Oscillation effective decrement δT against pressure initial peak A_0 for operation modes with various mixture ratio values in gas generator, experimental gas/liquid combustion chamber.

and was calculated from the following relationship with the relevant amplitude averaging:

$$(\delta T)_i = \ell n[(A)_i/(A)_{i-1}] \tag{11}$$

It can be seen from Fig. 4 that the curve for $\delta T_e(A)$ has the minimum value at disturbance amplitudes at which stable self-oscillations are realized on achieving $\delta T = 0$ (at the instability boundary). With a decrease in the mixture ratio in the gas generator, the decrement against amplitude curve moves down, which is evidence of the chamber process approaching the instability region.

III. Disturbance Devices

A device meeting the following requirements had to be developed for rating chamber process stability to hard excitation.

1) The artificial disturbance pulse should be equivalent (in its effect on the combustion chamber process) to most pressure disturbances observed in a combustion chamber.

2) The disturbance device (DD) should function reliably under the firing test condition of the LRE in the required range of combustion chamber and gas generator operating parameters (pressure, temperature, vibration level).

3) The disturbance should be applied to the system at a specified moment.

4) The shock wave amplitude at the device outlet should be constant. It is only in this case that stability quantitative evaluation can be carried out.

5) The disturbance being applied should have a certain form and duration.

In the early work of the design bureaus in Russia and of foreign firms, explosive disturbance devices, in which either an explosive charge was detonated or a powder charge ignited, were used. The large scatter of pulse characteristics was the main drawback of those devices. There were attempts to reduce the scatter by mounting a separating diaphragm with a calibrated burst pressure at the chamber inlet.

Disturbance Device Type	Operating Pressure MPa	Disturbance Pressure Pa	Number of Disturb. Device on Engine	Years of Application
Shock Tube (gas) a) 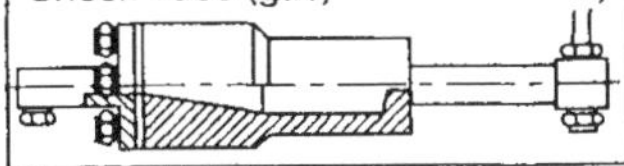	≤ 10	50–200	4	1966–1977
In-Chamber b) 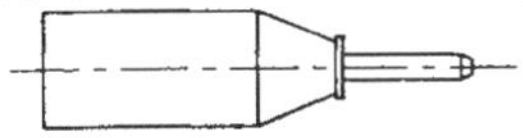	≤ 20	100–600	1	1971–1979
Four-Charge c) 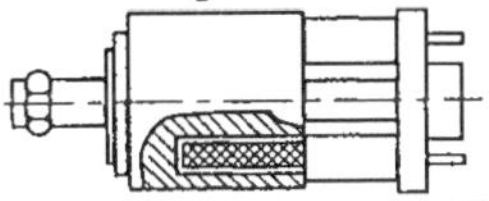	< 20	20–200	1	1974–1979
Two-Three-Charge d) 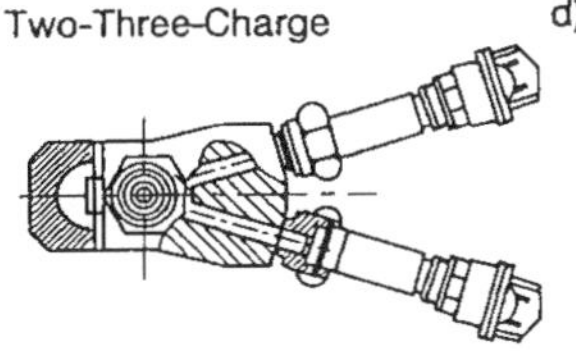	≤ 70	≤ 1,000	1	1980
Five-Charge e) 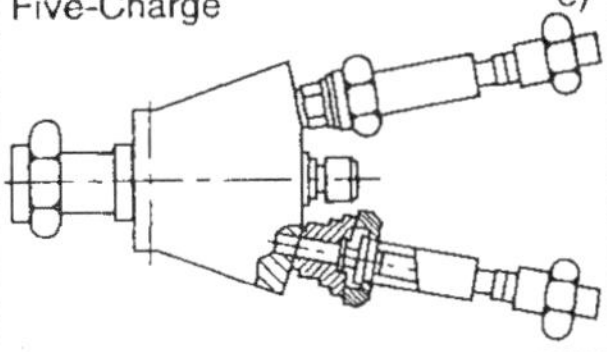	≤ 40	≤ 1,000	1	1980

Fig. 5 Disturbance devices.

Experts of NIICHIMMASH and the Physical Chemistry Institute of the Academy of Science developed a series of disturbance devices, a shock tube, shown in Fig. 5a being one of them. The shock tube includes a case having a cavity filled with a stoichiometric mixture of methane and oxygen that is separated from the combustion chamber by a burst diaphragm. On initiating detonation in the cavity, the diaphragm bursts, and a disturbance is introduced into the combustion chamber. The cavity filled with the detonable mixture has an "acceleration region" along which the combustion is accelerated after the ignition of the mixture, and then the combustion transition to detonation occurs.

The in-chamber disturbance device shown in Fig. 5b is an explosive charge placed in a case of special heat insulating material; the charge is ignited by a blasting cap (TAT-1-1) with a powder ignitor. The blasting cap is initiated on flame contact after the burnout of the protecting shell, the thickness of which defines the time delay of the device initiation. The possibility of placing the device in the chamber without breaking the chamber integrity and the device independence are its important advantages; the impossibility of introducing disturbances at any desired moment is its disadvantage.

A four-charge DD is shown in Fig. 5c. Four charges that are initiated at prescribed moments are placed in one case. Further improvement of the DD resulted in the development of multicharge devices with slower detonation modes.

A two-pulse DD for gas generators is shown in Fig. 5d. Charge chambers by way of tilted case channels are connected with a channel opening into a trap. At the channel outlet, there is a conical extension immersed into the trap that acts as a shield for repulsing diaphragm fragments.

In a three-charge version of the disturbance device for a gas generator, the charges are separated from the case channels by copper diaphragms to eliminate any oxygen gas effects. To prevent fragments from getting into the gas generator chamber, the body of this disturbance device has a separator with guides made in the form of ribs that prevent the fragments from entering the chamber while the disturbance "flowing over" the ribs enters the gas generator. The diaphragm fragment trap is made in the form of a Y-shaped channel with a decreasing diameter.

A five-charge DD for combustion chambers (Fig. 5e) is a system of disturbance devices placed in one case and connected with tilted channels to form a single outlet channel. Each charge in this device, as in the device for gas generators, is separated from the case channel with a diaphragm. For combustion chambers of small diameter (80–120 mm) a grid is mounted in the device to provide pulse control over a wider range and to trap the diaphragm fragments. The described devices were studied with the aim of analyzing the characteristics of the disturbances being applied. The study results show that the disturbance intensity and uniformity are defined mostly by the device characteristics, the disturbance introduction conditions, and the parameters of the volume being disturbed. Classification of the factors influencing the disturbance amplitude and form is given in Fig. 6. When we speak of DD characteristics, we mean its physicochemical parameters, including its dimensions (channel diameter and length) and configuration features (one charge, multicharge, the existence of damping elements, the explosive mass, and its combustion mode).

The existence of a reactive medium in the combustion chamber often results in an additional replenishing energy effect connected with additional energy release following the moving pressure front. This causes the disturbance amplitude

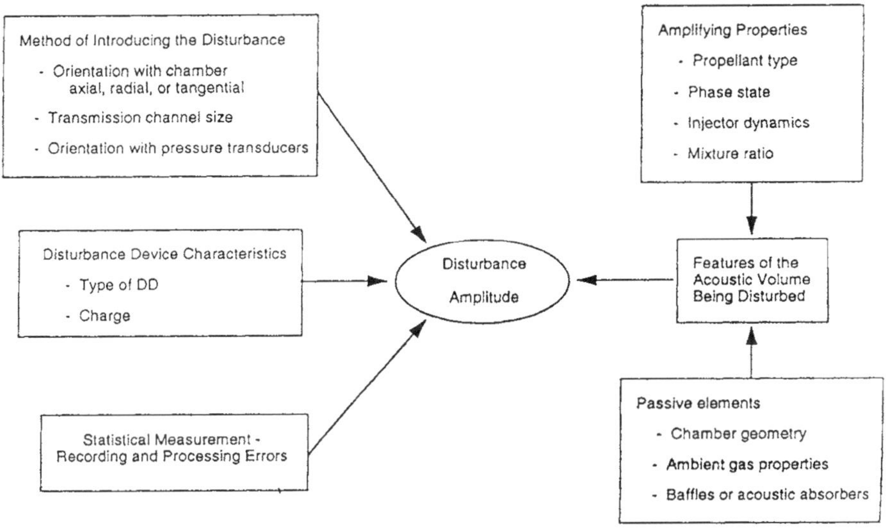

Fig. 6 Factors influencing the functional coupling of the disturbance amplitude in the combustion chamber with the charge mass of the disturbance device.

increase in the combustion chamber. The effect of the volume being excited, however, is defined not only by the intensifying properties of the combustion zone in the chamber but by such parameters as pressure, density, sound velocity, geometry, and other features of the object being analyzed.

To evaluate the influence of the aforementioned factors on the disturbance level $A_{\rm col}$, laboratory tests were carried out that gave some quantitative and qualitative relations.

1) For shock wave damping along the channel with a diameter of 4–12 mm and a length of up to 650 mm with various charge masses, the experiment showed that with length-to-diameter ratio increase up to $l_{\rm ch}/d_{\rm ch} \approx 10$ the disturbance amplitude declines by 20%. With a real channel diameter decrease, $d_{\rm ch} < 6$ mm, the damping along the channel is greater.

2) For evaluating the chamber pressure, medium density and sound velocity influences, experiments were carried out in volumes filled with air, helium, and pyroxylin combustion products. The experiments showed that the relationship between the medium initial pressure P_c and the sound velocity is the main factor defining the disturbance characteristics.

3) For the method of introducing the disturbance into the chamber, the experiments showed that at chamber pressures greater than 8 MPa, the influence of introduction angle on the disturbance form is small. The relationships obtained allow one to evaluate the requisite disturbance amplitude value at the chamber inlet for an inert medium. A five-charge DD provides the prescribed value of the disturbance with an accuracy of +10%.

The number of the LRE tests carried out to evaluate the operating process stability to hard excitation of high-frequency oscillations permits one to refine the results of cold tests. When comparing the results of cold tests and tests performed under actual conditions, disturbance amplitude in firing tests was found to be higher than in cold tests. From the performed studies it is evident that the

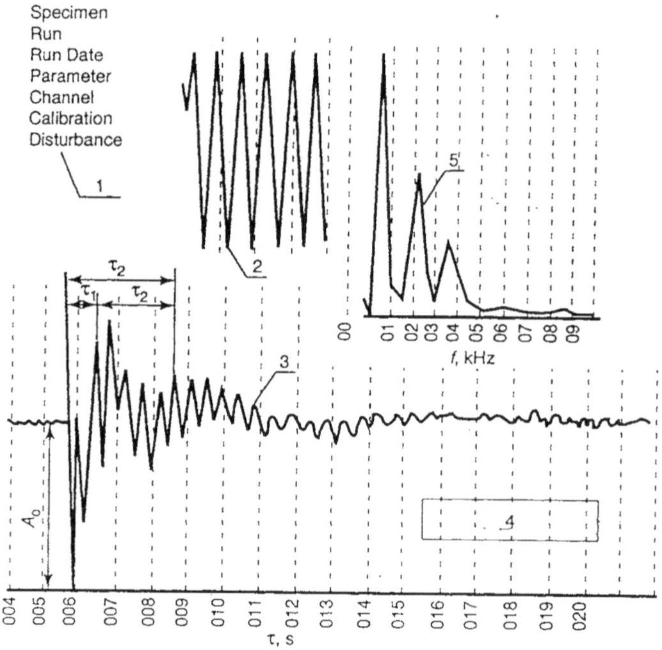

Fig. 7 Example of information processing results obtained after applying an artificial pressure disturbance. 1 information about test and parameter, 2 calibration, 3 total signal, 4 values of parameters being calculated, 5 energy spectrum; τ_r oscillation relaxation time and τ_1 time of pressure pulse direct action on chamber process.

ratio of firing test disturbance amplitude to cold test disturbance amplitude increased with chamber diameter increase. With chamber diameters changing from 0.08–0.1 to 0.26 m this ratio increases by 15–20%. For a combustion chamber running with liquid/liquid propellants (nitrogen tetroxide + UDMH) the disturbance amplitude in firing tests is a factor of 1.5–2 higher than that in cold tests. For a combustion chamber running with gas/liquid propellant this factor amounts to 1.1–1.2.

The experimental results allow one to define the expected value of the disturbance as a function of explosive mass with high accuracy. The results of processing the information obtained after introducing artificial disturbances into a combustion chamber are shown in Fig. 7. An evaluation of the results of the stability to pulse disturbances applied to a hydrogen/oxygen engine combustor (nominal $P_c = 5.8$ MPa, thrust = 75 kN, and $D_c = 0.2$ m) are given in Fig. 8.

IV. Decrement Determination from Natural Disturbances

In evaluating LRE combustion chamber process stability, combustion chamber noise can also be used as a natural disturbance source. It is natural to suppose that these disturbances are small enough so that the pulsations caused by them can be described with linearized equations. These disturbances are described with random functions, whose spectral densities are of a larger bandwidth than the frequency characteristics of the combustion chamber in the neighborhood of the natural frequencies being studied. Then the chamber pressure pulsations caused

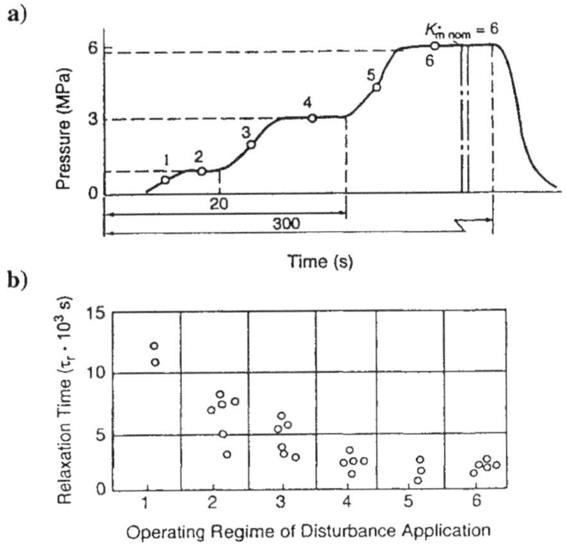

Fig. 8 Disturbance relaxation time τ_r variation with oxygen/hydrogen engine operating mode change: a) engine operation cycle where $i = 1$–6 indicate number of disturbance applications and b) values of various relaxation times.

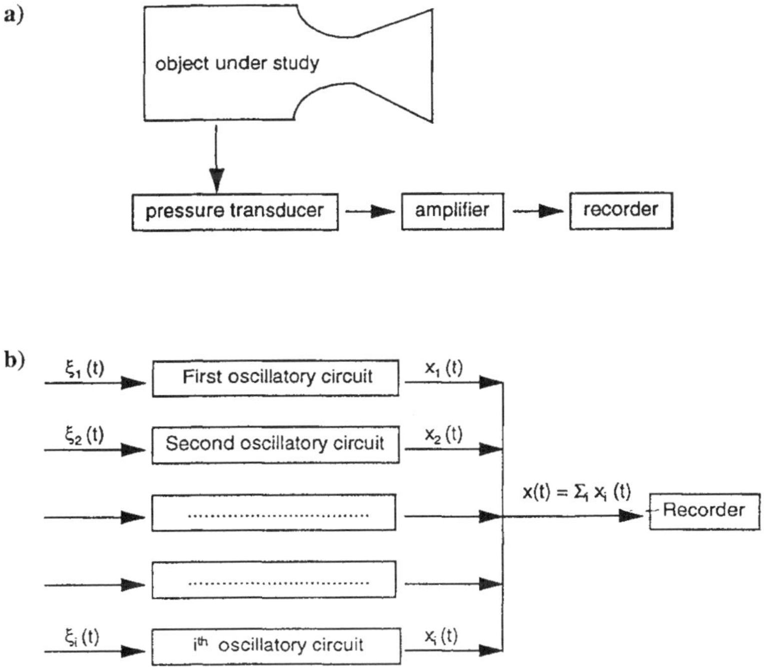

Fig. 9 Block diagram of a) pressure pulsation recording of the object under study and b) equivalent electrical block diagram of signal recording.

by these disturbances in a general case can be described with a system of linear equations and presented by the block diagram shown in Fig. 9. The combustion chamber, being an oscillatory system with distributed parameters, is reduced to the sum of oscillatory systems of second order.

A series of methods is devised for determining the decrement from natural disturbances. The following ones will be considered: spectral, correlational, amplitude, and instantaneous period method.

1) The spectral method defines the decrement from the width of the noise power spectrum (Fig. 10). In an actual case, the filter width Δ_f of the spectrum analyzer should be taken into account. For a simple filter, $\delta T = \pi(\Delta - \Delta_f)/f_0$, where δT is the oscillation decrement, Δ the width of power spectrum $S(\omega)$ at a level of half that of the maximum amplitude, and f_0 the resonant frequency.

An algorithm was devised for taking Δ_f into account for other types of filters as well. In an actual algorithm, more complicated formulas are used to account for neighboring (close) peaks, background, and other interferences (three-point methods). An exhaustive search of point pairs lying to the left and to the right of the maximum is performed, and a pair is selected that gives the minimum sum of the decrement value and its theoretical statistical spread.

2) The correlation method defines the decrement from the rate of decay of the auto- or cross-correlation function $B(\tau)$ (Fig. 11)

$$\delta T = [1/(m - n)] \cdot \ln[B(\tau_n)/B(\tau_m)] \tag{12}$$

where τ_n and τ_m correspond to maxima of $B(\tau)$ for nth and mth oscillation periods of $B(\tau)$.

3) The amplitude method (Fig. 12) defines decrement δT as a function of the ratio of two amplitudes at the outlet of two different filters tuned to a resonant frequency of the oscillatory process being analyzed: $\delta T = F(A_1/A_2)$. In a particular case the decrement is defined from the ratio of amplitudes at the inlet and the outlet of a filter (preferably a rejection one). Discrete values of the function F are determined by calibration on a simulator that is an oscillatory circuit with a varying decrement. The decrement is calculated from the discrete values by linear interpolation.

4) The instantaneous period method (Fig. 13) defines the decrement from the mean module value $|\bar{T}'|$ of instantaneous period fluctuations of the oscillations

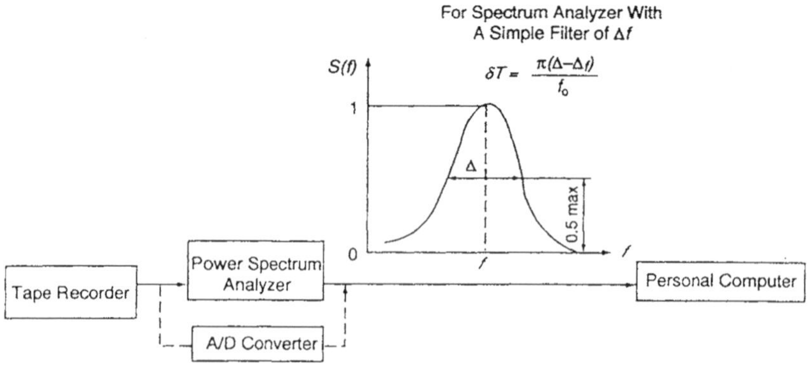

Fig. 10 Block diagram of oscillation decrement δT determined by a spectral method.

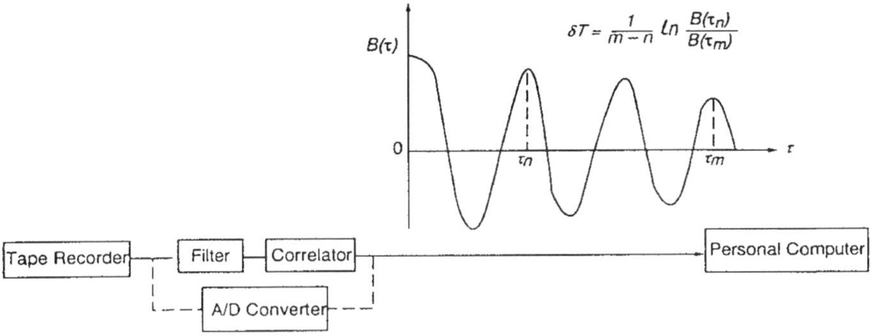

Fig. 11 Block diagram of oscillation decrement δT determined by a correlation method.

being analyzed: $\delta T = F(|\bar{T}'|)$. The instantaneous period of a random signal with a zero mean value is the sum of two adjacent intervals between the signal successive zeros. Discrete values of the function F are defined by calibration on a simulator, as in the amplitude method. The decrement is found from these values by linear interpolation.

There should be a definite idea of the decrement accuracy before using it for quantitative estimation of stability margins. The error of the decrement estimation consists of its statistical error and the error due to interfering factors, such as close frequencies and background noise.

The theoretical estimate of statistical error requires tedious calculations and results in awkward expressions. It is connected with the necessity to account for the relationship between the spectrum points and the correlation function. The results of these theoretical studies are, therefore, given in a simplified form.

For the spectral method, the value of the relative statistical error $\varepsilon_{\delta T}$ of the decrement estimate is approximately equal to the statistical error ε_{s_0} of the power

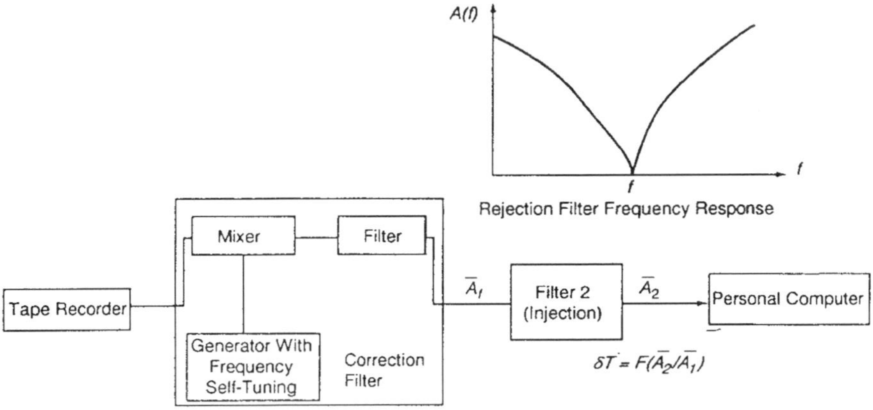

Fig. 12 Block diagram of oscillation decrement δT determined by an amplitude method.

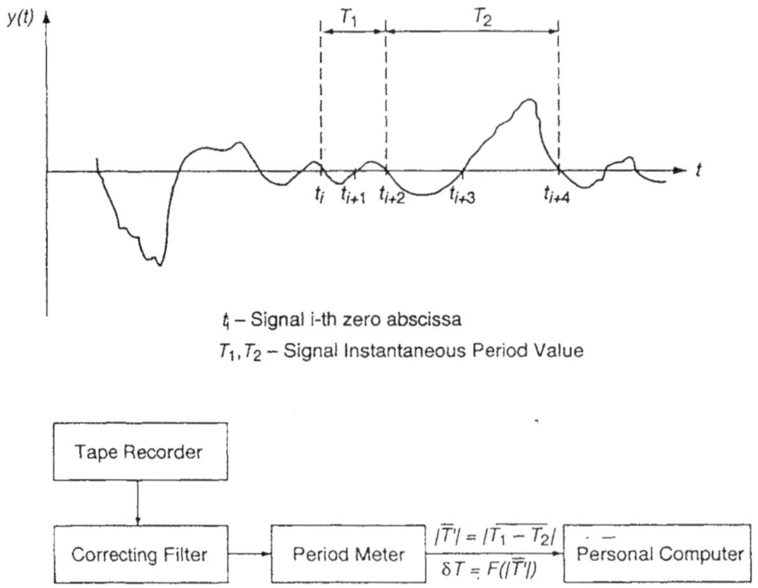

Fig. 13 Block diagram of oscillation decrement δT determined by a method of instantaneous period.

spectrum estimate $S(\omega_0)$ (the square of the filtered amplitude) at the resonant frequency ω_0,

$$\varepsilon^2_{\delta T} \simeq \varepsilon^2_{s_0} \approx 1/(\Delta_e \cdot T_a) \tag{13}$$

where

$$\varepsilon^2_{\delta T} = \overline{(\delta T - \overline{\delta T})^2}/(\delta T)^2$$

is the square of the relative statistical error of the estimate of the decrement δT,

$$\varepsilon^2_{s_0} = \overline{[\tilde{S}(\omega_0) - S(\omega_0)]^2}/S^2(\omega_0)$$

is the square of the relative statistical error of the estimate $\tilde{S}(\omega_0)$ of the power spectrum at the resonant frequency ω_0, T_a is the averaging time, Δ'_f the effective width of the filter amplitude-frequency characteristic raised to the fourth power $[A^4(\omega)]_f$. Also Δ'' is the effective width of the chamber plus filter amplitude-frequency characteristic raised to the fourth power $[A^4(\omega)]_{f+c}$, and $\Delta_e = (\Delta'_f)^2/\Delta''$. The effective width of the characteristic is the ratio of the area under the curve $A^4(\omega)$ to the maximum of $A^4(\omega)$.

Let us consider two extreme cases of very wide and very narrow filters, $(\Delta'_f \gg \delta)$ and $(\Delta'_f \ll \delta)$, respectively. For the wide filter case, we have minimum statistical error, $\varepsilon^2_{\delta T} \approx 1/\delta T_a$; however, the frequency resolution is poor and the error due to the existence of closely spaced frequencies, background, etc. is large. For the narrow filter case, we have good frequency resolution and large statistical error, $\varepsilon^2_{\delta T} \approx 1/\Delta'_f T_a$. From the extreme cases just analyzed, it follows that there exists an optimum filter width that gives the least total error.

For the correlation method, the formula of decrement error estimation for the simplest case when the decrement is determined from two points $B(t_1)$ and $B(t_2)$ with width $t_1 = 0$ is $\varepsilon^2_{\delta T} \simeq [1 + f(\alpha)]/\delta T_a$, where $\alpha = \delta T_n$, $n = \tau_2/T$ (the number of periods), and $f(\alpha) = (e^{2\alpha} - 1 - 2\alpha - 2\alpha^2)/2\alpha$ The minimum statistical error of the decrement estimate is achieved with $n = 1$ (adjacent maxima). In this case, $f(\alpha) \simeq 0$ and $\varepsilon^2_{\delta T} \simeq 1/\alpha T_a$.

To process an actual signal using the initial part of the correlation function is unreasonable, however, because of the background influence and the filter existence. Using adjacent maxima is unreasonable because of the instrument error influence and closely spaced frequencies; thus, the initial part $B(\tau)$ is eliminated, and the remaining part $B(\tau)$ is used more completely by proper averaging of some successive values of $B(\tau_n)/B(\tau_{n+1})$.

For the amplitude method, the relative statistical error consists of relative estimates of both filters' squared amplitudes and, as the relative error for the rejection filter is an infinitesimal value, the relative statistical error is actually equal to the relative error of the squared amplitude at the first filter outlet: $\varepsilon^2_{\delta T} \simeq \varepsilon_{A_1^2}$. Thus, everything that was said about the spectral method relative statistical error also holds true for the amplitude method. As for the instantaneous period method, no theoretical study of the statistical error was performed because of the complexity of the task. Only experimental error estimation with the help of a simulator was carried out. Actual statistical errors for all described methods amounted to about

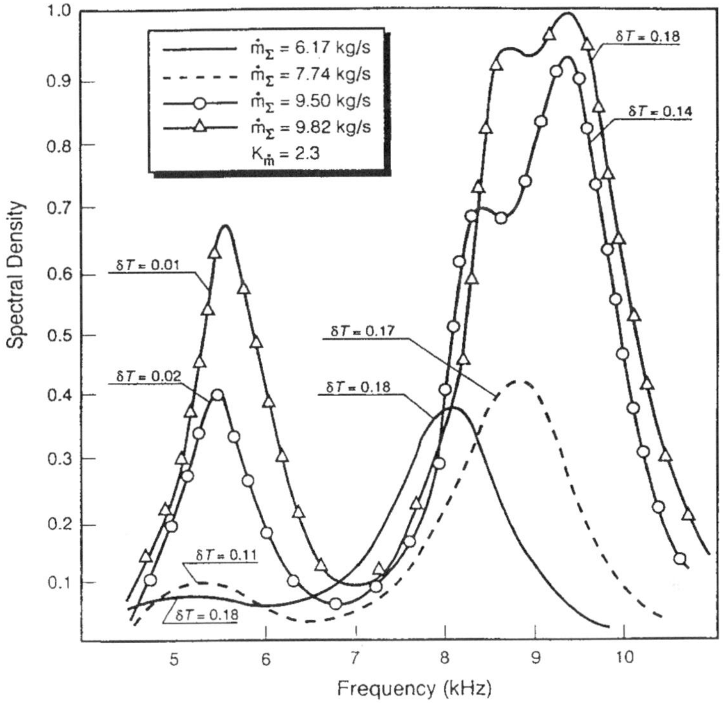

Fig. 14 Variation of combustion chamber pressure oscillation spectrum and decrement with propellant flow rate change with f = 5.5 kHz, first tangential mode, D_c = 120 mm, P_c = 80 bar, and nitrogen tetroxide and UDMH propellants.

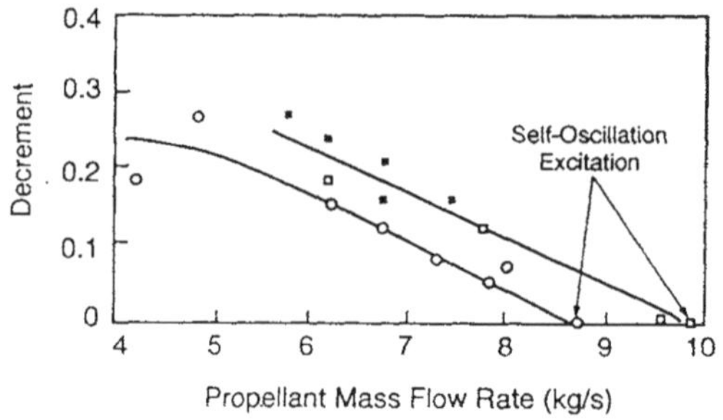

Fig. 15 Pressure oscillation decrement value as a function of propellant flow rate, first tangential mode: ○ engine tests and □ component tests.

10% and 30% with a realization time of 1 and 0.08 s, respectively. Thus with a realization time of 1 s, the statistical error is not a governing factor, and the total accuracy is defined by interfering factors already mentioned. Statistical error is often governing when the averaging time is 0.08 s.

The results of decrement determination from natural disturbances in component tests of a liquid/liquid engine combustion chamber running nitrogen tetroxide and UDMH are shown as an example in Figs. 14 and 15. The chamber diameter is 0.12 m, pressure 80 bar, and propellant total flow rate $\dot{m}_\Sigma = 6.2$ kg/s. Tests revealed (Figs. 14 and 15) that with flow rate increases the decrement corresponding to the first tangential mode decreases, approaching zero at a flow rate of 9.8 kg/s. It should be noted that the spectral density maximum moves to the right while approaching the combustion instability boundary (Fig. 14), which is indicative of an increase in volume-averaged sound velocity at the expense of a decrease in the cold zone volume.

The decrement against the flow rate while testing the chamber as part of the engine is shown for comparison in Fig. 15. The oscillation decrement found from natural noises substantially decreased and reached zero at a flow rate of $\dot{m}_\Sigma =$ 8.8 kg/s. This decrement change can be explained by the change of the combustion chamber operation conditions in the engine. The vibration level increased approximately by a factor of 1.5.

V. Flame Stabilization Effects

In the course of LRE development, some cases revealed that the scatter of the stability characteristics greatly exceeded their determination accuracy. While evaluating combustion chamber stability with pulse disturbances, differences between pressure oscillation noise spectra and values of the decrement δT (determined from noise) obtained before and after the application of the disturbance were found. The ambiguity appears in the occurrence of spontaneous random transitions from a state of small oscillations (noise) to a state of regular self-oscillations under otherwise fixed conditions. In such cases, an understanding of the physical processes causing the inconstancy of the stability characteristics is necessary for proper modification of combustion chambers and gas generators.[5,6]

The results of an experimental study[7] of the liquid kerosene/gaseous oxygen combustion chamber process bistability nature are presented next. The chamber injector had four emulsifying injection elements. A temperature transducer was mounted in the region where the fuel jet was introduced into the oxidizer flow in the

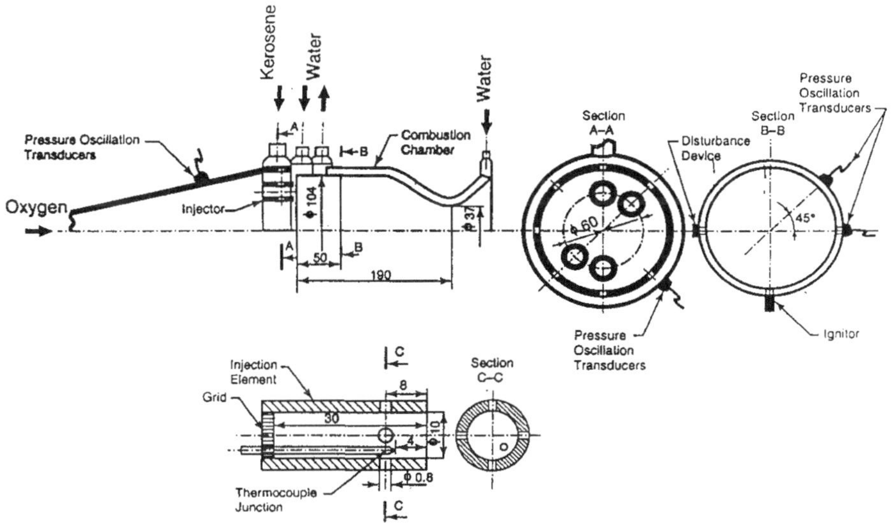

Fig. 16 Schematic of combustion instability study: DD indicates disturbance device and I ignitor; $\lambda r = 604$, $\lambda r = 608$, high-frequency pressure transducers.

injection element gas passage. The chamber was equipped with a self-contained ignitor and a device for applying pulse disturbances to the combustion zone. The dynamic state of the chamber process was monitored with high-frequency pressure transducers mounted in the reaction zone and in the cavities upstream of the injection elements (Fig. 16).

In the course of the experiment, the following states were recorded depending on the chamber pressure and equivalence ratio: the state of regular self-oscillations at one of the combustion chamber natural frequencies (6300 Hz); the state characterized by alternating combustion modes with noise waves and with self-oscillations, the instant of occurrence and the durability of each dynamic state being random in the series of similar realizations of the operating parameters; and the acoustically stable noise state.

The effect of an introduced disturbance is dependent on the chamber operation regime. The disturbance could have no effect, or organized oscillations could grow from chamber noise. If organized oscillations are present prior to the disturbance, the mode of these oscillations could be changed to random noise by the action of the pulse disturbance. The organized oscillations could also recover their initial amplitude after the pulse.

An analysis of test results revealed that the alteration of the dynamic state was accompanied by transients in the temperature of the gas in the injector passages. At stable operating conditions, with some amount of combustion noise, the measured gas temperature in the injector passage was about 325 K. At the onset of spontaneous instability with a frequency of ~6500 Hz and peak-to-peak amplitudes of about 20 kPa, the temperature of the gas in the injector passage increased to approximately 525 K within 0.01 s. A disturbance was introduced, that induced a transition in operation from unstable back to the combustion noise state. With the transition to the combustion noise state, the injector gas temperature went back down to about 325 K. Further analysis of the experimental data leads to a

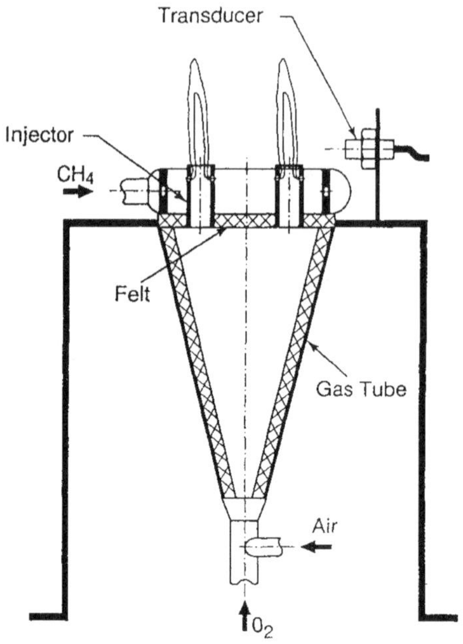

Fig. 17 Elements of gas burner setup.

conclusion that these temperature transients were caused by the variations in the flame stabilization mode of the injectors, which were either spontaneous or were induced by pause disturbances.

The whole complex of the results allows one to explain the observed phenomena if one allows the possibility of multiple combustion zone stationary states within every injection element, which differ in flame stabilization mode, as well as the possibility that the stationary states change, either spontaneously or inductively, by the action of artificial disturbances. The type of dynamic transition that occurs depends on whether the stationary states are stable or unstable with small acoustic disturbances.

In connection with the possibility of combustion zone multiple states occurring under fixed conditions, a question of the nature of dynamic transition arises. Some observations of an open diffusion flame formed by an injector were carried out to reveal physical processes preceding the transition to vibratory combustion. The injector, with a configuration identical to the injector depicted in Fig. 16, had four injection elements with a radial introduction of methane into the flow of gaseous oxygen and air (Fig. 17). The parameter range corresponding to random alternation of flame stabilization modes accompanied by simultaneous wide-band noise level change was studied. Random change of flame stabilization modes occurred at separate fuel holes, both within one injection element and within a group of elements. Flame stabilization could be controlled by varying the air flow rate.

The following two stabilization modes were observed.

1) The flame envelops the whole base of the fuel jet over the periphery: the flame has a distinct pattern and a low level of wide-band noise (Fig. 18a).

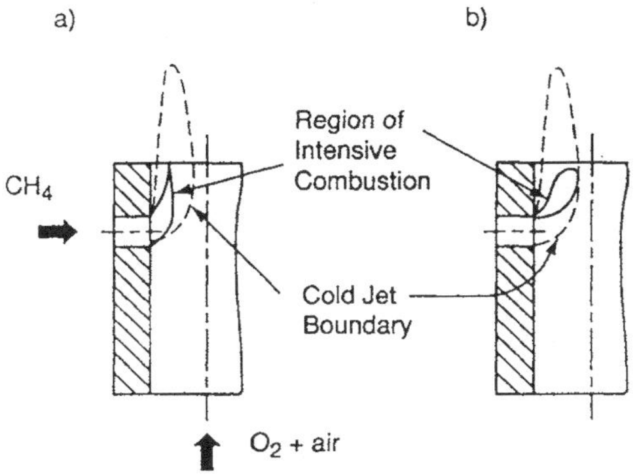

Fig. 18 Modes of flame stabilization in injection element: a) flame envelops base of fuel jet entirely and b) flame partly separates from fuel jet.

2) The flame partly separates from the fuel jet, being stabilized in back streams behind the jet: the flame has no definite contour and is characterized by an increase in noise level (Fig. 18b).

It was discovered that after some period of the system being in the regime of random alternation of flame stabilization modes with fixed propellant flow rates, a catastrophic change of the flame pattern takes place on all injection elements simultaneously, and this change is accompanied with a sharp (by an order of magnitude) increase of noise. This combustion mode is characterized by a hysteresis zone in the region of the governing parameters (oxygen, air and methane flow rates). High-level noise has a self-oscillation nature.

Many observations showed that oscillation self-excitation occurs after the transition of the flame stabilization at the jet base to the stabilization in back streams, at least on two opposite jets of one of the four injection elements. Thus, the obtained experimental results demonstrate one of the causes of the irreproducible nature of the stability characteristics.

In practice, chamber process stability can be determined by estimating stability characteristics before and after introducing artificial disturbances. A substantial excess of decrement scatter over the accuracy of its determination at fixed conditions is also indicative of combustion process instability.

VI. Conclusions

1) Liquid-propellant rocket engine reliability depends largely on the level of combustion chamber and/or gas generator process stability. This level should be rated with characteristics that give objective information about combustion stability. The application of pressure oscillation decrement and oscillation relaxation time as stability characteristics is reasoned theoretically. The decrement is defined from inherent combustion chamber noise at natural acoustic frequencies, and the relaxation time is defined from the decay of pressure oscillations induced by an artificial pulse disturbance at a certain relative maximum pressure peak. The effectiveness of these stability characteristics has been validated by experimental studies and the experience gathered in LRE development.

2) Some methods of oscillation decrement determination from random noise that is always present in the combustion chamber of an operating engine are devised and a theoretical and experimental estimation of their accuracy is made.

3) Disturbance devices were developed and investigated that allow application of pulse disturbances of a prescribed strength to the combustion chamber at a prescribed moment. Up to five disturbances can be applied to the chamber during one test.

4) Problems associated with stability evaluation were studied. It was illustrated by an example that the statistical character of self-oscillation excitation in LRE combustors is caused by the ambiguity of combustion zone stationary states at fixed feed system parameters. The tendency of the combustion zone to change stationary states can be revealed by applying pressure pulse disturbances to the combustion chamber.

Acknowledgments

The authors wish to acknowledge Mark Fisher and the NASA Marshall Space Flight Center for their help in preparing the figures.

References

[1]Harrje, D. T., and Readon, F. H. (eds.), *Liquid Propellant Rocket Instability,* Russian ed. Mir, Moscow, 1975.

[2]Oefelein, J. C., and Yang, V., "A Comprehensive Review of Liquid Propellant Combustion Instabilities in F-1 Engines," Dept. of Mechanical Engineering, Propulsion Engineering Research Center, Pennsylvania State Univ., University Park, PA, 1992.

[3]Awad, E., and Culick, F. E. C., "On the Existence and Stability of Limit Cycles for Longitudinal Acoustic Modes in a Combustion Chamber," *Combustion Science and Technology,* Vol. 46, pp. 195–222, 1986.

[4]Melchior, A., "A New Bipropellant Rocket Engine for Orbital Maneuvering," AIAA Paper 90-2052, June 1990.

[5]Bendat, J. S., *Principles and Applications of Random Noise Theory,* Russian ed., Nauka, Moscow, 1965.

[6]Dranovsky, M. L., Derevjanko, A. L., Ivanov, V. N., and Manchenkov, I. E., "About Dynamic Transients of Combustion Process in a Through-Flow Chamber," *Collected Abstracts of the (National) Tenth Symposium on Combustion and Explosion* (Chernogolovka, Russia, 1992), Russian Academy of Sciences, Moscow, 1994.

[7]Agarkov, A. F., and Ivanov, V. N., "Some Peculiarities of Gas Diffusion Flame Transition to Vibrational Combustion," *Collected Abstracts of the (National) Tenth Symposium on Combustion and Explosion* (Chernogolovka, Russia, 1992), Russian Academy of Sciences, Moscow, 1994.

Chapter 11

Liquid-Propellant Droplet Vaporization: A Rate-Controlling Process for Combustion Instability

W. A. Sirignano,* J.-P. Delplanque,† C. H. Chiang,‡ and R. Bhatia‡
University of California, Irvine, Irvine, California 92717

I. Introduction

THE problem of combustion instability in liquid-propellant rockets is a continuing challenge. Any enclosed or restricted chamber has an infinite number of modes for resonant oscillation. The heat release associated with the combustion processes can potentially provide a driving mechanism for the acoustical oscillation. The acoustical pressure and velocity oscillations tend to cause an oscillation in the local burning rate. If the amplitude of the burning rate oscillation is large enough and if the phase between the burning rate and pressure oscillations is small enough, the instability can be driven. Often the amplitudes are sufficiently large to cause such increased heat transport rates that a structural failure occurs at the injector face or at the nozzle throat. Combustion instability research was very active during the two decades of the 1950s and 1960s, especially in relation to the Apollo program. Experience with liquid oxygen- (LOX-) kerosene propellants and impinging injectors indicated severe instability problems. A general rule is that the instability is more likely and becomes more severe in terms of damage as the energy release rate per unit volume increases or, in other words, as performance increases. The behavior is more complicated than this simple rule; instability probability depends on spatial distribution of the energy release rate, combustion time lag, damping forces, and chamber dimensions (or natural frequencies). The rule, however, is a useful guide.

A. Historical Perspective

During the 1950s and 1960s, some limited research was performed on the basic driving mechanism of the instability. Several researchers[1-12] examined fuel droplet vaporization as the rate-controlling phenomenon and as the basic driving

* Professor, Department of Mechanical and Aerospace Engineering.

† Research Associate, Department of Mechanical and Aerospace Engineering.

‡ Research Assistant, Department of Mechanical and Aerospace Engineering.

mechanism. Because of limitations of their analyses (explained later in the discussion on spray combustion, Sec. VIII), they could not successfully explain the cause of combustion instability. The most widely used description of the combustion process in analyses was the Crocco time lag description.[1,13–15] This was an ad hoc representation of the combustion process through the use of a small number of parameters (usually two). One of these parameters was the sensitive time lag, which was based on the concept that the combustion process had a time lag so that the burning rate at some instant depended on conditions experienced by the reactants for some short time duration prior to that instant. The time duration or time-lag magnitude itself depended on the conditions prior to the instant of burning. The average time lag and other ad hoc parameters were considered to be functions of propellant combinations and injector design only, yielding the hope for straightforward scaling of results from small combustors to large combustors. This method had limited success.

The most reliable method for preventing large-amplitude oscillations in liquid propellant rocket engines was to damp the oscillations through the use of baffles or acoustic liners. That is, energy was taken from the most unstable modes of oscillation. The science of preventing the addition of energy to the oscillations by modifying or controlling the basic combustion process was never really developed. There is reason to believe, however, that this science now can be rapidly developed and that new technology on the control of combustion instability can be advanced. There is still a need for this new technology since the engineering fixes through damping mechanisms lead to increased volume, increased weight, increased cooling requirements, and decreased performance.

The resurgence in interest in space and the renewed interest in liquid-propellant systems creates the need to re-examine the combustion instability problem in an organized and intense manner. There is recent interest in LOX-methane systems whose instability characteristics are not well understood. LOX-kerosene systems are under active consideration again, and their instability problems are well-known. The LOX-hydrogen systems have not experienced severe instability problems. There is no inherent reason, however, why LOX-hydrogen should be more stable than other systems, and our experience with LOX-hydrogen has been with a very narrow range of designs. New designs intended to improve performance could easily lead to serious instability problems.

Because of limitations in the methodologies of those times, theoretical analyses were performed mostly with small perturbation techniques for linearized or weakly nonlinear oscillations. Other nonlinear approaches involved gross and sometimes nonrealistic simplification of the basic governing equations. On the experimental side, pressure was the only instantaneous and localized point measurement that could be made; even there pressure was measured only at outer walls. Because of the high temperature, high-frequency behavior, velocity, and scalar point measurements could not be made instantaneously in the field.

B. Combustion Instability

In this section, a brief outline of the important concepts is given for reference. The sensitive time-lag model is also mentioned for completeness. Details, however, are provided elsewhere.[1]

Combustion instabilities are more likely to occur in systems designed for propulsion because of the high-energy densities and low losses associated with such systems.[1,13] Combustion instability can be viewed as excitation of acoustical motions by heat addition. They are essentially unsteady motions of compressible gases. There are several types of combustion instabilities, usually classified using the range of frequency of the oscillations created. Of main interest are the high-frequency instabilities (or acoustic instabilities) because they are known to be the most destructive by grossly enhancing heat transfer. Note that the so-called POGO instabilities (coupling between low-frequency pressure oscillations, structural vibrations, and propellant feed system) have also been a major concern in the recent past but will not be discussed here.[1,13] There are other types of instability in a combustion chamber: hydrodynamic, chemical, thermodiffusive, and shock waves to cite just a few. These will not be treated here.[14] Combustion instabilities can be approximated by perturbations of the acoustic modes of the combustion chamber. Therefore, acoustical phenomena must be studied first. Acoustics is quite an ancient science; Lindsay[16] gives a historical perspective. Morse and Ingard give the necessary background of such phenomena.[17]

The issue at stake is to determine which of the many physical processes taking place in a combustion chamber organize into a mechanism that drives the instability. First, one must agree on the definitions of process and mechanism. There seems to be some disagreement in the combustion community regarding these two terms.[2] Our definitions follows. A mechanism is the organized sequence of events whereby a perturbation in the gas field affects a process that results in a change of the energy release rate, which, in turn, modifies the original perturbed quantity. A process is an event or a sequence of events by which a change is produced in at least one of the dependent variables (e.g., pressure, temperature, density, velocity). Vaporization, mixing, and chemical reaction are examples of processes.

The criterion used to assess if a mechanism can drive the instability is the so-called Rayleigh criterion, stated by Lord Rayleigh in 1894 in his pioneering work.[18] This criterion may be summarized as follows: A mechanism can drive the instability if the burning rate associated with it oscillates with a large enough amplitude and with a small enough phase lag with the pressure oscillation. That is, the mechanism's characteristic time is of the same order as the period of pressure oscillations. However, Lord Rayleigh was working on systems much simpler than liquid rocket engines, and the adaptation of his criterion to a more general system with chemical and phase changes is still the object of controversy. First, the response factor can be evaluated at each location in chamber or averaged over the entire combustor. The averaged approach is easier to apply and does provide an evaluation of the stability margin of the engine. The obtained response factor, however, is extremely dependent on the mode and frequency of oscillation and the engine operating conditions.[2]

The candidate mechanisms for liquid-propellant rocket combustion instability involve a coupling of the combustion chamber acoustics with any or some of the following processes: liquid injection, primary atomization, secondary atomization, chemical kinetics, vaporization and liquid heating, and mixing. A discussion of the relative importance of these mechanisms is provided in the next section.

The early attempts at describing analytically combustion instabilities in liquid-propellant engines were of two types: the time-lag model of Crocco and co-workers, and the vaporization model[1,3–6,15] of Priem and co-workers. The time-lag model probably originated in von Kármán's group in the early 1950s. Nevertheless, Crocco and his co-workers at Princeton University applied it to a liquid fuel system and introduced the idea of a sensitive part of the time lag, which constitutes the highly successful and widely used sensitive time-lag theory. The sensitive time-lag model is heuristic. The basic idea spawns from the observation that, among the exchanges taking place in a combustion chamber, the type most relevant to instability are energy exchanges. Then, the continuous gasification process is modeled as a discontinuous occurrence; it is assumed that there is a finite time interval between the injection of an element of propellant and the release of its chemical energy. The perturbations in the heat release rate at t are largely due to the pressure and velocity perturbations which occurred at $t - \tau$ in the reaction zone and are proportional to the intensity of the pressure (or velocity) heat release oscillation coupling modeled by the interaction index n (Refs. 1 and 15). The time lag is made of two parts: one that is constant and the other sensitive to pressure fluctuations. The existence of a time lag in the combustion process implies that instability is more likely to occur in the frequency range where the oscillation period is of the same order as the characteristic time for combustion. Bypassing the mathematical details, just note that the n–τ model accommodates longitudinal and transverse modes and uses both a small perturbation method (linear analysis) and nonlinear analysis for finite (although small) amplitudes.[1] The complete model is presented in a monograph by Crocco and Cheng.[15] The time lag and the interaction index can be regarded as empirical parameters. Hence, experimental data provided the needed parameters (average sensitive time lag and interaction index), yielding quantitatively accurate predictions on the stability of quite a few combustion systems.[1] The stability maps and the behavior of solutions of the n–τ model for nonlinear wave motion show how stability depends on the interaction index, the time lag, the type of instability (transverse, longitudinal, with or without shock), and the spatial distribution of combustion. Linear and nonlinear stability limits differ for extreme values of the time lag. This model allowed the treatment of the combustion instability problem without determining a physical link between the combustion rate and the fundamental processes involved.[13,19] This approach is outdated, however, since today's goal is to understand the instability mechanism in order to model it as well as the resulting instability[2] more accurately. Instead, theoretical models taking proper account of the vaporization process and other physiochemical processes must be developed.

II. Candidate Driving Mechanisms for Combustion Instability

Many potential mechanisms for combustion instability in liquid-propellant rockets have been identified. They involve coupling of the combustion acoustics with injection coupling, primary atomization, secondary atomization, vaporization, mixing, chemical kinetics, or any combination of these processes.

One may crudely view combustion in a liquid-propellant rocket motor as consisting of a sequence of processes. This sequence would be followed by an infinitesimally small element of propellant. In sequence, injection, primary atomization, secondary atomization, liquid heating, vaporization, gas-phase mixing and heating, gas-phase chemical reaction, and loss of some heat to fresh propellants are the processes. There are some caveats that should be added. The combustion is continuous, so that propellant elements injected at different times would not be necessarily in the same process at a given time. There is also a spatial distribution so that different elements injected at the same time can be in different processes at a given time. The details of any process for a given element can depend on both the preceding processes for that element and all of the processes for other elements. For example, the vaporization rate for an element depends on the atomization process for that element which determines the droplet size. Vaporization rate also depends on the heat transferred from other propellant elements that have already burned.

Each of the processes has a duration or characteristic time associated with it. The time can vary substantially. If one of the times for the sequential processes is much longer than any of the rest and, in fact, much longer than the sum of the rest, the time for burning of that propellant element is essentially the characteristic time for the longest process. We refer to that particular process as rate controlling because the burning rate approximately equals the rate for that process.

Variation of pressure, temperature, and velocity in the chamber will affect these processes and, in particular, can affect the characteristic times and rates associated with these processes. Dynamic or frequency-dependent phenomena are expected. Any driving mechanism must pass certain tests. In particular, the chemical conversion rate (or burning rate) associated with the mechanism must respond adequately to fluctuations of pressure, velocity, and temperature. That is, for a given amplitude of pressure oscillation, the conversion rate must oscillate with a sufficiently large amplitude and a sufficiently small phase relative to the pressure oscillation. Even if the conversion rate responds to some other quantity, such as temperature or velocity, it is the amplitude and phase relative to the pressure amplitude and phase that is important, since the feedback of energy to the oscillation depends on the relationship to pressure. Because maximum amplitudes of pressure and velocity can occur at different positions in the chamber, this results in a complex relationship for feedback of energy to the combustor oscillation. A change in a parameter generally not only affects the amplitude and phase of the burning rate oscillation but also modifies the spatial distribution of the burning. The most important basic parameters are those related to the injector design or the propellant choice. Three types of mechanisms can be important in combustion instability. The first type is associated with a combustion process that is rate controlling and has a characteristic time of the same order as the period of oscillation. This process can respond strongly to oscillations of the pressure and velocity in the chamber. The question still remains whether the amplitude is sufficient and the phase is proper to overcome the damping mechanisms and to drive the oscillation.

If the process is not rate controlling, oscillations in its rate cannot significantly affect the burning rate and the feedback to the combustor oscillation. If the process has a characteristic time much longer than the period of oscillation, it generally cannot respond to the oscillation. That is, the frequency of the oscillation is too high for the process; there is insufficient time for the particular process to adjust to the oscillations of pressure and velocity. When the characteristic time of the process is very short compared to the oscillation, it can respond. Processes with short characteristic times, however, are generally not rate controlling. If they still are rate controlling, there is the additional difficulty that the process would only respond in a quasisteady manner. The dynamic response could result in larger amplitudes of burning rate than the quasisteady response. The second type of process is one that affects the characteristic time of some other rate-controlling process. An example of this type would be atomization, which has a characteristic time much shorter than the typical period of oscillation; however, pressure and velocity variations produce variations in the initial droplet size distribution which, in turn, affects the characteristic time of rate-controlling processes. This type of process can have a dramatic indirect effect on the oscillatory process.

A third type of mechanism also has an indirect influence. This type of mechanism affects the spatial distribution of the combustion. Therefore, it produces variation in the local response that affects the coupling with a particular mode of oscillation. The second and third types of mechanisms are actually concurrent actions of the same mechanism since a change in characteristic time of a combustion process as viewed from a Lagrangian frame has the impact of a change in combustion distribution as viewed from an Eulerian frame.

Each of the major processes related to combustion in liquid propellant rocket engines is discussed in the following.

A. Liquid Injection

The liquid feed system involves storage tasks, pumps, feed lines, and injection parts into the combustion chamber. The lengths and volumes (or components) of the feed system tend to be large so that characteristic response times tend to be much longer than 1 ms.

Injection coupling is defined to mean that oscillations in the combustion chamber and oscillations in the feed system are coupled and reinforcement occurs. If there is no significant coupling, the combustion chamber will oscillate with essentially constant mass flow through the injector ports.

Injector coupling tends to be most significant for low-frequency chamber oscillations because of the large characteristic time for the feed system. That is, the feed system time matches better with the residence time in the combustion chamber than with the period of oscillation for any of the acoustical oscillatory modes. In the low-frequency modes, the chamber pressure is approximately spatially uniform but time varying.

These low-frequency modes have been well understood for decades. Summerfield analyzed these modes[1] using a constant time-lag concept. Injector coupling is relatively easy to eliminate by isolation of the feed system. Increasing the pressure drop across the injection port is a standard process that isolates the feed system since, for large-pressure drops, a given pressure fluctuation in the chamber produces a small variation in pressure drop and, therefore, in the mass flow.

The fluctuation in the feed system could result in modifications of other processes involved in combustion of the propellants. For example, the jet velocity at the port would fluctuate, thus modifying the primary atomization process of one or both propellants. Initial droplet size and droplet velocity fluctuations would vary. This would produce variations in the characteristic times of other processes. Also, it would affect the spatial distribution of combustion in the chamber. Since the characteristic times for these other processes are much shorter than the period of oscillation in the low-frequency cases and since little sensitivity to spatial distribution of combustion occurs in those low-frequency cases, these effects are generally not significant in terms of instability. Of course, some exceptions might exist.

B. Primary Atomization

Various atomization schemes are employed in liquid-propellant systems. In one scheme called impinging jets, injection ports are designed to work as pairs or triplets with the jets colliding in order to enhance breakup into droplets. Upon impingement, sheets of liquid in a conical fanlike shape emerge. Here, the conical cross sections are not circular but rather more like ellipses with a very large ratio of semimajor to semiminor axes. Usually, the impinging streams consist of "like" (identical) propellants so that liquid-phase mixing does not occur. This scheme is commonly used with LOX/RP1 propellants.

In another scheme involving coaxial jets, fuel and oxidizer are injected in parallel streams. This is commonly used for LOX/hydrogen and LOX/methane propellant combinations. In practice, LOX is always placed inside with the fuel on the outside.

At the high-injection velocities employed, the Rayleigh breakup mechanism is not of interest. Rather, the Kelvin–Helmholtz instability is the interesting mechanism. Surface and interface waves form and grow, causing distortion and ultimate breakup of the surface in this mechanism. Very little fundamental experimental and theoretical work[20,21] has been done on this mechanism. The characteristic wavelengths and periods of oscillation for these surface waves on the propellant streams are very small. An estimated frequency for these waves can be more than 10^5 cps. This characteristic time can be more than one order of magnitude lower than typical periods of oscillation for chamber instabilities. Variations in chamber conditions due to acoustic oscillations can be treated as quasisteady as far as primary atomization is concerned. The atomization process will not respond in a resonant fashion to chamber oscillations.

Strahle and Crocco[22] have shown by a linearized analysis that, in the absence of a resonant behavior and at low mean flow Mach numbers, fluctuations in droplet size and droplet number density do not cause the instability. The droplets move through the chamber at velocities much less than the speed of sound and carry the effect of initial size and number density. The kinematic wave resulting from the droplet initial conditions undulates more through the chamber than the pressure wave, which travels with an acoustic speed. This results in more undulation of the amplitude of the fluctuating source of burned gases that drives the instability. Because of the substantial undulation of the kinematic wave, there are many regions of both positive and negative reinforcement of the pressure wave. As a result, without a resonance effect, this phenomenon cannot produce a significant global effect since the positive contributions roughly equal the negative contributions.

Primary atomization is not projected to have a major direct effect on the driving mechanism of the instability. If the processes following it in the sequence do not respond significantly to oscillations of pressure and velocity, there would not be any sustained oscillation, according to linear theory. It is possible that a nonlinear perturbation could modify the conclusion. Since the issue relates more to the phase relationship between the source term and the pressure term rather than to an amplitude effect, however, it is not clear how nonlinearities would change the conclusion.

Note that atomization processes are quasisteady and, thus, do not respond differently to different frequencies in the range below 10^4 cps; therefore, this mechanism would not explain observed phenomena even if other factors, discussed earlier, were not present.

Atomization does affect initial droplet size distribution thereby influencing substantially droplet heating and vaporization. In this manner, it can affect combustion instability.

C. Secondary Atomization

Droplets that are formed by the primary atomization process can undergo further distortion and breakup. When smaller droplets result from a larger droplet, the process is called secondary atomization.

The surface tension tends to keep a droplet intact with a spherical shape. Aerodynamic pressure drag resulting from relative droplet-gas velocity tends to strain the droplet so that it assumes an oval shape with a decrease in the length parallel to the main flow direction and an increase in the transverse lengths. This can result in the "pinching off" of droplets.

Another mechanism involves a Kelvin–Helmholtz instability along the droplet surface. This results in the stripping of smaller droplets from the surface of the larger droplet.

Droplet distortion can significantly affect droplet drag and trajectory since the drag coefficient is modified. The likelihood for distortion, Kelvin–Helmholtz instability, and the pinching-off mechanism increases as the droplet Weber number increases. Secondary atomization tends to occur downstream, away from the dense spray regime.

The characteristic times associated with secondary atomization are much smaller than droplet vaporization and heating times. Also, they are much smaller than the typical periods of oscillation. Therefore, the same general conclusions can be made for secondary atomization as were made for primary atomization. It does not emerge as a strong candidate for direct feedback to the oscillation as a mechanism for combustion instability. There are, however, very important indirect ways in which the instability can be affected by secondary atomization.

D. Vaporization and Liquid Heating

Droplet vaporization and droplet heating should be considered as an integrated process. Heat from the high-temperature gaseous environment is transferred to the droplet largely by convective heating. The heat to the surface partially passes to the liquid interior raising the droplet temperature in the interior as well as at the surface. The remainder of the transferred heat sustains the phase change at the surface and increases the temperature of the vapors flowing away from the surface.

In practice, there is generally a relative velocity between the (LOX or fuel) droplet and the surrounding gas. Because of convection, the droplet heating rate and vaporization rate will depend on the Reynolds number based on this relative velocity. There will be a laminar boundary layer over the surface of the droplet. The shear stress at the surface will cause an internal circulation in the liquid droplet. This affects the transient heating of the liquid and the vaporization rate.

Below the critical pressure, a clear distinction can be made between the liquid droplet and the surrounding gas. The surface tension tends to keep the droplet spherical. The characteristic time for the liquid heating tends to be higher than the time for boundary-layer heating. For kerosene-type fuel droplets, the heating time and droplet lifetime are of the same order, whereas for LOX droplets, the heating time would be longer than the droplet lifetime. In liquid-propellant rocket motors, periods of acoustic oscillation tend to be of the order of the droplet heating time, suggesting a potential for coupling.

The situation is more complicated at near-critical or supercritical pressures. The reduced surface tension allows deviation from spherical shapes. The increased gas density decreases or removes the distinction between the phases and increases the gas-phase heating time. A coupling with the acoustic oscillations can still be expected, although less is known about the details.

Several investigators[1,3–11,23] have pursued the vaporization as the key process in the driving mechanism for combustion instability in liquid-propellant rocket engines. Generally, vaporization is the slowest of the processes, and, thus, is the rate-controlling process. A quasisteady response of the vaporization to combustion oscillations cannot provide sufficient feedback to drive the instability. Therefore, a dynamic or resonant response must occur if vaporization were to drive the instability. Strahle[7–11] has examined the gas-phase film and boundary layer surrounding a droplet to determine its oscillatory response. At the frequencies of interest, the heat and mass transport through the gas layer tend to oscillate in a quasisteady manner, without a resonance effect. Transient heating of the liquid interior was examined in Refs. 3–6 and 12. There was a dynamic response with the model[3–6, 12] in the frequency range of interest; however, the amplitude of the vaporization rate was not sufficient to drive the instability. It was assumed that liquid temperature was uniform spatially through the interior, albeit time varying. The model effectively employed an infinite conductivity so that the heat transferred through the liquid surface immediately was distributed throughout the liquid interior. This kept the surface temperature unrealistically low; the amplitude of surface temperature and, consequently, the amplitude of vaporization are kept at reduced levels because of this assumption.

In recent years, transient droplet heating and vaporization has been studied[23–37] with a better modeling of heat and mass diffusion in the liquid phase. In particular, spatial gradients of temperature and concentration through the liquid are taken into account. In these more accurate situations, it is found that amplitudes of surface temperature and vaporization rate can be large enough to drive an oscillation.[23, 38] The characteristic time for droplet heating is of the order of 1 ms and, therefore, commensurate with the practical periods of oscillation resulting in a strong dynamic response. Droplet heating time is, of course, dependent on initial droplet size. Some recent results[38] for liquid-fueled ramjets indicate that, along a stability limit, initial droplet diameter increases with the combustion chamber length (or the period of the natural oscillation).

The vaporization rate responds to fluctuations of both the ambient pressure and the relative gas-droplet velocity. It tends to be more sensitive to the latter. For transverse oscillations, the pressure and velocity tend to be in phase so that the pressure sensitivity and the velocity sensitivity can be reinforcing.

The theory for oscillatory droplet vaporization can explain the driving of combustion instabilities in situations where hydrocarbon fuels are employed and chamber operating pressures are subcritical. The theory must be extended to understand more fully situations where the pressure is supercritical or the vaporization of the liquid oxygen is rate controlling (hydrogen or methane fuel, for example).

In the supercritical-pressure case, the temperature will vary significantly from the ambient supercritical value to a subcritical value in the droplet. The energy of vaporization is zero in the supercritical-pressure domain. Properties will vary substantially over the droplet, and its surrounding gas film because of the wide variation in temperature; characteristic diffusion times can, therefore, be very different in different zones. This means that the supercritical-pressure case shares certain generic characteristics with the subcritical pressure case with regard to the response to ambient oscillations. That is, transport processes within certain regions of the field could potentially resonate at some frequency. Again, the theory here is not fully developed. Note that the supercritical case allows a greater likelihood for secondary atomization due to the reduced value of the surface tension.

In the case of oxygen droplets at subcritical pressures, the latent heat of vaporization is very low and the transfer number is high so that the droplet vaporization is very fast. The magnitudes of the temperature gradients in the liquid phase remains to be determined. Potentially, they can be large. The sensitivity to fluctuations also remains to be determined.

In summary, we can model an oscillatory vaporizing spray flow reasonably well for standard hydrocarbon type fuels at subcritical conditions. For supercritical conditions or for oxygen vaporization, the models have greater uncertainties. The vaporization rate is typically rate controlling and responds in a resonant fashion for frequencies of practical interest. Therefore, vaporization is a most prominent candidate as a driving mechanism for instability.

E. Mixing

After the two propellants vaporize, they must mix with each other before reaction can occur. The mixing process will be turbulent since Reynolds numbers based on distance between injector elements and propellant velocities will be in the range of 10^4–10^6. This distance is relevant for impinging like-on-like injectors. For like-on-unlike impingement or coaxial injectors, the diameter of the injection orifice or port is more relevant. In that case, we see a one order of magnitude reduction in the Reynolds number.

A characteristic time for turbulent mixing can be estimated to be less than 1 ms at its upper limit and as small as 10 μs. Some very small fraction of the propellant requires a time to mix that is substantially longer. That small fraction accounts for only a very small amount of released energy. Mixing, then, is generally faster than vaporization so that it cannot be rate controlling. It has minor impact on the distribution of the energy release through the chamber.

In this discussion, only mixing and transport on a scale larger than the droplet size or than the size of the boundary layer around the droplet is considered. The mixing and transport in the boundary layer and near wake of the droplet has been considered as part of the vaporization process. Mixing, as defined here, includes both the process of change of local composition and the transfer of heat from burned propellants to the fresh propellants. Because of the short characteristic time, the mixing process would tend to resonate at higher frequencies than are of practical interest. Mixing is not a promising candidate for a combustion instability driving mechanism.

F. Chemical Kinetics

After the propellants are well mixed and heated, they can react. At the high pressures and high temperatures associated with rocket combustion chambers, the chemical reaction rate will be large. The characteristic time for the chemical reaction will be at least one order of magnitude smaller than the vaporization time.

Variations in the chemical reaction time due to fluctuations in pressure and temperature will not be significant compared to variations in the vaporization time. Oscillations in the chemical rate will not, therefore, produce significant oscillations in the overall burning rate. Furthermore, there is no potential for a resonant response of the chemical kinetics in the frequency range of interest.

For these reasons, chemical kinetics is not viewed as an interesting driving mechanism for combustion instability in liquid-propellant rocket motors. Note that, for certain experimental configurations where propellants are introduced as premixed gases, we can see the importance of the chemical reaction as a driving mechanism. The chemistry, however, would respond in a quasisteady manner over a wide range of frequencies. No resonant phenomenon has been observed; for example, an increase in the chamber length does not modify the stability of the combustion process.

III. Early Models for Oscillatory Droplet Vaporization

Contrary to the sensitive time-lag approach, these models are based on a tentative description of the actual physicochemical processes. Strahle[7–11] investigated unsteady droplet combustion linearly. He neglected the thermal wave in the liquid, and his results did not provide sufficient evidence to indicate a potential for vaporization to drive the instability. Priem and Heidmann[12] assumed finite thermal inertia and assumed a uniform liquid temperature that was equivalent to an infinite conductivity assumption. Their model was quasisteady and one dimensional and, since the solution procedure was numerical, nonlinearities were less of a challenge. Priem and Guentert[3] and Priem[4] first showed that the stability was affected by vaporization and could be controlled by varying the characteristics of the vaporization process. This model did not yield response factors high enough to reach a clear conclusion on the potential driving of the instability by the vaporization process.[1] Priem and Guentert[3] compared the stability maps obtained by assuming that 1) the local instantaneous burning rate is equal to the local vaporization rate and 2) it is equal to the chemical reaction rate. Their results showed that for large burning rate parameters (which is the case in rocket engines) the vaporization model yields instabilities for smaller disturbances. Of course, due to the limitations of digital computers in the early 1960s, ad hoc assumptions had to be made in order to render the solution tractable. Heidmann and Wieber[5] showed that the n–τ model is in good agreement with the response obtained from the Priem–Heidmann model in the region of high response (low frequency). Crocco[19] noted that the conclusions of Priem and Guentert[3] depend more on the ad hoc simplifications than on the original physics of their combustion configuration and that much of the background flow nonlinearities had to be omitted in the computations. Nevertheless, addition of transverse velocity effects to this model could lead to nonlinear instability.

Recently, Nguyen and Muss[39] investigated the open-loop response of the burning rate of a droplet array for specified acoustic oscillations in the combustion chamber. They used an extension of the Agosta–Hammer[40] vaporization response model to mixed radial (R) and tangential (T) modes up to 8R and 8T. This model was basically an evolution of the model used by Priem. Infinite rate chemical kinetics and fast gas-phase mixing were assumed; hence, vaporization and burning rate were identified. Uniform droplet temperature, quasisteady heat and mass transfer processes between the droplet and the surrounding and constant liquid density were also assumed. The droplet temperature is estimated from an overall energy balance. Viscous effects of the gas on the droplets were included, but the reciprocal was neglected. They tested various assumptions concerning heat transfer within the droplet: finite conductivity with spherically symmetric droplet temperature, zero conductivity, and infinite conductivity (uniform temperature). Then they compared their results[39] with experimental measurements of droplet radius history in the case of a droplet vaporizing in a hot cross flowing airstream. The uniform temperature model gave the best fit.

Usually, when the "film" theory is used to model droplet combustion, the film surrounding the droplet is assumed to be quasisteady. To a large extent, this assumption is legitimate, since the characteristic time for heat diffusion through the film, τ_H, is typically two orders of magnitude smaller than the droplet lifetime. This assumption is even more accurate at high Reynolds numbers. However, Williams[41] proved that neglecting the transient of the droplet heating may lead to a 20% error in the evaluation of the total vaporization time in a quiescent environment. Many investigators[42–46] studied transient effects on droplet vaporization and burning. The main conclusion was that although the quasisteady theory gave a good estimate of droplet burning rates, it failed to predict correctly the evolution of the flame-to-droplet diameter ratio. Meanwhile, the unsteady theory predicted that the flame radius increased first and then decreased to zero,[42] thus corroborating experimental observations.[44] Furthermore, at high pressure and temperature, properties in the gas film were significantly modified, and τ_H was comparable to the droplet lifetime, which reinforced the intrinsic unsteadiness. The transient character of droplet vaporization and combustion under these conditions cannot be neglected.[47] Some investigators indicated that the quasisteady theory was not valid for reduced pressures above 0.1 in a stagnant environment.[48]

On the other hand, the very high pressures and temperatures encountered in a liquid-rocket combustion chamber imply that both the gas and the liquid phase may deviate substantially from their usual behavior. Specifically, the ideal gas assumption is not valid under these conditions, and the liquid-vapor equilibrium is not correctly represented by the classical laws and approximations. These phenomena complicate significantly the modeling of droplet vaporization and burning in such regimes but may have a critical importance in the liquid-rocket combustion instability phenomenon. As discussed by Wieber[49] over 30 years ago, near the critical point a small change in pressure might lead to violent boiling of the oxygen droplet, thereby supporting the combustion instability. Wieber also noticed that if the chamber pressure is so high that the droplet reached a supercritical state and became a vapor puff, the rate-controlling mechanism for combustion instability might be the vapor diffusion into the gaseous environment. Therefore, determining accurately whether the liquid drop gets close enough to the critical state and exhibits such a behavior is primordial, and a large amount of work was devoted during the past two decades by the combustion community to investigate the vaporization of a spherically symmetric drop in a quiescent environment under high pressure and temperature.

Various researchers have improved on Wieber's work, most of them in the domain of hydrocarbon fuel.[26,27,48,50–52] Lazar and Faeth[50] noted that Wieber did not account for high-pressure phase equilibrium, nonideal gas effects, effects of pressure on physical properties, and gas solubility in the liquid phase. They included these phenomena but considered a simplified droplet model neglecting all transient phenomena and convection. The agreement between the computed equilibrium temperatures and experimentally obtained data on suspended droplet in a microgravity environment was good. Manrique and Borman[51] used a somewhat more sophisticated quasisteady model and showed that if the ambient pressure was high enough, steady-state conditions were not attained. In 1972, Matlosz et al.[48] combined both transient theory and high-pressure effects on the phase equilibrium for a hexane/nitrogen system and provided experimental measurements supporting their results.[51] They[48] concluded that both phenomena must be included for correct predictions. They also noted that accurate methods to compute physical properties, and more specifically mass diffusion coefficients, under such regimes were crucial.

Once the whole droplet has reached a supercritical state, however, modeling is possible. The theory of supercritical combustion for such gas pockets was initially developed by Spalding.[53] He assumed that the droplet can be replaced by a point source of vapor and showed the importance of unsteadiness. He concluded that under supercritical conditions, the burning time increased with the cube root of the pressure. Rosner[54] extended this theory by considering a distributed source. More detailed numerical simulations were performed[55,56] showing that the burning time under supercritical conditions was shorter than under subcritical conditions.

IV. Opportunities of Modern Times

Now in the 1990s, we can benefit from the methodologies of computational fluid dynamics and of nonintrusive laser diagnostics. Fully nonlinear coupled systems can be analyzed. Measurements of velocity, temperature, and concentration can be made locally under high-temperature, oscillatory conditions. Clearly, the potential exists to learn much more about the driving mechanisms and other characteristics of combustion instability.

As numerical methods and computing power are improving, the mechanistic approaches seem to be able to yield more interesting and insightful results. The limitations of approaches such as the n–τ model reside mainly in their heuristic character. They do not provide a description of the combustion process but represent it with a time-lag concept.[15] These approaches are linear so that their response factor really applies to cases where the linear approximation is valid. When nonlinearities are included, the effects of the combustion response nonlinearity are excluded.[1,15] Vaporization models, however, allow the inclusion of vaporization and combustion nonlinearities and, therefore, yield a nonlinear response factor. Furthermore, in the simple version of the n–τ model, the interaction index and the sensitive time lag are uniform in the chamber. Crocco and his co-workers extended this analysis[1] to include velocity and displacement sensitivities (in some limited configurations). Because of the phenomenological nature of this approach, however, the values of all of the parameters involved have to be derived from experimental data, which precludes applications.[57] Vaporization models do not suffer from this limitation.

Since the 1960s, there has been substantial advancement in the understanding of basic combustion phenomena. Understanding of spray combustion, including transient droplet heating and vaporization, has advanced tremendously. Mixing of coflowing streams of fluids is now much better understood. Our knowledge of atomization processes and chemical kinetics has also advanced substantially. The research that led to these advancements addressed automotive combustion, gas turbine combustion, ramjet combustion, and industrial combustion problems rather than liquid rocket combustion. Nevertheless, liquid-propellant combustion research can build upon the recently established foundations.

V. Spherically Symmetric Models

The first step in droplet vaporization and combustion modeling is the simulation of the spherically symmetric vaporization of a droplet in a quiescent environment, with no gas motion other than that created by the Stefan flow and thermal expansion for the reaction zone. These models are considered next with an emphasis on high-pressure and supercritical environments. It should be noted in advance that velocity coupling is the major influence on droplet vaporization so that spherically symmetric vaporization models have little or no value in explaining combustion instability.

Recent studies[52] concluded that the droplet vaporizes completely before the liquid surface reaches its critical temperature, but the enthalpy of vaporization was approximated by the latent heat of vaporization, which was proven to be quite inaccurate.[48,51] Extension to the case of a multicomponent hydrocarbon droplet was contributed by Hsieh et al.[26] in a detailed numerical analysis that is too computer time consuming to be used in a larger spray combustion code. Umemura[58] showed that the mass diffusion coefficient must be zero at the interface, whereas the mass flux remains finite. This should affect greatly vaporization rate predictions.

Although liquid oxygen (LOX)/hydrogen (H_2) systems are currently being used for the Space Shuttle main engine and developed for the European launcher Arianne V because of the higher thrust they provide, only recent contributions[28,59] deal with the case of a LOX droplet vaporizing at high pressure. Indeed, the large molecular weight and temperature differences between the liquid surface and the ambient gas, their consequences on thermophysical property variations, and the disparity of the characteristic times result in increased numerical problems and make LOX/H_2 cases even more complex. Litchford and Jeng[59] used the quasisteady film theory developed for subcritical droplet vaporization and the appropriate high-pressure phase-equilibrium relations. They found that the droplet surface can reach the critical state for reduced ambient pressure slightly above a value of unity.

The singular behavior occurring when a thermodynamic system goes from a subcritical state to a supercritical state has not been thoroughly investigated yet, and the necessary knowledge to model the vaporization and burning of a drop going through the critical point is still lacking.[58]

The case of an isolated liquid oxygen droplet vaporizing in a supercritical quiescent environment was studied.[29] Effects of the gas-phase unsteadiness and effects of near-critical conditions were studied. First, unsteady spherically symmetric combustion of a single droplet in a quiescent surrounding gas was modeled, which corresponds to a film extending from the droplet surface to infinity. Combustion was included here and, although we were really interested in the unsteady effects on LOX vaporization in gaseous hydrogen, the case of a hexane droplet vaporizing in air was also treated as a reference case to provide a validation of the method used by comparison with classical results. In the cases studied, pressure and temperature were high enough so that unsteadiness in the film had to be considered, but the values near the surface were sufficiently below the mixture critical point for the real-gas mixture effects on phase-equilibrium and latent heat of vaporization to be neglected and for the usual liquid density expression to be used. Nonuniform and time-varying properties were considered. The density of the liquid, however, was assumed to remain constant. The effects of near-critical conditions were predicted using phase-equilibrium relations and thermophysical property expressions valid over the considered range. The pertinence of including each phenomenon specific to near-critical conditions in droplet vaporization modeling was evaluated. Here, relative motion between the droplet and the ambient gas was not considered. Thereby, spherical symmetry resulted.

Yang et al.[27] reported pressure-coupled vaporization results at supercritical conditions for hydrocarbons burning in air. Their model, however, did not allow for relative velocity effects that are known to be predominant. They found the vaporization response changed significantly when the critical mixing conditions were reached. Nevertheless, it should be noted that they did not indicate clearly how the mass flux was computed at the critical interface.

The exchange of mass, momentum, and energy across the critical interface becomes an important issue in spray calculations. Supercritical sprays can be conveniently described as two-phase fluids. The supercritical fluid can be considered as the continuous phase whereas the subcritical pockets of fluids can be considered as the discrete phase or droplets.

VI. Convective Droplet Studies

Classical droplet vaporization theory presents the well-known d^2 law. This analysis, however, assumes that the droplet is traveling at the velocity of the gas and that there is spherical symmetry with no transient behavior. It is known that this theory has rather limited applications. The relative motion between the droplet and the ambient gas (which is nonzero in most cases) creates the convection that increases the rates of heat and mass transfer over the values for the spherically symmetric droplet. Furthermore, the shear force on the liquid surface causes an internal circulation that enhances the heating of the liquid. As a result, vaporization rate increases with increasing Reynolds number. For droplets with initial radii of 25 μm and greater, the time for velocity relaxation due to drag is comparable or greater than the droplet lifetime.[24] These and other effects, like internal circulation, neglected by the classical droplet vaporization theory, lead to considerably smaller characteristic lengths and times for liquid-phase heat and mass transfer. The strong coupling among the drag, heat transfer, mass transfer, relative droplet velocity, and internal circulation necessitates a numerical solution.

A. Subcritical Environment

Tong and Sirignano[23] examined the effect of an oscillatory field (pressure and velocity) on droplet vaporization. They used an isolated droplet model but combined the vaporization rates, including the effects of the phase at injection, to obtain the equivalent response of the spray. They showed that the oscillatory part of the vaporization rate in phase with pressure is sufficient to sustain instability. Their version of the Rayleigh criterion is that "the growth or decay of the wave is determined by the net in-phase or out-of-phase energy or mass addition." They note that earlier models[3–12] could not prove that vaporization was the driving mechanism for instability, but these models either neglected thermal inertia or assumed a time varying but uniform droplet temperature. In a hot environment, the droplet vaporization rate is large; thus, the heating time and the droplet lifetime are of the same order. Tong and Sirignano's model is one-dimensional and isentropic, and proper account of internal circulation is taken through a vortex model.[25,30] Using this model showed that droplet heating and vaporization are unsteady and that internal temperature profiles are nonuniform during most of the droplet lifetime,[25] thereby refuting the assumptions used by Strahle[7] and Priem and Guentert,[3] Priem,[4] and Priem and Heidmann.[12] Gas-phase heat and mass transfer rates are modeled by an average between the stagnation point and shoulder region. This model was shown to be in good agreement with results of the exact solution of the axisymmetric Navier–Stokes equations. The continuous droplet size distribution is discretized using a finite number of droplet groups.[23] The fuel is hexane (10 atm, 1000 K, $\Delta V = 25$ m/s, and $D \simeq 95$ μm). The frequency range investigated is

100–1000 Hz; hence, the droplet lifetime and period of oscillation are of the same order and, therefore, the droplet behavior is significantly affected by the phase point at which the droplet is injected. This is taken into account here when computing the overall response. To estimate the importance of the component of the vaporization rate in phase with the pressure Tong and Sirignano[23] evaluated

$$G \equiv \int_0^{2\pi} m' p' \, d\theta \bigg/ \int_0^{2\pi} p'^2 \, d\theta \tag{1}$$

For vaporization to be able to drive longitudinal mode instability, $G \geq (\gamma + 1)/2\gamma$ ($\gamma \simeq 1.2$) is needed.[15] (Note that this assumes concentrated combustion. The minimum value of G to drive instability with distributed combustion will be higher.)

It is assumed by Tong and Sirignano[23] that vaporization is rate controlling; i.e., it occurs much slower than the mixing and chemical reactions, so that the burning rate is identified with the vaporization rate. The model is quasisteady in the gas phase. This is legitimate for frequencies corresponding to an oscillation period much smaller than the diffusion time across the gas-phase boundary layer, hence for frequencies lower than a few kilo hertz. Five types of oscillatory fields were investigated: pressure oscillations only, velocity oscillations only, traveling wave, standing wave, and a one-dimensional shock-wave model.[60] They[23] showed that the effects of velocity oscillations are predominant and that the oscillatory vaporization rate increases as the Mach number of the mean flow decreases. In the traveling wave case, vaporization is found to be driving at all frequencies investigated. In the standing wave case, however, vaporization is driving (marginally) only at low frequencies. Important nondimensional groups were identified in the search for the resonant frequencies: $f R_0^2/\alpha$, the Reynolds number based on initial droplet diameter and relative velocity, and the relative pressure and velocity fluctuation amplitudes. The main conclusion of Tong and Sirignano's work[23] is that if droplet vaporization (without coupling to atomization) is the controlling mechanism, self-sustained acoustical longitudinal mode oscillation can occur in the combustion chamber. The oscillations are almost always self-sustained for traveling waves, seldom for standing sinusoidal waves, and sometimes for the shock-wave oscillation.

Culick[13] challenges Tong and Sirignano's conclusions[23] that unsteady rate-controlling droplet vaporization provides a potential mechanism for driving combustion instabilities. Culick argues that if all terms are included and if combustion is ignored, vaporizing droplets cause damping and not driving. We believe that Culick has a semantical difficulty. Certainly, he is correct that vaporization absorbs heat rather than releases heat. Vaporization, however, can control the burning rate so that the characteristic times associated with the energy release are the vaporization times. In this sense, vaporization can be a rate-controlling process that is key to the driving mechanism.

The model used by Tong and Sirignano is valid only at high Reynolds number for the droplet. This precludes its use in situations where the relative velocity might change sign, which is the case in most situations. Assumptions of constant gas-phase thermophysical properties are made in the gas phase, and correction factors are used to account for surface blowing effects on the laminar boundary layer. Since then, Abramzon and Sirignano[31] defined a model that does not have this limitation. This model differs significantly from the one used by Tong and

Sirignano.[23] Internal heating is approximated using the effective diffusivity concept. Influence of Stefan convection on temperature and species profiles in the gas film as well as variable thermophysical properties and nonunitary Lewis number are included. Hence, significantly different results can be expected. Abramzon and Sirignano's model has not been used in a liquid rocket combustion instability study, but the authors could show in a simplified case that the vaporization rate is significantly affected by gas velocity oscillations. Bhatia and Sirignano[38] applied this model to a one-dimensional analysis of ramjet instability. They found that vaporization is rate controlling and identified the ratio of oscillation period to droplet heating time and the ratio of residence time to droplet heating time as important parameters. Turbulent mixing and transverse transport were not included in this study. Molavi[61] and Molavi and Sirignano[62] modeled combustion instability in a two-dimensional dump combustor using this droplet model. Turbulent mixing was included, but due to high numerical diffusion no precise conclusion could be drawn concerning the controlling processes. Note that for transverse oscillations in cylindrical geometries, pressure and velocity tend to be in phase and thus can be reinforcing.

Patnaik[32] and Patnaik et al.[33] developed an axisymmetric model that used a stream function and vorticity approach in the liquid phase and could accommodate variable density in the gas phase. The constant properties used in the calculation, however, were evaluated at ambient conditions and, as a result, the computed quantities were overestimated. Chiang et al.[34] developed an advanced droplet vaporization model with fewer assumptions and with the intention of creating a reference model against which simpler models could be measured. This axisymmetric, single-droplet model includes effects of surface blowing, transient heating, and internal circulation. Variable gas-phase properties are taken into account and viscosity variations in the liquid phase are also calculated. The formulation for the gas and the liquid phases of the axisymmetric droplet problem complete with boundary, interface, and initial conditions can be found in Chiang.[35] A complete primitive variable approach for both gas and liquid phases is given in Chiang and Sirignano.[36] The results given in this chapter are based on that approach.

B. Supercritical Environment

Most rocket engine combustors are operated in the very high-pressure domain where fuel droplets may vaporize at near- or even supercritical conditions. Hence, an understanding of fundamental transport processes underlying high pressure fuel-droplet vaporization is needed.

There are several mandatory considerations for both the spherically symmetric and the convective cases to extend the subcritical droplet vaporizing model to deal with the ambient pressure in the range from the near-critical region to the supercritical region. They are summarized in the following. Note that because of various approximations the quantitative accuracy of the model is limited.

1) A comprehensive calculation of variations of thermophysical properties with respect to temperature for each species component is necessary. The high-pressure correction, the mixing rules to evaluate the thermodynamic properties, and the quantum gas behavior of hydrogen vapor must be considered.

2) The real gas effect in the high-pressure case can be included in the gas-phase equation of state by incorporating a compressibility factor. A modified Redlich–Kwong[63] equation of state with the mixing rules of Chueh and Prausnitz[64] will serve

this purpose. The enthalpy of vaporization is derived from the equation of state by using the fugacity equation. Since the swelling effect of the droplet vaporization at the critical state may be very significant, a primitive variable approach is required to analyze the internal motion and thermal and mass transport within the droplet.

3) The solubility of the fuel vapor in the liquid phase must be considered, and it makes a multicomponent droplet formulation necessary. The principle of equal temperature, pressure, and the fugacity of both phases for phase equilibria is used to compute the surface mass fraction of species for gas and liquid. A brief description of the new phase equilibrium conditions at the droplet interface is given next.

The thermodynamic equilibrium of a mixture gives

$$T^{\ell} = T^{g}, \qquad P^{\ell} = P^{g}, \qquad f_i^{\ell} = f_i^{g} \tag{2}$$

which ultimately presents the following system of equations to be solved:

$$x_{O_2}^{\ell} \times \phi_{O_2}^{\ell}\left(x_{O_2}^{\ell}, x_{H_2}^{\ell}, T, P\right) = x_{O_2}^{g} \times \phi_{O_2}^{g}\left(x_{O_2}^{g}, x_{H_2}^{g}, T, P\right) \tag{3}$$

$$x_{H_2}^{\ell} \times \phi_{H_2}^{\ell}\left(x_{O_2}^{\ell}, x_{H_2}^{\ell}, T, P\right) = x_{H_2}^{g} \times \phi_{H_2}^{g}\left(x_{O_2}^{g}, x_{H_2}^{g}, T, P\right) \tag{4}$$

A phase equilibrium diagram and enthalpy of vaporization from this analysis is given in Fig. 1. The following interface equations for a multicomponent fluid are taken.

Conservation of species:

$$\left[\dot{m}_v Y_i - \rho D_{ij}\frac{\partial}{\partial r} Y_i\right]_{r^+} = \left[\dot{m}_v Y_i - \rho D_{ij}\frac{\partial}{\partial r} Y_i\right]_{r^-} \tag{5}$$

Conservation of energy:

$$-k\frac{\partial}{\partial r}T\bigg|_{r^-} = k\frac{\partial}{\partial r}T\bigg|_{r^+} + \left[\dot{m}_v Y_{O_2} - \rho D_{ij}\frac{\partial}{\partial r} Y_{O_2}\right]_{r^+}$$
$$\times\left(\Delta\bar{H}_{v,O_2} - \Delta\bar{H}_{v,H_2}\right) + \dot{m}_v \Delta\bar{H}_{v,H_2} \tag{6}$$

where $\dot{m}_v = \rho(V - \dot{r})_{r^+}$

4) A technique is required to locate the gas/liquid interface and to compute the regression rate due to vaporization as well as phase change across the critical surface with continuous density and temperature gradients at the surface.

Convective effects were considered by Delplanque and Sirignano.[65] The extended film model of Abramzon and Sirignano[31] was used here. This model was built on the classical film model to include the effect of blowing on boundary-layer thickness and profile distortion. Internal circulation effects were included using the effective diffusivity model.[31] The high-pressure phase equilibrium computation method already described was used here unmodified. Because the droplet surface temperature may reach the critical mixing value, the surface tension coefficient may become vanishingly small. This allows fluid removal from the droplet surface by shearing, thereby impeding the heating of the droplet surface temperature to its critical mixing value.

The enhanced heat transfer and mass transfer due to convection cause the surface of a LOX droplet vaporizing in hot gaseous hydrogen at a pressure supercritical for

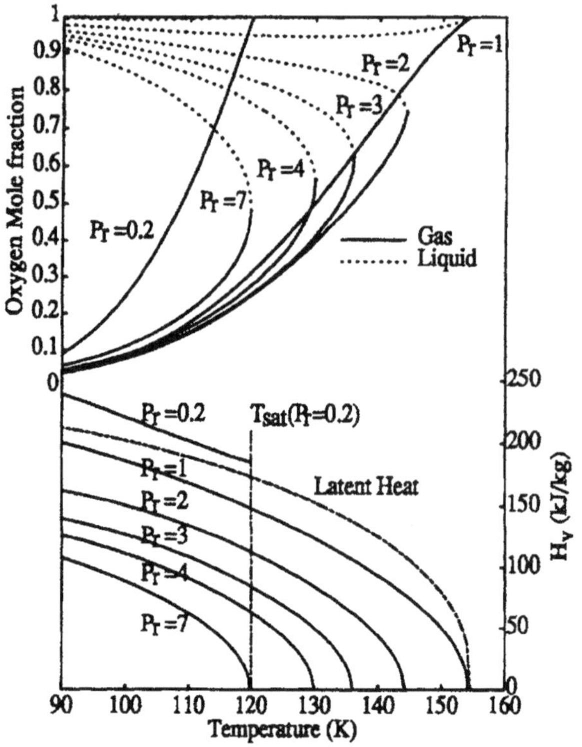

Fig. 1 Oxygen/hydrogen system, thermodynamic equilibrium.

oxygen to reach the critical conditions extremely quickly (less than 30 μs in the base case investigated). At such conditions, the predicted lifetime of the droplet is about 4.6 ms for a 100-μm droplet.

The reviewed literature indicates that the stripping breakup mode would indeed greatly affect the droplet behavior and may not be omitted in the model definition. In addition, the breakup times for the stripping mode found in the literature are much smaller than the droplet lifetime estimated with a simple vaporization and burning model; this could have significant effects on liquid rocket combustion instability study.

Delplanque and Sirignano[65] numerically investigated this phenomenon. The physical properties of liquid oxygen and a surrounding gaseous mixture of oxygen and hydrogen were carefully evaluated to compute the relevant nondimensional numbers and assess if the boundary-layer stripping mode was likely to be dominant according to the criteria found in the literature. Next, a simple analysis[66,67] was used to include, at least roughly, the boundary-layer stripping mechanism in a simple one-dimensional model without increasing unrealistically computation times. They also modified the boundary-layer analysis to incorporate Stefan flow effects.

Their results showed that, in an environment typical for liquid rocket engines, a LOX droplet undergoes boundary-layer stripping, which had two important effects on the droplet behavior. First, stripping significantly reduced the droplet lifetime (to 0.24 ms in the base case or more than one order of magnitude less than the value predicted without stripping). Hence, the smaller characteristic time introduced by this phenomena must be considered in liquid rocket combustion instability studies. Furthermore, stripping precluded, in most cases, heating of the droplet surface to the critical mixing conditions. Blowing reduced the stripping rate and, therefore, increased the predicted droplet lifetime ($\simeq$13%). Order of magnitude comparisons with experiment indicated that the stripping rate was controlled mechanically, by the size of the viscous layer, rather than thermally.

In the pressure range investigated ($Pr = 0.8$–2.5), the computed droplet lifetime was shown to exhibit a minimum at $Pr = 1$ if stripping was not included, whereas $Pr = 1$ corresponded to a maximum of the droplet lifetime with stripping.

In a recent experimental study,[68] n-heptane droplets gasifying at supercritical conditions with forced convection exhibited behaviors similar to those described by theory.[65] At low-convection intensities, the gasification regime was a continuous regression of the droplet surface. At higher convection intensities the droplets underwent what the authors[68] qualified as an "enhanced supercritical" gasification regime where the droplets deformed significantly and were "partially convected away as a dense vapor cloud." This was not quite a secondary breakup in the stripping mode as is described in the present study, but the largest relative velocities considered in these experiments[68] were one order of magnitude smaller than those typical of rocket engine conditions. More experimental studies of this kind are needed to validate the present model. Experimental data are needed to validate the values used for the effective mass flux quality and the mist temperature in the evaluation of the heat transfer effect of stripping.

In the preliminary unpublished work conducted by Chiang and Sirignano,[36] a calculation is performed to study vaporization behavior before the droplet reaches the critical mixing state. The changes of properties at the interface occur very rapidly during the droplet vaporization. Diffusion length scales are reduced at the elevated pressure. The transport rates are faster than those in the low-pressure case; hence, the droplet lifetime is shorter. As a result, a very small time step is mandatory to ensure accuracy. Figure 2 indicates that radius increases with time due to the swelling effect of droplet as well as the condensation of vapor. The reduced enthalpy of vaporization also contributes to the fast vaporization of the droplet.

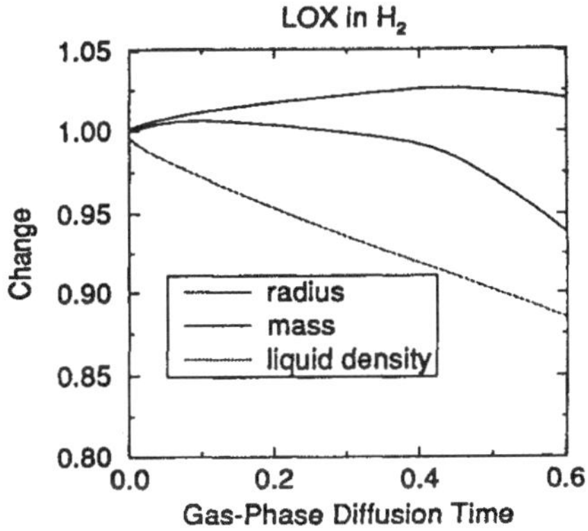

Fig. 2 Fractional change in oxygen droplet radius, mass, and density due to heating and vaporization.

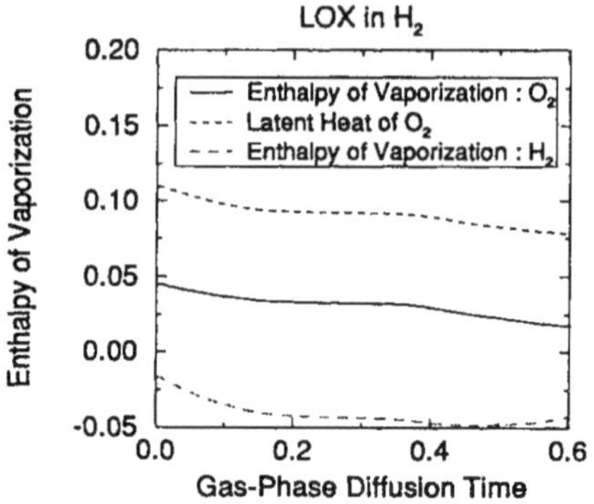

Fig. 3 Enthalpy of vaporization for hydrogen/oxygen mixture.

The enthalpy of vaporization shown in Fig. 3 is significantly reduced from the conventional "latent heat." A remarkable condensation of gas-phase vapor occurs at the early droplet lifetime. The vaporization begins as the surface temperature increases.

The compressibility factors (averaged around the azimuthal direction) for the gas and liquid phases at the interface become equal when the critical mixing state is reached. The two phases merge into one continuous phase at that point. Since the critical mixing state highly depends on temperature and mass fraction, the droplet surface at different locations will reach the critical mixing state at different times. The diffusivity for the liquid phase is, in general, two orders of magnitude less than that of the gas phase. As a result, most of the diffusing vapors are concentrated in the boundary-layer region at the interface.

To advance this high-pressure droplet model to deal with the vaporization at supercritical conditions, two major difficulties in the calculation must be addressed: 1) the formulation of boundary conditions, especially to trace the location of the droplet interface and 2) more accurate calculation of the near-critical properties for the fluid near the droplet surface.

VII. Response to an Oscillatory Environment

The model[65] for LOX-droplet vaporization is applied to prototypical cases pertinent to liquid rocket combustion instabilities.[69] The response of a LOX droplet to oscillatory ambient conditions over a wide range of frequencies is computed. The effects of stripping, pressure fluctuation, and mean pressure on the response are evaluated.

In an oscillatory environment, the effects of relative velocity fluctuations are predominant and cannot be overlooked in combustion instability studies. Effects of pressure fluctuations on the critical interface location appear to be minor in the cases investigated.

It was shown that the gasification process, with or without stripping, has enough potential to drive longitudinal mode combustion instabilities, under the simplifying assumptions used in this work. Furthermore, the peak frequency for the computed response factor is correlated to the droplet lifetime. Therefore, secondary atomization, which causes a substantial reduction of the droplet lifetime (one order of magnitude), results in a significant shift of the peak frequency, away from the common modes for standard rocket engines. This phenomenon could explain the observed better stability of cryogenic rocket engines relative to storable propellant rocket engines because, for cryogenic propellants, most droplets are likely to undergo secondary atomization in the stripping regime. A secondary peak in the response function was identified and linked to both the magnitude of the droplet velocity to sound velocity ratio and the possible occurrence of flow reversal. The computed response factors could be fitted using the simplified expression from the n–τ model within a reasonable margin. Various wave patterns (traveling wave, sawtooth wave, and pressure oscillations only) were considered. Results confirm the predominance of velocity effects and show that in a sawtooth wave pattern, when stripping is included, the droplet lifetime is too short for the gasification process to contribute effectively to combustion instability. Supercritical mean pressures have a stabilizing effect if stripping is not included, whereas higher pressures are destabilizing when stripping is included.

All of the previous spherically symmetric oscillatory work did not consider many of the vital effects that would significantly modify the characteristic time or length scales of droplet vaporization. The one-dimensional model[31] was also limited and, therefore, vital information such as the drag coefficient could not be obtained from the Abramzon and Sirignano study. A detailed analysis that includes convection, the transient heating, and variable properties of the fuel droplet vaporization under oscillating conditions is required to identify the role of vaporization in combustion instability.

A preliminary axisymmetric calculation of oscillatory vaporization was made by Duvvur et al.[70] who examined the role of droplet vaporization under oscillating ambient conditions. This extended the previous work by Tong and Sirignano.[23] The advanced model of Chiang et al.[34] and Chiang[35] was employed to study the effects of an oscillating gas phase on the heat and mass transfer processes in the liquid phase. Variable properties were calculated in both the gas and liquid phase. The oscillations of temperature, pressure, and velocity in the gas phase showed significant changes in the heat and mass transfer rates and also in drag coefficients. Droplets were introduced at various times in order to evaluate the combined effect. The response factor of the system was calculated in order to determine if the vaporization process is capable of driving instability.

The problem considered in this study[70] is that of a single n-octane droplet, injected into the hot gases of a combustor. Gas-phase perturbations are introduced on the velocity, pressure, and temperature. The velocity and pressure oscillations are dictated by the classical standing wave pattern, and the corresponding temperature oscillation is calculated by assuming an isentropic flow. This study[70] examined the effects of variation of critical parameters such as the initial phase of injection, frequency, and amplitude of oscillation on the vaporization history of the droplet. The effects of decreasing relative velocity, varying thermophysical properties, transient heating, internal circulation of liquid, boundary-layer blowing, and moving interface were incorporated into the calculations.

For the governing equations in the liquid and gas phases, one should refer to work by Chiang[35] and Chiang et al.[34] The initial conditions corresponding to the injection of a droplet into an unstable combustor depend on the particular time and location of injection. T, p and V_z are therefore dependent on the phases ϕ_t and ϕ_{po}. V_r and Y_f are set to zero on the inflow side and the gradients are set to zero on the downstream side. When the flow reverses, the boundary conditions are, in effect, specified downstream with the gradient set to zero on the upstream side.

The inlet conditions of the gas phase are as follows.
Velocity condition:

$$U'_\infty(z', t') = \bar{U}'_\infty + \epsilon'_u \sin(2\pi z'/\lambda')\sin(2\pi c' t'/\lambda') \tag{7}$$

Pressure condition:

$$p'_\infty(z', t') = \bar{p}'_\infty + \epsilon'_p \cos(2\pi z'/\lambda')\cos(2\pi c' t'/\lambda') \tag{8}$$

where

$$\epsilon'_u = (a'/\gamma)(\epsilon'_p/\bar{p}'_\infty) \tag{9}$$

Temperature condition:

$$\frac{T'_\infty}{\bar{T}'_\infty} = \left(\frac{p'_\infty}{\bar{p}'_\infty}\right)^{(\gamma-1)/\gamma} \approx 1 + \frac{\gamma - 1}{\gamma}\frac{\epsilon'_p}{\bar{p}'_\infty}\cos\frac{2\pi z'}{\lambda'}\cos\frac{2\pi c' t'}{\lambda'} \tag{10}$$

On the outlet side, the pressure is specified in a similar manner, but the gradients of the temperature and velocity are set to zero.

It is known that the driving force behind combustion instability is the energy input from the combustion process. The Rayleigh criterion is the most commonly used criterion to determine whether a particular mechanism is capable of driving the combustor into an unstable mode or not. As previously stated, the Rayleigh criterion for a mechanism to drive the instability is that the associated burning rate oscillates with the proper combination of a large enough amplitude and a small enough phase lag with the pressure oscillation. For the vaporization to be able to drive instability, it is necessary to evaluate the component of the vaporization rate that is in phase with the pressure. Duvvur et al.[70] approached this by evaluating the response factor G as suggested by the Rayleigh criterion[18]:

$$G = \oint\!\!\int m P \, \mathrm{d}\tau_{Hg} \, \mathrm{d}\zeta \Bigg/ \oint\!\!\int P^2 \, \mathrm{d}\tau_{Hg} \, \mathrm{d}\zeta \tag{11}$$

where ζ is a constant for each droplet during its lifetime. The ζ value represents the phase at the time of injection. G is a nondimensional measure of the integrated (over time and space) product of burning rate fluctuation and pressure fluctuation. This form is consistent with the form for the combustion response developed by linear theory.[15]

The definition of G just given involves integration in an Eulerian frame of reference. In order to apply this to the present model, it is adapted to a Lagrangian frame of reference. Figure 4 shows the droplets born at various phase values. In an Eulerian time frame (represented by solid lines), droplets injected at different phases are present at different axial locations at a given time. By the preceding definition, G is calculated by integrating in horizontal sweeps over axial location and then over the Eulerian time. Lagrangian time varies along the dotted lines of constant initial phase value (constant ζ) in the figure. G can, therefore, be evaluated by integrating in diagonal sweeps over Lagrangian time and then by integrating over the phase value at the time of injection.

Note that the droplets injected into the chamber of a liquid rocket combustor experience varied histories of ambient conditions depending on the phase of the

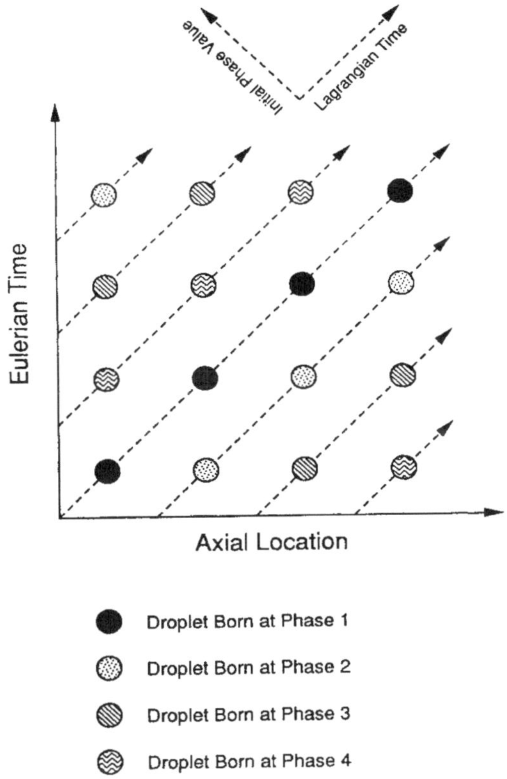

Fig. 4 Evaluation of *G* in a Lagrangian frame of reference.

oscillation at the time of injection. Figure 5 shows the vaporization rate histories (schematic) of the droplets present in a combustor. Droplets injected at phases of -2π, $-3\pi/2$, $-\pi$, and $-\pi/2$ have contributions to the area of interest (marked as A, B, C, and D, respectively). This area, however, corresponds to the area marked A, B, C, and D on the vaporization curves of droplets injected at zero, $\pi/2$, π, and $3\pi/2$, respectively. Thus, integrating over the vaporization curves of all droplets born between zero and 2π would, in effect, give the contributions of all droplets present in the combustor.

To be able to overcome the losses and to drive combustion instability, G should be above a certain minimum positive value. For the case of distributed combustion, this value corresponds to 3.72 for longitudinal mode instability.[15]

Four different parametric studies have been conducted or are underway. These involved 1) frequency comparison, 2) amplitude comparison, 3) droplet configuration comparison, and 4) variation in the liquid fuel. Duvvur et al.[70] have not yet completed their effort, but certain conclusions can be drawn from their preliminary analysis.

The imposed oscillations have significant effects on the heat and mass transfer process and on the liquid properties. Droplet vaporization produces enough energy in phase with the pressure to drive longitudinal mode combustion instability. An investigation of the frequency domain showed that droplet vaporization was only capable of driving combustion instability in certain frequency domains. The fre-

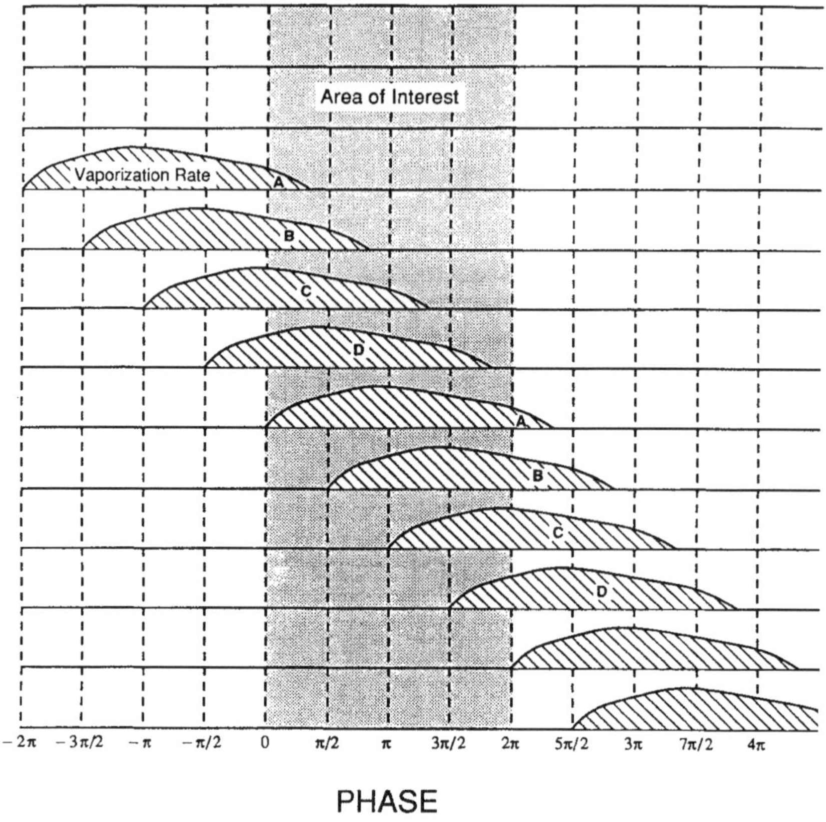

Fig. 5 Vaporization rate (schematic) distribution for various droplets present in the combustor.

quency of the oscillation is a key parameter in determining the fraction of the total energy that is in phase with the pressure oscillation. An evaluation of the effect of the disturbance amplitude on the response factor indicates that the effects of amplitude variation are not very significant.

Droplet vaporization studies related to combustion instability have neglected to emphasize whether the wake of the droplet is ahead of the droplet or behind it; there is a significant change in the response factor for the two situations. When the droplet is faster than the gas, the system has a high damping, whereas when the gas is faster than the droplet, the response factor is high enough to drive the instability. This is primarily due to the diverse effect the velocity oscillation has on the relative velocity. A variation of the volatility of the fuel has a fairly significant effect on the response factor, though it is not as significant as the orientation of the wake.

VIII. Spray Combustion Instabilities

A. One-Dimensional Analysis

A one-dimensional computational analysis[38] was performed to investigate the liquid-fueled combustion instability with the objective of gaining fundamental understanding of the effects of various design parameters on the stability of the combustion process in liquid-fueled systems. Spray characteristics, mixture ratio profiles, and combustor geometry were among the more interesting parameters investigated. Effort concentrated on the unsteady solution for the gas and liquid phases, evaluation of the effects of inlet gas velocity perturbations, and the coupling between gas and liquid phases under perturbed conditions.

The physical model studied was a simplified one-dimensional approximation of a ramjet combustion chamber in which vaporization and combustion of liquid spray occurs. Monodispersed, monocomponent, and subcritical liquid spray (*n*-decane) was assumed. The gas phase (air) was taken as unsteady, subsonic, compressible, and exchanging mass, momentum, and heat with the liquid phase. The basic set of equations in the Eulerian approach included continuity, momentum, chemical species, energy, and an equation of state for the gas phase.

The Lagrangian formulation of the liquid-phase equations was based on the effective conductivity model.[31] This droplet model assumes quasisteadiness in the gas film and employs the one-third-averaging rule to evaluate the gas-mixture properties. The assumptions of unity Lewis and Prandtl numbers have been removed. The effect of Stefan flow on the heat and mass transfer is modeled by a correction on the film thickness. For efficient computational time and accuracy, an effective value of the liquid thermal conductivity is used to account for the internal circulation within the droplet. The vaporization model also accounts for the droplet drag and the heat and mass transfer with the gas phase. One-step oxidation kinetics were employed. No assumption was made about a rate-controlling process.

Dirichlet boundary conditions were used at the combustor inlet, where mass flow (time-dependent), gas temperature, and species concentration were specified. The inlet pressure was determined by taking the inlet stagnation pressure to be constant. Overall equivalence ratio was specified at the inlet. At the outlet, Neumann boundary conditions were used, where zero gradients of the density, temperature, and species concentration were employed. The short-nozzle approximation at the combustor outlet (entrance to the nozzle) allowed the exit Mach number to be constant. Initially, the gas-phase temperature, velocity, pressure, and species concentration are specified. For the liquid phase, the droplet radius, location, velocity, and temperature were specified. The numerical solution was based on the modified implicit continuous-fluid (ICE) Eulerian scheme.[71,72]

Results from several cases covering a wide range of parameters were obtained. It was found that when the combustor is free to oscillate, the frequency of oscillations is independent of the frequency of the perturbation function and is solely determined by the length of the combustor, the droplet size, and the overall equivalence ratio. Also, this frequency is found to be very near the fundamental resonant frequency of the combustion chamber.

The results of this study are summarized in Figs. 6–9, which map the domains of stable/unstable combustion with the droplet radius, the combustor length, and the overall equivalence ratio (OER) as the parameters. Figure 6 shows the domain as determined by the droplet radius and the combustor length. With increase of the droplet size (and decrease in droplet number density) for a fixed combustor length and equivalence ratio, the combustion passes from stable to unstable and back to stable. Also evident is the effect of the combustor length. At the lower droplet radius of 50 μm, increasing length moves the combustion process from unstable to stable, whereas the opposite is true at the higher droplet radius of

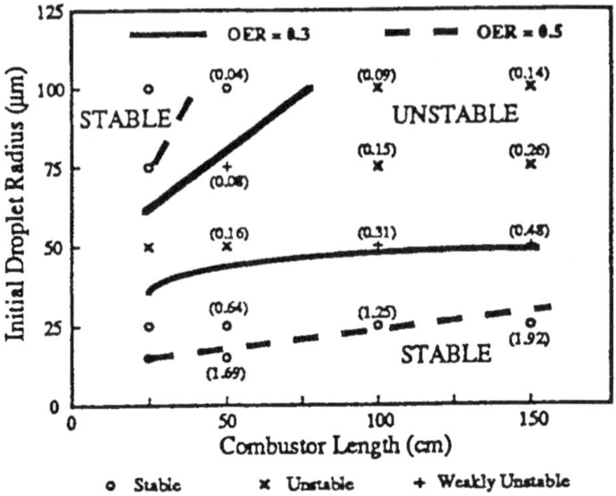

Fig. 6 Domain of stable/unstable operation: droplet radius and combustor length points shown correspond to OER = 0.3, numbers in parentheses indicate τ_p/τ_{dh}.

100 μm. The characteristic time for droplet heating for the n-decane droplets burning in 1000 K air at 10-atm pressure is estimated as $\tau_{dh} \sim (\rho_l C_{p_g}/\lambda_g) r_l^2 \approx d_l^2/1.22 \times 10^{-6}$ (m$^2 \cdot$ s/m^2). With fixed combustor length and increasing droplet radius, the characteristic time associated with the droplet heating increases. In some intermediate droplet radius range, it matches with the characteristic acoustic period so that the droplet vaporization can excite the acoustic modes. At higher droplet radii (75 and 100 μm), the low-frequency oscillations begin to get excited. This phenomenon is explained later. The characteristic time associated with the acoustic frequency, τ_a, is proportional to combustor length. This explains the increase in the stable region resulting from an increase in the combustor length. Also shown in Fig. 6 is the ratio of characteristic times for the oscillations to that

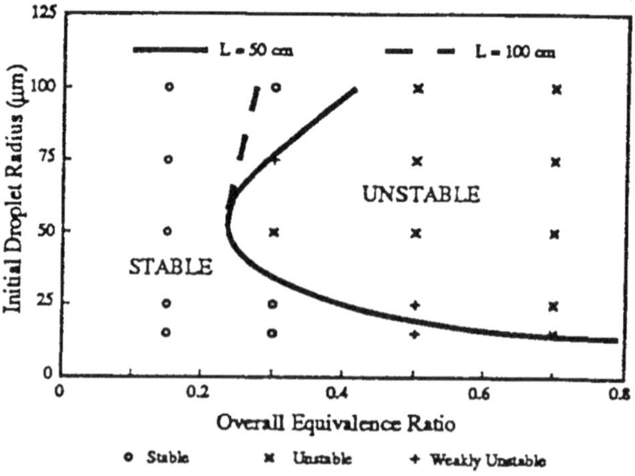

Fig. 7 Domain of stable/unstable operation: droplet radius and equivalence ratio points shown correspond to L = 50 cm.

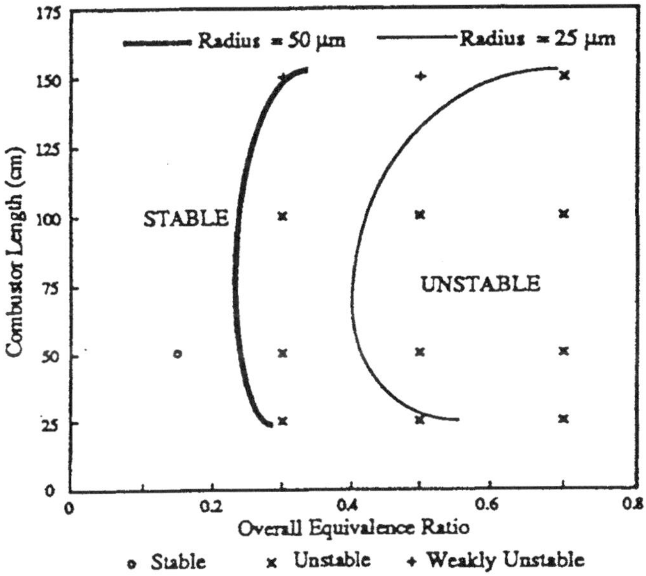

Fig. 8 Domain of stable/unstable operation: combustor length and equivalence ratio points shown correspond to $r_i = 50$ µm.

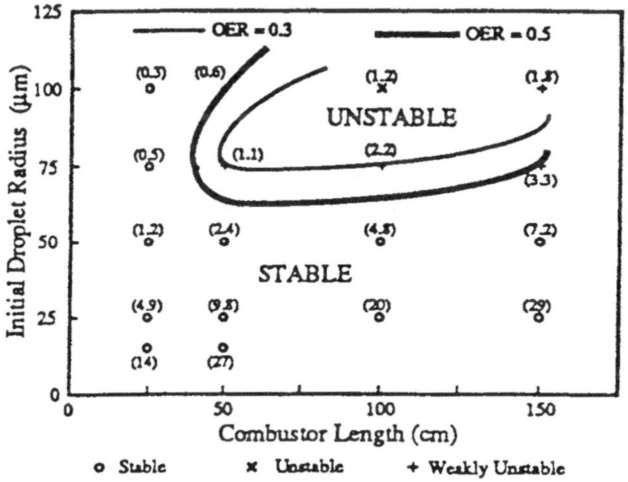

Fig. 9 Domain of stable/unstable operation for low-frequency oscillations only: droplet radius and combustor length points shown correspond to OER = 0.3, numbers in parentheses indicate τ_f/τ_{dh}.

for the droplet heating (τ_p/τ_{dh}). For unstable high-frequency cases, τ_p/τ_{dh} is in a range from 0.08 to 0.48. The points in the center of the unstable range vary between 0.14 and 0.16. Therefore, we take 0.15 as the most unstable value of the ratio. Tong and Sirignano,[23] in their study of the response of a vaporizing droplet in an unstable combustor, report a maximum gain occurring for oscillations of frequency 200 Hz for 100-μm droplets. Using a characteristic frequency of 30 Hz for 100-μm droplet heating, we obtain $\tau_p/\tau_{dh} = 0.15$ for the Tong–Sirignano calculation, too.

Figure 7 shows the domain as determined by the droplet radii and the overall equivalence ratio. The domain for unstable operation becomes larger at higher equivalence ratios, as the overtones get excited. The situation is further complicated by the interaction of the simulated injection process. For a fixed rate of injected fuel, the combustion process passes from stable to unstable at lower droplet radius, and again becomes stable at higher droplet radius. Figure 8 shows the stability limits as a function of the combustor length and the overall equivalence ratio. For a given droplet radius, the combustion passes from a stable to an unstable region with increase in the equivalence ratio. In Fig. 9, the low-frequency oscillations are isolated from the high-frequency ones. The figure shows that the low-frequency oscillations are important only in the cases of higher droplet radius. Also shown in the figure is the ratio of characteristic times for residence, taken as $\tau_f = 2L/u_{g,\text{inlet}}$, to the droplet heating time. It is inferred that to excite the low-frequency modes the characteristic time for the liquid droplet heating has to correspond to the residence time, or $\tau_f/\tau_{dh} \approx 1.0$.

Certain general statements can be made. Vaporization is rate controlling in the one-dimensional simulation where turbulent mixing and transverse transport are not considered. The combustion process can occur in either a stable or unstable mode. In the unstable mode, limit-cycle behavior of pressure, temperature, velocity, and gas-phase mass flow rate will occur. The magnitude of these limit-cycle oscillations is determined by the combustor geometry, spray characteristics, and the overall equivalence ratio. Low-frequency or entropy oscillations result when the characteristic time for droplet heating is close to that for the gas residence. High-frequency oscillations result when the ratio of characteristic time for period of oscillation to droplet heating is close to 0.15. At lower equivalence ratios, the frequency of free oscillations is very close to the fundamental resonant frequency of the combustor. At higher equivalence ratios, overtones are present. The frequency of the initial perturbation function has a small effect on the combustor oscillations. When the forcing function is removed, the combustor oscillations return to their natural frequency within a few cycles of oscillation. In cases where the damping of excited acoustic oscillations occurs, the excitations tend to dampen in about 10 cycles. The variation of droplet size, for the same equivalence ratio and combustor length, has a profound effect on the combustor stability. A range of droplet radii for which the combustion is unstable is observed. Below and above this range of droplet size, no limit-cycle behavior is present. This indicates the importance of spray combustion. An intermediate range also exists for combustor length, like the droplet radii, where the combustor operation is unstable. Increasing the equivalence ratio enlarges the domain of unstable operation. In the range of the calculations, all droplet radii and combustor length cases are unstable for the equivalence ratio of 0.7. When the inlet gas velocity is continuously excited, resonance occurs for an excitation frequency close to the combustor natural frequency.

B. Idealized Rocket Engine

The single LOX-droplet model[65] was used to investigate the combustion of arrays of LOX droplets in a simplified combustor at supercritical conditions.[73] The simplified configuration considered consists of a two-dimensional rectangular channel closed at one end where discrete LOX droplets are injected in streams (Fig. 10). This configuration, also known as parallel droplet streams, has been extensively used to investigate flame propagation in hydrocarbon fuel droplet streams[74–76] and in hazardous waste incineration.[77–79] A comprehensive review was contributed by Rangel and Sirignano.[80] Using this configuration, the transient effects in the gas phase and the consequences of discrete droplet injection (intermittence) may be included in the model.

Velocity and pressure fields in the channel are prescribed, and species and energy equations are solved. Droplets are followed in a Lagrangian manner and are viewed as point sources (or sinks) from the gas phase. Effects of supercritical ambient conditions on the droplet convective vaporization are included using the model already described.

The LOX droplets are initially in a subcritical state; in the combustor, they heat, vaporize, and burn. Depending on the external conditions, their surfaces may reach the critical mixing point. Note that, hitherto, actual combustion had not been included. The droplets respond to the changing external gaseous field as they travel through the idealized combustor. No direct interaction between drops is included, but the droplets do affect each other through the impact they have on the gaseous fields. Investigation of the influence of these interactions on the droplet transcritical behavior is the main purpose of a recent study.[73]

In the particular case of the LOX droplet in hydrogen, however, the contribution of the mass vaporized to the overall mass flux is not negligible because of the disparities of the molecular weights. To include these effects in the present model while keeping the overall algorithm reasonably simple, the resulting departure of the longitudinal velocity from the uniform profile is evaluated in a quasisteady manner from mass flux conservation. This non uniformity may, in certain cases, affect significantly the computed droplet lifetime (Fig. 11).

The source terms due to vaporization are summed over all of the droplets. The possibility of stripping is included in their evaluation. It was shown[65] that although some of the stripped liquid vaporizes in the film surrounding the droplet (secondary vaporization), thus affecting directly heat and mass transfer to the droplet, most of the stripped liquid vaporizes outside that film. The assumption is made here that, given the high-pressure, high-temperature conditions in the gas phase, and given

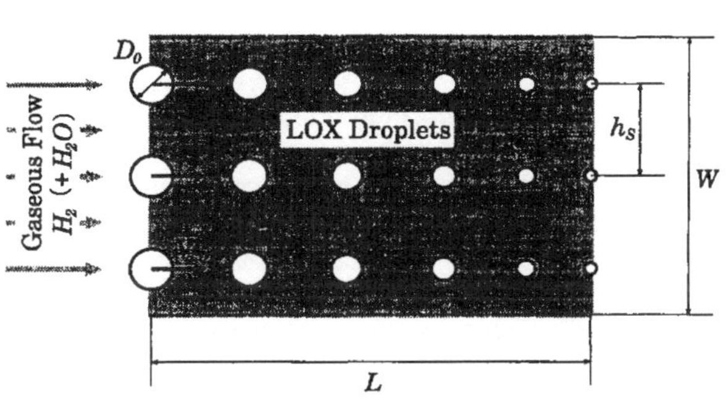

Fig. 10 Idealized parallel stream configuration.

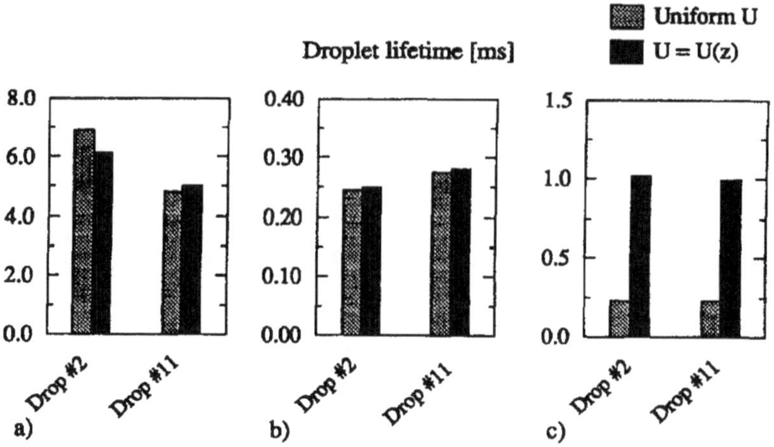

Fig. 11 Influence of nonuniform gas velocity on the droplet lifetimes: a) no stripping: $U_0 = 1$ m/s, $\Delta U = 1$ m/s, b) with stripping: $U_0 = 1$ m/s, $\Delta U = 20$ m/s, c) with stripping: $U_0 = 40$ m/s, $\Delta U = 20$ m/s.

the experimentally estimated size of the mist (1–10 μm), complete vaporization of the mist occurs within a small region around the parent droplet. Hence, as viewed from the gas flow, the mass source and energy sink due to the vaporization of the remaining mist may be taken as being a point source (sink) at the same location as the parent droplet.

In an attempt to simplify the chemical kinetics of the hydrogen/oxygen system while retaining the principal features of this reaction, the procedure described by Gutheil and Williams[81] and Chung and Williams[82] is followed. A one-step global mechanism is used. The overall reaction rate is approximated by the reaction rate of the first forward elementary reaction of the complete mechanism.

The most consequential result presented in the study[73] is that the combined effects of high temperatures in the reaction zone and reduced relative velocity cause the droplet surface temperature to reach the critical mixing conditions, even when stripping is occurring. This is a significant departure from the behavior predicted for an isolated vaporizing droplet.[65] Nevertheless, it was also demonstrated that stripping is the process controlling the droplet gasification and that the droplet lifetime is at least one order of magnitude smaller than that expected with primary vaporization only.

This simplified geometry was used to gain some insight concerning the consequences of these results on the response of a burning LOX droplet array to an oscillatory environment.[69] Note that the purpose of this comparison is to evaluate the influence of neighbor-induced gas field nonuniformities. High gas velocities, such as those considered in the cases already reported (Sec. VII), could not be used in the present case because they require a much larger domain and significantly more droplets and thus increase unrealistically the computational load. The gas injection velocity used here (1 m/s) is therefore not representative of real engine conditions (the droplet/gas relative velocity was kept unchanged at 20 m/s). Therefore, the values presented in Table 1 are much lower than the threshold value of the response factor for unstable operations.

An important difference between this type of computation and the single droplet evaluation reported earlier, comes from the fact that in the array configuration

Table 1 Comparison of the response factors obtained by the isolated droplet method and droplet in an array method with uniform response factor values shown for reference

Frequency, kHz	Isolated droplet	Droplet array	Uniform resp. factor
5	0.217 (16)[a]	0.425 (12)	0.1075
10	0.221 (16)	0.411 (6)	0.1077
15	0.273 (16) 0.739 (4)	0.726 (4)	0.1082

[a]Number of droplets per cycle in parentheses.

the mean velocity of the gas at infinity is affected by the gasification of the neighboring droplets. The stripping rate is, therefore, reduced, and the droplet lifetime increased.

The comparison of the values computed in both configurations (Table 1) indicates that the response spectra obtained with the vaporizing isolated droplet configuration underestimate the driving potential of the gasification process. In Table 1, the values labeled uniform response factor refer to the those obtained using an expression of the response factor resulting from a perturbation analysis as derived by Crocco and Cheng.[15] These values are much lower than those obtained by the other methods. This is attributed to the oversimplication of the vaporization process which such an analysis (linearization and perturbation methods) requires.

No simulation of the droplet-array configuration response could be performed with more realistic gas injection velocities. However, the foregoing argument indicates that such simulations would yield response factors higher than those obtained in the isolated droplet configurations[69] and, thus, confirm the conclusion drawn from these results (cf. Sec. VII).

IX. Conclusions

Consideration of the characteristic times for each of the major combustion processes in a liquid-propellant rocket motor and comparison with the natural periods for oscillations in typical rocket combustion chambers gives strong circumstantial evidence that vaporization is the major controlling mechanism for combustion instability. Atomization can be a vital coupling mechanism since it affects droplet size distribution that, in turn, directly relates to the characteristic vaporization time. More direct evidence about the prominence of the vaporization mechanism comes from several analyses. Theoretical/computational studies of individual fuel droplets in an oscillatory field indicate that the response of the vaporization to an ambient velocity oscillation is sufficiently large and the phase with the pressure oscillation is sufficiently small to drive combustion instability. Another computational study of combustion of liquid hydrocarbon fuel and air in a chamber with a choked nozzle at the exit confirms the importance of the vaporization process. Calculations by Tong and Sirignano[23] and Bhatia and Sirignano[38] give evidence that the rate-controlling vaporization for hydrocarbon fuels is a critical factor explaining the combustion instability mechanism.

Theoretical studies of oxygen propellant droplets at supercritical ambient pressure indicate substantial differences in the vaporization mechanism from subcritical vaporization. The immediate gas film can become a subcritical, multicomponent region. A secondary atomization mechanism identified as stripping can occur intermittently resulting in an increase in the frequency range for maximum response. Analyses of idealized combustion chamber configurations produced decreasing mass stripping rates with increased droplet lifetimes and a greater potential to drive combustion instability as compared to the isolated oxygen droplet studies.

To a substantial extent, there is need for calculations at more frequency values and over a wider range of other parameters in order to obtain a better picture of the role of vaporization in liquid rocket combustion instability.

Acknowledgments

The authors acknowledge critical programmatic support from the Société Européenne de Propulsion with P.-A. Baudart as liaison, the Office of Naval Research with Gabriel Roy as Scientific Officer, the Air Force Office of Scientific Research with M. Birkan as the Scientific Officer, and the Air Force Phillips Laboratory with J. Levine as Technical Monitor through a subcontract from Pratt and Whitney with G. Cox as Principal Investigator. Computational resources provided by the San Diego Supercomputer Center and the UCI Office of Academic Computing were also critical.

References

[1]Harrje, D. T., and Reardon, F. H. (eds.), *Liquid Propellant Rocket Combustion Instability,* NASA SP-194, Washington, DC, 1972, Chaps. 2–6.

[2]Jensen, R. J., "A summary of the JANNAF Workshop on Liquid Rocket Engine Combustion Driven Instability Mechanisms," *Proceedings of the 26th JANNAF Combustion Meeting,* (Pasadena, CA), Chemical Propulsion Information Agency, Laurel, MD, Oct. 1989.

[3]Priem, R. J., and Guentert, D. C., "Combustion Instability Limits Determined by a Nonlinear Theory and a One-Dimensional Model," NASA, Rept. TN D-1409, Oct. 1962.

[4]Priem, R. J., "Theoretical and Experimental Models of Unstable Rocket Combustors," *Ninth Symposium (International) on Combustion,* Combustion Inst., Pittsburgh, PA, 1963, pp. 982–992.

[5]Heidmann, M. F., and Wieber, P. R., "Analysis of *n*-heptane Vaporization in Unstable Combustor with Travelling Transverse Oscillations," NASA TN D-3424, Nov. 1965.

[6]Heidmann, M. F., and Wieber, P. R., "Analysis of Frequency Response Characteristics of Propellant Vaporization," NASA TM X-52195, June 1966.

[7]Strahle, W. C., "A Theoretical Study of Unsteady Droplet Burning: Transients and Periodic Solutions," Ph.D. Dissertation, Princeton Univ., Dept. of Aeronautical Engineering, Princeton, NJ, Dec. 1963.

[8]Strahle, W. C., "Periodic Solutions to a Convective Droplet Burning Problem: the Stagnation Point," *Tenth Symposium (International) on Combustion,* Combustion Inst., Pittsburgh, PA, 1964, pp. 1315–1325.

[9]Strahle, W. C., "High Frequency Behavior of the Laminar Jet Flame Subjected to Transverse Sound Waves," *Eleventh Symposium (International) on Combustion,* Combustion Inst., Pittsburgh, PA, 1966, pp. 747–754.

[10]Strahle, W. C., "Unsteady Laminar Jet Flames at Large Frequencies of Oscillation," *AIAA Journal,* Vol. 3, No. 5, 1965, p. 957.

[11]Strahle, W. C., "Unsteady Reacting Boundary Layer on a Vaporizing Flat Plate," *AIAA Journal,* Vol. 3, No. 6, 1965, p. 1195.

[12]Priem, R. J., and Heidmann, M. F., "Propellant Vaporization as a Design Criterion for Rocket Engine Combustion Chambers," NASA TR-R67, 1960.

[13]Culick, F. E. C., "Combustion Instabilities in Propulsion Systems," *Combustion Instabilities in Liquid-fuelled Propulsion Systems,* Propulsion and Energetics Panel 72nd B Specialists' Meeting (Bath, UK), AGARD, Neuilly-sur-Seine, France, NATO, Oct. 6–7, 1988.

[14]Williams, F. A., *Combustion Theory,* Benjamin-Cummings, Menlo Park, CA, 1985, Chap. 9.

[15]Crocco, L., and Cheng, S. I., *Theory of Combustion Instability in Liquid Propellant Rocket Motors,* AGARD Monograph No. 8, Butterworths, London, 1956, Chaps. 1–3.

[16]Lindsay, R. B. (ed.), *Acoustics: Historical and Philosophical Development,* Benchmark Papers in Acoustics, Dowden, Hutchinson and Ross, Stroudsburg, PA, 1973, pp. 5–20.

[17]Morse, P. M., and Ingard, K. U., *Theoretical Acoustics,* McGraw-Hill, New York, 1968, Chaps. 6, 10, 11, 14.

[18]Rayleigh, Lord, *The Theory of Sound,* Macmillan, London, 1894; 2nd ed. rev. and enlarged, Vol. 2, Dover, New York, 1945, p. 226, Sec. 322.

[19]Crocco, L., "Research on Combustion Instability in Liquid Propellant Rockets," *Eleventh Symposium (International) on Combustion,* Combustion Inst., Pittsburgh, PA, 1966, pp. 85–99.

[20]Rangel, R. H., and Sirignano, W. A., "Nonlinear Growth of Kelvin-Helmholtz Instability: Effect of Surface Tension and Density Ratio," *Physics of Fluids,* Vol. 31, No. 7, 1988, pp. 1845–1855.

[21]Rangel, R. H., and Sirignano, W. A., "The Linear and Nonlinear Shear-Instability of a Fluid Sheet," *Physics of Fluids A,* Vol. 3, No. 10, 1991, pp. 2392–2400.

[22]Strahle, W. C., and Crocco, L., "Analytical Investigation of Several Mechanisms of Combustion Instability," 5th Propulsion Meeting (Tampa, FL), Chemical Propulsion Information Agency, Laurel, MD, Nov. 1963.

[23]Tong, A. Y., and Sirignano, W. A., "Oscillatory Vaporization of Fuel Droplets in an Unstable Combustor," *Journal of Propulsion and Power,* Vol. 5, No. 3, 1989, pp. 257–261.

[24]Sirignano, W. A., "Fuel Droplet Vaporization and Spray Combustion Theory," *Progress in Energy and Combustion Science,* Vol. 9, No. 4, 1983, pp. 291–322.

[25]Tong, A. Y., and Sirignano, W. A., "Analysis of a Vaporizing Droplet with Slip, Internal Circulation, and Unsteady Liquid-Phase and Quasi-Steady Gas-Phase Heat Transfer," *ASME-JSME Thermal Engineering Joint Conference,* Vol. 2, American Society of Mechanical Engineers, New York, 1983, pp. 481–487.

[26]Hsieh, K. C., Shuen, J. S., and Yang, V., "Droplet Vaporization in High-Pressure Environments: 1. Near Critical Conditions," *Combustion Science and Technology,* Vol. 76, No. 1–3, 1991, pp. 111–132.

[27]Yang, V., Hsiao, C. C., and Shuen, J. S., "Pressure-Coupled Vaporization and Combustion Responses of Liquid Fuel Droplets in High-Pressure Environments," 27th AIAA/SAE/ASME/ASEE Joint Propulsion Conf., AIAA Paper 91-2310, Sacramento, CA, June 24–26, 1991.

[28]Yang, V., and Lin, N. N., "Vaporization of Liquid Oxygen (LOX) Droplets in Supercritical Hydrogen Environments," *Combustion Science and Technology,* Vol. 97, No. 4–6, 1994, pp. 247–270.

[29]Delplanque, J.-P., and Sirignano, W. A., "Numerical Study of the Transient Vaporization of an Oxygen Droplet at Sub- and Supercritical Conditions," *International Journal of Heat and Mass Transfer,* Vol. 36, No. 2, 1993, pp. 303–314.

[30]Tong, A. Y., and Sirignano, W. A., "Analytical Solution for Diffusion and Circulation in a Vaporizing Droplet," *Nineteenth Symposium (International) on Combustion,* Combustion Inst., Pittsburgh, PA, 1982, pp. 1001–1020.

[31]Abramzon, B., and Sirignano, W. A., "Droplet Vaporization Model for Spray Combustion Calculations," *International Journal of Heat and Mass Transfer,* Vol. 32, No. 9, 1989, pp. 1605–1618.

[32]Patnaik, G., "A Numerical Solution of Droplet Vaporization with Convection," Ph.D. Dissertation, Dept. of Mechanical Engineering, Carnegie-Mellon Univ., Pittsburgh, PA, April, 1986.

[33]Patnaik, G., Sirignano, W. A., Dwyer, H. A., and Sanders, B. R., "A Numerical Technique for the Solution of a Vaporizing Fuel Droplet," *Dynamics of Reactive Systems,* edited by J. R. Bowon, J. C. Leyer, and R. I. Solovkhin, Progress in Astronautics and Aeronautics, Vol. 105, AIAA, New York, 1986, pp. 253–266.

[34]Chiang, C. H., Raju, M. S., and Sirignano, W. A., "Numerical Analysis of Convecting, Vaporizing Fuel Droplet with Variable Properties," 27th AIAA Aerospace Sciences Meeting and Exhibit, AIAA Paper 89-0834, Reno, NV, Jan. 9–12, 1989; also *International Journal of Heat and Mass Transfer,* Vol. 35, No. 5, 1992, pp. 1307–1324.

[35]Chiang, C. H., "Isolated and Interacting, Vaporizing Fuel Droplets: Field Calculation with Variable Properties," Ph.D. Dissertation, Dept. of Mechanical Engineering, Univ. of California, Irvine, CA, July 1990.

[36]Chiang, C. H., and Sirignano, W. A., "Axisymmetric Vaporizing Oxygen Droplet Computations," 29th AIAA Aerospace Sciences Meeting and Exhibit, AIAA Paper 91-0281, Reno, NV, Jan. 1991.

[37]Chiang, C. H., and Sirignano, W. A., "Interacting, Convecting, Vaporizing Fuel Droplets with Variable Properties," *International Journal of Heat and Mass Transfer,* Vol. 36, 1993, pp. 875–886.

[38]Bhatia, R., and Sirignano, W. A., "One-Dimensional Analysis of Liquid-Fueled Combustion Instability," *Journal of Propulsion and Power,* Vol. 7, No. 6, 1991, pp. 953–961.

[39]Nguyen, T. V., and Muss, J. A., "Modification of the Agosta-Hammer Vaporization Response Model for the Prediction of High-Frequency Combustion Instabilities," 24th JANNAF Combustion Subcommittee Meeting, Chemical Propulsion Information Agency, Laurel, MD, Oct. 5–9, 1987.

[40]Agosta, V. D., and Hammer, S. S., "Vaporization Response of Evaporating Drops with Finite Thermal Conductivity," NASA CR-2510, Jan. 1975.

[41]Williams, F. A., "On the Assumption Underlying Droplet Vaporization and Combustion Theories," *Journal of Chemical Physics,* Vol. 33, No. 1, 1960, pp. 133–144.

[42]Chervinsky, A., "Transient Burning of Spherically Symmetric Fuel Droplets," *Israel Journal of Technology,* Vol. 7, No. 1–2, 1969, pp. 66–73.

[43]Nuruzzaman, A. S. M., and Beér, J. M., "On the Non-Steady State Nature of Droplet Combustion," *Combustion Science and Technology,* Vol. 3, 1971, pp. 17–24.

[44]Crespo, A., and Liñan, A., "Unsteady Effects in Droplet Evaporation and Combustion," *Combustion Science and Technology,* Vol. 11, 1975, pp. 9–18.

[45]Botros, P., Law, C. K., and Sirignano, W. A., "Droplet Combustion in a Reactive Environment," *Combustion Science and Technology,* Vol. 21, 1980, pp. 123–130.

[46]Saitoh, T., and Nagano, O., "Transient Behavior of Combustion and Evaporation of a Fuel Droplet with Finite Rate of Chemical Reaction and Variable Properties," *Technology Reports of the Tohoku University,* Vol. 51, No. 2, 1986, pp. 129–141.

[47]Rosner, D. E., and Chang, W. S., "Transient Evaporation and Combustion of a Fuel Droplet Near its Critical Temperature," *Combustion Science and Technology,* Vol. 7, 1973, pp. 145–158.

[48]Matlosz, R. L., Leipziger, S., and Torda, T. P., "Investigation of a Liquid Drop Evaporation in a High Temperature and High Pressure Environment," *International Journal of Heat and Mass Transfer,* Vol. 15, No. 4, 1972, pp. 831–852.

[49]Wieber, P. R., "Calculated Temperature Histories of Vaporizing Droplet to the Critical Point," *AIAA Journal,* Vol. 1, No. 12, 1963, pp. 2764–2770.

[50]Lazar, R. S., and Faeth, G. M., "Bipropellant Droplet Combustion in the Vicinity of the Critical Point," *Eleventh Symposium (International) on Combustion,* Combustion Inst., Pittsburgh, PA, 1966, pp. 801–811.

[51]Manrique, J. A., and Borman, G. L., "Calculation of Steady State Droplet Vaporization at High Ambient Pressures," *International Journal of Heat and Mass Transfer,* Vol. 12, No. 9, 1969, pp. 1081–1095.

[52]Curtis, E. W., and Farrell, P. V., "Droplet Vaporization in a Supercritical Microgravity Environment," *Astronautica Acta,* Vol. 17, No. 11/12, 1972, pp. 1189–1193.

[53]Spalding, D. B., "Theory of Particle Combustion at High Pressures," *ARS Journal,* Vol. 29, 1959, pp. 828–835.

[54]Rosner, D. E., "On Liquid Droplet Combustion at High Pressures," *AIAA Journal,* Vol. 5, No. 1, 1967, pp. 163–166.

[55]Polymeropoulos, C. E., and Peskin, R. L., "Combustion of Fuel Vapor in a Hot, Stagnant Oxidizing Environment," *Combustion Science and Technology,* Vol. 5, No. 4, 1972, pp. 165–174.

[56]Sanchez-Tarifa, C., Crespo, A., and Fraga, E., "A Theoretical Model for the Combustion of Droplets in Super-Critical Conditions and Gas Pockets," *Astronautica Acta,* Vol. 17, 1972, pp. 685–692.

[57]Culick, F. E. C., "Combustion Instabilities in Liquid-Fueled Propulsion Systems—An Overview," AGARD 72B Propulsion and Energetics Panel Meeting (Bath, England, UK), NATO, 1988.

[58]Umemura, A., "Supercritical Liquid Fuel Combustion," *Twenty-First Symposium (International) on Combustion,* Combustion Inst., Pittsburgh, PA, 1986, pp. 463–471.

[59]Litchford, R. J., and Jeng, S.-M., "LOX Vaporization in High-Pressure Hydrogen-Rich Gas," AIAA/SAE/ASSME/ASEE 26th Joint Propulsion Conf., AIAA Paper 90-210, Orlando, FL, 1990.

[60]Sirignano, W. A., and Crocco, L., "A Shock Wave Model of Unstable Rocket Combustors," *AIAA Journal,* Vol. 2, No. 7, 1964, pp. 1285–1296.

[61]Molavi, K., "Acoustic Oscillations in Dump Combustor Configurations," Ph.D. Dissertation, Dept. Mechanical Engineering, Univ. of California, Irvine, CA, April 1990.

[62]Molavi, K., and Sirignano, W. A., "Spray Combustion Driving Mechanism for Ramjet Instability," 25th JANNAF Combustion Meeting (Huntsville, AL), Chemical Propulsion Information Agency, Laurel, MD, Oct. 1988.

[63]Redlich, O., and Kwong, J. N. G., "On the Thermodynamics of Solutions," *Chemical Reviews,* Vol. 44, 1949, pp. 233–244.

[64]Chueh, P. L., and Prausnitz, J. M., "Vapor–Liquid Equilibrium at High Pressures—Vapor–Phase Fugacity Coefficients in Nonpolar and Quantum-Gas Mixtures," *Industrial and Engineering Chemistry Fundamentals,* Vol. 6, No. 4, pp. 493–498, 1967.

[65]Delplanque, J.-P., and Sirignano, W. A., "Boundary Layer Stripping Effects on Droplet Transcritical Convective Vaporization," *Atomization and Sprays,* Vol. 4, No. 3, 1994, pp. 325–349.

[66]Taylor, G. I., "The Shape and Acceleration of a Drop in a High-Speed Air Stream," *The Scientific Papers of G. I. Taylor,* edited by G. K. Batchelor, Vol. 3, Cambridge, Univ. Press, Cambridge, England, UK, 1963, pp. 457–464.

[67]Ranger, A. A., and Nicholls, J. A., "Aerodynamic Shattering of Liquid Drops," *AIAA Journal,* Vol. 7, No. 2, 1969, pp. 285–290.

[68]Litchford, R. J., Parigger, C., and Jeng, S.-M., "Supercritical Droplet Gasification Experiments with Forced Convection," AIAA/SAE/ASSME/ASEE 28th Joint Propulsion Conf., AIAA Paper 92-3118, Nashville, TN, July 1992.

[69]Delplanque, J.-P., and Sirignano, W. A., "Stability Influence of Transcritical LOX Droplet Vaporization in an Idealized Rocket Combustor," 31st AIAA Aerospace Sciences Meeting and Exhibit, AIAA Paper 93-0231, Reno, NV, Jan. 11–14, 1993; also "Influence of Transcritical LOX Droplet Vaporization on the Stability of an Idealized Rocket Combustor," *Journal of Propulsion and Power* (submitted for publication).

[70]Duvvur, A., Chiang, C. H., and Sirignano, W. A., "Oscillatory Fuel Droplet Vaporization: Driving Mechanism for Combustion Instability" (submitted for publication).

[71]Harlow, F. H., and Amsden, A. A., "A Numerical Fluid Dynamics Calculation Method for All Flow Speeds," *Journal of Computational Physics,* Vol. 8, No. 2, 1971, pp. 197–213.

[72]Westbrook, C. K., and Chase, L. L., "A One-Dimensional Combustion Model," Tech. Rept. UCRL-52297, Lawrence Livermore Lab., Univ. of California, Berkeley, CA, 1977.

[73]Delplanque, J.-P., and Sirignano, W. A., "Transcritical Vaporization and Combustion of LOX Droplet Arrays in a Convective Environment," 14th International Colloquium on the Dynamics of Explosions and Reactive Systems, Univ. of Combria, Portugal, Aug. 1993; also, *Combustion Science and Technology* (to be published).

[74]Rangel, R. H., and Sirignano, W. A., "Rapid Vaporization and Heating of Two Parallel Fuel Droplet Streams," *Twenty-First Symposium (International) on Combustion,* Combustion Inst., Pittsburgh, PA, 1986, pp. 617–624.

[75]Rangel, R. H., and Sirignano, W. A., "Unsteady Flame Propagation in a Spray with Transient Droplet Vaporization," *Twenty-Second Symposium (International) on Combustion,* Combustion Inst., Pittsburgh, PA, 1988, pp. 1931–1939.

[76]Rangel, R. H., and Sirignano, W. A., "Combustion of Parallel Fuel Droplet Streams," *Combustion and Flame,* Vol. 75, No. 3–4, 1989, pp. 241–254.

[77]Delplanque, J.-P., "Multicomponent Spray Combustion Computations: Application to Hazardous—Waste Incineration," Master's Thesis, Dept. of Mechanical Engineering, Univ. of California, Irvine, CA, Dec. 1989.

[78]Delplanque, J.-P., Rangel, R. H., and Sirignano, W. A., "Liquid-Waste Incineration in a Parallel-Stream Configuration: Parametric Study," *Twenty-Third Symposium (International) on Combustion,* Combustion Inst., Pittsburgh, PA, 1990, pp. 887–894.

[79]Delplanque, J.-P., Rangel, R. H., and Sirignano, W. A., "Liquid-Waste Incineration in a Parallel-Stream Configuration: Effect of Auxiliary Fuels," *Dynamics of Deflagrations and Reactive Systems: Heterogeneous Combustion,* edited by A. L. Kuhl, J. C. Leyer, A. A. Borisov, and W. A. Sirignano, Progress in Astronautics and Aeronautics, Vol. 132, AIAA, Washington, DC, 1991, pp. 164–186.

[80]Rangel, R. H., and Sirignano, W. A., "Spray Combustion in Idealized Configurations: Parallel Droplet Streams," *Numerical Approaches to Combustion Modeling,* edited by E. S. Oran, and J. P. Boris, Progress in Astronautics and Aeronautics, Vol. 135, AIAA, Washington, DC, 1990, pp. 585–613.

[81]Gutheil, E., and Williams, F. A., "A Numerical and Asymptotic Investigation of Structures of Hydrogen-Air Diffusion Flames at Pressures and Temperatures of High-Speed Combustion," *Twenty-Third Symposium (International) on Combustion,* Combustion Inst., Pittsburgh, PA, 1990, pp. 513–521.

[82]Chung, S. H., and Williams, F. A., "Asymptotic Structure and Extinction of CO-H_2 Diffusion Flames with Reduced Kinetic Mechanisms," *Combustion and Flame,* Vol. 82, No. 3–4, 1990, pp. 389–410.

Chapter 12

High-Frequency Injection-Coupled Combustion Instability

John J. Hutt* and Marvin Rocker*
NASA Marshall Space Flight Center, Huntsville, Alabama 35812

Nomenclature

A = admittance
a = admittance of a computational element
d = actual or hydraulic diameter of computational segment
F = force
f = friction factor
$\hat{f}$ = frequency
$\dot{m}$ = mass flow rate
$\boldsymbol{n}$ = normal unit vector
p = pressure
R = response
S = average cross-sectional area of control volume
S_I = cross-sectional area at the inlet of a computational segment
S_E = cross-sectional area at the exit of a computational segment
u = axial velocity
$\boldsymbol{V}$ = velocity vector
ρ = density
υ = volume
τ = combustion time lag
τ_w = wall shear stress
ω = angular frequency

Subscripts
c = combustion chamber
f = fuel

*Aerospace Engineer.

i = index of computational segment
o = oxidizer

I. Introduction

LIQUID-PROPELLANT rocket engines involve the injection of propellants, usually a separate fuel and oxidizer, into a cylindrical combustion chamber. The propellants have a high-energy density and react at a finite rate to release thermal and kinetic energy. The energy release rate includes a variety of time scales due to the many subprocesses that must occur before chemical reaction is completed. Because these subprocesses are dependent on the combustion-chamber pressure and gas velocity, a feedback mechanism for amplifying combustion chamber natural acoustic modes is often encountered. The feedback mechanisms are commonly subclassified into intrinsic and injection-coupled mechanisms. Injection coupling implies that the change in flow rate through the injector due to chamber pressure variation is a significant contributor to amplification. Often times, variation in subprocesses that occur after the propellant has been injected are the dominant mechanisms, and the injector is only passively involved. This is what is commonly referred to as the intrinsic mechanism. This chapter is focused on the unique aspects of the injection-coupled mechanism.

Because of the strong coupling among all of the subprocesses and sparse available instrumentation, it is difficult to clearly determine if injection coupling has occurred when an engine has gone unstable. There is one classic example that can be used to demonstrate a clearly injection-coupled mechanism, the J-2S.[1] The J-2S engine was a liquid-oxygen (LOX)/gaseous-hydrogen engine throttleable between 200-kN and 1140-kN vacuum thrust at mixture ratio from 4.4 to 6.0. The engine was pump fed by pumps driven from a hot-gas tapoff from the combustion chamber. The main injector was made up of shear coaxial injection elements. The initial LOX post design was orificed at the head end with an orifice-to-post area ratio of 0.30, giving it characteristics similar to an opened-end/closed-end acoustic resonator tube (Fig. 1). Although the LOX post lengths varied from 5.59 to 7.87 cm, a combustion instability occurred at 4400 Hz and was attributed to injection coupling between injector flow oscillations and combustion chamber acoustics.

Several unique features of this instability initially flagged this as different from a typical intrinsic instability. 1) The mode shape corresponded to a third tangential mode rather than a fundamental mode. 2) The growth rate of the oscillations was very slow, less than 10/s in contrast to a typical growth rate of an intrinsic instability, which is on the order of 100/s. 3) The oscillations limited at relatively low amplitudes without causing chamber damage. The injection elements used in this engine had staggered LOX post lengths ranging from 5.59 to 7.87 cm that served to spread the bandwidth of the injector response. Injector response calculations reported in Ref. 1 show a resonance at 4600 Hz for the 7.87-cm-long posts, which is very near the observed 4400-Hz frequency. The longer elements make up 237 of the 614 total elements, accounting for a significant percentage of available energy. The remaining elements, all shorter, have higher frequency resonances and contributed less to the instability. This may at least partially explain why the growth rates were slow. A larger contributor to this phenomenon lies in the behavior of the combustion itself. It is often approximated that the

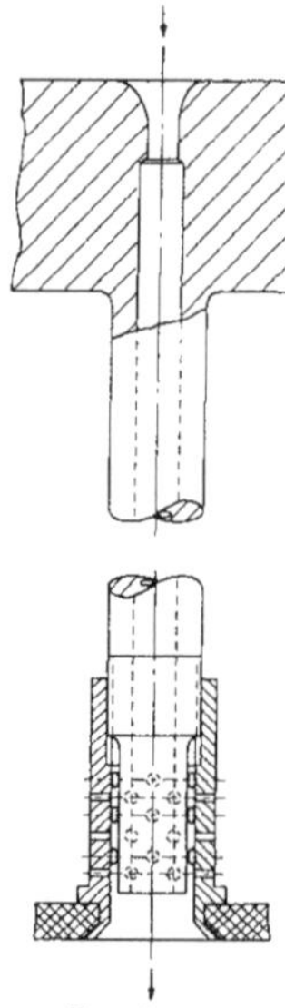

Fig. 1 Initial J-2S injector element.

combustion occurs at a fixed axial plane in the chamber (concentrated combustion), allowing a single characteristic time lag to determine when the energy of the injected propellants is released. This time lag controls the phase relationship between the injector oscillations and chamber oscillations. In an actual case the combustion is distributed with some fraction of the burning complete at each instant in time. With this being the case, only a fraction of the available energy may be released in phase. It was believed by the original investigators that the portion of the reaction going on in the cup region provided the proper phase relationship, resulting in a weakly coupled system with the small growth rate. The problem was corrected by altering the resistance of the flow passage allowing most of the loss to occur at the exit of the element. Figure 2 shows the final configuration of the LOX posts, designed to have a considerable pressure drop near the post exit.

The frequency response predictions of these respective elements is shown in Figure 3. The solid line represents the gain of propellant flow rate oscillation as a result of chamber pressure oscillation for the initial J-2S design. The curve has a substantial peak at 4600 Hz, indicative of a resonance condition. The dashed line represents the frequency shift and magnitude reduction that resulted when the post was redesigned. Testing of this design verified that the change was sufficient to stabilize the injector. The methodology used to generate these response curves is described in Sec. II. In this type of instability, injector effects can be isolated to the relatively simple acoustics of the LOX post. In general, however, both the fuel and oxidizer feed systems, including the propellant manifolds, may be coupled.

Another unique aspect of the combustion instability phenomenon with LOX/hydrogen propellants is the strong effect of the fuel injection temperature. Increasing the fuel temperature tends to stabilize the engine. This effect has been studied at least as far back as 1964, but a general agreement does not exist on the

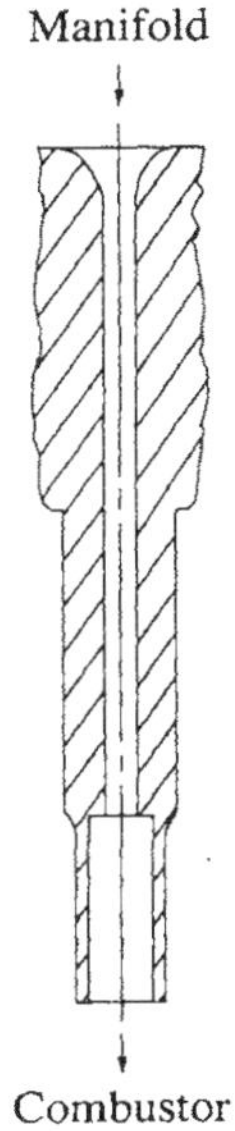

Fig. 2 Final J-2S LOX post.

controlling physical mechanisms. The effect is commonly attributed to either a momentum ratio or velocity ratio effect, which implies an intrinsic instability independent of injector flow rate oscillations. The results of the investigations in the 1960s, however, indicate exactly the opposite argument. The following paragraphs are a brief summary of the studies that lead to this conclusion.

Under NASA contract, an experimental program was conducted to investigate this phenomena.[2] Results of tests involving both a 2.54 × 12.7 cm slab motor and a 25.4-cm-diam cylindrical combustor were evaluated. Fuel injection temperature ranged from 25 to 139 K with chamber pressures ranging from 2050 to 5790 kPa. This program demonstrated high-frequency instability near and below the critical temperature of hydrogen in many tests. The effect was found to be essentially

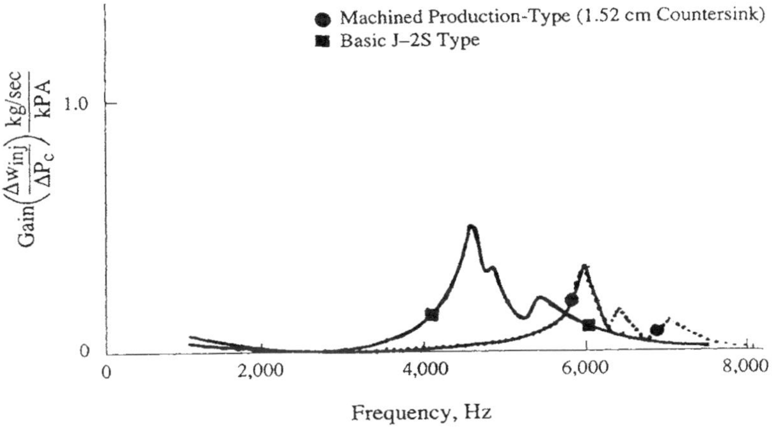

Fig. 3 Oxidizer post calculated frequency response.

independent of chamber pressure. An attempt was made to correlate the effect of momentum ratio, and no correlation was found. A trend was found that related stability to the inlet velocity ratio of the fuel and oxidizer.

In 1966 a study was conducted at NASA Lewis Research Center (LeRC) using a 89-kN thrust LOX/hydrogen rocket engine with a 27.36-cm-diam chamber.[3] The fuel temperature was varied from 33 to 107 K and the chamber pressure was held at 2070 kPa. For this engine, the stability boundary could always be defined in terms of a critical velocity ratio, regardless of the fuel temperature. Velocity ratio was controlled by varying the fuel flow areas. It was shown in a later study that velocity ratio correlation does not hold if the element size is changed.[4] A study was conducted to examine the case of varying oxidizer pressure drop independently of the exit velocity. The high-frequency instability limit was found to be independent of oxidizer pressure drop, supporting the argument of an intrinsic mechanism. It was found that in this combustor a LOX post recess of 0.25 cm increased the stability margin by 27.8 K but no conclusions were drawn as to why this was the case. The effects of hydrogen pressure drop were not examined, leaving open the possibility of an injector-coupled instability involving the hydrogen side of the injector. A continued study has shown that the hydrogen velocity ratio correlation actually can be explained by changes in the fuel pressure drop.[5] This information led to the understanding of one of the principal mechanisms of this type of instability. It was concluded that the change in fuel temperature results in a change in injection pressure drop across the fuel annulus allowing oscillations in the fuel annulus/manifold to tune with the chamber resonant oscillations. An empirical correlation was developed to predict the transition temperature. This relation empirically accounts for the closed-looped system that includes the fuel side of the injector. This correlation was found to be rather accurate for a fixed element density. The major shortcoming in this correlation is that the proportionality constant is dependent on the injector/combustor design. It was further shown that the change in hydrogen pressure drop due to LOX post recess accounts for the improved stability that results.

In 1974, NASA LeRC released a report that verified this injector-coupled phenomena.[6] This program involved tests using the same chamber as already discussed operated at 2070 kPa. This study demonstrated that a stability limit dependence on impingement angle and oxidizer tube blunt base thickness may be explained by changes in fuel side pressure drop. These characteristics were later shown also to apply to LOX/methane combustors using shear coaxial injection.[7]

To date a complete analysis of the data has not been conducted with a complete model that includes the fuel side manifold and fuel annulus acoustics. This analysis would be helpful in settling some of the controversy over the controlling mechanism.

It should be noted that injection coupling is never independent of the intrinsic subprocesses, such as atomization, propellant heatup, vaporization, and mixing because these processes determine the phase relationship between the injector response and the chamber response. Also, in the case of an intrinsic instability, where injector flow oscillations do not feed back energy to the chamber acoustic wave, the injector will be passively involved, with the chamber pressure oscillations observable in the propellant manifolds. Careful parametric evaluation and analysis are needed to determine the degree to which the injector is involved in any particular instability.

II. Modeling

A complete analysis of high-frequency injection coupling involves an evaluation of the interaction between the chamber pressure oscillations and the propellant burning rate oscillations including those that result from flow rate oscillations. The procedure used to perform a linear analysis of this problem is the same as that used for low-frequency, feed system stability with the injector response replacing the feed system response and the chamber acoustic response replacing the bulk chamber response. The model that describes the injector acoustics is unique to this chapter and will be described in the following paragraphs.

For many cases an acceptable level of modeling is to assume one-dimensional flow of a real, single-phase, compressible fluid in an injector element component such as a LOX post or fuel sleeve. The computational domain consists of the propellant flow path from the element inlet to the combustion chamber face. To account for axial variations in the flowfield, the domain is divided into $n - 1$ computational segments with n segment faces. The number of computational segments needed is chosen such that the length of a computational element is much shorter than the wavelength of the highest frequency acoustic mode to be considered. The control volume forms of the governing equations of mass and axial momentum can be written as follows.

Mass:

$$0 = \frac{\mathrm{d}}{\mathrm{d}t}\left(\iiint_{cv} \rho \, \mathrm{d}v\right) + \iint_{cs} \rho(\boldsymbol{V} \cdot \boldsymbol{n}) \, \mathrm{d}s \tag{1}$$

Axial momentum:

$$\sum F_x = \frac{\mathrm{d}}{\mathrm{d}t}\left(\iiint_{cv} \boldsymbol{V}\rho \, \mathrm{d}v\right) + \iint_{cs} \boldsymbol{V}\rho(\boldsymbol{V} \cdot \boldsymbol{n}) \, \mathrm{d}s \tag{2}$$

where the forces are pressure and friction forces. Assuming fully developed plug flow Eq. (1) can be integrated across a computational segment. Decomposing the result into mean and perturbed parts results in

$$0 = \bar{\dot{m}}_i - \bar{\dot{m}}_{i+1} \tag{3}$$

$$v_{i+1}\frac{\mathrm{d}\rho'_{i+1}}{\mathrm{d}t} = \dot{m}'_i - \dot{m}'_{i+1} \tag{4}$$

where barred quantities represent mean variables and primed quantities represent the perturbed variables. Assuming small, isentropic perturbations, the density is a function of pressure only. Incorporating this assumption and normalizing by the mean flow quantities, Eq. (4) becomes

$$C_{i+1}\frac{\mathrm{d}}{\mathrm{d}t}\left(\frac{p'_{i+1}}{\bar{p}_c}\right) = \frac{\dot{m}'_i}{\bar{\dot{m}}} - \frac{\dot{m}'_{i+1}}{\bar{\dot{m}}} \tag{5}$$

where

$$C_{i+1} = \frac{\bar{p}_c v_{i+1}}{\bar{\dot{m}}}\left(\frac{\partial \bar{\rho}}{\partial \bar{p}}\right)_{i+1}$$

It should be noted here that it is not assumed that the mean flowfield is isentropic, only that an isentropic acoustic field may be superimposed on a real, nonisentropic mean flow. To apply the momentum equation, the wall shear stress is represented empirically as

$$\tau_w = \frac{f}{4}\frac{\rho u^2}{2} \tag{6}$$

Integrating Eq. (2) across the computational element and decomposing into mean and perturbed variables gives, for the mean,

$$f_i\left(\frac{l_i}{d_i}\right)\left(\frac{1}{\bar{\rho}_i + \bar{\rho}_{i+1}}\right)\left(\frac{\bar{\dot{m}}}{S_i}\right)^2 = \frac{(S_I)_i}{S_i}\bar{p}_i - \frac{(S_E)_i}{S_i}\bar{p}_{i+1} + \frac{\bar{\dot{m}}^2}{S_i}\left\{\frac{1}{\bar{\rho}_i (S_I)_i} - \frac{1}{\bar{\rho}_{i+1}(S_E)_i}\right\} \tag{7}$$

Dropping all higher order terms and using the mean equation to eliminate f results in a manageable, linearized equation for the perturbed axial momentum,

$$L_i \frac{\mathrm{d}}{\mathrm{d}t}\left(\frac{\dot{m}'_i}{\bar{\dot{m}}}\right) + R_i\frac{\dot{m}'_i}{\bar{\dot{m}}} + \xi_i\frac{\dot{m}'_{i+1}}{\bar{\dot{m}}} - \phi_i\frac{\dot{m}'_{i-1}}{\bar{\dot{m}}} = (1 + D_i - G_i)\frac{p'_i}{\bar{p}_c} - (1 - U_i - H_i)\frac{p'_{i+1}}{\bar{p}_c} \tag{8}$$

where the constant coefficients are defined as

$$L_i = \frac{l_i\bar{\dot{m}}}{S_i\bar{p}_c} \qquad R_i = \frac{2(\bar{p}_i - \bar{p}_{i+1})}{\bar{p}_c} + \phi_i - \xi_i \qquad \xi_i = \frac{2\bar{\dot{m}}^2}{\bar{p}_c\bar{\rho}_{i+1}(S_E)_i S_i}$$

$$\phi_i = \frac{2\bar{\dot{m}}^2}{\bar{p}_c\bar{\rho}_i(S_I)_i S_i} \qquad D_i = \frac{R_i\bar{p}_c}{2(\bar{\rho}_i + \bar{\rho}_{i+1})}\left(\frac{\partial\bar{\rho}}{\partial\bar{p}}\right)_i \qquad U_i = \frac{R_i\bar{p}_c}{2(\bar{\rho}_i + \bar{\rho}_{i+1})}\left(\frac{\partial\bar{\rho}}{\partial\bar{p}}\right)_{i+1}$$

$$G_i = \frac{\phi_i\bar{p}_c}{2\bar{\rho}_i}\left(\frac{\partial\bar{\rho}}{\partial\bar{p}}\right)_i \qquad H_i = \frac{\xi_i\bar{p}_c}{2\bar{\rho}_{i+1}}\left(\frac{\partial\bar{\rho}}{\partial\bar{p}}\right)_{i+1}$$

To evaluate these coefficients the mean pressure and density field as well as the local speed of sound are needed. These may be evaluated using standard one-dimensional engineering equations for internal flow. This generalized formulation allows for the treatment of variable flow area and heat transfer through the element walls.

Combining Eqs. (5) and (8) and taking the Laplace transform gives

$$\left(\phi_i s + \frac{1 + D_i - G_i}{C_i}\right)a_{i-1}(s) - \left(L_i s^2 + R_i s + \frac{1 + D_i - G_i}{C_i} + \frac{1 - U_i - H_i}{C_{i+1}}\right)a_i(s) + \left(\xi_i s - \frac{1 - U_i - H_i}{C_{i+1}}\right)a_{i+1}(s) = 0 \tag{9}$$

where the local admittance is defined as

$$a_i(s) = \bar{p}_c \dot{m}_i'(s)/\bar{\dot{m}} p_c'(s) \qquad \text{and} \qquad s = j\omega \tag{10}$$

Equation (9) applies to all of the interior segments. The first and last segments require boundary conditions and can be written,

$$\left(L_1 s^2 + (R_1 - \phi_1)s + \frac{1 + U_1 - G_1}{C_2}\right)a_1(s) + \left(\xi_1(s) - \frac{1 - U_1 - H_1}{C_2}\right)a_2(s)$$

$$= (1 + D_1 - G_1)s\frac{p_i'(s)}{p_c'(s)} \tag{11}$$

$$\left(\phi_{n-1}s + \frac{1 + D_{n-1} - G_{n-1}}{G_{n-1}}\right)a_{n-2}(s) - \left(L_{n-1}s^2 + (R_{n-1} + \xi_{n-1})s\right.$$

$$\left. + \frac{1 + D_{n-1} - G_{n-1}}{C_{n-1}}\right)a_{n-1}(s) = (1 + U_{n-1} - H_{n-1})s = 0 \tag{12}$$

The local admittance at the $n-1$ segment represents the response in injector flow rate as a result of flow rate oscillations. For clarity we will, therefore, define the injector admittance for either a fuel or oxidizer flow rate as

$$A_o = a_{o,n-1}(s) = \frac{\bar{p}_c \dot{m}'_{o,n-1}(s)}{\bar{\dot{m}} p_c'(s)} \qquad A_f = a_{f,n-1}(s) = \frac{\bar{p}_c \dot{m}'_{f,n-1}(s)}{\bar{\dot{m}} p_c'(s)} \tag{13}$$

To obtain this admittance the simultaneous solution of the tridiagonal matrix formed from Eqs. (9–12) is required with boundary conditions required at the upstream node. The simplest condition is to set the upstream pressure oscillation to zero, which occurs when manifold effects are decoupled from the injector oscillations.

The goal of this analysis is to predict a coupling between the response of chamber pressure oscillations to burning rate oscillations and the response of burning rate oscillations to chamber oscillations. The burning rate response in this case is a function of the injector flow rate oscillations and the combustion process. The simplest approach for including the effects of the combustion process is to apply the well-known sensitive time-lag model resulting in the following equation:

$$\frac{\dot{m}'_B(s)}{\bar{\dot{m}}} = \frac{\dot{m}'_o(s)}{\bar{\dot{m}}}e^{-s\bar{\tau}_o} + \frac{\dot{m}'_f(s)}{\bar{\dot{m}}}e^{-s\bar{\tau}_f} + A_B(s)\frac{p'(s)}{\bar{p}} \tag{14}$$

where the burning admittance can be used as derived by Crocco and Cheng,[8]

$$A_B = n(1 - e^{-s\bar{\tau}})$$

Equation (14) can be rewritten as

$$\frac{\dot{m}'_B(s)\bar{p}_c}{p_c'(s)\bar{\dot{m}}} = \frac{\bar{r}A_o(s)e^{-s\bar{\tau}_o} + A_f(s)e^{-s\bar{\tau}_f}}{\bar{r} + 1} + A_B(s) \tag{15}$$

where the mixture ratio is defined as

$$\bar{r} = \frac{\bar{\dot{m}}_o}{\bar{\dot{m}}_f}$$

An overall injector admittance can now be written

$$A_I(s) = \frac{\bar{r}A_o(s)e^{-s\bar{\tau}_o} + A_f(s)e^{-s\bar{\tau}_f}}{\bar{r}+1} \tag{16}$$

The derivation for the injector admittance was based on a single-injector element. If all of the elements in the injector are identical with the same mean mass flow rates, this equation applies directly simply by using the total injector mean flow rates. The phase relationship due to the element spatial distribution is implicitly accounted for because all of the element oscillations must be synchronized to feed energy into the same acoustic wave if an instability is to be supported. However, if the elements are different, as in the case of variable length elements, the admittance must be written as a mass-weighted summation of the admittance of each individual element.

With these relations the total response of the combustion process to chamber pressure oscillations can now be written as a function of only the injector and burning admitttances,

$$[\dot{m}'_B(s)\bar{p}_c/p'_c(s)\bar{\dot{m}}] = A_I(s) + A_B(s) \tag{17}$$

To evaluate the potential of an unstable feedback of this oscillatory energy back into the chamber acoustic modes, a chamber response function is needed. This function can be obtained from the methods described in Chapter 15 and written as

$$R_c(s) = [p'_c(s)\bar{\dot{m}}/\dot{m}'_b(s)\bar{p}_c] \tag{18}$$

If the proper magnitude and phase relationship exits between the burning and chamber response, the linear theory results in an unbounded growth in the chamber pressure. The resulting characteristic equation is formed by combining Eqs. (17) and (18),

$$0 = 1 - R_c(s)\{A_I(s) + A_b(s)\} = 1 + F(s) \tag{19}$$

To satisfy Eq. (19), the following must be true:

$$|F(j\omega)| = 1 \tag{20}$$

$$[\arg\{F(j\omega)\}] = \pm\pi \tag{21}$$

The phase relationship depends on the values of the combustion time lag. Since these values are the weakest part of the analysis, a conservative approach is usually taken where it is assumed that Eq. (21) is always satisfied. To gain insight into the relative contribution of the chamber response and combustion response, a common approach is to graph

$$\left|\frac{1}{R_c(j\omega)}\right| \quad \text{vs} \quad \hat{f} = \frac{\omega}{2\pi} \tag{22}$$

$$|A_I(s) + A_b(s)| \quad \text{vs} \quad \hat{f} = \frac{\omega}{2\pi} \tag{23}$$

on the same plot. Points where the curves cross indicate frequencies that satisfy Eq. (20) and that indicate an instability. Because of the infinite number of mixed modes and harmonics that are possible from the acoustics, experience must be used to determine which modes should be considered. These are typically limited to the chamber fundamental modes and those up to the second harmonic. Historically, frequencies higher than these are not usually a problem.

The model described was implemented in a stability program conducted at the NASA Marshall Space Flight Center. Tests were conducted in a rectangular chamber with a single row of shear coaxial injection elements. The propellants were liquid oxygen and gaseous hydrogen with mixture ratios of 4.0 to 6.0 running at chamber pressures from 690 to 2760 kPa. The chamber measures 34.29 cm in width and 26.67 cm in length with a thickness of 2.54 cm. The results of plotting expressions (21) and (22) are shown in Fig. 4. The element LOX posts were 11.86 cm in length and 0.35 cm in diameter with a head orifice with a diameter of 0.22 cm. Each of the parabola-shaped curves represents a responsive band around a natural acoustic mode of the chamber. For example, the first widthwise mode has a natural frequency of about 3000 Hz. The bottom curve represents the vectored sum of the injector acoustics and the combustion response. The dominating feature of these curves is the peaks representing the natural acoustic modes of the LOX post. For example, the natural frequency of the first mode of the LOX post is about 2750 Hz. This example is typical of what is often seen when analyzing engines with shear coaxial injectors. By strict interpretation of the equations, an instability only can occur when the curves cross. In this plot, the only such points are at the very high-frequency first width-third longitudinal (1W-3L) and 1W-4L modes. Typically, these higher mixed modes do not go unstable, and an instability would not be expected to occur. The 1W mode curve, however, comes very close to

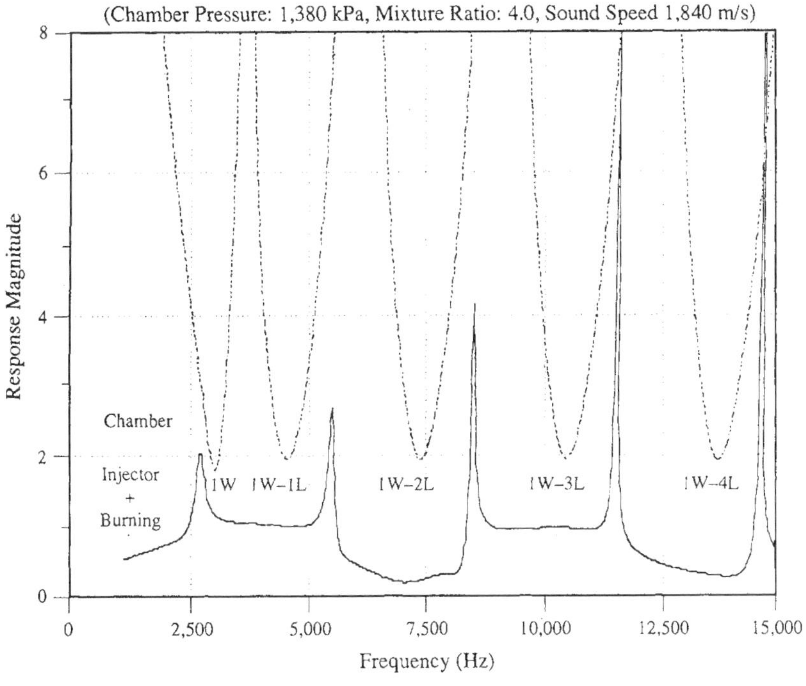

Fig. 4 Two-dimensional stability research combustor response plot.

touching the peak from the fundamental mode of the injector, and an instability in this mode would not be unexpected. As mentioned earlier no attempt is typically made to evaluate the phase relationship due to the general lack of confidence in this number.

III. Summary

High-frequency combustion instability in liquid-propellant engines often occurs due to the geometry of the elements that are often used. Shear coaxial injector elements are particularly prone to this type of instability because the resonant acoustic frequencies of the LOX post are often close to the resonant frequencies of the combustion chamber. Linearized acoustic response models of the injector acoustics, chamber acoustics, and the combustion response can be used to evaluate an engine's susceptibility to this phenomena. The preferred frequencies are controlled primarily by the natural acoustic modes of the chamber and injector, and the phase relationship is determined by the combustion process. Since the combustion process is difficult to adequately model, the conservative approach is usually taken that resonant frequency matches between the two components are to be avoided.

References

[1]Anon., "4400-Hz Vibration Investigation Final Report," Rocketdyne Engineering R-8742, Rockwell International, Rocketdyne Div., Canoga Park, CA, June 1971.

[2]Anon., "Summary Report on an Investigation of Combustion Instability for Liquid Oxygen and Liquid and Cold Gas Hydrogen Propellants," Pratt and Whitney Aircraft, United Aircraft Corp., PWA FR-1005, Aug. 1964.

[3]Wanhainen, J. P., Parish, C. P., and Conrad, E. W., "Effect of Propellant Injection Velocity on Screech in 20,000-pound Hydrogen-Oxygen Rocket Engine," NASA TN D-3373, April 1966.

[4]Hannum, N. P., and Conrad, E. W., "Performance and Screech Characteristics of a Series of 2500-Pound-Thrust-Per-Element Injectors for a Liquid-Oxygen-Hydrogen Rocket Engine," NASA Lewis Research Center, E-3283, Cleveland, OH, June 1967.

[5]Conrad, E. W., Bloomer, H. E., Wanhainen, J. P., and Vincent, D. W., "Interim Summary of Liquid Rocket Acoustic-Mode—Instability Studies at a Nominal Thrust of 20,000 pounds," NASA TN D-4968, April 1968.

[6]Hannum, N. P., Russell, L. M., Vincent, D. W., and Conrad, E. W., "Some Injector Element Detail Effects on Screech in Hydrogen-Oxygen Rockets," NASA TM X-2982, Feb. 1974.

[7]Jensen, R. J., Dodson, H. C., and Claflin, S. E., "LOX/Hydrocarbon Combustion Instability Investigation," NASA CR-182249, July 1989.

[8]Crocco, L., and Cheng, S. I., *Theory of Combustion Instability in Liquid Propellant Rocket Motors,* Butterworths, London, 1956, pp. 19–24.

Chapter 13

Acoustic Waves in Combustion Chambers

Vigor Yang,* Josef M. Wicker,† and Myong Won Yoon‡
Pennsylvania State University, University Park, Pennsylvania 16802

Nomenclature

$\bar{a}$ = speed of sound in mixture
D = diameter of cylindrical chamber
$\mathcal{E}$ = acoustic mechanical energy
f = source term in Eq. (2)
H = height of chamber
h = source term in Eq. (1)
J_m = Bessel function of order m
k = wave number
L = length of chamber
L_b = length of baffle blade
M = Mach number
$\dot{m}_b$ = propellant burning rate
$\boldsymbol{n}$ = unit outward normal vector
$\boldsymbol{r}$ = position vector
r = amplitude of acoustic mode
R = radius of cylindrical chamber
S = cross-sectional area
V = volume
α = linear growth constant
$\bar{\gamma}$ = mass-averaged specific heat ratio
η = amplitude of acoustic mode
θ = azimuthal coordinate or linear frequency shift
ϕ = phase angle of acoustic mode
ω = radian frequency

---

*Professor, Mechanical Engineering.
†Graduate Research Assistant, Aerospace Engineering.
‡Graduate Research Assistant, Mechanical Engineering.

Subscripts
NC = nonlinear combustion response
0 = limit cycle quantity

Superscripts
$'$ = fluctuation
$-$ = mean
$\cdot$ = time derivative

I. Introduction

HIGH-FREQUENCY combustion instabilities are generally characterized by mutual coupling between combustion processes and acoustic motions within a chamber. The study of acoustics in combustion chambers, therefore, constitutes a necessary step in approaching the complete problem. It is the purpose of this chapter to address various important aspects of acoustics in combustion chambers with respect to stability. Issues involving combustion processes and the detailed interactions between combustion and acoustics are treated in other chapters.

The fact that some combustion instabilities are intimately tied to acoustics was recognized early in the development of liquid rocket engines due to striking similarities between the unsteady motions of instabilities and the natural or normal acoustic modes of the chamber. Sometimes called screaming, screeching, or squealing modes, the frequencies of acoustic-coupled instabilities are generally higher than those of feed-system or injector-coupled instabilities, although delineation by frequency alone is not always adequate. An instability is usually classified in terms of the acoustic mode it most closely resembles, with the acoustic motion being either longitudinal or transverse. In rectangular geometries, transverse modes are all basically similar, whereas in cylindrical chambers the transverse modes can be either radial or tangential. Tangential modes can be standing or spinning, depending on whether the pressure node line is stationary or rotating. Besides these simple modes, there are also mixed modes with characteristics of multiple pure modes, each having discrete natural frequencies of oscillation.

Much of the previous theoretical work on liquid rocket combustion instabilities has been limited to linear analysis,[1–3] which is confined to treatment of infinitesimal (or at least sufficiently small) disturbances in the chamber. The global behavior predicted by a linear model will always yield one of three scenarios: the disturbance grows unbounded (linearly unstable), decays completely (linearly stable), or remains unchanged (neutrally stable). Because of the ubiquitous presence of small perturbations in any real system, the latter case is not physically realizable.

Nonlinear analysis deals with finite amplitude fluctuations, offering a wider scope for investigating oscillatory behavior.[4–6] Specifically, the phenomena of spontaneous[7–11] and pulse-triggered[12–14] oscillations may be treated. Spontaneous oscillations are observed in linearly unstable systems in which disturbances grow with no external forcing and eventually reach constant amplitudes, commonly known as limit cycles. Pulsed oscillations occur in linearly stable systems and require an initial excitation with amplitude above some threshold magnitude. For disturbance magnitudes less than the threshold value, the fluctuations decrease to zero; for those above the threshold, the fluctuations grow to a limit cycle value. Both types of nonlinear phenomena have practical manifestations. For example, the limiting amplitude of spontaneous or pulsed instabilities can determine if the instability warrants design modifications. Also, finding the disturbance threshold for pulsed instabilities is an important part of most engine design programs, since the commonly used bomb stability rating technique basically quantifies the "triggerability" of a system. A firm grasp on modeling, predicting, and analyzing nonlinear phenomena is essential for enhancing liquid rocket engine design and development.

In the remainder of this chapter, the general formulation of acoustic motions in a combustion chamber is explained, followed by a brief summary of normal modes. Several types of nonlinear acoustic waves are then examined in detail, with specific consideration of second- and third-order gasdynamics, mean flow/acoustic interactions, and nonlinear combustion responses.

II. Formulation

Analytically, a combustion chamber is a complex environment, and the equations which govern the behavior of acoustic fluctuations must be developed with care. The general formulation by Culick and Yang[3] outlined in Chapter 1 of this volume allows consideration of two-phase flow, combustion, and other processes pertinent to acoustic motions. After mass averaging the gaseous and condensed phases into one compressible medium, the conservation equations of mass, momentum, and energy are manipulated to give a nonlinear wave equation in terms of fluctuating pressure,

$$\nabla^2 p' - \frac{1}{\bar{a}^2}\frac{\partial^2 p'}{\partial t^2} = h \tag{1}$$

Boundary conditions on p' are specified in terms of the gradient of p' normal to the boundary surface

$$\boldsymbol{n} \cdot \nabla p' = -f \tag{2}$$

where $\boldsymbol{n}$ is the outward unit vector normal to the surface. The exact forms of the source terms h and f depend on the order of analysis and the types of processes being considered. Acoustic admittance functions may be conveniently employed to represent f in actual calculations.

The forcing functions h and f represent deviations from unperturbed acoustic modes and provide the mechanism for initiating and sustaining oscillations. Although exact analytical solution of Eqs. (1) and (2) is not feasible for all but the simplest forms of h and f, the forcing is typically so small that the unsteady motions in a chamber can be treated as a perturbation of the classical acoustics. Using the approximate method described in Chapter 1, the acoustic pressure and velocity fields are therefore most sensibly expressed as a series of the acoustic normal modes, with arbitrary time dependence accounted for by time-varying coefficients,

$$p'(\boldsymbol{r}, t) = \bar{p}\sum_{n=1}^{\infty} \eta_n(t)\psi_n(\boldsymbol{r}) \tag{3}$$

$$u'(r, t) = \sum_{n=1}^{\infty} \frac{\dot{\eta}_n(t)}{\bar{\gamma} k_n^2} \nabla \psi_n(r) \tag{4}$$

The normal modes ψ_n satisfy the homogeneous Helmholtz equation

$$\nabla^2 \psi_n + k_n^2 \psi_n = 0 \tag{5}$$

subject to the boundary condition

$$n \cdot \nabla \psi_n = 0 \tag{6}$$

Here $k_n = \omega_n / \bar{a}$ is the wave number and ω_n the frequency of the nth mode. After substituting these expressions into Eqs. (1) and (2), and applying a spatial-averaging technique similar to the Galerkin method, the following set of ordinary differential equations is derived to govern the temporal evolution of each acoustic mode:

$$\frac{d^2 \eta_n}{dt^2} + \omega_n^2 \eta_n = F_n \tag{7}$$

where

$$F_n = -\frac{\bar{a}^2}{\bar{p} E_n^2} \left[\iiint h \psi_n \, dV + \oiint \psi_n f \, dS \right] \tag{8}$$

and E_n is the Euclidean norm defined as

$$E_n^2 = \iiint \psi_n^2 \, dV \tag{9}$$

III. Normal Modes

Selection of the acoustic normal modes as basis functions for the unsteady pressure and velocity fields not only gives the series in Eqs. (3) and (4) good convergence properties, but also allows Eq. (7) to be written compactly, due to the orthogonality of the normal mode set. For an unbaffled rectangular chamber with length L, width W, and height H, the method of separation of variables may be used to easily find the normal mode shapes,

$$\psi_n = \cos[(\pi l / L) x] \cos[(\pi m / W) y] \cos[(\pi s / H) z] \qquad \begin{cases} l = 0, 1, 2, \cdots \\ m = 0, 1, 2, \cdots \\ s = 0, 1, 2, \cdots \end{cases} \tag{10}$$

with

$$k_n^2 = \omega_n^2 / \bar{a}^2 = \pi^2 [(l/L)^2 + (m/W)^2 + (s/H)^2] \tag{11}$$

Here the subscript n represents the triplet of indices (l, m, s), chosen so that $\omega_{n+1} \geq \omega_n$ for every mode; in two-dimensional problems, n would correspond to an index doublet. In an unbaffled cylindrical chamber with length L and radius R, azimuthal degeneracy leads to two possible normal modes for each frequency,

$$\psi_n = \cos[(\pi l / L) x] J_m(\kappa_{ms} r) \begin{cases} \cos(m\theta) \\ \sin(m\theta) \end{cases} \qquad \begin{cases} l = 0, 1, 2, \cdots \\ m = 0, 1, 2, \cdots \\ s = 0, 1, 2, \cdots \end{cases} \tag{12}$$

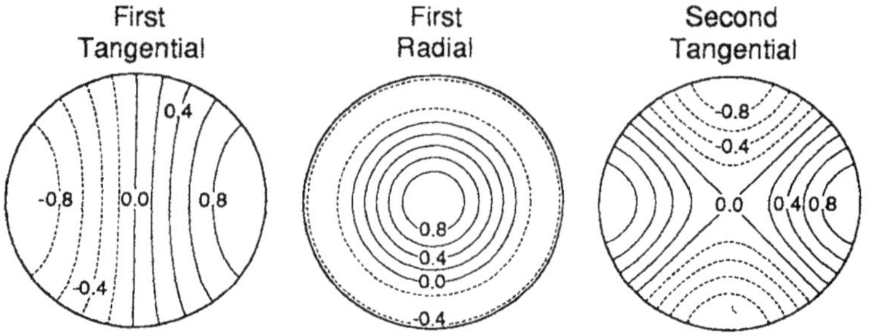

Fig. 1 Normalized acoustic pressure contours for first three transverse modes in an unbaffled cylindrical chamber.

with

$$k_n^2 = \omega_n^2/\bar{a}^2 = (\pi l/L)^2 + (\kappa_{ms})^2 \tag{13}$$

where $(\kappa_{ms} R)$ is the sth positive zero of the function $\mathrm{d}J_m(r)/\mathrm{d}r$ for $s \neq 0$ and $\kappa_{ms} = 0$ for $s = 0$. Figure 1 shows the first three transverse modes of acoustic pressure in a cylindrical chamber. The existence of two independent solutions at each frequency also allows for spinning normal modes in cylindrical chambers.

In combustors with complicated geometry, as with injector face baffles, the eigenfunction expansion technique explained in Refs. 15 and 16 may be used to find normal mode solutions. The acoustic fields in each baffle compartment and in the main chamber are represented separately using a series expansion in terms of transverse eigenfunctions for each region. An equation governing the axial dependence is subsequently developed. By matching acoustic pressure and the axial gradient of acoustic pressure at the baffle/main chamber interface, a set of equations for the series coefficients in each region can be solved, with the normal mode frequency treated as an eigenvalue. Figure 2 presents the normal mode shape

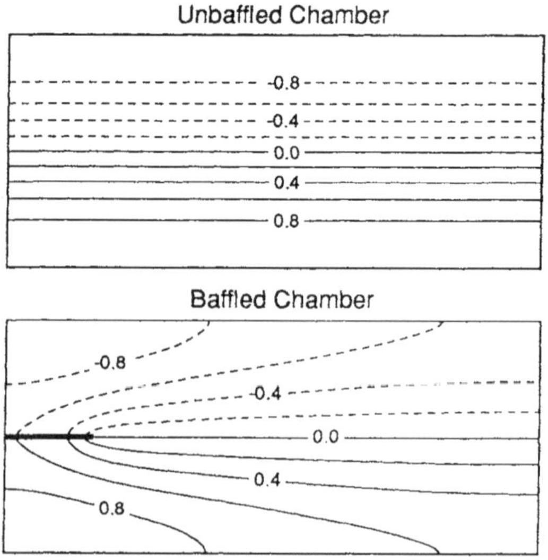

Fig. 2 Normalized acoustic pressure contours for first transverse mode in unbaffled and baffled (L_b = 0.2H) two-dimensional chambers, L/H = 1.25.

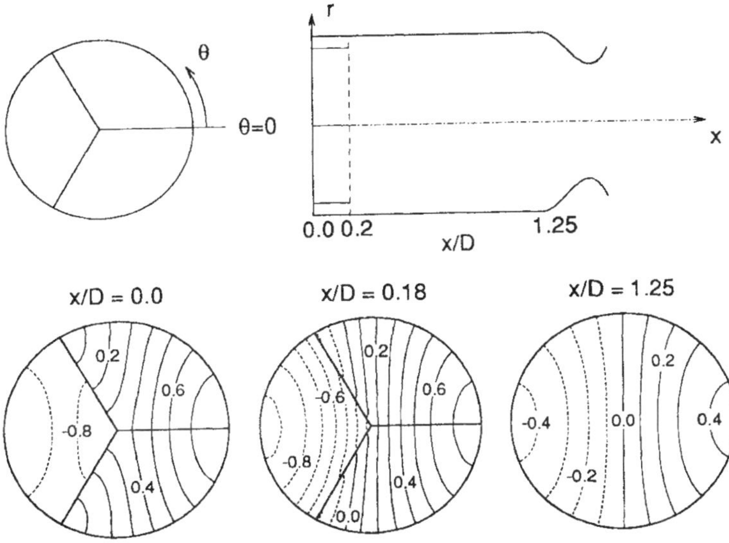

Fig. 3 Normalized acoustic pressure contours at various cross sections for first tangential mode in baffled cylindrical chamber, $L_b = 0.2D$ and $L/D = 1.25$.

for the first transverse mode in a two-dimensional rectangular chamber without baffles and with one baffle. The first tangential mode shape in a three-dimensional cylindrical chamber with three radial baffles is given in Fig. 3.

Examination of Figs. 2 and 3 shows that baffles can have a significant effect on the spatial structures of acoustic waves. Whereas a transverse mode shape in an unbaffled chamber varies only in the transverse direction, the baffle introduces longitudinal variation, especially near the head end. Indeed, for longer baffle lengths, the motions within the baffle compartments become almost totally longitudinal for certain transverse modes. Often accompanying this longitudinalization is a concentration of the acoustic pressure extremes toward the injector face. If the combustion processes (atomization, evaporation, reactions, etc.) are sensitive to pressure fluctuations, baffles may actually destabilize the system by increasing the pressure fluctuations in this sensitive region and enhancing combustion response.[15] Given any number of baffles and type of mode, the degree of acoustic pressure concentration may vary substantially among the baffle compartments. In some instances bulk-type pressure oscillations may be found in some compartments, whereas in other cases a baffle compartment may have almost no pressure fluctuations. Since these scenarios have very different implications for stability with respect to pressure-sensitive mechanisms, the global stability of baffled combustors must be considered on a case-by-case basis.

Because of the longitudinalization of the acoustic wave and the partitions between baffle compartments, the transverse component of acoustic velocity is substantially restricted in the compartments. For systems with velocity-sensitive combustion processes, baffles may therefore enhance stability with respect to transverse modes.[15] A region of large acoustic velocity exists just downstream of the baffle, corresponding to the strong acoustic pressure gradient found there. Since the no-slip condition must be physically satisfied at the baffle blade, viscous damping caused by the resulting shear stresses is proposed in Ref. 16 as one way in which baffles stabilize a chamber. The strong transverse acoustic flow may also result in periodic vortex shedding from the blade. Energy dissipated in these vortical structures or convected with them out of the chamber may be other damping mechanisms of baffles. The vortices could influence local mixing of propellants, affecting the combustion in a periodic fashion. Furthermore, near-wall coupling between fluctuating velocity and combustion processes should be considered in a rigorous stability analysis.

Another important effect of baffles is on the acoustic frequency. Numerous studies have concluded that the frequencies of transverse modes decrease after baffles are inserted.[15] Figure 2 shows how the wave "turns" around the baffle blade, yielding a longer acoustic particle path length and lower frequency. This effect may be stabilizing, destabilizing, or neutral, depending on the frequency response characteristics of the system.

IV. Second-Order Nonlinear Acoustic Waves

A linear treatment of unsteady motions, such as the stability analysis detailed in Chapter 1, is only meaningful for small disturbances. Investigation of finite amplitude phenomena, such as limit cycles and pulse-triggered instabilities, requires consideration of nonlinear physical mechanisms. The most fundamental nonlinear analysis is a second-order one in which terms having products of three or more fluctuating quantities may be safely ignored. If the mean flow Mach number is reasonably assumed to be small (of the same magnitude as a perturbational quantity itself), then the source terms in Eqs. (1) and (2) become[8]

$$\begin{aligned} h = &-\nabla\cdot\left[\bar{\rho}(\bar{\boldsymbol{u}}\cdot\nabla\boldsymbol{u}'+\boldsymbol{u}'\cdot\nabla\bar{\boldsymbol{u}})\right]+\frac{1}{\bar{a}^2}\left[\bar{\gamma}\frac{\partial p'}{\partial t}\nabla\cdot\bar{\boldsymbol{u}}+\bar{\boldsymbol{u}}\cdot\nabla\frac{\partial p'}{\partial t}\right] \\ &-\nabla\cdot\left[\bar{\rho}(\boldsymbol{u}'\cdot\nabla\boldsymbol{u}')+\rho'\frac{\partial\boldsymbol{u}'}{\partial t}\right] \\ &+\frac{1}{\bar{a}^2}\left[\frac{\partial}{\partial t}(\boldsymbol{u}'\cdot\nabla p')+\bar{\gamma}\frac{\partial}{\partial t}(p'\nabla\cdot\boldsymbol{u}')\right]+\nabla\cdot\mathcal{F}'-\frac{1}{\bar{a}^2}\frac{\partial\mathcal{P}'}{\partial t} \end{aligned} \tag{14}$$

and

$$f=\left\{\left[\bar{\rho}\frac{\partial\boldsymbol{u}'}{\partial t}+\bar{\rho}(\bar{\boldsymbol{u}}\cdot\nabla\boldsymbol{u}'+\boldsymbol{u}'\cdot\nabla\bar{\boldsymbol{u}})\right]+\left[\rho'\frac{\partial\boldsymbol{u}'}{\partial t}+\bar{\rho}(\boldsymbol{u}'\cdot\nabla\boldsymbol{u}')\right]-\mathcal{F}'\right\}\cdot\boldsymbol{n} \tag{15}$$

where $\mathcal{F}'$ and $\mathcal{P}'$ represent the fluctuating influences of mass, momentum, and energy transfer between the condensed and gaseous phases, heat release due to combustion, heat conduction, and viscosity. Substituting these expressions into Eq. (8), F_n clearly shows the basic physical processes affecting acoustic fluctuations,

$$\begin{aligned} F_n = -\frac{\bar{a}^2}{E_n^2\bar{p}}\Bigg[&\underbrace{\bar{\rho}k_n^2\iiint\psi_n(\bar{\boldsymbol{u}}\cdot\boldsymbol{u}')\,\mathrm{d}V}_{\mathrm{a}}\;\underbrace{-\bar{\rho}\iiint[\boldsymbol{u}'\times(\nabla\times\bar{\boldsymbol{u}})]\cdot\nabla\psi_n\,\mathrm{d}V}_{\mathrm{b}} \\ &\underbrace{-\frac{1}{\bar{a}^2}\iiint\frac{\partial p'}{\partial t}\bar{\boldsymbol{u}}\cdot\nabla\psi_n\,\mathrm{d}V}_{\mathrm{c}}\;\underbrace{+\frac{\bar{\gamma}-1}{\bar{a}^2}\iiint\psi_n\frac{\partial p'}{\partial t}\nabla\cdot\bar{\boldsymbol{u}}\,\mathrm{d}V}_{\mathrm{d}} \end{aligned}$$

$$\underbrace{+\bar{\rho}\iiint (\boldsymbol{u}'\cdot\nabla\boldsymbol{u}')\cdot\nabla\psi_n\,\mathrm{d}V}_{\mathrm{e}}\;\underbrace{-\frac{1}{\bar{a}^2}\iiint\frac{\partial p'}{\partial t}\boldsymbol{u}'\cdot\nabla\psi_n\,\mathrm{d}V}_{\mathrm{f}}$$

$$\underbrace{+\frac{\bar{\gamma}-1}{\bar{a}^2}\iiint\psi_n\frac{\partial p'}{\partial t}\nabla\cdot\boldsymbol{u}'\,\mathrm{d}V}_{\mathrm{g}}\;\underbrace{+\frac{\bar{\gamma}-1}{\bar{a}^2}\iiint\psi_n p'\nabla\cdot\frac{\partial\boldsymbol{u}'}{\partial t}\,\mathrm{d}V}_{\mathrm{h}}$$

$$\underbrace{+\oiint\psi_n\frac{\partial}{\partial t}(\bar{\boldsymbol{u}}\rho'+\boldsymbol{u}'\bar{\rho}+\boldsymbol{u}'\rho')\cdot\boldsymbol{n}\,\mathrm{d}S}_{\mathrm{i}}$$

$$\left.\underbrace{-\iiint(\mathcal{F}'\cdot\nabla\psi_n)+\left(\frac{1}{\bar{a}^2}\psi_n\frac{\partial\mathcal{P}'}{\partial t}\right)\mathrm{d}V}_{\mathrm{j}}\right] \tag{16}$$

The terms a–d are linear in the perturbation quantities and represent all influences of mean flow on unsteady motions. Terms e–h are the second-order effects of gas-dynamics, which give rise to cross- and self-coupling among modes of finite amplitude. Term i contains the integrated effects of boundary conditions due to injector and nozzle responses, acoustic liners, etc. The final term j includes any additional linear and nonlinear influences (to second order) of heat conduction, viscosity, interphase momentum and heat transfer, and volumetric combustion response.

By substituting Eqs. (3) and (4) into Eq. (16), F_n can be written as a function of η_n and $\dot{\eta}_n$. Although Eq. (7) could then be directly integrated with numerical techniques, some insight can be gained through application of the method of time averaging.[9] The method basically filters out high-frequency information from the results, leaving a simpler problem to be solved for only the slowly varying behavior, which is of most interest anyway. This filtering feature also decreases the stiffness of the system for numerical solution. To begin, $\eta_n(t)$ and $\dot{\eta}_n(t)$ are expressed in terms of $r_n(t)$ and $\phi_n(t)$, measuring the time-varying amplitude and phase of the oscillation, respectively,

$$\eta_n(t) = r_n(t)\sin[\omega_n t + \phi_n(t)] \tag{17}$$

$$\dot{\eta}_n(t) = r_n(t)\omega_n\cos[\omega_n t + \phi_n(t)] \tag{18}$$

where ω_n is the radian frequency of the nth normal mode as found in the linear analysis. The temporal development of the nth mode may be characterized by two disparate time scales: a fast time scale t_f, equal to ω_n^{-1}, and a slow time scale t_s, over which the gradual changes in nonlinear oscillations are observed. In Eq. (17) rapid variations of η_n are well captured by the $\omega_n t$ portion of the sinusoid, leaving only slow variations in the amplitude and phase. Substituting Eqs. (17) and (18) into Eq. (7) yields a set of first-order ordinary differential equations in terms of the new variables r_n and ϕ_n. If these equations for each mode are time averaged over a time interval τ, which is large compared to t_f but small with respect to t_s, then slowly varying quantities can be considered as constant during the integration. It can be shown that rapidly varying terms in the r_n and ϕ_n equations vanish exactly for longitudinal modes, since the mode frequencies are integer multiples of the first mode frequency. For transverse modes in cylindrical chambers, there is no such relationship between the modes. If, however, $t_f \ll \tau \ll t_s$, the effects of rapidly varying terms over many periods of oscillation are quite small, and the terms are neglected.[9]

A. Effect of Acoustic Mode Coupling

The linear coupling between acoustic modes is often weak due to disparity of the normal mode frequencies; this implies that linear modes behave independently of one another. For nonlinear waves, however, interactions between modes and self coupling can have profound influences. If only second-order effects of nonlinear acoustics [processes e–h in Eq. (16)] are considered, F_n can be written as

$$F_n = -\sum_{i=1}^{\infty}[D_{ni}\dot{\eta}_i + E_{ni}\eta_i] - \sum_{i=1}^{\infty}\sum_{j=1}^{\infty}[A_{nij}\dot{\eta}_i\dot{\eta}_j + B_{nij}\eta_i\eta_j] \tag{19}$$

where linear coefficients D_{ni} and E_{ni}, and nonlinear coefficients A_{nij} and B_{nij} are defined in Refs. 8 and 9. To maintain some degree of analytical tractability, only the first two modes (e.g., either longitudinal or transverse) are retained in the expansions of acoustic pressure and velocity. These common modes adequately capture the major features of nonlinear behavior, since the amplitudes of higher modes are usually small due to the effective viscous damping at higher frequencies. Time averaging then produces the following set for a two-mode system, for longitudinal as well as baffled and unbaffled transverse mode cases[7,9,15]:

$$\frac{dr_1}{dt} = \alpha_1 r_1 + a_1 r_1 r_2 \cos\Psi \tag{20}$$

$$\frac{dr_2}{dt} = \alpha_2 r_2 + b_1 r_1^2 \cos\Psi \tag{21}$$

$$\frac{d\Psi}{dt} = -2\theta_1 + \theta_2 + \Omega_1 - \left(2a_1 r_2 + b_1\frac{r_1^2}{r_2}\right)\sin\Psi \tag{22}$$

where

$$\Psi = 2\phi_1 - \phi_2 + \Omega_1 t$$
$$\alpha_n = -(D_{nn}/2), \qquad \theta_n = -(E_{nn}/2\omega_n), \qquad \Omega_1 = (2\omega_1 - \omega_2) \tag{23}$$

The coefficients a_1 and b_1 arise from the nonlinear gasdynamic coupling between modes. If nonlinear terms are ignored, Eqs. (20–22) become independent, with α_n and θ_n representing the linear growth constant and frequency shift of the nth mode, respectively.

The mode amplitudes r_1 and r_2 are constant for a limit cycle and cause all derivatives in Eqs. (20–22) to vanish. Limit-cycle amplitudes (denoted by the subscript 0) become

$$r_{10} = \left\{\frac{\alpha_1\alpha_2}{a_1 b_1}\left[1 + \left(\frac{2\theta_1 - \theta_2 - \Omega_1}{2\alpha_1 + \alpha_2}\right)^2\right]\right\}^{\frac{1}{2}} \tag{24}$$

$$r_{20} = -\frac{\alpha_1}{a_1}\left[1 + \left(\frac{2\theta_1 - \theta_2 - \Omega_1}{2\alpha_1 + \alpha_2}\right)^2\right]^{\frac{1}{2}} \tag{25}$$

Since $a_1 b_1$ is always negative in the present study of nonlinear gasdynamics, Eq. (24) requires that

$$\alpha_1 \alpha_2 < 0 \tag{26}$$

In other words, one mode must be stable and the other unstable for a limit cycle to exist. This obviously prevents any triggering to a limit cycle due to second-order gasdynamics alone since triggered instabilities can occur only in a system that is linearly stable. For a limit cycle to be observed, the amplitudes given by Eqs. (24) and (25) must be stable equilibrium points. Conditions for limit-cycle stability may be found by examining the behavior of an infinitesimal disturbance in the neighborhood of the limit-cycle solution.[9]

Analysis of phase in a limit cycle ϕ_{n0} also yields useful insight. For a constant value of Ψ required for an equilibrium point, the following form of ϕ_{n0} is found:

$$\phi_{n0} = \nu_n t + \zeta_n \tag{27}$$

where ν_n and ζ_n refer to frequency modulation and relative phase in the limit cycle, respectively. Because one phase is arbitrary, ζ_1 may be specified with no loss of generality. After several manipulations, expressions for the limiting mode oscillations become

$$\eta_{10}(t) = r_{10} \sin[(\omega_1 + \nu_1)t + \zeta_1] \tag{28}$$

$$\eta_{20}(t) = r_{20} \sin[2(\omega_1 + \nu_1)t + \zeta_2] \tag{29}$$

The second mode, therefore, oscillates at a modulated frequency exactly twice that of the first, even for transverse modes in cylindrical[9] or baffled[15] chambers. With the exception of $\nu_1 = 0$, the fundamental mode frequency is shifted from its linear value, even for systems originally having normal mode frequencies, which are already integer multiples of the first.

Nonlinear coupling via gasdynamics within the acoustic modes reveals an energy cascade, which explains the development of limit cycles. The rate of change of time-averaged acoustic energy $\langle \mathcal{E}_n \rangle$ within each mode is[15]

$$\frac{\mathrm{d}\langle \mathcal{E}_1 \rangle}{\mathrm{d}t} = \frac{E_1^2}{\bar{\gamma}^2}\left[\alpha_1 r_1^2 + a_1 r_1^2 r_2 \cos\Psi\right] \tag{30}$$

$$\frac{\mathrm{d}\langle \mathcal{E}_2 \rangle}{\mathrm{d}t} = \frac{E_2^2}{\bar{\gamma}^2}\left[\alpha_2 r_2^2 + b_1 r_1^2 r_2 \cos\Psi\right] \tag{31}$$

where

$$\langle \mathcal{E}_n \rangle = \left(r_n^2 E_n^2 / 2\bar{\gamma}^2\right) \tag{32}$$

Energy enters the system through linear processes of the unstable (e.g., first) mode and is transferred by nonlinear coupling mechanisms to the stable (e.g., second) mode where it is linearly damped. Substituting results from the limit-cycle analysis shows that a limit cycle is characterized by constant levels of energy within each mode and the entire system. For systems with $a_1 = -b_1$ (e.g., longitudinal modes, transverse modes in unbaffled rectangular two-dimensional chambers), the energy extracted in a limit cycle by the unstable mode exactly equals that dissipated by the stable mode; this is not exactly true, however, for cases with $a_1 \neq -b_1$ (e.g., transverse modes in cylindrical or baffled chambers). Perhaps energy is not directly transferred from mode to mode but, rather, is managed through a kind of energy repository, which may itself have the capacity to modulate energy. The physical manifestation of such an intermediate stage of energy transfer is presently unclear from the time-averaged or complete governing equations. Another strong possibility is that energy is truly conserved in the transfer mechanisms, but the definition in Eq. (32) does not adequately resolve the energy of the system. For example, the effect of frequency modulation is not explicitly accounted for. This question remains somewhat open and deserves further study.

Figure 4 shows the evolution of limit cycles for the first tangential and first radial modes in an unbaffled cylindrical chamber. This behavior is typical of limit-cycle development, with the unstable and stable modes initially showing exponential growth and decay, respectively. As the unstable mode grows, nonlinear coupling mechanisms become more important and eventually form the limit cycles.

Reference 9 also explored a three-mode expansion for a cylindrical chamber, considering the first tangential (1T), first radial (1R), and second tangential (2T) modes. The same basic type of mode coupling was found between the lowest frequency mode (1T) and each of the others, as observed for the two-mode expansion. Furthermore, the nonlinear coupling modulates the mode frequencies in such a way that the two higher modes oscillate at exactly twice the frequency of the first (see Fig. 5). It appears that the three-mode behavior is dominated by the coupling between the 1T and 1R mode due to more efficient nonlinear energy transfer from the 1T to the 1R mode than to the 2T mode.[9] For completeness, coupling between the unstable 1T mode and longitudinal or even mixed modes should be examined to see if other modes are important in the actual transfer of acoustic energy through nonlinear coupling.

It is advantageous to use time averaging and a two-mode expansion in studying nonlinear acoustics because sweeping results can be obtained due to the analytical tractability. If the governing equations were simply integrated with numerical

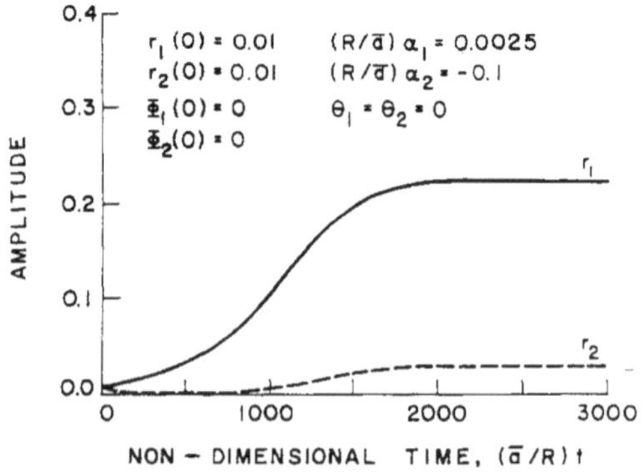

Fig. 4 Example of limit-cycle evolution for first tangential and first radial modes in unbaffled cylindrical chamber.

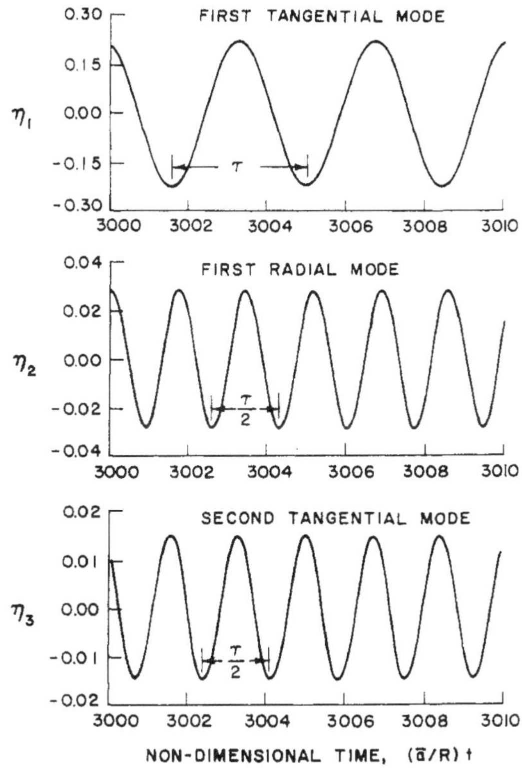

Fig. 5 Close up of limit-cycle oscillations in three-mode system (1T/1R/2T) showing frequency modulation.

techniques, results would be relevant only to specific cases. Jahnke and Culick[10,11] applied dynamical systems theory to nonlinear chamber acoustics and obtained fairly general results without resorting to time averaging or a few-mode expansion. The method utilizes a modified continuation technique to find limit-cycle solutions provided one solution is already known. The power of the technique lies in the ability to compute an entire locus of limit-cycle solutions as a function of system parameters (e.g., linear coefficients of each mode). Furthermore, solutions may be found for many modes in the expansion of Eqs. (3) and (4) only at the expense of increased computational time.

Time-averaged, two-mode solutions were compared with continuation results using different numbers of modes and considering only second-order gasdynamics in a longitudinal mode system.[10,11] Cases with unstable first modes or unstable second modes were treated, with all other modes assumed to be stable. Differences between the two approaches were observed primarily for very unstable systems (large values of α_1 or α_2); time averaging caused only quantitative differences, but truncation to two modes yielded qualitative differences. Specifically, in very unstable systems, higher modes can be responsible for a significant share of the total energy dissipation. If these modes are ignored, the amplitude of other stable modes must (unrealistically) increase to compensate for the loss of energy dissipation; in the extreme, the limit cycle in the two-mode system becomes artificially unstable. Practically, these results may not be too restrictive, since the disagreement occurs for mode amplitudes near the limit of applicability for second-order approximation anyway. Furthermore, the generality of the conclusions may be somewhat confined since only the unstable mode growth constant was varied. The two-mode expansion with an unstable second mode, however, was found to be physically limited, since the natural flow of energy from lower modes to higher modes through gasdynamics is not possible.[11]

B. Nonlinear Combustion Response

Although second-order gasdynamics alone are incapable of causing triggered instabilities, nonlinear combustion response to acoustic oscillations is a candidate mechanism in this important nonlinear phenomenon. Previous work[8,12,14] has emphasized combustion response in solid-propellant motors, which enters through the unsteady mass flux in the boundary effects term "i" in Eq. (16). Results thus obtained are still meaningful to liquid rocket engines in which nonlinear effects of the boundary are plausible, due to combustion at the boundary (e.g., "concentrated combustion" assumption or combustion of propellant film at chamber walls), acoustic response of the injector and nozzle, near-injector phenomena such as jet breakup, response of ablative surfaces, or response of acoustic dampers. More important, the effects of nonlinear response of surface combustion are expected to have structural analogs in systems with volumetric combustion (in the limit of concentrated combustion, volumetric and surface burning would have identical influences). In Refs. 8, 12, and 14, combustion response is not formulated from first principles; rather, ad-hoc response functions are used to obtain qualitative results for large classes of physically viable responses. For surface combustion, this amounts to finding a response function for the mass flow normal to the boundary.

In general, the combustion response function should depend on pressure and velocity fluctuations, since these quantities are the primary manifestations of acoustic waves. Two cases are treated here to illustrate the dependence of acoustic characteristics on the type of combustion response function. Case 1 assumes that the fluctuation of propellant burning rate $\dot{m}'_b$ is proportional to any linear combination of the following quantities:

$$\dot{m}'_b \propto p', \qquad u', \qquad p'^2, \qquad u'^2, \qquad p'u' \tag{33}$$

Substituting Eqs. (3) and (4) into the most general form of Eq. (33) will contribute to the forcing function as follows[14]:

$$(F_n)_{\rm NC} = \sum_{i=1}^{\infty} D^*_{ni}\dot{\eta}_i + \sum_{i=1}^{\infty} E^*_{ni}\eta_i + \sum_{i=1}^{\infty}\sum_{j=1}^{\infty} A^*_{nij}\dot{\eta}_i\dot{\eta}_j + B^*_{nij}\eta_i\eta_j + M^*_{nij}\dot{\eta}_i\eta_j \tag{34}$$

where all starred coefficients are constant values. This form of forcing function produces the same basic type of mode coupling as second-order gasdynamics alone, with governing equations similar to Eqs (20–22).[12] Therefore, this rather broad class of combustion response functions is incapable of causing pulse-triggered instability.

Another viable set of nonlinear combustion response functions involves the absolute value of oscillatory quantities. For example, response to $|u'|$ in liquid engines may be due to enhanced mixing of adjacent fuel and oxidizer streams, accelerated jet breakup, or droplet shattering caused by unsteady gas motions. For case 2 we assume that the fluctuating burning rate is proportional to any linear combination of the following quantities:

$$\dot{m}_b' \propto |u'|, \qquad |p'|, \qquad u'|p'|, \qquad p'|u'| \tag{35}$$

Of all of the possible forms of this dependence, consideration of response only to $|u'|$ (case 2a) and $p'|u'|$ (case 2b) is sufficient, since unsteady pressure can be related to velocity fluctuations through a simple phase shift. After expressing these functions in terms of η_n and $\dot{\eta}_n$ and following the method of time averaging, the equations for the amplitudes and phases of the first two modes become[14] for case 2a, $\dot{m}_b' \propto |u'|$,

$$\frac{dr_1}{dt} = \alpha_1 r_1 + a_1 r_1 r_2 \cos\Psi - 4c_{1a} r_2 \sin\Psi \tag{36}$$

$$\frac{dr_2}{dt} = \alpha_2 r_2 + b_1 r_1^2 \cos\Psi - c_{1a} r_1 \sin\Psi \tag{37}$$

$$\frac{d\Psi}{dt} = -2\theta_1 + \theta_2 + \Omega_1 - \left(2a_1 r_2 + b_1 \frac{r_1^2}{r_2}\right)\sin\Psi - \left(16\frac{r_2}{r_1} + \frac{r_1}{r_2}\right) c_{1a}\cos\Psi \tag{38}$$

For case 2b, $\dot{m}_b' \propto p'|u'|$,

$$\frac{dr_1}{dt} = \alpha_1 r_1 + a_1 r_1 r_2 \cos\Psi - c_{1b}\frac{16}{9}\left[r_1^2 - \frac{4}{25} r_2^2 \cos 2\Psi\right] \tag{39}$$

$$\frac{dr_2}{dt} = \alpha_2 r_2 + b_1 r_1^2 \cos\Psi - c_{1b}\frac{16}{15}\left[4 + \frac{7}{15}\cos 2\Psi\right] r_1 r_2 \tag{40}$$

$$\frac{d\Psi}{dt} = -2\theta_1 + \theta_2 + \Omega_1 - \left(2a_1 r_2 + b_1 \frac{r_1^2}{r_2}\right)\sin\Psi - c_{1b}\sin 2\Psi\left[2\left(\frac{16}{15}\right)^2 \frac{r_2^2}{r_1} - 7\left(\frac{4}{15}\right)^2 r_1\right] \tag{41}$$

where c_{1a} and c_{1b} are parameters (typically negative) measuring the strength of combustion response in each case. Although these equations were derived for longitudinal modes and two-dimensional transverse modes, extension to transverse modes in cylindrical or baffled chambers should present only quantitative differences.

The preceding set of equations may be compared with the case of pure nonlinear gasdynamic coupling, Eqs. (20–22). It has been shown that the linear cross coupling caused by combustion response to $|u'|$ in Eqs. (36) and (37) and the quadratic self-coupling caused by combustion response to $p'|u'|$ in Eq. (39) can produce triggering phenomena.[14] Both cases require strong linear damping of the second mode and weaker damping of the first mode for triggering to occur. For stronger combustion response, damping of both modes must increase for the system to trigger nonlinear instability.

The rates of mode energy are governed by the following equations for case 2a, $\dot{m}_b' \propto |u'|$,

$$\frac{\mathrm{d}\langle\mathcal{E}_1\rangle}{\mathrm{d}t} = \frac{E_1^2}{\bar{\gamma}^2}\left[\alpha_1 r_1^2 + a_1 r_1^2 r_2 \cos\Psi - c_{1a} 4 r_1 r_2 \sin\Psi\right] \tag{42}$$

$$\frac{\mathrm{d}\langle\mathcal{E}_2\rangle}{\mathrm{d}t} = \frac{E_2^2}{\bar{\gamma}^2}\left[\alpha_2 r_2^2 + b_1 r_1^2 r_2 \cos\Psi - c_{1a} r_1 r_2 \sin\Psi\right] \tag{43}$$

and for case 2b, $\dot{m}_b' \propto p'|u'|$,

$$\frac{\mathrm{d}\langle\mathcal{E}_1\rangle}{\mathrm{d}t} = \frac{E_1^2}{\bar{\gamma}^2}\left[\alpha_1 r_1^2 + a_1 r_1^2 r_2 \cos\Psi - c_{1b}\frac{16}{9}\left(r_1^3 - \frac{4}{25} r_1 r_2^2 \cos 2\Psi\right)\right] \tag{44}$$

$$\frac{\mathrm{d}\langle\mathcal{E}_2\rangle}{\mathrm{d}t} = \frac{E_2^2}{\bar{\gamma}^2}\left[\alpha_2 r_2^2 + b_1 r_1^2 r_2 \cos\Psi - c_{1b}\frac{16}{15}\left(4 + \frac{7}{15}\cos 2\Psi\right) r_1 r_2^2\right] \tag{45}$$

These expressions are similar to Eqs. (30) and (31) but with the addition of energy sources (or sinks) due to the nonlinear combustion response. In the case of linear cross coupling (case 2a), the combustion can add or detract energy from both modes, depending on the mode phases (specifically, the sign of $\sin\Psi$). For nonlinear self-coupling (case 2b), the explicit influence of nonlinear combustion response on mode energy is almost independent of mode phase, except for the relatively small $\cos 2\Psi$ terms in the energy flow of each mode. Assuming zero phase lag in the response function, nonlinear combustion in case 2b can only drive the second mode but can damp the first mode, if the amplitude of the second mode is sufficiently large and the amplitude of the first is sufficiently small.

Although modal energy transfer plays a key role in system behavior, triggering can occur with energy flow from lower modes to higher modes, or vice versa. For the linear cross-coupling case, the energy source term from nonlinear combustion involves the product of both mode amplitudes, i.e., the presence of one mode is required for the nonlinear combustion to feed (or subtract) energy to (or from) the other. Physically, this might be interpreted as transfer of energy between modes, but it is not the usual transfer associated with nonlinear gasdynamics. In the scenario of nonlinear self coupling, Eqs. (44) and (45) also contain nonlinear combustion response terms with the product of mode amplitudes, but the first mode has the additional r_1^3 term—the first mode can receive energy from nonlinear combustion independent of the second mode.

Examples of pulse-triggered instabilities for cases 2a and 2b are shown in Fig. 6. The typical behavior expected of triggering is observed, with small pulses damping away and larger pulses waxing to limit cycles. An issue of practical concern involves determination of the threshold amplitude, above which an initial pulse causes persisting oscillations and below which it damps out. The bomb test for rating the dynamic stability of an engine experimentally seeks to answer this question. Theoretically, further insight is available through examination of the equilibrium points of Eqs. (36–38) and (39–41). Based on studies of simplified one- and two-degree-of-freedom systems[13] and analysis of nonlinear combustion

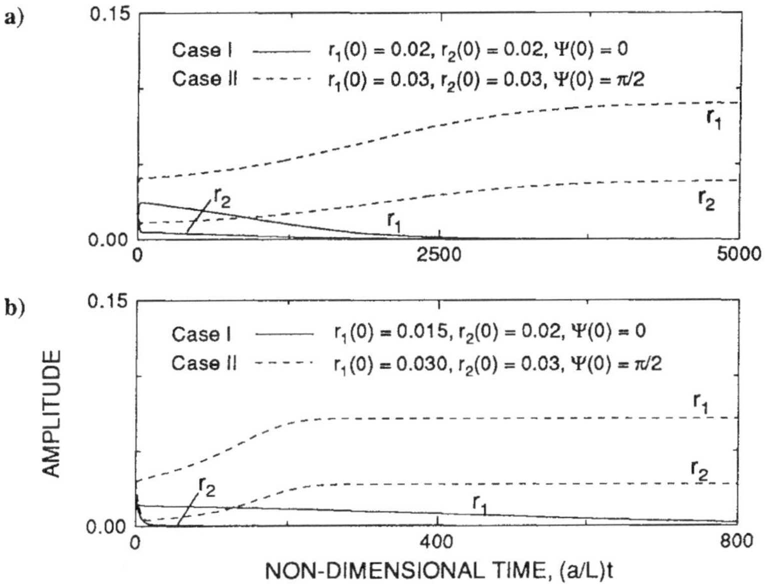

Fig. 6 Examples of pulse-triggered instabilities for cases of a) linear cross coupling and b) nonlinear self-coupling.

response,[12,14] the essence of triggering entails a stable trivial solution ($r_1 = r_2 = 0$) and a stable limit cycle of finite amplitude, with an unstable limit cycle of intermediate amplitude. The stable manifold of this intermediate limit cycle forms the boundary for the basin of attraction of the trivial solution (i.e., the region of initial disturbances which decay). It is reasonably hypothesized in Ref. 12 that initial pulses having energy greater than that of the intermediate limit cycle would grow to the stable limit cycle, but lower-energy pulses would decay. Although the amplitude of the intermediate limit cycle does appear to scale the region of stable pulses, the region is significantly distorted according to the phase and harmonic content (relative amplitudes of the modes) of the initial pulse.[14] In fact, for both cases 2a and 2b, certain stable pulses have initial amplitudes substantially larger than other pulses that trigger the system, but with suitable differences in initial phase and harmonic content. Practically, this means that the idea of a single "threshold" amplitude is physically incomplete—all aspects of the initial disturbance must be considered (e.g., traveling vs standing wave, "bomb" disturbance, harmonic disturbance, etc.).

V. Third-Order Nonlinear Acoustic Waves

The qualitative effects of third-order gasdynamic interactions are presented in this section. One motivation for originally pursuing third-order mechanisms was the faulty notion that these processes are vital in producing pulse-triggered phenomena.[5,6] It has already been shown that such effects are clearly not necessary.[13] Gasdynamics and mean flow/acoustics interactions to third order are not even completely sufficient for triggering. Investigation of these mechanisms is important, nonetheless, for complete understanding and for extending the analytical treatment to higher amplitude fluctuations.

A. Effect of Third-Order Acoustics

If the sources h and f are developed to third order (again, treating mean flow Mach number as a perturbation quantity), two new processes arise: third-order acoustics and interactions between mean flow and second-order acoustics.[13] By simply setting the mean flow equal to zero, the effects of third-order acoustics may be isolated for clearer interpretation. The time-averaged amplitude and phase difference equations for the first two longitudinal modes become

$$\frac{\mathrm{d}r_1}{\mathrm{d}t} = \alpha_1 r_1 + a_1 r_1 r_2 \cos\Psi \tag{46}$$

$$\frac{\mathrm{d}r_2}{\mathrm{d}t} = \alpha_2 r_2 + b_1 r_1^2 \cos\Psi \tag{47}$$

$$\frac{\mathrm{d}\Psi}{\mathrm{d}t} = -2\theta_1 + \theta_2 + \Omega_1 - \left(2a_1 r_2 + b_1 \frac{r_1^2}{r_2}\right)\sin\Psi + 2\beta_t\left(r_2^2 - r_1^2\right) \tag{48}$$

where $\beta_t = \omega_1(\bar{\gamma}+1)/(64\bar{\gamma}^2)$. It is interesting to note that comparison of the preceding expressions with the governing Eqs. (20–22) for the second-order acoustics shows that third-order acoustics influence only the phase angle of each acoustic mode, with basically identical coupling between the mode amplitudes. For a limit cycle to exist, one mode must still be linearly stable and the other linearly unstable–no triggering is possible due to third-order acoustics alone. The effects of second- and third-order gasdynamics are similar, providing a path for energy exchange between the modes. The only major effect of third-order acoustics is to modify the amplitudes and phases of each mode, with different conditions required for stable limit-cycle oscillations.

B. Mean Flow/Acoustics Interactions

The effects of mean flow interactions with second-order acoustics is investigated[13] separately by ignoring all third-order acoustic terms in h and f. The mean flow is assumed to be only axial and linearly varying,

$$\bar{u} = (\bar{u}_e/L)x \tag{49}$$

where L and $\bar{u}_e$ are the combustor length and mean velocity at the combustor exit, respectively. The equations governing the amplitudes and phase difference of the first two longitudinal modes become

$$\frac{\mathrm{d}r_1}{\mathrm{d}t} = \alpha_1 r_1 + r_1 r_2(a_1\cos\Psi - 7\beta_m \sin\Psi) \tag{50}$$

$$\frac{\mathrm{d}r_2}{\mathrm{d}t} = \alpha_2 r_2 + r_1^2(b_1\cos\Psi + \beta_m\sin\Psi) \tag{51}$$

$$\frac{\mathrm{d}\Psi}{\mathrm{d}t} = -2\theta_1 + \theta_2 + \Omega_1 - \left(2a_1 r_2 + b_1\frac{r_1^2}{r_2}\right)\sin\Psi + \beta_m\left(-14r_2 + \frac{r_1^2}{r_2}\right)\cos\Psi \tag{52}$$

where $\beta_m = \bar{u}_e/(32\bar{\gamma}L)$. The mode coupling in this scenario is similar to the second-order acoustics case, but with an important exception; due to the $\sin\Psi$

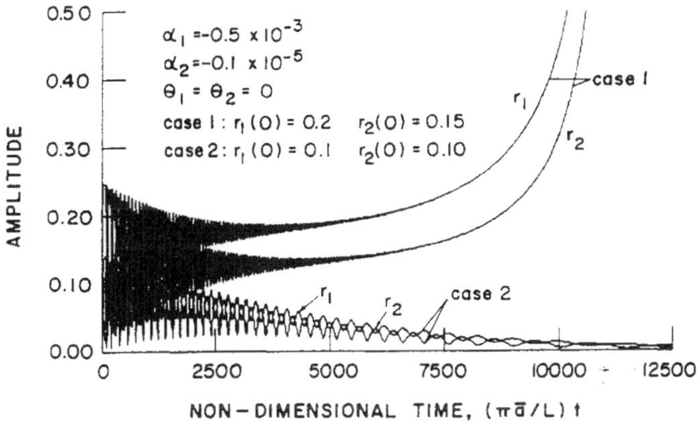

Fig. 7 Example of triggering to unstable limit cycle due to interactions between mean flow and second-order nonlinear acoustics.

terms in the amplitude equations, limit cycles can exist in a linearly stable chamber (both α_1 and α_2 negative). However, stability analysis shows that oscillations can trigger only to an unstable limit cycle, as shown in Fig. 7.Since this corresponds to unbounded growth, the ability of mean flow/acoustic interactions to cause triggering to a stable limit cycle is incomplete; other processes must act simultaneously to limit the unsteady motions.

VI. Conclusions

Because high-frequency combustion instabilities arise from interactions between combustion processes and acoustic motions, studies emphasizing chamber acoustics have added substantially to the general understanding of instabilities. The general two-phase formulation described in Chapter 1 of this volume has proven to be especially fruitful in such investigations. By treating unsteady motions in an integral fashion, it was shown in Chapter 1 that the problem of linear stability can be treated quite generally. Given knowledge of the normal modes of the chamber, the methodology also allows exploration of nonlinear phenomena as well.

The most fundamental of nonlinear processes in a thrust chamber is second-order gasdynamics. The intermodal coupling due to this mechanism allows energy transfer between modes, such that limit-cycle oscillations develop. For this to happen, one mode must be linearly stable and the other unstable, precluding the possibility of triggering due to second-order gasdynamics alone. By additionally considering appropriate nonlinear combustion responses (e.g., $\dot{m}_b'$ proportional to $|u'|$ or $p'|u'|$), triggering to stable limit cycles has been established, suggesting that this mechanism is an important player in pulse-triggered oscillations. The key to triggering is introduction of linear cross coupling between the modes, or nonlinear self-coupling in at least one mode. Results show that initial harmonic content and phase characteristics of the initial pulse (in addition to overall amplitude) are important in determining the nonlinear stability of a disturbance. Third-order gasdynamics have also been treated, yielding behavior basically similar to that of second-order gasdynamics alone (limit cycle may be observed, but no triggering). For assumed small mean flow, another third-order mechanism is the interaction between mean flow and second-order acoustics. Here, the intermodal energy transfer is modified by the mode phase difference so that triggering to an unstable limit cycle is possible. Since triggered oscillations grow unbounded in this case, other processes must be considered for complete modeling of chamber acoustics.

The majority of the work done on nonlinear acoustics has employed a few-mode expansion to represent the acoustic fields, along with the method of time averaging. Although this procedure has provided key insight into the basic physics of the problem by preserving analytical tractability, the quest for more detailed understanding promotes different solutions. With the current level of computational power available, it seems plausible to simply relax these constraints. A continuation method has been applied to the study of nonlinear acoustics, indeed revealing limitations of a few-mode expansion. Using bifurcation theory, global aspects of system stability and behavior may also be determined with such a methodology. Application of other aspects of modern dynamical systems theory, along with more numerically intensive approaches, is expected to offer further understanding of chamber acoustics.

Acknowledgments

This work was conducted under the sponsorship of the Air Force Office of Scientific Research, Grant AFOSR-91-0171, and by the Propulsion Engineering Research Center, Pennsylvania State University. The support of Mitat A. Birkan is greatly appreciated.

References

[1]Crocco, L., and Cheng, S. I., *Theory of Combustion Instability in Liquid Propellant Rocket Motors,* AGARDograph No. 8, Butterworths, London, 1956, pp. 76–147.

[2]Harrje, D. J., and Reardon, F. H. (eds.), *Liquid Propellant Rocket Instability,* NASA SP-194, 1972, pp. 167–231.

[3]Culick, F. E. C., and Yang, V., "Prediction of the Stability of Unsteady Motions in Solid-Propellant Rocket Motors," *Nonsteady Burning and Combustion Stability of Solid Propellants,* edited by L. DeLuca, E. W. Price, and M. Summerfield, Vol. 143, Progress in Astronautics and Aeronautics, AIAA, Washington, DC, 1992, pp. 719–779.

[4]Maslen, S. H., and Moore, F. K., "On Strong Transverse Waves Without Shocks in a Circular Cylinder," *Journal of the Aeronautical Sciences,* Vol. 23, No. 6, 1956, pp. 583–593.

[5]Sirignano, W. A., "A Theoretical Study of Nonlinear Combustion Instability: Longitudinal Mode," Ph.D. Thesis, Dept. of Aerospace and Mechanical Sciences, Princeton Univ., Princeton, NJ, March 1964.

[6]Zinn, B. T., and Powell, E. A., "Nonlinear Combustion Instability in Liquid Propellant Rocket Engines," *Proceedings of the 13th Symposium (International) on Combustion,* Combustion Inst., Pittsburgh, PA, 1971, pp. 491–503.

[7]Awad, E., and Culick, F. E. C., "On the Existence and Stability of Limit Cycles for Longitudinal Acoustic Modes in a Combustion Chamber," *Combustion Science and Technology,* Vol. 46, Nos. 3–6, 1986, pp. 195–222.

[8] Kim, S. I., "Nonlinear Pressure Oscillations in Combustion Chambers," Ph.D. Thesis, Pennsylvania State Univ., Dept. of Mechanical Engineering, University Park, PA, May 1989.

[9]Yang, V., and Culick, F. E. C., "On the Existence and Stability of Limit Cycles for Transverse Acoustic Oscillations in a Cylindrical Combustion Chamber. 1: Standing Modes," *Combustion Science and Technology,* Vol. 72, Nos. 1–3, 1990, pp. 37–65.

[10]Jahnke, C. C., and Culick, F. E. C, "An Application of Dynamical Systems Theory to Nonlinear Combustion Instabilities," *Journal of Propulsion and Power,* Vol. 10, No. 4, 1994, pp. 508–517.

[11]Jahnke, C. C., and Culick, F. E. C, "Nonlinear Energy Transport Between Longitudinal Acoustic Modes in Cylindrical Combustion Chambers," 30th AIAA/ASME/SAE/ASEE Joint Propulsion Conf., AIAA Paper 94-3190, June 1994.

[12]Greene, W. D., "Triggering of Longitudinal Combustion Instabilities in Rocket Motors," M.S. Thesis, Pennsylvania State Univ., Dept. of Aerospace Engineering, University Park, PA, Dec. 1990.

[13]Yang, V., Kim S. I., and Culick, F. E. C., "Triggering of Longitudinal Pressure Oscillations in Combustion Chambers. I: Nonlinear Gasdynamics," *Combustion Science and Technology,* Vol. 72, Nos. 4–6, 1990, pp. 183–214.

[14]Wicker, J. M., Greene, W. D., Kim, S. I., and Yang, V., "Triggering of Longitudinal Pressure Oscillations in Combustion Chambers. II: Nonlinear Combustion Response," *Journal of Propulsion and Power* (submitted for publication).

[15]Wicker, J. M., Yoon, M. Y., and Yang, V., "Linear and Nonlinear Pressure Oscillations in Baffled Combustion Chambers," *Journal of Sound and Vibration* (to be published.)

[16] Baer, M. R., and Mitchell, C. E., "Theoretical Evaluation of Rigid Baffles in the Suppression of Combustion Instability," NASA CR-134986, March 1976.

Chapter 14

Experimental Procedures Aiding the Design of Acoustic Cavities

E. Laudien,* R. Pongratz,† R. Pierro,† and D. Preclik‡
DASA-Deutsche Aerospace AG, Munich 81663, Germany

Nomenclature

A = cross-sectional area of combustion chamber, m^2, and amplitude factor in Eqs. (2) and (3)
c = speed of sound, m/s
d = diameter of resonators' opening, m
f = frequency, Hz
f_0 = resonant frequency, Hz
i = $\sqrt{-1}$
J = Bessel function of first kind
ℓ = orifice length of Helmholtz resonator, m, and longitudinal mode number and eigenvalue in Eqs. (1) and (4)
ℓ_R = orifice length of quarter-wave resonator, m
$\Delta\ell$ = orifice length correction of resonators
L_C = idealized cylinder length of combustion chamber, m
L_R = length of quarter-wave resonator, m
m = tangential mode number
n = number of resonators
p = pressure, Pa
p_1 = pressure measured in combustion chamber, Pa
p_2 = pressure measured in resonator volume, Pa
R = resistance, kg/m^2 s
R_C = combustion chamber radius, m
R_r = resistance radiation term, kg/m^2 s
R_i = internal resistance (viscous dissipation), kg/m^2 s
r = specific resistance; radial coordinate in Eq. (4)
S = open area of resonator, m^2

---

*Senior Research Engineer, Central Laboratory, Aeroacoustics.
†Research Engineer, Central Laboratory, Aeroacoustics.
‡Space Transport and Propulsion Systems, Rocket Engines.

t = time, s
x = specific reactance; axial coordinate in Eq. (4)
V = volume of the Helmholtz resonator, m^3
Z = impedance, kg/m^2 s
α = absorption coefficient
ε = resistance factor, $\varepsilon_\ell + \varepsilon_{n\ell}$ (linear and nonlinear term)
η = damping factor
ϑ = phase angle, deg
λ = eigenvalue
ν = dynamic viscosity, Pa s
ξ = conductance
ρ = density, kg/m^3
ρc = specific impedance of combustion chamber, kg/m^2 s
Φ = velocity potential function, m^2/s
ϕ = mode shape of potential function, m^2/s^2
χ = reactance, kg/m^2 s
ψ = phase angle, deg
ω = circular frequency, Hz
ω_0 = circular resonant frequency, Hz

Indices

C = combustion chamber
cold = cold gas condition
hot = hot gas condition
i = ith resonator
ℓ = longitudinal mode number
m = tangential mode number
n = radial mode number
R = resonator
0 = at resonance

I. Introduction

MANY rocket engine development programs have experienced high-frequency combustion instabilities. Elimination of these instabilities is often difficult, calling for a huge number of expensive full-scale tests, and eventually results in redesign of parts of the injector/combustor system at later stages of a project. Consequently, it is important from the very beginning to systematically exploit all possibilities that may help to avoid the occurrence of these highly destructive oscillations.

This chapter provides some experimental procedures that can be employed to aid the design process of acoustic resonators and cavities, i.e., elements that are used as passive means to suppress or limit the destructive potential of high-frequency oscillations. In fact, changing the damping characteristics of a combustion chamber by incorporating acoustic devices is a well-known technique in rocket engine development, even though the actual realization may be quite difficult at times as was the case during the development of the F1-rocket motor.

The mechanism of high-frequency instabilities in rocket engines has been investigated over a period of several decades; however, no general rules have yet been established for designing inherent stable injector/combustor systems. One reason for this is the enormous variety of existing injector configurations for the different propellant combinations used, all featuring specific atomization, vaporization, mixing, and burning behavior. In contrast to this, it is much easier to formulate general design rules for instability suppression devices since their variability is more limited, at least concerning the general types of baffles, Helmholtz, and quarter-wave resonators discussed here. These damping devices produce good results in stabilizing combustion oscillations. The latter two have the advantage of being less exposed to the adverse core flow conditions within a combustion chamber.

In the first part of this chapter, the role of a full-scale acoustic chamber model in gaining first acoustic data is explained. General baffles as well as cavity design principles are then recalled, and experimental procedures that aid the conventional resonator layout process are described. Experimental data are provided where important to highlight typical resonator characteristics under hot-firing conditions. Finally, several issues addressing the shape of the inlet orifice of an acoustic cavity and its impact on overall resonator absorption behavior are discussed.

II. Acoustic Tests with Model Chambers

Acoustic resonators and baffles are damping devices successfully used as suppression mechanisms for the reduction or elimination of combustion oscillations. Resonators are narrowband absorbers that must be tuned accurately to discrete acoustic eigenfrequencies of the combustion chamber. These frequencies must be known at the beginning of any preliminary layout of resonating cavities. A full-scale model chamber is a useful tool here for first tests under ambient conditions. Cold-test results can then be used to evaluate the chamber's eigenfrequencies during hot-firing tests.

In the current investigation, a model chamber was equipped with a number of small loudspeakers in the injector faceplate: one in the center and others equally spaced at two different radii. A microphone mounted on a thin rod was attached to a small tube, making it possible to reach any axial, radial, and azimuthal position in the chamber. Additional microphones were mounted flush at the chamber wall immediately behind the faceplate.

Random noise was used as the source for the loudspeakers. Resonant frequencies in the chamber were identified by registering microphone signals throughout the chamber. Once the frequencies were known, a sine generator was used to activate only one particular mode. The shape pattern of the mode was then determined by scanning the chamber in azimuthal and/or radial and axial directions for nodal lines. These lines of minimum pressure can be found, first, by observing the amplitude and second, by locating any phase shift in the chamber. The latter is accomplished by comparing the phase between the measured signal and a reference signal taken directly from the loudspeaker. Each time the microphone moves through a nodal line, the acoustic pressure will be at a minimum and the phase will shift by 180 deg. A schematic layout of the experimental setup is illustrated in Fig. 1.

Another possibility of generating broadband noise in a chamber without loudspeakers is by means of high-speed airflow through the injection holes. This was practiced with smaller liquid-propellant engines where the air pressure in a reservoir behind the faceplate was held constant for the test duration.

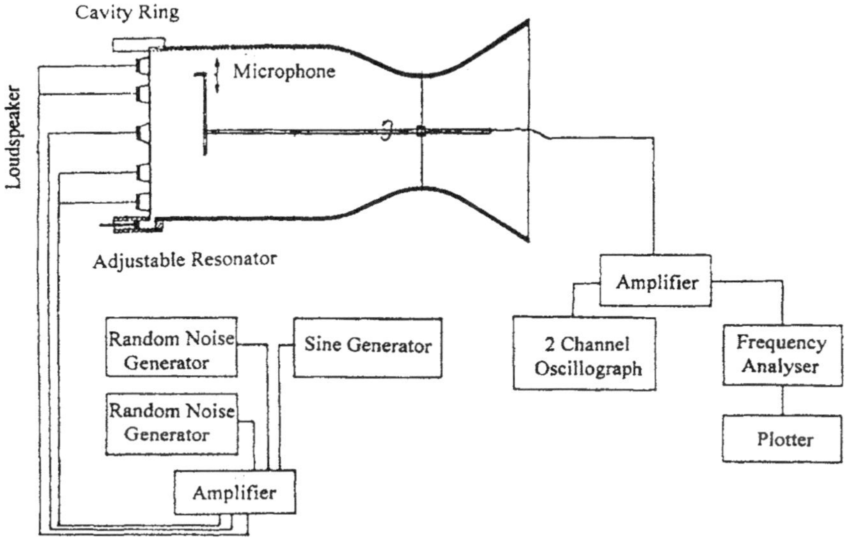

Fig. 1 Experimental setup.

A. Resonant Frequencies

A typical spectrum of a full-scale model test performed in air under ambient conditions is shown in Fig. 2. The notations 1T, 1L, and 1R stand for first tangential, first longitudinal, and first radial mode, respectively. The combined first longitudinal/first tangential mode is denoted 1L/1T. It exhibits distinct peaks, demonstrating that the chamber itself rarely has any damping effect for transversal modes (tangential and radial) or their combinations with longitudinal frequencies. Longitudinal (axial) modes were not identified during these measurements since they are damped by the convergent part of the nozzle. Only in those chambers with a long cylindrical part and/or short nozzle section were L modes measurable.

Figure 2 depicts two frequency scales: the upper one represents frequencies measured in the model under ambient condition, and the lower one represents frequencies expected in the actual chamber during hot firing. The difference is the speed of sound ratio between these two conditions. Whereas the wavelength of an eigenfrequency is determined by the geometry of the chamber, and thus fixed for a specific model chamber, the corresponding frequency changes with the speed of sound ratio as the hot-gas condition is approached.

Acoustic tests with models have shown that the combustion chamber and nozzle configuration must be treated theoretically as a closed/closed system even though the nozzle is not actually closed. The eigenfrequencies for a cylindrical chamber closed at both sides can be calculated with Eq. (1),

$$f_{\ell mn} = \frac{c}{2\pi}\sqrt{\left(\frac{\lambda_{mn}^2}{R_c^2} + \frac{\ell^2\pi^2}{L_c^2}\right)} \qquad \ell, m, n = 0, 1, 2, \ldots \tag{1}$$

where λ_{mn} is the transverse eigenvalue listed in Table 1 for the first few modes.

For rocket combustion chambers, L_c must be replaced by the effective acoustic length, i.e., the distance between the faceplate and nozzle throat less approximately one-half of the converging nozzle length. Table 2 compares calculated resonant

Table 1 Transversal eigenvalues

	m					
n	0	1	2	3	4	5
0	0	1.8412	3.0542	4.2012	5.3176	6.4156
1	3.8317	5.3314	6.7061	8.0152	9.2824	10.5199
2	7.0156	8.5363	9.9695	11.3459	12.6819	13.9872
3	10.1730	11.7060	13.1704	14.5859	15.9641	17.3128

Table 2 Natural frequencies of acoustic modes in a rocket chamber

Mode	Theoretical frequency, Hz	Experimental frequency, Hz	Deviation, %	Hot-gas frequency, Hz
1T	957	985	−2.8	3152
1L1T	1235	1240	−0.4	3968
2T	1588	1610	−1.3	5152
1L2T	1769	1826	−3.1	5843
1R	1992	2022	−1.5	6470

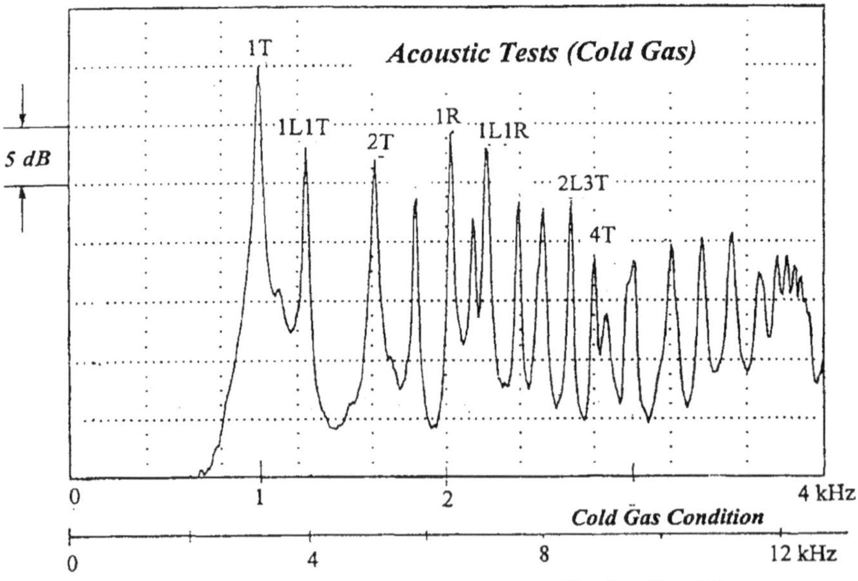

Fig. 2 Frequency spectrum measured in a combustion chamber model under cold-gas condition, random noise generation, see Fig. 1.

frequencies with those measured experimentally in a full-scale model chamber (see Ref. 1). The theoretical frequencies were computed from Eq. (1) using the effective length L_c of the combustor. One can see that the theoretical frequencies derived from the simple closed/closed tube model are in excellent agreement with experimental data. The expected frequencies under hot-gas conditions are also listed in Table 2. These latter frequencies were scaled from the experimental results using a speed of sound ratio of 3.2.

B. Mode Shape

The acoustic modes of a combustion chamber arise from pressure waves traversing the cavity at the speed of sound and reflecting from the chamber walls. If the fluid oscillates freely in the longitudinal ℓ, tangential m, and radial n mode, the dynamic component of the fluid pressure is simply the sum of the contributions of the individual modes, and can be expressed in terms of the total velocity potential function:

$$\Phi = \sum_{\ell mn} A_{\ell mn} \frac{c^2}{\omega_{\ell mn}} \varphi_{\ell mn} \sin(\omega_{\ell mn} t + \psi_{\ell mn}) \tag{2}$$

The dynamic variation in the acoustic pressure can be expressed in terms of this velocity potential

$$p = \rho \frac{\partial \Phi}{\partial t} = \sum_{\ell mn} A_{\ell mn} \rho c^2 \varphi_{\ell mn} \cdot \cos(\omega_{\ell mn} t + \psi_{\ell mn}) \tag{3}$$

where A is a constant that specifies the amplitude, ψ the phase angle with respect to the other modes, and φ the nondimensional mode shape of the potential function. For a closed/closed cylindrical chamber this mode shape is

$$\varphi_{\ell mn} = J_m(\lambda_{mn}(r/R_c)) \cdot \cos(\ell x \pi / L_c) \cdot \cos m\vartheta \tag{4}$$

J_m is the Bessel function of the first kind and is of order m.

The acoustic pressure amplitudes of the tangential modes reach their maximum at the chamber wall and are zero in the center, whereas those of the radial modes are largest in the center. This was also observed during measurements in the model chamber where a microphone placed in the combustor's center measured only radial and combined radial/longitudinal modes. From the acoustic pressure field distribution, one can conclude that the ideal placement of acoustic resonators for maximum effectiveness is at the circumference of the combustion chamber if tangential modes are to be damped, and at the center of the faceplate if radial modes are to be damped.

Experience with a number of liquid-propellant engines shows that if a chamber is susceptible to instability, it is usually sensitive to tangential modes. It is also this acoustic mode that may drive the most destructive type of instabilities, since its pressure oscillation shape can destroy the thermal boundary layer and cause immediate chamber wall damage.

Mode-shape measurements in a chamber model under ambient conditions are restricted to the location of nodal lines. Tangential modes are in general unsettled, taking any position in the chamber unless an irregularity in the injection locks them to a fixed location. Cold-gas tests with a limited number of loudspeakers have shown that the nodal pressure lines change their position corresponding to a change in loudspeaker location. A nodal pressure line always avoids the position of maximum acoustic excitation; i.e., under hot-gas conditions the nodal line spins away from regions of more intensified combustion. Cold tests also showed that the nodal line of a T mode always turns into the position of an acoustic resonator, thus, rendering the absorber ineffective. As a consequence, resonators tuned to a certain mode should be spaced nonsymmetrically around the circumference of an engine.

III. Baffles

Transverse wave motion in the combustion chamber can be altered by baffles at the injector side. Baffles modify the acoustic resonance properties and change the oscillatory flow pattern in the vicinity of the injector plate. It is further believed that the formation of vortices at the blade edges dissipates energy and, thus, increases damping. This, however, cannot be simulated in an acoustic chamber model.

A two-bladed baffle could, at the most, prevent a tangential mode from spinning. This mode would be fixed in its position, with pressure antinodes at the baffle blades and velocity antinodes displaced at 90 deg. A configuration that damps the first and second tangential mode would be a three-bladed design. A four-bladed concept has little or no effect on the second tangential mode because it can still develop within the baffle arrangement. The five-bladed design used in many liquid-propellant engines is expected to influence all T modes up to the fourth.

Experiments with acoustic model chambers confirm that a five-bladed baffle shifts the resonant frequency to lower values, especially the first and second tangential mode. The baffle used in these acoustic tests had a length-to-radius ratio of 0.33. The first tangential resonant frequency was reduced by 24%, which is in agreement with measurements reported by various investigators.[2] A convenient way to determine the amount of damping present in a system is to measure the decay rate of the oscillating pressure. This measurement is made by abruptly stopping the driving sources—in this case, the loudspeakers installed in the injector plate—which excite the chamber acoustically. Several microphones located in the cavity ring were used to detect the chamber's response. The measured amount of pressure decrease in decibels per second divided by the frequency of the particular tangential mode are plotted in Fig. 3 in line with the damping factor, the determining quantity of the decay rate.

Another useful response measure is the bandwidth method (Fig. 4), which is applicable only if the damping rate is relatively low. Both methods, decay rate and bandwidth measurement, were applied to determine the damping effect of the baffles; results are shown in Fig. 3. Damping increases by a factor of five for the first two modes; however, the five-bladed baffle did not affect the third and fourth tangential eigenfrequency. The diagram in Fig. 5 (taken from Ref. 2) shows the ratio of unbaffled to baffled decay rate vs baffle length for acoustic as well as hot-gas tests. The current experimental data have been added to this diagram. It is surprising that results from acoustic tests agree so well with hot-gas data since the effects of mixing and vaporizing propellant sprays, inhomogenities in the chamber's gas, and burning rate fluctuations are not present in an acoustic model. The good agreement between cold- and hot-gas decay rates reported in Ref. 2 encourages the use of acoustic chamber models as reasonably accurate approximations of combustion acoustics.

Hot-gas tests with a five-bladed plus hub baffle were performed with an engine operating with the hypergolic propellant combination N_2O_4 and MMH. The baffle

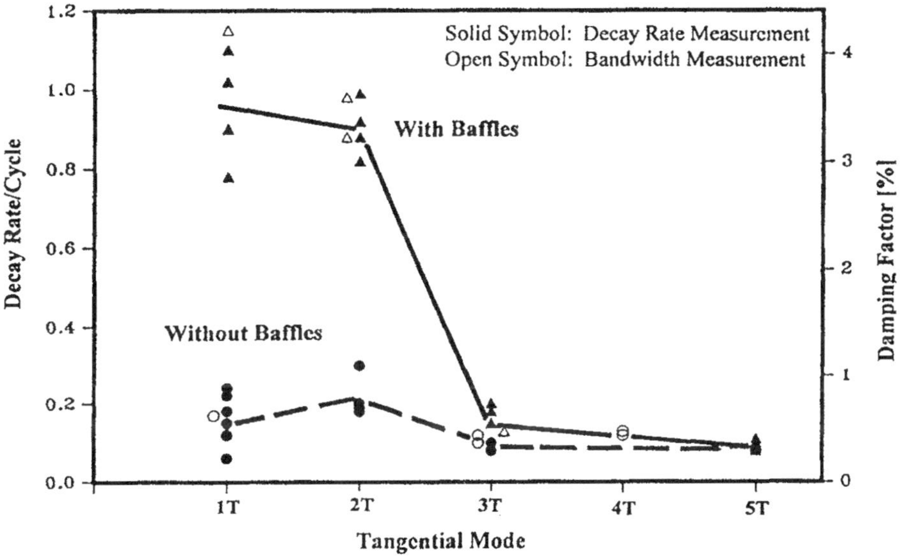

Fig. 3 Decay rate and damping factor measured in a baffled and unbaffled acoustic chamber model for tangential modes, length to radius rate = 0.33.

length was set at 40 mm. This corresponded to a length-to-radius ratio of 20%, which, according to Fig. 5, increased the expected decay rate by a factor of about 2.5. Implementation of the baffles in the chamber had a stabilizing effect on the engine's combustion behavior, and thus allowed reduction in the number of acoustic cavities[1] originally foreseen without the baffles.

IV. Acoustic Cavities

A. Theoretical Considerations

Acoustic cavities such as Helmholtz and quarter-wave resonators were successfully used as damping devices for the suppression of combustion oscillations. resonator consists of a small volume connected with the combustion chamber through an orifice. If the dimensions of the various resonator elements are small in comparison to the wavelength of the oscillation, the gas motion behavior in the resonator is analogous to a mass-spring-dashpot system. The undamped resonant frequency f_0 can be calculated from the geometrical dimensions (Fig. 6) as

$$f_0 = \frac{c}{2\pi}\sqrt{\frac{S}{V(\ell + \Delta\ell)}} \tag{5}$$

where S is the orifice area, ℓ the orifice length, V the cavity volume, and $\Delta\ell$ the mass correction which is about $0.85 \cdot d$ for lower noise levels. A quarter-wave resonator (Fig. 6) may also be treated as a mass-spring-dashpot system even if mass and spring must be evaluated differently. The equation for the frequency is

$$f_0 = c/(4(L_R + \Delta\ell)) \tag{6}$$

where L_R is the length of the quarter-wave resonator.

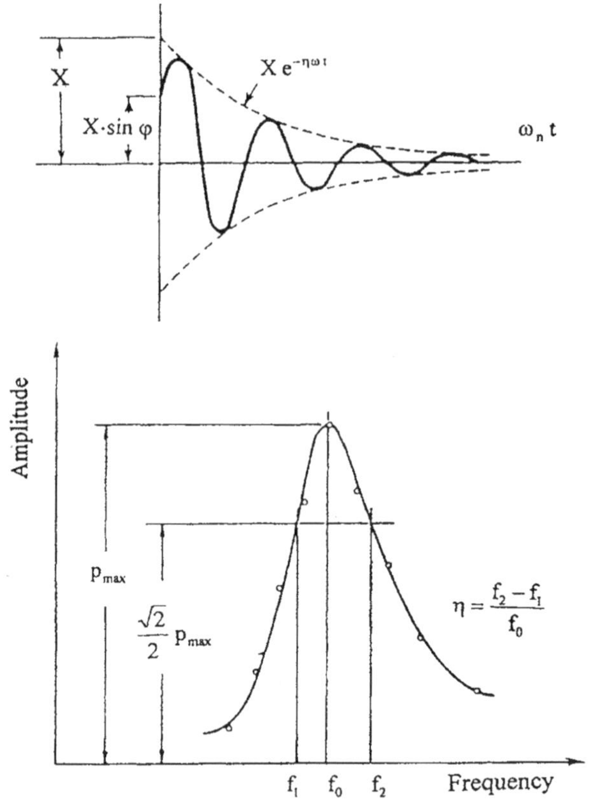

Fig. 4 Evaluation of the damping factor by the decay rate and bandwidth method.

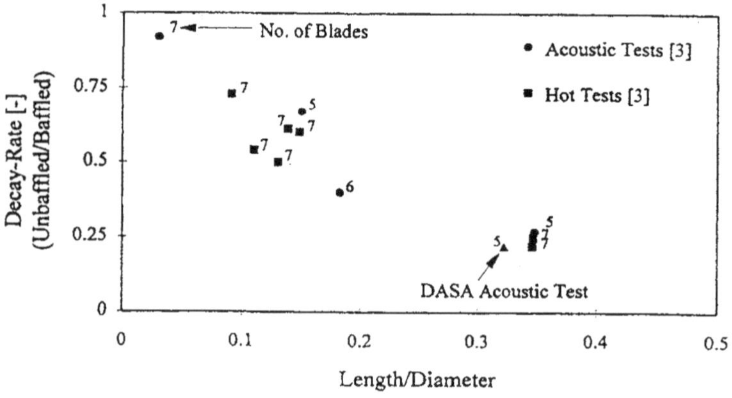

Fig. 5 Decay rate vs baffle length for the IT mode.

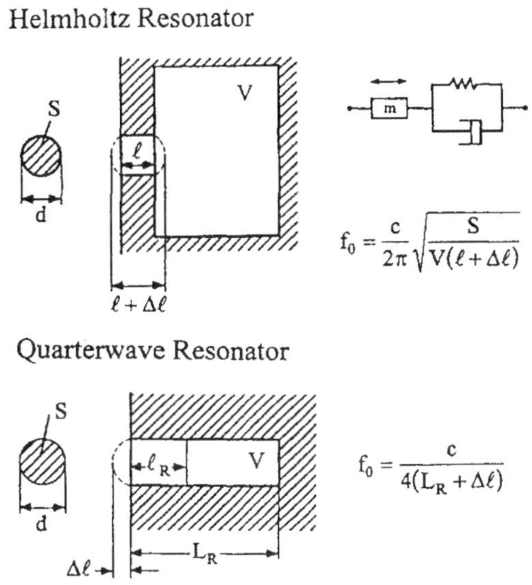

Fig. 6 Helmholtz and quarter-wave resonators.

To study the acoustical behavior of the resonators, and especially its absorption effect, impedance Z must be determined. This impedance is, in general, a complex unit consisting of the sum of an acoustic resistance R and an acoustic reactance χ,

$$Z = R + i\chi \tag{7}$$

The acoustic impedance is determined by the geometric and mechanical properties of the resonator, in particular the oscillating mass in the orifice, the spring stiffness of the volume, and the resistive element associated with viscous dissipation in the orifice. The reactance can be written in terms of the cavity dimensions as

$$\chi = 2\pi f \rho(\ell + \Delta\ell)\left(1 - \left(f_0^2/f^2\right)\right) \tag{8}$$

The total resistance is a combination of radiation and viscous dissipation at the orifice wall, i.e., internal resistance,

$$R = R_r + R_i \tag{9}$$

The radiation term can be written as

$$R_r = 2\pi\rho S f_0^2/c \tag{10}$$

According to Morse and Ingard,[3] however, this term is usually neglected for resonators in ducts. The internal resistance is a function of the orifice length and is defined as

$$R_i = 4(\varepsilon_\ell + \varepsilon_{n\ell} + \ell/d)\sqrt{\nu\rho\pi f} \tag{11}$$

where ν is the dynamic viscosity and $\varepsilon = \varepsilon_\ell + \varepsilon_{n\ell}$ is the resistance factor that depends on the quality of the orifice wall surface (ε_ℓ), as well as on a nonlinear correction term for high-sound-pressure levels ($\varepsilon_{n\ell}$). This equation also holds true for quarter-wave resonators, only ℓ has to be replaced by ℓ_R (the length of the effective acoustic mass). The reactance of a quarter-wave resonator is

$$\chi = -4\rho(\ell_R + \Delta\ell) f_0 \cdot \cot(\pi f / 2 f_0) \tag{12}$$

When these terms are multiplied by the area ratio A/S of the combustion chamber, one obtains the specific cavity resistance r and reactance x per unit area. For an arrangement of n, differently tuned absorbers around the circumference of a combustion chamber, the sum of the acoustic impedance is given by

$$Z = \left(\frac{1}{A} \sum_{i=1}^{n} \frac{S_i}{Z_i} \right)^{-1} \tag{13}$$

Once the impedance is known, the absorption coefficient and the conductance (real part of resonator admittance) can be evaluated:

$$\alpha = \frac{4r}{\rho c} \bigg/ \left[\left(1 + \frac{r}{\rho c} \right)^2 + \left(\frac{x}{\rho c} \right)^2 \right] \tag{14}$$

$$\xi = \frac{r}{\rho c} \bigg/ \left[\left(\frac{r}{\rho c} \right)^2 + \left(\frac{x}{\rho c} \right)^2 \right] \tag{15}$$

The reactance vanishes at resonance, thus simplifying this equation in which ρc is the specific resistance of the gas in the combustion chamber.

In practice, a cavity layout can be made by either optimizing the absorption coefficient or maximizing the real part of the acoustic boundary admittance (the conductance ξ is also called the power loss factor). According to Ref. 3, the loss in acoustic power for any surface of local reaction (orifice surface) is the conductance times the dynamic pressure at that surface. Whereas the absorption coefficient has its optimum at 100% (i.e., no reflection of sound intensity at the chamber wall), the conductance term is not limited by definition. Figure 7 may explain this difference in more detail by showing the spectral behavior of the absorption coefficient as well as the conductance for three different absorber arrangements in a closed tube. Optimizing the system's damping by means of the absorption coefficient would suggest, in this example, a three-resonator configuration, in contrast to maximizing the conductance, which suggests the configuration with 12 or even more absorbers. Acoustically speaking, more than three absorbers overdamp this system, which implies less than 100% absorption but increases the bandwidth. Some investigators prefer the use of the conductance term for the combustion process. According to Ref. 2, however, the usual approach attempts to optimize the absorption coefficient, despite the uncertainty of this model, which is the result of the inhomogenous sound pressure field.

The preceding discussion leads to the conclusion that once the basic design of a resonator has been evaluated from acoustic tests that consider the combustion chamber's hot-gas properties, maximization of the absorption coefficient for hot-firing conditions must be done by selecting the right number of resonators within the absorber arrangement. If an increase in bandwidth is desired (see also Fig. 7) because the combustion chamber hot-gas properties either cannot be precisely determined or are expected to shift during engine operation, the number of resonators within the arrangement may be increased.

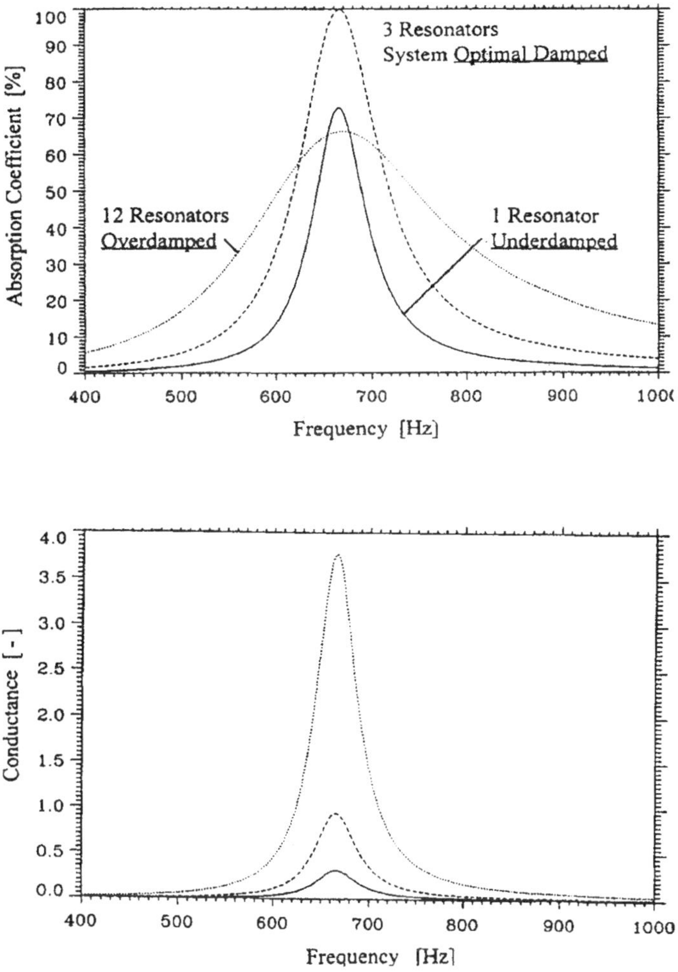

Fig. 7 Absorption behavior of an under-, optimized-, and overdamped system.

B. Impedance and Absorption Measurements

Impedance and absorption measurements with Helmholtz and quarter-wave resonators were done in a standing wave apparatus. Typical results for a quarter-wave cavity are presented in Fig. 8. Good agreement between calculated and experimental data was achieved by varying the resistance factor [see Eq. (11)]. The absorption maximum at resonance is about 80% for this particular cavity, with a normalized resistance $r/\rho c = 0.4$. This means that the system is overdamped. The number of cavities has to be reduced in this case in order to attain 100% absorption.

C. Nonlinear Effects

High-fluctuating-pressure levels in a combustion chamber will change the damping behavior of acoustic resonators. Under such circumstances, energy losses as

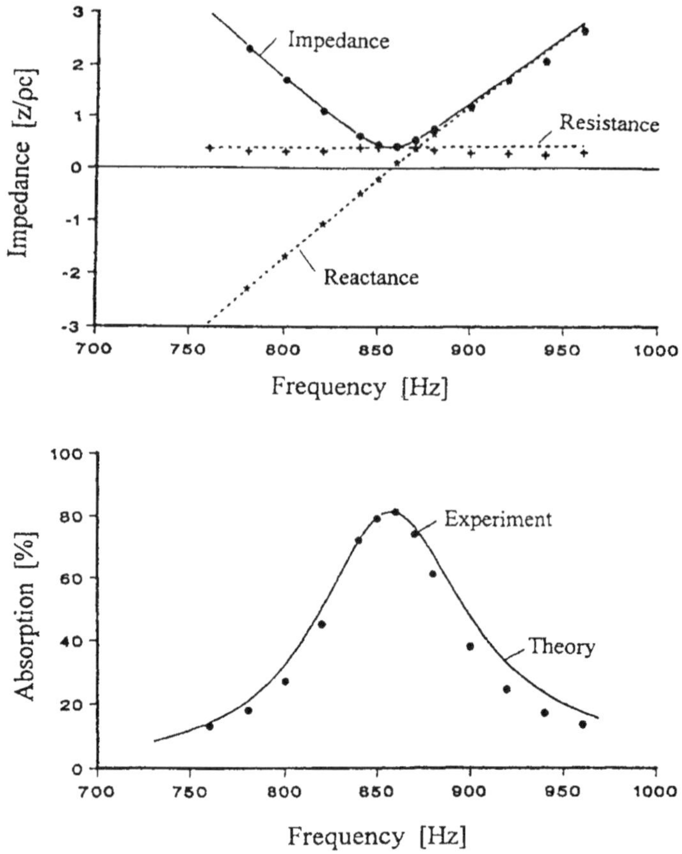

Fig. 8 Experimental and theoretical impedance and absorption.

a function of viscous dissipation at the orifice wall become negligible. The main contribution of these losses results from the formation of vortices at the orifice edges, which dissipate into turbulence. These losses are amplitude dependant and dominate the resistance term at high-sound-pressure levels. Consequently, a remarkable change in absorption and bandwidth can be expected.

Acoustic tests in a standing wave apparatus were performed with an increasing sound pressure level (SPL) up to nearly 160 dB. The measured ratio of nonlinear to linear resistance $(\varepsilon_\ell + \varepsilon_{n\ell})/\varepsilon_\ell$ vs SPL is shown in Fig. 9 for a Helmholtz and quarter-wave cavity. It can be seen that the resistance increases for both types of cavities, but with a remarkable difference in slope; the ratio for the Helmholtz resonator increases by a factor of 13 whereas only a factor of 2 was found at 155 dB for the quarter-wave resonator. There are two reasons for this. In the Helmholtz type, the cross-sectional jump between orifice and resonator volume implies a velocity gain and, hence, results in higher resistance values. Second, this cross-sectional jump causes additional turbulence which is not present in a quarter-wave resonator. The effect of such a rising resistance on absorption is that a chamber optimally damped for low SPLs becomes underdamped for high SPLs when Helmholtz resonators are used. There is, however, only a marginal

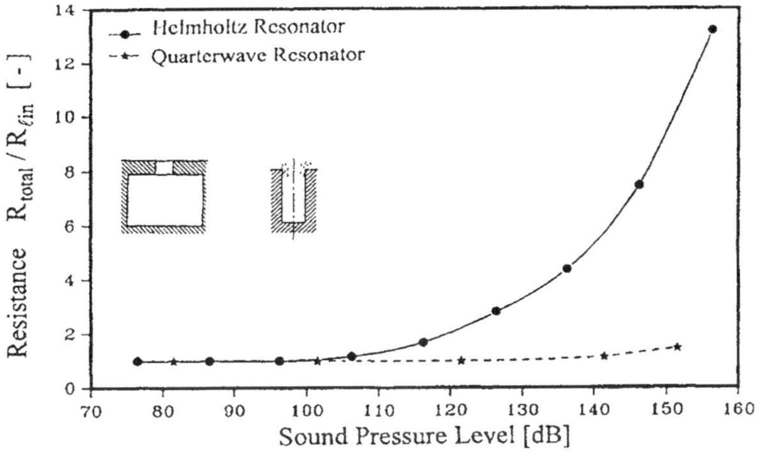

Fig. 9 Nonlinear resistance for Helmholtz and quarter-wave cavities at high-sound pressure levels.

shift in damping when quarter-wave resonators are used. The bandwidth increases continuously with higher SPLs for both types of resonators.

It should also be mentioned that a high, fluctuating, pressure level not only influences the resistance but also the resonant frequency. In all experiments, a frequency rise was observed with increasing SPL. This means that the oscillating mass is reduced by shortening the effective orifice length. Again, this effect is negligible in the case of quarter-wave resonators.

D. Orifice Shape

The fact that the resistance increases with rising dynamic chamber pressure makes absorption prediction very difficult. A steep increase in resistance provides the absorber with higher damping but with lower absorption. If the design criterion asks for a high-absorption coefficient, however, then the resonator must be designed in such a way that nonlinear effects are kept as low as possible even for high SPLs. Such a criterion usually exists if the eigenfrequency of the combustion chamber does not change during engine operation. On the other hand, if the frequency varies as a result of changing engine operation conditions (as is the case during transients), then the bandwidth should be large in order to supply sufficient damping at any time. Since energy is dissipated in the nonlinear regime due to the formation of vortices at the resonator entrance and the orifice to resonator volume transition area, it is clear that design modifications must focus on these locations.

Several variations of the orifice geometry were tested experimentally. The orifice edges for both types of cavities were rounded with different radii. In addition, the orifice of a quarter-wave resonator received a wedge-shaped constriction, reducing the open area to 50% of its original value. The various orifice geometries are shown in Fig. 10.

A significant reduction in resistance, even in the linear region, was observed with rounded edges for the Helmholtz resonators. The transition from rectangular to rounded edges was especially effective, whereas the radius of curvature had only

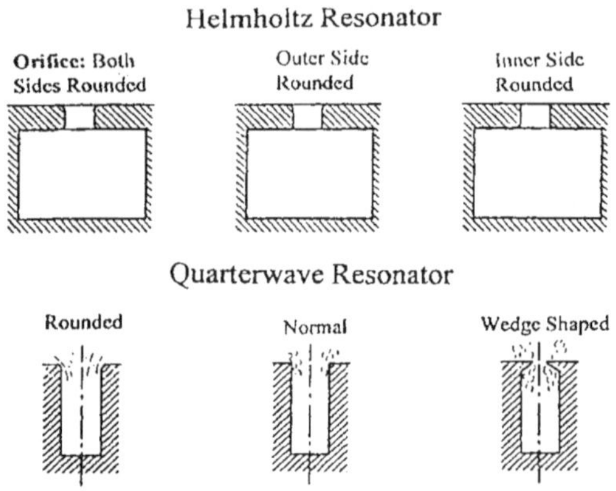

Fig. 10 Various orifice geometries for Helmholtz and quarter-wave resonators.

a minor influence. Furthermore, it was important to round off both the inner and outer throat edges in order to postpone the onset of nonlinearity. The experimental results are illustrated in Fig. 11a. One can see that the resistance increase is not as rapid for the rounded orifice as for the normal-edged orifice, especially if adjusted at both sides.

Quarter-wave resonators, as pointed out earlier, are less sensitive to nonlinearities. The increase in resistance for a normal-edged cavity is low; nevertheless, the rounding effect is measurable reducing the slope as expected. A greater effect was observed when the orifice-edge was sharpened by applying a thin wedge-shaped constriction at the cavity opening (Fig. 10). As expected, the formation of vortices increased the resistance remarkably, making the cavity more sensitive to nonlinear effects (Fig 11b).

V. Hot-Firing Test Results

With the information gained from acoustic tests, a first cavity layout was made. Turning toward hot testing, the engine's dynamic behavior, the cavity performance, and the following parameters need to be observed: dynamic chamber pressure signal, dynamic cavity pressure signal, and cavity gas temperature.

The transfer function between two pressure transducers, installed in the chamber and in the cavity, give the information needed to evaluate the performance of the cavity.

If the complex relation between the dynamic pressure inside the resonator and the excitation pressure at the wall of the combustion chamber are measured, the acoustic properties of the resonator can be calculated with the following equation:

$$\frac{p_2}{p_1} = \frac{-\rho c^2 S}{\omega V} \frac{\omega\rho(\ell + \Delta\ell)\left(1 - \left(\omega_0^2/\omega^2\right)\right) + iR}{\left[\omega\rho(\ell + \Delta\ell)\left(1 - \left(\omega_0^2/\omega^2\right)\right)\right]^2 + R^2} \tag{16}$$

At resonance, Eq. (16) reduces to

$$\frac{p_2}{p_1} = -i\frac{\omega_0\rho(\ell + \Delta\ell)}{R} \tag{17}$$

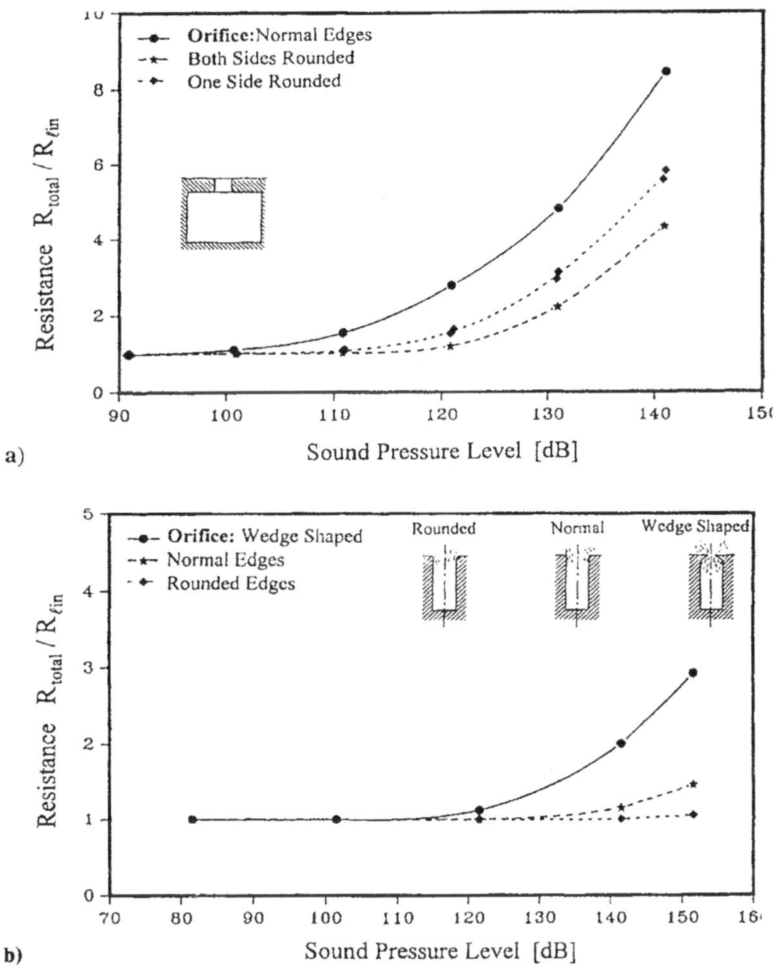

Fig. 11 Nonlinear resistance for different rounded Helmholtz orifices.

The resonant frequency and the resistance can now be determined by measuring the phase between the cavity and chamber signal and by the dynamic pressure amplification of the absorber. A typical curve of a transfer function with magnitude and phase taken during a stable combustion period is shown in Fig. 12. This procedure was applied to different tuned quarter-wave absorbers, leading to the curve in Fig. 13, where resonant frequencies are plotted against cavity length. The results are shown together with cold-gas data. From the frequency ratio between hot- and cold-gas data, an effective speed of sound in the cavities was calculated under the assumption of an effective length equal to that measured in the acoustic model chamber. Equation (6) leads to the following relation:

$$\frac{f_{0,\mathrm{hot}}}{f_{0,\mathrm{cold}}} = \frac{c_{\mathrm{hot}}}{c_{\mathrm{cold}}} \cdot \frac{(L_R + \Delta\ell)_{\mathrm{cold}}}{(L_R + \Delta\ell)_{\mathrm{hot}}} \tag{18}$$

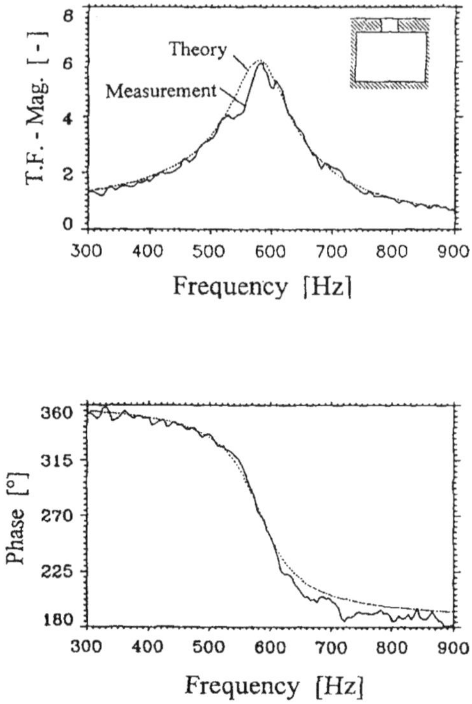

Fig. 12 Typical cavity/chamber transfer function with magnitude and phase of a stable combustion.

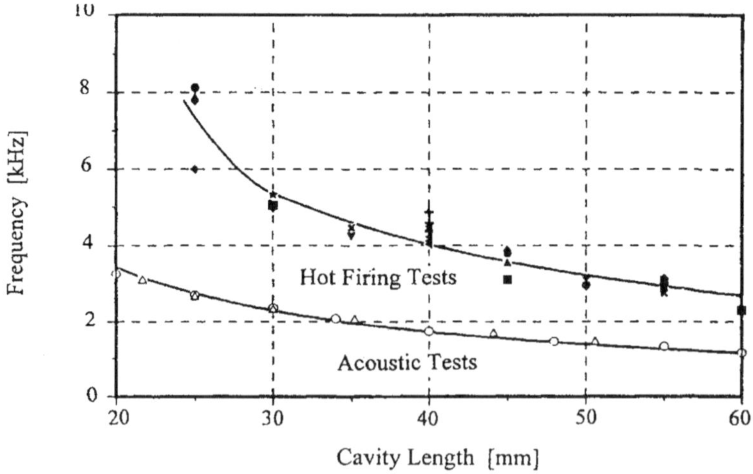

Fig. 13 Hot- and cold-gas cavity resonant frequencies for various cavity lengths.

Since this frequency ratio remains almost constant for quarter-wave cavities with lengths beyond 30 mm, one can assume that there is no major change in the average speed of sound of a resonator. On the other hand, if resonator tuning requires quarter-wave lengths below 30 mm, the assumption of a constant average speed of sound within those cavities will undoubtedly lead to mistuned resonators. To confirm this, the gas temperature was measured at different depths of a quarter-wave cavity. It was found that the gas temperature drops strongly within the first 15 mm from the orifice inlet but stays almost constant as the depth of 30 mm is exceeded. Hence, care must be taken in tuning quarter-wave resonators to higher modes that call for very short resonator lengths. These cavities are most sensitive to immediate changes of absorber gas properties, particularly when a local change occurs in the combustion process close to the resonator inlet.

An additional feature observed during hot-firing was that the eigenfrequency of a quarter-wave resonator grew with increasing distance of the absorber from the injection plane. According to the preceding statement, the temperature inside the resonators at greater distances from the injector were higher since they are positioned closer to the mean combustion zone. This explains the higher speed of sound and, consequently, the increasing eigenfrequency. With regard to resonator tuning, one can thus conclude that each axial cavity position should be tuned separately in order to optimize overall absorber efficiency.

As mentioned earlier, another important property of resonators is their impedance and, particularly, their resistance. With the measured transfer function and Eq. (16), it is possible to calculate resistance R as well as resistance factor ε at resonance. A necessary condition for applying the transfer function method is a good coherence between the chamber and cavity signal. This requirement will most often be met by placing the two pressure transducers together as closely as possible. If this is not practicable from the design point of view, it may be appropriate to work with the transmissibility rather than the transfer function under insufficient signal coherence. The transmissibility gives the absolute value of the pressure ratio without any correlation between the two signals.

The comparison between the quarter-wave resonators with rounded, normal, and wedge-shaped orifices during hot runs reveals much the same behavior as that found during the acoustic tests. The resistance and resistance factor rise with an increasing sharpness of the orifice edge for both cold- and hot-gas tests at aproximately the same SPL, as shown in Table 3.

Figure 14 presents the measured transfer functions for each of the three orifice configurations. The pressure ratio, which corresponds to the resistance magnitude, decreases with increasing sharpness of the resonator edge.

To confirm the slight increase in resistance for quarter-wave resonators at high-sound-pressure levels during a hot-firing test the measured hot-gas data of a

Table 3 Resistance R and resistance factor ε for various quarter-wave orifice types

	Cold gas			Hot gas		
	Round	Normal	Wedge	Round	Normal	Wedge
R, kg/m^2 s	15.7	17.9	35.1	117	137	207
ε	9.9	12.4	25.8	30	32	48
R/R_{round}	1.0	1.14	2.24	1.0	1.17	1.77

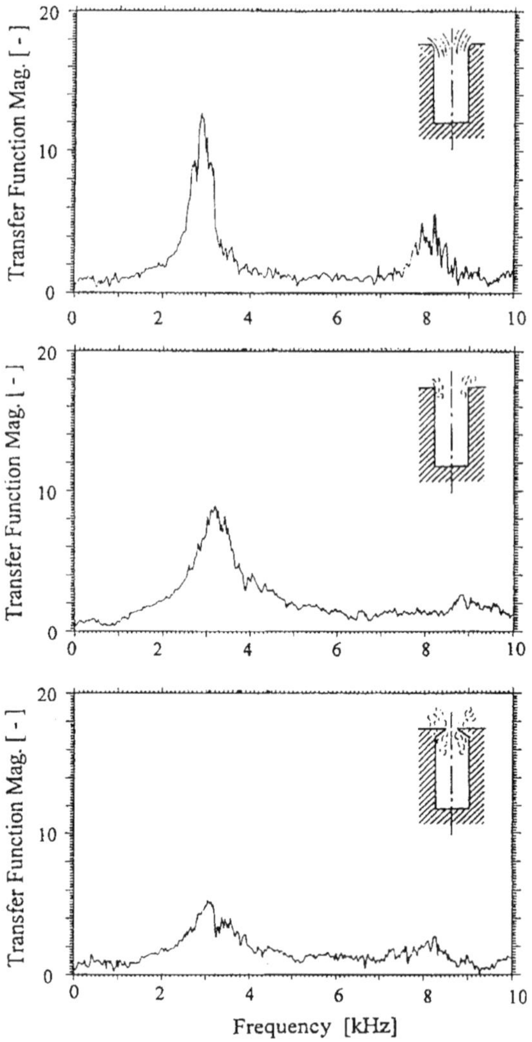

Fig. 14 Measured hot-run transfer functions for different quarter-wave cavities.

normal-edged resonator was used to calculate the linear behavior when using combustion gas parameters. Even when there is a difference in the nonlinear resistance term (Table 3), the linear term, i.e., energy losses as a function of the viscous dissipation at the wall, can be assumed to be constant. The comparison between this calculated result and the measured result showed a resistance increase by a factor of two as also measured under ambient conditions (Fig. 9). This result underlines the low-amplitude dependancy of quarter-wave resonators and, thus, their good predictable behavior under combustion conditions with huge pressure levels.

The same procedure was followed with the Helmholtz resonators, again with a normal-edged orifice. The comparison between cold- and hot-gas tests shows a significant reduction in the dynamic pressure ratio and a remarkable increase in bandwidth. The corresponding resistance increases by a factor of 20. This

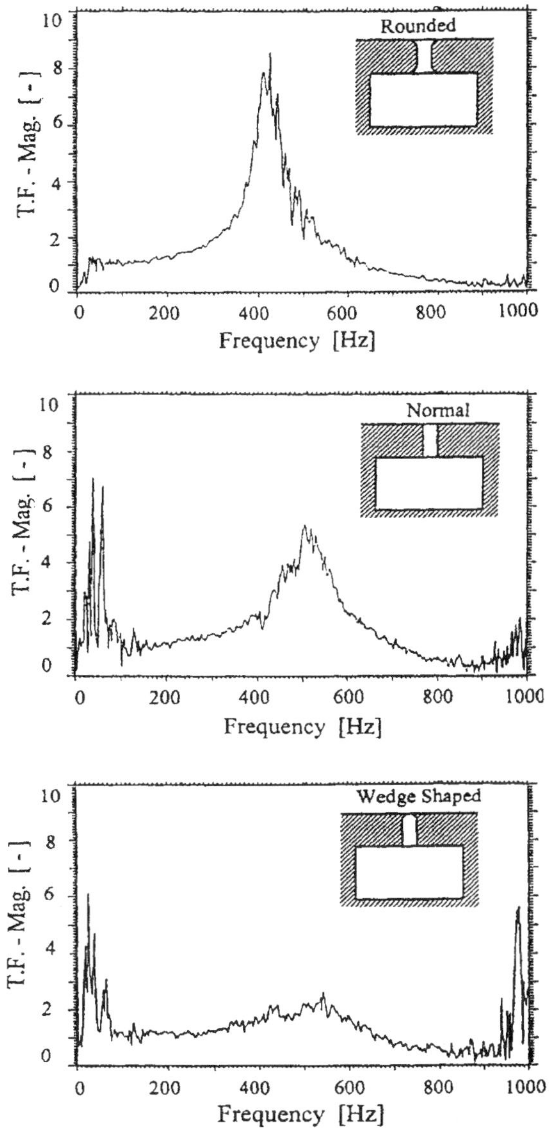

Fig. 15 Transfer function for different Helmholtz resonators measured with a laboratory combustor.

rise confirms the nonlinear behavior of the Helmholtz resonator measured during acoustic tests in air under ambient conditions.

Similar results were obtained at a small test facility at the Deutsche Aerospace Acoustic Laboratories (see Ref. 4). A methane/oxygen combustor was installed to make it possible to test resonators of different geometries under hot-gas conditions without the huge costs of a rocket test facility. Further advantages were quick realization and completion of tests and immediate availability of results. A great deal of valuable empirical data was gained through the use of this laboratory combustor. Figure 15 documents, for instance, the behavior of Helmholtz resonators

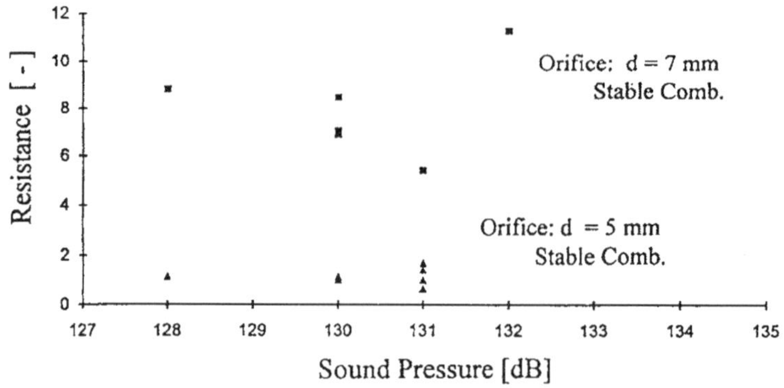

Fig. 16 Specific resistance for different resonator orifice diameters.

with the already described rounded, normal, and wedge-shaped edges. The results are qualitatively and quantitatively, similar to Fig. 14, which was measured in a full-scale engine.

Another series of tests of resonators with two different orifice diameters led to the interesting result that at a constant opening area fraction, a larger number of the cavities with a small orifice area have a lower resistance r than is the case with fewer cavities with large orifices. Figure 16 shows the specific resistance for resonators with a diameter of 5 and 7 mm. The increased resistance seems to result from the generation of more intense vortices for the larger diameter. Further measurements showed that the resistance of an absorber is highly influenced by the velocity pattern and the gas composition in the combustion chamber.

To judge the effectiveness of absorbers located at different axial chamber positions, a measurement series was made with an unstable combustor. The number of absorbers needed to stabilize the combustion was three, with the absorbers located at a distance of 23 mm from the faceplate. When this distance was increased to 39 mm, the combustor returned to an unstable state. Now, 30% more resonators were necessary to guarantee a stable combustion. A further increase of the distance to 55 mm destabilized the combustion again (see Fig. 17). The effectiveness of resonators in a combustion chamber is, therefore, considerably influenced by its location.

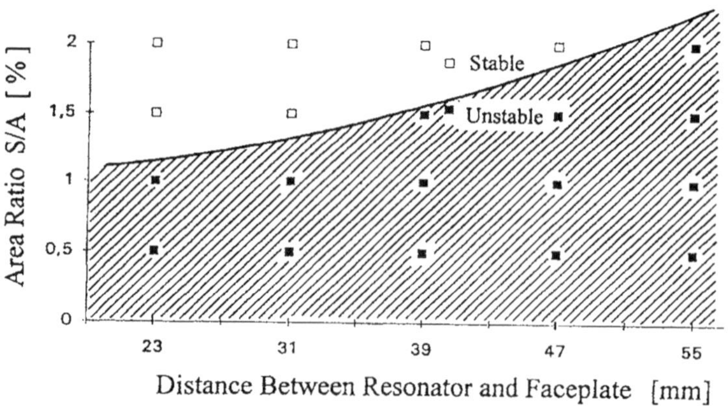

Fig. 17 Influence of the axial resonator position to the stability.

In addition, the position of the nodal lines of the acoustic modes, as well as the velocity, the gas properties, and the acoustical driving source must be taken into consideration. These parameters play an important role in the behavior of a resonator.

VI. Conclusion

Combustion instabilities in rocket engines can be eliminated by increasing the damping characteristics of the chamber. Baffles, quarter-wave, and Helmholtz resonators are ideal devices for suppressing combustion oscillations.

To accurately tune these suppression devices, acoustic tests in air under ambient conditions with full-scale chamber models and acoustic absorbers are essential. These tests provide a basic tool for predicting the damping behavior in a rocket engine under hot-gas conditions as well as a general survey of the characteristic quantities of various types of acoustic absorbers under linear as well as nonlinear conditions. With these data and an estimation of hot-gas parameters, a first cavity layout was made.

Transfer function measurements between cavity and chamber dynamic pressure during hot-firing tests yielded additional precise data regarding cavity resonances and impedances. These data confirmed that a high-absorption coefficient can be reached with low-resistance absorbers provided by rounding off the orifice edges. This chapter has further shown that the layout of quarter-wave resonators is more predictable than that of Helmholtz resonators, since the latter are very sensitive to variation of dynamic pressure levels prevailing in combustion chambers.

Further research was carried out through the use of a laboratory combustor. Results gained with this test facility proved that maximum absorption is obtained with a rounded resonator orifice. The test series with variation of the axial cavity position led to the conclusion that the flowfield and gas properties at the resonator opening have a major influence on damping behavior. With increased distance between resonator and faceplate, a decrease in the efficiency of the absorbers could be observed. Small laboratory combustors, such as those described in Refs. 5 and 6 are good tools for optimizing damping devices.

References

[1]Langel, G., Laudien, E., Pongratz, R., Habiballah, M., Grand-Perret, S., and Vingert, L., "Combustion Stability Characteristics of the Ariane 5 L7 Engine," 42nd Congress of the International Astronautical Federation, IAF-91-250, Oct. 1991.

[2]Harrje, D. T., and Reardon, F. H. (eds.), *Liquid Propellant Rocket Combustion Instability,* NASA SP-194, 1972, pp. 385–429.

[3]Morse, P. M., and Ingard, K. U., *Theoretical Acoustics,* 1st ed., Princeton Univ. Press, Princeton, NJ, 1968, Chap. 9.1.

[4]Laudien, E., and Pierro, R., "Auslegung von Helmholtz-Resonatoren für Brennkammern mit Hilfe eines Laborbrenners," Fortschritte der Akustik—DAGA '94, DPG GmbH, Bad Honnef, March 1994, pp. 669–672.

[5]Poinsot, T., Bourienne, E., Esposito, E., and Lang, W., "Suppression of Combustion Instabilities by Active Control," AIAA/SAE/ASME/ASEE 23rd Joint Propulsion Conf., AIAA Paper 87-1876, July 1987.

[6]Huynh, A., Ghafourian, A., Mahalingam, S., and Daily, J. W., "Combustor Design for Atomization Study in Liquid Rocket Engines," AIAA 30th Aerospace Science Meeting and Exhibit, AIAA Paper 92-0465, Jan. 1992.

[7]Crocco, L., and Cheng, S., *Theory of Combustion Instability in Liquid Propellant Rocket Motors,* AGARDograph No. 8, Butterworths, London, 1956.

[8]Fang, J., "Application of Combustion Time-Lag Theory to Combustion Stability Analysis of Liquid and Gaseous Propellant Rocket Engines," AIAA 22nd Aerospace Sciences Meeting, AIAA Paper 84-0510, Jan. 1984.

[9]Ingard, U., "On the Theory and Design of Acoustic Resonators," *Journal of the Acoustical Society of America,* Vol. 25, No. 6, 1953, pp. 1037–1061.

[10]Reardon, F. H., Crocco, L., and Harrje, D. T., "Velocity Effects in Transverse Mode Liquid Propellant Rocket Combustion Stability," *AIAA Journal,* Vol. 2, No. 9, 1964, pp. 1631–1641.

[11]Lewis, G. D., and Garrison, G. D., "The Role of Acoustic Absorbers in Preventing Combustion Instability," AIAA/SAE 7th Propulsion Joint Specialist Conf., AIAA Paper 71-0699, June 1971.

[12]Oberg, C. L., Wong, T. L., and Ford, W. M., "Evaluation of Acoustic Cavities for Combustion Stabilization," NASA Manned Spacecraft Center, Contract NASA-9866, July 1971.

Chapter 15

Analytical Models for Combustion Instability

Charles E. Mitchell*
Colorado State University, Fort Collins, Colorado 80523

Nomenclature

A_N = coefficient defined in Eq. (46)
a = sonic speed
F_1 = nonhomogeneous part of the linear wave equation, Eq. (16)
F_2 = surface boundary condition function, defined in Eq. (26)
G = Green's function, defined in Eq. (19)
i = imaginary unit, $\sqrt{-1}$
K = acoustic impedance
$\boldsymbol{k}$ = unit vector in axial direction
L = axial length of combustion chamber
M = Mach number
N = set of three integers defining a specific eigenfunction
n = interaction index, time lag model
$\boldsymbol{n}$ = unit outward normal
p = pressure
Q = volumetric mass source term defined in Eq. (1)
$\boldsymbol{q}$ = velocity vector
R = cylindrical combustion chamber radius
r = radial coordinate
S = surface area
s = entropy
T = temperature
t = time
u = axial component of velocity
V = volume
x = acoustic slot depth
z = axial coordinate
α = phase angle of combustion response factor

*Professor, Department of Mechanical Engineering.

β = acoustic admittance
Γ = absorber coefficient defined in Eq. (38)
γ = ratio of specific heats
ε = amplitude factor
η = combustion response factor
θ = azimuthal coordinate
λ = linear decay rate
μ_N = eigenfunction expansion coefficient matrix element
$\boldsymbol{\xi}$ = vector field due to vorticity
ρ = density
τ = time lag
Φ = time-dependent velocity potential function
ϕ = time-independent velocity potential function
Ω_N = acoustic eigenfunction specified by the integer set N
ω = complex frequency

Subscripts

a = acoustic absorber quantity
e = nozzle entrance quantity
inj = quantity at injector end of combustion chamber
l = liquid phase quantity
N = set of integers defining eigenfunction
$\hat{N}$ = set of integers defining dominant mode
R = real part of a complex quantity
I = imaginary part of a complex quantity

Superscripts

$'$ = space-dependent part of a perturbation variable
$\wedge$ = amplitude of an oscillatory variable
$\sim$ = approximate solution quantity
$-$ = steady-state quantity
$*$ = oscillatory variable

I. Introduction and Problem Formulation

THE complex oscillatory phenomena that characterize acoustic or high-frequency combustion instability in liquid-propellant rocket engines pose a challenging problem for those who would model them. In some instances the oscillations grow spontaneously, apparently from combustor background noise or some aspect of the startup sequence, and may eventually reach large amplitudes, of the order of the mean chamber pressure or more. Other engine configurations are spontaneously or intrinsically stable but can be excited into the execution of similar large-amplitude oscillations through the introduction of pulses or disturbances of a certain size and form. The time histories of these oscillations may be more or less sinusoidal, though for the large amplitude waves they are typically steep fronted. In almost all cases the frequency is observed to be within a few percent of that of one of the normal acoustic modes for the basic geometry of the combustion chamber. The stability of a given engine configuration is often found to be sensitive to relatively minor changes in injector design or sizing. Any comprehensive theory must account for all of these features as well as other, more specific details of the oscillatory fields in order to be a useful tool in designing stable engines, or even in interpreting data and guiding development.

In a general way it must be true that for instability to occur the driving supplied to any particular oscillation mode must exceed the damping available. An accounting and representation of all important driving and damping mechanisms must, thus, be included in a complete stability model. Moreover, their mutual interactions must be interconnected reasonably by satisfaction of the appropriate conservation and state equations. In all engine configurations, the major driving possibility is the combustion of the liquid propellants. Damping can be provided by the mean flow, by interphase damping, and possibly by the supercritical nozzle that must terminate the combustion chamber. There may be, as well, strongly nonlinear interactions between elements of the oscillatory field which produce driving or damping (for example, the presence of a shock wave discontinuity). In more complex engine designs, components such as acoustic liners and slots or injector face baffles may be added to increase damping. Clearly these must also be modeled if a complete stability picture is to be drawn.

Two general approaches to the stability modeling problem just posed can be identified. The first, the one that will be treated in this chapter, is to attempt a direct analytical solution of the partial differential equations and associated boundary conditions that describe the flowfield in the rocket engine. If direct solution is not possible (as it usually is not), then reduction of the system to ordinary differential equations or integral equations and their subsequent numerical integration is also considered to march under the banner of so-called analytical solutions. The second general approach, direct numerical integration of the partial differential equations, is usually termed a numerical analysis, and this type of approach is discussed elsewhere in this volume. The available analytical approaches can themselves be separated into two families. The first of these is devoted to the analysis of small-amplitude oscillations. Specifically, the nonlinear partial differential equations are simplified by assuming that the products of all oscillatory variables are negligibly small. We will place all approaches of this sort under the heading of linear analyses. All other treatments, which include at least some part of the nonlinear terms in the governing partial differential equations, will be called nonlinear analyses. We will classify models for engine components in the same way. For example, the convergent nozzle may be treated by either linear or nonlinear analysis.

A comprehensive review of the status of both linear and nonlinear analysis for liquid rocket engine instability as of about 1970 is available as part of Ref. 1. Chapter 4 of this reference, in particular, deals with the engine stability models available at that time. Analytical models for engine components, such as nozzles and acoustic absorbers, are also included in other chapters. In what follows frequent reference will be made to this book.

The purpose of this chapter is not only to provide a brief review of some of the progress made in liquid rocket instability modeling since 1970, but also to describe in a little detail approaches that represent the state of the art in analytical modeling for the liquid rocket instability problem. To do this, the following scheme of presentation has been adopted. First, the basic equations and assumptions generally used in liquid engine stability analyses will be set down. Following this, linear analysis and methods will be discussed. This discussion will include some results of calculations made using existing models. A brief presentation of approaches for the nonlinear instability problem will then be made. A few conclusions and a discussion of possible directions for future efforts will conclude the presentation.

II. Basic Equations and Assumptions

The equations used and assumptions made in analytical approaches since the 1970 review have been no more general than those given in that reference.[2] Consequently, they will be given here in a slightly different form but with the same essential content as presented there. The equations are given in nondimensional form with the characteristic values used in the nondimensional scheme being steady-state values in the gas phase at the injector face. Lengths are made nondimensional using either the chamber radius (circular cross section) or the smaller transverse direction (rectangular cross section).

Continuity:

$$\frac{\partial \rho}{\partial t} + \nabla \cdot (\rho \boldsymbol{q}) = Q \tag{1}$$

Momentum:

$$\rho \frac{\mathrm{D}\boldsymbol{q}}{\mathrm{D}t} + \frac{\nabla p}{\gamma} = Q(\boldsymbol{q}_l - \boldsymbol{q}) \tag{2}$$

Energy:

$$\rho T \frac{\mathrm{D}s}{\mathrm{D}t} = Q\left[1 - T\left(\frac{\gamma - 1}{2}\right)(2\boldsymbol{q}_l \cdot \boldsymbol{q} - \boldsymbol{q}^2)\right] \tag{3}$$

State:

$$p = \rho T \qquad \rho = p^{-1/\gamma} e^{-s} \tag{4}$$

In deriving these equations, it was assumed that a single-component calorically perfect product gas was produced with nondimensional volumetric rate Q from the liquid phase whose volume was taken to be negligible. It was also assumed that the stagnation enthalpy of the liquid phase is a constant. Finally it was assumed that drag between the droplets and the gas is negligibly small. Regarding this last assumption, the drag terms were included in the general development of Ref. 2. They were also shown there, however, to have a very minor impact on liquid rocket stability.

III. Linear Analysis

If the amplitude of the oscillations are small, then a linear analysis, in which products of the oscillatory variables are neglected, is an appropriate mathematical approximation. In a practical sense, a linear analysis can be used to assess the intrinsic stability of a combustor to small amplitude disturbances and to determine the frequency, growth rate, and spatial character of any resulting oscillation. The impact of design strategies, such as the addition of acoustic absorbers or baffles, modifications in the contraction ratio, or the alteration of the steady-state combustion profile on intrinsic stability, can also be ascertained using a linear analysis.

In this section an analytical framework for investigating and describing the linear stability of liquid rocket engines, which includes the flexibility necessary to include acoustic absorbers, baffles, small contraction ratios, and arbitrary axial combustion distributions, will be presented. A few results of calculations showing the impact and interaction of these combustion chamber features on intrinsic stability will also be given.

A. Governing Equations

To develop the appropriate linear forms of the conservation equations the dependent variables p, ρ, etc. are represented as sum of a steady component and an oscillatory component, thus,

$$p = \bar{p} + p^*, \qquad \rho = \bar{\rho} + \rho^*, \qquad \text{etc.} \tag{5}$$

Here the overbar indicates a steady-state (time-independent) variable whereas an asterisk denotes an oscillatory variable. The set of equations for the steady-state can be solved in a straightforward way if it is assumed that the steady-state flow is one dimensional (axial). The solution gives $\bar{p}$, $\bar{\rho}$, etc. in terms of $\bar{Q}(z)$, the steady-state mass source distribution, which can be specified in an arbitrary manner. Steady-state solutions of this type are given, for example, in Refs. 2 and 3 and will not be repeated here. Subtraction of these equations from the complete set leaves the nonsteady (oscillatory) equations which must be solved. These are linearized in the usual way by neglecting products of the perturbation quantities p', ρ', etc. In addition, all of the oscillatory components are represented through an exponential time dependence. Thus,

$$p^* = p'e^{i\omega t} \qquad \rho^* = \rho' e^{i\omega t} \qquad \text{etc.} \tag{6}$$

where

$$\omega = \omega_R + i\lambda \tag{7}$$

where ω_R is the nondimensional frequency and λ is the linear decay rate (positive implies exponential decay). The following set of equations results:

$$i\omega\rho' + \nabla \cdot (\rho'\bar{u}\boldsymbol{k} + \bar{\rho}\boldsymbol{q}') = Q' \tag{8}$$

$$\begin{aligned} &\bar{\rho}[i\omega\boldsymbol{q}' + \nabla(\bar{u}\boldsymbol{k}\cdot\boldsymbol{q}') - (\bar{u}\boldsymbol{k}x\nabla x\boldsymbol{q}')] + \rho'\bar{u}\frac{\mathrm{d}\bar{u}}{\mathrm{d}z}\boldsymbol{k} + \nabla\frac{p'}{\gamma} \\ &= Q'(\bar{u}_l - \bar{u})\boldsymbol{k} - \bar{Q}\boldsymbol{q}' \end{aligned} \tag{9}$$

$$\begin{aligned} &i\omega s' + \bar{u}\frac{\partial s'}{\partial z} = \frac{\bar{Q}}{\bar{p}}[-T' - (\gamma - 1)[\bar{u}_l\boldsymbol{k}\cdot\bar{\boldsymbol{q}}' - \bar{u}\boldsymbol{k}\cdot\boldsymbol{q}']] \\ &+ \left(\frac{Q'}{\bar{p}} - \frac{\bar{Q}p'}{\bar{p}}\right)(\gamma - 1)\bar{u}(\bar{u} - \bar{u}_l) \end{aligned} \tag{10}$$

$$p'/\bar{p} = (\rho'/\bar{\rho}) + (T'/\bar{T}) \tag{11}$$

$$\rho'/\bar{\rho} = (1/\gamma)(p'/\bar{p}) + s' \tag{12}$$

To simplify this system it is observed that the steady flowfield is irrotational and that any vorticity generated in the oscillatory field must be the result of interactions between the oscillations and the steady and unsteady source terms. Moreover, since the source terms should be of the order of the mean flow Mach number, the vorticity should also be relatively small. Because of this, $\boldsymbol{q}$ is represented as the sum of the gradient of a potential function and rotational vector field generated by source and mean flow interactions. Thus, $\boldsymbol{q}' = \nabla\phi + \boldsymbol{\xi}$. Substituting this into the perturbed momentum equation (9) results in the following expressions after a few manipulations:

$$\nabla\left[i\omega\phi + \bar{u}\frac{\partial\phi}{\partial z} + \frac{p'}{\gamma\bar{\rho}} + \frac{\bar{Q}}{\bar{\rho}}\phi\right] = 0 \tag{13}$$

$$\begin{aligned}\frac{\partial}{\partial z}(\bar{u}\xi_z) + \left(i\omega + \frac{\bar{Q}}{\bar{\rho}}\right)\xi_z &= \frac{Q'}{\bar{\rho}}(\bar{u}_l - \bar{u}) + \phi\frac{\mathrm{d}\bar{Q}/\bar{\rho}}{\mathrm{d}z} \\ &+ \frac{\rho'}{\bar{\rho}}\bar{Q}(\bar{u}_l - \bar{u}) - \left[\rho'\frac{\bar{u}}{\bar{\rho}}\frac{\mathrm{d}\bar{u}}{\mathrm{d}z} + \frac{p'}{\gamma\bar{\rho}^2}\frac{\mathrm{d}\bar{p}}{\mathrm{d}z}\right]\end{aligned} \tag{14}$$

and $\xi_r = \xi_\theta = 0$ (or $\xi_x = \xi_y = 0$ for a rectangular cross section). Equation (13) can be integrated directly, with the arbitrary constant set equal to zero to give

$$p' = -\gamma\bar{p}\left[i\omega\phi\bar{u}\frac{\partial\phi}{\partial z} + \frac{\bar{Q}}{\bar{\rho}}\phi\right] \tag{15}$$

It is seen immediately that p' depends directly only on ϕ and steady flow variables and not on s' or $\boldsymbol{\xi}$.

The vanishing of the components of $\boldsymbol{\xi}$ in directions perpendicular to the direction of mean flow implies that (as one might expect physically) the vorticity vector must lie in a plane perpendicular to the mean flow direction. Formal integration of Eqs. (14) and (10) can be carried out to give expressions for ξ_z and s', respectively, in terms of ϕ and steady-state variables. Since these integral expressions are somewhat complicated and their exact form is not critical to this presentation, they will not be given here; see Ref. 3 for examples of these expressions.

Substitution of the expression for p' along with the state equations into Eq. (8) eventually leads to an equation of the following form:

$$\nabla^2\phi + \omega^2\phi = F_1(\phi, \xi_z, \bar{Q}, Q') \tag{16}$$

where F_1 is quite lengthy and complicated. Equation (21) together with the two expressions for s' and ξ_z given next form the set of equations to be solved,

$$s' = s'(\phi, \xi_z, \bar{Q}, Q') \tag{17}$$

$$\xi_z = \xi_z(\phi, s', \bar{Q}, Q') \tag{18}$$

To complete the formulation, the perturbation mass source Q', must be given in terms of p', $\boldsymbol{q}'$, $\bar{Q}$, etc.; that is, some combustion response relationship must be used, though its form is arbitrary.

The approach used to solve this system of equations requires first that Eq. (16) be transformed to an integral equation. This is accomplished using standard techniques (for example, see Ref. 4). A Green's function $G(\boldsymbol{r}/\boldsymbol{r}_0)$ is introduced such that

$$\nabla^2 G + \omega^2 G = \delta(\boldsymbol{r} - \boldsymbol{r}_0) \tag{19}$$

where $\delta(\boldsymbol{r} - \boldsymbol{r}_0)$ is the Dirac delta function, and $\nabla G \cdot \boldsymbol{n} = 0$ on all bounding surfaces of the combustion chamber. The equation for G is multiplied by ϕ, whereas Eq. (16) is multiplied by G. Subtraction of the two resulting equations and application of Green's theorem gives

$$\phi(\boldsymbol{r}) = \int_{V_0} G(\boldsymbol{r})/(\boldsymbol{r}_0) F_1(\boldsymbol{r}_0)\, \mathrm{d}V_0 - \int_{S_0} G(\boldsymbol{r}/\boldsymbol{r}_0)(\nabla\phi \cdot \boldsymbol{n})\, \mathrm{d}S_0 \tag{20}$$

The Green's function is represented by an expansion in orthonormal eigenfunctions for the chamber geometry, thus,

$$G = \sum_N \frac{\Omega_N(\boldsymbol{r})\Omega_N(\boldsymbol{r}_0)}{\omega^2 - \eta_N^2} \tag{21}$$

where

$$\nabla^2\Omega_N + \eta_N^2\Omega_N = 0 \tag{22}$$

$$\nabla\Omega_N \cdot \boldsymbol{n} = 0 \tag{23}$$

on all surfaces, and N is, in general, a set of three integers that depends on the geometry of the chamber. Substitution of this representation for G into Eq. (20) results in an eigenfunction expansion representation of ϕ,

$$\phi = \sum \mu_N \Omega_N \tag{24}$$

where

$$\mu_N = \frac{1}{\omega^2 - \eta_N^2}\left[\int_V \Omega_N F_1(\phi, \xi_z, s')\, \mathrm{d}V - \int_S \Omega_N F_2(\phi, \xi_z, s')\, \mathrm{d}S\right] \tag{25}$$

In the last expression we have assumed a general form for the surface boundary condition, that is,

$$\nabla\phi \cdot \boldsymbol{n} = F_2(\phi, \xi_z, s') \tag{26}$$

where F_2 varies depending on the boundary surface and may, of course, be zero.

A successive approximation approach is used to solve Eq. (25) as well as the expressions for ξ_z and s'. This approximate technique is based on the assumption that the solution for ϕ is close (in some sense) to a known solution for a similar problem, say, $\tilde{\phi}$. In many cases the choice $\tilde{\phi} = \Omega_{\hat{N}}$ is appropriate, where $\Omega_{\hat{N}}$ is a particular eigenfunction (acoustic mode). If this choice is made, then

$$\phi = \Omega_{\hat{N}} + \sum_{N \neq \hat{N}} \mu_N \Omega_N \tag{27}$$

This implies $\mu_{\hat{N}} = 1$ or, using Eq. (25) with $N = \hat{N}$

$$\omega^2 = \eta_{\hat{N}}^2 + \int_V \Omega_{\hat{N}} F_1(\phi)\, \mathrm{d}V \int_S \Omega_{\hat{N}} F_2(\phi)\, \mathrm{d}S \tag{28}$$

At this point it should be remembered that ϕ, μ_N, etc. are complex quantities as is ω $(\omega_R + i\lambda)$. Thus, Eq. (28) when solved gives the frequency eigenvalue and the decay rate, whereas Eq. (25) yields a complex coefficient matrix (three dimensional, in general) for the eigenfunction expansion of ϕ. The solution of the system of Eqs. (17), (18), (27), and (28) is then attempted by the technique of successive approximation, using to lowest order $\phi = \Omega_{\hat{N}}$, $\omega = \eta_{\hat{N}}$, $\xi_z = 0$, and $s' = 0$. These are used in the integral terms of all four equations to produce first-order estimates of ϕ, ω, ξ_z, and s'. In turn these are used in the integral terms for the second order, and so on. If the procedure converges so that the jth approximation is arbitrarily close to the $(j - l)$th, then a formal solution to the integral equations has been found. Of course, there is no guarantee that the technique will converge. In fact, experience has shown that the simple procedure just given, which uses $\Omega_{\hat{N}}$ as the lowest order approximation, will only converge for restricted engine configurations. Specifically, it is necessary that the combustion be well-distributed axially (rather than concentrated), that the nozzle effect (admittance) be small, that the total acoustic absorber open area be small and well distributed along the walls of the chamber, that no injector face baffles be present, and that the contraction ratio be sufficiently large (nozzle entrance Mach number less than about 0.2).

When the combustion zone is concentrated, the nozzle effect is large, or the nozzle entrance Mach number is greater than about 0.2, a better approximating function than $\Omega_{\hat{N}}$ is required, primarily because of a strong axial variation in the oscillatory field. In theses cases a $\tilde{\phi}$ based on the solution for a chamber with mean flow and nozzle effects but without combustion terms has proved useful. If any $\tilde{\phi}$ other than $\Omega_{\hat{N}}$ is used, it is a straightforward matter to manipulate Eqs. (25) and (28) to produce the following forms:

$$\phi = \tilde{\phi} + \sum_N \mu_N \Omega_N \tag{29}$$

$$\mu_N = \int_V \Omega_N F_3(\phi, \tilde{\phi})\, \mathrm{d}V - \int_S \Omega_N F_4(\phi, \tilde{\phi})\, \mathrm{d}S \tag{30}$$

$$\omega^2 = \tilde{\omega} \int_V \Omega_{\hat{N}} F_5(\phi, \tilde{\phi})\, \mathrm{d}V - \int_S \Omega_{\hat{N}} F_6(\phi, \tilde{\phi})\, \mathrm{d}S \tag{31}$$

where F_3, F_4, F_5, and F_6 depend on the choice of $\tilde{\phi}$, and μ_N is a correction matrix giving the change in ϕ from $\tilde{\phi}$.

It must be pointed out that the nonconvergence of the successive approximation technique when $\Omega_{\hat{N}}$ is used as the approximating function for a particular engine implies that, for this engine configuration, $\Omega_{\hat{N}}$ is an inadequate representation of the wave motion for linear stability analysis. Calculations show that even in those situations in which convergence is obtained using $\Omega_{\hat{N}}$ as the lowest order approximation, the final values of ω_R and λ often differ from the first-order values (with $\Omega_{\hat{N}}$ used in the integrals) very significantly, and the coefficient matrix μ_N has considerable content.

Stability calculations require models for combustion response, acoustic damping devices, and the convergent nozzle in order to develop expressions for $Q'F_1$, and F_2 in Eqs. (25) and (28). These will be discussed next.

B. Combustion Response Models

The state of the art in modeling the response of combustion processes to oscillations in liquid rocket engines has advanced very little since 1970 (as described in Ref. 1). There are still no complete analytical, numerical, or empirical models that represent accurately the major combustion processes, injection, mixing, atomization, vaporization, and chemical kinetics, and their linear or nonlinear response to chamber oscillations. This situation does not result solely from a lack of interest or attention by modelers, though research activity in this area has certainly been very limited. Rather, the current reality is that the critical experiments necessary for such model development have yet to be performed. Until this is accomplished it is hard to imagine how any accurate mechanistic models of the combustion feedback process, suitable for inclusion in an analytical stability model, can be constructed. In the absence of usable mechanistic representations, liquid rocket stability modelers have continued to resort to very simplified two-parameter formulations for combustion response. In the case of linear oscillations, these models take the general form

$$Q' = \bar{Q}\eta p' \tag{32}$$

Q' and $\bar{Q}$ are the perturbed and steady-state volumetric mass generation rates at a given location, and η represents the so-called (complex) response factor. Since η can be written as $|\eta|e^{i\alpha}$, then $|\eta|$ represents the local gain and α the phase between mass generation fluctuations and the pressure oscillations. The two parameters $|\eta|$ and α depend on the model for Q and on the frequency of the oscillations. This formulation is quite general and values for $|\eta|$ and α must be obtained for a given combustion process as a function of frequency. This has been done in the past, for example, in the case of quasisteady droplet vaporization.[5] Even sophisticated mechanistic combustion response models of the future may well be representable in this form (at least for linear responses), since both of the model parameters can vary with location and their frequency dependence is arbitrary. A response model of this type is easily inserted into the volume integrals of the stability model presented earlier in this chapter.

A widely used simplified model for η that postulates a specific frequency dependence and introduces a characteristic time for the combustion response process was developed by Crocco and Cheng[6] in the 1950s. This so-called time-lag model has the following form:

$$Q' = \bar{Q}n(1 - e^{i\omega\tau})p'/\bar{p} \tag{33}$$

The major assumption in this model is that n, the interaction index (mainly the gain factor) and τ, the time lag (mainly the phase factor) are constants for the entire combustion process (not location dependent) and are independent of frequency. The model was originally proposed solely based on some physical arguments.[6] Subsequently, the existence of a characteristic time (time lag) in liquid rocket instability was demonstrated experimentally.[2] Additionally, for droplet vaporization controlled combustion, averaged values of η for the complete combustion region were shown to match the form of the time-lag model fairly well in terms of frequency dependence.[5,7] Finally, Reardon[8] found that the stability behavior of a range of liquid engine designs could be correlated qualitatively in terms of the time-lag model. Nevertheless, it remains unlikely that the complicated liquid rocket combustion response process can be represented completely and accurately by this simple two-parameter model. In the absence of more sophisticated formulations, however, the time-lag model can be used to predict and interpret stability trends at least qualitatively. It can be particularly helpful in assessing and quantifying the stabilizing impact of devices such as injector face baffles and acoustic absorbers, as will be done later in this chapter.

C. Nozzle Response Model

It has long been the conventional procedure to separate the oscillatory flow in the nozzle from that in the combustion chamber in performing linear stability analyses. The nozzle solution then serves to provide a boundary condition or nozzle admittance relationship for the chamber flow. The work of Crocco and Sirignano,[9] who treated axially symmetric and two-dimensional nozzles, remains the standard reference. Numerical integration was used by them to determine admittance coefficients relating the oscillatory pressure, velocity, and entropy at the nozzle entrance plane. In times past, tables of the admittance coefficients from the Crocco and Sirignano[9] work were used in stability analyses. More recently the increased speed of computers has made direct integration of their equations a common procedure as part of many linear stability analyses. Following this latter procedure really makes the separation between the nozzle and chamber problems mainly formal. In performing stability analyses, however, it is often possible to use a single nozzle calculation for several variations in chamber operating conditions. Consequently, the separation of chamber from nozzle remains a practical convenience. An example of a computer program for nozzle admittance calculations of this type can be found as part of the ROCCID code.[10]

Experiments designed to validate the nozzle admittance predictions at least in cold-flow situations have been performed by Zinn et al.[11] and Bell and Zin.[12] In general, the agreement between theory and experiment has been quite good, especially for longitudinal and mixed longitudinal-tangential mode oscillations. This would seem to indicate that the Crocco–Sirignano approach can be used with some confidence; however, it must be pointed out that actual hot-firing data are lacking.

Some characteristics of the nozzle admittance predictions should be mentioned. The first is that for some ranges of nozzle entrance Mach numbers the admittances are very sensitive to frequency. That is, for the same general mode (say, first tangential) small variations in ω away from the normal acoustic mode frequency for the chamber can cause significant changes in the admittance coefficients. If the admittance coefficients were always small (say, of the order of the mean chamber Mach number), this would probably not have a significant impact on stability. It is found, however, that the admittance coefficients frequently are closer to $\mathcal{O}(1)$ than to $\mathcal{O}(\bar{M})$. One last but very important prediction of the nozzle theory is that the admittance condition for pure tangential modes is found to be driving rather than damping in many cases. That is, the nozzle admittance tends to drive rather than damp the oscillation. In these driving situations the admittance is typically of $\mathcal{O}(M)$, but not vanishingly small. For these situations stability predictions are heavily dependent on the nozzle interaction. Physically the mechanism for producing a driving impact at the nozzle seems to be a conversion of mean flow energy to oscillatory energy in the nozzle convergent section.

The nozzle admittance condition just discussed is appropriate for all nozzles with convergence sufficiently gradual that a quasi-one-dimensional description of the mean flow is appropriate. This includes most nozzles of practical interest. For the special case of nozzles that are extremely short relative to the wavelength of the oscillation of interest, the well-known quasisteady or short nozzle approximation may be appropriate. For these instances the admittance condition becomes simply

$$u' = [(\gamma - 1)/2\gamma]\bar{M}_e p' \qquad (z = L) \tag{34}$$

Since $(\gamma - 1)/2\gamma$ is small itself, regardless of $\bar{M}_e$ this will provide a relatively small damping (only) effect. The short nozzle approximation thus provides a simple, frequency-independent boundary condition at the nozzle. Unfortunately, it can hardly be considered a plausible model for real engine nozzles, though it has been used often in the past.

D. Acoustic Absorber Models

Given the very incomplete models for combustion response in liquid rocket engines, there has been considerable effort and interest in the use of acoustic damping devices such as slots and liners. The general motivation for this is to provide sufficient damping so that exact knowledge of the driving supplied by the combustion processes to a given mode of oscillation is not a necessity. To effect this strategy, the actual damping provided by a particular absorber design in the complicated oscillatory flow environment of the engine must be assessed. This requires first that accurate models for the absorbers that predict their response to arbitrary oscillations be available. In addition, the impact of these absorbers on the other stability affecting components of the engine and, thus, on the stability of the chamber's complete driving and damping mechanisms must be determined accurately. The latter requirement, in the case of linear oscillations at least, generally can be satisfied using the integral technique given earlier. It remains to discuss response models for acoustic absorbers.

There are several types of acoustic dampers which have been either proposed or actually employed in liquid rocket engines. These include arrays of individual absorbers (generally of the Helmholtz resonator type), which are called acoustic liners. These liners may cover the complete periphery of the combustion chamber or only some fraction of it. In addition to liners, acoustic slots in a wide variety of designs have been used. These include quarter-wave slots, Helmholtz resonator slots, and slots with a design intermediate between the quarter-wave tube and Helmholtz resonators. Beyond these classical devices whose design permits a one-dimensional flow evaluation, L-shaped slots or slots with a more arbitrary two-dimensional geometry have also seen application.[13–18]

Models for acoustic liners and for the classical one-dimensional devices have long been available. These were reviewed in the liquid rocket combustion instability book[1] and also in a slightly later and more complete publication by NASA.[13] More recently slots with an arbitrary two-dimensional geometry and slots with nonuniform mean temperatures have been modeled.[3,19–21] For all configurations the analysis requires that the frequency dependence and damping be predicted. The former problem can be successfully attacked using linear analysis, the latter requires a nonlinear approach, at least for the oscillation amplitudes present in liquid rocket engines.

For use in the type of integral stability model being discussed here, the performance of acoustic absorbers is most conveniently given in terms of an acoustic admittance β, defined by

$$q' \cdot n = \beta p' \tag{35}$$

The condition is to be applied at the absorber–main chamber interface with the unit normal directed out of the combustion chamber. Also, q' is the local chamber oscillatory velocity, p' the local pressure oscillation, and β the (complex) admittance. The real part of β is mainly a measure of the damping provided by the absorber, the imaginary part affects the frequency directly and overall stability indirectly, as will be discussed later. In analyzing the behavior of the absorber by itself, it is somewhat more convenient and consistent with historical practice in acoustics to deal with the reciprocal of β, the acoustic impedance. Clearly

$$\beta = \frac{1}{K} = \frac{K_R - iK_I}{(K_R^2 + K_I^2)} \tag{36}$$

If K_I is zero then so is β_I. Also, at this point, β_R takes its maximum value and, of course, u' and p' are in phase. The absorber is said to be tuned at this point, and the frequency that gives $K_I = 0$ is called the tuning frequency. In modeling the performance of acoustic absorbers, it has usually been adequate to separate the problems of determining K_I and K_R, that is, the tuning and damping problems.

To determine the phasing between u' and p' (that is, K_I) experience indicates that solving the linear wave equation for the cavity gives a good approximation.[3,20] For one-dimensional geometries this is quite straightforward (i.e., quarter-wave tubes, Helmholtz resonators, intermediate absorbers) and the tuning formulas are well known and validated by much experimental data. For example, in the case of a quarter-wave slot with uniform mean temperature,

$$K_I = -\gamma \bar{\rho}_a \bar{a}_a \cot(\omega L_{\text{eff}}/\bar{a}_a) \tag{37}$$

The parameter L_{eff} is equal to the (nondimensional) depth of the cavity plus a small empirical correction, to account for the inertia of the gas in the slot. Somewhat more complicated results for the other one-dimensional cavity flows with uniform temperature are readily available.[13] When a two-dimensional analysis is required, and/or when the mean temperature in the cavity varies, the analysis becomes more complicated and simple closed-form results are not possible. In this case the integral approach described has been used with success for the absorber problem as well[3,20] and results in terms of eigenfunction expansions for u' and p' with ω as a parameter result.

In developing models to predict K_R it was recognized early on by Ingard[22] that the basic dissipation mechanism for amplitudes of the magnitude to be expected in liquid rocket engines was related to jet dissipation at the chamber-cavity interface. Quasilinearization of the flow in the interfacial region leads to predictions of K_R of the following form:

$$K_R = \Gamma \bar{\rho}_a \hat{u}_a \tag{38}$$

In this expression $\bar{\rho}_a$ is the mean density at the cavity entrance, $\hat{u}_a$ the amplitude of the oscillatory velocity at the cavity entrance, and Γ a constant dependent on the cavity geometry and the magnitude of crossflow at the cavity entrance. Γ has been predicted by analysis and measured in both cold- and hot-firing environments.[14,23]

With the expressions for K_R and K_I already given, the impact of an acoustic absorber on stability can be assessed by inserting these expressions into the integral Eqs. (25) and (28). Note that K (or β) has been defined as a strictly local variable, that is, $\beta = \beta(\theta, z)$. The design of the absorber can change from place to place on the surface of the chamber. Multiple slots, slots designed for multiple mode tuning, and liners of variable β can all be included. The impact of absorbers of arbitrary design can, thus, be assessed using the stability model. The answers to questions, such as the following, concerning the impact of acoustic absorbers on stability can also be obtained. What geometry tunes a given absorber when the influence of other engine components affecting stability are included? Are combustion zone–acoustic absorber interactions significant? What is the overall impact of a given absorber on linear stability? Does tuning an absorber provide maximum stability enhancement? A surprising result from stability predictions to be presented shortly is that the last question is not always answered in the affirmative.

E. Injector Face Baffle Modes

Because the presence of baffles alters the geometry of the combustion chamber to a complicated and frequently irregular form, the wave motions in the baffle cavities and main chamber can be very complex indeed. It follows that including the effects of these baffles is quite a challenging problem analytically. When the liquid rocket stability reference book[1] was written in 1970, only the simplest one-dimensional treatment was available, as given by Reardon.[24] Though this simple model did predict the frequency depression observed in experiments, it made no attempt to describe the complicated three-dimensional wave interactions between the baffle cavities and main chamber. Moreover, no mechanism for dissipation was included, so any damping that occurred could only be the result of indirect effects (such as the frequency depression just mentioned). Oberg et al.[25] attempted the problem of baffled combustors in the early 1970s. In this approach a variational integral method similar to the technique already presented was used. The bulk of this work did not include mean flow, nozzle, or combustion effects. In addition, no dissipation mechanism for the baffled chamber oscillations was included. A surprising and unrealistic prediction of this work was that baffles were found to provide linear driving rather than damping. The work of Oberg et al.[25] did, however, predict the frequency depression observed experimentally.

The first analysis of the baffle problem to provide a mechanism for baffle dissipation was performed by Baer and Mitchell.[26] In this work a finite mean flow, a concentrated combustion zone, and a short (constant Mach number) nozzle were included. The analysis separated the problem into an outer flow region away from the baffle tips and an inner region in the high-velocity region near the baffle blade tips. The outer solution was obtained using an eigenfunction expansion technique. Irrotational, isentropic flow was assumed to exist in this outer region, so the governing equation was simply

$$\nabla^2\phi + \omega^2\phi = 2\bar{M}i\omega\frac{\partial\phi}{\partial z} + \bar{M}^2\frac{\partial^2\phi}{\partial z^2} \tag{39}$$

with

$$p' = -\gamma\left[i\omega\phi + \bar{M}\frac{\partial\phi}{\partial z}\right] \tag{40}$$

and

$$\xi_z = s' = 0 \tag{41}$$

In the main chamber, ϕ was represented as an expansion in orthonormal eigenfunctions that satisfied no-flow conditions at the chamber walls and the short nozzle condition at the nozzle entrance plane. For each baffle cavity ϕ was represented as well by a similar expansion. In the cavities the eigenfunctions obeyed zero normal velocity conditions at the baffle blades and the concentrated combustion form of the sensitive time-lag combustion response model at the injector face. Requiring that pressure and axial velocity match at the main chamber–cavity interface then provided the conditions necessary to determine the coefficients in the series for both main chamber and baffle cavities. In addition, a value for the (complex) frequency eigenvalue was determined. The effects of dissipation were included first by performing an asymptotic analysis in a local coordinate system near the baffle blade tip. This inner solution was matched to the outer solution in order to provide an accurate description of the flowfield in this high-velocity region. This solution was, in turn, used to drive an oscillatory boundary layer on the baffle blades near the tips (away from the tips the velocity was small enough that viscous dissipation was negligible). The energy dissipated in the boundary layer was given by a surface integral over the baffle blades area of the form

$$E_{\text{dis}} = \int_{\text{blades}} \gamma \sqrt{\frac{\mu_{\text{eff}}\omega}{2}} \hat{u}^2 \, \mathrm{d}S \tag{42}$$

where $\hat{u}$ was the velocity amplitude at the edge of the boundary layer and μ_{eff} was a combined molecular-turbulent effective viscosity coefficient. This dissipation integral was used by Baer and Mitchell[26] in an integral stability determination based on the method of Cantrell and Hart.[27] The complete outer solution was used in volume and surface integrals, similar to Eq. (28), in order to obtain a value for the decay coefficient λ. It is important to note that this procedure assumed that the outer solution is not affected by the dissipation in the boundary layer, that is, no attempt was made to correct ϕ.

More recent work[28] has extended the Baer and Mitchell analysis to include arbitrary combustion distributions, finite length nozzle admittances, and acoustic absorbers. This later analysis is cast in the form of the integral method described in this chapter. The basic solution for the main chamber and cavities (including the inner solution at the tip region) is determined for the no mean flow, no combustion case. That is, the matched inner-outer, main chamber–baffle cavity solutions are all for the no-flow acoustic situation. Taken together they define $\tilde{\phi}$, the approximating function of Eqs. (30) and (31). To date only the second of these, Eq. (31), has been used to determine baffled chamber stability. Because $\nabla\tilde{\phi} \cdot n = 0$ on all boundaries in the present application this equation can be written

$$(\omega - \bar{\omega}) \int_V \tilde{\phi}\tilde{\phi} \, \mathrm{d}V = \int_V \tilde{\phi} F_1(\tilde{\phi}) \, \mathrm{d}V - \int_S \tilde{\phi} F_2(\tilde{\phi}) \, \mathrm{d}S \tag{43}$$

Using this equation, the effects of mean flow, combustion, nozzle, acoustic absorbers, and dissipation on ω_R and λ are determined with the wave represented by $\tilde{\phi}$. No attempt has yet been made to successively improve ϕ from $\tilde{\phi}$ by applying

Eq. (30) to include the mean flow, combustion, nozzle, absorber, and baffle dissipation effects. This remains as an interesting and, perhaps, formidable extension to the theory. It should also be noted that all analyses to date have considered baffles without radial hubs. That is, baffles designed to suppress only tangential and not radial modes have been analyzed. Work is in progress, however, which will treat both hubs and radial blades with hubs.

F. Some Predictions

The bulk of the calculations performed using the integral method described have been completed for somewhat simplified forms of the governing equations.[3, 19–21, 27] This was done both to reduce the cost of the calculations and the complexity of the algebraic manipulations and because the simplifications were found to have relatively minor effects on most predictions. Specifically, in Eq. (16) all terms of $\mathcal{O}(\bar{M}^3)$ or smaller were dropped and $\bar{Q}$ was taken to be approximately constant. The latter assumption implies that the steady-state volumetric mass generation rate was taken to be uniform in the combustion zone. The axial extent and location of the combustion zone were arbitrary, however, as was the corresponding magnitude of $\bar{Q}$. Finally, the time-lag model was used to represent the response of the combustion zone to oscillations in the results that will be discussed here.

1. Effects of Combustion Concentration

The impact of the axial extent (distribution) of the combustion zone on stability was examined in a series of calculations.[2] Some results for a particular cylindrical engine configuration are presented in Fig. 1. The two curves shown are neutral stability ($\lambda = 0$) curves for the first tangential mode of oscillation and two different

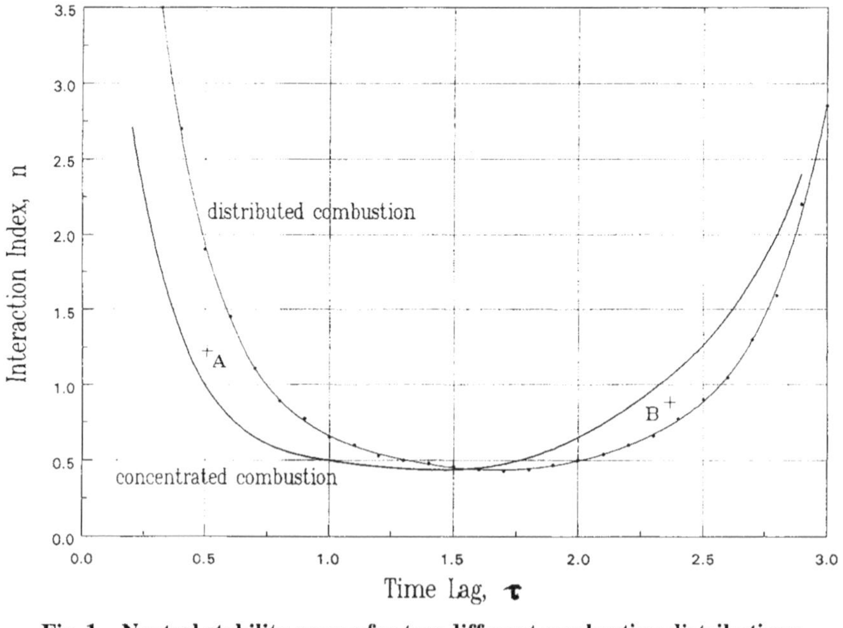

Fig. 1 Neutral stability curves for two different combustion distributions.

axial combustion distributions. The curve labeled "concentrated combustion" is for a combustion zone concentrated in the first 10% of the chamber length, next to the injector. The "distributed combustion" curve corresponds to a combustion zone which extends along the entire axial length of the combustion chamber. Points along both curves were generated by solving the integral equations described earlier with the real part of the frequency as the independent variable and the imaginary part of the frequency, λ, set equal to zero. The corresponding dependent variables for each solution point were the coefficient matrix μ_N (which determines the spatial form of the oscillation) and the combustion response eigenvalues, interaction index n and time lag τ, required to drive the oscillation. A low value of n for a point along either curve indicates that a relatively small gain is required from the combustion process in order to support the oscillation, whereas a large value of n requires a high gain.

To illustrate, suppose that the combustion process for a particular engine supplied an overall combustion response characterized by the n, τ values at point A in the figure. For this value of τ, the n value required for neutral stability in the case of concentrated combustion (as determined by the intersection of a vertical line through A and the appropriate neutral stability curve) is about 1.0. The n value supplied by the combustion process (point A) is about 1.25. Thus, more gain is supplied than necessary for neutral stability, and the engine with concentrated combustion is predicted to be unstable. Indeed, if the n and τ at point A were used as inputs to the stability model with complex frequency as output (the reverse of what was done to generate the stability curves), one would find that the imaginary part of the frequency, λ, would be negative, predicting exponential growth of the oscillation. (The calculated value found using the model for the point in question is $\lambda = -0.036$.) For an engine with distributed combustion having the same n, τ values supplied by the combustion process (point A), the n supplied is less than that required for neutral stability. The engine is predicted to be stable and λ is positive (exponential decay). In contrast, if the n and τ supplied by the combustion process corresponded to point B, the opposite situation would obtain: the engine with distributed combustion would be predicted unstable, the one with concentrated combustion predicted stable. Overall, an inspection of Fig. 1 indicates that, at least for the particular engine configuration considered, distributing the combustion zone has an effect on stability which depends on the characteristic time of the combustion process. If τ is relatively short (as at point A), distributing the combustion is stabilizing, whereas if τ is relatively long (as at B), distributing the combustion is destabilizing. Since there is essentially no upward shift in the minimum of the neutral stability curve with a change in combustion distribution, there is no change in the minimum gain required when phasing is optimal. These results are typical of those found in calculations for other engine configurations.[3]

2. *Acoustic Slot Effects*

The same engine configuration just discussed was modified by adding a quarter-wave slot oriented radially in the chamber wall at the injector end of the chamber. The open area of the slot at its interface with the main chamber was taken to be 10% of the area of the injector face. The depth of the slot was varied in the calculations. In Fig. 2 three neutral stability curves are shown. The lowest curve is for the baseline engine with no slot and axially distributed combustion. Above this is the neutral stability curve for the engine with a slot having a depth such that

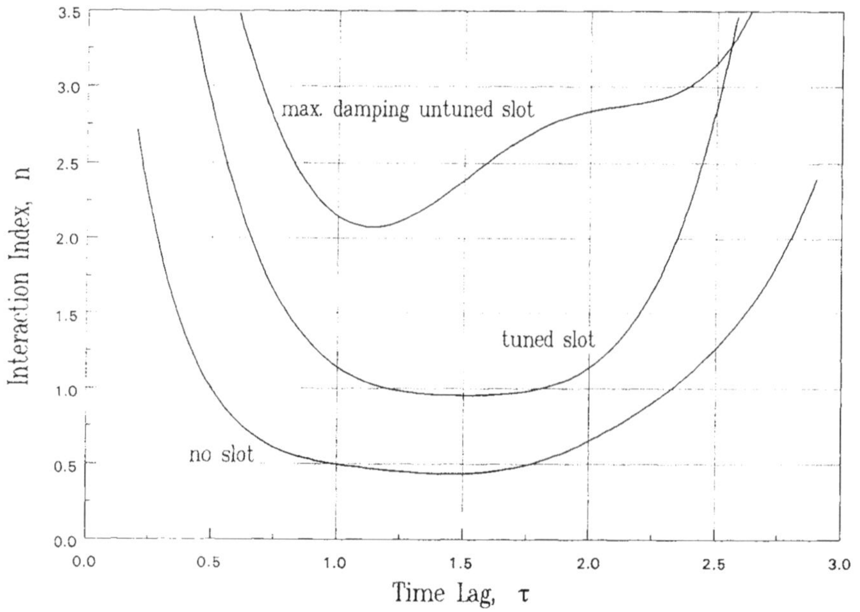

Fig. 2 Effect of quarter-wave slots on neutral stability curves.

it is tuned $\beta_I = 0$ for the frequency at the minimum point (most unstable) on the n, τ curve. The third curve is the neutral stability curve for the slot of the depth that produced the largest upward shift in the minimum n point compared with the baseline engine. This slot had a depth less than that for the tuned slot and was consequently detuned significantly. This was reflected in the fact that the value of β_R (real part of the slot admittance) was about half of the value for the tuned slot.

A more quantitative sense of the damping produced by the addition of slots was gained by making decay rate calculations. Values of n and τ were fixed at the values corresponding to the minimum n point of the baseline neutral stability curve in these calculations while the depth of the slot was varied. Results of calculations for λ, expressed in terms of percent decay per cycle, are shown in Fig. 3 for two different slot open areas. The large increase in decay rate away from the tuning point is evident in the sharply peaked 10% open area curve. A maximum increase in decay rate of about a factor of three over the tuned slot is predicted.

Addition of a slot was also found to cause a very substantial change in the spatial form of the pressure oscillation. Evidence of this is shown in Fig. 4 where the pressure amplitude is plotted as a function of axial distance. For the baseline engine with no slot a significant decrease in amplitude (about 50%) between the injector and nozzle can be seen. In the case of the tuned slot the opposite is true; injector end amplitude is considerably smaller. In contrast, for the maximum damping untuned slot the amplitude is almost uniform.

3. *Interpretation of Slot Results*

It is worthwhile to consider the main implications of the results of the calculations just presented. Of primary importance is the fact that the spatial waveform

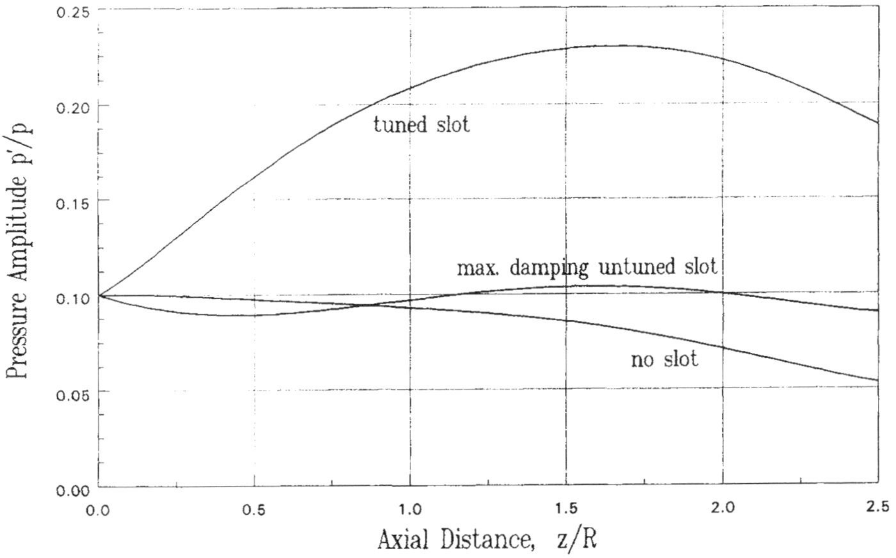

Fig. 3 Axial variation of pressure amplitude with and without slots.

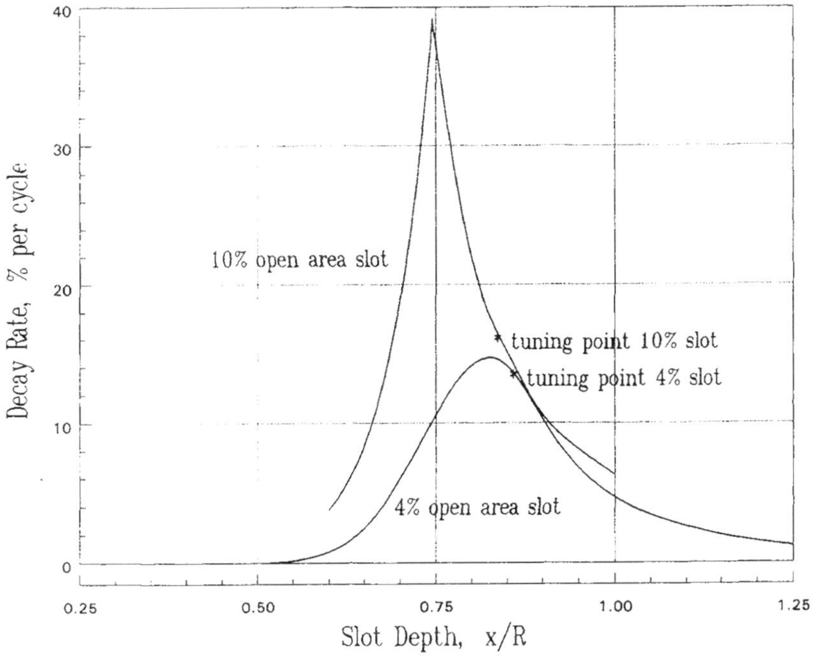

Fig. 4 Decay rate dependence on depth of quarter-wave slots.

of the oscillations discussed is quite different from that of the corresponding simple acoustic mode $\Omega_{\hat{N}}$. This implies that for the significant mean flow, nozzle, and absorber influences present in typical liquid rocket engine configurations it is not an adequate approximation to use $\Omega_{\hat{N}}$. as the approximating function in an integral stability equation such as Eq. (28). The impacts of these components can only be included correctly if an accurate representation of the actual waveform is generated. Examination of Eq. (28) shows why. Both nozzle and absorber provide damping through terms like

$$\int_S \Omega_{\hat{N}} \beta p' \, \mathrm{d}S \tag{44}$$

where S is the active area over the nozzle or absorber entrance. On the other hand, the leading term in combustion driving is

$$\int_V \Omega_{\hat{N}} \bar{Q} \eta p' \, \mathrm{d}V \tag{45}$$

All three integrals are dependent on the local pressure amplitude. Since this varies significantly, according to the calculations just presented, its form must be well approximated in order to give the true stability picture. Certainly if the relative amplitudes of the pressure oscillations at the nozzle and injector change by a factor of four upon the addition of a 10% open area slot as predicted, then a model which assumes that the amplitude is in fact the same at both locations is quite inaccurate.

An important implication of this requirement for a true picture of actual linear wave shape is that it makes no sense to try to separate the complete stability problem into independent calculations for the λ contribution of one component (say, the slot) while ignoring the presence of the others (combustion and nozzle). Adding up the λ values for these problems cannot give an accurate measure of the overall decay coefficient. The fact that a tuned absorber does not provide maximum overall damping is ample evidence of this.

This is not to say that there are no instances in which $\Omega_{\hat{N}}$ can be used as an adequate approximating function. It must, of course, be true that if the impacts of combustion, absorber, and nozzle are small enough (that is, $\bar{M}$, $\bar{Q}$, β_{nozzle}, and β_{absorber} are all small), then $\Omega_{\hat{N}}$ will serve. Also, if the impact of one driving or damping element is small (say, the absorber), then the description of the waveform will not be much affected by that element. As an example of this consider a chamber oscillation that is neutrally stable with no absorber present. If an absorber tuned to the frequency of the oscillation without the absorber present is added, then as long as its open area is small enough it should provide optimal damping. At some point, however, the absorber begins to affect the form of the oscillation and the tuning depth may no longer be best. The results of calculations that demonstrate this are given in Fig. 5. Curves are shown for a tuned slot and also for a variable depth slot designed to give maximum λ as the slot width is increased. For small enough open areas (up to about 3%) the curves coincide. For large open areas there is considerable divergence. Note that the effectiveness of the tuned absorber actually decreases with increasing open areas above about 6%. The tuned absorber provides an increase in damping by itself, but also affects the other components so that the net damping decreases.

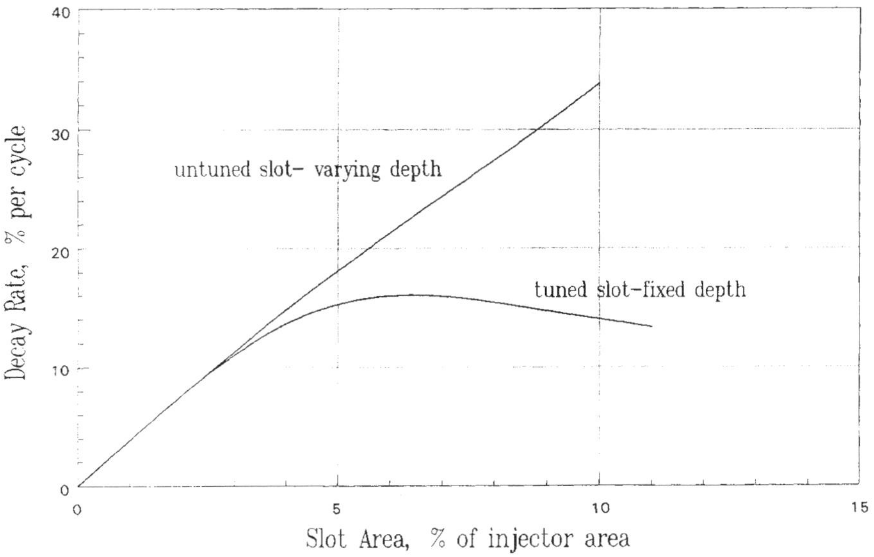

Fig. 5 Decay rate dependence on open area for tuned and untuned slots.

4. *Baffled Chamber Predictions*

The calculations to be discussed briefly here were performed for a combustor with a variable length combustion zone and a conical nozzle with admittance predicted by the Crocco–Sirignano[9] theory. As discussed earlier, the approximating function for baffled chamber analysis was described by the solution to the baffled chamber problem without mean flow, combustion, or nozzle. These latter stability components are only included to first order through Eq. (31). A more complete description of these results and the computer code which was used to generate them is given in Ref. 28.

Neutral stability curves in terms of n and τ for a short, three-bladed baffle, concentrated combustion, and the first tangential (1T) mode of oscillation (main chamber) are presented in Fig. 6. The blade length is 33% of the chamber length. It can be observed immediately that the unbaffled chamber is very unstable; that is, the minimum point on the neutral stability curve for the unbaffled chamber is very low, with a minimum n around 0.25. A major reason for this is that for the particular nozzle design and contraction ratio of this engine, the nozzle admittance has a large, negative real part; it is strongly destabilizing. All combustion zone locations and concentrations are predicted to give identical stability behavior in unbaffled engine. Enhanced stability as reflected by the upward shift of the neutral stability curves is predicted for the same engine with the three-bladed baffle added. Results for two different combustion distributions are shown. The curve labeled "con. combustion" is for a concentrated combustion zone at the injector end of the chamber occupying just 12% of the total axial length. The curve labeled "dist. combustion" is for a uniform distribution occupying the entire chamber length. Comparing the two curves it is seen that the stability enhancement predicted is almost twice as large for distributed combustion as for concentrated combustion.

Why? The plots of the axial dependence of the oscillatory pressure amplitude shown in Fig. 7 give some insight. For the unbaffled engine the pressure profile is just the horizontal line to be expected for the acoustic solution without baffles.

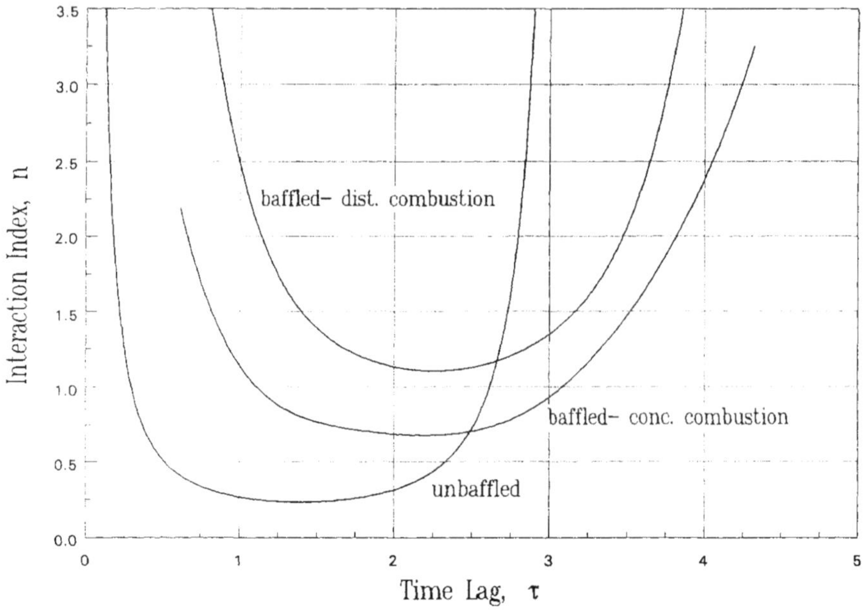

Fig. 6 Axial variation of pressure amplitude with and without a baffle.

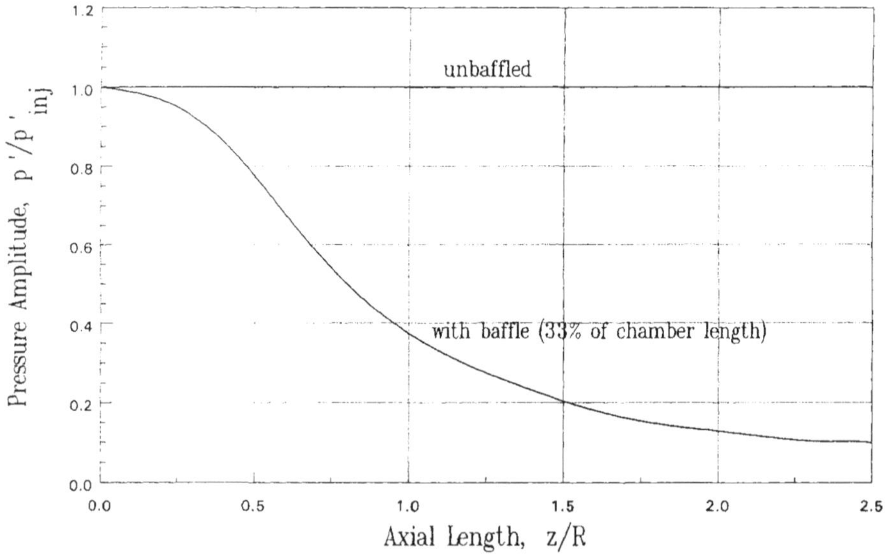

Fig. 7 Neutral stability curves for baffled and unbaffled engine designs.

(Recall that the plot is for p based on $\tilde{\phi}$, and $\tilde{\phi} = \Omega_{\hat{N}}$ in the absence of a baffle.) The 33% baffle length, on the other hand, causes a dramatic increase in the pressure amplitude at the injector end relative to the nozzle end. This means that for combustion sources located near the nozzle end, at a location with relatively low amplitude, the contribution of these sources to the combustion driving integral is relatively small. On the other hand, the baffle dissipation integral occurs in a region of relatively high amplitude. The net effect is a considerable stability enhancement. With the combustion zone located at the injector (in the baffle cavities) both dissipation and combustion occur at high amplitude, and the net stabilizing effect is smaller. The general conclusion to be drawn from these calculations is that baffles can provide stabilization both through viscous damping at the blade tips and through a reduction in combustion driving caused by distorting the spatial form of the oscillation.

IV. Nonlinear Analysis

Though the linear analyses just discussed can be used to assess the intrinsic stability of a liquid-propellant rocket engine including the impact of the major driving and damping mechanisms and components, the linearization of the governing equations that makes the analysis relatively straightforward also precludes the possibility of describing at least two important features of the instability problem. The first of these is a prediction of the finite amplitude waveform (in space and time) that oscillations finally assume. Having a model that can predict these limit cycles is important, not only for a complete description of instability phenomena, but also because of the implications for validation of the theory. Experiments for fully developed oscillations are easier to perform. Measurements of amplitude and waveform for finite oscillation in engines are more straightforward to make than the growth rate or stability boundary measurements for spontaneous oscillations that are generally required for validation of linear theories. The second well-known characteristic of combustion instability that only a nonlinear theory can treat is the phenomenon of triggering. Ample experimental results indicate that in many cases liquid rocket engines which are intrinsically stable to small-amplitude background disturbances can exhibit finite amplitude oscillations upon the introduction of disturbances of sufficient amplitude and of the correct form.

To treat the nonlinear combustion instability problem completely, nonlinearities in all phases of the problem must be included: gas dynamics, combustion response, nozzle admittances, and damping mechanisms including slots and baffles. The lack of a mechanistic combustion response model or even of a viable nonlinear heuristic model which characterizes the nonlinear response of the burning processes to oscillations in pressure and velocity has generally limited analytical work to the treatment of nonlinear gas dynamics and the resulting nonlinear wave motion. In most treatments the combustion response model has remained linear. This is generally true for the early work of Sirignano and Crocco,[29] Mitchell et al.,[30] Crocco and Mitchell,[31] and Zinn.[2] These analyses were reviewed in Ref. 2, and a detailed recapitulation is not needed here. It should be remarked, however, that the analyses were quite successful in predicting the general form of finite amplitude waves, including both shock waves and steep fronted waves for the cases of axial and annular combustion chamber geometries. Triggering was also predicted in this work but only for very narrow ranges of the linear combustion response parameters (n and τ) close to the neutral stability limit in the region of linear stability. These ranges were, in fact, so narrow that discriminating between the triggered instability predicted and spontaneous instability in an experimental engine would hardly be possible. Thus, in essence these results indicated that no viable triggering mechanism was to be found in the nonlinear gas dynamics for any geometry common in liquid rocket engines. A few words of caution, however, are appropriate. First, this work was limited to moderate amplitude waves, with terms of the order of the pressure amplitude cubed generally ignored in the equations. Second, the nonlinear form of the short nozzle, constant Mach number boundary condition was used in all of the work except Zinn's, which in turn was limited by an irrotational assumption.

Subsequent work by Zinn and Powell[33] and Powell and Zinn[34] attacked the nonlinear instability problem using a different technique. This approach was an approximate integral analysis based on the method of weighted residuals that they termed a modified Galerkin technique. The basic governing equation used by them can be derived directly from the conservation equations and assumptions presented earlier, if, in addition, an assumption of irrotational motion is made. Noting that the steady-state flow is taken to be one dimensional and that for linear oscillations the influence of combustion generated vorticity is small, this assumption is not as restrictive as it might seem at first glance. Moreover, the early nonlinear work of Crocco and Mitchell,[31] which included vorticity effects in axial and annular mode oscillations, showed that only a modest effect on stability resulted from its inclusion. The velocity potential Φ was represented as a (truncated) expansion in terms of appropriate orthogonal eigenfunctions for the geometry in question, $\Omega_N(\boldsymbol{r})$, thus,

$$\Phi = \sum_N A_N(t)\Omega_N(\boldsymbol{r}) \tag{46}$$

The coefficients $A_N(t)$ were in turn determined by applying the Galerkin orthogonality condition to require approximate satisfaction of the governing equation and boundary conditions. Substitution of the series representation for Φ into these integral conditions resulted in a system of nonlinear coupled ordinary differential equations for the $A_N(t)$ coefficients. These were then integrated numerically. Results of calculations for both longitudinal and transverse (tangential and radial) mode oscillations showed that limit cycles qualitatively similar to those observed experimentally were predicted. Triggering predictions were consistent with the earlier work just discussed in that triggering was found to exist only for very restricted areas in the linearly stable region of the n, τ plane lying very close to the neutral stability curves. No triggering at all for 1T and axial modes was predicted if only terms to second order in pressure amplitude were retained. A third-order theory was required to predict any triggering. In their work Zinn and Powell[33] found that only a few terms in the series for Φ needed to be retained (five terms for axial modes and three terms, 1T, 2T, 1R, for the 1T mode).

There are distinct advantages to the Galerkin approach. It is straightforward, relatively simple, and gives a time history for developing oscillations. In view of the linear results presented earlier, however, one must wonder if it can model adequately the complex interactions between modes that occur for engines with concentrated combustion, strong nozzle effects, and damping components such as slot absorbers. A very complex spatial structure is predicted in these engines for linear oscillations. How can a few modes suffice for the nonlinear case? Certainly for linear oscillations, the Galerkin approximation used by Zinn and Powell[33] leads to a spatial representation through $\Omega_{\hat{N}}$. As has been shown in the preceding section on linear stability this is rarely adequate.

From the discussion up to this point it can be seen that nonlinear treatments of the gas dynamics in the combustion chamber that include linear combustion response models have not been adequate to predict the observed triggering characteristics of many engine configurations. It is reasonable to suspect then that inclusion of a sufficiently realistic nonlinear combustion response model is required in order to predict triggered instabilities. An attempt was made by Jotiban[35] in the mid-1970s to do this. He pursued an analytical investigation that posited a nonlinear vaporization-based combustion response model as the driver in a liquid rocket engine. Both the combustor model and the response model were quite simplified and, perhaps, quite unrealistic. Nevertheless, some fairly strong triggering was predicted, and the method may bear re-examination for future application. The engine model included a concentrated combustion zone at the injector end, a short nozzle at the nozzle end, a moderate contraction ratio (terms of order $\bar{M}^3$ were dropped in the analysis) and an acoustic slot with a Helmholtz resonator geometry. Irrotational flow was assumed so that the governing equation was essentially the same as that of Zinn and Powell.[33] The nonlinear short nozzle boundary condition was used as was a nonlinear form of the absorber admittance for the Helmholtz resonator. The concentrated combustion assumption was made for analytical simplicity. It is, of course, inconsistent with a droplet vaporization model requiring a finite droplet lifetime. Consequently, a characteristic time (essentially the droplet lifetime) was artificially included in a monodisperse droplet vaporization model. Though the physical reality of the vaporization model is questionable, it was certainly strongly nonlinear.

To simplify calculations, only terms through second order in ε, an amplitude parameter, were kept. Also it was assumed that only traveling tangential (θ direction) modes were to be considered. The velocity potential was represented as a harmonic series truncated to two terms, thus,

$$\Phi = Re[\Phi_1 + \Phi_2] \tag{47}$$

where

$$\Phi_1 = \phi_1(r)e^{i\alpha}, \qquad \Phi_2 = \phi_2(r)e^{2i\alpha} \tag{48}$$

and

$$\alpha = \theta + \omega t, \qquad \omega = \omega_R + i\lambda \tag{49}$$

Substitution into the governing equation and boundary conditions led to two partial differential equations, one for the first harmonic component and one for the second. Following the integral transformation method given earlier, these were, in turn, cast a two integral equations of the form

$$\phi_i = \int_{V_0} G_i F_{1i}(\phi_1, \phi_2)\,\mathrm{d}V_0 - \int_{S_0} G_i F_{2i}(\phi_1, \phi_2)\,\mathrm{d}S_0 \tag{50}$$

where $i = 1, 2$ and F_{1i} and F_{2i} are nonlinear functions of ϕ_1 and ϕ_2. F_{2i} was determined by application of the boundary conditions and included combustion driving terms. The equations were solved by successive approximation. To lowest order $\phi_1 = \varepsilon\Omega_{\hat{N}}$ and $\phi_2 = 0$. Substitution in the integral equations gave improved approximation for ϕ_1 and ϕ_2 (in terms of eigenfunction expansions) and so on. The requirement that the amplitude of $\Omega_{\hat{N}}$ be ε also gave a successively improved value for $\omega = \omega_R + i\lambda$. It is to be noted that F_{1i} and F_{2i} contain both the linear

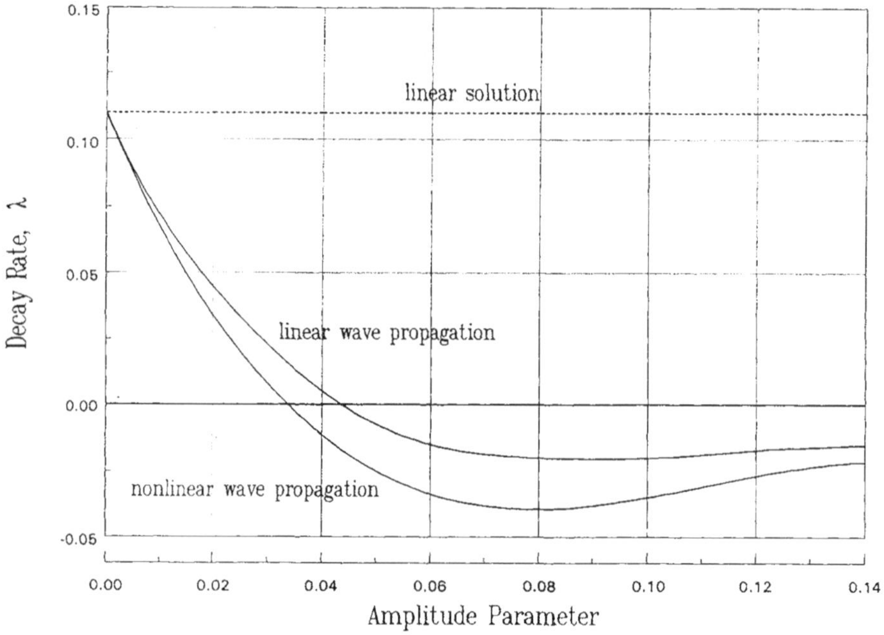

Fig. 8 Nonlinear stability predictions for engine model with vaporization controlled combustion response.

terms due to mean flow, combustion, etc., as well as the nonlinear terms. Thus, if ε is set identically to zero, the usual linear solution results.

Results of calculations for one set of model parameters are shown in Fig. 8. To construct this figure, calculations of the type just described were completed for many values of ε, starting at $\varepsilon = 0$. For each value of ε, the functional forms of ϕ_1 and ϕ_2 were determined and values of ω_R and λ were obtained. At $\varepsilon = 0$, the linear solution gives a prediction of λ of approximately 0.11. That is, linear stability with a large linear decay rate is predicted. As ε is increased, λ decreases, passes through zero, and becomes negative. Negative values of λ indicate growing oscillations. What is shown is, indeed, a prediction of triggered oscillations. For the complete nonlinear solution labeled as "nonlinear wave propagation" in the figure, the curve passes through $\lambda = 0$ at an amplitude parameter value of about 0.032. For values of ε below this the engine is stable, for values larger than this the engine is unstable. Thus $\varepsilon \cong 0.032$ represents the minimum triggering amplitude parameter required, and the oscillation that occurs at this point represents an unstable limit cycle. Physically the calculation procedure represents a prediction of how the engine configuration would respond to the imposition of a 1T traveling wave of a certain amplitude ε. The prediction takes into account all of the driving and damping mechanisms and the nonlinear interactions up to $\mathcal{O}(\varepsilon^2)$. It is a "snap shot" at a particular time; no time history is possible with the technique. The fact that the curve does not recross the axis $\lambda = 0$ for larger values of ε means that stable limit cycles were not predicted. There is, however, a gradual tendency toward the axis. Calculations were not carried out beyond $\varepsilon = 0.14$ because higher order terms would become important. (Indeed, at $\varepsilon = 0.14$ the peak-to-peak amplitude of the oscillation was predicted to be more than 50% of the mean.) A third-order analysis might predict stable limit cycles.

The second curve labeled "linear wave propagation" in the figure was generated by repeating the calculation procedure when all of the nonlinear terms in the governing equation (gas dynamics) were neglected. Nonlinearities in the driving and damping boundary conditions were retained. It can be seen that there is a qualitative agreement between the complete nonlinear prediction and that the quantitative differences are not that large, with a slightly higher minimum ε for triggering being predicted. Clearly, for the engine model being used, the boundary conditions, and not the nonlinear gas dynamics, dominate. This provides evidence that the required triggering mechanism must reside in the nonlinear combustion response and not in nonlinear gas dynamics.

It is interesting to contrast the Jotiban approach with the Galerkin method of Zinn and Powell[33] (or with the mainly equivalent but more recent and more elegant approach described by Culick[36]). In the Jotiban[35] approach a simple representation of the temporal behavior of the oscillation is combined with a complete description of spatial interactions. For the Zinn and Powell[33] or Culick[36] techniques, a simple spatial representation is combined with a complete investigation of the time-dependent behavior of the oscillation. It would seem that both approaches provide partial pictures of the complete nonlinear situation. Both are of interest; both warrant future application to the liquid rocket instability problem. Finally, both require accurate models for nonlinear combustion response processes before meaningful practical application is possible.

V. Concluding Remarks

A major conclusion from the review just presented is that, as was the case in the 1970s, the most fundamental need in successful analytical modeling of combustion instability in liquid rocket engines is for mechanistic, realistic models for the combustion process, and for the response of these processes to both linear and nonlinear oscillations. Such models are not currently available. In their absence the other important physical contributors to the stability picture have received the greatest attention. For the case of small-amplitude oscillations especially, a quite complete analytical framework exists, which lacks only the insertion of realistic response models to become a useful predictive tool. The work on these other aspects of the linear stability problem, however is not all done. As mentioned earlier, a more complete description of instability in baffled combustion chambers is required. The mutual influences of baffle dissipation, combustion, and acoustic absorbers on the basic wave pattern generated by the baffle must be modeled. Also baffle hubs to suppress radial instability must be included in the baffled chamber analysis. The profound destabilizing influence of the convergent part of nozzles, which is predicted in some cases by the widely used nozzle admittance analysis of Crocco and Sirignano,[9] needs to be verified. The magnitude of the nozzle driving is predicted to be of the same size as the damping provided by large open area acoustic slots, at least for pure transverse modes in some engine configurations. An extension of the steady-state combustion description to include arbitrary radial and azimuthal variations is a generalization which will better model some real combustion situations and will more readily accept mechanistic combustion response models as they become available.

The state of analytical modeling for large, amplitude spontaneous or triggered oscillations is less well developed. Given the considerable increase in the speed at which computations can be done today as compared to what was possible years ago when the original work was done, a much more general and at the same time accurate modeling framework should be possible. Both mode history (Galerkin type) and the harmonic analysis approaches should be re-examined and extended as appropriate. Here again, the goal is to develop an analytical framework which can accept quite general nonlinear combustion response models and determine through direct calculation if their driving mechanisms result in stability predictions consistent with the observed phenomena in liquid-propellant rocket engines.

References

[1]Harrje, D. T., and Reardon, F. H. (eds.), *Liquid Propellant Rocket Combustion Instability,* NASA SP-194, 1972.

[2]Crocco, L., "The Sensitive Time Lag Theory," *Liquid Propellant Rocket Combustion Instability,* NASA SP-194, 1972, Chap. 4.

[3]Mitchell, C. E., *Final Report: Stability Design Methodology,* Final Rept. Air Force Contract FO4611-86-K0020, AF-TR-89-041, Vols. 1 and 2, Oct. 1989.

[4]Morse, P. M., and Ingard, K. U., *Theoretical Acoustics,* McGraw–Hill, New York, 1968, pp. 285–291.

[5]Harrje, D. T., and Reardon, F. H. (eds.), "Use of Analytical Models in Design and Development," *Liquid Propellant Rocket Combustion Instability,* NASA SP-194, 1972, Chap. 6.

[6]Crocco, L., and Cheng, S. I., *Theory of Combustion Instability in Liquid Propellant Rocket Motors,* AGARD Monograph No. 8, Butterworths, London, 1956, pp. 13–24.

[7]Nguyen, T. V., and Muss, J. A., "Modification of the Agosta-Hammer Vaporization Response Model for the Prediction of High-Frequency Combustion Stability," *24th JANNAF Combustion Meeting Proceedings,* CPIA Pub. No. 476, Vol. 3, Chemical Propulsion Information Agency, Laurel, MD, 1987, pp. 576–591.

[8]Reardon, F. H., "The Sensitive Time Lag Model," *Liquid Propellant Rocket Combustion Instability,* NASA SP-194, 1972, Chap. 6.3, pp. 277–286.

[9]Crocco, L., and Sirignano, W. A., *Behavior of Supercritical Nozzles Under Three Dimensional Oscillatory Conditions,* AGARDograph No. 117, Nevilly Sur Seine, 1967.

[10]Muss, J. A., and Nguyen, T. V., *User's Manual for Rocket Combustor Interactive Design (ROCCID) and Analysis Computer Program,* NASA CR-1087109, Vols. 1 and 2, May 1991.

[11]Zinn, B. T., Bell, W. A., Daniel, B. R., and Smith, A. J., Jr., "Experimental Determination of Three-Dimensional Liquid Rocket Nozzle Admittances," *AIAA Journal,* Vol. 11, No. 3, 1973, pp. 267–272.

[12]Bell, W. A., and Zinn, B. T., "The Prediction of Three Dimensional Liquid Propellant Rocket Nozzle Admittances," NASA CR-121129, Feb. 1973.

[13]Anon., "Liquid Rocket Engine Combustion Stabilization Devices," NASA SP-8113, 1974.

[14]Oberg, C. L., Wong, T. L., and Ford, W. M., "Evaluation of Acoustic Cavities for Combustion Stabilization," Rocketdyne Rept. R-8757, Canoga Park, CA, 1971.

[15]Harrje, D. T., and Reardon, F. H. (eds.), "Design Factors Affecting Damping," *Liquid Propellant Rocket Combustion Instability,* NASA SP-194, 1972, Chap. 8.

[16]Oberg, C. L., "Space Shuttle Maneuvering Engine Reusable Thrust Chamber Program," NASA CR-141676, 1975.

[17]Moriarty, M. P., and Nestlerode, J. A., "Acoustic Damping Characteristics of the Space Shuttle Main Engine Pre-Burner Acoustic Cavity," *10th JANNAF Combustion Meeting,* CPIA Pub. No. 243, Vol. 2, Chemical Propulsion Information Agency, Laurel, MD, 1973, pp. 271–284.

[18]Kesslring, R. C., and Huebner, A. W., "Combustion Stability Aspects of a Lox/JP-4 Gas Generator," *15th JANNAF Combustion Meeting,* CPIA Pub. No. 297, Vol. 2, Chemical Propulsion Information Agency, Laurel, MD, 1979, pp. 485–500.

[19]Mitchell, C. E., and Dodd, F. E., "Two Dimensional Acoustic Cavity Models," *24th JANNAF Combustion Meeting,* CPIA Pub. No. 476, Vol. 3, Chemical Propulsion Information Agency, Laurel, MD,1987, pp. 555–566.

[20]Mitchell, C. E., Dodd, F. E., Acker, T. L., and Howell, D. J., "Two Dimensional and Temperature Effects on Acoustic Cavity Tuning and Performance in Rocket Combustors," *Combustion Instabilities Driven by Thermo-Chemical Acoustic Sources,* ASME Heat Transfer Div., Vol. 128, American Society of Mechanical Engineers, New York, 1989, pp. 11–16.

[21]Acker, T. L., and Mitchell, C. E., "Combustion Zone–Acoustic Cavity Interactions in Rocket Combustors," *Journal of Propulsion and Power,* Vol. 10, No. 2, 1994, pp. 235–243.

[22]Ingard, U., "On the Theory and Design of Acoustic Resonators," *Journal of the Acoustical Society of America,* Vol. 25, No. 6, 1953, pp. 1037–1067.

[23]Tang, P. K., Harrje, D. T., and Sirigano, W. A., "Experimental Verification of the Energy Dissipation Mechanisms in Acoustic Dampers," *Journal of Sound and Vibration,* Vol. 26, No. 2, 1973, pp. 263–267.

[24]Reardon, F. H., "Available Theory," *Liquid Propellant Rocket Combustion Instability,* NASA SP-194, 1972, Chap. 8.2, pp. 386–389.

[25]Oberg, C. L., Evers, W. H., and Wong, T. L., "Analysis of the Wave Motion in Baffled Combustion Chambers," NASA CR-72897, 1971.

[26]Baer, M. R., and Mitchell, C. E., "A Theoretical Evaluation of Rigid Baffles in Suppression of Combustion Instability," *AIAA Journal,* Vol. 15, No. 2, 1977, pp. 212–217.

[27]Cantrell, R. H., and Hart, R. W., "Interaction Between Sound and Flow in Acoustic Cavities: Mass, Momentum and Energy Consideration," *Journal of the Acoustical Society of America,* Vol. 36, No. 4, 1964, pp. 697–706.

[28]Mitchell, C. E., Howell, D. J., Dodd, F. E., and Acker, T. L., "Users Manual for the Multidimensional Baffle Model Computer Programs," Mechanical Engr. Rept. No. 87-5, Colorado State Univ., Fort Collins, CO, Sept. 1987.

[29]Sirignano, W. A., and Crocco, L., "A Shock Wave Model of Unstable Rocket Combustion," *AIAA Journal,* Vol. 2, No. 7, 1964, pp. 1285–1296.

[30]Mitchell, C. E., Crocco, L., and Sirignano, W. A., "Nonlinear Longitudinal Instability in Rocket Motors with Concentrated Combustion," *Combustion Science and Technology,* Vol. 1, 1969, pp. 35–64.

[31]Crocco, L., and Mitchell, C. E., "Nonlinear Periodic Oscillations in Rocket Motors with Distributed Combustion," *Combustion Science and Technology,* Vol. 1, 1969, pp. 147–169.

[32]Zinn, B. T., "A Theoretical Study of Nonlinear Transverse Combustion Instability in Liquid Propellant Rocket Motors," Ph.D. Thesis, Dept. of Aerospace and Mechanical Sciences, Princeton Univ., Princeton, NJ, May 1966.

[33]Zinn, B. T., and Powell, E. A., "Nonlinear Combustion Instability in Liquid Propellant Rocket Engines," *13th Symposium (International) on Combustion,* Combustion Inst., Pittsburgh, PA, 1970, pp. 491–503.

[34]Powell, E. A., and Zinn, B. T., "Theoretical Investigation of Nonlinear Three-Dimensional Instabilities in Liquid Rockets with Real Nozzles," *Acta Astronautica,* Vol. 1, 1974, pp. 1051–1073.

[35]Jotiban, Y., "Nonlinear Analysis of Wave Distortion Response," Ph.D. Thesis, Colorado State Univ., Fort Collins, CO, 1977.

[36]Culick, F. E. C., "Combustion Instabilities in Propulsion Systems," *Combustion Instabilities Driven by Thermo-Chemical Acoustic Sources,* ASME Heat Transfer Div., Vol. 128, American Society of Mechanical Engineers, New York, 1989, pp. 33–52.

Chapter 16

Effects of Turbulence on Linear Acoustic Instability: Spatial Inhomogeneity

Jong Soo Kim*
University of California, San Diego, La Jolla, California 92093-0411

Nomenclature

A_n = amplitude of the *n*th eigenmode
a_f = overall frozen isentropic sound speed
a_0 = reference sound speed
$c_{p,f}$ = overall frozen specific heat at constant pressure
d = rocket chamber diameter
$\boldsymbol{F}$ = nondimensional rate of momentum transfer by viscous stress and peculiar velocity
G = Green's function
K_f = overall frozen adiabatic compressibility
k = wave number
ℓ_c = nondimensional turbulence correlation distance
ℓ_t = characteristic turbulence length scale
M = Mach number based on the reference axial velocity as defined in Eq. (18)
$\wp$ = nondimensional pressure defined in Eq. (20)
p = pressure
$\mathcal{Q}$ = nondimensional rate of heat release
R_p = nondimensional linear acoustic pressure response of $\mathcal{Q}$
$\boldsymbol{R}_v$ = nondimensional linear acoustic velocity response of $\mathcal{Q}$
T = temperature
t = temporal coordinate
$\boldsymbol{u}$ = nondimensional velocity defined in Eq. (20)
$\vec{u}$ = velocity
u_t = characteristic turbulence velocity scale
u_0 = reference axial velocity averaged a transverse cross-sectional area
$\boldsymbol{x}$ = nondimensional spatial coordinate defined in Eq. (19)

*Assistant Research Scientist, Center for Energy and Combustion Research.

$\vec{x}$ = spatial coordinate
Y = mass fraction
α = volume fraction
β_f = overall frozen volumetric thermal expansion coefficient
γ = ratio of the specific heat at constant pressure to that at constant volume
ε = intensity of the turbulent temperature fluctuations
ζ = longitudinal coordinate in the ξ coordinate
η = stochastic linear growth rate
κ = nondimensional adiabatic compressibility defined in Eq. (21)
λ = deterministic linear growth rate
ξ = inversely mass weighted nondimensional spatial coordinate defined in Eq. (33)
ρ = density
ϱ = nondimensional density defined in Eq. (21)
τ = nondimensional temporal coordinate normalized by the characteristic acoustic time $t_a = d/a_0$ as defined in Eq. (19)
τ' = nondimensional temporal coordinate normalized by the characteristic instability growth time as defined in Eq. (69)
φ = Helmholtz eigenfunction
ψ = acoustic potential
ω = acoustic frequency

Subscripts

f = chemically frozen state
g = gas phase
s = variables of the mean field
0 = stationary part of the mean field
1 = fluctuating part of the mean field

I. Introduction

THE purpose of the analysis described in this chapter is to provide a suitable framework for studying the linear instability of acoustic waves in liquid-propellant rocket engines driven mainly by interactions with combustion processes. Most of the discussion here is devoted to the formulation of the problem in a manner that enables us to take into account the effects of the turbulence and spray as well as the chemical reaction. Emphasis will be placed on the mathematical principles for treating wave propagation in random media characterized by stochastic distributions of state variables, velocity, and chemical reaction.

Liquid-propellant rocket acoustic instability was intensively studied during the 1950s and 1960s, the decades of rocket development. Since then, it has undergone a period of de-emphasis during which the fundamental understanding of the related physics remained almost unchanged. Consequently, most of the ideas about acoustic instability can be found in reviews based on information available at that time.[1–3] Because knowledge in turbulent combustion was lacking then, predictions of acoustic instability had to rely on highly simplified models. The most important feature of those models is that the medium of wave propagation is homogenized by taking averages of state variables and assuming homogeneous and empirical chemical reaction rates dependent only on the averaged variables. The problems, thus, involve only one length scale, i.e., the wavelength. Although from the viewpoint of the acoustic wave such models could give a first approximation, the effects of phenomena taking place at much smaller length scales, such as those of turbulence and influences of interactions with sprays, had to be significantly compromised or ignored. The greatest shortcoming, however, is that the models did not properly include results of the flame-structure analysis. The flame is not only the main energy source for the wave field but also must be located where the actual amplification mechanisms of the acoustic field occur. By ignoring flame-structure analyses, results of the identified amplification mechanisms[4–7] cannot be incorporated into the predictions; therefore, the models had to rely on empiricisms such as time-lag descriptions.[8]

To describe the effects of turbulence and of detailed local flame structure, the basic idea is patterned after the laminar flamelet model of turbulent combustion,[9–11] in which the disparity of the length scales enables us to assume that a turbulent flame is an ensemble of laminar flamelets. Consideration of the length scales involved in the problem indicates that the largest length scale will be the wavelength, whose order of magnitude is the diameter of rocket chamber, denoted by d. The important length scale for turbulence, ℓ_t, is the integral scale, measuring the distance that a turbulent fluctuation can travel on average. In this analysis, the wavelength is always asymptotically larger than the turbulent length scale. This long-wave assumption allows us to adopt a hypothesis of local homogeneity, which states that the wave is considered to be homogeneous over a distance of the order of ℓ_t. The length scale for the flame is even smaller than the turbulence length scale in the flame-sheet regime, and modeling is needed to resolve all of these length scales.

The modeling in this analysis can occur at two levels, the macroscopic and microscopic levels. At the macroscopic level, the acoustic wave is considered to propagate in a multiphase, turbulent reactive medium. In this case, the resulting wave equation as well as the wave field are stochastic. Using a mathematical method called smooth perturbation,[12–14] the mean wave field is calculated in terms of moments of the stochastic variables. Flame structures are considered at the microscopic level. Since the flame is dominantly a diffusion flame, the mixture-fraction variable can be adopted to simplify the flame-structure analysis. The main objective of this later analysis is to obtain the dependence of variables such as the rate of heat release on acoustic fluctuations as functions of the mixture fraction. The results of the microscopic-level analyses are introduced into the macroscopic level to obtain the moments of stochastic variables through the mixture fraction that is available from the turbulent-flame calculation.[9]

Time scales are relevant to the phenomena. Since the velocity scale of the wave is a representative sound speed, its time scale can be taken to be given by d/a_f where a_f is frozen sound speed. On the other hand, the characteristic turbulence time scale is ℓ_t/u_t, where u_t is the turbulent velocity scale in large eddies. It will be assumed that the time scale of the wave is small compared with that of turbulence, i.e., $\ell_t/u_t \gg d/a_f$, which is true in most rocket engines. The wave is then considered to propagate in a quasistationary random medium so that the time dependence of the turbulent fluctuations is decoupled from the acoustic field. Another important time scale is the chemistry time scale, denoted by t_c, which affects the flame-structure analysis. If the chemistry time scale is shorter than the acoustic time scale, then chemical equilibrium prevails everywhere, so that flame structures can be calculated by considering quasisteady mixing only. The rate of heat release, however, still fluctuates in response to the acoustic field since the rate of mixing is modified by the acoustic field. As the chemistry time scale becomes equal to or longer than the characteristic turbulence time scale, the dependence of the finite chemical reaction rate on the acoustic fluctuations becomes important for calculating the mean and oscillatory rates of heat release. As an extreme case, flames may be extinguished due to excessively high strain rates exerted by turbulent fluctuation if t_c is much larger than the turbulent time scale. The amplification mechanisms for the acoustic field can be identified according to these limiting cases.

The amplification mechanisms in acoustic instability can be divided in two groups, boundary amplification and medium amplification. Boundary amplification is more common in solid-propellant rocket engines in which the propellant burns at its surface. In this case, acoustic instability can be analyzed in terms of the acoustic admittance of the burning propellant. Instead, attention is focused on medium amplification driven by combustion, vaporization, atomization, etc. Since the approach is analytic, additional simplifications are introduced. We consider the linear stability only. In later analyses, generalizations to nonlinear stability need to be addressed. With these assumptions, a stochastic wave equation arising from spatial inhomogeneity flow is derived from conservation equations in Sec. II. The stochastic wave equation is averaged by using the method of smooth perturbation in Sec. III. A dispersion relation is obtained to yield turbulence corrections to acoustic frequency and linear-growth rate of acoustic pressure. In Sec. IV, other effects of turbulence on acoustic instability are briefly discussed.

II. Formulation

A. Conservation Equations

Prior to introducing the conservation equations, it is important to clarify the definitions. We let ρ_{ji} denote the mass of chemical species i in phase j per unit volume of phase j and α_j the volume fraction of phase j. Generally $j = g, f, o$ denotes gas, fuel, and oxidizer streams, respectively. With N chemical species in the system, the density of phase j is

$$\rho_j = \sum_{i=1}^{N} \rho_{ji}$$

and the total density is $\rho = \sum_j \alpha_j \rho_j$. From these definitions, we are able to define the mass fractions, namely,

$$Y_j = \frac{\alpha_j \rho_j}{\rho}, \qquad Y_{ji} = \frac{\alpha_j \rho_{ji}}{\rho} \qquad \text{with} \qquad \sum_j Y_j = 1, \qquad \sum_i Y_{ji} = Y_j \tag{1}$$

where Y_j is the mass fraction of phase j and Y_{ji}, that of species i in phase j. If $\vec{u}_j$ is the mass-weighted average velocity of phase j and $\rho\vec{u} \equiv \sum_j \alpha_j \rho_j \vec{u}_j$, i.e., $\vec{u} = \sum_j Y_j \vec{u}_j$, then we can define the peculiar velocity of phase j as

$$\vec{U}_j = \vec{u}_j - \vec{u} \qquad \text{with} \qquad \sum_j Y_j \vec{U}_j = 0 \tag{2}$$

Overall mass conservation is

$$\frac{\partial \rho}{\partial t} + \nabla \cdot (\rho \vec{u}) = 0 \tag{3}$$

Mass conservation for phase j is

$$\frac{\partial \rho Y_j}{\partial t} + \nabla \cdot \{\rho Y_j(\vec{u} + \vec{U}_j)\} = -w_j \tag{4}$$

where w_j is the rate of mass loss per unit volume for phase j. In terms of the diffusion velocity $\vec{U}_{ji}$ for species i in phase j, conservation of chemical species i in phase j is

$$\frac{\partial \rho Y_{ji}}{\partial t} + \nabla \cdot \{\rho Y_{ji}(\vec{u} + \vec{U}_j + \vec{U}_{ji})\} = \alpha_j w_{ji} + \sum_{j' \neq j} w_{j'} \beta_{jij'} \tag{5}$$

where w_{ji} is the mass rate of production of chemical species i in phase j, and $\beta_{jij'}$ is the mass of species i added to phase j from phase j' per unit mass loss of phase j'. Vaporization and droplet combustion laws determine w_j and $\beta_{jij'}$, whereas diffusion equations determine $\vec{U}_{ji}$ and chemical kinetics w_{ji}. Overall momentum conservation is

$$\frac{\partial \rho \vec{u}}{\partial t} + \nabla \cdot \left\{ \sum_j \rho Y_j (\vec{u} + \vec{U}_j)(\vec{u} + \vec{U}_j) \right\} + \nabla p = \nabla \cdot \left(\sum_j \alpha_j \overleftrightarrow{T}_j \right) \tag{6}$$

where $\overleftrightarrow{T}_j$ is the viscous stress tensor in phase j.

Overall energy conservation written in terms of the total (thermal plus chemical) enthalpy per unit mass, denoted by h, is

$$\frac{\mathrm{D}h}{\mathrm{D}t} = \frac{1}{\rho}\frac{\mathrm{D}p}{\mathrm{D}t} + \frac{1}{\rho}\left[\overleftrightarrow{T} : \nabla \vec{u} + \nabla \cdot \left(\sum_j \alpha_j \vec{U}_j \cdot \overleftrightarrow{T}_j \right) - \nabla \cdot \left(\sum_j \alpha_j \vec{q}_j \right) \right]$$

$$- \frac{1}{2}\frac{\mathrm{D}U^2}{\mathrm{D}t} - \frac{1}{\rho}\left[\nabla \cdot \left(\sum_j \rho Y_j H_j \vec{U}_j \right) + \vec{u} \cdot \nabla \cdot \left(\sum_j \rho Y_j \vec{U}_j \vec{U}_j \right) \right] \tag{7}$$

where $h \equiv \sum_j Y_j h_j$, $\overleftrightarrow{T} = \sum_j \alpha_j \overleftrightarrow{T}_j$, $\vec{q}_j$ is the heat flux vector for phase j, and the material derivative is defined as

$$\frac{\mathrm{D}}{\mathrm{D}t} \equiv \frac{\partial}{\partial t} + \vec{u} \cdot \nabla$$

In addition, the stagnation enthalpy per unit mass for phase j, denoted by H_j, and the overall stagnation enthalpy per unit mass H are defined as

$$H_j \equiv h_j + u_j^2/2 \qquad H = h + u^2/2 + U^2/2$$

respectively, in which $u^2 = \vec{u} \cdot \vec{u}$ and $U^2 = \sum_j Y_j \vec{U}_j \cdot \vec{U}_j$.

Focus on the mechanical phenomena that occur in acoustic waves can be achieved by invoking two thermodynamic identities[10]

$$T\frac{\mathrm{D}s}{\mathrm{D}t} = \frac{\mathrm{D}h}{\mathrm{D}t} - \frac{1}{\rho}\frac{\mathrm{D}p}{\mathrm{D}t} - \sum_{j,i} \frac{\mu_{ji}}{W_i}\frac{\mathrm{D}Y_{ji}}{\mathrm{D}t} \tag{8}$$

and

$$\frac{Dp}{Dt} = \left(\frac{\partial p}{\partial \rho}\right)_{s,Y_{ji}} \frac{D\rho}{Dt} + \left(\frac{\partial p}{\partial s}\right)_{\rho,Y_{ji}} \frac{Ds}{Dt} + \sum_{j,i} \left(\frac{\partial p}{\partial Y_{ji}}\right)_{s,\rho,Y_{kl(\neq ji)}} \frac{DY_{ji}}{Dt} \tag{9}$$

where s is the entropy per unit mass, W_i the molecular weight of species i, and μ_{ji} the chemical potential per unit mass of species i in phase j. Equation (7) may be used to eliminate $Dh/Dt - \rho^{-1}Dp/Dt$ from Eq. (8), which may then be employed to remove Ds/Dt from Eq. (9), leaving many source terms, a number of which involve DY_{ji}/Dt, which is eliminated in favor of other source terms by use of Eq. (5). Employing Eq. (3) for $D\rho/Dt$ in this reduced form of Eq. (9) then yields

$$\frac{1}{\rho a_f^2}\frac{Dp}{Dt} = -\nabla \cdot \vec{u} + \frac{\beta_f}{c_{p,f}} Q \tag{10}$$

where a_f, β_f, and $c_{p,f}$ are defined by

$$a_f^2 \equiv \left(\frac{\partial p}{\partial \rho}\right)_{s,Y_{ji}}, \qquad \beta_f \equiv -\frac{1}{\rho}\left(\frac{\partial \rho}{\partial T}\right)_{p,Y_{ji}}, \qquad c_{p,f} \equiv \left(\frac{\partial h}{\partial T}\right)_{p,Y_{ji}} \tag{11}$$

It is usually physically more relevant to use the overall adiabatic compressibility defined by

$$K_f \equiv \frac{1}{\rho}\left(\frac{\partial \rho}{\partial p}\right)_{s,Y_{ji}} = \frac{1}{\rho a_f^2} \tag{12}$$

instead of a_f^2. A simplified recipe to estimate the molecular properties in a mixture of the ideal gases and liquid spray is presented in the Appendix.

The composite term Q denotes the rate of heat exchange per unit mass and for convenience is taken to be separated into four terms, Q_v, Q_l, Q_c, and Q_e, which represent, respectively, various contributions from molecular transport, contributions from peculiar velocities, contributions from homogeneous chemical reaction, and contributions from phase change. They are expressed as

$$Q_v = \frac{1}{\rho}\left[\overleftrightarrow{T} : \nabla\vec{u} + \nabla \cdot \left(\sum_j \alpha_j \vec{U}_j \cdot \overleftrightarrow{T}_j\right) - \nabla \cdot \sum_j \alpha_j \vec{q}_j + \sum_{j,i} h_{ji} \nabla \cdot (\rho Y_{ji} \vec{U}_{ji}) \right]$$

$$Q_l = -\frac{1}{2}\frac{DU^2}{Dt} - \frac{1}{\rho}\left[\nabla \cdot \left(\sum_j \rho Y_j H_j \vec{U}_j\right) + \vec{u} \cdot \nabla \cdot \left(\sum_j \rho Y_j \vec{U}_j \vec{U}_j\right) - \sum_{j,i} h_{ji} \nabla \cdot (\rho Y_{ji} \vec{U}_j) \right] \tag{13}$$

$$Q_c = -\sum_{j,i} h_{ji} \alpha_j \frac{w_{ji}}{\rho}, \qquad Q_e = -\sum_{j,i} h_{ji} \sum_{j' \neq j} \frac{w_{j'}}{\rho} \beta_{jij'}$$

An applicable wave equation may be derived by linearization of Eqs. (6) and (10). For convenience, expressing Eq. (6) in a form similar to that of Eq. (10), conservation equations for the overall energy and momentum are

$$K_f \frac{\mathrm{D}p}{\mathrm{D}t} + \nabla \cdot \vec{u} = \mathcal{Q}, \qquad \rho \frac{\mathrm{D}\vec{u}}{\mathrm{D}t} + \nabla p = \vec{F} \tag{14}$$

where

$$\mathcal{Q} = \frac{\beta_f}{c_{p,f}} Q, \qquad \vec{F} = \nabla \cdot \left\{ \sum_j \alpha_j (\overleftrightarrow{T}_j - \rho_j \vec{U}_j \vec{U}_j) \right\}$$

Here $\vec{F}$ represents the momentum transfer resulting from the viscous stresses and from the peculiar velocities of the different phases. It also can be viewed as a kind of external force applied to the ideal multiphase fluid. Analyses at the microscopic level will provide the most useful avenues for modeling of the source terms $\mathcal{Q}$ and $\vec{F}$.

B. Scaling and Derivation of Wave Equation

It is important to recognize that variations of the mean field and of the acoustic field scale differently, since changes in the mean field are scaled by a characteristic flow velocity u_0, whereas the acoustic field is scaled by a characteristic sound speed a_0. The fact that the Mach number $M = u_0/a_0$ is small provides a scaling that ultimately yields a much simplified wave equation.[15,16]

To nondimensionalize the variables, characteristic values for the length, density, and velocity need be selected. First, reference values for the density and adiabatic compressibility are obtained by volumetric and time averages to be

$$\rho_0 = \lim_{t' \to \infty} \frac{1}{Vt'} \int_0^{t'} \int_{\mathcal{V}} \rho \, \mathrm{d}\vec{x} \, \mathrm{d}t, \qquad K_0 = \lim_{t' \to \infty} \frac{1}{Vt'} \int_0^{t'} \int_{\mathcal{V}} K_f \, \mathrm{d}\vec{x} \, \mathrm{d}t \tag{15}$$

where $\mathcal{V}$ denotes the integration domain and V is the total volume of the integration domain $\mathcal{V}$. The mean chamber pressure is similarly defined as

$$p_0 = \lim_{t' \to \infty} \frac{1}{Vt'} \int_0^{t'} \int_{\mathcal{V}} p \, \mathrm{d}\vec{x} \, \mathrm{d}t \tag{16}$$

From ρ_0 and K_0, the characteristic velocity scales for the mean and acoustic fields are found to be

$$u_0 = \dot{m}/\rho_0, \qquad a_0 = 1/(K_0 \rho_0)^{1/2} \tag{17}$$

where $\dot{m}$ is the area-averaged axial mass flux and is independent of the axial coordinate. Then the Mach number M is defined as

$$M \equiv u_0/a_0 = \dot{m}/\rho_0 a_0 = t_a/t_r \tag{18}$$

which is also the ratio of the characteristic acoustic time $t_a = d/a_0$ to the characteristic flow time $t_r = d/u_0$, where the chamber diameter d is taken to be the characteristic geometric length.

The nondimensional spatial coordinates $\boldsymbol{x}$ and temporal coordinate τ are

$$\boldsymbol{x} \equiv \vec{x}/d, \qquad \tau \equiv t a_0/d \tag{19}$$

The nondimensional dependent variables are

$$\wp \equiv \frac{p}{\rho_0 a_0^2}, \qquad \boldsymbol{u} \equiv \frac{\vec{u}}{a_0}, \qquad \mathcal{Q} \equiv \frac{Q\beta_f d}{a_0 c_{p,f}}, \qquad \boldsymbol{F} \equiv \frac{\vec{F} d}{\rho_0 a_0^2} \tag{20}$$

and the density and the adiabatic compressibility are scaled by

$$\varrho \equiv \rho/\rho_0, \qquad \kappa \equiv \rho_0 a_0^2 K_f \tag{21}$$

Scalings for variations of the mean field variables are

$$\begin{aligned} \wp_s &\equiv \frac{p_s}{\rho_0 u_0^2} = M^{-2}\frac{p_s}{\rho_0 a_0^2}, \qquad & \boldsymbol{u}_s &\equiv \frac{\vec{u}_s}{u_0} = M^{-1}\frac{\vec{u}_s}{a_0} \\ \mathcal{Q}_s &\equiv \frac{Q_s\beta_f d}{u_0 c_{p,f}} = M^{-1}\frac{Q_s\beta_f d}{a_0 c_{p,f}}, \qquad & \boldsymbol{F}_s &\equiv \frac{\vec{F}_s d}{\rho_0 u_0^2} = M^{-2}\frac{\vec{F}_s d}{\rho_0 a_0^2} \end{aligned} \tag{22}$$

where the subscript s denotes variations of the mean field. In addition, the mean chamber pressure is scaled by

$$\wp_0 \equiv \frac{p_0}{\rho_0 a_0^2} \tag{23}$$

We now decompose the nondimensionalized variables into sums of the mean and acoustic components. The acoustic field can be sought in order of ϵ where the small parameter ϵ measures a small amplitude of the acoustic pressure relative to the mean chamber pressure. Although ϵ is, in general, a function of Mach number M, an arbitrary small number can be taken as ϵ if only linear instability is considered. Then we seek a solution in the form

$$\begin{aligned} \wp &= \wp_0 + M^2\wp_s + \epsilon\wp', \qquad & \boldsymbol{u} &= M\boldsymbol{u}_s + \epsilon\boldsymbol{u}', \\ \mathcal{Q} &= M(\mathcal{Q}_s + \epsilon\mathcal{Q}'), \qquad & \boldsymbol{F} &= M^2(\boldsymbol{F}_s + \epsilon\boldsymbol{F}') \end{aligned} \tag{24}$$

where $\wp'$ and $\boldsymbol{u}'$ are acoustic fluctuations of the pressure and velocity, and $\mathcal{Q}'$ and $\boldsymbol{F}'$ are the acoustic responses of $\mathcal{Q}$ and $\boldsymbol{F}$, respectively. Here the orders of magnitude for the acoustic responses are assumed to be the same as the orders of magnitude for the corresponding mean field variations. All of the quantities in the right-hand side of Eq. (24) except the small parameters M and ϵ are of order unity.

Since $\wp'$ and $\boldsymbol{u}'$ are the dependent variables, the acoustic responses $\mathcal{Q}'$ and $\boldsymbol{F}'$ have to be expanded as linear functions of $\wp'$ and $\boldsymbol{u}'$ such that

$$\mathcal{Q}' = \varrho R_p\wp' + \varrho\boldsymbol{R}_v\cdot\boldsymbol{u}', \qquad \boldsymbol{F}' = \boldsymbol{B}_p\wp' + \overleftrightarrow{\boldsymbol{B}}_v\cdot\boldsymbol{u}' \tag{25}$$

where R_p, $\boldsymbol{R}_v$, $\boldsymbol{B}_p$, and $\overleftrightarrow{\boldsymbol{B}}_v$ are defined as

$$\begin{aligned} R_p &= \frac{1}{\varrho M}\frac{\partial\mathcal{Q}}{\partial\wp} = \frac{\rho_0^2 a_0^2 d}{u_0}\frac{1}{\rho}\frac{\partial Q}{\partial p}, \qquad & \boldsymbol{R}_v &= \frac{1}{\varrho M}\frac{\partial\mathcal{Q}}{\partial\boldsymbol{u}} = \frac{\rho_0 a_0 d}{u_0}\frac{1}{\rho}\frac{\partial Q}{\partial\vec{u}} \\ \boldsymbol{B}_p &= \frac{1}{M^2}\frac{\partial\boldsymbol{F}}{\partial\wp} = \frac{a_0^2 d}{u_0^2}\frac{\partial\vec{F}}{\partial p}, \qquad & \overleftrightarrow{\boldsymbol{B}}_v &= \frac{1}{M^2}\frac{\partial\boldsymbol{F}}{\partial\boldsymbol{u}} = \frac{a_0 d}{\rho_0 u_0^2}\frac{\partial\vec{F}}{\partial\vec{u}} \end{aligned} \tag{26}$$

Here R_p, $\boldsymbol{R}_v$, $\boldsymbol{B}_p$, and $\overleftrightarrow{\boldsymbol{B}}_v$ are, in general, complex functions of $\wp'$ or $\boldsymbol{u}'$, and subjects of analyses in the microscopic level.

Substituting Eqs. (15–25) into Eq. (14), and collecting the terms of $\mathcal{O}(\epsilon)$ and $\mathcal{O}(\epsilon M)$ only, conservation equations for the acoustic pressure and velocity are found to be

$$\begin{aligned} \kappa\frac{\partial \wp'}{\partial \tau} + \nabla\cdot\boldsymbol{u}' &= -M(\kappa\boldsymbol{u}_s\cdot\nabla\wp' - \varrho \mathrm{R}_p\wp' - \varrho\boldsymbol{R}_v\cdot\boldsymbol{u}') \\ \varrho\frac{\partial \boldsymbol{u}'}{\partial \tau} + \nabla\wp' &= -M(\varrho\boldsymbol{u}_s\cdot\nabla\boldsymbol{u}' + \varrho\boldsymbol{u}'\cdot\nabla\boldsymbol{u}_s) \end{aligned} \tag{27}$$

If the density and adiabatic compressibility are constant, and the terms on the right-hand side are not present, then Eqs. (27) reduce to a Helmholtz equation upon eliminating $\boldsymbol{u}'$ from the equations. It is important to note that the nonhomogeneous terms are responsible for amplification (or attenuation) and refraction of the acoustic pressure and that they are only of order of M. Therefore, growth or decay of the acoustic pressure amplitude can occur only in the characteristic residence time t_r that is longer than the characteristic acoustic time t_a by an order of M^{-1} (Ref. 16). Such multiple time-scale behavior is found in evolutions of the sustaining acoustic oscillations in rocket engines. From time to time, it has been questioned whether the order of magnitude for the acoustic velocity response $\boldsymbol{R}_v$ is M^{-1}. This question stems from the observation that relative amplitude of the acoustic velocity to its mean velocity is larger than that of the acoustic pressure by an order of M^{-1}, so that the acoustic response can be much larger than assumed in the present analysis. If such ordering for the $\boldsymbol{R}_v$ term exists, then the acoustic oscillations will grow or be damped in less than a few cycles, so that no sustaining acoustic oscillations can be observed. Since our interest is restricted to the sustaining acoustic oscillations in liquid-propellant rocket engines, $\boldsymbol{R}_v$ must be of order unity.

To investigate acoustic waves, it is more convenient to introduce complex representations of the acoustics variables, in terms of the complex wave number k, as

$$\wp' = \psi\exp(-ik\tau), \qquad \boldsymbol{u}' = \boldsymbol{v}\exp(-ik\tau) \tag{28}$$

where the wave number k is an eigenvalue of the problem. If the imaginary part of k is positive, then the acoustic oscillation grows, thereby leading to the instability. The wave number is related to the dimensional frequency ω by

$$\omega = \frac{k}{d}\sqrt{\frac{1}{K_0\rho_0}} \tag{29}$$

From the second of Eqs. (27), $\boldsymbol{v}$ is found in terms of ψ as

$$\begin{aligned} \boldsymbol{v} &= \frac{1}{i\varrho k}[\nabla\psi + M(\varrho\boldsymbol{u}_s\cdot\nabla\boldsymbol{v} + \varrho\boldsymbol{v}\cdot\nabla\boldsymbol{u}_s)] \\ &= \frac{\nabla\psi}{i\varrho k} - \frac{M}{k^2}\left[(\boldsymbol{u}_s\cdot\nabla)\left(\frac{\nabla\psi}{\varrho}\right) + \left(\frac{\nabla\psi}{\varrho}\right)\cdot\nabla\boldsymbol{u}_s\right] + \mathcal{O}(M^2) \end{aligned} \tag{30}$$

Substituting Eq. (30) into the first of Eqs. (27), a single scalar wave equation is obtained as

$$\nabla \cdot \left(\frac{\nabla\psi}{\varrho}\right) + \kappa k^2 \psi$$
$$= \frac{iM}{k}\left\{ -k^2 \kappa \boldsymbol{u}_s \cdot \nabla\psi + \nabla \cdot \left[(\boldsymbol{u}_s \cdot \nabla)\left(\frac{\nabla\psi}{\varrho}\right) + \left(\frac{\nabla\psi}{\varrho}\right) \cdot \nabla \boldsymbol{u}_s \right]\right\}$$
$$+ M\left[ik\varrho R_p \psi + \varrho \boldsymbol{R}_v \cdot \left(\frac{\nabla\psi}{\varrho}\right)\right] \tag{31}$$

Here the term involving the acoustic pressure response R_p is usually the main amplification effect, whereas the terms involving the mean field velocity $\boldsymbol{u}_s$ on the right-hand side of Eq. (31) are responsible for the possible damping effect arising from the convection in the longitudinal direction. It appears from Eq. (31) that all of the perturbations in the right-hand side vanish as the Mach number M approaches zero. However, the acoustic source terms that do not involve $\boldsymbol{u}_s$ explicitly may be rescaled in an alternative way at zero M, since the mean axial convective transport is no longer the dominant transport mechanism for the acoustic source terms. If the turbulent mixing becomes more important than the mean axial convective transport, a new scaling can be introduced by a new Mach number based on the characteristic turbulent velocity, and the resulting equation will be identical to Eq. (31).

Wave equation (31) is applicable even for more realistic nonuniform acoustic media. We limit our interest, however, to the case in which the acoustic medium is uniform in the leading order, the assumption that is commonly adopted in most of the acoustic instability analyses. This simplification may give rise to significant error in evaluating the eigenfunctions and eigenvalues, but such error estimates have not yet been made, although it might be simple to do so. Since our main interest is to analyze the effects of the turbulence, the nonuniformity of the acoustic medium is assumed to be only of order of ε for the purpose of circumventing the lack of knowledge about acoustics in nonuniform deterministic media. Here the new small parameter ε measures the intensity of the turbulent fluctuations. Also the wave equation is subject to the simplest boundary condition, namely, the Neumann condition,

$$\boldsymbol{n} \cdot \nabla\psi = 0, \qquad \text{for} \quad x \in B \tag{32}$$

where $\boldsymbol{n}$ is the unit normal vector at the boundary.

C. Introduction of Turbulent Fluctuations

This analysis is based on the limit in which $u_t/a_0 \ll \ell_t/d$, so that the characteristic acoustic time is shorter than the characteristic turbulence time. The acoustic medium is then considered to be quasi-stationary, and the time dependence of the turbulent fluctuations is decoupled from the acoustic field. Only the spatial inhomogeneity of the mean field needs be considered. In liquid-propellant rocket engines, the spatial inhomogeneity arises from a variety of sources, such as turbulent combustion, or fluctuations of phase or velocity fields. Although the spatial fluctuations associated with these effects have the same characteristic time and length scales, their orders of magnitude are different. Since the order of magnitude for the temperature fluctuations associated with turbulent combustion far exceeds those associated with the phase and flow turbulence, only the spatial inhomogeneity arising from turbulent combustion needs to be considered in order to obtain the first approximation to influences of the turbulence.

Not all of the mean-field variables are directly coupled with turbulent combustion. For instance, if the medium is assumed to be a mixture of the ideal gases and spray, κ is proportional to the volume fraction of the gas phase, since only the gas phase is compressible. The volume fraction of the liquid phase is usually small, so that inhomogeneity of κ can be ignored. The density ϱ and the acoustic responses, R_p and $\boldsymbol{R}_v$, however, are directly influenced by turbulent combustion, and their fluctuations must be included. On the other hand, the velocity field is predominantly longitudinal with a constant mass flux. Since the longitudinal velocity is inversely proportional to the density, fluctuations of the longitudinal velocity also needs be included.

Prior to introducing the expansions for the fluctuation fields, it is convenient to introduce a coordinate transformation that will significantly simplify the algebra involving the variable density. This can be achieved by introducing new nondimensional coordinates that are weighted by the nondimensional density. The new coordinates $\boldsymbol{\xi}$ are then defined by

$$\xi_i = \int_0^{x_i} \rho \mathrm{d}x_i', \qquad i = 1, 2, 3 \tag{33}$$

where the origin of the coordinate is placed at the center of the injector assembly. The wave equation in the new coordinate system is

$$\nabla^2\psi + \frac{\kappa}{\varrho}k^2\psi = \frac{iM}{k}\{-k^2\kappa\boldsymbol{u}_s\cdot\nabla\psi + \nabla\cdot[(\varrho\boldsymbol{u}_s\cdot\nabla)\nabla\psi + \varrho\nabla\psi\cdot\nabla\boldsymbol{u}_s]\}$$
$$+ M[ikR_p\psi + \boldsymbol{R}_v\cdot\nabla\psi] \tag{34}$$

where ∇ now denotes the gradient operator in the $\boldsymbol{\xi}$ coordinate.

Letting ε be the intensity of the temperature-field fluctuations, we now expand the mean-field variables into sums of their mean and fluctuating values,

$$\begin{aligned} &\varrho = 1 + \varepsilon\varrho_1, \qquad \kappa = 1 + o(\varepsilon), \qquad R_p = R_{p0} + \varepsilon R_{p1} \\ &\boldsymbol{R}_v = \boldsymbol{R}_{v0} + \varepsilon\boldsymbol{R}_{v1}, \qquad \varrho\boldsymbol{u}_s = \boldsymbol{e}_\zeta, \qquad \boldsymbol{u}_s = \boldsymbol{e}_\zeta(1 - \varepsilon\varrho_1) \end{aligned} \tag{35}$$

where ζ is the longitudinal coordinate in the $\boldsymbol{\xi}$-coordinate system and $\boldsymbol{e}_\zeta$ denotes the unit longitudinal vector. Substituting the expansions (35) into Eq. (34), we then obtain

$$\nabla^2\psi + k^2\psi = M\left[ikR_{p0}\psi + \boldsymbol{R}_{v0}\cdot\nabla\psi - 2ik\frac{\mathrm{d}\psi}{\mathrm{d}\zeta}\right] + \varepsilon\varrho_1 k^2\psi$$
$$+ \varepsilon M\left[ikR_{p1}\psi + \boldsymbol{R}_{v1}\cdot\nabla\psi + 2ik\varrho_1\frac{\mathrm{d}\psi}{\mathrm{d}\zeta} - i\frac{\mathrm{d}}{\mathrm{d}\zeta}(\nabla\varrho_1\cdot\nabla\psi)\right] \tag{36}$$

where the first line corresponds to the mean-field part of the wave equation, whereas the second line is the fluctuating part. Since $\xi_i \to x_i + o(\varepsilon)$ for $x_i \gg \varepsilon$ with the density profile given in Eq. (35), the boundary condition remains the same as Eq. (32) even in order of ε.

III. Mean Wave Field

A. General Consideration for Stochastic Averaging

Since the wave equation in Eq. (36) is a random field, the resulting wave is also a random field. Then the terms in the second line of Eq. (36) involve multiplicative randomness, which is not necessarily vanishing by averaging. Therefore, the resulting mean wave field is different from the wave field without randomness. An averaging technique is discussed to obtain the mean wave field.[14]

Let us consider a simple linear stochastic wave equation given by

$$L_r(\boldsymbol{r})u(\boldsymbol{r}) + X(\boldsymbol{r}, \mu)u(\boldsymbol{r}, \mu) = 0 \tag{37}$$

where L_r is a given deterministic differential operator in the $\boldsymbol{r}$ coordinate, $X(\boldsymbol{r}, \mu)$ is a given random-field operator arising from a random process μ, and $u(\boldsymbol{r})$ is the resulting random wave field. Averaging Eq. (37) yields

$$L_r\langle u\rangle + \langle Xu\rangle = 0 \tag{38}$$

where the average operator $\langle\ \rangle$ can be understood in an ensemble-average sense. The equation includes not only $\langle u\rangle$, but also the second moment $\langle Xu\rangle$. To find the equation for $\langle Xu\rangle$, one may multiply Eq. (37) by $X(\boldsymbol{r}_1, \mu)$ and take averages, giving

$$L_r\langle X(\boldsymbol{r}_1, \mu)u\rangle + \langle X(\boldsymbol{r}_1, \mu)X(\boldsymbol{r}, \mu)u\rangle = 0 \tag{39}$$

which now contains the third-order moment. To describe the subsequently occurring higher order moments, an infinite set of hierarchy equations is necessary. Clearly, it is not possible to obtain a finite and closed system of equations just as has been encountered in the averaged Navier–Stokes equations. One is then forced to introduce closure procedures by postulating that the higher order moments can be expressed in terms of the lower order moments.

One of the closure schemes is the correlation-discard approximation, namely,

$$\langle X(\boldsymbol{r}_1, \mu)X(\boldsymbol{r}, \mu)u\rangle = \langle X(\boldsymbol{r}_1, \mu)X(\boldsymbol{r}, \mu)\rangle\langle u\rangle \tag{40}$$

This hypothesis is equivalent to

$$\langle X(\boldsymbol{r}_1, \mu)X(\boldsymbol{r}, \mu)\Delta u\rangle = 0 \qquad \text{with} \quad \Delta u = u - \langle u\rangle$$

which means that the product of the field fluctuation with a random coefficient is independent of the random coefficient at a different location. This approximation is also known as the local-independence hypothesis. Substitution of Eq. (40) into Eq. (39) gives

$$\langle X(\boldsymbol{r}_1, \mu)u\rangle = -L_r^{-1}\langle X(\boldsymbol{r}_1, \mu)X(\boldsymbol{r}, \mu)\rangle\langle u\rangle = -\big\langle X(\boldsymbol{r}_1, \mu)L_r^{-1}X(\boldsymbol{r}, \mu)\big\rangle\langle u\rangle \tag{41}$$

where L_r^{-1} denotes the inverse of the operator L_r. Substituting Eq. (41) into Eq. (38), the resulting equation becomes

$$\big[L_r - \big\langle X(\boldsymbol{r}_1, \mu)L_r^{-1}X(\boldsymbol{r}, \mu)\big\rangle\big]\langle u\rangle = 0 \tag{42}$$

which describes the mean wave field.

If the stochastic operator X is of order of ε as found in the present analysis, a more systematic closure scheme, called smooth perturbation,[12–14] can be adopted. There exists a solution u_r obeying

$$L_r u_r = 0 \tag{43}$$

Then, a solution of a linear stochastic wave equation $L_r u + \varepsilon X u = 0$ can be iterated, in a form of a perturbation about u_r,

$$u = u_r - \varepsilon L_r^{-1} X u_r + \varepsilon^2 L_r^{-1} X L_r^{-1} X u_r + \cdots \tag{44}$$

Taking the average and solving for u_r, we find

$$\begin{aligned} u_r &= \left\{1 + \varepsilon^2 L_r^{-1} \langle X L_r^{-1} X \rangle\right\}^{-1} \langle u \rangle + \cdots \\ &= \langle u \rangle - \varepsilon^2 L_r^{-1} \langle X L_r^{-1} X \rangle \langle u \rangle + \cdots \end{aligned} \tag{45}$$

Application of L_r to the both sides of Eq. (45) yields an equation for $\langle u \rangle$ as

$$\left\{L_r - \varepsilon^2 \langle X L_r^{-1} X \rangle\right\} \langle u \rangle = 0 \tag{46}$$

which is identical to Eq. (42) considering that X in Eq. (42) is now scaled by a small parameter ε.

In the present analysis, the operators L_r and X are the composite terms of order unity and of order of M. When a solution is sought in the neighborhood of an eigensolution of the Helmholtz equation, the operators can be decomposed as

$$L_r = L_0 + M L_1, \qquad X = X_0 + M X_1 \tag{47}$$

where the Helmholtz operator L_0 is

$$L_0 \equiv \nabla^2 + k_n^2 \tag{48}$$

whereas the other higher order operators are

$$\begin{aligned} L_1 &= \frac{(k^2 - k_n^2)}{M} - i k_n R_{p0} - \boldsymbol{R}_{v0} \cdot \nabla + 2 i k_n \frac{\mathrm{d}}{\mathrm{d}\zeta} \\ X_0 &= -\varrho_1 k_n^2 \\ X_1 &= -i k_n R_{p1} - \boldsymbol{R}_{v1} \cdot \nabla - 2 i k_n \varrho_1 \frac{\mathrm{d}}{\mathrm{d}\zeta} + i \frac{\mathrm{d}}{\mathrm{d}\zeta} (\nabla \varrho_1 \cdot \nabla) \end{aligned} \tag{49}$$

Here the subscript n denotes the nth eigenmode ordered from the smallest wave number. The difference of the eigenvalue k from k_n is of order of M, so that the term $(k^2 - k_n^2)/M$ is of order unity. Substituting Eq. (47) into Eq. (46), we obtain a new symbolic equation for the mean wave field as

$$\begin{aligned} &\left\{L_0 + M L_1 - \varepsilon^2 \langle X_0 L_0^{-1} X_0 \rangle \right. \\ &\left. \quad - \varepsilon^2 M \left[\langle X_1 L_0^{-1} X_0 \rangle + \langle X_0 L_0^{-1} X_1 \rangle - \langle X_0 L_0^{-1} L_1 L_0^{-1} X_0 \rangle\right]\right\} \langle u \rangle = 0 \end{aligned} \tag{50}$$

To obtain the mean wave equation, it is necessary to evaluate the terms involving the inverse Helmholtz operator L_0^{-1}. The inverse operator L_0^{-1} may be represented by an integral operator of Green's function defined by

$$L_0 G(\xi, \xi') \equiv \delta(\xi - \xi') \tag{51}$$

where $\delta(\xi)$ is the Dirac delta function. Green's function for the Helmholtz equation in cylindrical coordinates[17,18] is given by

$$G(\xi, \xi') = \sum_{i \neq n} \frac{\varphi_i(\xi)\varphi_i(\xi')}{k_n^2 - k_i^2} \tag{52}$$

where φ_i are the orthonormal eigenfunctions for $(\nabla^2 + k_i^2)\varphi_i = 0$. As seen from the summation, which excludes the eigenmode φ_n, a role of Green's function is to resolve the particular solution whose functional form can be decomposed into a sum of the eigenmodes that differ from the homogeneous solution φ_n. In terms of Green's function, L_0^{-1} becomes

$$L_0^{-1} f(\xi) \equiv \int_{\mathcal{V}} G(\xi, \xi') f(\xi')\, \mathrm{d}\xi'$$

Then a representative term in Eq. (50) involving L_0^{-1} becomes

$$\left\langle X_0 L_0^{-1} X_0 \right\rangle \langle u(\xi) \rangle = \left\langle X_0(\xi) \int_{\mathcal{V}} G(\xi, \xi') X_0(\xi') \langle u(\xi') \rangle \mathrm{d}\xi' \right\rangle \tag{53}$$

The other functional operations involving L_0^{-1} can be evaluated in the same manner.

B. Mean Wave Equation

First, we evaluate the second-order moments of the random operators. Considering $\langle X_0 L_0^{-1} X_0 \rangle$, it is found to be

$$\begin{aligned} \langle X_0 L_0^{-1} X_0 \rangle \langle \psi(\xi) \rangle &= k_n^4 \left\langle \varrho_1(\xi) \int_{\mathcal{V}} G(\xi, \xi') \varrho_1(\xi') \langle \psi(\xi') \rangle \, \mathrm{d}\xi' \right\rangle \\ &= k_n^4 \int_{\mathcal{V}} \langle \varrho_1(\xi) \varrho_1(\xi') \rangle G(\xi, \xi') \langle \psi(\xi') \rangle \, \mathrm{d}\xi' \end{aligned} \tag{54}$$

which brings in two-point correlation of two scalar fluctuations by using commutativities of the $X_0(\xi)$ and average operators with the integral in the ξ' coordinates. Since the turbulent length scale ℓ_t is much shorter than the typical wave length, i.e., $k_n \ell_t \ll d$, the correlation $\langle \varrho_1(\xi) \varrho_1(\xi') \rangle$ behaves like a delta function. Green's function $G(\xi, \xi')$ and the mean wave $\langle \psi(\xi') \rangle$ can then be approximated by their values at $\xi' = \xi$. We find $\langle X_0 L_0^{-1} X_0 \rangle$ as

$$\left\langle X_0 L_0^{-1} X_0 \right\rangle \langle \psi(\xi) \rangle = k_n^4 G_n(\xi) \langle \psi(\xi) \rangle \int_{\mathcal{V}} \langle \varrho_1(\xi) \varrho_1(\xi') \rangle \, \mathrm{d}\xi' \tag{55}$$

where the modified Green's function $G_n(\boldsymbol{\xi})$ is

$$G_n(\boldsymbol{\xi}) = G(\boldsymbol{\xi}, \boldsymbol{\xi}) = \sum_{i \neq n} \frac{\varphi_i^2(\boldsymbol{\xi})}{k_n^2 - k_i^2} \tag{56}$$

To complete the averaging procedure, the two-point correlation is introduced as a function of the distance vector $\boldsymbol{r} = \boldsymbol{\xi}' - \boldsymbol{\xi}$,

$$\langle \varrho_1(\boldsymbol{\xi}) \varrho_1(\boldsymbol{\xi}') \rangle = \langle \varrho_1^2 \rangle C(\boldsymbol{r}) \tag{57}$$

where $\langle \varrho_1^2 \rangle$ is the correlation value at $\boldsymbol{r} = 0$ and measures the intensity of the fluctuations. The functional form of the correlation function for two arbitrary scalar fluctuations may be approximated as

$$C(\boldsymbol{r}) = \exp\left(-r_a^2/\ell_a^2 - r_b^2/\ell_b^2\right) \tag{58}$$

where r_a and r_b denote the nondimensional longitudinal and transverse distance between two points, respectively, and where ℓ_a and ℓ_b are their corresponding nondimensional correlation distances normalized by the chamber diameter d. If the turbulence is isotropic, then the correlation distances ℓ_a and ℓ_b are the same and can be denoted by ℓ_c. Also, if the turbulence is homogeneous, the intensity $\langle \varrho_1^2 \rangle$ is not a function of $\boldsymbol{\xi}$. Since these approximations affect the algebraic results by only a small degree, the turbulence is assumed to be homogeneous and isotropic in order to simplify the algebra. Then the two-point correlation for the density fluctuations where $r \equiv |\boldsymbol{\xi} - \boldsymbol{\xi}'|$ is expressed by

$$\langle \varrho_1(\boldsymbol{\xi}) \varrho_1(\boldsymbol{\xi}') \rangle = \langle \varrho_1^2 \rangle \exp\left(-r^2/\ell_c^2\right) \tag{59}$$

The volumetric integration of the correlation is found to be

$$\begin{aligned} \int_{\mathcal{V}} \langle \varrho_1(\boldsymbol{\xi}) \varrho_1(\boldsymbol{\xi}') \rangle \, \mathrm{d}\boldsymbol{\xi}' &= \langle \varrho_1^2 \rangle \int_0^\infty \exp\left(-r^2/\ell_c^2\right) 4\pi r^2 \, \mathrm{d}r \\ &= \langle \varrho_1^2 \rangle \sqrt{2\pi^3} \ell_c^3 \end{aligned} \tag{60}$$

The final expression for $\langle X_0 L_0^{-1} X_0 \rangle$ then becomes

$$\langle X_0 L_0^{-1} X_0 \rangle \langle \psi(\boldsymbol{\xi}) \rangle = \sqrt{2\pi^3} \ell_c^3 k_n^4 \langle \varrho_1^2 \rangle G_n(\boldsymbol{\xi}) \langle \psi(\boldsymbol{\xi}) \rangle \tag{61}$$

A similar procedure can be applied to yield

$$\begin{aligned} \langle X_0 L_0^{-1} X_1 \rangle \langle \psi(\boldsymbol{\xi}) \rangle &= i\sqrt{2\pi^3} \ell_c^3 k_n^3 G_n(\boldsymbol{\xi}) \left[\langle \varrho_1 R_{p1} \rangle + 2\langle \varrho_1^2 \rangle \frac{\mathrm{d}}{\mathrm{d}\zeta} \right] \langle \psi(\boldsymbol{\xi}) \rangle \\ \langle X_0 L_0^{-1} X_1 \rangle \langle \psi(\boldsymbol{\xi}) \rangle &= i\sqrt{2\pi^3} \ell_c^3 k_n^3 \langle \psi(\boldsymbol{\xi}) \rangle \left[\langle \varrho_1 R_{p1} \rangle + 2\langle \varrho_1^2 \rangle \frac{\mathrm{d}}{\mathrm{d}\zeta} \right] G_n(\boldsymbol{\xi}) \end{aligned} \tag{62}$$

where the correlation function $C(\boldsymbol{r})$ is identical for each of the two-point scalar correlations because the fluctuations arise from a single random process, i.e., the turbulent flame. The second-order moments associated with the random vector fields $\boldsymbol{R}_{v1}$ and $(\mathrm{d}/\mathrm{d}\zeta)\nabla\varrho_1$ vanish, because their correlations with ϱ_1 are first-order tensors that are in general odd functions with zero volumetric integrals. Finally, the last second-order moment $\langle X_0 L_0^{-1} L_1 L_0^{-1} X_0 \rangle$ is not important in linear analyses because the contributions that can be decomposed by φ_n are filtered out after a successive application of Green's function. This term thus cannot affect the mean wave $\langle \psi \rangle$ that is extremely close to the eigenmode φ_n, although it may be important in nonlinear analyses.

By collecting the terms in Eqs. (61) and (62), the equation for the mean stochastic wave field is found to be

$$
\begin{aligned}
[\nabla^2 + k^2]\langle \psi(\boldsymbol{\xi}) \rangle &= iM\left[k_n R_{p0} - i\boldsymbol{R}_{v0}\cdot\nabla - 2k_n \frac{\mathrm{d}}{\mathrm{d}\zeta}\right]\langle \psi(\boldsymbol{\xi})\rangle \\
&+ \varepsilon^2 \sqrt{2\pi^3}\ell_c^3 k_n^4 \langle \varrho_1^2 \rangle G_n(\boldsymbol{\xi})\langle \psi(\boldsymbol{\xi})\rangle \\
&+ i\varepsilon^2 M\sqrt{8\pi^3}\ell_c^3 k_n^3 \left[\langle \varrho_1 R_{p1}\rangle + \langle \varrho_1^2 \rangle \frac{\mathrm{d}}{\mathrm{d}\zeta}\right]\{G_n(\boldsymbol{\xi})\langle \psi(\boldsymbol{\xi})\rangle\}
\end{aligned} \tag{63}
$$

Here the term of order of ε^2 describes the effective variation of the acoustic frequency arising from the random density profile, whereas the terms of order of $\varepsilon^2 M$ give the correction for the growth rate. Since the mean linear-growth rate is of order of M, the turbulent corrections to the acoustic frequency and linear-growth rate are both of order of ε^2. Taking the average of Eq. (32) gives the appropriate boundary condition as

$$
\boldsymbol{n}\cdot\nabla\langle \psi \rangle = 0 \qquad \text{for} \quad \boldsymbol{\xi} \in B \tag{64}
$$

C. Dispersion Relation for a Cylindrical Chamber

To calculate the wave number k, we consider a mean wave that is extremely close to an eigenmode φ_n. Now multiply Eq. (63) by φ_n and the Helmholtz equation $(\nabla^2 + k_n^2)\varphi_n = 0$ by $\langle \psi \rangle$, subtract the results, and integrate over the chamber to obtain

$$
\begin{aligned}
&\int_{\mathcal{V}} \{\varphi_n \nabla^2 \langle \psi \rangle - \langle \psi \rangle \nabla^2 \varphi_n\}\,\mathrm{d}\boldsymbol{\xi} + (k^2 - k_n^2)\int_{\mathcal{V}} \langle \psi \rangle \varphi_n \,\mathrm{d}\boldsymbol{\xi} \\
&= iM\int_{\mathcal{V}}\left[k_n R_{p0} - i\boldsymbol{R}_{v0}\cdot\nabla - 2k_n\frac{\mathrm{d}}{\mathrm{d}\zeta}\right]\langle \psi(\boldsymbol{\xi})\rangle \varphi_n\,\mathrm{d}\boldsymbol{\xi} \\
&+ \varepsilon^2\sqrt{2\pi^3}\ell_c^3 k_n^4 \int_{\mathcal{V}} \langle \varrho_1^2\rangle G_n(\boldsymbol{\xi})\langle \psi(\boldsymbol{\xi})\rangle \varphi_n\,\mathrm{d}\boldsymbol{\xi} \\
&+ i2\varepsilon^2 M\sqrt{2\pi^3}\ell_c^3 k_n^3 \int_{\mathcal{V}}\left[\langle \varrho_1 R_{p1}\rangle + \langle \varrho_1^2\rangle\frac{\mathrm{d}}{\mathrm{d}\zeta}\right]\{G_n(\boldsymbol{\xi})\langle \psi(\boldsymbol{\xi})\rangle\}\varphi_n\,\mathrm{d}\boldsymbol{\xi}
\end{aligned} \tag{65}
$$

Green's integral theorem shows that the first integral of Eq. (65) vanishes. Integrating the leftover terms in Eq. (65), the dispersion relation is obtained, namely,

$$
\begin{aligned}
k &= k_n\left\{1 + \frac{1}{2}\varepsilon^2\ell_c^3\,\Xi\langle \varrho_1^2\rangle\right\} + \frac{M}{2k_n}\int_{\mathcal{V}} (\boldsymbol{R}_{v0}\cdot\nabla\varphi_n)\varphi_n\,\mathrm{d}\boldsymbol{\xi} \\
&+ iM\left\{\frac{1}{2}\int_{\mathcal{V}} R_{p0}\varphi_n^2\,\mathrm{d}\boldsymbol{\xi} + \varepsilon^2\ell_c^3\,\Xi\langle \varrho_1 R_{p1}\rangle\right\}
\end{aligned} \tag{66}
$$

where the factor Ξ, of order unity, is

$$\Xi = \sqrt{2\pi^3} k_n^2 \int_{\mathcal{V}} G_n(\boldsymbol{\xi}) \varphi_n^2(\boldsymbol{\xi}) \, \mathrm{d}\boldsymbol{\xi} \tag{67}$$

The first term on the right-hand side of Eq. (66) is the main contribution to the real part of the wave number k which corresponds directly to the acoustic frequency by Eq. (29). The last term carrying iM is usually the main amplification effect, whereas the second term comes from the acoustic velocity response. Since $\boldsymbol{R}_{v0}$ is a complex vector function, it is not clear whether its dominant consequence will be amplification or refraction.

Whether turbulence will increase or decrease the linear-growth rate and acoustic frequency depends on the signs of $\langle \varrho_1^2 \rangle$, $\langle \varrho_1 R_{p1} \rangle$ and $G_n(\boldsymbol{\xi})$. Whereas $\langle \varrho_1^2 \rangle$ is positive definite, $\langle \varrho_1 R_{p1} \rangle$ tends to be negative. The acoustic pressure response occurs mainly in relatively thin reactive layers, and its real part is generally positive because reaction rates increase with pressure. Since the density is lower in the reaction zone, the correlation $\langle \varrho_1 R_{p1} \rangle$ is likely to have a negative real part. The sign of the factor Ξ will be negative for a few lowest eigenmodes. Putting these results into Eq. (66), it can be seen that turbulence tends to increase the amplification rate and decrease the acoustic frequency for lower eigenmodes, and otherwise for higher eigenmodes.

The turbulent corrections to the acoustic frequency and amplification rate are both of order of $\varepsilon^2 \ell_c^3$. Although the corrections have a fifth-order dependence on the small parameters ε and ℓ_c, their actual values can be much larger. First, a typical value for ε is about one-half, since temperature fluctuations in the turbulent flame are sometimes as large as the mean temperature. If the average distance between the injectors is taken as the characteristic turbulent length, a typical value of ℓ_c is of order 10^{-1}. On the other hand, the actual value of Ξ is about 10^2, because k_n^2 is usually larger than 10, and an additional factor 10 comes from $\sqrt{2\pi^3}$. Combining all of these numbers, a realistic turbulence correction can be a few percent of the mean value. Moreover, the turbulence correction increases with the cubic power of ℓ_c, so that it can be much more important in flows with larger turbulent length scales. It is also worthy of note that the dispersion relation in Eq. (66) is still valid even if ε is a value of order of unity, because the mean equation obtained from the local-independence hypothesis is identical to that obtained from the smooth perturbation.

A few percent correction of the acoustic frequency may not be considered of great importance; however, such a correction to the linear-growth rate is found to be significant in some circumstances. Acoustic instabilities in rocket engines can be seen in general as a Hopf bifurcation, which occurs when the imaginary part of the wave number begins to be positive. For systems that are far from the onset condition of Hopf bifurcation, effects of the turbulence will be qualitatively insignificant. Near the critical condition of Hopf bifurcation in which the mean linear-growth rate is of the same order as the turbulent correction or smaller, however, the turbulence correction can have a dominant amplification contribution. It is the stochastic acoustic pressure evolution regime in which Clavin et al.[16] found that, combined with a subcritical Hopf bifurcation, evolution of the acoustic pressure is qualitatively different from what has been obtained from deterministic analyses.

IV. Other Effects of Turbulence on Acoustic Instabilities

Turbulence effects on acoustic instabilities in rocket engines manifest themselves in a variety of ways, such as the spatial and temporal randomnesses, and engine variation of test results. Up to now, this chapter has been devoted to describing the averaging technique that is capable of treating spatial randomness. Other consequences of turbulence on acoustic instabilities are briefly discussed in order to help the readers understand the current status of development concerning acoustic instabilities in turbulent flows.

In general, turbulent flows in rocket engines are statistically stationary but only on time scales much longer than the characteristic turbulence time. The linear-growth rate is expected to fluctuate about its mean value with a typical fluctuation time corresponding to the characteristic turbulence time. The spatial averaging presented here is still valid when the characteristic turbulence time is longer than the characteristic acoustic time. Considering an acoustic mode φ_n, the time-dependent mean wave can now be expressed as

$$\langle\psi\rangle = A_n(\tau')\varphi_n(\xi)\exp(-ik_n\tau) \tag{68}$$

where A_n is the acoustic pressure amplitude and the new temporal coordinate τ' is defined by

$$\tau' = M\tau \tag{69}$$

where τ' is normalized by the characteristic instability growth time that is equivalent to the characteristic flow time. Nonlinear evolution of the acoustic pressure amplitude is described by a Landau–Stewart equation, namely,

$$\frac{\mathrm{d}A_n}{\mathrm{d}\tau'} = (\lambda + \eta)A_n \pm A_n^3 + \text{higher order terms} \tag{70}$$

where the deterministic and stochastic linear-growth rates in the τ' coordinate, i.e., λ and η, can be, respectively, given by

$$\lambda = \frac{1}{2}\mathrm{Im}\left\{\int_{\mathcal{V}}(\boldsymbol{R}_{v0}\cdot\nabla\varphi_n + i R_{p0}\,\varphi_n)\varphi_n\,\mathrm{d}\xi\right\}, \qquad \eta = \mathrm{Im}\left\{i\varepsilon^2\ell_c^3\Xi\langle\varrho_1 R_{p1}\rangle\right\} \tag{71}$$

This nonlinear amplitude evolution equation is a single-mode approximation of general acoustic amplitude evolution equation.

Comparing the characteristic turbulence time with the characteristic instability growth time, Clavin et al.[16] identified two distinct interaction regimes, depending on whether the turbulence time is or is not much smaller than the characteristic instability growth time. Larger instabilities are associated with the second of these classes in which evolution of the acoustic pressure during each fluctuation exhibits a deterministic behavior. The transition behavior in the first class involves random walks of the stochastic linear-growth rate η, requiring a statistical description that was performed by approximating the stochastic linear-growth rate by a white noise. A key feature of Eq. (70) is that influences of the stochasticity on the acoustics appear as multiplicative noise rather than additive noise.

In a white-noise approximation, the Stratonovich rather than Ito interpretation of the stochastic integral must be employed, and the Markov property of the white-noise process then enables the evolution of the probability density of the pressure amplitude to be described by a Fokker–Planck equation. The stationary solutions of the Fokker–Planck equation lead to the results shown in Fig. 1, which demonstrates that the transition behavior depends on the character of the bifurcation. The results shown in Fig. 1 were highlighted by the bimodal behavior seen in the metastable region of subcritical bifurcation, thereby suggesting that resurging patterns of the pressure amplitude signal may occur near the transition regime of subcritical Hopf bifurcation. In other cases, the probabilities of the acoustic pressure amplitude are unimodal, and the most probable pressure amplitudes are close to the deterministic pressure amplitudes.

Results of the analysis can be related to engine variation of acoustic instability characteristics. Suppose that we run a series of rocket engine tests for different chambers which are all built in the same way and working under the same condition. Some of them may exhibit strong instability which could lead to the destruction of the engines, whereas others maintain only a weak level of acoustic instability. In such cases, a statistical description is necessary to investigate the chance of the strong instability occurring. Once random evolution of the acoustic instability becomes stationary, the resulting statistics are equivalent to that measured in ensemble average sense due to the ergodicity. Then the bimodal probability density function could explain strong engine variations of test results.

Similar analyses were performed by different authors. Culick et al.[19] wrote a two-mode approximation different from Eq. (70) and identified a corresponding

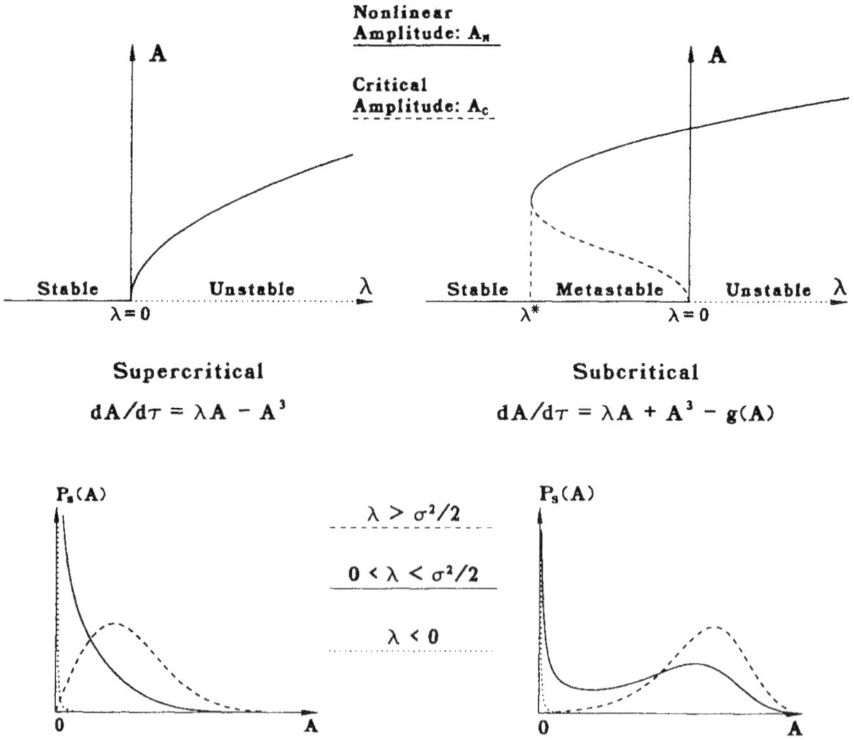

Fig. 1 Supercritical and subcritical bifurcations, illustrating the resulting unimodal and bimodal probability-density functions for the pressure amplitude.

Fokker–Plank equation. They reported results of a Monte Carlo simulation that exhibited only unimodal probability density functions. Because no subcritical bifurcation case was considered, they were not able to obtain the bimodal probability density functions. In an analysis of Roudakov,[20] a bimodal probability density function associated with subcritical bifurcation was obtained from an electrical analog. Although a number of his observations are similar to those of Clavin et al.,[16] there is a fundamental difference in that Roudakov employed an additive (rather than multiplicative) noise arising from random turbulent pressure fluctuations. Since the additive noise associated with turbulent pressure fluctuations is smaller than the multiplicative noise coming from the noisy linear-growth rate by order M, his results may not represent the main route to the bimodal probability density function.

Another promising avenue to study nondeterministic behaviors of acoustic instabilities[3,16,19] is chaotic transition in dynamical systems, which developed considerably during the past two decades. As is well known, nonlinear terms of relatively simple deterministic equations can lead to chaotic evolutions, and turbulence in nonreacting flows is considered to be an example of such transitions. The noisy behavior is here associated with a strange attractor (which can occur already in low-dimensional systems) that introduces, as a consequence, a strong sensitivity to initial conditions. Thus far there is no conclusive evidence that acoustics in rocket engines exhibit chaotic behaviors, although Jahnke and Culick[21] reported a Poincaré intersection map suggesting the possible existence of a toroidal attractor arising from a quasiperiodic bifurcation. If chaotic acoustic oscillations ever occur in rocket engines, the resulting dynamic behaviors can provide different routes to complicated evolution patterns of acoustic instabilities. Since chaos can occur usually in sufficiently nonlinear systems, bimodal probability density functions might be produced without involving the metastability of the subcritical Hopf bifurcation, the necessary condition from which Clavin et al.[16] obtained the bimodal probability density functions. Therefore, a more extensive and systematic search for chaotic acoustic oscillations is warranted.

Until now, we have discussed the turbulence effects on the dynamic behaviors of acoustic pressure amplitude evolution. However, the most important turbulence effect on acoustic instability lies in combustion by which the actual amplification mechanisms occur. Difficulties in including turbulent combustion effects come from the fact that no general theory describing such flows is currently available. To make progress, it is therefore necessary first to limit attention to specific types of combustion and to introduce simplifications of the combustion models. One of the simplified models is the reaction-sheet model[9–11] of turbulent combustion, in which the turbulent flame is considered to be an ensemble of laminar flamelets. Provided that the statistics regarding distributions of the elementary laminar flamelets throughout the chamber and their response functions to acoustic excitation are known, the linear growth rates can be calculated from the spatial averaging procedure demonstrated in this chapter. Although a few models[9] exist that enable us to describe turbulent combustion statistically, in many respects the acoustic response functions of laminar flamelets are still unknown and deserve more attention.

Pioneering work concerning laminar-flame acoustic response was performed by Strahle.[4–7] However, all of these works postulated infinite rate chemistry, thereby precluding study of the interaction of finite-rate chemical kinetics of the flame with the acoustic waves. Since the dominant nonlinear term in combustion processes arises from the chemical reaction, neglecting finite rate chemical-reaction effects may lead to significant underestimation of amplification rates under some conditions. In earlier attempts to address acoustic-chemistry interaction, applications of activation-energy asymptotics (AEA)[22] can be found in the works of McIntosh[23] and Clavin et al.[24] in which the stabilities of one-dimensional premixed flames were studied. Recently, Kim and Williams[25] and Im et al.[26] extended AEA to problems related to amplification mechanisms in rocket engines. In their analyses, particular attention was focused on near-extinction flames, a condition that can be caused by strong turbulent fluctuations of the velocity field. The amplification rates near extinction are found to be larger by an order of magnitude than those near equilibrium, thereby once again suggesting the significant underestimation of the amplification rate obtained from equilibrium flames.

Analytical improvements for AEA have recently been developed for alkane-air flames[27–29] and for hydrogen-air flames[30] by use of reduced chemical-reaction models systematically deduced from the detailed chemistry, followed by application of the methods of rate-ratio asymptotics (RRA). With sufficient effort, these methods can be applied to analyses of acoustic amplification. Since phenomenological chemistry parameters, such as the overall activation energy and the overall frequency factor, are no longer needed in these approaches, the RRA methods may provide more fundamentally based, quantitative predictions for acoustic amplification mechanisms.

V. Concluding Remarks

The present chapter deals with effects of spatial inhomogeneities in the acoustic medium associated with two-phase flow and with turbulence. Attention is restricted to a particular class of interaction in which the characteristic time scale of turbulence is asymptotically large compared with that of acoustic waves, so that the acoustic medium can be considered to be a quasistationary random medium. In addition, it is assumed that the acoustic wavelength is long compared with any scale of inhomogeneity within the chamber, motivating a homogenized description, in which averages of the state variables become appropriate. On the basis of these assumptions, the Navier–Stokes equations for two-phase flows were reduced to a nonhomogeneous stochastic Helmholtz equation. Source terms of the equation were identified as arising mainly from homogeneous chemical reaction and from the resulting spatial variations of molecular properties such as density.

To handle the stochastic wave equation, the method of smooth perturbation was used and resulted in a Helmholtz equation for an equivalent deterministic acoustic medium. With use made of Green's integral theorem, the dispersion relation for transverse acoustic modes in cylindrical chambers was obtained in the form $k = k_n\{1 + \varepsilon'\langle \varrho_1^2\rangle/2\} + iM\{\langle R_{p0}\rangle/2 + \varepsilon'\langle \varrho_1 R_{p1}\rangle\}$ where $\langle \varrho_1^2\rangle$, $\langle R_{p0}\rangle$, and $\langle \varrho_1 R_{p1}\rangle$ are calculated from appropriate averages of the source terms. In particular, the $\varepsilon'\langle \varrho_1 R_{p1}\rangle$ term corresponds to the main stochastic contributions to the linear-growth rate arising from the spatial correlation of the density with the pressure response function. Compared with the deterministic contributions to the linear-growth (or damping) rate and the acoustic frequency, the ratios of the stochastic contributions, ε', are both found to be of the order of $\varepsilon^2\ell_c^3$ where ε is the intensity of the temperature fluctuations and ℓ_c is the ratio of the characteristic length scale for turbulence to the chamber diameter, thus implying that most of the stochastic contributions comes from the larger turbulent eddies. The stochastic effects are expected to be significant in the transition regime in which the deterministic growth rate is close enough to zero for transition between stable and unstable domains to occur reasonably frequently by stochastic variations of the growth rate.

A number of other aspects associated with turbulence are discussed in the preceding section regarding dynamic behaviors of nonlinear acoustic instability. Recent analyses by Clavin et al.[16] and Culick et al.[19] and Jahnke and Culick[21] suggested that extremely complicated evolution patterns of acoustic instability can occur if the acoustic oscillatory systems are situated in a metastable region of subcritical Hopf bifurcation or chaotic transition regime. However, the most important and challenging aspect of turbulence on acoustic instability is the turbulent combustion. Since turbulence can enhance nonequilibrium chemistry, the dominant nonlinear process of flows in rocket engines, the presence of strong turbulence may lead to an order-of-magnitude increase in the overall amplification rate. In order to predict the turbulence effects on amplification mechanisms, emphasis must be placed on two aspects of turbulent combustion: 1) what type of flame element produces the largest flame response to acoustic oscillations and 2) how many of these flame elements exit in the rocket engine. The first problem is currently being investigated by a model chemical mechanism employing AEA.[25,26] However, detailed chemical kinetic mechanisms, perhaps simplified by reduced chemical kinetics, need to be introduced to obtain quantitative predictions of amplification rates coming from elementary flames. The second problem can be addressed by statistical description of turbulent flames in rocket engines, which is in turn dependent on progress of modeling for general turbulent flames. In these studies, special emphasis on multiphase effects is warranted since they introduce the most pronounced difference in statistical description of turbulent flames in rocket engines.

Appendix: Molecular Properties in Multiphase Mixture

In this Appendix, we discuss a way to estimate the molecular properties of a multiphase system consisting of a number of liquid phases dispersed in an ideal-gas mixture. To consider a_f, we first write the overall density as

$$\rho = \frac{m}{V} = \sum_j m_j \Big/ \sum_j V_j \tag{A1}$$

where m_j is the mass of phase j in a small control volume V and V_j is the volume of phase j in that control volume. Since the differential operator acts only on the V_j,

$$\frac{1}{a_f^2} = \left(\frac{\partial \rho}{\partial p}\right)_{s,Y_{ji}} = -\sum_j \frac{m}{V^2}\left(\frac{\partial V_j}{\partial p}\right)_{s,Y_{ji}} = \sum_j \frac{\alpha_j^2}{Y_j}\frac{1}{a_{f,j}^2} \tag{A2}$$

where the frozen sound speed of phase j is

$$\frac{1}{a_{f,j}^2} = -\frac{m_j}{V_j^2}\left(\frac{\partial V_j}{\partial p}\right)_{s,Y_{ji}} \tag{A3}$$

In liquid-propellant rocket engines, the volume fraction of each liquid phase is usually small, and the sound speed is higher in the liquid phase than in the gas phase, so that the overall sound speed can be approximated as being determined by that of the gas phase only. Since the gas phase is assumed to be an ideal-gas mixture, the overall frozen sound speed is expressed as

$$a_f^2 \approx \frac{Y_g a_{f,g}^2}{\alpha_g^2} \approx \frac{Y_g \gamma_g R T_g}{\alpha_g^2} \tag{A4}$$

where R is the gas constant per unit mass. Then the adiabatic compressibility K_f is correspondingly approximated as

$$K_f = \frac{1}{\rho a_f^2} \approx \frac{Y_g}{\alpha_g} \frac{1}{\rho_g a_f^2} \approx \frac{\alpha_g}{\gamma_g p_0} \tag{A5}$$

which shows that K_f is proportional mainly to the volume fraction of the gas phase.

Similarly, β_f is expressible as

$$\beta_f = \sum_j \alpha_j \beta_{f,j} \approx \frac{\alpha_g}{T_g} \tag{A6}$$

On the other hand, $c_{p,f}$ is given by

$$c_{p,f} = \sum_j Y_j c_{p,f,j} \tag{A7}$$

which shows that $c_{p,f}$ is weighted by the mass fraction rather than by the volume fraction, so that contributions from liquid phases are not generally negligible.

Acknowledgments

This research has been supported by the U.S. Air Force Office of Scientific Research through AFOSR Grant F49620-94-1-0166. The author is indebted to F. A. Williams for his guidance during the course of this work and also would like to thank Paul Clavin for his clarification on the stochastic effects on acoustics and Vigor Yang for many helpful suggestions on the first manuscript of this chapter.

References

[1]Harrje, D. J., and Reardon, F. H. (eds.), *Liquid Propellant Rocket Instability,* NASA SP-194, 1972.

[2]Culick, F. E. C., "Combustion Instabilities in Liquid-Fueled Propulsion Systems—An Overview," AGARD 72B PEP Meetings, Oct. 1988.

[3]Culick, F. E. C., "Some Recent Results for Nonlinear Acoustics in Combustion Chambers," *AIAA Journal,* Vol. 32, No. 1, 1994, pp. 146–169.

[4]Strahle, W. C., "Unsteady Laminar Jet Flame at Large Frequencies of Oscillation," *AIAA Journal,* Vol. 3, No. 6, 1965, pp. 957–960.

[5]Strahle, W. C., "Unsteady Reacting Boundary Layer on a Vaporizing Flat Plate," *AIAA Journal,* Vol. 3, No. 6, 1965, pp. 1195–1198.

[6]Strahle, W. C., "Periodic Solutions to Convective Droplet Burning Problem: The Stagnation Point," *Tenth Symposium (International) on Combustion,* Combustion Inst., Pittsburgh, PA, 1965, pp. 1315–1325.

[7]Strahle, W. C., "High-Frequency Behavior of the Laminar Jet Flame Subjected to Transverse Sound Waves," *Eleventh Symposium (International) on Combustion,* Combustion Inst., Pittsburgh, PA, 1967, pp. 747–754.

[8]Crocco, L., and Cheng, S. I., *Theory of Combustion Instability in Liquid Propellant Rocket Motors,* AGARD Monograph No. 8, Butterworths, London, 1956.

[9]Libby, P. A., and Williams, F. A. (eds.), *Turbulent Reacting Flows,* Academic, San Diego, CA, 1994.

[10]Williams, F. A., *Combustion Theory,* Addison-Wesley, Menlo Park, CA, 1985.

[11]Peters, N., "Laminar Flamelet Concepts in Turbulent Combustion," *Twenty-First Symposium (International) on Combustion,* Combustion Inst., Pittsburgh, PA, 1986, pp. 1231–1250.

[12]Karil, F. C., and Keller, J. B., "Elastic, Electromagnetic, and Other Waves in a Random Medium," *Journal of Mathematical Physics,* Vol. 5, No. 4, 1964, pp. 537–547.

[13]Chow, P. L., "Thermoelastic Wave Propagation in a Random Medium and Some Related Problems," *International Journal of Engineering Science,* Vol. 11, 1973, pp. 953–971.

[14]Sobczyk, K., *Stochastic Wave Propagation,* Elsevier, Amsterdam, 1985.

[15]Margolis, S. B., "Nonlinear Stability of Combustion-Driven Acoustic Oscillations in Resonance Tubes," *Journal of Fluid Mechanics,* Vol. 253, 1993, pp. 67–103.

[16]Clavin, P., Kim, J. S., and Williams, F. A., "Turbulence-Induced Noise Effects on High-Frequency Combustion Instabilities," *Combustion Science and Technology,* Vol. 96, 1994, pp. 61–84.

[17]Stakgold, I., *Boundary Value Problems of Mathematical Physics,* Macmillan, New York, 1968.

[18]Mitchell C. E., "Stability of Combustors with Partial Length Acoustic Liners," *Combustion Science and Technology,* Vol. 6, 1972, pp. 61–70.

[19]Culick, F. E. C., Paparizos, L., Sterling, J., and Burnley, V., "Combustion Noise and Combustion Instabilities in Propulsion Systems," *Proceedings of the AGARD Conference on Combat Aircraft Noise,* AGARD CP 512, Paris, 1992.

[20]Roudakov, A. S., "Combustion Instability in LRE," personal communication, June 1993.

[21]Jahnke, C. C., and Culick, F. E. C., "An Application of Dynamical Systems Theory to Nonlinear Combustion Instabilities," AIAA Paper 93-0114, Jan. 1993.

[22]Liñán, A., "The Asymptotic Structure of Counterflow Diffusion Flames for Large Activation Energies," *Acta Astronautica,* Vol. 1, 1974, pp. 1007–1039.

[23]McIntosh, A. C., "Pressure Disturbances of Different Length Scales Interacting with Conventional Flames," *Combustion Science and Technology,* Vol. 75, 1991, pp. 287–309.

[24]Clavin, P., Pelcé, P., and He, L. T., "One-Dimensional Vibratory Instability of Planar Flames Propagating in Tubes," *Journal of Fluid Mechanics,* Vol. 216, 1990, pp. 299–322.

[25]Kim, J. S., and Williams, F. A., "Contribution of Strained Diffusion Flames to Acoustic Pressure Response," *Combustion and Flame,* Vol. 98, No. 3, 1994, pp. 279–299.

[26]Im, H. G., Law, C. K., Kim, J. S., and Williams, F. A., "Response of Counterflow Diffusion Flames to Oscillating Strain Rates," *Combustion and Flame,* Vol. 100, No. 1, 1995, pp. 21–30.

[27]Peters, N., and Williams, F. A., "The Asymptotic Structure of Stoichiometric Methane-Air Flames," *Combustion and Flame,* Vol. 68, No. 2, 1987, pp. 185–207.

[28]Seshadri, K., and Peters, N., "Asymptotic Structure and Extinction of Methane-Air Diffusion Flames," *Combustion and Flame,* Vol. 73, No. 1, 1988, pp. 23–44.

[29]Card, J. M., "Asymptotic Analysis for the Burning of *n*-Heptane Droplets Using a 4-Step Reduced Mechanism," *Combustion and Flame,* Vol. 93, No. 4, 1993, pp. 375–390.

[30]Balakrishnan, G., Treviño, C., and Mauss, F., "The Asymptotic Structure of Hydrogen-Air Diffusion Flames," *Combustion and Flame,* Vol. 91, No. 3-4, 1992, pp. 246–256.

Chapter 17

Calculations of Combustion Response Profiles and Oscillations

Richard J. Priem* and Kevin J. Breisacher†
NASA Lewis Research Center, Cleveland, Ohio 44135

Nomenclature

A_f = injector face admittance
A_n = nozzle admittance
A_w = wall admittance
g = conversion constant
J = Bessel function
n = pressure interaction index
p = pressure
R_p = combustion response
r = radial coordinate
t = time
v = velocity
v_{liq} = velocity of injected liquid
w = local burning rate
z = axial coordinate
γ = ratio of specific heats
θ = tangential
ρ = density
τ = sensitive time lag
Φ = potential function
ω = complex frequency

Subscripts

n = spatial dimension, mode number
$0, t$ = time averaged
$1, 2, \ldots$ = harmonic component

---

*Private Consultant, Priem Consultants, Inc.
†Aerospace Engineer, Space Propulsion Technology Division.

I. Introduction

ANY review of liquid rocket combustion instability model development in the United States prior to 1972 inevitably leads to Ref. 1. This publication is a compendium of the combustion instability modeling efforts by industrial, academic, and government teams from across the nation. Their efforts were aided by the considerable financial support these groups received during the 1960s due to the perceived threat of combustion instabilities to achieving national space goals. The work of these groups also benefitted from the considerable experimental studies of combustion instability phenomena including liquid rocket combustion tests.

After 1972, financial support for combustion instability research was drastically reduced and very little effort was devoted to the subject. In the interim, major advances were made in the computational capability of computers and in numerical solution techniques for nonlinear partial differential equations. This increase in capability has permitted investigators to make calculations that were only dreams in the early 1970s. Increased computational capabilities also allowed the development of new quantitative spray measurement techniques.

A. Status as of 1972

The majority of the modeling efforts prior to 1972 used linear or nonlinear acoustics but very cursory combustion models. Modeling efforts representative of this era were made by Crocco and Cheng,[2] Crocco et al.,[3] Crocco and Mitchell,[5] Culick,[6] and Priem and Rice.[7] These models were used to gain an understanding of instabilities and guide engine design changes intended to achieve stability. In general, the differences between the chamber response obtained with linear and nonlinear acoustic models were small.[3,4] Some models for chamber acoustics included the separate effects of baffles (radial baffles or hub baffles, but not combinations of hub and radial baffles) and acoustic absorbers but not the combined effects of baffles and absorbers. With these models of chamber acoustics, designers had a greater ability to predict the need for and proper design of acoustic absorbers.

The combustion models used in the frequency-domain codes related the local instantaneous combustion rate to the local pressure using two constants. Reardon[8] did correlate time-lag parameters with engine design and operating parameters. These correlations were also used as a guide to obtaining stable engine configurations. Individual combustion processes such as injection, atomization, and vaporization were too computationally intensive to model, and in the case of atomization detailed models were not available. Movies of the combustion zone in experimental combustors[9–11] have shown the highly nonlinear nature of the combustion process during an instability. During a full-blown instability, it almost appeared as if a shock wave was traversing the combustion zone. It is not surprising that the simple two-constant combustion models could not provide a detailed understanding of instability or a quantitative evaluation of the influence of design and operating changes on stability.

Numerical studies[12,13] were conducted to examine the response of individual combustion processes (injection, atomization, vaporization, and chemical reaction) to idealized chamber acoustic waves. These acoustic waves corresponded to the resonant frequency of cylindrical chambers with no stability aids, axial velocity profiles, and ideal nozzles. The studies demonstrated that all of the individual processes were capable of driving an instability. The results of these studies were correlated to provide values for the two constants, n and τ, used in the simple acoustic models. None of the frequency-domain models of this era attempted a comprehensive model that tracked the propellant from the injector until it was completely burned. Simple one- and two-dimensional time-domain combustion instability models[14–16] were developed prior to 1972. Although these codes typically modeled injection, a mechanistic treatment of atomization was omitted. The combustion rate was usually modeled as an empirical function of pressure. None of these early time-domain codes had a "complete" combustion model. The codes were able to demonstrate the existence of a minimum disturbance amplitude necessary to initiate an instability. The wave characteristics calculated with these codes were similar to those observed from dynamic stability tests. These codes had considerable problems with numerical stability, particularly the two-dimensional codes.

B. Design Tools for the 1990s

The goal of current stability code development efforts is to produce a collection of analysis tools to allow a designer to assess the stability margin of candidate combustor designs. This collection of codes may include fast, acoustic n–τ codes for preliminary design screening. Fourier time expansion and simple time-domain codes could be utilized for stability margin evaluation and stability aid design for candidate combustor designs. Rigorous time-domain codes could be used for final design evaluation. The features of these codes are summarized in Table 1 and are discussed next.

Table 1 Three approaches to instability modeling

Acoustics: simple combustion	Time domain CFD	Fourier time expansion
Features		
Linear amplitude or mode series expansion	Time history all parameters	Fourier expansion in time
Continuity and momentum conserved	Continuity, momentum, and energy conserved	Continuity and momentum conserved
Simple two-parameter combustion model	Mechanistic combustion model	Mechanistic combustion model
Stability aids modeled	Stability aids modeled	Stability aids modeled
Advantages		
Identifies problem modes	Includes all nonlinear effects	Determines stability margin
Results obtained quickly	Direct comparison to stability tests	Simulates resulting oscillation
Disadvantages		
Poor modeling of combustion	Computationally intense	No real time history generated
Difficult to determine instability source	Different runs for each disturbance level, type	No interaction between modes

Any analysis of combustion instability in rocket engines is only as accurate as the specification of the boundary conditions (in particular, the injection and atomization rates at the injector face) for the problem. Unfortunately, the boundary conditions for stability analyses in liquid rocket engines are highly nonlinear and not very well understood. The rate and location at which injected propellant is atomized and the resulting size and velocity of the droplets is crucial input to any instability analysis. The subsequent behavior of the propellant as it burns or undergoes further breakup is critically linked to the atomization boundary condition. This boundary condition is also part of a highly nonlinear feedback loop in which the combustor gas field influences atomization and atomization influences burning and the combustor velocity profile. Accurately modeling this phenomena is the paramount challenge of rocket combustor modeling.

Typically, a shear coaxial element is modeled using correlations developed from cold-flow tests.[17,18] These correlations provide a relationship between global parameters (gas velocity, liquid flow rate, jet diameter, etc.) and drop size or atomization rate. Unfortunately, in many models, these global correlations have been applied in a subgrid fashion. The overall significance of droplet coalescence and secondary breakup phenomena has not been demonstrated. Secondary breakup is typically modeled using a spring-mass oscillator analogy or linear-surface wave growth rates even though the majority of experimental evidence suggests a violent stripping phenomena.[10] One can imagine the difficulty in trying to model the drop size produced by an impinging element subject to a high-amplitude transverse wave.

The biggest advantage of time-domain codes is that they can provide a complete time history of any parameter of interest. These time histories allow for a direct comparison with engine test results. Physical submodel development and code validation is also facilitated by use of these time histories. Interaction between different modes of oscillations is handled automatically, within the filtering limits of the finite difference method. Updating of one- and two-dimensional time-domain codes to take advantage of the increase in knowledge of the stability characteristics of various finite differencing techniques is underway.[19–23] However, most instabilities due to injector design or the wave dynamics are inherently three-dimensional. A fully three-dimensional simulation of the Space Shuttle main engine (SSME) (manifolds, baffles, absorbers, and 600 injection elements) is beyond current computational capabilities, even if atomization could be modeled adequately. Computational time remains a major drawback of the time-domain approach.

The Fourier time-expansion (FTE) approach is something of a hybrid between the numerical time-domain approach and the semianalytical acoustic-modeling approach. In the FTE approach, the thermodynamic variables and velocity components are expanded in a Fourier time series and substituted in the appropriate conservation equations. The resulting equations are differenced and solved numerically to determine the gas-phase oscillations. All of the basic combustion processes are modeled as they would be in a time-domain approach and then converted into complex responses for use with the FTE approach. Solutions are obtained at different oscillation amplitude levels for the complex frequency (observed frequency and decay rate). The oscillation amplitude at which the decay rate changes from negative to positive is the minimum disturbance amplitude required to excite that mode of instability. The oscillation amplitude at which the decay rate changes back from positive to negative is the equilibrium or limit cycle amplitude of the instability. A disadvantage of the FTE approach is that it does not permit having oscillations present with different basic frequencies, i.e., chug and a high-frequency oscillation or two transverse modes with different frequencies that beat. Another disadvantage is that no time history of flow parameters is calculated directly for comparison with experiments.

II. Theory for the Fourier Time Expansion

In this chapter, the theory for the Fourier time expansion model will be discussed and calculations made using the method will be presented for an SSME combustor. The results of the FTE approach were developed for incorporation into the high-frequency injection coupled combustion instability program (HICCIP).[24–26] The code described in Ref. 24 was restricted to using wave characteristics derived from a pseudo-three-dimensional theory. The code assumed uniform combustion response at all radial and tangential positions for a given axial position. Also, the wave model used could not account for the presence of baffles and discrete absorbers. Therefore, a three-dimensional model for the chamber gas oscillations was required for HICCIP to predict the characteristics in a typical rocket engine.[25,26]

The gas-phase oscillations calculated with the three-dimensional FTE approach are used to determine burning-rate oscillations throughout the combustor. This is an open-loop calculation where burning-rate oscillations are assumed and used to calculate gas-phase oscillations. The gas-phase oscillations are then used to calculate burning rate oscillations. If the final calculated burning-rate oscillations agreed with the assumed values, a consistent, complete solution would have been obtained. To close the loop and obtain the complete solution would require iterating the procedure described earlier. Work is continuing toward completing the closed-loop model.

The basic conservation equations used to define flow in a cylindrical chamber are as follows.

Conservation of mass:

$$\frac{\partial \rho}{\partial t} + \nabla \cdot (\rho v) = w \tag{1}$$

Conservation of momentum:

$$\frac{\partial (\rho v)}{\partial t} + \nabla \cdot \rho v v + g \nabla p = w v_{\text{liq}} \tag{2}$$

Adiabatic ideal gas:

$$\frac{p}{p_0} = \left(\frac{\rho}{\rho_0}\right)^{\gamma} \tag{3}$$

It is assumed that viscous and gravitational forces are negligible. To simplify the analysis, it is further assumed that the radial and azimuthal steady-state flow are minimal and are, therefore, zero. This avoids requiring a three-dimensional steady-state solution to determine radial and azimuthal flow and pressure profiles. As in HICCIP, all of the variables (p, ρ, v_z, v_r, v_θ, w) are expanded in a Fourier time series given by

$$F = F_0 + F_1 e^{i\omega t} + F_2 e^{2i\omega t} + \cdots + F_n e^{ni\omega t} + \cdots \tag{4}$$

where F is a dependent variable in the conservation equations.

Consistent with the preceding assumptions, ρ_0 and p_0 vary only in the axial direction, v_{r0} and $v_{\theta 0}$ are zero (no radial or tangential average flow), and v_{z0} and w_0 are variable in all three space dimensions.

In the remaining portion of this chapter, only the terms through F_1 are used. When Eq. (4) is substituted into the conservation equations the terms with the ωt dependence are the only ones with F_1 coefficients, and these terms are, therefore, not influenced by F_2 and higher harmonic terms. After determining the F_1 values, the equations could again be solved using the known F_0 and F_1 values to find the F_2 values. This process could be continued to determine as many F_n values as required to completely and rigorously define the oscillations. Solutions for the F_1 values, however, define the complex frequency of the oscillations (both the observed frequency and the damping or growth rate for a specified pressure amplitude at a single point and combustion response). Substituting the Fourier time series into the conservation equations and retaining only the terms containing F_0 and F_1 yields the following set of equations.

Continuity:

$$i\omega\rho_1 + \frac{1}{r}\frac{\partial(r\rho_0 v_{r_1})}{\partial r} + \frac{1}{r}\frac{\partial(\rho_0 v_{\theta_1})}{\partial \theta} + \frac{\partial(\rho_0 v_{z_1})}{\partial z} + \frac{\partial(\rho_1 v_{z_0})}{\partial z} = w_1 \tag{5}$$

Axial momentum:

$$i\omega\rho_0 v_{z_1} + i\omega\rho_1 v_{z_0} + 2\frac{\partial(\rho_0 v_{z_0} v_{z_1})}{\partial z} + \frac{\partial(\rho_1 v_{z_0} v_{z_0})}{\partial z} + \frac{1}{r}\frac{\partial(r\rho_0 v_{r_1} v_{z_0})}{\partial r}$$

$$+ \frac{1}{r}\frac{\partial(\rho_0 v_{\theta_1} v_{z_0})}{\partial \theta} + \gamma\frac{p_0}{\rho_0}\frac{\partial \rho_1}{\partial z} = w_1 v_{\text{liq}} \tag{6}$$

Radial momentum:

$$i\omega\rho_0 v_{r_1} + \frac{\partial(\rho_0 v_{z_0} v_{r_1})}{\partial z} + \gamma\frac{p_0}{\rho_0}\frac{\partial \rho_1}{\partial r} = 0 \tag{7}$$

Azimuthal momentum:

$$i\omega\rho_0 v_{\theta_1} + \frac{\partial(\rho_0 v_{z_0} v_{\theta_1})}{\partial z} + \gamma\frac{p_0}{r\rho_0}\frac{\partial \rho_1}{\partial \theta} = 0 \tag{8}$$

Adiabatic ideal gas:

$$p_1 = \gamma\frac{p_0}{\rho_0}\rho_1 \tag{9}$$

The boundary conditions at the chamber wall for the combustor are

$$v_{r_1} = \gamma A_w p_1 \tag{10}$$

where the wall admittance A_w is zero for hard walls and is an input value for wall absorbers.

At the centerline of the chamber ($r = 0$) all radial fluxes are zero as the area is zero.

The boundary conditions at the injector face are

$$v_{z_1} = \gamma A_f p_1 \tag{11}$$

where A_f is zero for a solid injector face and is input for absorbers on the injector face.

At the end of the combustion chamber the boundary conditions are

$$v_{z_1} = \gamma A_n \rho_1 \tag{12}$$

where A_n is the nozzle admittance. In this chapter, it is assumed to be zero (short compact ideal nozzles).

For the two baffle surfaces, the radial baffle $v_{\theta 1} = 0$, and the baffle is at an azimuthal momentum cell face; the hub baffle $v_{r1} = 0$, and the baffle is at a radial momentum cell face.

It is also possible to model the oscillations as one- or two-dimensional phenomena. This is accomplished by assuming that the oscillations in the radial and/or azimuthal directions can be specified by an analytic wave equation solution in the desired direction. With this assumption, the conservation equations in that direction are not necessarily satisfied. The solution can be obtained in terms of the potential function given as

$$\Phi = f(z)e^{i\omega t}e^{in\theta}J_r(mr) \tag{13}$$

and the variables are determined by[27]

$$\rho_1 = C_1 \frac{\partial \Phi}{\partial t} \tag{14}$$

$$v_{z_1} = C_2 \frac{\partial \Phi}{\partial z} \tag{15}$$

$$v_{r_1} = C_3 \frac{1}{r}\frac{\partial \Phi}{\partial r} \tag{16}$$

$$v_{\theta_1} = C_4 \frac{\partial \Phi}{\partial \theta} \tag{17}$$

For one cell in the radial direction, the radial derivatives in Eqs. (5–8) are given as follows:

$$\frac{1}{r}\frac{\partial}{\partial r}(r\rho_0 v_{r_1}) \equiv \frac{\rho_0}{r}\left\{\frac{\partial^2}{\partial r^2}[rJ_n(mr)] \Big/ \frac{\partial}{\partial r}[J_n(mr)]\right\} v_{r_1} \tag{18}$$

$$\frac{1}{r}\frac{\partial}{\partial r}(r\rho_0 v_{z_0} v_{r_1}) \equiv \frac{\rho_0 v_{z_0}}{r}\left\{\frac{\partial^2}{\partial r^2}[rJ_n(mr)] \Big/ \frac{\partial}{\partial r}J_n(mr)\right\} v_{r_1} \tag{19}$$

$$\gamma \frac{p_0}{\rho_0}\frac{\partial \rho_1}{\partial r} = \frac{\gamma p_0}{\rho_0}\left\{\frac{\partial}{\partial r}[rJ_n(mr)] \Big/ J_n(mr)\right\} \rho_1 \tag{20}$$

where the derivatives of the Bessel functions are determined from Bessel function tables (at a radius equal to half the chamber radius) for different Bessel function indices m and n, which describe the mode in the radial direction and the tangential mode (n). The calculations then correspond to a very thin annular ring at half the distance to the chamber wall.

Similarly for only one cell in the azimuthal direction the azimuthal derivative in Eqs. (5–8) are given as follows:

$$\frac{1}{r}\frac{\partial \rho_0 v_{\theta_1}}{\partial \theta} = i\frac{n}{r} v_{\theta_1} \tag{21}$$

$$\frac{1}{r}\frac{\partial (\rho_0 v_{z_0} v_{\theta_1})}{\partial \theta} = \frac{\rho_0 v_{z0}}{r} i n v_{\theta_1} \tag{22}$$

$$\frac{1}{r}\frac{\partial (\rho_0 \rho_1)}{\partial \theta} = \frac{\gamma \rho_0}{r} i n \rho_1 \tag{23}$$

where n specifies the number of modes in the azimuthal direction. The calculations then correspond to a very thin "pie" section cut out of the chamber.

III. Sample Calculations for the Space Shuttle Main Engines

To demonstrate the capability of the FTE approach, calculations were made for the SSME at nominal operating conditions with and without baffles and absorbers for different oscillation modes. The configuration used to represent the SSME is presented in Table 2.

A time-averaged combustion profile was used similar to that obtained with a vaporization limited model (50% combustion completed 0.058 m from the injector, 90% in 0.191 m, and 99% completion within the chamber). The time-averaged combustion profile remained constant. This is not a realistic assumption as the gas oscillations do influence the time-averaged burning rate. The influence of these oscillations will be evaluated when the FTE approach is combined with the combustion models in HICCIP.

The definition for combustion response is

$$R_p = \frac{W_1/W_0}{p_1/p_0} \tag{24}$$

The combustion response is a vector in the complex plane. The real portion is the magnitude of the nondimensional burning rate oscillation that is in phase with the pressure. The imaginary portion indicates the phasing of combustion relative to pressure. A positive imaginary response indicates that the combustion is leading the pressure oscillation, a negative response indicates combustion is lagging.

In the present work, the combustion response is assumed to be spatially invariant. The simplifications of the burning process, although not rigorous, allow a demonstration of the FTE approach and indicate how some of the combustor design variables influence stability of the SSME.

The combustion response needed to drive various waves in the SSME were calculated iteratively. For a given frequency, a combustion response was assumed

Table 2 SSME Configuration

Chamber diameter, m	0.451	Chamber pressure, MPa	19.6
Cylindrical length, m	0.383	Oxidizer flow rate, kg/s	378.2
Hub baffle length, m	0.051	Fuel flow rate, kg/s	102.2
Hub baffle diameter, m	0.225	Absorber wall gap, m	0.009
Radial baffles	5 blades	Absorber tuned mode	3T
Radial baffle length, m	0.051	Speed of sound, m/s	1579.8

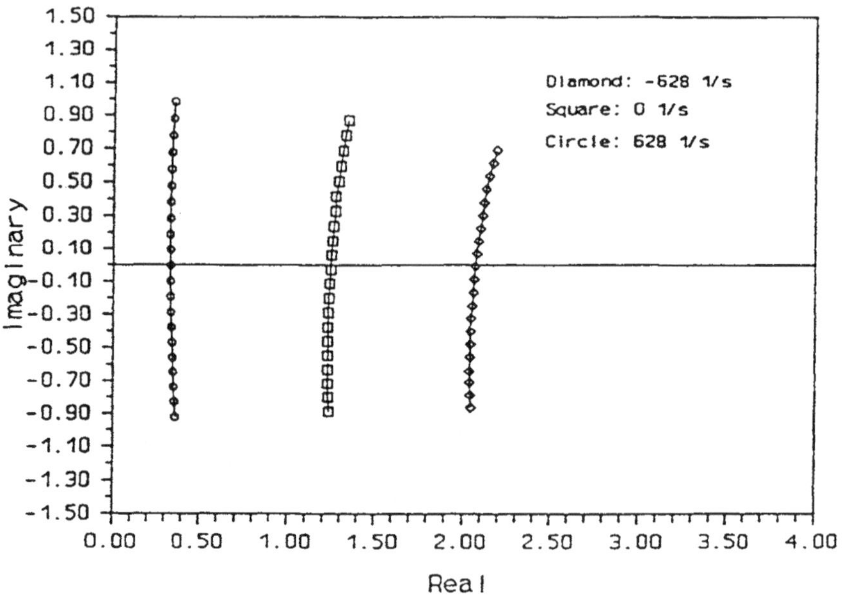

Fig. 1 SSME response curves.

and a burning rate oscillation was approximated using an analytical acoustic wave solution. This initial guess at the burning rate oscillation was used to determine oscillating pressure and velocity profiles in the combustor. With the calculated pressure profile and assumed burning rate oscillations, an average combustion response was calculated (based on the fraction of propellant burned at each location) which was then used to calculate a new burning rate oscillation. This process was repeated until the change in the average combustion response between successive iterations was minimal. With the assumptions and procedure outlined, a combustion response for any frequency can be determined. The combustion response was calculated (using a grid size of 4 radial, 10 azimuthal, and 8 axial cells) for the first tangential (1T) mode and the results are shown in Fig. 1. The complex combustion response is plotted as a function of frequency between 1860 and 2060 Hz. Three curves are shown in Fig. 1, corresponding to damping rates (imaginary portion of the complex frequency) of +628, 0, −628 /s. As the frequency is increased, the combustion response changes from combustion lagging pressure to leading pressure and crosses the in-phase (1.39 real and zero imaginary) plane at 1960 Hz with neutral damping. Normally, this is considered the natural frequency for the 1T mode in this engine. Depending on the combustion response, however, oscillations in the engine can occur at other frequencies as shown in Fig. 1. Larger combustion responses make the engine unstable; lower values result in a stable engine. Negative imaginary response lowers the frequency from the natural, and positive imaginary responses increase the frequency. Similar plots can be obtained for all of the modes possible in the chamber. The mode type is determined by the initial assumption for the pressure profile or burning rate oscillations in the chamber.

To illustrate how the FTE approach can be used to determine the influence of design parameters on stability and stability modes, calculations were made to determine the frequency and combustion response for neutral stability at the natural frequency, i.e., where the combustion response has no imaginary value and the oscillations have zero growth or decay rate. The results of these calculations for different stability aid configurations and modes are presented in Table 3.

Table 3 illustrates how the acoustic absorber improves stability in the SSME. The original absorber design is tuned to the 3T mode and increases the required combustion response of the 3T mode for neutral stability from 1.64 to 2.63, which is equivalent to going from neutral stability to a damping rate of −628 1/s in Fig. 1. This damping rate would decrease the amplitude of a 100% chamber pressure peak-to-peak wave to the allowable 10% amplitude in 3.6 ms. The damping time as required by Ref. 28 is 18 ms. Therefore, the SSME absorber design provides damping five times that required for the 3T mode.

The original absorber design provides a very small change in frequency and response for the 1T and first longitudinal (1L) modes because the absorber is tuned to a much higher frequency. The 2T mode is actually made more unstable by the absorber because it reduces the frequency and decreases the chamber response. Retuning the SSME absorber to the 1T and 1L modes by decreasing the absorber opening in the chamber from 0.009 to 0.003 m and increasing the aperture length from 0.013 to 0.025 m will produce a large increase in stability, more than enough to meet the damping requirements of Ref. 28. The increase in stability at the 1T and 1L frequencies is less in comparison to the stability of the original design at the 3T because the absorber open area has been reduced to obtain tuning at these lower frequencies.

Adding the baffle to the SSME changes the tuned frequencies and decreases the response required for neutral stability. The frequency change is 105 Hz for the 1T mode and 380 Hz for the 2T mode. The reduction in response is also greater for the 2T mode.

Table 3 Neutral stability conditions for the SSME

Combustor configuration	Mode	Frequency, Hz	Response
Unbaffled, no absorber	1L	1966	1.69
	1T	2057	1.69
	2T	3407	1.65
	3T	4686	1.64
Unbaffled, original SSME absorber	1L	1955	1.73
	1T	2045	1.63
	2T	3366	1.50
	3T	4680	2.63
Unbaffled, retuned absorber	1L	1966	1.94
	1T	2052	1.90
	2T	3403	2.51
	3T	4680	2.63
Baffled, original SSME absorber	1L	1955	1.73
	1T	1957	1.35
	2T	3023	1.08
	3T	4305	2.26

These calculations demonstrate that oscillations can occur in baffled chambers. Oscillations with either standing or spinning wave characteristics are possible. With the baffle, the wave is distorted within and near the baffles, but an acousticlike wave does exist. Downstream of the baffles, the wave is similar to the acoustic wave for a pure cylinder. Near the baffles, they become distorted and within the baffle compartment, the oscillations are predominantly axial with small variations in the radial and azimuthal directions. The decrease in response occurs because of the wave distortion produced by the baffles coupled with the steady flow profile within the chamber. The baffles reduce the loss associated with the steady flow convecting spinning oscillations out of the chamber by producing axial oscillations.

To test the capability of the FTE approach to predict observed acoustic phenomena, calculations were performed to compare with the results published by Wieber[29] and the GEMSIP program.[30] Good agreement with this data was obtained. Calculated results for the SSME were somewhat dependent on grid size, especially for the baffled chambers.[25,26]

IV. Space Shuttle Main Engine Burning Rate Oscillations

Local burning rate oscillations resulting from the gas-phase oscillations just described were calculated using the HICCIP model described in Ref. 24. These calculations were performed at the resonant frequency (zero imaginary part of the combustion response and no damping rate) for the SSME chamber and for the SSME with stability aids removed. The HICCIP model includes: 1) fuel flow oscillations, 2) oxidizer flow oscillations, 3) atomization in the recess and chamber,[17] and 4) vaporization using an onion skin model.[20] With these models, the propellant is followed from the time it enters the fuel or oxidizer dome, through the injection element, recess, and combustion chamber (as a function of time and space). Mass is always conserved throughout the combustion process. The processes are dependent on the local conditions as required by the available equations.

To make the burning rate oscillation calculations, a proper selection of the constants in the atomization models for the rate of atomization and drop size is required. The atomization rate constant was selected so that 50% of the oxidizer was atomized in the recess. This resulted in an atomization plane (the effective location where vaporization starts) at the injector face. The drop size constant was selected to produce a steady-state gas velocity profile similar to that used in the gas-phase oscillation calculations already described (99% vaporized in the chamber). This resulted in a steady-state drop size of 507 μ being formed at the injector face.

The oscillations in injector flow rates as given by a flow response for different rows in the injector are shown in Table 4. The large fuel response is due to the very low fuel-injection pressure drop (2.1 MPa). The variation in oxidizer response is due to the different lengths of the oxidizer tubes connecting the oxidizer manifold to the combustion chamber in the SSME injector. Oxidizer tube length varied from 0.127 m near the center to 0.228 m at the chamber wall. Local combustion responses are shown in Fig. 2 as a function of axial position for five different radial and tangential positions. To compare the local combustion rate oscillation at different locations they are nondimensionalized as follows:

$$R_c = \frac{p_0 \sum_t w_1 p_1}{p_{i1}^2 \sum_t w_0} \tag{25}$$

Table 4 Injector flow responses[a]

Row	Oxidizer response
1	$-0.42 - 0.30i$
2	$-0.53 - 0.40i$
3	$-0.70 - 0.46i$
4	$-0.93 - 0.44i$
5	$-1.16 - 0.25i$
6	$-1.22 + 0.08i$
7	$-1.06 + 0.34i$
8	$-0.83 + 0.44i$

[a]Fuel response $-5.9 + 1.2i$

where w_1 is the local oscillating burning rate, w_{01} the average burning rate in the first axial cell, p_0 the average chamber pressure, and p_{i1} the oscillating pressure in the first axial cell. Thus, cells with very low burning rates and/or pressure amplitude do not contribute to the oscillations and are reduced proportionally in combustion response. The responses are referred to as normalized local combustion responses in Fig. 2. This weighting eliminates large values of the combustion response at the end of the combustor where only a few drops remain to burn, and the combustion response is very dependent on when in the pressure cycle these remaining drops are vaporized.

The gas oscillations and combustion response field repeat around the chamber with each baffle compartment. Therefore, only the first four tangential locations, corresponding to one baffle compartment, are shown in Fig. 2 and discussed herein. Profiles inside the hub baffle are presented in Fig. 2 as well as within a baffle compartment.

The responses for radial positions near the centerline and within the hub and all tangential positions are very similar as the oscillations are spinning. Next to the hub (radius = 0.098 m) the influence of the radial baffles outside the hub are noticed, and the responses corresponding to the cells near the radial baffles (radius = 0.098 m) are shown. Within the baffle compartment, the responses at the corners and the center show the extreme variations calculated.

At the first axial position, the real part of the response is always a large positive number, and the imaginary part is a larger negative number. This is a result of the large variation in drop size produced by the oscillation in fuel flow during a cycle. In the baffle compartments, the drop size varied between 270 and 950 μ during a cycle. With the large negative fuel response listed, the axial velocity is very low when the pressure is high. Thus, large drops are being produced when the pressure is high, resulting in low burning rates for these drops. Within the first cell 40–50% of the oxidizer is burned (depending on the oscillation amplitudes). The average time for the oxidizer to travel from the injector face (atomization plane) to the center of the first axial cell is 0.40 of a cycle. Therefore, on average the large drops produced at the high-pressure portion of the oscillations are burning at the low-pressure portion of the oscillation in the first cell. This produces the large positive combustion response in the first cell. The fact that the combustion response has a larger negative imaginary response indicates that the time delay has resulted in the large drops burning closer to the period when the pressure is near zero and increasing.

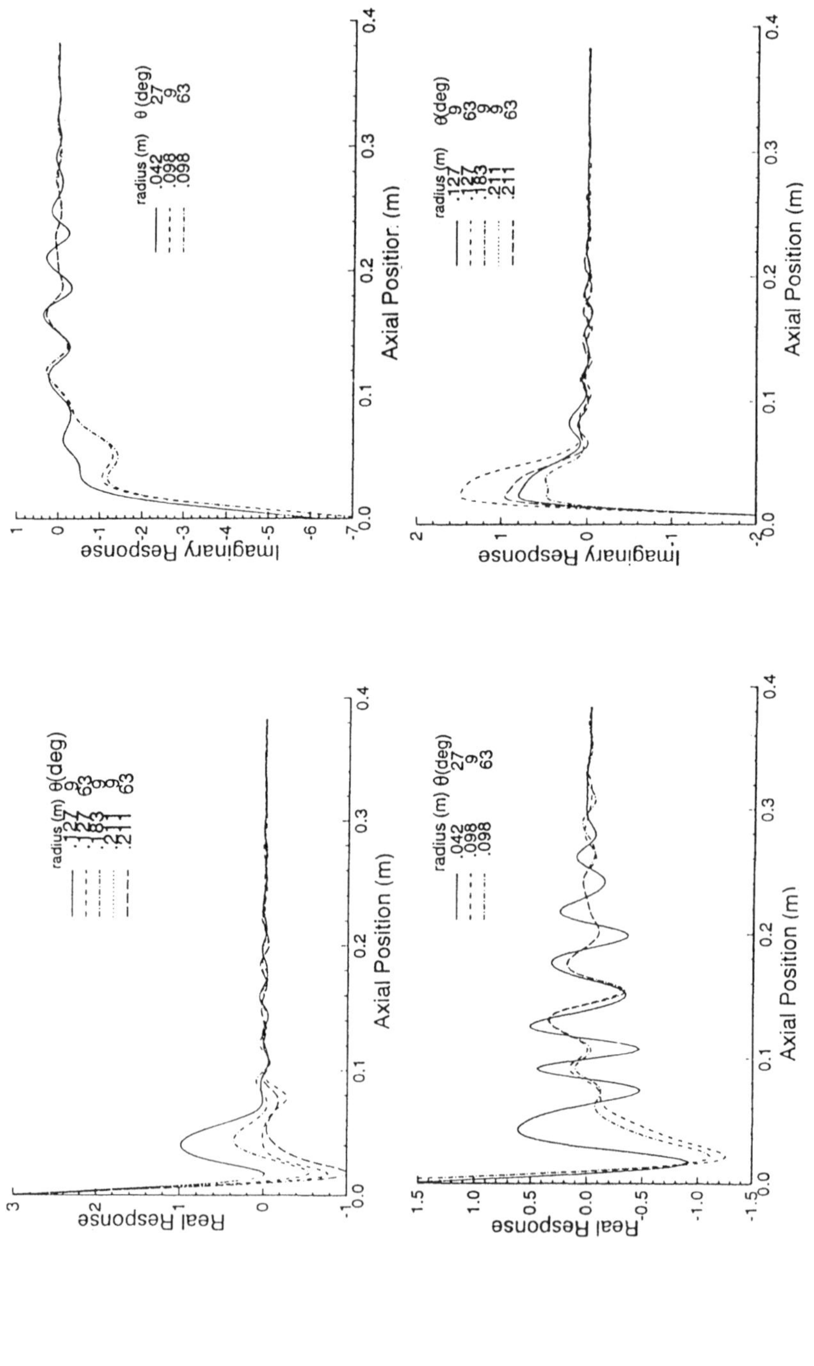

Fig. 2 Normalized local combustion responses at various radial and tangential locations.

The real part of the combustion response in the first cell varies between 1.78 and 3.41 whereas the imaginary part of the combustion response varies between −5.30 and −7.67. It is evident that these combustion responses cannot be considered constant as is assumed in the simple combustion models discussed earlier.

The normalized combustion response drops very fast after the first cell, due to the decrease in burning rate axially and to the fact that the drops take different amounts of time to arrive at downstream cells. The effects of the variation in drop size with time are being smeared by the drops having different life times. It takes several cycles for the drops to arrive at these downstream cells. As a result, the contribution of the oscillating burning downstream is minimized. However, the variation in combustion response with radial and tangential directions is very large. Again, this violates the assumption used in the simple combustion models that the combustion response is uniform throughout the chamber.

To determine the best representation of a combustion response for the entire chamber, the burning rates were summed over the full chamber volume. For the baffled chamber, the normalized full chamber combustion response was (1.358, −2.084i). This is in contrast to the value of (1.35, +0.0i) determined using the FTE approach to drive the 1T mode. Obviously, the combustion response will not drive the oscillations at the natural frequency since the phasing of the two vectors are not the same (the tuning frequency would be lower as indicated by Fig. 1 for negative imaginary response). In addition, the local combustion oscillations calculated in the two stages are very different, which means we do not have the correct oscillation profiles.

Similar local combustion response calculations were made for the unbaffled SSME with no absorber. For the 1T mode in an unbaffled chamber, all of the angular positions see the same oscillation. Therefore, all of the combustion responses are the same for a given radial position. The normalized stream tube (an axial row of cells at a fixed radial location) combustion responses are shown in Table 5 for the SSME with stability aids removed.

Table 5 again shows the large variation in combustion response with position with an unbaffled chamber. Since the wave is spinning, there are variations in only the radial and axial directions. The first axial position again has a very large value (twice the normalized stream tube value) and then rapidly drops, as seen in Fig. 2 for the baffled chamber. The largest response values are located near the chamber wall where they are most effective in driving an oscillation. The

Table 5 Stream tube combustion response for the unbaffled chamber[a]

Radius, m	Streamtube response	Response at injector face
0.014	0.78 − 2.57i	1.67 − 3.95i
0.042	0.29 − 3.23i	0.87 − 5.61i
0.070	0.79 − 3.12i	1.68 − 5.81i
0.098	1.14 − 2.90i	2.24 − 5.52i
0.126	1.30 − 2.66i	2.57 − 5.05i
0.154	1.34 − 2.46i	2.74 − 4.73i
0.183	1.33 − 2.30i	2.81 − 4.59i
0.211	1.34 − 2.19i	2.89 − 4.58i

[a]Full chamber combustion response 1.272 − 2.45i.

calculated response for the full chamber has a slightly lower real value for the unbaffled chamber compared to the baffled chamber. This would indicate that the unbaffled chamber is more stable than the baffled chamber. With the SSME, the large combustion response is due to the varying atomization produced by the large injector fuel flow response.

The results and discussions indicated that the fuel response is responsible for a large portion of the combustion response in the SSME. This result was checked by performing calculations with different portions of the combustion model turned off. Calculations were also performed with different oscillations amplitude levels to show the nonlinear characteristics of the combustion process. The results of these calculations are shown in Table 6.

Table 6 confirms the importance of fuel flow oscillations on the combustion response. It is the phenomena making the largest contribution to combustion response at all amplitudes. The influence is very nonlinear, decreasing with increasing amplitude. At higher oscillation amplitudes the velocity of the fuel in the injector element reaches sonic velocity during a portion of the oscillation. Since the velocity cannot exceed sonic velocity, this reduces the fuel flow oscillation. To account for this fact, the fuel response in HICCIP is decreased by the square root of the oscillation amplitude at which sonic velocity is first achieved divided by the actual amplitude. For the 20% peak-to-peak chamber pressure oscillations, sonic condition is reached in the four rows within the baffle compartment and the response is reduced to 0.8 of its value at low amplitude. At 80% Pc peak-to-peak rows 2–10 reach sonic velocity and the fuel response in the baffle compartment area is reduced to 0.4 of its value at low amplitude (values in Table 6). As a result of this feature, the contribution of the fuel flow response decreases with increasing amplitude. The other terms are relatively insensitive to the oscillation amplitude at low amplitudes but do change at very high amplitudes.

Axial velocity oscillations are the next most important influence on combustion response after the fuel response. With the presence of the baffles, an axial wave is present in the baffle compartments, thereby producing large axial velocity oscillations. With a steady axial velocity in all cells, the axial velocity produces a sinusoidal oscillation in burning rate (the burning is Reynolds number dependent). The baffles also reduce the frequency of the oscillation. Within the hub, the reduced frequency combined with the smaller diameter results in a large amplitude increase

Table 6 Factors influencing combustion response

Oscillation amplitude	5% of Pc peak-to-peak	20% of Pc peak-to-peak	80% of Pc peak-to-peak
Fuel flow oscillation	$1.453 - 3.505i$	$0.803 - 2.268i$	$0.166 - 0.909i$
Axial velocity oscillation	$0.025 + 0.652i$	$0.025 + 0.655i$	$0.020 + 0.424i$
Density oscillation	$0.419 - 0.046i$	$0.417 - 0.069i$	$0.405 - 0.085i$
Oxidizer flow oscillation	$0.242 - 0.322i$	$0.277 - 0.341i$	$0.417 - 0.232i$
Pressure oscillation	$0.070 + 0.010i$	$0.064 + 0.011i$	0.045
Radial and axial velocity oscillation	0.0	$0.001 - 0.001i$	$-0.008 - 0.002i$
Linear sum of all oscillations	$2.20 - 3.201i$	$1.41 - 2.04i$	$1.05 - 0.804i$
Computed response with all oscillations	$2.22 - 3.026i$	$1.36 - 2.08i$	$0.646 - 1.29i$

from the injector to the tips of the baffle. This is accompanied by large axial velocity oscillations within the hub. The contribution of the axial velocity is out of phase (imaginary response) as the axial velocity oscillations are generally 120-deg out of phase with the pressure.

Density oscillations always produce a 0.4 contribution to the combustion response. This is associated with the density term in the Reynolds number, which influences the burning rate in proportion to the square root of the Nusselt number. Since density is proportional to pressure divided by the ratio of specific heats, the predicted influence of density on response is about 0.4 and is independent of amplitude, frequency, drops size, etc.

Oxidizer flow oscillations have a nominal influence on the combustion response in the SSME. Oxidizer injector pressure drop is not excessive, and as Table 4 shows, the oxidizer flow response is as large as −1.2 at some rows. The negative flow response results in a positive combustion response because the time lag between injecting the oxidizer and burning is close to a half-cycle. The oxidizer flow response is reduced due to the smearing of the flow oscillations through many cycles in the combustion process. The influence of the oxidizer flow oscillations increases with amplitude due to the change in drop size and velocity, which changes the time lag between injection and burning, as well as the burning rates.

Radial and tangential gas velocity and pressure oscillations have little influence on the combustion response. Without steady state or time-averaged radial and tangential velocities, the effect of these oscillations on burning rate cancel (burning rate is a function of the magnitude of velocity). The radial and tangential gas velocities would contribute to the harmonic terms in the Fourier expansion. The pressure oscillation also has little influence as it changes the surface temperature of the drop, which slightly changes the heat required to vaporize the drop. Pressure also slightly influenced the drop size and atomization rate via the density influence in the atomization process, which was considered as a pressure oscillation influence. These factors were all small and not very dependent on oscillation amplitude.

Contributions of the various factors toward the total combustion response is additive at the lower amplitudes. The entry "linear sum of all" in Table 6 is the value obtained by adding all of the individual contributions. At the 5 and 20% peak-to-peak levels, the actual calculated response using all of the terms is within 5% of the individual sums. At higher amplitudes, the factors interact, as one would expect, and the sum appears to be quite different than the actual. The major influence, however, has been to change the phasing of the combustion response (larger imaginary and lower real values). Basically, this means that the time lag has changed.

All of the results presented are very dependent on the constants used in the atomization process to determine drop size and atomization rates.[17,24,25] The atomization constants are premulitipliers for the drop size and mass stripping rate equations. The drop size and stripping rate are functions of the relative velocity of the fuel and oxidizer, the densities and viscosities of the fuel and oxidizer, and the surface tension of the oxidizer. These constants will change the steady-state profiles as well as the time lags between injection and burning. To show this influence, calculations were made with the same gas-phase oscillations but with different atomization constants. These results are shown in Table 7.

Table 7 shows the strong tuning influence of the atomization constants. Both the size and rate constants show optimum values at which the response is either a

Table 7 Influence of atomization constant on combustion response

Size const.	Rate const.	Dropsize, μ	Length to 50% vaporized, m	Combustion response, baffled	Combustion response, unbaffled
18,800	0.266	268	0.01	$0.46 - 4.7i$	$0.21 - 4.7i$
26,320	0.266	375	0.13	$1.1 - 3.3i$	$0.96 - 3.6i$
37,600	0.266	507	0.023	$1.4 - 2.1i$	$1.3 - 2.5i$
52,640	0.266	711	0.041	$1.4 - 1.0i$	$1.4 - 1.6i$
75,200	0.266	1072	0.071	$1.2 - 0.06i$	$1.2 - 0.79i$
37,600	0.133	1558	0.145	$2.6 - 1.3i$	$2.6 - 1.3i$
37,600	0.187	712	0.038	$1.5 - 2.2i$	$1.4 - 2.8i$
37,600	0.267	507	0.023	$1.4 - 2.1i$	$0.96 - 2.5i$
37,600	0.533	480	0.020	$1.4 - 1.8i$	$1.4 - 2.1i$

maximum or minimum. The size constant of 37,600 (used in all of the calculations presented) results in a response for both the baffled and unbaffled chamber that is almost a maximum. Increasing or decreasing the size would improve stability. The rate constant of 0.2666 (also used) results in a minimum response for both the baffled and unbaffled chamber. Drop size and location of the zone where 50% of the oxidizer is vaporized is also shown. Changing the rate constant changes the atomization plane and the drop size. The change in drop size is due to a change in the gas velocity field surrounding the liquid when the atomization rate is changed.

Table 7 also shows the importance of the atomization constants in comparing the stability of the baffled to the unbaffled chamber. With either larger or smaller rate constants the baffled chamber is more stable than the unbaffled. At the rate constant used in the previous results, the baffled chamber is more stable than the unbaffled. At smaller size constants, the unbaffled chamber becomes much more stable than the baffled.

V. Summary

A brief review of instability modeling prior to 1972 was made. Current approaches to instability modeling were outlined, and the advantages and disadvantages of each approach were discussed. The theory for the Fourier time-expansion approach was outlined in detail. Calculations were made for the combustion response profiles in the SSME combustor (with a five-bladed hub baffle and absorbers). The contributions of various physical phenomena (atomization, injection, etc.) to the combustion response profiles were assessed.

References

[1]Harrje, D. T., and Reardon, F. H. (eds.), *Liquid Propellant Rocket Combustion Instability,* NASA SP-194, 1972.

[2]Crocco, L., and Cheng, S. I., *Theory of Combustion Instability in Liquid Propellant Rocket Motors,* AGARD Monograph No. 8, Butterworths, London, 1956.

[3]Crocco, L., and Harrje, D. T., et al., "Nonlinear Aspects of Combustion Instability in Liquid Propellant Rocket Motors," Princeton Univ., NASA CR 72426, Princeton, NJ, June 1968.

[4]Crocco, L., and Mitchell, C. E., "Nonlinear Periodic Oscillations in Rocket Motors with Distributed Combustion," *Science and Technology,* Vol. 1, No. 2. 1969.

[5]Zinn, B. T., and Savell, C. T., "A Theoretical Study of Three-Dimensional Combustion Instability in Liquid Rocket Motors," *Twelfth Symposium (International) on Combustion,* Combustion Inst., Pittsburgh, PA, 1969, pp. 139–147.

[6]Culick, F. E. C., "Stability of High Frequency Pressure Oscillations in Rocket Combustion Chambers," *AIAA Journal,* Vol. 1, No. 5., 1963, pp. 1097–1104.

[7]Priem, R. J., and Rice, E. J., "Combustion Instability with Finite Mach Number Flow and Acoustic Liners," *Twelfth Symposium (International) on Combustion,* Combustion Inst., Pittsburgh, PA, 1969, pp. 149–159.

[8]Reardon, F. H., "Correlation of Sensitive-Time-Lag-Theory Combustion Parameters with Thrust Chamber Design and Operating Variables," *Fifth Interagency Chemical Rocket Propulsion Group Combustion Conference,* CPIA Pub. 183, Chemical Propulsion Information Agency, Laurel, MD, Dec. 1968, pp. 237–244.

[9]Lawhead, R. B., "Photograhpic Studies of Combustion Processes in Liquid Propellant Rockets," *Eighth Symposium on Combustion,* Combustion Inst., Pittsburgh, PA, 1962.

[10]Heidmann, M. F., "Visualization Studies of Combustion Instability in a Hydrogen-Oxygen Model Combustor," NASA Lewis Research Center, Film C-226, 1972.

[11]Burrows, M. C., "Thermal Radiation as a Tool in Analyzing Unstable Combustion," *First Interagency Chemical Rocket Propulsion Group Combustion Instability Conference,* CPIA Pub. 68, Vol. 1, Chemical Propulsion Information Agency, Laurel, MD, 1964, pp. 301–309.

[12]Heidmann, M. F., and Wieber, P. R., "Analysis of Frequency Response Characteristics of Propellant Vaporization," NASA TN D-3749, Dec. 1966.

[13]Heidmann, M. F., and Groeneweg, J. F., "Analysis of the Dynamic Response of Liquid Jet Atomization to Acoustic Oscillations," NASA TN D-5339, July 1969.

[14]Priem, R. J., and Guentert, D. C., "Combustion Instability Limits Determined by a Nonlinear Theory and a One-Dimensional Model," NASA TN D-1409, Oct. 1962.

[15]Hoffman, R. J., Beltran, M. R., Breen, B. P., and Wright, R. O., "Extension of the Priem-Guentert Annular Combustion Instability Model to a Bi-Propellant System," *Third Interagency Chemical Rocket Propulsion Group Combustion Conference,* CPIA Pub. 138, Vol. I, Chemical Propulsion Information Agency, Laurel, MD, Feb. 1967, p. 309.

[16]Burstein, S. Z., and Chinitz, W., "Nonlinear Combustion Instability in Liquid-Propellant Rocket Motors," Quarterly Repts. to Jet Propulsion Lab., Contract NAS7-100, Mathematical Applications Group, White Plains, NY, 1967–1968.

[17]Wu, P.-K., Ruff, G. A., and Faeth, G. M., "Primary Breakup in Liquid/Gas Mixing Layers," AIAA Paper 91-0285, Jan. 1991.

[18]Wu, P.-K., Tseng, L.-K., and Faeth, G. M., "Primary Breakup in Gas/Liquid Mixing Layers for Turbulent Liquids," AIAA Paper 92-0462, Jan. 1992.

[19]Giridharan, M. G., "A Computer Model for Liquid Jet Atomization in Rocket Thrust Chambers," NAS8-38425, Aug. 1991.

[20]Priem, R. J., "Calculating Vaporization Rates in Liquid Rocket Engines Above Critical Pressure-Temperature Conditions," 27th JANNAF Combustion Subcommittee Meeting, Nov. 1991.

[21]Merkle, C. L., "Application of Computational Fluid Dynamics (CFD) Techniques to Engine Instability Study," First International Symposium on Liquid Rocket Combustion Instability, Pennsylvania State Univ., University Park, PA, Jan. 18–20, 1993.

[22]Yang, H. Q., Wang, Z. J., Pindera, M. Z., and Przekwas, A. J., "Advanced CFD Methodology for Fast Transients Encountered in Nonlinear Combustion Instability Problems," NAS8-38489, July 1992.

[23]Habiballah, M., and Dubois, I., "Numerical Analysis of Engine Instability," First International Symposium on Liquid Rocket Combustion Instability, Pennsylvania State Univ., University Park, PA, Jan. 18–20, 1993.

[24]Priem, R. J., and Bresacher, K. J., "High Frequency Injection Coupled Combustion Instability Program (HICCIP) Status and Plans," 1990 Conf. on Advanced Earth-To-Orbit Propulsion Technology, NASA Marshall Space Flight Center, CP 3092, July 1990.

[25]Priem, R. J., and Breisacher, K. J., "3D Combustor Acoustic Analysis," 1992, Conference on Advanced Earth-To-Orbit Propulsion Technology, NASA Marshall Space Flight Center, CP-3174.

[26]Priem, R. J., and Breisacher, K. J., "3D Rocket Combustor Acoustics Model," 28th AIAA/ASME/SAE/ASEE Joint Propulsion Conf., AIAA Paper 92-3228, June 1992.

[27]Maslen, S., and Moore, F. K., "On Strong Transverse Waves Without Shocks in a Circular Cylinder," *Journal of the Aeronautical Sciences,* Vol. 23, No. 6, 1956, pp. 583–593.

[28]Anon., "Guidelines for Combustion Stability Specifications and Verification Procedures for Liquid Propellant Rocket Engines," CPIA Pub. 247, Chemical Propulsion Information Agency, Laurel, MD, 1973.

[29]Wieber, P. R., "Acoustic Decay Coefficients of Simulated Rocket Combustors," NASA TN D-3425, 1969.

[30]Anon., "Gemini Stability Improvement Program Final Report," Aerojet General, Vol. 5, Aug. 1965.

Chapter 18

Numerical Analysis of Engine Instability

M. Habiballah* and I. Dubois†
ONERA, Châtillon, 92322 France

Nomenclature

b = chamber width
D = species diffusivity
D_m = diameter of group m droplets
e = gas-phase specific internal energy
h_k = specific enthalpy of species k
M_k = molar mass of species k
n_m = number of group m droplets per unit volume
p = pressure
R = perfect gas constant
T = temperature
t = time
u = axial gas velocity
u_m = axial droplet velocity
$\boldsymbol{V}$ = gas velocity
$\boldsymbol{V}_m$ = droplet velocity
v = transverse gas velocity
v_m = transverse droplet velocity
x = axial direction
y = transverse direction
z = radial direction
λ = thermal conductivity
μ = viscosity
ρ = gas-phase density
ρ_k = partial density of species k
ρ_L = liquid density

---

*Division Head, Liquid Propellant Engines and Launchers.
†Research Engineer, Liquid Propellant Engines and Launchers Division.

ρ_m = partial density of group m droplets
$\sigma_{xx}, \sigma_{xy}, \sigma_{yy}$ = viscous stress tensor components

Subscripts and Superscripts

c = combustion
i = injection
k = gaseous species k
l = laminar
lf = lateral flow
m = group m
t = turbulent
v = vaporization
1 = gaseous fuel
2 = gaseous oxidizer
3 = burnt products

I. Introduction

COMBUSTION instability results from the coupling between combustion and the fluid dynamics of the system. Several types of instability have been observed. They are all characterized by chamber pressure oscillations, although the frequency and amplitude of the oscillations and their external manifestations normally vary with the type of instability.

Oscillatory operation of an engine is undesirable for many reasons. One of the most important effects is the very high vibration levels that may impair the operation of the guidance system and severely affect payloads. Another detrimental effect of oscillatory operation is the highly increased heat transfer. This increase often leads to destruction of the engine. Several different classes of instability have been identified and studied experimentally: low-frequency (chug, pogo), intermediate-frequency (buzz), and high-frequency instability (screaming, screeching). High-frequency instability, also called resonant combustion or acoustic instability, is the most destructive type of instability.

Combustion instability problems have been encountered in almost all of the propulsion system development programs, for example, during the development programs of the F-1, J-2, and J-2S in the United States, and the Viking engine in Europe. Although a considerable effort has been given to understanding and controlling this undesirable phenomenon over the last 40 years, it continues to present significant challenges in reliable engine design. An unforeseen instability problem can cause engine loss and facility damage, and it can result in cost over-run and program schedule slippage. Stabilizing devices often used to solve the instability problem impose additional cost, weight, and complexity to engine technology.

A significant amount of work on combustion instability was performed in the United States in the 1950s and the 1960s during the Apollo development program. Much of this work was compiled by Harrje and Reardon.[1] Since then, due to limited development programs, research on combustion instability has been reduced significantly. Culick[2] has reported more recent work on the subject.

This chapter focuses on the numerical analysis of engine instability. After a short review of the numerical analysis of combustion instability, we describe the research we have been carrying out on combustion instability in the framework of the Ariane European launcher development program.[3]

Until now, stability assessment of rocket engines has relied on qualitative tools and full-scale testing. The complex nature of the combustion instability phenomenon and the lack of computational capabilities precluded any attempt to describe it in the past. Most of the effort put into understanding, predicting, controlling, and eliminating combustion instability in rocket motors has been devoted to linear analyses.[4–6] These analyses, which must make assumptions such as small-amplitude oscillations and single-frequency sinusoidal oscillations to solve the problem, are not always consistent with observed instability phenomena. Finite disturbances are often required to excite instability, and waves are usually steep fronted. Linear analyses can predict neither the large-amplitude oscillations often observed in case of instability nor the limiting cycle behavior.

Comparatively limited work has addressed the solution of nonlinear combustion instability problems in liquid rockets. Over the past decade, substantial applications of computational fluid dynamics (CFD) techniques to solid rocket and ramjets instabilities have been reported,[7–11] whereas these techniques have received only limited attention for liquid rocket combustion instabilities.[12,13]

With the limited resources available today, more economical quantitative analysis tools and subscale testing must be used to rate the stability of an engine. The advent of large supercomputers, CFD modeling capabilities, and progress in diagnostic techniques make numerical analysis of combustion instability a promising new approach to the problem.

Numerical analysis of combustion instability requires the solution of the differential conservation equations that describe the flow in a typical liquid rocket engine. Thus, it is necessary to describe in detail potential mechanisms involved in the instability phenomenon. Such mechanisms are recognized as injection, atomization, vaporization, mixing, and combustion of liquid propellants.

Priem and Guentert[14] were the first to address combustion instabilities in liquid rockets by solving the conservation equations numerically. Their approach was later adopted by others, like Hoffman et al.[15] Only one- or two-dimensional problems were treated. The two-dimensional problem was formulated for an annular ring in the combustion chamber. The nonlinear conservation equations were solved with source terms representing vaporization and combustion. Hoffman et al.[15] improved the Priem model; they treated variable oxidizer/fuel mixture ratio, included the droplet drag and momentum transfer between the liquid and gas phases, and adapted the vaporization model by including the Reynolds number dependence.

In these works, numerical analysis was initiated by a pressure wave having sinusoidal amplitude distribution around the annular ring. Its time evolution was calculated as it propagated around the ring, which could be placed anywhere along the axis of the chamber. Axial propagation of the wave was not treated.

The approach just described was motivated by the analysis of tangential acoustic modes, the most destructive kind of instabilities. Burstein et al.[16] carried out a similar numerical analysis. They also treated wave propagation in an annulus, including the influence of axial baffles, and were able to treat wave propagation in the axial direction. This appears to be the last work reported in the 1970s.

More recently, some research effort on numerical analysis of combustion instability was initiated in a fewlaboratories, either by adapting existing numerical models to the problem of combustion instability[17–19] or by developing specific models (See Chapter 19).

II. Approach to the Numerical Analysis of Combustion Instability

In the early 1980s, as part of the European Ariane rocket development program, a research activity on high-frequency combustion instability in liquid-propellant engines was begun. At that time, computational resources were sufficiently advanced to make it possible to address the problem numerically. The complex processes involved in the combustion chamber of a rocket engine that needed to be modeled were many. Emphasis was placed on those processes recognized to be involved in combustion instability phenomenon: atomization, droplet vaporization, and burning. Submodels resulting from atomization and droplet vaporization studies were implemented in the simultaneously developed numerical model.

Although the approach is general and may be used for any liquid propellant engine, the analysis presented herein was carried out with the Ariane liquid rocket engines in mind. Two engine types were considered: the Viking engine using UDMH/N_2O_4 propellants and a radial impinging injector system, and the cryogenic Vulcain engine using LOX/H_2 and a coaxial element injector system.

A. Atomization Model

The intent of the research on atomization was to describe the spray characteristics resulting from the injector system. More precisely, attention was focused on the measurement of droplet size and velocity distributions. The lack of a theoretical model to describe droplet formation for an impinging or coaxial injector dictated an experimental approach. Measurements were performed with a Malvern sizer. In most cases, simulant fluids were used to replace actual propellants. Different parameters were investigated, such as fluid properties, working pressure, and mass flow rate. The injector element geometry was also investigated. The spray characterization output was given in a compact form which described drop size distribution as a function of working conditions and geometry. Both impinging and coaxial injector elements were considered, and the results were described in several papers[20–23] (see also Chapter 6).

B. Vaporization Model

Droplet vaporization and burning have been recognized as playing a fundamental role in combustion instability phenomenon. Research was undertaken to provide insight into this process and to determine how it affects combustion instability. The approach taken dealt with detailed numerical analysis of the flowfield of vaporizing and burning droplets. It included both one-dimensional (burning droplet in a quiescent atmosphere) and two-dimensional (burning droplet in a convective field) analyses.[24] The mathematical formulation includes multispecies gas-phase equations and liquid-phase equations, which are solved numerically. Special attention was paid to the analysis of vaporization and burning of droplets in a high-pressure and high-temperature atmosphere that represented the actual working conditions of rocket engines. The analysis was applied to storable (UDMH, N_2O_4)and cryogenic (LOX) propellant droplets. The effects of pressure and velocity waves on vaporization and burning were investigated for both UDMH and N_2O_4 droplets. For liquid oxygen droplets, vaporization has been considered under subcritical and supercritical conditions.[25] Results of these analyses were summed up in analytical laws that were implemented in the numerical model.

C. Phedre Numerical Model

Despite progress made in computing capabilities, it was not realistic at the time the present research started to deal with the development of a three-dimensional code that would describe the flowfield in a typical rocket engine. Limitations resulted from both computational time and memory requirements. The alternative was to develop a two-dimensional annular model, Phedre, to analyze combustion instability. This approach was suggested by the concern for describing acoustic tangential modes, which are by far the most critical ones. This numerical code is discussed in the following section.

III. Numerical Analysis of Combustion Instability

The two-dimensional numerical model Phedre was developed to describe the unsteady, turbulent, two-phase reacting flow in a liquid rocket engine combustion chamber such as that represented in Fig. 1a. The model formulation, presented in this section, enables the treatment of different engine types. It includes axial or radial propellant injection, baffles, and acoustic resonator modeling. The injected propellants may be either both liquid or one liquid and one gas.

The axisymmetric chamber geometry is transformed into an equivalent annular geometry as shown in Fig. 1b. This transformation consists of spreading out the actual geometry according to the following rules. The axial lengths (x axis), such as the chamber length L and the baffle length L_b, are kept unchanged. The chamber transverse length l (y axis) is adjusted so that the nth tangential mode frequencies of the real and equivalent chambers are the same (n being the order of the mode being analyzed). The width in the third direction (z axis), which varies only with x, is determined so that the cross-sectional areas of the two chambers match, enabling the axial flow evolution to be the same in both chambers. The flowfield is not resolved in the z direction.

An n-bladed baffle is represented by placing evenly spaced n plates in the transverse direction. To represent acoustic resonators distributed at the chamber periphery, surface boundary conditions are applied at the chamber-resonator interface. These conditions, simulating a gas flow incoming or leaving the chamber through the cavity aperture, consist of solving a set of equations which describe the resonator response to the chamber flow disturbances.

To model radial injection and acoustic resonators, the two-dimensional model equations were written in a form that enables the injection of gas and/or liquid through the chamber lateral walls. Later, these flows will be referred to as lateral flows.

With this formulation, it is also possible to describe two-dimensional planar flows useful for pure longitudinal-mode analyses, provided the relevant boundary conditions are implemented. Then, the transverse length is typically chosen equal to the chamber diameter, and the width is determined as for the annular geometry.

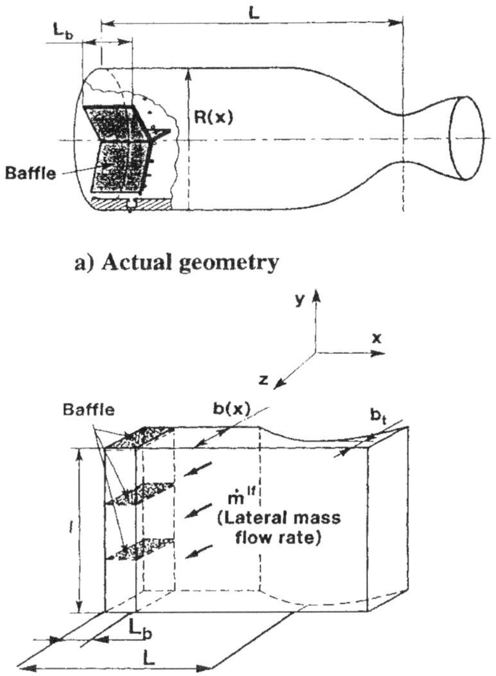

Fig. 1 Definition of equivalent annular chamber.

A. Governing Equations

1. Gas-Phase Equations

The gas phase is described by solving the Favre-averaged Navier–Stokes equations with specific source terms. These terms are the result of the interaction of the gas phase with the liquid phase and the injection of a lateral gas flow when present. The gas phase consists of three species: fuel, oxidizer, and burnt products. Additional transport equations for the fuel and the oxidizer are also solved. The gas-phase equations are written in the following two-dimensional form.

Conservation equation of mass:

$$\frac{\partial \rho}{\partial t} + \frac{1}{b}\frac{\partial \rho u b}{\partial x} + \frac{\partial \rho v}{\partial y} = \dot{W}^v + \dot{W}^{lf} \tag{1}$$

where $\dot{W}^v$ and $\dot{W}^{lf}$ are the mass source terms due to droplet evaporation and lateral flow, respectively.

Conservation equations of momentum:

$$\frac{\partial \rho u}{\partial t} + \frac{1}{b}\frac{\partial \rho u^2 b}{\partial x} + \frac{\partial \rho u v}{\partial y} = -\frac{\partial p}{\partial x} + \frac{\partial \sigma_{xx}}{\partial x} + \frac{\partial \sigma_{xy}}{\partial y} + f_x + S_x^v + S_x^{lf} \tag{2}$$

$$\frac{\partial \rho v}{\partial t} + \frac{1}{b}\frac{\partial \rho u v b}{\partial x} + \frac{\partial \rho v^2}{\partial y} = -\frac{\partial p}{\partial y} + \frac{\partial \sigma_{xy}}{\partial x} + \frac{\partial \sigma_{yy}}{\partial y} + f_y + S_y^v + S_y^{lf} \tag{3}$$

where f_x and f_y are the drag force components, S_x^v and S_y^v are the components of the momentum exchange $\boldsymbol{S}^v$ between the two phases due to droplet evaporation, and S_x^{lf} and S_y^{lf} are the components of the momentum source term $\boldsymbol{S}^{lf}$ due to lateral gas flow.

Conservation equation of energy:

$$\begin{aligned}\frac{\partial \rho e}{\partial t} + \frac{1}{b}\frac{\partial \rho e u b}{\partial x} + \frac{\partial \rho e v}{\partial y} &= -p\left(\frac{1}{b}\frac{\partial u b}{\partial x} + \frac{\partial v}{\partial y}\right) \\ &+ \sigma_{xx}\frac{1}{b}\frac{\partial u b}{\partial x} + \sigma_{yy}\frac{\partial v}{\partial y} + \sigma_{xy}\left(\frac{\partial u}{\partial y} + \frac{1}{b}\frac{\partial v b}{\partial x}\right) \\ &- \frac{1}{b}\frac{\partial q_x b}{\partial x} - \frac{\partial q_y}{\partial y} + S_e^v + S_e^{dw} + S_e^{lf}\end{aligned} \tag{4}$$

where S_e^v, S_e^{dw}, and S_e^{lf} represent the energy source terms due to droplet evaporation, drag force, and lateral flow, respectively; q_x and q_y are the heat flux components due to both conduction and species enthalpy diffusion expressed according to Fick's law,

$$\begin{aligned} q_x &= -\lambda\frac{\partial T}{\partial x} - \rho D \sum_k h_k \frac{\partial}{\partial x}\left(\frac{\rho_k}{\rho}\right) \\ q_y &= -\lambda\frac{\partial T}{\partial y} - \rho D \sum_k h_k \frac{\partial}{\partial y}\left(\frac{\rho_k}{\rho}\right)\end{aligned} \tag{5}$$

Conservation equations of species concentrations:

$$\begin{aligned}\frac{\partial \rho_k}{\partial t} + \frac{1}{b}\frac{\partial \rho_k u b}{\partial x} + \frac{\partial \rho_k v}{\partial y} &= \frac{1}{b}\frac{\partial}{\partial x}\left[\rho b D \frac{\partial}{\partial x}\left(\frac{\rho_k}{\rho}\right)\right] \\ &+ \frac{\partial}{\partial y}\left[\rho D \frac{\partial}{\partial y}\left(\frac{\rho_k}{\rho}\right)\right] + \dot{W}_k^v + \dot{W}_k^c + \dot{W}_k^{lf}\end{aligned} \tag{6}$$

where $\dot{W}_k^v$, $\dot{W}_k^c$, and $\dot{W}_k^{lf}$ are the source terms of species k due to droplet evaporation, gas-phase combustion, and lateral flow, respectively.

This system of equations is closed by the perfect gas equation of state,

$$p = RT \sum_k \frac{\rho_k}{M_k} \tag{7}$$

where the temperature is derived from the internal energy using

$$e(T) = \sum_k \frac{\rho_k}{\rho}\left[h_k(T) - \frac{R}{M_k}T\right] \tag{8}$$

Turbulence model equations: Turbulence effects are taken into account using the Boussinesq approximation. Thus, the transport coefficients are expressed as the sum of a laminar coefficient and a turbulent one. The turbulent viscosity is given by the Deardoff subgrid scale (SGS) turbulence model[26]:

$$\mu_t = \rho c_t \delta^2 \left[2\left(\frac{\partial u}{\partial x}\right)^2 + 2\left(\frac{\partial v}{\partial y}\right)^2 + \left(\frac{\partial u}{\partial y} + \frac{\partial v}{\partial x}\right)^2 \right]^{\frac{1}{2}} \tag{9}$$

where c_t is a dimensionless constant $c_t = 0.04$ and δ the minimal cell size.

Turbulent thermal conductivity and species diffusivity are, thus, obtained from the turbulent Prandtl and Schmidt numbers,

$$\lambda_t = \mu_t C_p / Pr_t, \qquad D_t = \mu_t / \rho Sc_t \tag{10}$$

2. *Liquid-Phase Equations*

The liquid phase is represented by droplet groups. Each group includes droplets having the same composition, diameter, velocity, and temperature at injection. As the droplets move through the chamber, they remain in their initial group and, as a result, there is no mass transfer between groups. For each group, equations describing the Eulerian motion of the droplets are solved. Only the drag forces are considered. Coalescence, droplet breakup, and interaction forces between droplets are neglected. The liquid-phase equations are, thus, written in the following two-dimensional form.

Conservation equation of mass: For each group m, partial density ρ_m is defined as the mass per unit volume of droplets belonging to group m,

$$\frac{\partial \rho_m}{\partial t} + \frac{1}{b}\frac{\partial \rho_m u_m b}{\partial x} + \frac{\partial \rho_m v_m}{\partial y} = \dot{W}_m^v + \dot{W}_m^{lf} \tag{11}$$

where $\dot{W}_m^v$ and $\dot{W}_m^{lf}$ represent the mass loss of group m caused by evaporation and the mass source due to lateral injection, respectively.

Conservation equations of momentum:

$$\frac{\partial \rho_m u_m}{\partial t} + \frac{1}{b}\frac{\partial \rho_m u_m^2 b}{\partial x} + \frac{\partial \rho_m u_m v_m}{\partial y} = f_{mx} + S_{mx}^v + S_{mx}^{lf} \tag{12}$$

$$\frac{\partial \rho_m v_m}{\partial t} + \frac{1}{b}\frac{\partial \rho_m u_m v_m b}{\partial x} + \frac{\partial \rho_m v_m^2}{\partial y} = f_{my} + S_{my}^v + S_{my}^{lf} \tag{13}$$

where f_{mx} and f_{my} are the group m drag force components, S_{mx}^v and S_{my}^v the momentum exchange terms between the two phases due to group m evaporation, and S_{mx}^{lf} and S_{my}^{lf} the momentum source terms due to lateral injection.

Conservation equation of droplet number:

$$\frac{\partial n_m}{\partial t} + \frac{1}{b}\frac{\partial n_m u_m b}{\partial x} + \frac{\partial n_m v_m}{\partial y} = S_{m,n_m}^{lf} \tag{14}$$

where S_{m,n_m}^{lf} represents the rate of change of the number of droplets resulting from lateral injection.

The droplet diameter is thus derived from the following expression, assuming spherical droplets:

$$D_m = \left(\frac{6\rho_m}{\pi \rho_L^m n_m} \right)^{1/3} \tag{15}$$

No droplet energy equation is solved as we neglect the transient heating time of the droplets and consider that the droplet surface temperature immediately reaches the vaporization temperature at local pressure and remains constant. This assumption is consistent with those made to establish the droplet vaporization model. It assumes that the heat transfer within the droplet is slow, which is valid for a vaporizing droplet in a high-temperature quiescent atmosphere. In the presence of convection where internal circulation tends to homogenize temperature inside the droplet, however, this assumption is not satisfactory. Indeed, some limited numerical analyses, including transient heatup and convection, showed that the surface temperature and the minimum temperature, which correspond to the vortex center, increase continuously. The analyses also showed that even with convection, a temperature gradient still exists inside the droplet. This suggests use of an effective conductivity model to account for this gradient. Improvement of the droplet heating model will be addressed in future work.

B. Source Terms due to Liquid and Gas-Phase Interaction

1. *Mass Source*

Vaporization. As mentioned earlier, the one-dimensional droplet combustion model provides an analytical expression for the unsteady vaporization mass flow rate of a nonmoving droplet. The Ranz–Marshall correction is then applied to account for convection effects. This finally gives an expression of the vaporization mass flow rate $\dot{m}$ for one droplet as a function of the droplet diameter, temperature, and local gas flow conditions. The mass loss term for group m may be given by

$$\dot{W}_m^v = -n_m \dot{m} \tag{16}$$

Consequently, the mass source terms are

$$\dot{W}^v = -\sum_m \dot{W}_m^v = \sum_m n_m \dot{m} \tag{17}$$

for the gas phase, and

$$\dot{W}_k^v = -\sum_{m_k} \dot{W}_{m_k}^v = \sum_{m_k} n_{m_k} \dot{m} \tag{18}$$

for those gaseous species $k = 1, 2$. The index m_k refers only to those groups of droplets with the same chemical composition as the kth species.

Combustion. A one-step chemical reaction is assumed to describe gas phase combustion: υ_1 fuel $+\ \upsilon_2$ oxidizer $\rightarrow$ burnt products.

The chemical reaction rate $\dot{\omega}$ is given by the Arrhenius law,

$$\dot{\omega} = CCe^{-\frac{T_a}{T}} \prod_{k=1,2} \left(\frac{\rho_k}{M_k} \right)^{\alpha_k} \tag{19}$$

where CC, T_a, and α_k are coefficients to be given later for each propellant combination.

The species mass source terms due to gas-phase combustion are expressed as

$$\dot{W}_k^c = -\upsilon_k M_k \dot{\omega} \tag{20}$$

2. *Momentum Source*

Vaporization. Assuming the vapor has the same velocity as the droplet when leaving the droplet surface, the momentum loss components for a group m may be written as

$$S_{mx}^v = \dot{W}_m^v u_m \qquad S_{my}^v = \dot{W}_m^v v_m \tag{21}$$

and the momentum source components for the gas phase as

$$\begin{aligned} S_x^v &= -\sum_m S_{mx}^v = -\sum_m \dot{W}_m^v u_m \\ S_y^v &= -\sum_m S_{my}^v = -\sum_m \dot{W}_m^v v_m \end{aligned} \tag{22}$$

Droplet drag force. The total droplet drag force appearing in the gas momentum conservation equations is given by

$$f = -\sum_m f_m \tag{23}$$

and the droplet drag force for a group m is

$$f_m = (\pi/8)\mu_l n_m D_m C_{Dm} Re_m (V - V_m) \tag{24}$$

where Re_m is the Reynolds number based on the relative gas-droplet velocity,

$$Re_m = \frac{\rho D_m}{\mu_l}|V - V_m| \tag{25}$$

The expression of the droplet drag coefficient C_{Dm} was corrected using the detailed droplet vaporization and burning analyses. Results obtained with the two-dimensional combustion model indicated that the drag coefficient of an evaporating droplet was approximately one-half the drag coefficient of a nonevaporating droplet. This is due to blowing of the boundary layer by the vaporized mass flow, which results in reduction of the friction forces.

The following relations have thus been used for the drag coefficient:

$$\begin{aligned} C_{Dm} &= 12/Re_m\left(1 + 0.15 Re_m^{0.687}\right) \qquad \text{if} \qquad Re_m \le 1000 \\ C_{Dm} &= 0.22 \qquad \text{if} \qquad Re_m > 1000. \end{aligned} \tag{26}$$

3. Energy Source

Vaporization. To derive the internal energy source term, first the exchange of total energy $E = e + V^2/2$ between the two phases is evaluated. The gas-phase energy source is a balance between the heat loss due to heatup and vaporization of the droplets and the energy increase due to the vaporized gas. The total energy source term is given by

$$S_E^v = S_E^{\dot{W}^v} - \Phi = \sum_{m_k} S_{E,m}^{\dot{W}^v} - \Phi_m \tag{27}$$

where Φ is the heat flux absorbed by droplets.

According to the assumptions relative to droplet temperature and vapor velocity, the total energy produced by a group m may be written as

$$S_{E,m}^{\dot{W}^v} = -\dot{W}_m^v\left[e\left(T_m^v\right) + \left(V_m^2/2\right)\right] \tag{28}$$

The heat flux Φ_m needed to increase the droplet temperature from the injection temperature to the vaporization temperature and vaporize the droplet is given by

$$\Phi_m = -\dot{W}_m^v\left(\int_{T_m^i}^{T_m^v} Cp_m^L \mathrm{d}T + L_m^v\right) \tag{29}$$

where Cp_m^L is the specific heat at constant pressure and L_m^v is the latent heat of vaporization for the liquid constituting the group m.

The following expression is then obtained for S_E^v:

$$S_E^v = -\sum_m \dot{W}_m^v\left[h_m^i - r_m T_m^v + \left(V_m^2/2\right)\right] \tag{30}$$

where h_m^i is the specific enthalpy of group m at injection temperature and $r_m = R/M_m$, with M_m being the molar mass of the group m vaporized gas.

Deriving the internal energy conservation equation from the total energy and momentum conservation equations results in the following expression of the internal energy source term:

$$S_e^v = S_E^v + \dot{W}^v\left(V^2/2\right) - V \cdot S^v \tag{31}$$

This relation and the foregoing results lead to the following expression for S_e^v:

$$S_e^v = -\sum_m \dot{W}_m\left[h_m^i - r_m T_m^v + \tfrac{1}{2}(V - V_m)^2\right] \tag{32}$$

Drag force. The work of the drag force S_e^{dw} appearing in the energy equation is expressed as

$$S_e^{dw} = \sum_m f_m \cdot (V - V_m) \tag{33}$$

C. Source Terms due to Lateral Flow

Because the model cannot resolve the flow in the third direction (z), a lateral mass flow (due to injection or the presence of an acoustic resonator) is handled as source terms (superscript lf) in the conservation equations. This is achieved by integrating the liquid- and gas-phase equations in an elementary volume of the computational domain. Note that lateral flow assumes no tangential (y axis) velocity component.

The expressions of the source terms are written next in the general case of a lateral flow which may be gaseous (mass flowrate $\dot{m}^{lf}$) or liquid (mass flow rate $\dot{m}_m^{lf}$ for each group m). V_{cell} refers to the computational cell volume.

The source terms for the gas phase are, for mass source,

$$\dot{W}^{lf} = \dot{m}^{lf} / V_{\text{cell}} \tag{34}$$

and, for momentum source,

$$S_x^{lf} = \dot{m}^{lf} u^{lf} / V_{\text{cell}} \qquad S_y^{lf} = 0 \tag{35}$$

where u^{lf} is the longitudinal component of the lateral flow velocity V^{lf}.

The energy source is

$$S_e^{lf} = \frac{\dot{m}^{lf}}{V_{\text{cell}}} \left(h^{lf} + \frac{(V^{lf})^2}{2} - u u^{lf} + \frac{V^2}{2} \right) \tag{36}$$

where h^{lf} is the enthalpy of the lateral gas flow. The k-species mass source is

$$\dot{W}_k^{lf} = y_k^{lf} \left(\dot{m}^{lf} / V_{\text{cell}} \right) \tag{37}$$

where y_k^{lf} is the mass fraction of k species in the lateral flow.

The source terms for the liquid phase are for the mass source

$$\dot{W}^{lf} = \dot{m}^{lf} / V_{\text{cell}} \tag{38}$$

for the momentum source

$$S_{mx}^{lf} = \dot{m}_m^{if} u_m^{lf} / V_{\text{cell}} \qquad S_{my}^{lf} = 0 \tag{39}$$

and for the droplet number source

$$S_{m,n_m}^{lf} = \frac{1}{V_{\text{cell}}} \frac{6 \dot{m}_m^{lf}}{\pi \left(D_m^{lf} \right)^3 \rho_L^m} \tag{40}$$

where D_m^{lf} is the diameter of the group m droplets injected laterally.

Acoustic resonator modeling is discussed next. When lateral flow is due to gas flow exchange between the chamber and its acoustic resonators, the mass flow rate is given by a system of differential equations. This system describes the inflow and outflow of gas at the resonator orifice in response to chamber pressure fluctuations, according to the Helmholtz theory.[1] This theory assumes that the motion of the gas in the resonator is analogous to that of a mass-spring-dashpot system. The mass of the system is the gas in the orifice and its immediate vicinity, which is considered to move as a unit. This theory also assumes that the dimensions of the acoustic resonators are small enough compared to the wavelengths of the chamber acoustic modes.

Therefore, the lateral mass flow rate $\dot{m}^{lf}$ is given by the following ordinary differential equations:

$$l_{\text{eff}}\frac{\mathrm{d}\dot{m}^{lf}}{\mathrm{d}t} + R_0\frac{\dot{m}^{lf}}{\bar{\rho}} + \frac{\bar{\rho}\bar{a}^2 A_{\text{ori}}^2}{V_{\text{cav}}}\tilde{z} = A_{\text{ori}}\,p' \tag{41}$$

$$\dot{m}^{lf} = \bar{\rho}A_{\text{ori}}\frac{\mathrm{d}\tilde{z}}{\mathrm{d}t} \tag{42}$$

where $\tilde{z}$ is the gas displacement along the orifice axis, $\bar{\rho}$ and $\bar{a}$ are the mean gas density and sound velocity in the resonator, p' is the chamber pressure fluctuation, A_{ori} is the cross-sectional area of the resonator orifice, and V_{cav} is the resonator cavity volume.

The effective orifice length l_{eff} is the sum of the orifice length l_{ori} and a length correction l_c that accounts for the flow effects in the vicinity of the orifice ends. An empirical equation for l_c, the most suitable for all conditions when the ratio $A_{\text{ori}}/A_{\text{cav}}$ is less than 0.15, is

$$l_c = 0.96\sqrt{A_{\text{ori}}}\left(0.5 - 0.7\sqrt{A_{\text{ori}}/A_{\text{cav}}}\right) \tag{43}$$

The resistance per unit orifice cross-sectional area R_0 accounts for nonlinear effects. Its expression is obtained from nonlinear analyses of Helmholtz resonator behavior,

$$R_0 = \Gamma\bar{\rho}\bar{U} \tag{44}$$

where Γ is a constant and $\bar{U}$ the amplitude of the orifice gas velocity oscillations. In our model the instantaneous gas velocity is used instead of the amplitude. Γ is taken equal to 1, a commonly used value.

The quantity p' appearing in the acoustic resonator equations is taken as the chamber pressure fluctuation, and couples the resonator flow to the chamber flow.

Some assumptions have to be made with regard to the temperature and chemical composition of the flow in acoustic resonators.

D. Boundary Conditions

Because of the hyperbolic nature of the convective part of the gas-phase equations, the number of physical boundary conditions to be specified is based on the method of characteristics. At a boundary, this number is thus equal to the number of characteristics running into the computational domain. For the running-out characteristics, numerical boundary conditions, which depend on the computed flow, must be formulated.

Let us first consider the case of a gas flow injected at the inlet boundary. The subsonic conditions at this boundary imply that only one characteristic is running out of the computational domain. Thus, we chose to specify the gas momentum components, the species mass fractions, and the temperature, whereas the pressure is extrapolated from downstream flowfield. When no propellant or only droplets are injected at the inlet boundary, this boundary is considered as a solid wall for the gas.

At the outlet boundary, the flow being supersonic, all characteristics are out running, so that boundary conditions need not be set. All variables are extrapolated from the computational domain. When the exit flow is subsonic at the outlet of the computational domain, which was the case for some computations, the Mach number is imposed.

Considering now droplets at the inlet boundary, a complete set of independent parameters is specified. For each group m, we impose the velocity components, volume concentration, and diameter.

Solid walls may be present at the inlet, inside the combustor for baffle modeling, or at the lower and upper boundaries for planar two-dimensional analysis. At these boundaries, the normal component of gas and droplet velocity is set to zero, along with an adiabatic wall assumption. Wall shear stress is determined by a wall function matching approach as

$$\tau_w = \beta_f \mu_l u_T^0 / \Delta \tag{45}$$

where Δ is the distance to the wall at which the shear stress is evaluated and β_f is a friction coefficient given by an analytical law as a function of the local Reynolds number:

$$\beta_f = 1 + 0.0236 \frac{(Re/80) + (Re/240)^2}{1 + (Re/80) + (Re/240)^2} (Re/4)^{0.811} \tag{46}$$

The Reynolds number is expressed as

$$Re = \rho u_T^0 \Delta / \mu_l \tag{47}$$

where u_T^0 is determined directly or by an iterative method, depending on whether or not Δ is in the viscous sublayer.

For tangential modes analysis using the annular geometry, the lower and upper boundary conditions are irrelevant in that it is only necessary to ensure periodicity.

E. Numerical Method

The partial differential equations for the gas and the liquid phases are solved by means of a finite difference method using the explicit predictor-corrector MacCormack[27] scheme. This is combined with a staggered grid; fluxes are defined at the cell sides, whereas scalar quantities are defined at the cell centers.

F. Instability Triggering Methods

To simulate unsteady behavior of the combustor, the model just described must be completed by a method of instability triggering. Different methods, based on industrial combustion stability rating techniques, have been investigated. They are all applied once the chamber steady-state flow has been computed.

The most commonly used method to rate engine dynamic stability is using explosive bombs. To model the finite amplitude pressure wave generated by a bomb, a sinusoidal pressure disturbance is initially imposed in a limited zone. Thus, the pressure distribution in this zone is given by

$$p(x, y) = \bar{p}(x, y)[1 + \alpha_p \sin(\beta_1 x + \beta_2) \sin(\gamma_1 y + \gamma_2)] \tag{48}$$

where $\bar{p}$ is the steady-state pressure and α_p, β_i, and γ_i are constants to be set. Similar equations are used for densities ρ and ρ_k and internal energy ρe, assuming that the bomb-supplied mass has the same temperature and composition as the gas in the chamber.

Since this excitation method simultaneously triggers longitudinal, tangential, and combined acoustic modes of the chamber, an alternative version has been used to trigger exclusively tangential modes. In this case, the disturbed pressure applied on the chamber generatrix is given by

$$p(x, y) = \bar{p}(x, y)[1 + \alpha_p] \tag{49}$$

Another triggering method intended to simulate siren tests, which were performed in small divergentless chambers, was implemented in the numerical model. During such siren tests, longitudinal acoustic modes are triggered by successive partial closing up of the throat by means of a rotating wheel. This excitation method was implemented in the code as a time variation of the throat width (b_t) at the longitudinal mode frequency during a few cycles.

IV. Assessment of Engine Stability

In this section, numerical analysis of the stability of a cryogenic engine, including the effect of baffle and acoustic resonator, is presented. The investigation concerned the gas generator and the thrust chamber of the Ariane V main-stage engine. Both the gas generator and the thrust chamber use liquid oxygen and gaseous hydrogen injected through coaxial injector elements.

A. Physical Submodels

Let us first describe the atomization and vaporization models used for the analyses.

1. *Atomization Model*

A cold-flow test program has been carried out to characterize coaxial injector sprays for both the gas generator and thrust chamber (see Chapter 6). Hardware configuration and test conditions were selected to make cold-flow simulant testing correspond as closely as possible to the engine operating conditions. Results of this program have led to the following correlation for the Sauter mean diameter D_{32} (in micrometers):

$$D_{32} = 127\left(\frac{V_r}{160}\right)^{-0.93}\left(\frac{D_l}{2.10^{-3}}\right)^{+2.25}\left(\frac{D_g}{3.10^{-3}}\right)^{-2.65}\left(\frac{rec}{D_l}\right)^{-0.26} \tag{50}$$

where V_r is the relative gas-droplet velocity (in meters per second), and D_l, D_g, and *rec* are the LOX post inner diameter, hydrogen annulus diameter, and recess, respectively (in meters).

This correlation, established with water/nitrogen and water/helium simulant combinations, is corrected for actual fluids (according to Ferrenberg et al.[28]), by multiplying the obtained diameter by the following coefficient K:

$$K = \frac{\left(\mu_L^{1/3}\sigma_L^{1/2}\rho_L^{-1/6}\right)_{\text{actual propellant}}}{\left(\mu_L^{1/3}\rho_L^{1/2}\sigma_L^{-1/6}\right)_{\text{simulant fluid}}} \tag{51}$$

where σ_L is the liquid surface tension.

2. *Vaporization Model*

Analyses of oxygen droplet vaporization and burning in a quiescent atmosphere were carried out under supercritical conditions. Different ambient conditions (pressure, temperature, and hydrogen mass fraction) were investigated. An analytical law for the droplet instantaneous diameter evolution was derived from numerical results. As the droplet surface reaches supercritical conditions, droplet diameter was defined as that of the sphere bounded by the critical isotherm. This law is similar to those obtained in subcritical combustion

$$D^2(t) = D_i^2 - (K_v t)^{\alpha_v} \left(D_i^2\right)^{1-\alpha_v} \tag{52}$$

Coefficients α_v and K_v depend on local pressure, temperature, and hydrogen mass fraction.

The vaporization rate in the absence of convection is then derived as

$$\dot{m}_0 = \rho_L \frac{\pi D(t)}{4} \left(D_i^2\right)^{(1-\alpha_v)/\alpha_v} \alpha_v K_v \left[D_i^2 - D^2(t)\right]^{(\alpha_v - 1)/\alpha_v} \tag{53}$$

The Ranz–Marshall correction is then applied to take into account the convection effects.

3. *Combustion Model*

To describe the H_2/O_2 combustion, the parameters appearing in the Arrhenius law were taken as[29]

$$CC = 1.210^6, \qquad T_A = 800\,\mathrm{K}, \qquad \alpha_1 = 1, \qquad \alpha_2 = 1 \tag{54}$$

B. Effect of Damping Devices on Engine Stability

Numerical analysis of combustion stability of both the Vulcain thrust chamber and gas generator was carried out for different operating conditions. We illustrate these results by presenting those corresponding to nominal working conditions.

The reference operating points in terms of total mass flow rate and overall LOX/H_2 mixture ratio are as follows. For the gas generator $\dot{m}_T = 8.25$ kg/s, and $r_m = 0.9$. For the thrust chamber $\dot{m}_T = 232$ kg/s, and $r_m = 5.9$.

The analysis was performed, using only one group of droplets to represent the spray. Injection droplet diameter and velocity were chosen as the Sauter mean diameter and the liquid-jet velocity, respectively. Droplets and gaseous hydrogen were uniformly injected at the chamber inlet.

To trigger instabilities, the computation was initiated by a bomb explosion as described earlier, the pressure perturbation having a sinusoidal form. The α_p constant, which represents the maximum overpressure, was $\alpha_p = 0.8$ for the gas generator and $\alpha_p = 0.5$ for the thrust chamber.

To illustrate the results, the time evolutions of pressure in a cell close to the inlet boundary are given. Reduced values are used; the reference pressure is the steady-state pressure, and the reference time is of the order of the millisecond.

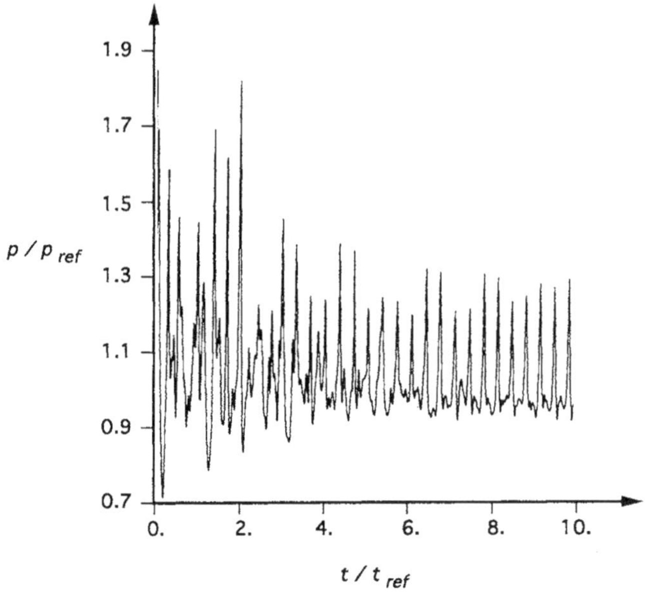

Fig. 2 Pressure response to bomb explosion; gas generator without a damping device.

1. *Gas Generator Results*

Figure 2 shows undamped pressure oscillations in the gas generator without any damping device. The corresponding Fourier analyses are presented in Fig. 3.This figure shows that all the first harmonics of the chamber acoustic modes are initially excited, but the longitudinal and combined modes are rapidly damped. Unstable modes are the tangential ones, especially the first tangential (1T) and second tangential (2T). Conversely, when the gas generator is equipped with a baffle (Fig. 4) or acoustic resonators tuned to the first tangential mode frequency (Fig. 5), its behavior is stable.

2. *Thrust Chamber Results*

The thrust chamber was found to be stable even without a damping device. A stable behavior was obtained at this operating point with overpressure intensities up to $\alpha_p = 2$.

C. Effect of Initial Droplet Size on Engine Stability

In the earlier analyses the gas generator without a damping device appeared to be the most sensitive to combustion instability. Thus, this chamber configuration was used to point out the effects of the spray injection parameters on the results. This problem was studied by varying the droplet injection diameter.

Computations were performed at the following operating point: $\dot{m}_t = 4.13$ kg/s, and $r_m = 0.94$. Three different values of injection diameter were used. The largest one, $D_i = 130\,\mu$m, was determined according to the correlation law [Eq. (50)] without the Ferrenberg and Varma correction; the smallest one, $D_i = 27.6\,\mu$m, was calculated taking into account this correction; and the third one, $D_i = 50\,\mu$m,

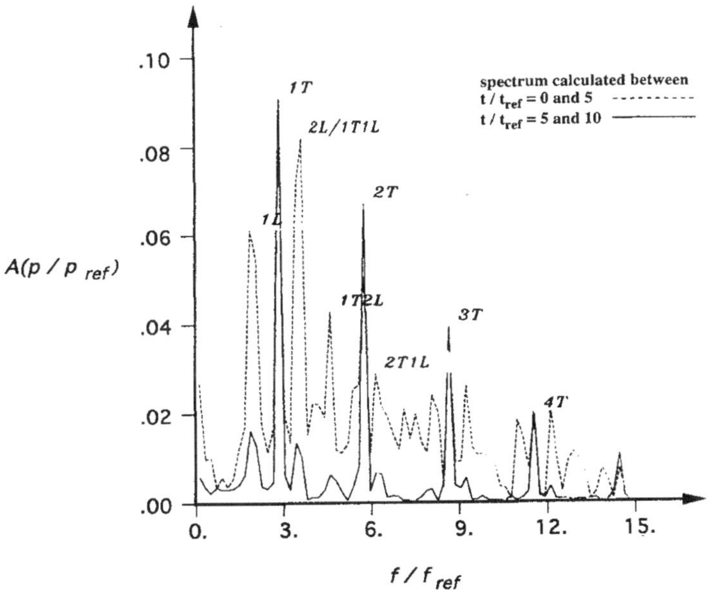

Fig. 3 Spectra of Fig. 2 curve.

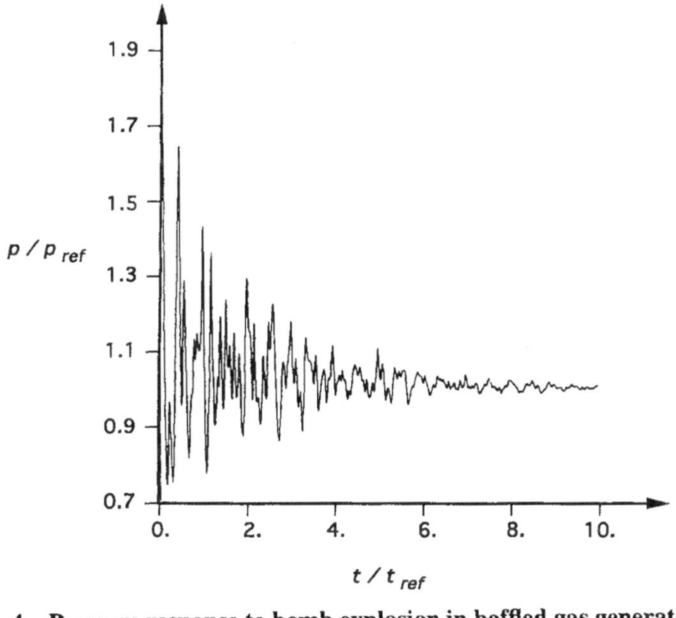

Fig. 4 Pressure response to bomb explosion in baffled gas generator.

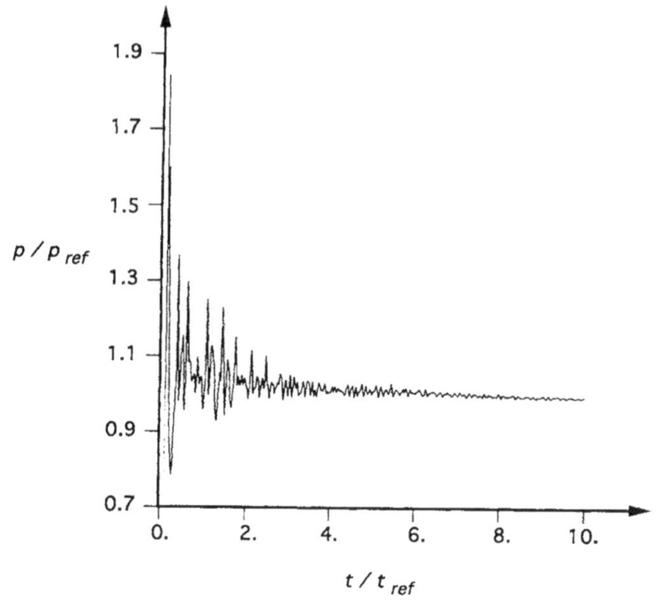

Fig. 5 Pressure response to bomb explosion; gas generator with acoustic resonators.

is an intermediate value. The unsteady computations were initiated using the bomb explosion method with a coefficient $\alpha_p = 1.5$.

Figures 6–8 show the pressure evolutions in a cell close to the inlet boundary using reference values as first defined. The two larger drop diameters lead to unstable operation whereas in the third case the fluctuations are damped. The larger the droplet size, the higher the amplitude of the pressure fluctuations. In the three cases, a large number of acoustic modes are initially excited, but most of them are quickly damped since the pressure signal rapidly appears well organized on one frequency. Spectral analysis of this signal indicates the presence of the second tangential mode and its harmonics. A second tangential combustion instability is thus associated with the largest droplet diameter, and to distributed vaporization. Injection of smaller droplets, which corresponds to high-vaporization rates in the injection zone, does not sustain instabilities. This result was confirmed by simulations for which the vaporized mass flow rate was artificially increased or decreased. An increase of this rate favors engine stability with regard to tangential modes.

V. Comparison of Test Results with Numerical Analysis

In this section, we discuss comparisons between experimental data and numerical results. Experimental data were obtained for the storable propellant Viking engine, a one element LOX/H_2 combustor, and the Vulcain gas generator.

A. Viking Engine

After a combustion instability problem occurred during the operation of the Viking engine, a campaign was waged to rate the engine stability for different operating conditions, propellant nature, and injector geometries. These tests, carried out by Sociéte Européenne de Propulsion (SEP),[30] enabled the characterization of

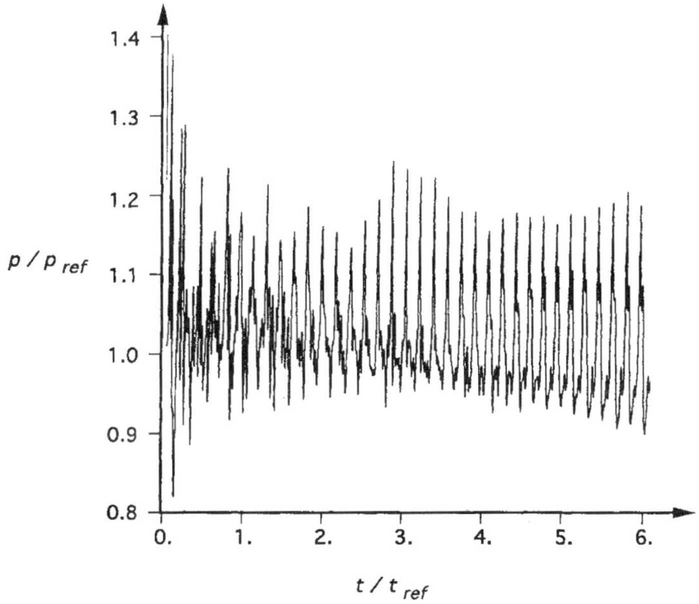

Fig. 6 Effect of droplet diameter on gas generator stability; pressure response to bomb explosion, D_i = 130 μm.

engine stability. The engine was more stable with the Viking injector type 2 (VI2) than with the Viking injector type 1 (VI1), the stability margin decreased as the working pressure increased, and the engine was more stable with the UH25 (75% of UDMH + 25% of hydrazine hydrate) than with UDMH.

Numerical computations were carried out to check whether these features could be recovered by the code.

The spray was represented by one group size of droplets with the Sauter mean diameter D_{32} derived from spray characterization studies.[12] Droplet injection velocity was set equal to the liquid-jet velocity. Droplet vaporization was described by analytical laws obtained from detailed droplet combustion analysis.[13] Gas-phase combustion was represented by the Arrhenius model with the following set of coefficients:

$$\begin{gathered} \upsilon_1 = 1, \qquad \upsilon_2 = 1.203, \qquad CC = 10^5, \qquad T_A = 0 \\ \alpha_1 = 1, \qquad \alpha_2 = 1.203 \end{gathered} \tag{55}$$

Different operating conditions were investigated, including injector geometry and working pressure. Instabilities were triggered by the bomb method. In the present analysis the disturbance had the following form:

$$p(x, y) = \bar{p}(x, y)[1 + \alpha_p] \tag{56}$$

Depending on the strength of the overpressure (value of α_p), the engine may or may not recover a stable operation. For given operating conditions, numerical analysis showed stable operation for small values of α_p and unstable operation

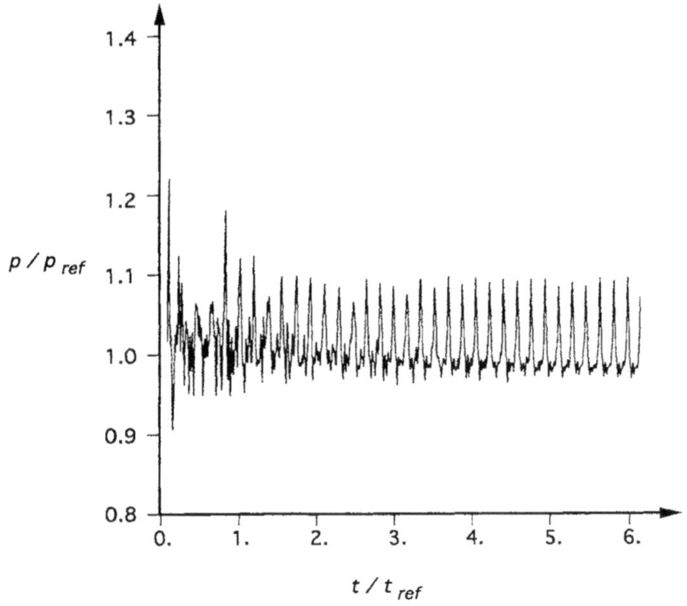

Fig. 7 Effect of droplet diameter on gas generator stability; pressure response to bomb explosion, D_i = 50 μm.

for high values. A critical value noted $\alpha_{p\text{lim}}$ corresponding to the transition from stable to unstable regime could be thus defined.

Numerical simulation shows that, for unstable operation, oscillations develop and reach a limiting cycle regime. The resulting peak-to-peak pressure level $\tilde{P}$ depends on the working pressure P_S.

Figures 9 and 10 present the final results of the engine stability assessment. Figure 9 shows the critical value $\alpha_{p\text{lim}}$ vs the reduced working pressure P_S/P_{ref}, and indicates that the stability margin decreases as the operating pressure increases. Figure 10 presents the pressure level $\tilde{P}/P_{\text{ref}}$ vs P_S/P_{ref}. Peak-to-peak pressure level is increased with increasing working pressure. What is quite clear from these figures is the stabilizing effect of the VI2 injector. In another study, (not reported here) the propellant (UH25 instead of UDMH) caused the stability margin to increase. These results fit quite well with the experimentally observed engine behavior.

B. Vulcain Engine Gas Generator

During the development program of the gas generator of the Vulcain engine, SEP conducted a stability rating program, including subscale and full-scale tests. Subscale tests were performed using a one-element LOX/H_2 combustion chamber designed to operate at the same conditions as the gas generator. Full-scale tests were performed with the gas generator.

In this section we discuss the numerical analysis carried out for both combustors and compare the analysis with experimental data.

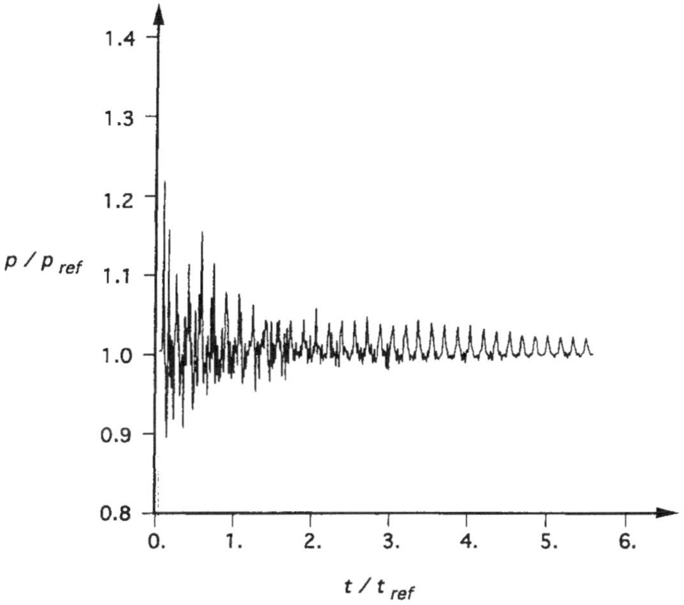

Fig. 8 Effect of droplet diameter on gas generator stability; pressure response to bomb explosion, $D_i = 27.6$ μm.

1. Subscale Tests

These tests, called siren tests, excite an acoustic longitudinal mode of the chamber by means of a cogwheel located at the chamber exit. The wheel speed rotation is adjusted so that the cogs partially close the exit section of the chamber at a frequency near the mode frequency and produce a pressure wave. An example of the experimental filtered pressure fluctuations recorded during a cycle is given in Fig. 11

These tests were simulated using the two-dimensional planar version of the numerical model. Excitation was achieved numerically by varying the throat width b_t, at the relevant frequency. The time variation of b_t is such that the throat area

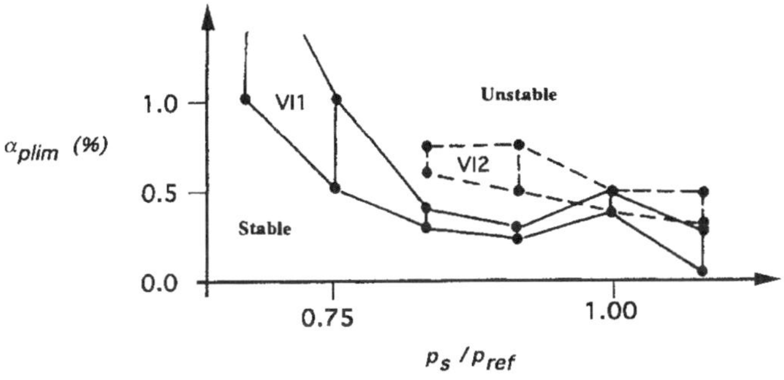

Fig. 9 Viking engine stability assessment; stability limit curve vs reduced working pressure.

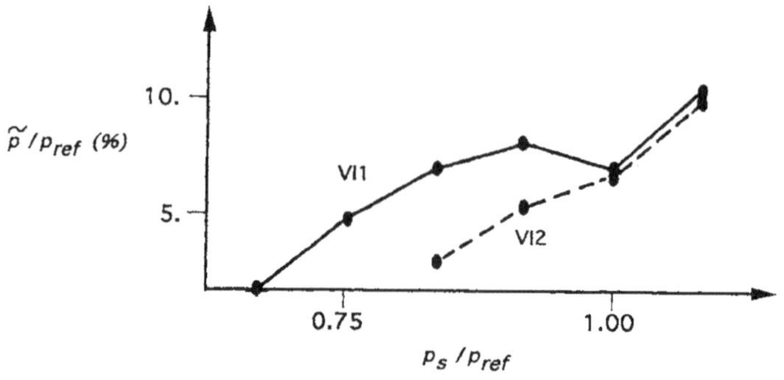

Fig. 10 Viking engine stability assessment; pressure level $\tilde{p}$ vs reduced working pressure.

of the annular chamber and the exit area of the real chamber remain equal at each time step.

An example of pressure evolution during a cycle (at the same operating conditions as those of Fig. 11) is given in Fig. 12. Pressure recorded near the injector face plate is filtered around the first longitudinal mode frequency. Reference time and reference pressure are defined as before. The figures indicate a growth of the oscillations as the throat is being closed, followed by an alleviation. The maximum pressure level corresponds to the maximum blocking of the throat. A damping ratio may be derived from these curves. The figures show a similar behavior between experimental and numerical results. The predicted maximum level, however, is higher than the experimental level.

Different operating conditions investigated during this numerical analysis included variable LOX/H_2 mixture ratio and working pressure. Both first and second longitudinal modes were triggered. The effect of the closed area on stability was also analyzed. Numerical results indicated that for given operating conditions of

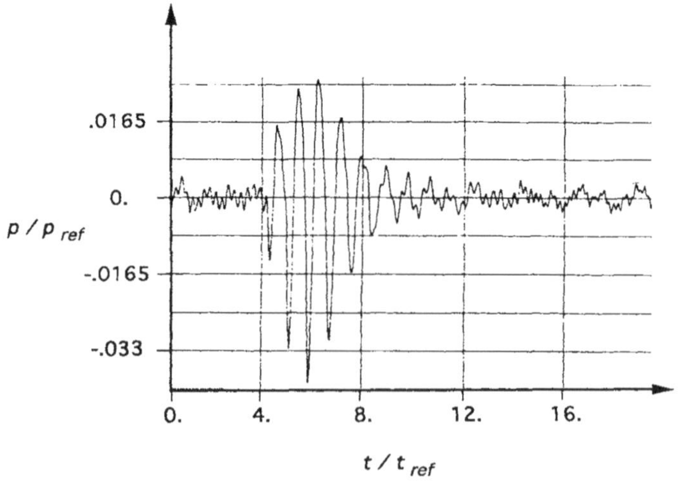

Fig. 11 Example of pressure evolution recorded during siren test, filtered signal.

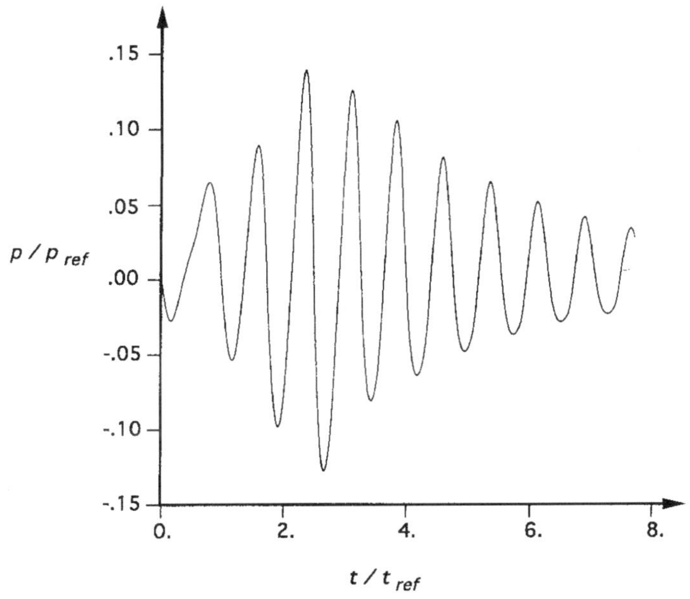

Fig. 12 Filtered pressure evolution during siren test numerical simulation, obturation = 50%.

the combustor, the maximum level of the pressure oscillations depends on the closed area, but the damping ratio does not.

Comparison between predicted and experimental results indicated a higher experimental damping ratio than had been calculated but relatively good agreement on the evolution of this damping with the mixture ratio. The damping ratio of both first and second longitudinal modes decreased with increased mixture ratio. Even if absolute values were not recovered, the results were useful in comparing the stability of the engine under different working conditions.

Differences between predicted and experimental damping ratios may be caused by nozzle geometry representation, atomization, and vaporization models that must still be improved.

2. *Full-Scale Tests*

These tests concern the assessment of dynamic stability of the gas generator. A series of bomb tests was carried out with a baffled and an unbaffled gas generator for this purpose. Different bomb sizes were used. The stability rating was determined for different working conditions. The tests were simulated with the Phedre numerical model, and results were compared with those of experimental tests.

The bomb is modeled using a group of cells called the bombed area, where the initial overpressure is imposed. The center of this area corresponds to the actual position of the bomb center.

Figure 13 is an example of the computed pressure evolution in a cell corresponding to the actual location of a pressure transducer. Figure 14 presents the same curve, filtered around 9000 Hz with a band width of 2000 Hz. The actual frequency of the second tangential mode was approximately 9600 Hz.

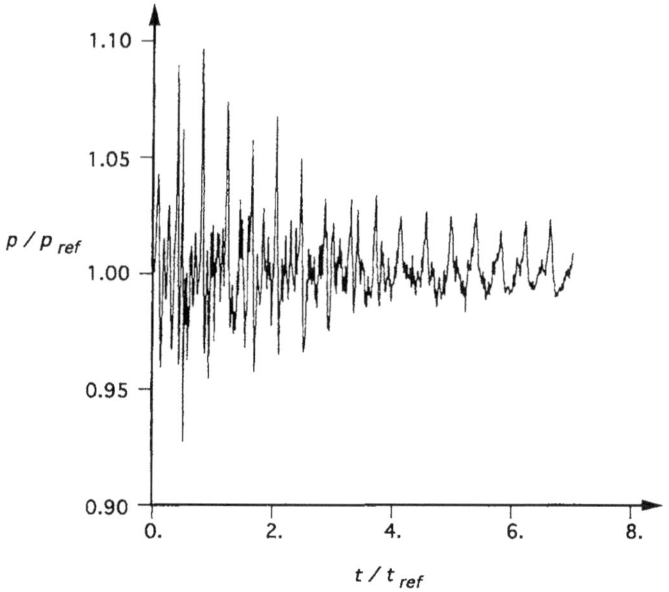

Fig. 13 Numerical simulation of bomb test in baffled gas generator; example of pressure evolution.

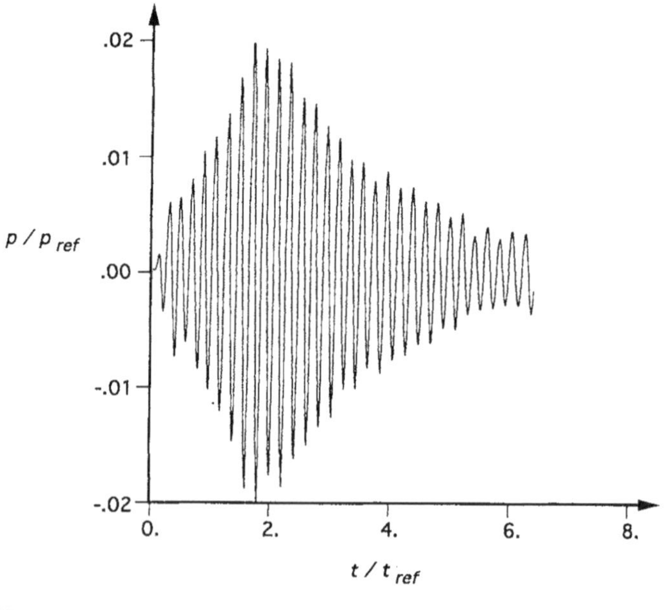

Fig. 14 Numerical simulation of bomb test in baffled gas generator; pressure evolution filtered around 9000 Hz.

After adjusting the initial overpressure at the center of the bombed area on one experimental operating point to recover the initial level recorded on the pressure transducers, the entire pressure-mixture ratio domain was explored, i.e., all operating points where a bomb test was performed were numerically analyzed.

The conclusion reached from this work was that although the absolute value of the experimental damping ratio was not recovered by the numerical model, relative trends were well predicted. Differences between the numerical model and test results may be due to the vaporization and atomization models. Moreover, the Vulcain gas generator has a three-dimensional outlet, which cannot be modelized by the Phedre code. Only the evolution of the section along the curvilinear abscissa on the generator axis could be represented.

VI. Conclusion

During the last four decades, liquid-propellant rocket engine development programs have been hampered by combustion instability problems. These problems remain a major concern for any rocket engine development program, particularly when a new design is involved.

Despite efforts to control and understand this phenomenon, rating the stability of an engine still relies on full-scale testing and semiempirical tools. Stabilizing devices such as baffles and acoustic resonators are often required to ensure stability.

The need for engines with high reliability and efficiency continues to increase. With limited funding for development programs, the instability problem cannot be solved by a trial and error method alone. Increased testing and more quantitative tools are needed.

Recent progress in diagnostic methods and computational resources offer new opportunities to develop tools that would help solving the problem. To control combustion instability, it is necessary to understand basic mechanisms; this requires extensive experimental research. Such efforts should involve both laboratory facilities accustomed to dealing with small-scale testing and intermediate- and full-scale test facilities. The latter are necessary to validate and complete data obtained by subscale testing, especially when laboratory experiments alone cannot cover the entire range of engine operating conditions.

Numerical analysis appears to be the proper direction toward better control of combustion instability. Current and future computers offer the means to achieve a better physical understanding of the detailed processes involved in liquid rocket operation from injection to combustion. Close contact must be maintained between research on basic mechanisms, numerical analysis, and full-scale testing. With this in mind, it should be possible to obtain predictive tools beneficial to the design of efficient and reliable engines.

The approach developed for liquid rocket combustion instability analysis involves development of a numerical model and research on physical modeling, with emphasis on spray characterization and detailed description of droplet vaporization and burning. A two-dimensional numerical model, Phedre, which incorporates various physical submodels (atomization, vaporization, and chemical reaction) is now available. It also includes baffles, acoustic resonators, and triggering methods.

Numerical analysis of combustion instability was performed for different combustors using storable and cryogenic propellants. Numerical results were compared with available data obtained from subscale or full-scale firing tests. Comparison between experiments and numerical analysis almost always revealed/ good qualitative agreement and identical general trends. The fact that quantitative agreement is not always satisfactory highlights the need for physical modeling improvement and more realistic geometries (i.e., a three-dimensional numerical model).

Currently efforts are being made to develop a three-dimensional numerical model for combustion instability analysis.[30] Investigations on basic processes to obtain more relevant physical submodels are also being conducted on the ONERA cryogenic bench MASCOTTE.

Acknowledgments

This work is supported by the Centre National d'Etudes Spatiales (CNES).

References

[1]Harrje, D. T., and Reardon, F. H. (eds.), *Liquid Propellant Rocket Combustion Instability,* NASA SP-194, 1972.

[2]Culick, F. E. C., "Combustion Instabilities in Liquid-Fueled Propulsion Systems—An Overview," AGARD-CP-450, Oct. 1988.

[3]Habiballah, M., Dubois, I., Vingert, L., Lourme, D., and Scherrer, D., "High Frequency Combustion Instability in Liquid Propellant Rocket Engines—A Review of Studies Carried out at ONERA for the Ariane Launcher," AIAA Paper 92-0774, Jan. 1992.

[4]Nguyen, T. V., "An Improved High-Frequency Combustion Stability Model," AIAA Paper 88-2853, 1988.

[5]Priem, R. J., and Breisacher, K. J., "Nonlinear Combustion Instability Model in Two-to Three-Dimensions," NASA-TM-102381, Oct. 1989.

[6]Philippart, K. D., and Moser, M. D., "Stability Analysis of Liquid Oxygen/Methane Injectors Using Currently Available Analytical Tools," AIAA Paper 88-2851, 1988.

[7]Baum, J. D., and Levine, J. N., "Numerical Techniques for Solving Instability Problems in Solid Rocket Motors," *AIAA Journal,* Vol. 20, No. 7, 1982, pp. 955–961.

[8]Levine, J. N., and Baum, J. D., "A Numerical Study of Nonlinear Instability Phenomena in Solid Rocket Motors," *AIAA Journal,* Vol. 21, No. 4, 1983, pp. 557–564.

[9]Baum, J., Levine, J., and Lovine, R., "Pulsed Instability in Rocket Motors: A Comparison Between Predictions and Experiments," *Journal of Propulsion and Power,* Vol. 4, No. 4, 1988, pp. 308–316.

[10]Lupoglazoff, N., and Vuillot, F., "Two-Dimensional Numerical Simulation of the Stability of a Solid Propellant Rocket Motor," AIAA Paper 91-0205, Jan. 1991.

[11]Vuillot, F., and Kuentzmann, P., "Flow Turning and Admittance Correction: An Experimental Comparison," *Journal of Propulsion and Power,* Vol. 2, No. 4, 1986, pp. 345–353.

[12]Habiballah, M., Lourme, D., and Pit, F., "PHEDRE—Numerical Model for Combustion Stability Studies Applied to the Ariane Viking Engine," *Journal of Propulsion and Power,* Vol. 7, No. 3, 1991, pp. 322–329.

[13]Dubois, I., and Habiballah, M., "Numerical Simulation of High Frequency Instability in an Oxygen/Hydrogen Rocket Engine," AIAA Paper 91-1860, 1991.

[14]Priem, R. J., and Guentert, D. C., "Combustion Instability Limits Determined by a Nonlinear Theory and a One-Dimensional Model," NASA-TN-D 1409, Oct. 1962.

[15]Hoffman, R. J., Wright, R. O., and Breen, B. P., "Combustion Instability Prediction Using a Bipropellant Vaporization Model," NASA-CR-920, 1968.

[16]Burstein, S. A., Chinitz, W., and Schechter, H. S., "A Nonlinear Model of Combustion Instability in Liquid Propellant Rockets Engines," AIAA/SAE 8th Joint Propulsion Specialist Conf., AIAA Paper 72-1146, 1972.

[17]Kim, Y. M., Chen, C. P., and Ziebarth, J. P., "Numerical Simulation of Combustion Instability in Liquid-Fueled Engines," 30th Aerospace Sciences Meeting, AIAA Paper 92-0775, Reno, NV, Jan. 1992.

[18]Kim, Y. M., Chen, C. P., Ziebarth, J. P., and Chen, Y. S., "Prediction of High Frequency Combustion Instability in Liquid Propellant Rocket Engines," AIAA/SAE/ASME/ASEE 28th Joint Propulsion Conf., AIAA Paper 92-3763, Nashville, TN, July, 1992.

[19]Liang, P., and Ungewitter, R., "Multi-phase Simulations of Coaxial Injector Combustion," 30th Aerospace Sciences Meeting, AIAA Paper 92-0345, Reno, NV, Jan. 1992.

[20]Lourme, D., Schmitt, D., and Brault, F., "Experimental Characterization of the Spray Originating from an Impinging-Jet Injector in a Liquid Rocket Engine," *Acta Astronautica,* Vol. 11, No. 7–8, 1984, pp. 469–482.

[21]Brault, F., and Lourme, D., "Experimental Characterization of the Spray Formed by Two Impinging Jets in Liquid Rocket Injector," *ICLASS-85, 3rd International Conf. on Liquid Atomization and Spray Systems,* Inst. of Energy, London, July 1985.

[22]Lourme, D., "Like-on-like Injector Spray Characterization for the Ariane Viking Engine," AIAA Paper 86-1444, 1986.

[23]Vingert, L., "Coaxial Injector Spray Characterization for the Ariane 5 Vulcain Engine," 6th Annual Conf. Liquid Atomization and Spray Systems in Europe, Pisa, Italy, July 1990.

[24]Scherrer, D., "Effet de la Convection sur l'Evaporation et la Combustion d'une Goutte de Combustible," AGARD-CP-422, Oct. 1987.

[25]Lafon, P., "Modélisation et Simulation Numérique de l'Evaporation et de la Combustion de Gouttes à Haute Pression," Ph.D. Thesis, Université d'Orléans, Orléans, France, 1994.

[26]Deardorff, J. W., "On the Magnitude of the Subgrid Scale Eddy Coefficient," *Journal of Computational Physics,* Vol. 7, 1971, pp. 120–133.

[27]Mac Cormack, R. W., "The Effect of Viscosity in Hypervelocity Impact Craterting," AIAA Paper 69-354, 1969.

[28]Ferrenberg, A., Hunt, K., and Duesberg, J., "Atomization and Mixing Study," NASA-CR-178751, Dec. 1985.

[29]Przekwas, A. J., Edwards, J., and Gross, K., "SSME Thrust Chamber Modeling With Navier–Stokes Equations," AIAA Paper 86-1517, 1986.

[30]Souchier, A, Lemoine, J. C., and Dorville, G., "Résolution du Probléme des Instabilités sur le Moteur Viking," *Proceedings of the 33rd International Astronautical Federation Congress,* IAF Paper 82-363, 1982.

[31]Dubois, I., Habiballah, M., and Lecourt, R., "Numerical Analysis of Liquid Rocket Engine Combustion Instability," AIAA Paper 95-0607, 1995.

Chapter 19

Application of Computational Fluid Dynamics Techniques to Engine Instability Studies

J. M. Grenda,* S. Venkateswaran,* and C. L. Merkle†
Pennsylvania State University, University Park, Pennsylvania 16802

I. Introduction

OUR present understanding of liquid rocket combustion instability has been achieved through a combination of experimental investigations and approximate analytical models. Computational fluid dynamics (CFD) has now advanced to the point where it can potentially complement these existing methods. The purposes of this chapter are to assess the potential role of CFD in combustion instability research and to explore the issues that are involved in bringing CFD methods to bear on combustion instability problems.

One of the advantages of computational analyses is that they permit very detailed physical descriptions of the liquid- and gas-phase processes to be coupled with the acoustics of the combustion chamber. Consequently, a most promising role of CFD is to serve as a computational testbed for demonstrating the effect of new analytical models of important subprocesses such as atomization and vaporization on the stability characteristics of a combustor. Similarly, such a computational testbed can be used to investigate the impact of experimental findings on stability behavior.

The development of a computational testbed for combustion instability requires careful validation against both analytical solutions and experimental results. Validation against analytical solutions is necessary to ensure that the numerical results represent accurate solutions to the equations and submodels that are chosen to represent the physics of the instability process. Validation against experimental data is necessary to ensure that the equations and submodels are physically meaningful.

In the present chapter, we discuss some of the fundamental characteristics that computational instability models must possess and identify some computational problems that are encountered in modeling the instability process. We also discuss the necessity for ensuring that computational models are sufficiently accurate and present methods for validation. Finally, to demonstrate current capabilities of

*Research Associate, Department of Mechanical Engineering.

†Professor of Mechanical Engineering.

computational instability methods, we present some representative results based on currently available physical submodels.

Computational instability models for liquid propellant engines are currently being developed in several groups. The model developed by Habiballah et al.[1] and Refs. 2–4 at ONERA, which has been applied to real engines perhaps more extensively than any other, solves the Navier–Stokes equations for the gas phase plus Eulerian formulation for the liquid phase. Their model considers annular geometries and includes comprehensive submodels for atomization, vaporization, turbulence, and finite rate chemistry. A similarly detailed model by Liang et al.[5,6] likewise includes atomization, vaporization, secondary breakup and droplet agglomeration, turbulence, species diffusion, and chemical kinetics. This model has been applied to both steady-state combustion and instability simulations.

In more recent work, Kim et al.[7] have reported solutions of the reacting Navier–Stokes equations for both axisymmetric and annular geometries, and Litchford and Jeng[8] have coupled an unsteady vaporization model based on a simple pressure-sensitive response with a two-dimensional annular geometry. In addition to these CFD models, Priem and Breisacher[9] have developed a three-dimensional, frequency-domain analysis that incorporates injection, atomization, and vaporization submodels. Their model provides the linear response to single-frequency disturbances although the physical subprocesses are incorporated as full nonlinear models.

The CFD analysis of the present authors[10–12] is philosophically similar to the Priem–Breisacher approach except that it is a time-domain method, and all nonlinear terms are retained. The time-domain solution enables the direct simulation of experimental events such as spontaneous disturbance growth and the response to bombs, but requires larger CPU requirements than does the linearized frequency-domain method. To circumvent this and to provide validation of our fully nonlinear analysis, we also have developed a companion linear time-domain analysis that is competitive with the Priem–Breisacher[9] analysis in CPU requirements. Wang et al.[13] have recently presented similar CFD results for the Euler equations utilizing a two-parameter collapsed combustion zone assumption analogous to that previously used by the current authors.

In the following section, we present the governing equations for the two-phase flowfields in liquid rocket engine combustors and outline appropriate models for the contributing physical subprocesses. Following this discussion, the important issues and numerical aspects of computational combustion instability are addressed.

II. Governing Equations

Computational analyses of combustion instability require equation sets for the gas and the liquid phases along with appropriate interaction terms. The formulation for the gas phase entails standard equations although there may be some argument as to which terms should be retained. The analysis of the liquid phase requires a description of the continuous liquid as it enters the chamber, the ensuing atomization process, and the dynamics of the droplet cloud as it traverses the combustor, as well as expressions for droplet vaporization, mixing, and combustion. The submodels for these processes entail substantial physical modeling. Here, we concentrate upon employing available descriptions of the atomization and vaporization processes.

A. Selection of the Gas-Phase Equations

The first major issue in modeling gas-phase phenomena is whether to start from the Navier–Stokes equations or the Euler equations. The Reynolds-averaged Navier–Stokes equations coupled with species diffusion and chemical kinetics represent the most complete description of combustion instability, but the simpler Euler equations accurately describe the wave dynamics while omitting only the effects of viscous wave dissipation. The primary diffusive effects occur in the boundary layers along the combustor walls where they have little effect on the wave dynamics. Consequently, boundary layers may profitably be ignored and replaced by slip boundary conditions regardless of whether diffusive and turbulence effects are included elsewhere in the flowfield.

The dominant diffusion processes in the instability problem are those associated with the coupling between the energy release process and the gas dynamics. These include microscale mixing on the droplet level and mean flow recirculation effects near the injector face on the scale of individual injector elements. The microscale mixing phenomena between the vaporizing droplets and the surrounding hot gases most certainly affect the instability mechanisms, but their scale is too small to be resolved with either an Euler or a Navier–Stokes formulation. Consequently, their effect must be included globally in either representation so that their presence does not detract from the choice of an Euler formulation.

The mixing on the scale of injection elements may, however, be compromised by the choice of the Euler equations. Individual elements introduce oxidizer/fuel (O/F) striations into the flow whose convection and redistribution by acoustic disturbances could provide driving mechanisms. In addition, the recirculation zones near the injector face on the individual element scale could impact both steady and unsteady atomization. Unfortunately, these scales are also probably too small to be resolved in unsteady Navier–Stokes solutions.

In view of the preceding arguments and the computational requirements needed for the spray dynamics, we choose here to focus on the Euler equations. We emphasize that an accurate, validated Euler solution is a necessary precursor to a reacting Navier–Stokes solution. This extension requires few additional new concepts in terms of numerics or physics although the required computational resources are larger.

The Euler equations in cylindrical coordinates take the form

$$\frac{\partial \boldsymbol{Q}}{\partial t} + \frac{\partial \boldsymbol{E}}{\partial x} + \frac{1}{r}\frac{\partial \boldsymbol{F} r}{\partial r} + \frac{1}{r}\frac{\partial \boldsymbol{G}}{\partial \theta} = \frac{1}{r}(\boldsymbol{H}_{\mathrm{gas}} + \boldsymbol{H}_{\mathrm{liq}}) \tag{1}$$

where x, r, and θ are the axial, radial, and azimuthal coordinates. The vectors, $\boldsymbol{Q}, \boldsymbol{E}, \boldsymbol{F}$, and $\boldsymbol{G}$, are given by

$$\boldsymbol{Q} = \begin{bmatrix} \rho \\ \rho u \\ \rho v \\ \rho w \\ e \end{bmatrix} \quad \boldsymbol{E} = \begin{bmatrix} \rho u \\ \rho u^2 + p \\ \rho u v \\ \rho u w \\ (e+p)u \end{bmatrix} \quad \boldsymbol{F} = \begin{bmatrix} \rho v \\ \rho u v \\ \rho v^2 + p \\ \rho v w \\ (e+p)v \end{bmatrix} \quad \boldsymbol{G} = \begin{bmatrix} \rho w \\ \rho u w \\ \rho v w \\ \rho w^2 + p \\ (e+p)w \end{bmatrix} \tag{2}$$

Here, u, v, and w are the velocity components in the axial, radial, and azimuthal directions, respectively. The density and pressure are given by ρ and p, and the total energy is defined in terms of the internal energy $e = \rho\varepsilon + 1/2\rho(u^2 + v^2 + w^2)$, where $\varepsilon = RT/\gamma - 1$ and R and γ represent the gas constant and specific heat ratio. For simplicity, we use constant specific heats and close the system by means of the perfect gas formulation $p = \rho RT$. Although written here in cylindrical coordinates, it is assumed that generalized coordinates are used throughout.

The source terms appearing on the right-hand side of Eq. (1) include radial and azimuthal effects, $\boldsymbol{H}_{\text{gas}}$, that arise from placing the governing equations in cylindrical coordinates, and the coupling terms, $\boldsymbol{H}_{\text{liq}}$, between the liquid and gas phases. These vectors are given as

$$\boldsymbol{H}_{\text{gas}} = \begin{bmatrix} 0 \\ 0 \\ p + \rho w^2 \\ -\rho v w \\ 0 \end{bmatrix} \qquad \boldsymbol{H}_{\text{liq}} = \begin{bmatrix} S_c \\ S_{mx} \\ S_{mr} \\ S_{m\theta} \\ S_e \end{bmatrix} \tag{3}$$

The terms in $\boldsymbol{H}_{\text{gas}}$ are standard. The quantities in $\boldsymbol{H}_{\text{liq}}$ represent mass, momentum, and energy transfer from the liquid to the gas. It is through these terms that effects such as droplet vaporization and combustion energy release are coupled to the gas phase. These terms are defined subsequently.

B. Liquid-Phase Equations

For the liquid phase, we divide the droplet size distribution into bins of class k, partition the entering liquid among each droplet class at each instant of time, and track each class in Lagrangian fashion. As the droplets vaporize in the chamber, they transfer mass, momentum, and energy to the gas phase, thereby determining the source term $\boldsymbol{H}_{\text{liq}}$ in Eq. (3).

The dynamic equations that govern each droplet class include an appropriate force balance and companion mass and energy conservation statements, written in vector form as

$$\frac{\mathrm{d}\boldsymbol{Q}_\ell}{\mathrm{d}t} = \boldsymbol{H}_\ell \tag{4}$$

where

$$\boldsymbol{Q}_\ell = [m_{d,k}, \boldsymbol{v}, T_d]^T \qquad \boldsymbol{H}_\ell = \left[-\dot{m}_{\text{vap},k}, \frac{\Sigma \boldsymbol{F}}{m_{d,k}}, \Delta\dot{E}_s\right]$$

Here, $\boldsymbol{v}$ is the droplet velocity, $\boldsymbol{F}$ the drag forces associated with the relative velocities between the phases, $m_{d,k}$ the mass of a droplet in class k (kilogram per droplet), T_d the droplet temperature, and $\dot{m}_{\text{vap},k}$ the vaporization rate of the droplet (kilogram per second per droplet).

The temperature distribution within the droplet is obtained from an energy balance whose form is dependent on the particular approach. Here we use $\Delta\dot{E}_s$ to represent an appropriate difference between the energy arriving at and leaving the droplet surface. Options for treating the droplet temperature are considered in more detail subsequently.

C. Coupling Between the Liquid and Gas Phases

The source term $\boldsymbol{H}_{\text{liq}}$ in Eqs. (1) and (3) is determined by dividing the combustor into finite volumes and summing the mass and energy generation rates and the forces in Eq. (4) over all classes of particles in each volume. When expressed in the vector form used for the gas-phase equations, Eq. (1), these relations become

$$
\boldsymbol{H}_{\text{liq}} = r \begin{bmatrix} \sum_k n_k \dot{m}_{\text{vap},k} / \Delta V \\ \sum_k \left(n_k \dot{m}_{\text{vap},k} u_k - \frac{1}{6} \pi \rho_k \text{d}_k^3 n_k \frac{\text{d}u_k}{\text{d}t} \right) \Big/ \Delta V \\ \sum_k \left(n_k \dot{m}_{\text{vap},k} v_k - \frac{1}{6} \pi \rho_k \text{d}_k^3 n_k \frac{\text{d}v_k}{\text{d}t} \right) \Big/ \Delta V \\ \sum_k \left(n_k \dot{m}_{\text{vap},k} w_k - \frac{1}{6} \pi \rho_k \text{d}_k^3 n_k \frac{\text{d}w_k}{\text{d}t} \right) \Big/ \Delta V \\ \sum_k n_k \dot{m}_{\text{vap},k} \left[\overline{c_{p,f}} (T_k - T_g) - \lambda + h_{\text{comb}} \right] / \Delta V \end{bmatrix} \tag{5}
$$

Here, ΔV is the local cell volume, u_k, v_k, and w_k are the droplet velocities, d_k is the droplet diameter, n_k is the number of droplets, and T_k and T_g are the entering vapor temperature and the ambient gas temperature. Remaining quantities in the energy source are the appropriately averaged specific heat of the vapor/gas mixture $\bar{c}_{p,f}$, the latent heat λ, and the heat of combustion h_{comb}.

D. Phenomenological Models for Physical Subprocesses

The evaluation of the sources in Eqs. (4) and (5) requires an atomization model and a vaporization model. Here, we take these models from the available literature.

1. *Atomization Models*

Typical atomization models include both a mean drop size correlation and a droplet distribution function. Some representative mean drop size correlations[14–16] for coaxial injectors are provided in Table 1. In these expressions, u_{rel} is the relative velocity between the velocity and the gas phase; the bulk properties include the liquid viscosity μ_ℓ, the surface tension σ, and the gas and liquid densities ρ and ρ_ℓ. In addition, D_1 and D_2 are post and injector diameters, and B is an empirical constant. From these mean drop size correlations and the liquid density, the mass per droplet and the number of droplets formed can be found.

Table 1 Summary of atomization models for collapsed liquid zone

Model	Mean dropsize correlation
Ingebo[14]	$d = 0.123 D_1 \left[\left(\frac{\rho}{\rho_\ell} \right) \frac{\rho^2 D_1^2 u_{\text{rel}}^2}{\sigma \mu_\ell} \right]^{-0.44}$
Mayer[15]	$d = 9\pi \sqrt[3]{16} B \left[\frac{\mu_\ell \sigma^{\frac{1}{2}}}{\rho u_{\text{rel}} \sqrt{\rho_\ell}} \right]^{\frac{2}{3}}$
Nurick[16]	$d = 3.3 \times 10^{-5} \left[\frac{\mu_\ell^{0.61}}{(u - \mu_\ell)^{1.4}} \right] \left[\frac{D_1}{D_2} \right]$

The application of any steady-state atomization model in a quasisteady sense in an unsteady CFD code gives time-varying drop sizes, but this quasisteady treatment is only a stop-gap measure. Unsteady transverse disturbances are expected to have an important effect on droplet size, and models that include unsteady effects are urgently needed. Research aimed at providing more detailed models of the atomization process from first-principles calculations is currently underway. Although such calculations offer promise of providing improved atomization models, the effects of crossflow velocities and unsteady gas-phase conditions are beyond near-term computer capabilities, and they will probably have to be obtained from experiments. In the interim, the present quasisteady implementations must serve as a logical starting point for CFD-based instability analyses.

2. *Droplet Vaporization*

Once the liquid droplets have been formed by atomization, they are convected through the combustor by the high-temperature gases while they simultaneously vaporize and burn. Substantial effort has gone into developing phenomenological models of spray vaporization. Most models are based on global heat and mass transfer balances over an isolated droplet in a quiescent environment that are later adjusted for convective effects. As a representative model, we consider the Priem and Heidmann model[17] that expresses the vaporization rate $\dot{m}_{\text{vap}}$ and the heat feedback q_v as

$$\dot{m}_{\text{vap}} = \pi d \bar{\rho} D Sh \, \ln[p/(p - p_{v,s})] \tag{6}$$

$$q_v = \left[\dot{m}_{\text{vap}} \bar{c}_{p,v} (T_g - T_{d,s})/(e^z - 1)\right] \tag{7}$$

For clarity, the bin subscript k has been dropped. The parameters in Eqs. (6) and (7) include the mass diffusivity D, the partial pressure of the vapor at the droplet surface $p_{v,s}$, and the Sherwood number for mass transfer Sh. Additional predictions given later also include the Abramzon–Sirignano model.[18]

The expressions for the vaporization rate and the heat flux in Eqs. (6) and (7) require a knowledge of the droplet temperature. There are several levels of sophistication that can be used to model the droplet temperature variation. These range from an unsteady, two-dimensional treatment of the internal droplet fluid dynamic processes to the simple assumption of a uniform fixed droplet temperature. For a multidimensional computation with a large number of droplets, a full solution of the flowfield within each droplet is computationally prohibitive. At the other extreme, however, it is likely that a fixed droplet temperature will not allow appropriate stability predictions. A tractable compromise is the rapid-mixing limit in which the droplet temperature remains uniform because of strong liquid convection effects, but its time rate of change varies according to the difference between the energy leaving the surface due to vaporization and the energy arriving due to heat transfer,

$$\frac{\mathrm{d}T_d}{\mathrm{d}t} = \frac{q_v - \dot{m}_{\text{vap}} \lambda}{m_d c_{p,\ell}} \tag{8}$$

where T_d corresponds to the droplet temperature. In an unsteady gas phase, this model allows temporal oscillations in the droplet temperature that induce a phase difference between the vaporization and gas oscillations. This is an important instability driving mechanism as is demonstrated later.

3. *Other Potential Coupling Mechanisms*

In addition to atomization and vaporization mechanisms, other phenomena including injector coupling, droplet-turbulence interactions, chemical kinetics, and secondary droplet breakup can also drive combustion instability. As with the work discussed here, algebraic correlations for any of these phenomena could, in principle, be incorporated in numerical simulations of combustion instability.[19]

III. Computational Fluid Dynamics Aspects of Combustion Instability

A. Separation of the Equations into Mean and Disturbance Components

The computational resources required for solving the Navier–Stokes or Euler equations in three-dimensional unsteady flowfields restricts the number of solutions that can be practically computed. Consequently, it is important to be able to rely on some simpler analyses for obtaining initial qualitative understanding and for providing a guide and validation procedure for the full numerical solutions. We obtain a simple analysis by splitting the flow variables into mean and fluctuating quantities and linearizing the results. Accordingly, expand the vectors $\boldsymbol{Q}$, $\boldsymbol{E}$, $\boldsymbol{F}$, $\boldsymbol{G}$, and $\boldsymbol{H}$ as

$$\begin{aligned} \boldsymbol{Q}(x, r, \theta, t) &= \bar{\boldsymbol{Q}}(x) + \boldsymbol{Q}'(x, r, \theta, t) \\ \boldsymbol{E} &= \bar{\boldsymbol{E}} + \boldsymbol{E}' = \bar{\boldsymbol{E}} + \bar{A}\boldsymbol{Q}' \end{aligned} \tag{9}$$

and similarly for the flux vectors $\boldsymbol{F}$, $\boldsymbol{G}$, and $\boldsymbol{H}$. In these expressions, overbars refer to the mean and primes to unsteady fluctuations, and $\boldsymbol{H}$ indicates the sum of $\boldsymbol{H}_{\text{liq}}$ and $\boldsymbol{H}_{\text{gas}}$. The Jacobians are given by $A = \partial \boldsymbol{E}/\partial \boldsymbol{Q}$, $B = \partial \boldsymbol{F}/\partial \boldsymbol{Q}$, $C = \partial \boldsymbol{G}/\partial \boldsymbol{Q}$, and $\boldsymbol{D} = \partial \boldsymbol{H}/\partial \boldsymbol{Q}$.

Substituting Eq. (9) into Eq. (1) gives separate equation sets for the mean flow and the unsteady portion. The mean flow equations are uncoupled from the unsteady part and can be solved before the perturbation solution. The perturbation equation takes the linear three-dimensional form,

$$\frac{\partial \boldsymbol{Q}'}{\partial t} + \frac{\partial \bar{A}\boldsymbol{Q}'}{\partial x} + \frac{1}{r}\frac{\partial \bar{B}\boldsymbol{Q}'r}{\partial r} + \frac{1}{r}\frac{\partial \bar{C}\boldsymbol{Q}'}{\partial \theta} = \frac{1}{r}\bar{D}\boldsymbol{Q}' \tag{10}$$

Although this equation is linear, its solution takes nearly as much effort as solving the entire nonlinear equations.

To obtain an equation whose numerical solution can be obtained very rapidly, we restrict the mean flow to one-dimensional (axial) variations and take the radial and azimuthal mean velocities $\bar{v}$ and w as zero so that the equations admit a separable solution,

$$\boldsymbol{Q}' = \begin{bmatrix} \hat{\rho}(x, t)RZ \\ \hat{\rho u}(x, t)RZ \\ \hat{\rho v}(x, t)R'Z/\lambda \\ \hat{\rho w}(x, t)RZ'/\lambda r \\ \hat{e}(x, t)RZ \end{bmatrix} \tag{11}$$

where R is a Bessel function of order μ, $R = C_3 J_\mu(\lambda r)$, and Z is the trigonometric function, $Z = C_1 \cos(\mu\theta) + C_2 \sin(\mu\theta)$. (Superscript primes denote differentiation with respect to the arguments.) The wave number λ is determined by requiring that the normal velocity fluctuations vanish at the wall, $J_\mu(\lambda r_w) = 0$, where r_w denotes the wall radius.

After substituting Eq. (11) into Eq. (10) and separating variables, we obtain the equation for the amplitude function,

$$\frac{\partial \hat{Q}}{\partial t} + \frac{\partial \bar{A}\hat{Q}}{\partial x} + \bar{\bar{B}}\hat{Q} = \bar{D}_{\text{liq}}\hat{Q} \tag{12}$$

where the vector $\hat{Q} = (\hat{\rho}, \hat{\rho}u, \hat{\rho}v, \hat{\rho}w, \hat{e})^T$, and the matrices $\bar{A}, \bar{\bar{B}}$, and $\bar{D}_{\text{liq}}$ are given elsewhere.[19] Note that the solution requires that $\hat{\rho}v$ and $\hat{\rho}w$ be identical.

As Eq. (12) suggests, the three-dimensional perturbation relation has been reduced to an unsteady equation in one space variable, even though the resulting solutions represent full three-dimensional disturbances. This equation can be solved at a small fraction of the effort of the fully three-dimensional set in Eq. (10) and can be used very effectively as a diagnostic tool. Further, it provides an efficient means of performing exploratory three-dimensional stability surveys prior to attempting nonlinear computations. The full nonlinear calculations can then be validated against these small amplitude results in the small disturbance limit.

B. Constant Coefficient Equation and Analytical Solution

By requiring the mean flow and the atomization, vaporization, and combustion models to be independent of x, Eq. (12) can be solved analytically. These solutions are restricted in the physical problems they simulate, but they are useful for qualitative understanding and numerical validation. The restrictions on the analytical solution can later be removed in the numerical code to test the impact of simplifying assumptions on the overall physics. Representative comparisons of this type are given in Sec. IV. The analytical solution also requires a formal linearization of the physical submodels for atomization, vaporization, and injector coupling.

If no mean axial variations are present and the liquid-phase source term is taken as constant, Eq. (12) admits solutions of the form

$$p' = \text{Re}\left[\left(P_1 e^{-k^+ x} + P_2 e^{-k^- x}\right) J_\mu(\lambda r) e^{i\mu\theta} e^{i\omega t}\right] \tag{13}$$

with similar relations for the remaining variables. Here, ω is the frequency, $k^\pm$ are the axial wave numbers of the disturbance, P_1 and P_2 denote the amplitude of the pressure, and Re refers to the real part.

Because our reason for obtaining an analytical solution is to understand the physical aspects of the instability problem and to validate the numerical algorithm, we choose a two-zone, distributed combustion model as depicted in Fig. 1, and linearize the atomization and vaporization submodels in this region. We start from the atomization relations given in Table 1 that show that the droplet size depends on the liquid- and gas-phase velocities, the physical properties of the liquid, and the thermodynamic state of the gas. To represent the gas, we choose its pressure and entropy s. We then express the atomization relations in the functional form, $d = d(p, u, s, T_\ell, u_\ell)$ where u here represents the velocity magnitude, and the subscript ℓ refers to the liquid. For simplicity, we ignore variations in the liquid and droplet velocity and consider an isentropic gas-phase fluctuation. These are

Table 2 Summary of linearized collapsed atomization models

Model	α_d	β_d
Ingebo[14]	$\dfrac{-1.32}{\gamma}$	$-0.88\dfrac{\bar{u}}{\bar{u}_{\text{rel}}}$
Mayer[15]	$-\dfrac{2}{3\gamma}$	$-\dfrac{4}{3}\dfrac{\bar{u}}{\bar{u}_{\text{rel}}}$
Nurick[16]	0	$-1.4\dfrac{\bar{u}}{\bar{u}_{\text{rel}}}$

retained in the full numerical solutions. With these approximations, the droplet size fluctuations depend on the pressure and velocity only and can be expressed as

$$\frac{d'}{\bar{d}} = \alpha_d \frac{p'}{p} + \beta_d \frac{u'}{\bar{u}} \tag{14}$$

These expressions can easily be incorporated into the linearized formulation presented in the preceding section. A summary of the linearization constants for the three models for a known mean drop size presented in Table 1 is given in Table 2.

A similar expansion of the vaporization correlation in Eq. (6) gives the linearized vaporization expression

$$\frac{\dot{m}'_{\text{vap}}}{\bar{\dot{m}}_{\text{vap}}} = \alpha_v \frac{p'}{\bar{p}} + \beta_v \frac{u'}{\bar{u}} + \delta_v \frac{T'_\ell}{\bar{T}_\ell} \tag{15}$$

where $\dot{m}'_{\text{vap}}$ indicates the fluctuation in the vaporization rate. Here, we have expressed the generic vaporization model as a function of the flow quantities,

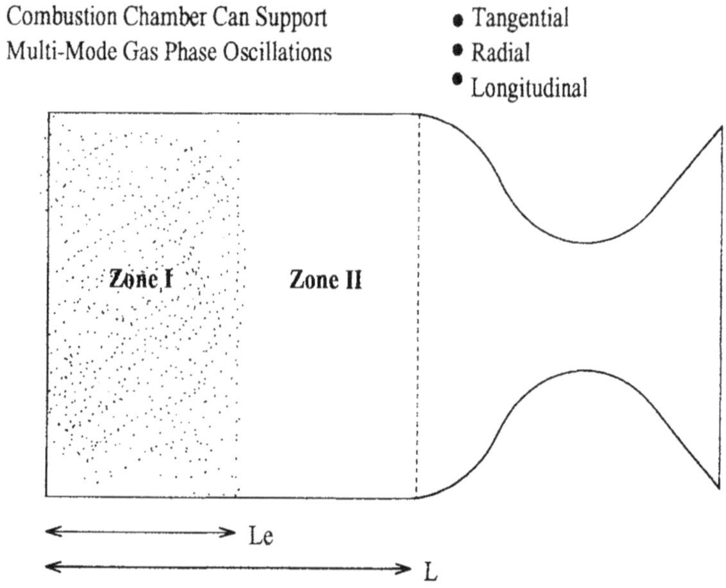

Fig. 1 Schematic of the two-zone computational problem.

$\dot{m}_{\text{vap}} = (p, u, s, T_\ell)$, and here again, we omitted variations in the droplet velocity and are restricted to isentropic gas fluctuations. Note that the vapor-pressure dependence in Eq. (6) can be expressed implicitly as a dependence on the liquid temperature through the Clausius–Clapeyron equation, $p_v = f(T_\ell)$.

Equation (15) can be reduced to two variables by combining it with an analogous linearized expression for the heat feedback to the droplet [Eq. (7)],

$$\frac{q_v'}{\bar{q}_v} = \left[\alpha_q \frac{p'}{\bar{p}} + \beta_q \frac{u'}{\bar{u}} + \delta_q \frac{T_\ell'}{\bar{T}_\ell}\right] \tag{16}$$

where the constants α_q, β_q, and δ_q represent the partial derivatives for the heat transfer relations. Expressions for these coefficients are given in Table 3 for the Priem–Heidmann model. Here, Re is the droplet Reynolds number, and Pr and Nu are the Prandtl and Nusselt numbers, respectively.

To eliminate the temperature dependence in Eq. (16), we linearize the energy balance in Eq. (8) to give

$$\frac{\mathrm{d}T_\ell'}{\mathrm{d}t} = \frac{1}{\bar{m}_d \bar{c}_{p,\ell}}\left[q_v' - \dot{m}_{\text{vap}}' \bar{\lambda} - \overline{\dot{m}_{\text{vap}}'} \frac{\mathrm{d}\lambda}{\mathrm{d}T_\ell} T_\ell'\right] \tag{17}$$

where the derivative of the heat of vaporization with respect to the liquid temperature is here retained to allow expressions that depend on both T_ℓ' and its time rate of change. Then, combining Eqs. (15–17) and using complex notation, we obtain the two-parameter vaporization expression,

$$\dot{m}_{\text{vap}}' = \bar{\dot{m}}_{\text{vap}}\left[\alpha_v^* \frac{p'}{\bar{p}} + \beta_v^* \frac{u'}{\bar{u}}\right] \tag{18}$$

Table 3 Summary of Priem–Heidmann linearization constants

	$\alpha_v(X_1/\gamma) - (p_{v,s}/Y)$
	$\beta_v(uX_1/\lvert u - u_d\rvert)$
	$\delta_v = \dfrac{T_\ell}{Y}\dfrac{\mathrm{d}p_{v,s}}{\mathrm{d}T_\ell}$
	$\alpha_q = \alpha_v + \dfrac{\gamma - 1}{\gamma}\dfrac{T}{T - T_\ell} + \dfrac{Z^2 p_v}{Y} - \dfrac{Z^2}{\gamma}(X_1 - X_2)$
	$\beta_q = \beta_v - Z^2\left[\dfrac{X_1}{\lvert u - u_d\rvert} - \dfrac{X_2}{\lvert u - u_d\rvert}\right]$
	$\delta_q = \delta_v - [T_\ell/(T - T_\ell)] - Z^2\delta$
	$Z = e^z/(e^z - 1)$
Note:	$X_1 = \dfrac{0.3Re^{1/2}Sc^{1/3}}{2 + 0.6Re^{1/2}Sc^{1/3}}$
	$X_2 = \dfrac{0.3Re^{1/2}Pr^{1/3}}{2 + 0.6Re^{1/2}Pr^{1/3}}$
	$Y = (P - P_{v,s})\ell n[P/(P - P_{v,s})]$
	$z = \dot{m}_{\text{vap},k} c_{p,\ell}/\pi dkNu$

where the complex coefficients α_v^* and β_v^* are given by

$$\alpha_v^* = \alpha_v + \frac{\delta}{\phi}\left[\alpha_q - \bar{\lambda}\frac{\bar{\dot{m}}_{\text{vap}}}{\bar{q}_v}\alpha_v\right] \tag{19}$$

$$\beta_v^* = \beta_v + \frac{\delta}{\phi}\left[\beta_q - \bar{\lambda}\frac{\bar{\dot{m}}_{\text{vap}}}{\bar{q}_v}\beta_v\right] \tag{20}$$

and the quantity ϕ is given by

$$\phi = \frac{\bar{\dot{m}}_{\text{vap}}\bar{T}_\ell}{\bar{q}_v}\frac{d\lambda}{dT_\ell} + \frac{\bar{\dot{m}}_{\text{vap}}\bar{\lambda}}{\bar{q}_v}\delta_v - \delta_q + iw\frac{m_d c_{p,\ell}\bar{T}_\ell}{\bar{q}_v} \tag{21}$$

Equation (18) provides a relationship for the rate of vaporization from a liquid droplet as a function of pressure and velocity fluctuations. The expressions α_v^* and β_v^* are functions of the complex frequency of the gas-phase oscillations. Because Eq. (17) is a differential rather than an algebraic expression, the vaporization response is not in phase with the pressure and velocity oscillations. This out-of-phase component arises from the temperature fluctuations within the liquid droplet and has a major impact on stability characteristics.

In the absence of droplet temperature fluctuations, Eq. (18) reduces to Eq. (16), with the T_ℓ' variations set to zero. In this case, α_v^* and β_v^* become real and reduce to the linearization coefficients α_v and β_v. The values of α_v^* and β_v^* along with corresponding values for α_d and β_d directly determine the stability characteristics of a combustor.

Finally, we note that the effects of atomization and vaporization can be combined with other physical phenomena such as injector coupling or secondary breakup to obtain a complete instability model. The overall linearization coefficients relating unsteady responses to disturbances in the chamber can be combined as

$$\alpha = \alpha_v^* - 3\alpha_d + \alpha_{inj} + \cdots \tag{22}$$

with an analogous expression for β where the factor of three multiplying the droplet fluctuation changes the droplet size fluctuation to a vaporization fluctuation.

C. Nonlinear Finite Difference Spectral Procedure

The solution of nonlinear three-dimensional transient flowfields remains nontrivial from the standpoint of computational resources. Finite difference solutions of the three-dimensional unsteady equations given in Eq. (1) coupled to the liquid phase may become prohibitively expensive. Because of limitations in grid size and difficulties caused by the presence of the axis of symmetry, economical solutions to the three-dimensional Euler equations are perhaps most efficiently obtained through a spectral decomposition of the tangential direction. Traditional finite differences are used in the axial and radial directions. This mixed character approach utilizes a Fourier expansion of the primary variable $\boldsymbol{Q}$ represented by truncated series of the form

$$\boldsymbol{Q} = \sum_{m=0}^{M}\left[\hat{\boldsymbol{Q}}_{m,c}(x,r,t)\cos m\theta + \hat{\boldsymbol{Q}}_{m,s}(x,r,t)\sin m\theta\right] \tag{23}$$

where the amplitude coefficients are

$$\hat{Q}_{m,c} = (\hat{\rho}_c, \hat{\rho u}_c, \hat{\rho v}_c, \hat{\rho w}_c, \hat{e}_c, 0, 0, 0, 0, 0)^T \tag{24}$$

$$\hat{Q}_{m,s} = (0, 0, 0, 0, 0, \hat{\rho}_s, \hat{\rho u}_s, \hat{\rho v}_s, \hat{\rho w}_s, \hat{e}_s)^T \tag{25}$$

These coefficients corresponding to the Fourier mode m are time-dependent functions of both the axial and radial directions. The general series formulation permits both spinning and standing mode wave solutions to be calculated.

By substituting these expressions into Eq. (1), the general dynamic equation governing the development of the flowfield in Fourier space can be written as

$$\frac{\partial \hat{Q}_{m,T}}{\partial t} + \frac{\partial \hat{E}(\hat{Q}_{m,T})}{\partial x} + \frac{1}{r}\frac{\partial \hat{F}(\hat{Q}_{m,T})^r}{\partial r} = \mp \frac{m}{r}\hat{G}(\hat{Q}_{m,T}) + \frac{1}{r}\hat{H}_{\text{tot}}(\hat{Q}_{m,T}) \tag{26}$$

Here, the vectors $\hat{E}$, $\hat{F}$, $\hat{G}$, and $\hat{H}_{\text{tot}}$ are written as functions of the amplitude coefficients $\hat{Q}_{m,T}$. In this expression, the variable $\hat{Q}_{m,T}$ is defined as the sum of the amplitude coefficients $\hat{Q}_{m,c}$ and $\hat{Q}_{m,s}$, and $\hat{H}_{\text{tot}}$ is a combination of the gaseous- and liquid-phase source term vectors. This relationship reduces the three-dimensional equations for the continuum gas phase to $2M + 1$ sets of two-dimensional equations. For standing mode calculations, the equations can be recast in a form in which each of the dynamic equations decouple in the tangential direction. This results in only $M + 1$ sets of two-dimensional equations.

In both the standing and spinning mode computations, the calculation of the shape functions in the Fourier domain provide the flow variables in three-dimensional physical space. These are defined by the summation of the truncated series of the primary vector $\hat{Q}$. In either situation, the spectral decomposition procedure should result in a decrease in computational time required over three-dimensional finite difference calculations since the Fourier series representation in $M + 1$ or $2M + 1$ modes replaces the discretized azimuthal direction. The solution of the general dynamic equation given in Eq. (26) is obtained through an explicit four-step Runge–Kutta time integration. A more detailed discussion of the mixed finite difference/Fourier spectral scheme can be found in Ref. 19.

IV. Representative Results

In this section, we present representative results to give concrete examples of the concepts just discussed and to illustrate how available models of atomization and vaporization can be combined with CFD analyses to study instability characteristics. The use of existing physical submodels enables instability characteristics to be tied to specific hardware configurations such as injector type and size, engine operating conditions, and combustor geometry. In addition, the identification of the most sensitive parameters in the models points the way for further diagnostic experiments, whereas the computational tools provide a testbed in which the effects of these findings on stability characteristics can be determined. We begin with an investigation of the global topology of the unstable regions of combustion instability models based on representative atomization–vaporization descriptions. Chamber conditions for the following predictions are: chamber diameter and length (chamber of length 0.36 m and radius 0.23 m), mean flow Mach number, 0.10 (150 m/s), droplet injection velocity 100 m/s and temperature 125 K, andmean pressure 40 atm and temperature 3000 K.

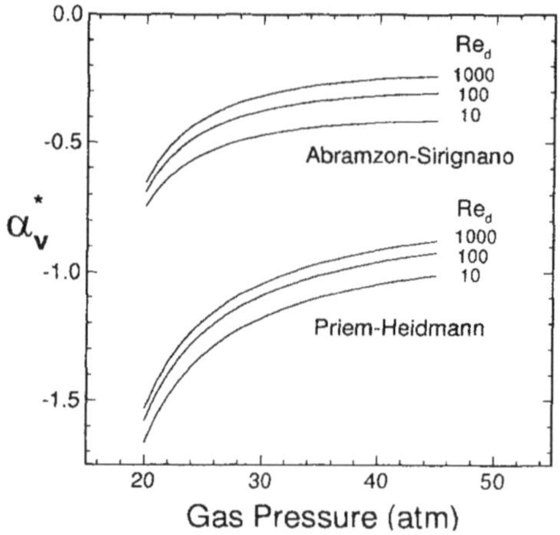

Fig. 2 Comparison of linearized pressure coefficients for Priem–Heidmann and Abramson–Sirignano vaporization models, $\bar{T}_{gas} = 3000$ K and $d = 110$ μm.

A. Global Stability Characteristics of Atomization-Vaporization-Based Models

The nondimensional coefficients α^* and β^* represent a two-parameter combustion response function that defines a complete stability map. To apply such two-parameter analyses, both the range of α^* and β^* values that lead to instability and the range of values that is physically realizable must be determined. The stability characteristics of an engine are then defined by the overlap between these two α^*–β^* regions. To facilitate understanding, we begin by considering the case where the droplet temperature is fixed so that the parameters α and β become real so that the analysis can be presented on the α^*–β^* plane.

The magnitude of the atomization parameters can be inferred directly from Table 2, but the vaporization parameters require some calculation. To demonstrate their magnitude, representative ranges are given in Figs. 2 and 3 for both the Priem–Heidmann[17] and Abramzon–Sirignano[18] models. Figure 2 shows that α_v^* is negative for both models ($\dot{m}'_{vap}$ decreases as pressure increases) and is a weak function of droplet Reynolds number, whereas Fig. 3 shows that β_v^* is positive and decreases with Reynolds number. The rapid decrease in β_v^* at low Reynolds numbers reflects the increased sensitivity as the mean relative velocity approaches zero.

Given these physically realizable ranges of α_v^* and β_v^*, we now compare them with the companion stability characteristics obtained by solving the two-parameter model. Results for the most unstable (or least stable) mode selected from the first tangential (1T), first radial (1R), and arbitrary longitudinal (nL)(typically first or second) mode are shown in Fig. 4. These results are obtained from a combination of the analytical and linearized numerical models.

The two-parameter formulation shows an unstable region in the upper left hand corner of the α–β plane as indicated by the neutral stability curve (solid line). Note that $\alpha = \alpha_v$ and $\beta = \beta_v$ for this constant droplet temperature case. The dashed lines show contours of constant growth (decay) rate, with positive values indicating growth and negative values attenuation. The magnitude of these numbers indicates the percent amplitude change per cycle. Realistic ranges of the α–β parameters corresponding to the two vaporization models (Figs. 2 and 3)

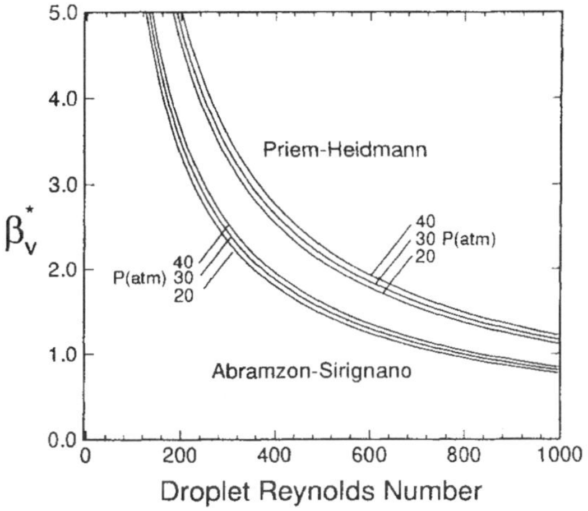

Fig. 3 Comparison of linearized velocity coefficients for Priem–Heidmann and Abramson–Sirignano vaporization models, $\bar{T}_{gas}$ = 3000 K and d = 110 μm.

are also shown on Fig. 4. As can be seen, these values lie well inside the stable region. The vaporization-only results are more stable than the combination of atomization plus vaporization, but neither the Abramzon–Sirignano model nor the Priem–Heidmann model predicts instability for the constant temperature droplet case. The α–β parameter range for combined atomization-vaporization results is, however, close enough to the neutral stability curve to suggest that the addition of a third effect (for example, injection coupling) could lead to instability in this constant droplet temperature case.

Before leaving this initial example, we pause to reflect on the difficulty of verifying that a given instability model is absolutely stable while using only a CFD code. Proof of absence of instability without analytical guidelines is essentially impossible, even though it is critical to establishing instability mechanisms by turning various effects "on" and "off" as is so easy to do with CFD analyses. This observation is the primary reason for pursuing the present synergistic modeling philosophy.

As a second example, we consider the effects of temperature fluctuations in the droplets. Because the coefficients α and β now become complex, the stability diagram cannot be shown on a plane as in Fig. 4. Instead, we present the results as a function of the temperature and size of the entering droplets. Again, we consider the 1T, 1R, nL modes. Initial droplet temperatures between 118 and 145 K are considered for initial droplet diameters ranging from 25 to 400 μm. The vaporization length for the smallest (25 μm), warmest (145 K) droplets was some 20% of the chamber length whereas the larger, colder drops did not completely vaporize within the chamber as the triangular upper bound on the results in Fig. 5 suggests.

Two neutral stability curves are shown on Fig. 5. The solid line represents the effects of vaporization alone (Priem–Heidmann[17]), whereas the curve with long and short dashes depicts the neutral stability curve for a combination of atomization (Ingebo[14]) and vaporization (Priem–Heidmann[17]). The stable region is to the right of these curves while the unstable region is to the left. The dashed lines show contours of constant amplification (decay) rate for the vaporization-only

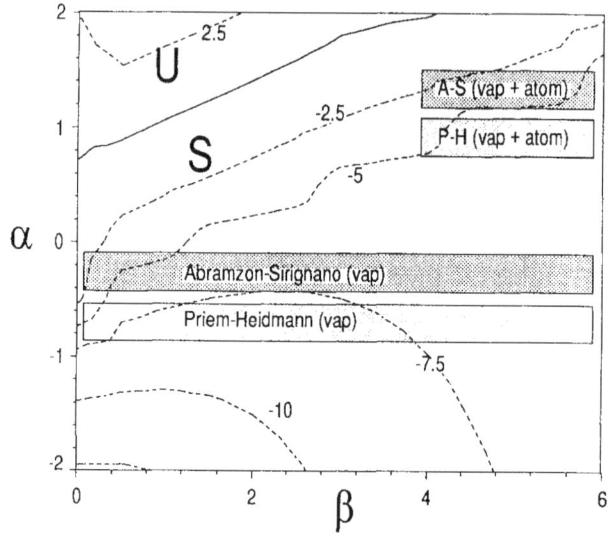

Fig. 4 Stability map as a function of the linearized pressure and velocity coefficients for constant temperature droplets, showing parameter range for Priem–Heidmann and Abramzon–Sirignano models for vaporization alone and atomization plus vaporization; $\bar{T}_{liq}$ = 125 K, d = 110 μm, $\bar{T}_{gas}$ = 3000 K, and $\bar{P}_{gas}$ = 40 atm.

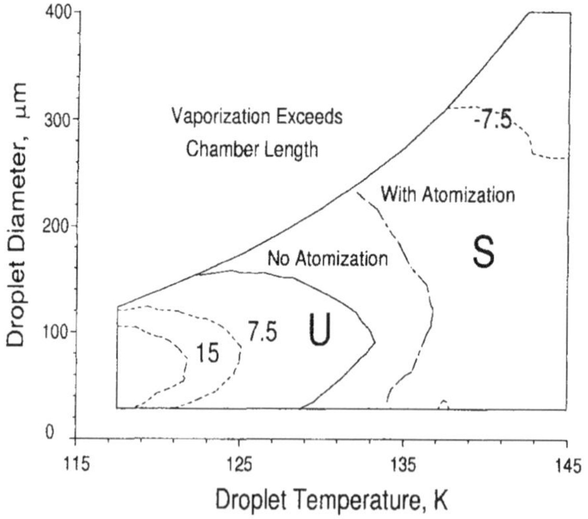

Fig. 5 Stability map as a function of droplet diameter and liquid temperature for fluctuating temperature droplets using Priem–Heidmann vaporization model, $\bar{T}_{gas}$ = 3000 K and $\bar{P}_{gas}$ = 40 atm.

case with numbers again indicating the percent growth per cycle. As can be seen, the temperature lag introduced by the presence of temperature fluctuations in the droplets moves the realistic parameter range into the unstable region. Colder drops tend to drive instability whereas warmer drops do not. Note that the supercritical properties range is not considered in these atomization–vaporization models.

B. Validation and Error Estimation of Numerical Solutions by Comparison with Analytical Solutions

The analytical solution described can be used to validate both the linearized numerical procedure and the full nonlinear analysis in the small amplitude limit. The validation referred to here has only to do with ensuring that the numerical results represent the correct solutions to the stability model. The issue of verifying the accuracy of the stability model must be done by comparison with experiment.

In validating the numerical procedures, we use an initial condition corresponding to a single-mode analytical solution. The resulting amplitude growth and phase error can then be determined by comparing with the exact solution. We consider an instability model based on vaporization effects alone. We also include the effects of temperature fluctuations. Inlet conditions for the LOX droplets are 120 K and a monodisperse droplet diameter of 90 μm. These conditions result in a vaporization length of about half the combustor. This operating condition is subject to an unstable 1T, 1R, 1L mode that grows approximately 20% per cycle as shown in Fig. 5.

Comparisons of the numerical and analytical pressure oscillations are given in Fig. 6 for two angular positions near the wall and the injector face for the linearized numerical analysis with 201 grid points in the axial direction. As can be seen, the numerical and analytical solutions are nearly indistinguishable. Quantitative evaluation of the difference indicates an amplitude error of 0.1% per cycle and a

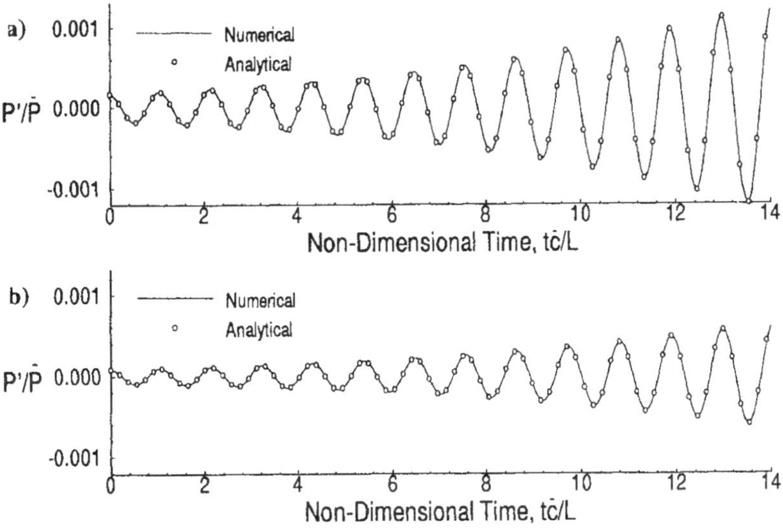

Fig. 6 Comparison of linearized numerical and analytical pressure fluctuation vs time traces at injector face/wall location; fluctuating temperature droplets, $\bar{T}_{liq}$= 120 K and d = 89 μm; three-dimensional gas phase, pure mode initial conditions: a) 0 deg and b) 60 deg.

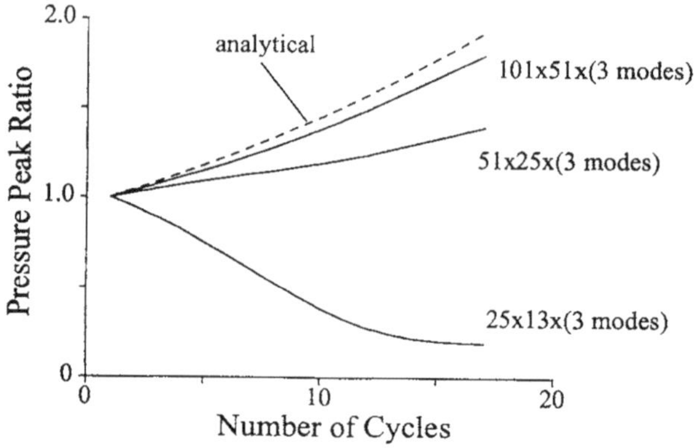

Fig. 7 Comparison of nonlinear numerical and analytical pressure fluctuation envelopes vs oscillation cycle at injector face/wall/zero-deg location; pure mode initial conditions.

phase error of 0.3 deg per cycle. Similar accuracies are obtained for the velocity fluctuations but are not shown here. Computations of the same case but with constant temperature droplets (not shown) predict that the disturbances decay with time and give accuracies similar to those in Fig. 6.

An analogous study of numerical accuracy in the nonlinear code is presented in Fig. 6. Here we show the envelope of the pressure peaks for three grid sizes, $25 \times 13 \times 3, 51 \times 26 \times 3$, and $101 \times 51 \times 3$. (Note the three Fourier modes in the azimuthal direction give quite accurate representation of disturbances with low θ wave numbers.) The amplitude results in Fig. 7 show that the sparse ($25 \times 13 \times 3$) grid calculation predicts wave decay rather than wave growth, whereas the intermediate ($51 \times 26 \times 3$) grid predicts growth but at a rate that is too slow. The finest ($101 \times 51 \times 3$) grid gives quite accurate results that are certainly acceptable for engineering use. Similar results for the phase error[19] show that the finest grid again gives reliable predictions.

These grid refinement studies show that second-order algorithms require some 40 grid points per wavelength. This grid density requirement can be reduced by switching to fourth-order systems that require only about 10 points per wavelength.[20]

C. Response to Arbitrary Disturbances

Unstable disturbances in liquid rocket engines generally arise from broadband disturbances within the chamber. With a broadband initial condition, each mode grows or decays according to its individual stability character, and the long-time solution will be dominated by the most unstable mode. The analytical solution predicts the most unstable mode (apart from nonlinear effects) for any set of operating conditions and can again be used as a check on the numerical results. Long-term nonlinear calculations should differ from linear calculations thereby displaying effects of finite amplitude.

Solutions obtained with the linearized numerical model to verify that it will pick out the most unstable mode are given in Fig. 8. Figure 8a is for the fluctuating droplet temperature case whereas Fig. 8b is for the constant droplet temperature case. The initial condition for the fluctuating temperature droplet is an arbitrary

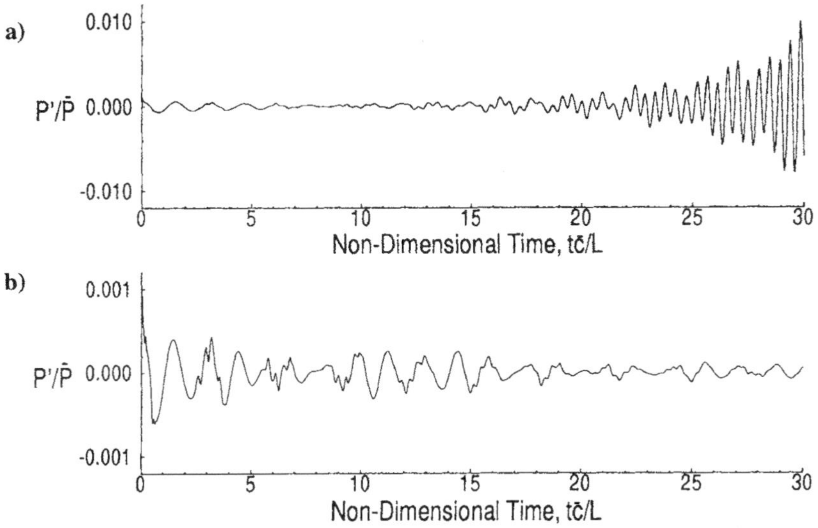

Fig. 8 Comparison of pulsed linearized numerical pressure-time traces at injector face/wall location 0 deg; three-dimensional gas phase: a) variable temperature droplets $\bar{T}_{liq}$ = 125 K and d = 56 μm and b) constant temperature droplets $\bar{T}_{liq}$ = 125 K and d = 56 μm.

pressure disturbance of 0.1% magnitude applied near the injector face and superimposed on an otherwise steady flow. The short-term response to this disturbance contains low-frequency waves that slowly propagate through the flowfield. As the disturbances interact with the distributed combustion process, the oscillations grow, and a beat frequency appears as the selective wave amplification by the instability model begins to take over. When carried out for long times, the final growth rate agrees with the analytically predicted maximum growth rate to within 1/2%, thus giving further validation of the analysis.

Corresponding calculations for the constant temperature droplet result in stable response over the entire spectrum as can be seen from Fig. 8b. At large times, the disturbance in the numerical solution decays to zero, thereby indicating that the numerical procedure is likewise globally stable to all disturbances. To obtain a nontrivial problem, this case is started with a combination of a single-mode solution plus an arbitrary disturbance.

The preceding examples have relied on the two-zone, constant mean flow conditions that were used to obtain analytical solutions. A uniform mean flow assumption is not necessary for the linearized numerical solutions (or the full Euler equations), and we now relax this restriction. Because a variable mean flow does not allow analytical solutions, we lack an appropriate initial condition for single-mode solutions, and so all results are limited to the response to arbitrary disturbances.

A variable mean flow solution obtained with the Priem–Heidmann[17] vaporization model is given in Fig. 9a for an initial LOX droplet size of 120 μm. The chamber pressure at the upstream is 35 atm and the mass flow rate increases with axial distance to a final value of 7.7 kg/s. The corresponding species mass fractions of hydrogen and water indicate the rapid change in the flow character near the injector face as the LOX droplets vaporize and burn. The inlet Mach number starts at 0.05 and increases.

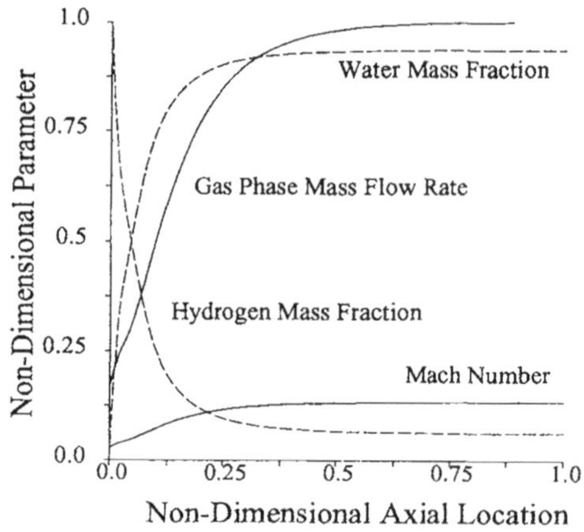

Fig. 9a Steady-state mean flow solution for coupled gas/liquid phases; gas-phase mass flow rate, Mach number, and species mass fractions, flow rate normalized with respect to exit value of 7.8 kg/s; O/F=7, $\bar{P}_{gas}$= 35 atm, upper-limit droplet distribution with $\bar{d}$ = 120 μm, σ = 2, injection temperature of 90 K with u_d = 100 m/s.

Starting from these nonuniform mean flow conditions, a transient solution was initiated by providing a pressure pulse of 1% amplitude near the injector face. The combustion stability model was here taken as the Priem–Heidmann[17] vaporization model (without atomization) with constant temperature drops. Results of the pressure fluctuations at the head end of the motor are given in Fig. 9b.

D. Finite Amplitude Effects

Extension of these linear computations to the nonlinear regime are shown in Figs. 10 and 11. Figure 10 shows representative solutions that contain wave steepening. These results are for a single-mode initial condition at two different initial amplitudes, 2.0% (Fig. 10a), and 20% (Fig. 10b). The disturbance corresponds to a weakly decaying mode. The numerical predictions for the 2.0% initial amplitude

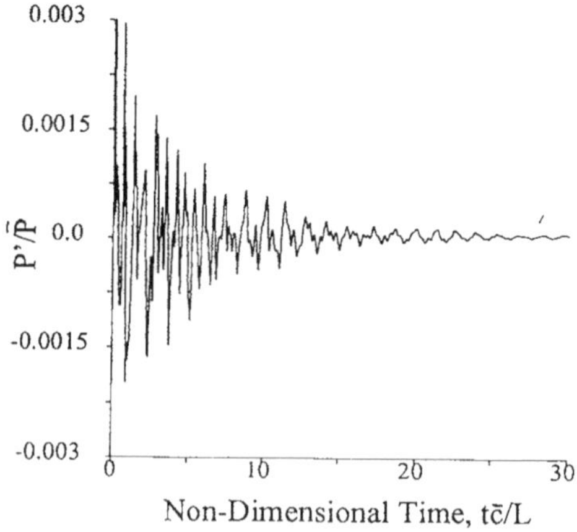

Fig. 9b Pulsed linearized numerical pressure-time traces at injector face/wall location 0 deg for previous mean flow conditions.

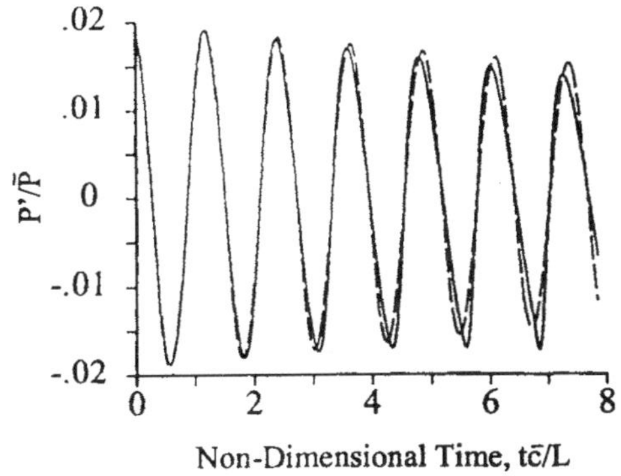

Fig. 10a Comparison of nonlinear numerical and linear analytical solutions of pressure oscillation vs time, (initial amplitude 2% $\bar{p}$; numerical solution exhibits slight wave steepening and shock formation.

exhibit slight wave steepening as can be seen by comparing the nonlinear solution in Fig. 10a (solid line) with the linear solution (dashed line). At 20% amplitude (Fig. 10b), the response shows strong wave steepening that quickly produces a triangular wave structure. The associated entropy increases also provide added damping so that the decay rate of this nonlinear disturbance is faster than the linear rate. Representative solutions of exponentially growing disturbances (not shown) verify that at long times, nonlinear effects limit growth and a limit cycle is reached.

As a final example, we consider the response to a bomb like those used in engine stability rating. This initial disturbance gives rise to a rich variety of initial modes, each of which grows according to its individual characteristics. For this test case, vaporization effects are taken to be uniform over the first half of the chamber. Computations based on variable temperature droplets and constant temperature droplets are compared in Figs. 11a and 11b. The variable temperature droplet case (Fig. 11a) shows an initially rapid decay in which the stable modes are attenuated, followed by a steady growth as the unsteady modes begin to dominate. The droplet temperature case (Fig. 11b) shows continued decay indicating that the constant

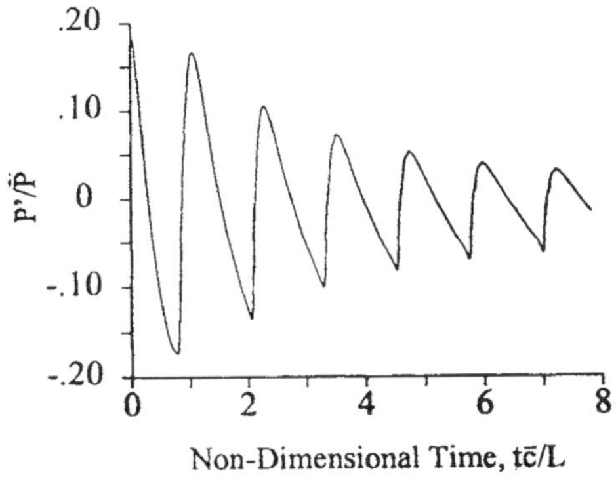

Fig. 10b Nonlinear numerical solution pressure-time trace, initial amplitude 20% $\bar{p}$; numerical solution exhibits strong wave-steepening and nonlinear effects.

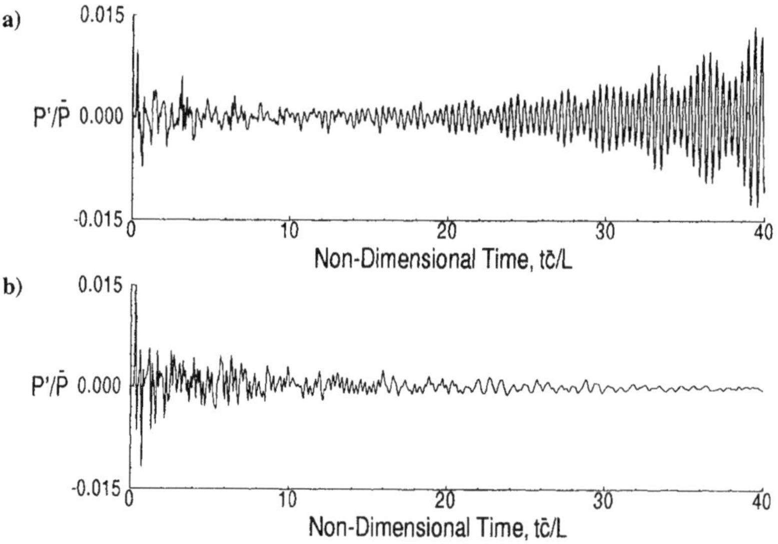

Fig. 11 Comparison of nonlinear numerical pulsed-mode pressure-time traces at injector face/wall location 0 deg: a) variable temperature liquid droplets = 110 μm and b) constant temperature liquid droplets = 110 μm.

temperature model is stable. This effect of droplet temperature on the global stability of the model is in agreement with that noted earlier.

V. Conclusions

It would appear that the most promising role for computational fluid dynamics in combustion instability research is to serve as a testbed for determining the effect of various physical mechanisms on the instability process. This testbed would then evaluate analytical models of individual processes that had been developed through diagnostic experiments and approximate analytical procedures to ascertain their effect on stability. The capability for turning various mechanisms on and off in the CFD testbed should be particularly useful in understanding fundamental mechanisms.

For a CFD analysis to be used in combustion instability, it is imperative that it be validated carefully. Validation at small amplitude is most effectively made by comparing with analytical results from simplified, closed-form analytical expressions, including linearizations of the atomization and vaporization models. Additional validation against experimental measurements is then needed to verify submodel predictions.

In addition to using linearized models for validation purposes, these models are also useful for identifying the basic instability features of a given instability model. These features are difficult to identify with a full, multidimensional, nonlinear computer model, but a simplified small amplitude numerical procedure that permits full three-dimensional disturbances in a one-dimensional mean flow is given that can rapidly provide a complete map of instability solutions to define an entire engine operating regime. The full nonlinear analysis can then be used to ascertain finite amplitude effects, more realistic mean flow variations on more detailed submodels in specific locales of the parameter space that makes up the instability hypersurface, rather than having to compute the entire multidimensional surface.

As a demonstration of the types of results that can be obtained from a CFD testbed, this hierarchy of synergistic models has been used to compute the instability characteristics as determined by classical atomization and vaporization models. This choice of submodel ties the combustion model directly to observable physical phenomena and ensures that the parameters used in the model are realistic. The results again point out the importance of phase lags introduced by unsteady vaporization effects in the droplets. The results also point out the necessity for developing models that describe unsteady atomization effects and more accurate dynamic models of droplet vaporization.

References

[1]Habiballah, M., Lourme, D., and Pit, F., "A Comprehensive Model for Combustion Stability Studies Applied to the Ariane Viking Engine," AIAA Paper 88-0086, Jan. 1988.

[2]Dubois, I., and Habiballah, M., "Numerical Simulation of High Frequency Instability in an Oxygen/Hydrogen Rocket Engine," AIAA Paper 91-1860, June 1991.

[3]Habiballah, M., Lourme, D., and Pit, F., "PHEDRE-Numerical Model for Combustion Stability Studies Applied to the Ariane Viking Engine," *Journal of Propulsion and Power,* Vol. 7, No. 3, 1991, pp. 322–329.

[4]Habiballah, M., Dubois, I., Vingert, L., Lourme, D., and Scherrer, D., "High Frequency Combustion Instability in Liquid Propellant Rocket Engines— A Review of Studies Carried Out at ONERA for the Ariane Launcher," AIAA Paper 92-0774, Jan. 1992.

[5]Liang, P., Fisher, S., and Chang, Y., "Comprehensive Modeling of a Liquid Rocket Combustion Chamber," *Journal of Propulsion and Power,* Vol. 2, No. 2, 1986, pp. 97–104.

[6]Liang, P., Fisher, S., and Chang, Y., "Numerical Analysis of SSME Preburner Injector Atomization and Combustion Processes," *Journal of Propulsion and Power,* Vol. 3, No. 6, 1987, pp. 508–514.

[7]Kim, Y., Chen, C., and Ziebarth, J., "Prediction of High Frequency Combustion Instability in Liquid Propellant Rocket Engines," AIAA Paper 92-3763, July 1992.

[8]Jeng, S., and Litchford, R., "Liquid Rocket Spray Combustion Stability Analysis," AIAA Paper 92-3227, July 1992.

[9]Priem, R., and Breisacher, K., "Nonlinear Combustion Instability Model in Two- and Three-Dimensions," Tech. Rept., NASA TM-102381, Oct. 1989.

[10]Grenda, J., Venkateswaran, S., and Merkle, C., "Three-Dimensional Analysis of Combustion Instabilities in Liquid Rocket Engines," AIAA Paper 93-0235, Jan. 1993.

[11]Grenda, J., Venkateswaran, S., and Merkle, C., "Liquid Rocket Combustion Instability Analysis by CFD Methods," AIAA Paper 91-0285, Jan. 1991.

[12]Grenda, J. M., Venkateswaran, S., and Merkle, C. L., "Analyses of Liquid Rocket Instabilities Using a Computational Testbed," *Twenty-fifth Symposium (International) on Combustion* (Irvine, CA), Combustion Inst. Pittsburgh, PA, 1994, pp. 1619–1625.

[13]Wang, Z., Yang, H., and Przekwas, A., "Assessment of Total Variation Diminishing Schemes in Combustion Instability Problems," AIAA Paper 93-0111, Jan. 1993.

[14]Ingebo, E. D., "Dropsize Correlation for Cryogen Liquid-Jet Atomization," NASA TM-102432, 1990.

[15]Mayer, E., "Theory of Atomization in High Velocity Gas Streams," *ARS Journal,* Vol. 31, No. 12, 1961, pp. 1783–1785.

[16]Nurick, W. M., "A Review of Shear Coaxial Element Atomization," W. J. Shafer Assoc., Memo No. 082, Los Angeles, CA, Feb. 1992.

[17]Priem, R., and Heidmann, M., "Propellant Vaporization as a Design Criterion for Rocket-Engine Combustion Chambers," Tech. Rept., NASA TR-67, Sept. 1960.

[18]Abramzon, B., and Sirignano, W., "Droplet Vaporization Model for Spray Combustion Calculations," *International Journal of Heat and Mass Transfer,* Vol. 32, No. 9, 1989, pp. 1605–1618.

[19]Grenda, J. M., "Computational Analyses of Combustion Instabilities in Liquid Rocket Engines," Ph.D. Thesis, Dept. of Mechanical Engineering, Pennsylvania State Univ., University Park, PA, May 1995.

[20]Yu, S.-T., Hultgren, L. S., and Liu, N.-S., "Direct Calculations of Waves in Fluid Flows Using High-Order Compact Difference Schemes," *AIAA Journal,* Vol. 32, No. 9, 1994, p. 1766.

Chapter 20

Full-Scale Component and Engine Stability Testing Using Spectral Analysis

K. P. Denisov,* A. S. Kadishevich,†
and J. D. Povolotzky‡
Scientific Research Institute of Chemical Machine Building, NIICHIMMASH, Moscow, Russia
and
F. G. Sarychev§
Production Association Southern Machinery Plant, JUZMASH, Dnepropetrovsk, Ukraine

Introduction

COMBUSTOR process high-frequency stability refers to the ability of a process property to continue without developing oscillations that correspond to acoustic frequencies. The spectral analysis of pressure pulsations upstream of the injector and combustor shell vibrations is a common way of studying stability in liquid rocket engines (LREs) both in stand-alone tests and in tests where the LRE is part of a stage. Measurements of pressure pulsations upstream of the injector and measurement of combustor shell vibration, rather than direct measurement of pressure pulsations in combustors and gas generators, are done primarily for two reasons. 1) The measurements of pulsations and vibrations upstream of the injector are sensitive enough to various phenomena in the combustor, the engine fluid systems (including the cooling system), and others that influence engine high-frequency stability to provide information with which to estimate stability margin. 2) Provision of engine test safety is easier while using these types of measurement, as they do not require any additional holes in the combustion zone.

There are various ways of applying results from the spectral analysis of these measurements to the study of engine high-frequency stability. Several features of a spectral analysis are 1) the number of measured parameters being analyzed (it may be only pressure pulsations upstream of the injectors, only shell vibration,

*Vice-General Director.

†Head of Engine Development and Analysis Division.

‡Retired Head of Engine Development and Analysis Division.

§Head of Engine Development and Analysis Division.

or both); 2) the bandwidth of frequencies being analyzed, i.e., a narrow band of frequencies specified a priori in an interval not more than $+/-10\%$ of natural acoustic frequencies, or a wide band of several thousand hertz within which all narrow bands mentioned are analyzed; 3) single amplitude-frequency spectrum parameters, e.g., amplitude and oscillation decrement; 4) characteristics that are used for subsequent solutions, single measurement results (maximum values when amplitude measurements are used for solutions, minimum values if decrements are analyzed), or statistical estimates of these parameters; and 5) the possibility of comparing the reduced results from the formal analysis.

Combinations of these features form a matrix of conceptually possible ways of using spectral analysis data to study LRE high-frequency instability. Such a matrix is shown in Table 1. Of course, the matrix does not include all possible situations; for example, there is no division by the number of similar measurements (pressure pulsations upstream of injection elements of one or both propellants, one- or many-directional vibrations). The matrix, however, does contain the main methodological peculiarities connected with obtaining initial information for subsequent analysis and the engine state qualification as acceptable or nonacceptable from the standpoint of high-frequency stability. Generalized data on the application of spectral analysis in the practice of engine development are given in Table 1. This chapter describes one of the most frequent methods, which has the following features: 1) spectral analysis of pressure pulsations upstream of the injector, and vibrations of the combustion chamber and gas generator shell; 2) investigation of a wide frequency band (0.2–15 kHz) by successive analysis of narrow intervals composing the band; 3) use of statistical characteristics that are obtained by spectral analysis; and 4) comparability of information found in narrow intervals.

This last feature, the comparability of information found in separate narrow intervals, is one of the main advantages that distinguishes the spectral analysis method from others given in Table 1. This method relies not on a priori models of processes, which may finally result in a loss of combustor stability, but rather on the basis of newly established relations of amplitude-frequency spectrum parameters to create models characteristic of specific situations.

The spectral analysis method shares certain features with other methods presented in Table 1, but the use of quantities characterizing the relations between the results obtained in relatively narrow frequency intervals is a peculiarity of the spectral analysis method. The method's main drawback is the amount of initial information associated with a narrow band that will appear unnecessary after finding the cause of the anomaly. It will be particularly evident if the same effect is obtained within one or two frequencies defined a priori. However, the spectral analysis method tends to be most reliable, and the superfluous information that is obtained may be useful for other cases, including other engine designs.

The main methodological principles used by the authors in LRE high-frequency stability analysis on the basis of information obtained in full-scale tests of these engines may be of some interest to specialists in other fields as well. With this in mind, it is appropriate to expound on the following issues: 1) the procedure that comprises primary processing of the pressure pulsation and the shell vibration amplitude-frequency spectrum, 2) the procedure of using data from the primary processing step to estimate LRE high-frequency stability, 3) examples of the procedure application in practice, and 4) recommendations on ways to improve the method.

Table 1 Matrix of possible methods for the application of spectral analysis to LRE high-frequency stability data

Parameters obtained	Characteristics used	Formalized signs application	Application estimate of parameters measured: Pulsations, frequency band		Vibrations, frequency band		Pulsations and vibrations, frequency band	
			narrow	wide	narrow	wide	narrow	wide
Amplitude	Single	Yes	2[a]	2	1[b]	1	3[c]	3
		No	2	2	1	1	3	3
	Statistical	Yes	2	2	1	1	3[d]	3[d]
		No	2	2	1	1	2[d]	2[d]
Oscillation decrements	Single	Yes	3	—[e]	—	—	—	—
		No	—	—	—	—	—	—
	Statistical	Yes	2	—	—	—	—	—
		No	—	—	—	—	—	—
Amplitude oscillation decrements	Single	Yes	2	—	—	—	—	—
		No	3	3	—	—	—	—
	Statistical	Yes	2	2	—	—	—	—
		No	2	2	—	—	—	—

[a] Applied often indicated by 2. [b] Applied seldom indicated by 1. [c] Applied very often indicated by 3. [d] Method discussed in this chapter. [e] Not applied.

Amplitude-Frequency Spectrum Primary Processing

The amplitude-frequency spectrum is the expansion of the total signal into its components in which certain amplitudes correspond to certain frequency values. The expansion into a series (in this case a Fourier series) is carried out anywhere from about 200 Hz to 15 kHz. The primary processing is performed in the following sequence.

1) At each LRE operating regime, from 10 to 100 magnetogram (magnetic tape record) data samples are taken of pressure pulsations and/or shell vibrations. Each sample is expanded into a Fourier series within the stated frequency band.

2) The frequency band under study is divided into subranges (usually not less than three, e.g., 200–1000 Hz, 1000–5000 Hz, 5000–10,000 Hz). Each subrange is analyzed further separately.

3) While investigating a subrange, the following information is recorded for each separate spectrum: the amplitude maximum value A_i and the corresponding frequency intervals $f_i \pm \Delta f_i$, where Δf_i is approximately 2–3% of the interval of the frequency subrange Δf_s taken for analysis.

4) From the spectra obtained for the engine operation regimes the following determinations are made. Estimates are made of the mathematical expectation of amplitude $\bar{A}_{fi}$ for each narrow frequency intervals $(f_i \pm \Delta f_i)$

$$\bar{A}_{fi} = \sum^{N_{fi}} A_{fi}/N_{fi} \tag{1}$$

where A_{fi} is the current value of the maximum amplitude in the frequency interval $f_i \pm \Delta f_i$, and N_{fi} is the number of samplings taken for the operation regime under study in which maximum amplitudes were realized in the frequency range $f_i \pm \Delta f_i$. The rate is ascertained at which a maximum amplitude appeared in each narrow frequency interval that will be later referred to as regularity R_{fi},

$$R_{fi} = N_{fi}/n \tag{2}$$

where n is the number of samplings for the engine given operation regime. The estimate of mathematical expectation of amplitude $\bar{A}$ is made in the narrow frequency interval being studied

$$\bar{A} = \sum^{n} A_i/n \tag{3}$$

Finally, the maximum regularity, $R_{\max}$ is determined in the frequency band under study and in the narrow frequency interval $(f_i \pm \Delta f_{i,\max})$ corresponding to this regularity.

Special attention should be paid to the necessity of studying each subrange of the overall frequency band separately to define all of the mentioned characteristics for each subrange. Special attention should also be paid to the significance of regularity in the whole analysis. It is regularity that defines the contribution of various frequencies to the overall process or, rather, the level at which various frequencies affect the process. The examination of Eq. (2) shows that changes in regularity with engine transition from one regime to another, even without changes in amplitude of pressure pulsations or shell vibrations, represent changes in the spectrum structure that are often indicative of physical changes in the system.

Regularity to some extent may be treated as a characteristic of the process "band-narrowness," i.e., of the oscillation energy concentration in a relatively narrow interval of frequencies.

It is useful to compare regularity with another LRE process stability characteristic, the oscillation decrement. The following interpretation of these characteristics seems reasonable: oscillation decrement is a characteristic uniquely related to the probability of sustained self-oscillations of a certain frequency without any connection to its amplitude; regularity is an estimate of the probability of self-oscillations appearing at a certain frequency at an amplitude maximum for the frequency band under study. The differences between these two characteristics, decrement and regularity, are most apparent at relatively low-amplitude oscillatory conditions. It is interesting to define possible regions in which the mentioned characteristics can be applied most effectively.

The use of decrement as a primary chamber process stability criterion is quite well justified for the analysis of combustion chamber and gas generator component tests where there is no connection to the engine state, in which case stability is defined to a high degree by pressure pulsation and vibration amplitude level. When the engine stability is analyzed, however, then both the amplitude value characteristic of self-oscillations and the probability of self-oscillation excitation at the maximum amplitude level are of importance. If the engine stability is evaluated without amplitude value being taken into account, the so-called producer risk will be, as a rule, too high.

Thus, in evaluating LRE high-frequency stability the authors give preference to regularity. At the same time it should be emphasized that in selecting criteria for defining LRE high-frequency stability, the variety of mechanisms that can lead to instability and the lack of theoretical knowledge of the problem makes a priori preference of a specific characteristic unwarranted. The main methodological principles used in the choice of high-frequency stability criteria and the procedures used to process these parameters will be discussed in the next section.

Estimating High-Frequency Stability Using Spectral Processing: Systematic, Essential, and Preventative Conditions

The problem is reduced to the consideration of the following two situations.

1) The combustion chamber process stability loss was accompanied by an increase in pressure pulsation amplitude and vibratory overload and by the appearance of some other indications of the intensification of instability.

2) The process stability loss was not confirmed by available information.

The same problems arise in both cases; that is, which characteristics of those obtained in primary spectral processing are most suitable for the estimation of high-frequency stability of the design. It is evident that the task is easier in the first case.

Theoretically any characteristic (estimate of mathematical expectations of amplitudes, or regularity of certain frequencies and maximum regularities corresponding to them) obtained may be used as an effective means of engine stability control if this characteristic can be used to identify a disposition toward stable or unstable operation, and has the properties of being systematic, essential, and preventative; these three terms will be explained in detail.

The term systematic refers to a situation whereby features found in certain series of tests are systematically (without any exclusion) observed in other independent test series, including experiments with different specimens carried out simultaneously. For example, if a multichamber engine is tested and high-frequency instability is found with a given rate in each of its chambers, then being systematic implies that the character of the parameter chosen for high-frequency stability control changes for successful and failed tests for each chamber in a reproducible fashion.

If the parameter is chosen to be the estimate of the mathematical expectation of amplitude in a certain frequency band, the requirement of being systematic is fulfilled if, for each chamber, the estimate value in successful engine tests is lower than in failed tests. Attention should also be paid to the systematic correspondence between the character of the parameter change and the change of certain design, manufacturing, and/or operation factors.

Another feature of the parameters that may be used for high-frequency stability evaluation is essentiality. One may speak of an essential difference if its value can be accounted for with an acceptable significance level on the basis of statistical data. Here, the significance level value is equal to $1-\gamma$, where γ is the fiducial probability specified by the user for evaluating the lower boundary of the confidence interval of failure-free operation probability. In the authors' experience the significance level is usually about 5–10%. Let $\bar{X}_1$ and $\bar{X}_2$ be the estimates of mathematical expectation of the parameters to be compared and $S_{\bar{X}_1}$ and $S_{\bar{X}_2}$ be the mean square errors of the mathematical expectations. The requirement of essentiality is fulfilled if the following relationship holds:

$$\frac{|X_1 - X_2|}{\sqrt{S_{\bar{X}_1}^2 + S_{\bar{X}_2}^2}} - t_{\text{st}}|1-\gamma;\, N_1 + N_2 - 2| \geq 0 \tag{4}$$

where t_{st} is the quantile of the student distribution, corresponding to $(N_1 + N_2 - 2)$ degrees of freedom at a significance level of $1-\gamma$; N_1 and N_2 are the number of measurements in the samples being compared.

In cases where the parameter X_i used for high-frequency stability estimation has a certain quantifiable dependence on some test condition or design factor, e.g., $X_i = f(j_i)$, the estimate of essentiality is virtually an estimate of the certainty of the dependence and is based on the comparison of an estimate of the variance of separate parameter measurements relative to the relationship $[S_X = f(j_i)]$ and an analogous estimate made with the assumption that there is no such relation (S_{X_0}). The requirement of essentiality is met if

$$\frac{S_{X_0}^2}{S_{X=f(\varphi_i)}^2} - t_F(N_2;\, N_1;\, 1-\gamma) \geq 0 \tag{5}$$

where t_F is the quantile of the Fisher distribution corresponding to the number of experiments in each sample and the chosen significance level.

It should be noted that unlike the systematic condition estimate, negative results of the essentiality estimate do not mean that the proposed parameter is inadequate for checking the engine state; such results indicate that there is insufficient information to state for certain that the observed difference is not a random realization of the same relationship. These results do not necessarily eliminate the proposed parameter as being adequate for high-frequency stability control. However, it should be known to what extent the search is promising, i.e., what is the chance that the experiment continuation will not finally result in the loss of being systematic. Here the following statement may be helpful: Further search (statistics accumulation) is not promising if relationship (6) holds and Eq. (4) does not hold for the samples being compared,

$$[N_2/(N_1 + N_2)]M_2 + (\gamma - 1) \leq 0 \tag{6}$$

where N_2 is the total number of samples, N_1 is the number of samples in which no cases of high-frequency instability were observed, and M_2 is the number of observations of high-frequency instability.

Equation (6) is an estimate based on an alternative characteristic ("success or failure") of the product state. The application of parametric methods instead of alternative ones is justified only in cases where the former are more informative. The number of engines or tests required for defining the effect of various factors on the engine high-frequency stability is the informativeness measure. This means that application of a certain parameter is justified only in a case in which it allows a reduction in the number of engines required for finding certain relationships. When relationships (4) and (5) are not met and Eq. (6) is met, two versions are possible: 1) the parameter is not informative, 2) the proposed parameter does not characterize the process under study.

In both cases any attempts to confirm the parameter adequacy will lead nowhere.

If Eqs. (4) or (5) hold, and the requirement of being systematic is met, the parameter may be reasonably considered as a basis for further application for an estimate of engine instability.

Here a question arises: at what engine state (how far from the stability boundary) does the amplitude-frequency spectrum of pressure pulsations or/and vibrations allow the systematic and essential distinction of negative changes of the hardware state and engine processes? It is most important that the requirement of being preventative mentioned at the beginning of this section is met. The preventative condition is the parameter's ability to reveal negative tendencies before they are found by more simple means, i.e., before the occurrence of serious damage.

It is only with these parameters that a probability (or improbability) of high-frequency engine instability occurrence can be estimated before the engine failure and without testing a large number of specimens. This possibility of being able to estimate the probability of ensuring high-frequency stability with reasonable accuracy and certainty is the main motive stimulating the application of parametric methods.

It should be underlined that only parameters directly connected to and governed by design, technological, and some other engine features can be used for estimating the probability of ensuring high-frequency stability.

If an estimate is based on the measurement of parameters that are not direct characteristics of the engine, the extension of the estimate to subsequent products is valid only in the case that the features defining the high-frequency stability are constant. If direct engine properties are checked, such confirmation is easy and does not require failure tests (in this case a failure caused by disturbance of combustion process). If out-of-engine characteristics (external and test conditions) are used as parameters of an operable state, the guarantees may be confirmed only by systematic tests of driving the combustion chamber and gas generator process to an unstable state or by tests carried out under conditions promoting such anomalies. This makes the application of amplitude-frequency spectrum parameters more attractive than other parameters even when the latter are better in terms of being systematic and essential. The application of out-of-engine parameters for the probabilistic prediction of high-frequency stability may be accepted only as a temporary, forced measure and are not suitable for either high-frequency stability control in the course of full-scale production or for the estimate of high-frequency stability margins.

Any amplitude frequency spectrum parameter may be used for checking high-frequency stability without restrictions. Theoretically it may be a parameter related to pressure pulsations upstream of a single injection element or analogous parameters related to combustion chamber or gas generator shell vibrations. Complex parameters, i.e., parameters practically duplicating each other, to some extent can be used as well. Common parameter combinations will be considered presently. The requirements of being systematic, essential and preventative are to be met in all situations. If these requirements are fulfilled for several parameters, the one that best meets the essentiality requirement should be chosen. Because state differences are expressed by complexes of a type

$$\frac{|\bar{X}_1 - \bar{X}_2|}{\sqrt{S^2_{\bar{X}_1} + S^2_{\bar{X}_2}}}$$

where two states are analyzed or

$$S^2_{X_0}/S^2_{X=f(\varphi)}$$

where relationships are analyzed, one may (designating parameters to be compared by indices a and b) express the conditions under which one of the parameters, a for example, is more preferable than the other by

$$\left\{\frac{|\bar{X}_1 - \bar{X}_2|}{\sqrt{S^2_{\bar{X}_1} + S^2_{\bar{X}_2}}}\right\}_a - \left\{\frac{|\bar{X}_1 - \bar{X}_2|}{\sqrt{S^2_{\bar{X}_1} + S^2_{\bar{X}_2}}}\right\}_b > 0 \tag{7}$$

$$\left\{\frac{S^2_{X_0}}{S^2_{X=f(\varphi)}}\right\}_a - \left\{\frac{S^2_{X_0}}{S^2_{X=f(\varphi)}}\right\}_b > 0 \tag{8}$$

Thus, the choice of amplitude-frequency spectrum parameters for subsequent control of high-frequency stability is a formalized problem. However, this normalization prevents only the rough mistake of choosing a parameter not adequate for the problem solution. The choice in the most simple case (one of the parameters is suitable) is a laborious task, i.e., a complete exhaustive search of all parameter possibilities. The selection of a complex parameter on the basis of theoretical concepts may be a very difficult task.

In the simplest case when enough experience has been gathered in the study of the unstable state of a certain engine configuration, a parameter obtained from the primary processing of the amplitude-frequency spectrum may be chosen. It may be the regularity of pressure pulsations in a narrow interval of frequencies upstream of an injection element R_f, the estimate of the mathematical expectation of amplitudes in the narrow interval upstream of an element $\bar{A}_f$, or the estimate of the mathematical expectation of amplitude in one of the frequency bands being studied $\bar{A}$.

The regularity of pressure pulsations upstream of an injection element is a frequent choice due to the fact that the underlying mechanisms of high-frequency instability are unknown a priori. Therefore, a frequency band in which a certain instability mechanism manifests itself should be found. As a rule the regularity R_{fi} parameter is most suitable for stability investigation when its value is less than

0.1. In cases qualified as being disposed to high-frequency instability, its value increases.

The estimate of the mathematical expectation of amplitude in a certain narrow frequency interval A_{fi} is not always suitable for stability study because of its low preventativity. The same may be said about the estimate of the mathematical expectation of amplitudes in the frequency band under study $\bar{A}$. In the case when the investigated process is accompanied by a regularity greater than 0.7 in the narrow frequency interval characterized by the chosen values of $\bar{A}_{fi}$, methods based on $\bar{A}_f$ and $\bar{A}$ are practically the same. These methods are widely used in investigating a possibility of high-frequency instability appearing under the existence of pressure self-oscillations in combustion chambers and fluid lines. A regularity level close to 1.0 is characteristic of such situations, and failure-free operation depends on the ability not to permit anomalous amplitude increase. In cases when self-oscillations are not systematic in time and have a tendency to increase in duration, the most simple complex parameter, the product $\bar{A}_{fi} \cdot R_{fi}$ or $\bar{A} \cdot R_{fi}$ is often preferable. This quantity may arbitrarily be called a mean integral amplitude in a narrow frequency interval, $f_i \pm \Delta f_i$. The arbitrariness of the definition is because in the first case ($\bar{A}_{fi} \cdot R_{fi}$), for samplings where no maximum amplitudes were observed, their values were equal to zero, and in the second case ($\bar{A} \cdot R_{fi}$), their values were equal to those observed in other frequency intervals. Mean integral amplitudes are applied to study processes characterized by self-oscillations developing in time and, as a consequence, by developing amplitude. If the regularity value at a qualified stable state is taken as a reference, then the development of regularity from the range 0.2–0.4 to the range 0.5–1.0 is characteristic of a state being inclined to instability.

For the cases just discussed when only one parameter may be taken for the stability study, the following expression defining the amplitude value vs the engine operation durability and the relevant parameter (the observed regularity, for example) is typical:

$$A_\tau = A_0 + B_0\tau^\beta\left[1/\left(\nu - R_{fi}^\alpha\right)\right] \tag{9}$$

where A_τ is the amplitude value at time τ, A_0 is the amplitude value at the start, and B_0, β, ν, and α are constants characteristic of the development of certain mechanisms of instability.

With Eq. (9) in mind, guidelines for applying each of the abovementioned amplitude-frequency spectrum parameters given in Table 2 where R is the regularity value characteristic of the situation when no failures are caused by combustion instability. It is clear that in the first case a quick process stability loss occurs, whereas in the others a relatively slow development of the event occurs.

Table 2 Guidelines for applying amplitude-frequency spectrum parameters

Initial conditions			Most informative parameter
$\nu \to 1$	$\alpha \leq 1$	$R_0 = 0$–0.05	Regularity R_{fi}
$\nu \ll 1$	$\alpha \gg 1$	$R_0 \to 1$	$\bar{A}_{fi}$; $\bar{A}$
$\nu \ll 1$	$\alpha \gg 1$	$R_0 = 0.2$–0.4	$\bar{A}_{fi} R_{fi}$; $\bar{A}_i R_{fi}$

Complex Parameters

It is clear that after determining the dependence of constants A_0, B_0, β, α, and ν at certain test conditions one may define the character of the development of processes leading to instability under various conditions. However, it should be kept in mind that the possibilities of the scheme described with Eq. (9) are rather limited. It defines the self-development of the process. At the same time experiment shows that in many cases the interaction of processes in the engine fluid systems and mutual influence of hydraulic, thermal, and mechanical effects are the governing factors of high-frequency instability. It is clear that for investigating such processes the parameters describing the interaction of these factors should be used. These complex parameters are based on parameters determined during primary processing.

The following notion of a possible mechanism of high-frequency instability is the basis of this complex parameter: various factors defining chamber and gas generator high-frequency stability act independently according to the laws of random independent events before the oscillations reach a certain level. When each of the factors reaches a certain limit value they influence each other unfavorably, thereby promoting high-frequency instability. Since high-frequency instability results in high levels of pressure pulsation and vibration amplitudes, the search and qualification of various engine states should be based on the investigation of the connection of the amplitude-frequency spectrum and engine parameters as a function of increases in pressure pulsations and shell vibration amplitude.

Complex parameters may be used for high-frequency stability analysis if the following relationships are satisfied (while meeting all other conditions already formulated):

$$\frac{\partial \bar{A}_i}{\partial X_i} \geq 0 \tag{10}$$

$$P\left[\bar{r} A_i = \varphi\left(A_{i_i}\right)\right]_{\text{lim}} \tag{11}$$

where $(\partial \bar{A}_i / \partial X_i)$ is a derivative of pressure pulsations or vibration amplitude with respect to any component (X_i) of the complex parameter. The derivative of the amplitude of pressure pulsations upstream of an injection element and shell vibrations with respect to each other should be positive. Equation (11) is the probability that the estimate of the mathematical expectation of the correlation coefficient of the interdependence of pressure pulsation and vibration amplitude differs from zero at the limit and higher values of amplitude.

Approximate equations, implying that full-size tests do not always give enough information for full correlation analysis, may be used for calculations. With the minimum required amount of information (two intervals, i.e., relatively low and higher level of amplitude), the following operations should be carried out for determining pressure pulsation or vibration limit values.

1) For each of the amplitudes being analyzed the difference between their values ΔA_i should be analyzed,

$$\Delta A_i = A_{i_1} - A_{i_2} \tag{12}$$

where index i_1 denotes the amplitude of any measured parameter (pressure pulsation upstream of an injection element or vibration) at an initial time interval and index i_2 is the value of the same parameter at the next interval.

2) The relation K_i of the amplitude changes of engine parameter in each separate test is determined,

$$K_i = \Delta A_{j_1} / \Delta A_{j_2} \tag{13}$$

where the indices 1 and 2 denote different engine parameters.

3) The mathematical expectation $\bar{K}_i$ and its mean square error $S_{\bar{K}_i}$ are evaluated from well-known relations. If $\bar{K}_i \leq 0$ the calculation is ceased as the relationship (10) is not satisfied. If $\bar{K}_i > 0$, the probability of the characteristic lying in the region of positive values is estimated; it is this condition that allows the results to be a consequence of certain relationships,

$$P = \varphi_{\text{st}}\left(K_i / S_{k_i}; N\right) \tag{14}$$

where φ_{st} is student's function.

At $P > \gamma$ the estimated expectation of the amplitude used in the calculations may be taken as a value at which various parameters mutually increase the tendency to development of high-frequency instability. The case $P > \gamma$ suggests that the estimated expectation lies in the superlimit range, i.e., in the range where the relationship is certain. If $P < \gamma$, the validity of the relationship is not confirmed. With enough data the dependence of the probability on the estimated expectation of amplitude could be found, and the estimated limit value ensuring $P = \gamma$ could be determined. Since this is not always possible, amplitude limit values may be found from relations of the form in Eq. (15) that imply that the probability asymptotically approaches unity when the value of the relevant amplitude tends to infinity, and the probability approaches zero when the amplitude tends to zero:

$$1 - P = e^{-\lambda A_i} \tag{15}$$

where λ is a constant.

With the preceding condition an equation may be written for the limit value of the amplitude of a parameter A_{lim}:

$$A_{\text{lim}} = A_p \frac{\log(1-\gamma)}{\log(1-P)} \tag{16}$$

where A_p is the amplitude value satisfying the probability $P(1 - P)$.

It is clear that the limit amplitude value A_{lim} can be easily found if the probability P is known. In the same way the limit values of each amplitude-frequency spectrum characteristic for each parameter connected to an instability may be found. A complex parameter connected with the characteristics of the engine's amplitude-frequency spectrum may be based on various possible combinations: pressure pulsations upstream of one injection element and shell vibrations in one direction, the same pulsations and vibrations in another direction, the same combinations for an injection element of another propellant, or combinations of the same parameter (pulsation-pulsation, vibration-vibration).

In contrast to situations characteristic of simple cases when the analysis is based on one parameter, in the more complex cases when the analysis is based on the relationships of several amplitude frequency characteristics, the parameter combinations do not compete with each other. There is no preferable combination. Any combination should be taken into account if it satisfies the requirements of being systematic and essential and it satisfies the condition $P > \gamma$ at the amplitude values that took place. It is clear that if this condition was satisfied the possibility of finding an anomaly was confirmed.

While studying high-frequency stability in full-size tests the same problems are solved by means of simple and complex parameters. The engine state is qualified as stable, disposed to instability, or unstable. The effectiveness of measures taken to increase stability and the probability of stability assurance are evaluated.

Qualification criteria are actually the same: no failures due to chamber process instability, and the values of the parameter used for stability analysis must remain within a certain so-called sublimit range. These are considered sufficient to qualify the engine as stable; maintaining the operable state with control parameters lying in the superlimit range qualifies the engine as being disposed to instability; conditions when the engine is qualified as unstable are evident. However, there are some differences between the approaches used, for example, to define the limit level of parameters being controlled. In the simplest case, this is a quantity characterizing an amplitude-frequency spectrum parameter. For a complex parameter the limit value is an aggregate of limit values of a parameter combination. For example, if a parameter combination is made of the pressure upstream of an injection element and one value of shell vibration, parameters of the amplitude-frequency spectrum should include the estimated computed expectation in the frequency band being studied $\bar{A}$ and their maximum regularity R, the limit values of the estimated expectation of pressure pulsation and vibration amplitudes, and the limit value of the maximum regularity of these parameters. If it turns out that a high correlation of these parameters (e.g., pressure pulsations upstream of an injector element and vibration) appears beginning from some minimum frequency difference ($f_i \pm \Delta f$) corresponding to maximum regularities, then the complex parameter limit should include the minimum allowable frequency difference as well.

If, in the example just given, all complex parameter components except one reach or exceed the relevant limit value, the engine still should be qualified as stable if it maintains its operable state in the absence of other faulty circumstances.

Some peculiarities of the complex parameter appear when the stability margin is estimated. Such calculations are necessary for each new measure introduced to ensure engine stability. In the simplest case, i.e., when the stability is analyzed with an amplitude-frequency spectrum parameter found in primary processing or with a parameter of mean-integral amplitude type, the margin is determined as the difference between the parameter's current and limit values from the estimated expectation of the parameter's critical value. For the versatile complex parameter it is clear that its margin can not be estimated the same way. First of all, separate parameters comprising the complex one should be made commensurable. There are several ways for solving the problem; the one applicable in this case follows.

1) All limit values of parameters that make the complex parameter are assigned the same value, that is, zero.

2) Value intervals are set for each parameter, and each interval is assigned a value that increases by a unit with a departure from the limit value toward the direction of favorable stability.

3) The dimensionless characteristic of each parameter making the complex parameter is the sum of intervals separating the current value from zero in the direction of favorable stability.

This way to make the parameters forming the complex parameter commensurable with each other can not be considered certain as it is based on arbitrary intervals taken for each parameter. However, any possible detrimental effects due to this approach can be partly compensated by the commensurability of intervals for similar parameters.

The estimate of the margin of the complex stability parameter is defined as the sum of dimensionless estimates of all parameters that comprise the complex parameter. Using margins calculated in such a way, the effectiveness of measures taken for assuring high-frequency stability and the probability of its realization are estimated. Both problems are solved in accordance with reliability assurance and control documents. The dimensionless estimate of the complex parameter is formally the estimate of the probability of this estimate lying above zero.

The validity of using dimensionless complex parameters to calculate stability probability may be confirmed by comparing these results with estimates based on assumptions about the character of the sublimit range of the complex parameter values. In the sublimit range, factors governing stability act independently and, consequently, the probability q_0 of reaching the limit state is defined by

$$q_0 = \prod^{Z} q_{0_i} \tag{17}$$

where q_{0_i} is the probability of achieving the limit state of the single amplitude-frequency spectrum parameter, and Z is the number of complex parameter components.

The value of q_0 is defined from well-known equations of probability parametric estimates. The estimated probability of stability using dimensionless complex parameters may be considered acceptable if the following relation is met:

$$P_{l_A} < P_{l_{cp}} \leq 1 - \prod^{Z} q_{u P_{0_i}} \tag{18}$$

where $P_{l_{cp}}$ is the lower confidence interval boundary of the probability of the dimensionless complex parameter lying in the range above zero (all confidence interval boundaries being defined with the fiducial probability set for engine reliability estimate), P_{l_A} is the lower confidence interval boundary of the probability of stability determined by the alternative method, and $q_u q_{0_i}$ is the upper confidence interval boundary of the probability of reaching the limit value of the single amplitude-frequency spectrum parameter.

Failure to satisfy the left-hand side of Eq. (18) means that the proposed parameter is not informative enough and, consequently, it cannot be used. Failure to satisfy the right-hand part of the relation means that the initial preconditions of the process development model do not conform to each other or, rather, the significance of the intervals of dimensionless estimates is chosen incorrectly. Such situations are possible for states close to the limit. In this case a review of the scale of the dimensionless estimates is necessary. In conclusion it should be said that when there are several possible combinations of single parameters for making a complex parameter, the margin of such a parameter is estimated from the least sum of the dimensionless combination of the estimates of the single parameters.

Application of the Described Methods

In the application of the described methods two basic problems should be considered: 1) actual situations when two parameters that had not been widely used before turned out most effective: pressure pulsation regularity in a certain narrow frequency interval and a complex (versatile) parameter; 2) some information about the reproducibility of high-frequency stability level obtained with the application of complex parameters.

As an example, consider the logic of selecting the regularity and a complex parameter for estimating the stability margin in two different cases. In the first case, an amplitude-frequency spectrum parameter must be found that conforms to the following observations: until a certain propellant temperature at the engine inlet was reached, no high-frequency instability was observed; instability was observed many times on reaching this temperature; and the rate of instability differed for two chambers of a two-chamber engine. In the second case the preferable parameter conforms to the following observation: instability appeared when the chamber pressure was changed. In this case, as in the first one, the instability appeared when the propellant at the engine inlet achieved a certain temperature. In both cases, the essentiality of statistical characteristic difference for different conditions was evident enough, i.e., the influence of the conditions on the engine stability seemed well grounded. In both cases the regularity of pressure pulsations in a narrow frequency interval upstream of an injection element was the most effective amplitude-frequency spectrum parameter. In the first case fuel injection elements were used, and the frequency interval was identified with the first modes of longitudinal and transverse oscillations. In the second case the oxidizer injection elements were used, and the interval differed from frequencies that usually accompany high-frequency instability.

In the first case the regularity of pressure pulsations upstream of the oxidizer injection elements competed with the chosen parameter. The competing parameter was rejected as inconsistent with differences of instability frequencies.

In the second case no other parameter of the amplitude-frequency spectrum of pressure pulsations and vibrations could compete with the chosen one as no other parameter conformed to statistical requirements. It should be noted that the frequency interval in which pressure pulsation regularity was most effective coincided with the interval of maximum vibration regularity through the engine's operation cycle.

Some data on stability reproducibility are given in Fig. 1. The rms scatter of the measured pressure pulsation and vibration spectrum at engine start and during steady-state operation are presented. The ordinate is the ratio of the observed estimate of the rms scatter of the complex parameter obtained in tests of multiple

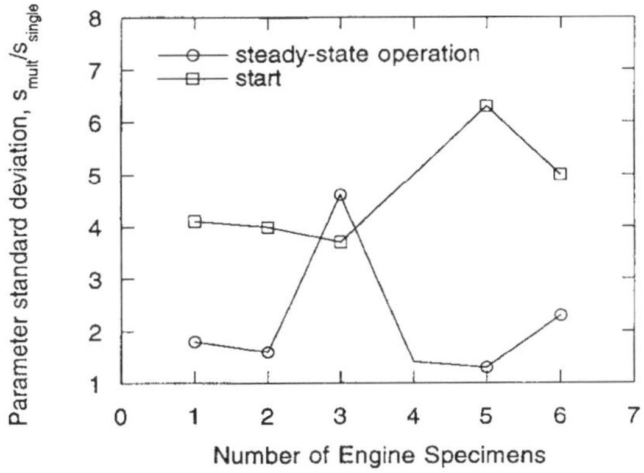

Fig. 1 Reproducibility of complex parameter stability estimates in repeated tests with a single combustor and with multiple combustors.

combustors to the observed estimate of the rms scatter of the parameter in repeated tests of a single combustor. The estimate was based on the analysis of statistical data of versatile parameters composed of amplitude-frequency spectrum parameters.

In connection with the data presented in Fig. 1 it should be noted that there are two equally important aspects in the problem of assuring stability reproducibility. The first is stability reproducibility in a single specimen (engine, combustor, gas generator). It may be expected that the engine that successfully (from the standpoint of stability) underwent acceptance tests will reproduce satisfactory characteristics in operation. The second problem is the reproducibility of satisfactory levels of stability in all engine specimens. The solution of the first problem is a necessary condition of the solution of the second problem.

The following observations can be made from Fig. 1.

1) The reproducibility of stability characteristics at the start and during the steady-state operation is higher for repeated tests of the same specimen than for tests of different specimens.

2) The differences observed in steady-state operation are well founded by statistical data; the differences observed at the engine start are found only in approximately half of the cases.

3) The ratio of the unreproducibility characteristic from specimen to specimen to that for repeated tests of the same specimen is higher in steady-state operation than at the start. This is attributed to the fact that repeated tests, which were accompanied by a change of elements affecting the start, were also taken into account.

The presented data could be grounds for a review of the design and technology documentation and the production process with the objective of providing higher reproducibility of design characteristics from specimen to specimen. It should be remembered, however, that the presented results cannot be taken as a generalization but serve to illustrate the application of complex parameters for the solution of a particular problem.

Conclusions

In conclusion, the following points should be stated:

1) Various mechanisms of high-frequency instability excitation are observed in full-size tests and their investigation is not always possible with a priori established parameters.

2) The procedure of applying parameters of the amplitude-frequency spectrum of pressure pulsations upstream of injection elements and combustor shell vibrations may promote the ability to find the most informative parameters for specific cases of high-frequency stability.

3) The effectiveness of the procedure may be increased by using amplitudes lower than the maximum amplitude. Though the experience in using the method for the analysis of high-frequency stability can be considered as a whole, some drawbacks of the method are evident. In particular, the method uses only a limited part of the spectral information. This limits the potential of the method for early detection of peculiarities leading to high-frequency instability. The potential of the method may be extended by using not only the largest amplitudes (the so-called major maxima) but also second-order amplitudes in the analysis. This should present no major problems considering the computational power presently available. The amount of information will be enlarged considerably, and the criteria used for qualifying liquid rocket engine stability will be modified. The modification of the current method by using second-order amplitudes may be the subject of a significant study.

Chapter 21

Scaling Techniques for Liquid Rocket Combustion Stability Testing

Steven C. Fisher,* Fred E. Dodd,† and Robert J. Jensen‡
Rockwell International, Canoga Park, California 91303

I. Introduction

THE development of liquid rocket combustion devices has historically been a costly effort when combustion instability is encountered. To prevent this unnecessary cost, typical programs are structured to deal with the issue of combustion stability through a balance of analysis and stability rating testing. If properly incorporated into the development methodology, subscale testing can be quite beneficial in eliminating configuration and design problems early in the development program. Subscale development efforts will never result in complete confidence in the full-scale design, but with a properly structured program full-scale development risks can be reduced. Stability modeling is an important tool in establishing low-cost and low-risk stability development programs. A variety of different subscale combustors are discussed in this chapter. The combustors reviewed include single-element, three-dimensional round, and two-dimensional configurations. The information that can be gained from each type of combustor is included in the discussion along with examples. Typically, various mechanisms of combustion instability can be studied more effectively with specific subscale combustors. Finally, suggestions are provided relative to maximizing the amount of information that can be extracted from a subscale program in order to minimize the cost of the full-scale program.

II. Injector Combustion Stability Development Issues

Combustion instability has proven to be a formidable issue in liquid rocket engine development programs. As demonstrated by many incidents, combustion

*Associate Director, Advanced Combustion Devices, Rocketdyne Division.
†Member of the Technical Staff, Combustion Devices Analysis, Rocketdyne Division.
‡Manager, Combustion Devices Analysis, Rocketdyne Division.

instability can cause significant damage or loss of the combustor, the test facility, or even the vehicle and launch pad within fractions of a second. One example is a liquid oxygen (LOX) and methane injector[1] that was severely damaged after about 0.8 s of unstable operation at 2000 psia (Fig. 1). Figure 2 shows an F-1 engine injector with major damage following an instability during development testing. In back-to-back incidents, two early Atlas vehicles equipped with baffleless MA-2 booster engines were lost together with their respective launch pads due to unstable combustion. From the standpoint of an engine or engine combustor component development program, combustion instability is a critical issue along with performance and injector/chamber compatibility. As a result, significant resources are typically set aside to analyze and experimentally demonstrate the stability of the injector design selected.

Combustion instability has been defined as the occurrence of pressure fluctuations of a well-defined frequency (or set of frequencies) within a combustion chamber that are coupled with the combustion process.[2,3] Often these frequencies can be readily identified with characteristic acoustic modes of the combustion chamber. However, the amplitudes of these oscillations are typically orders of magnitude above that of environmental acoustic noise sources. Peak-to-peak pressure wave amplitudes as high as 150% of the average chamber pressure have been measured during fully developed combustion instabilities.[1] For the LOX/methane injector shown in Fig. 1, the peak-to-peak amplitude during unstable combustion was as high as 3000 psi (Fig. 3). For certain types of instabilities the propellant feed system is also coupled with the chamber oscillations, such that the dynamics of the entire propulsion system may dictate the oscillation growth rate, mode, and frequency. Combustion processes such as transient atomization, vaporization, and propellant mixing may occur causing locally enhanced burning rates and pressure-wave amplification. As a result of such mechanisms combustion instability can grow to full amplitude within milliseconds. Chamber and injector damage associated with combustion instability result from the high localized heating rates caused by the oscillatory motion of the combustion gases. Often minor changes in configuration (such as LOX post length) or operating conditions may cause a previously stable injector to exhibit an instability.

Because of these concerns, the design process within industry has been structured to address the issue of stability in a stepwise manner which is fully integrated with other design objectives. As indicated in the process outlined in Fig. 4 a full-scale injector concept is developed based on available data and analyses. Although the conceptual design is often relatively similar to the final configuration, many details must be defined that can significantly affect combustion stability. These details are dependent on characteristics that can only be established with confidence through hot-fire testing. For example, cavity design is driven by the instability frequencies that must be suppressed. This frequency is typically calculated based on the assumption that the chamber acoustic velocity is equal to that of combustion products at the injected mixture ratio based on equilibrium chemistry.[4,5] In some cases, instabilities have occurred at frequencies significantly above or below that assumed frequency. There are several reasons why the measured frequencies may not correspond to the equilibrium speed of sound of the product gases. Since data indicate that most of the acoustic energy and damage is localized near the injector face, it is reasonable that the frequency of oscillation of the transverse modes is dependent on the speed of sound at the injector face. The near injector speed of

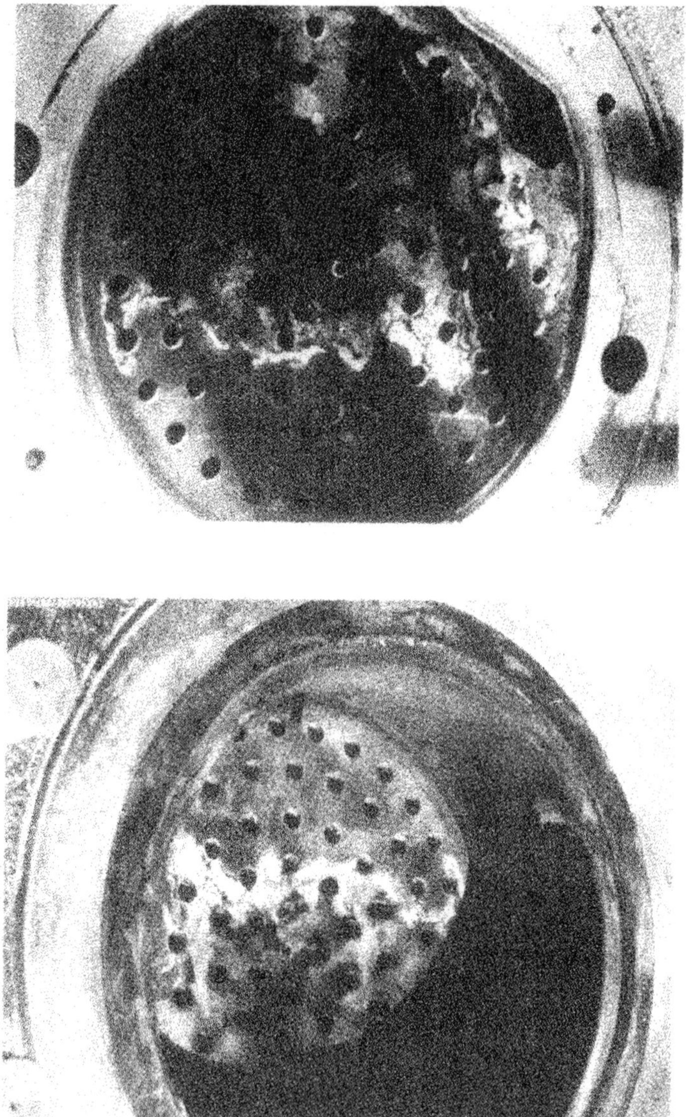

Fig. 1 **LOX/methane injector damaged as a result of 800 ms of combustion instability.**

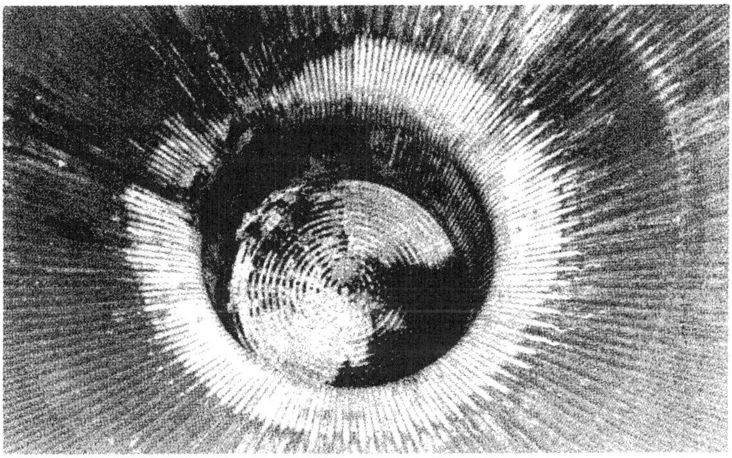

Fig. 2 F-1 injector damaged by a combustion instability incident.

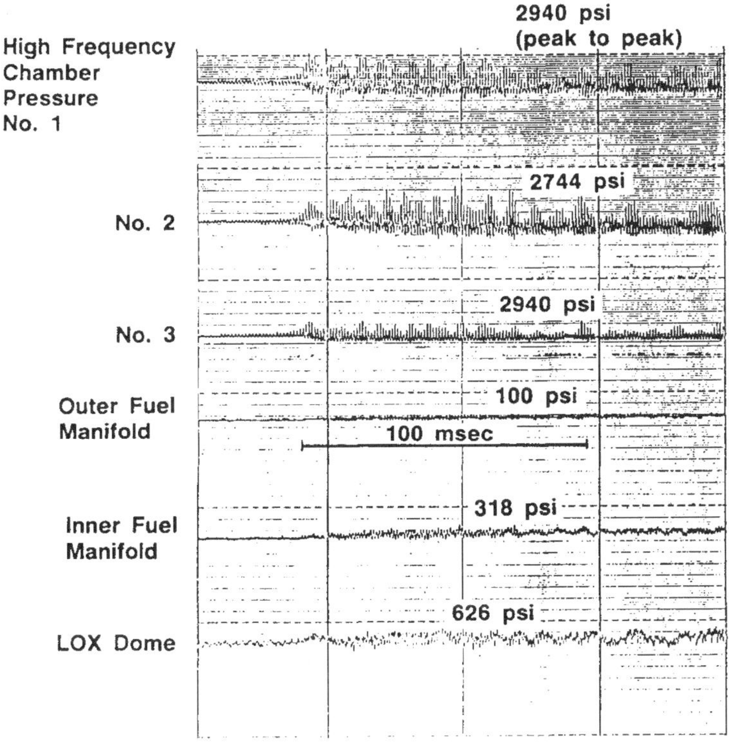

Fig. 3 Brush chart of high-frequency pressure transducer data for a first tangential instability; peak-to-peak oscillation levels indicated.

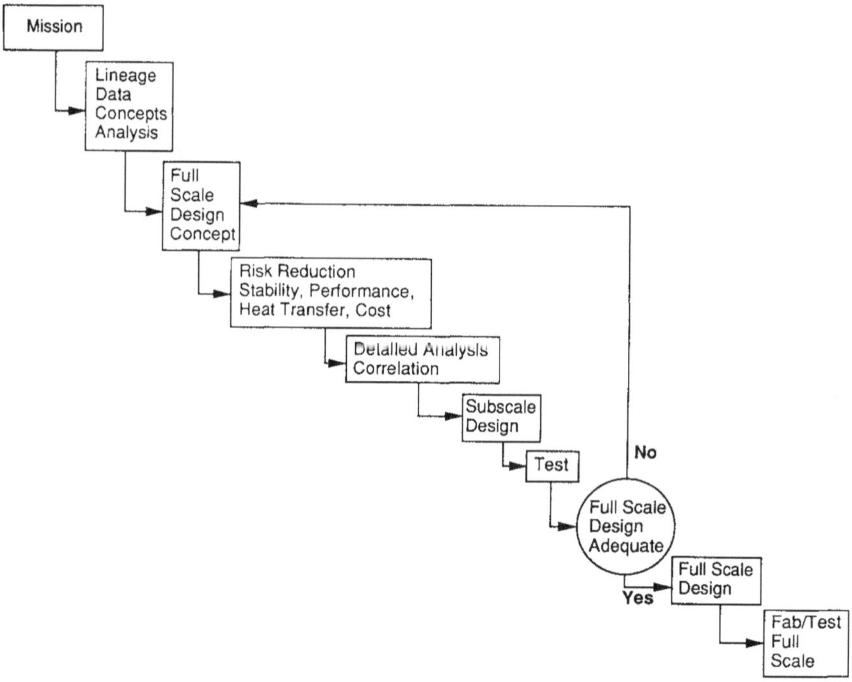

Fig. 4 Booster engine development logic.

sound may vary from the equilibrium product gas sound speed for reasons including a two-phase composition at the face, unreacted gaseous propellants that have a substantially different speed of sound than the combustion gas, and heat transfer to the wall, for example. Although the exact causes of a shift in frequency may not be immediately obvious from the data, hot-fire testing is critical to confirm what frequencies the cavities should be designed for.

As indicated earlier some occurrences of combustion instability can be traced to minor changes in operating condition or hardware design that may occur during combustor development. Ideally the impact on stability of such changes should be anticipated through analyses. However, whereas modeling techniques have proven useful in describing stability trends, these analyses are by no means fully matured and anchored. One of the key current limitations of analytical stability tools is the inability to predict the magnitude of the combustion instability gain. Often it is not even possible to determine which mechanisms are responsible for a given combustion instability, much less model it. Such experiences reinforce the need to conduct stability rating hot-fire testing with representative hardware.

It would be desirable to conduct all such testing at the full-scale level to match the stability-controlling characteristics of the final engine configuration (including the chamber acoustic modes and frequencies and the injector pattern) as closely as possible. For small (for example, upper stage class) engines and thrusters, testing is typically conducted at the full-scale size and thrust level. It is difficult if not impossible, however, to conduct all of the desired stability development testing for booster class engines at full-scale because of cost and schedule considerations. Because of test operations and hardware costs, only a limited number of full-scale injector configurations and tests can be conducted with actual engines. Most often, initial testing is accomplished using pressurized facility propellant tanks rather than actual engine pumps to minimize test hardware costs and to facilitate testing early in the engine development program. In addition, the potential for failure of a large test article and the associated safety, cost, and schedule issues if such a failure were to occur must be considered. Finally, other combustion related characteristics such as combustor performance and the impact of the injector flowfield on chamber life and cooling (injector/chamber compatibility) may be of greater concern than stability for a given program. Like stability, these concerns motivate the engine developer to seek some measure of confidence in his design prior to committing to full-scale booster testing.

To provide an intermediate, cost-effective evaluation of full-scale booster stability and performance characteristics, lower thrust subscale combustor hardware is typically hot-fire tested as part of the full-scale design effort. As shown in Fig. 4 the subscale effort is embedded within the detailed full-scale design effort and follows the conceptual design phase. The thrust level is often dictated by the maximum flow rates available at a given (subscale) facility. Typical industrial facility thrust levels are on the order of 20,000–100,000 lbf (89,000–445,000 N) at sea level.

III. Development of a Stability Program Utilizing Subscale Testing

Fundamental to the entire stability development effort is the fact that the injector and its associated stability aids (baffles and cavities) are the primary stability and performance controlling components in the engine. Hence, subscale test programs must be structured to evaluate the stability related characteristics of the full-scale injector. In particular, to be successful, subscale test programs must preserve those characteristics to the greatest extent possible. However, certain subscale design features are often the result of tradeoffs between stability, performance, and heat flux characterization objectives. Depending on the subscale approach selected, these tradeoffs may result in adjustments to the subscale injector pattern, the combustion chamber constant diameter section length, nozzle inlet geometry, and the amount of wall film cooling, for example. In addition, the desire to have a modular thrust chamber assembly to facilitate a wide range of injector and chamber parameters may result in baffles and cavities that are geometrically dissimilar to those in the full-scale design. For instance, the modular subscale thrust chamber hardware may utilize a removable spool with radial aperture cavities whereas the full-scale engine may have axial aperture cavities that are integral with the injector body. If possible, subscale chamber dimensions are selected such that full-scale baffle compartment and/or chamber frequencies are matched. Even if no pertinent mode of the full-scale combustor is matched, the occurrence of a combustion instability in subscale testing would indicate that further stability improvements are necessary before committing to a full-scale injector. The probability of an instability occurring in a full-scale combustor with an injector pattern from an unstable subscale combustor is very high. The higher frequency transverse modes of a full-scale combustor may match the range of frequencies in the subscale hardware. In addition, depending on the bandwidth of mechanisms relevant to subscale instabilities, these same mechanisms may easily support growth of lower frequency tangential oscillations characteristic of the larger chamber diameter hardware.

Although the subscale combustor diameter (or transverse dimensions) are typically smaller than the full-scale design, operating conditions (mixture ratio and chamber pressure) are usually close to that of the full-scale design for both performance and stability considerations. Injector element dimensions, injector to wall spacing, combustor length, and contraction ratio also usually match full-scale values for similar reasons. Throat heat flux simulation requirements dictate that the convergent section wall radii be smaller than full scale, potentially resulting in a chamber contour quite unlike the full-scale design. To correctly make such trades and accommodate subscale data requirements, a systematic design approach is necessary.

To fill this need, a judicious selection of stability and performance modeling techniques is required. Modeling can be employed to establish both appropriate (including off-nominal) ranges of operating conditions and hardware configurations to evaluate performance and stability characteristics of the subscale hardware. This is particularly true for cases where some tradeoffs between stability and performance design requirements are being considered. The results of a parametric modeling study can be used to structure a test plan and a test logic in parallel to establishing the subscale design and the required instrumentation. In addition, the results of subscale testing can later serve to better anchor the stability and performance models for conditions and configurational details typical of the full-scale design. Stability codes can be incorporated to evaluate spontaneous stability characteristics and acoustic cavity effectiveness. These codes can also be used in a correlative technique to quantify the combustion response (combustion induced gain and characteristic frequency). Theoretically, the combustion response is independent of the combustor scale. Hence, when anchored to subscale data these models can then be applied to estimate the performance and stability of a full-scale combustion device, provided that the injector and operating conditions tested at the subscale level are similar to that of the full-scale hardware.

Together with the subscale database, anchored stability and performance models supply a basis for confidence that the full-scale combustor will meet the design requirements. Far from eliminating the need to verify these characteristics at full scale, subscale testing and associated analyses provide an ability to cost effectively eliminate unsatisfactory designs early in the development process and to partially characterize the remaining candidates.

Critical to the subscale test program and the analytical efforts is an understanding of the key stability parameters and mechanisms. Subscale testing that is conducted at conditions similar to that of the full-scale design will exhibit similar combustion physics. For example, atomization rates, droplet size distribution, vaporization, and mixing should be reasonably similar between subscale and full-scale combustor conditions. If full-scale chamber or baffle compartment acoustic frequencies are matched by the subscale hardware, the interactions (including the relative characteristic times of the various processes) between the combustion and the oscillations in the full-scale design should be reasonably simulated in the subscale testing. With respect to stability modeling, although exact functional forms describing instability mechanisms are not required, predicting the preferred coupling frequencies and gain exhibited by injection and combustion processes as a function of geometry and operating conditions is essential to structuring a meaningful subscale stability test plan. Furthermore, the degree to which the key instability mechanisms are represented in the modeling will be reflected in the a priori accuracy of stability predictions for an untested full-scale combustor. A code that is well anchored to the subscale data across a range of operating conditions and configurations will be more likely to correctly model the effect of key stability physics for the full-scale design. Hence, part of the subscale test effort is typically aimed at gaining understanding of the controlling stability mechanisms and representing them in the codes.

An understanding of the mechanisms (feed system dynamics, atomization, vaporization, mixing, or chemistry) associated with instabilities may be isolated by performing tests at selected operating points. Mixture ratio excursions and throttling may expose feed system mechanisms. Chamber pressure excursions typically aid in the evaluation of the atomization/vaporization process as a controlling mechanism. Measured performance (and sometimes heat flux profiles) are compared with trends predicted by performance codes to attempt to determine the relative influence of each of the physical mechanisms modeled. At times, however, a single controlling mechanism cannot be isolated.

Examples and discussion of the more common subscale test approaches and modeling tools that are matched to the data obtained from such testing follow. Emphasis is placed on current industrial practice. Recommendations for future test and analysis technology improvements are also presented.

IV. Subscale Experimental Techniques

During most rocket engine development programs, series of subscale tests are typically performed to mitigate risk. As discussed earlier, typical subscale rocket engine development tests include cylindrical chamber hardware that is geometrically smaller with respect to the full-scale hardware, has similar contraction ratio and injector elements, and operates at similar pressures and mixture ratios. More generally, subscale combustors include single-element, subscale three-dimensional cylindrical combustors, two-dimensional "slab" combustors, and a variety of more obscure combustors, such as wedge-shaped combustors.

The experimental testing must be carefully planned in order to efficiently obtain as much information as possible. The stability considerations of a given full-scale design are not the only reasons for performing subscale development tests. Typically, performance information, heat transfer data, and injector-chamber compatibility data can also be generated during subscale testing depending on the combustor. The next section discusses the various subscale combustors and the information that can be obtained from each to reduce full-scale development risk.

A. Single-Element Combustors

Single-element combustors are also called labscale devices. The injector is composed of one or a few elements. Single-element combustor testing represents the most convenient and lowest cost form of subscale testing. However, only a very limited amount of stability information (relating to full-scale stability trends) can be generated. The most obvious information that can be obtained from single-element testing is the flow and combustion characteristics of an isolated injector element (intraelement flow characteristics). Several design features influence on the intraelement flow properties of the combustor. For a typical coax injector these include the fuel annulus dimensions, the oxidizer post tip thickness, and the oxidizer post inner diameter. The combustion dynamics may be evaluated by disturbing the flow with a high-explosive charge or by periodically interrupting the flowfield. Some of the physical results that can be obtained include approximations of the relative combustion time lag (using advanced photographic diagnostics), an approximation of the average combustion distribution, the propensity of the single-element injector to react to artificially generated high-amplitude disturbances, and single-element spontaneous stability characteristics. One example of the use of artificial disturbances was a set of experiments conducted in France with a rotating wheel to interrupt the combustor exhaust.[6] The specific results of these tests were not scaled to larger liquid propellant hardware. It should also be noted that element-to-element interactions that affect performance and stability are not simulated in single-element experiments. Specific evidence of a single-element spontaneous instability has not been obtained, but a 5 lb, three-element Marquardt engine was found to be spontaneously unstable.[7]

Single-element testing has limitations. Because the single-element chamber residence time and the propellant manifolding are quite different from the full-scale combustor, very little is offered with respect to chug stability information even though the element hydraulic response is the same as for the full-scale combustor. The chamber length can be increased with subscale hardware to represent full-scale chamber residence time and heat transfer aspects, but the increase in length of the combustor is destabilizing with respect to the longitudinal mode. Another disadvantage of single-element testing is that the tangential acoustic modes associated with the combustion chamber are not usually representative of the full-scale combustor. The tangential mode acoustic frequencies of a single-element combustor are higher than typical full-scale baffle compartment and main chamber frequencies. In addition, the injection density is typically lower than full scale which may affect the instability physical process relevant to full-scale instability behavior.

Instrumentation is critical with single-element hardware as with all combustors. Single-element chambers, however, can offer additional instrumentation opportunities not easily available to full scale. Typical instrumentation includes chamber pressure, steady flow rate, propellant and chamber temperatures, and some nonintrusive diagnostics. Windowed combustors facilitate a variety of both qualitative and quantitative combustion measurements. Qualitative visual information regarding general combustion zone structure, ignition and flame holding characteristics, and intraelement mixing characteristics may be obtained. Quantitative information includes droplet sizing and velocity data, liquid stream length, and temperature measurements. Such information can be useful for the determination of combustion performance characteristics that are typically needed to conduct an analysis of the combustion stability of a particular injector. Current single-element combustors allow uncooled windowed chamber sections to facilitate optical diagnostics that include Raman spectroscopy, planar laser-induced fluorescence, laser schlieren, droplet sizing measurements, and other optical techniques that characterize the combustion process. Window cooling can be a significant issue especially at high pressure. The aforementioned diagnostics have been applied to the single-element windowed combustor[8] shown in Fig. 5. Key objectives of this program were to collect data for use in evaluating Space Shuttle main engine (SSME) preburner ignition processes and validating a mechanistic [computational fluid dynamics (CFD)] combustion model.

B. Three-Dimensional Cylindrical Subscale Chambers

Unlike a true single-element combustor, a typical three-dimensional chamber has multiple injector elements of the same basic element size, geometry, and pattern as the full-scale. The number of elements, however, is reduced to match the smaller injector face area and lower overall flow rate. Three-dimensional round

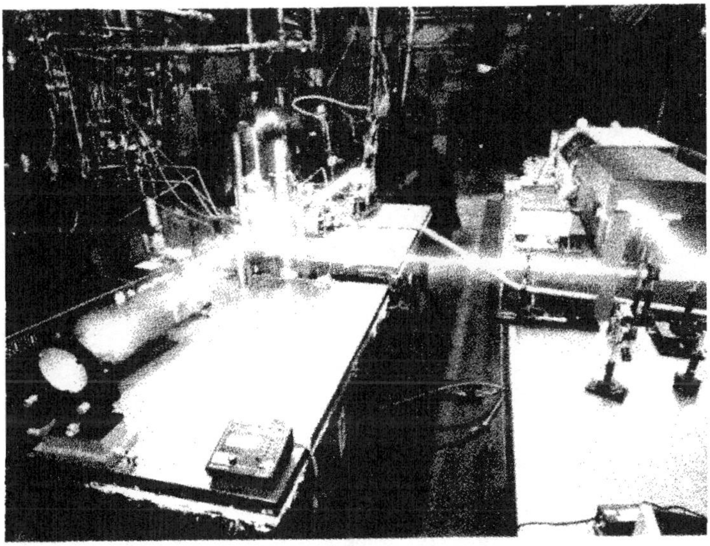

Fig. 5 Turbine drive single-element combustor with Raman and schlieren instrumentation.

subscale chambers can provide information on inter-element interaction that is characteristic of the full-scale design. In addition, the transverse wavelength of subscale hardware may approximate that of a full-scale baffle compartment. The full-scale mass flux, contraction ratio, and operating conditions (chamber pressure and mixture ratio) are specified for the three-dimensional subscale test article. Performance and heat flux information (using water cooled calorimeter chambers) may be obtained to aid in the design of the full-scale combustor. Care should be taken, however, in the design of the subscale hardware in order to not jeopardize the stability development program.

There are several advantages to using subscale three-dimensional round hardware. These advantages include reduced flow rates (and thrust) relative to full scale, a large hardware experience base from which to draw, and a large test database to compare with. All of these factors make it relatively easy to plan and execute a successful development program. Similar to single-element combustors, combustion characteristics can be obtained experimentally and later applied to the analysis of a full-scale combustor. The response of the combustion process to high-amplitude steep-fronted waves can be examined with three-dimensional round chambers via stability rating bomb tests similar to those typically used in full-scale tests. This rating technique has been successfully incorporated on many full-scale rocket engine programs including the J-2, F-1, and the SSME. During bomb tests, a small charge of high explosive is installed in the chamber combustion zone and detonated during combustor operation. The response of the combustor to the bomb pressure pulse has been shown to be a reproducible stability rating technique. The amplitude of the induced pressure wave can be controlled by the amount of explosive placed in the bomb.

Three-dimensional round chamber subscale testing also has some limitations. The most prominent limitation is that the hardware is geometrically smaller than the full-scale hardware. As a result, frequencies observed are not the same as the lower frequency transverse modes of the full-scale combustors. Another limitation is that the subscale combustor may have a larger length-to-diameter ratio than the full scale. The reason is that to preserve the same combustion residence time and allow for complete droplet vaporization, the subscale length is made identical to that of the full-scale engine. Combustion chambers with relatively large length-to-diameter ratios are susceptible to longitudinal instabilities.

Key stability results obtained in three-dimensional cylindrical hardware include basic stability data such as fluctuating pressure and acceleration levels, cavity temperature measurements, high-frequency time-lag data (which is calculated from the ring-down of a pressure perturbation), and the dynamic response (pulse induced pressure wave amplitude growth) associated with a specific injector pattern. Other data typically include chamber pressure measurements, calorimeter temperature measurements (heat flux), propellant flow rates, and external high-speed motion pictures of the plume. If the framing speed of the film is fast enough, some low-frequency (e.g., chug stability) information may be obtained from plume motion pictures. Plume spectroscopy is sometimes incorporated with subscale three-dimensional round hardware testing to detect the onset of combustor erosion or compatibility problems. Propellant flow rates are typically measured with cavitating venturis, ASME type venturis, or turbine flow meters. Typically, high-frequency pressure measurements in the manifolds and combustion chamber are made with piezoelectric pressure transducers. Since acoustic mode high-frequency instability is defined as oscillations in chamber pressure at a specific frequency and amplitude, chamber high-frequency pressure measurements should be made on all subscale and full-scale hardware. Typically, combustion instability pressure oscillations are at their maximum amplitude at the injector face. Hence, high-frequency pressure transducers should be located at various circumferential locations close to the injector face to provide accurate information on the mode and peak amplitude of those oscillations.

Many examples exist with respect to three-dimensional cylindrical subscale hardware. The 82-element LOX/CH_4 hardware[1] is displayed in Fig. 6. The combustor diameter was 5.66 in., the main stage operating pressure was approximately 2000 psia, and the mixture ratio ranged from 1.9 to 3.7. The nominal propellant flow rate was approximately 90 lb/s. During the testing of the 82-element LOX/CH_4 injector, two separate types of stability rating tests were performed in order to evaluate the stability margin of a combustor with this particular injector. The first tests performed were fuel temperature ramp testing. The combustor was found to be unstable at methane temperatures ranging from -16 to -23°F. During the J-2 program, it was discovered that instabilities would occur with the engine when the nominal fuel temperature was lowered to a specific value for a given injector. As a result, this technique was established as a stability rating approach for liquid/gas coaxial injectors. Many such subscale ramping tests were conducted at NASA Lewis Research Center in the 1960s and 1970s.[9] Two mechanisms can account for the temperature ramp instabilities observed. The first is that the gaseous fuel density decreases appreciably with temperature and results in a loss in hydraulic resistance across the injector. The loss of resistance results in increased combustion chamber sensitivity to fuel side feed system oscillations. The second mechanism is that the fuel injection velocity decreases linearly with temperature and results in poor atomization, improving the potential for combustion coupled acoustic instability.[10]

Fig. 6 Santa Susana test facility 82-element LOX/CH_4 subscale hardware.

Figure 7 shows the varying LOX dome frequency spectrum that was typical of these tests. Each spectrum trace corresponds to a "slice" (short sampling period) of time during the test. Note that the characteristic combustion frequency shifts as the test proceeds and the fuel temperature continues to decrease. The start of the instability that is noted corresponds to the onset of high-amplitude chamber oscillations. Subscale fuel temperature ramp tests can provide direct input into the development of the start sequence of a full-scale engine.

Some programs that have implemented three-dimensional subscale hardware include the SSME 40,000-lb test article,[11] the heavy hydrocarbon main injector technology program,[4] Marshall Space Flight Center RP-1 and methane tests[12] and Lewis Research Center RP-1 hot-fire tests.[13] A large amount of useful data has been collected with three-dimensional cylindrical subscale hardware.

C. Two-Dimensional Subscale Combustors

Two-dimensional slab combustors represent perhaps the most representative means by which full-scale acoustical stability can be evaluated with subscale hardware. Two-dimensional subscale combustors are typically sized such that the first width or "slosh" mode of the two-dimensional combustor matches the first tangential mode of the full-scale hardware. This is done in order to expose the combustion process to pressure perturbations of the same frequency that occur in the full-scale hardware. It should be noted that the two-dimensional tests do not yield exact combustion stability simulation to the full-scale hardware, because the spatial pressure waveform differs between two-dimensional and full scale due to the differences in geometry of each type of combustor. Furthermore, a spinning type full-scale instability cannot physically occur in the two-dimensional hardware.

The main advantage of two-dimensional subscale testing is that it is possible to have the same acoustical frequencies as the full-scale hardware at much

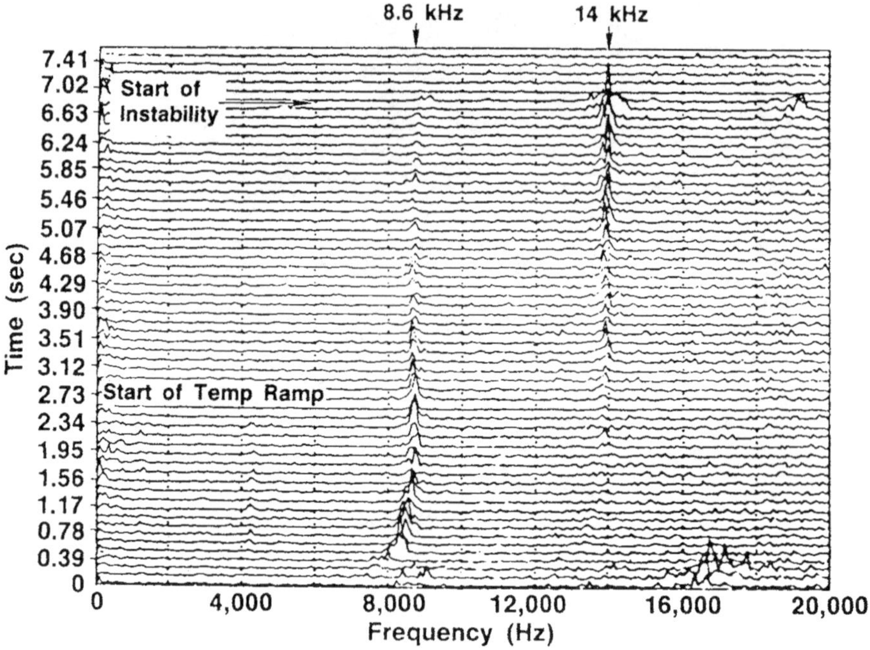

Fig. 7 Oxidizer manifold high-frequency pressure transducer isoplot for a LOX/methane subscale fuel temperature ramping test.

lower flow rates (compatible with a smaller test facility) than the full-scale design. The injector elements match the full-scale hardware. The two-dimensional hardware can also be used for acoustic cavity sizing configuration evaluation efforts.[14] Baffle length studies can also be performed with two-dimensional subscale combustors. The damping rates and frequencies of combustors can be evaluated as a function of baffle length.[15] The damp time of bomb disturbances typically decreases as the baffle length increases, but longer baffles require more coolant which detracts from performance. Two-dimensional combustors afford the developer an opportunity to optimize baffle length for full-scale hardware by means of relatively lower cost testing. Two-dimensional subscale hardware has also been manufactured with transparent chamber walls that facilitate flow and combustion visualization. The transparent wall J-2S two-dimensional combustor is shown in Fig. 8.

Two-dimensional combustors have some limitations. Besides the facts that spinning tangential instabilities are precluded and that the waves exhibit different spatial pressure distributions relative to full scale, the other drawback is that the surface area to volume ratio of two-dimensional hardware is much larger than the corresponding full-scale combustor. The large surface area results in somewhat greater acoustic damping relative to full-scale hardware. However, the amount of damping occurring at the wall is thought to be small in comparison to the convective damping through the combustor throat.

Typical instrumentation with the two-dimensional combustor includes high- and low-frequency pressure transducers, thermocouples, accelerometers, flow meters, and high-speed motion picture cameras. Various low-pressure two-dimensional

Fig. 8 J-2S transparent two-dimensional thrust chamber installed at Rocketdyne Santa Susana field laboratory.

combustors have incorporated transparent walls.[10,14,16] Nonintrusive optical instrumentation such as schlieren, shadowgraph, or direct combustion photography can be used if a transparent wall is incorporated in the hardware. One particular advantage is that two-dimensional combustors have two large surface area regions that are easily accessible for multiple instrumentation ports and windows if needed.

A significant amount of information was generated during two-dimensional testing that directly supported the design of the F-1 engine. In particular, the effectiveness of various baffle lengths, the occurrence of the resurge instability mode, and the stability of innovative injector concepts were all studied in two-dimensional combustors as part of Project First, which supported F-1 engine development.[15] Additionally, good corroboration was noted between two-dimensional and full-scale chambers in a study of the dynamic stability of an injector using various fuels.[16] J-2S slab motor testing was conducted to study the flow characteristics in the region of a hot-gas tap-off port.

A recent example of a two-dimensional subscale device is the two-dimensional stability research combustor (SRC). The SRC was designed to aid in the understanding of predominant physical mechanisms associated with unstable combustion. Either coaxial or impinging injectors may be installed. The combustion chamber is highly modular and permits combustor widths up to 13.5 in. long to be tested. The combustion chamber includes instrumentation ports for various test measurements and also incorporates a wall with windows to facilitate optical diagnostic techniques discussed earlier. The hardware is shown installed at NASA Marshall Space Flight Center in Fig. 9.

Fig. 9 Stability research combustor installed at Marshall test stand 115.

D. Other Subscale Hardware

Other types of subscale hardware have also been used to a limited degree to aid in understanding how full-scale hardware should be designed to maximize stability margin. Three subscale configurations that have been tested in the past are annular, barrel, and wedge combustors. Examples of all three combustors are displayed schematically in Fig. 10.

As depicted, the annular chamber has an outer chamber, a centerbody, and a spud or ring type injector pattern located at the head end of the annular gap. This design is intended to facilitate the study of full-scale frequencies in a reduced thrust combustor. The annular combustor can be used to model spinning instabilities in the outer zone of the combustion chamber. Assuming the annular chamber has the same diameter as the full-scale hardware, the spinning or racetrack mode frequencies are typically lower than the full-scale combustor. Annular combustor radial modes will be at significantly higher frequencies. The annular chamber, however, poses significant fabrication issues (high cost and design complexity for thermal load management on the centerbody) that should be considered in the subscale program planning. In addition, like the two-dimensional combustor, wall losses are higher than in cylindrical chambers.

The barrel (or pulse type) chamber is another subscale alternative that offers full-scale acoustics with reduced flow rates. The barrel chamber is similar to the annular chamber but without the centerbody. As before, spud or ring type injectors are located near the chamber wall to study the influence of the near-wall elements on stability. Like the annular chamber, this is a relatively low-thrust

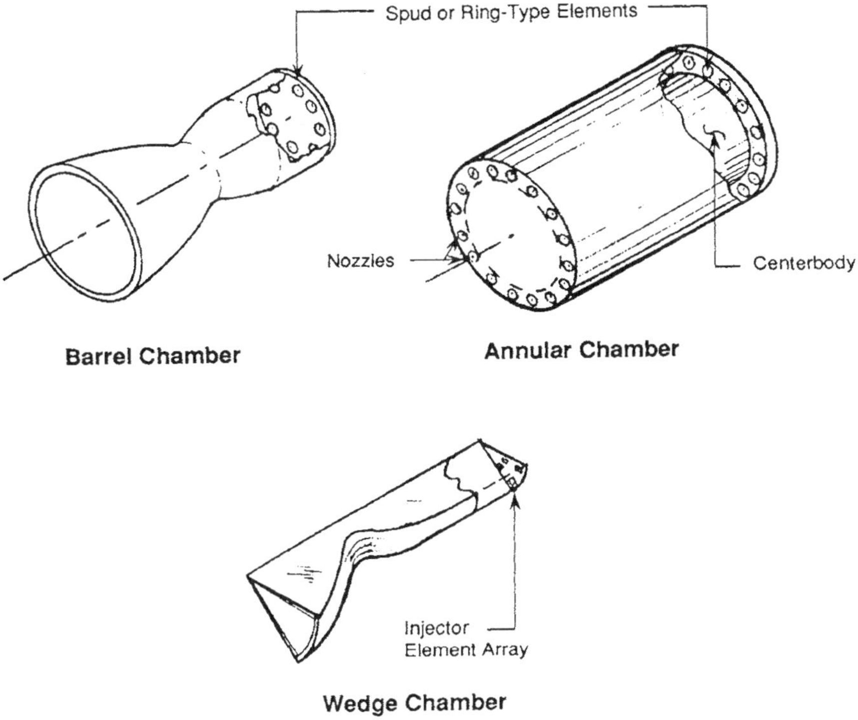

Fig. 10 Annular, barrel, and wedge combustors (from NASA SP-194).

device. Because the chamber diameter is essentially full scale, the combustion instability frequencies approximately match those of the full scale. The drawback to the barrel chamber is that the mass flux in the combustion chamber is significantly lower than a full-scale chamber. This leads to contraction ratio (or chamber pressure) discrepancies that can affect the convective loading and diffusion characteristics of the propellant sprays. Such compromises may result in misleading information with respect to the combustion response. A high-contraction ratio may reduce convective damping of the nozzle and result in a significantly different stability margin than the corresponding full-scale hardware. Major differences in stability characteristics have been noted between barrel and full-scale hardware in the study of various propellants.[16]

Wedge chambers have also been previously used to investigate baffle compartment acoustics. Previous programs have displayed baffle compartment modes and ring down frequencies that match the predicted compartment acoustics. The wedge-shape chambers give an exact representation of the full-scale baffle compartment but fail to give information on the interaction between the compartment and chamber.

One final method of testing that may be considered is cold-flow acoustic response testing. Information can be gained from these tests, but careful extrapolation must take place to utilize cold-flow subscale test results to project hot-fire combustor characteristics. The information that can be gained from these tests includes the complex acoustic modes of the full-scale geometry, the hydraulic resistance of the injector, and visual flow representation. Additionally cold-flow droplet vaporization and breakup studies have been conducted with the purpose of obtaining a better understanding of combustion instability mechanisms and developing a database for code anchoring efforts. However, combustion effects can be substantial such that cold-flow mixing and atomization test results must be interpreted analytically to properly extrapolate to hot-fire conditions.

V. Analytical Approaches

As stated earlier, subscale engine component testing is used to mitigate the risk associated with full-scale combustor development. A key application of the subscale test results is to validate analytical tools that will be used in the extrapolation to full scale. Several analytical approaches have been used to model the aspects of combustion instability in liquid-propellant rocket engines. The approaches include the correlative sensitive time-lag analysis and multidimensional mechanistic time-domain numerical models. All of the analyses lend credibility to the full-scale design when properly validated.

The sensitive time lag approach was first introduced in 1956 by Crocco and Cheng.[17] The approach incorporates mechanistic modeling of the gas acoustic field in the combustion chamber and correlative modeling of the gain and characteristic frequency (combustion response) of the propellants. The time-lag approach has been used successfully on a number of rocket engine programs. Because of the mechanistic approach to the acoustics modeling, combustion stability aids such as acoustic cavities and injector face baffles can be evaluated analytically. The time-lag theory assumes that all of the combustion processes can be correlated to the unsteady pressure in the combustor. The drawback to the time-lag theory is that any new rocket engine design analysis must be based on extrapolated data. If subscale tests are run and combustion response growth or decay data are obtained such that the time-lag database can be extended, the amount of extrapolation is minimized.

The combustion response can be calculated from the pressure oscillation decay rate (ring-down) of an artificial or naturally occurring disturbance during hot-fire testing. Alternatively, the combustion response can also be related to changes in instability margin if the instability damping characteristics are parametrically varied, say, by varying the cavity open area. Using a linear stability model the decay rate measured from the ring-down of the pressure disturbance is then used to calculate the gain and overall combustion response.[17,18] Nonlinear combustion response stability models have also been developed which predict the steep-fronted pressure perturbation amplitude required to drive a given combustion process unstable.[19] Both modeling techniques require inputs that can be obtained from subscale testing and/or combustion performance modeling, such as the combustion profile and droplet sizes.

Time-lag theory was used to predict the available stability margin as a function of cavity tuning and open area on a high-pressure LOX/RP-1 subscale test program and on the full-scale (upper stage) XLR-132 storable propellant engine.[4,20] Linearized theory was also used in scaling the XLR-132 to a higher thrust storable upper stage (SUS) engine. Quantitative results were generated based on the scaling technique. When tested, the SUS combustor was spontaneously and dynamically stable without need for changing the predicted stable acoustic cavity configuration.[21]

The second approach that has been taken to model unstable combustion in liquid propellant rocket engines is mechanistic numerical analysis. This includes a limited amount of transient CFD analyses of the combustion process.[22,23] The relevant combustion processes include injection hydraulics, atomization (both primary and secondary), mixing, vaporization, and chemical reaction. Recent efforts have concentrated on the development of submodels that describe these processes based on first principles. These analyses also have to be anchored with detailed spatial data (both time averaged and transient) which are used to validate the predicted values. Subscale testing (especially single-element tests and cold-flow tests) provides an excellent means to collect the necessary data to accomplish the anchoring process. Advances in combustion diagnostic instrumentation (particularly nonintrusive optical diagnostics) facilitate acquisition of the required data. Key data that can now be measured in reacting flows include droplet sizes and velocity and gas-phase species, temperature, and velocity.

All of the data also aid in the validation of current and future steady-state performance codes. These codes also yield information that can aid in the evaluation of the stability characteristics of a combustion chamber. An example discussed earlier is the steady near injector acoustic environment for a given injector. Parametric performance design analysis codes, such as the standardized distributed energy release model (SDER), can be used to approximate the conditions. However, three-dimensional CFD analyses in principle can provide the most accurate and detailed near injector acoustic profile necessary for this calculation.

VI. Technology Improvements

There are many areas of rocket engine scaling that can be improved. The most important aspect of a stability development or a scaling program is to establish the direction from the beginning of the program. A well-defined subscale test methodology will enable the program to progress to the desired results on time and within budget. The current design methodology should be enhanced so that the most advanced analytical and instrumentation technologies are incorporated early in the test planning and hardware design. If properly structured, the design methodology related to combustion stability scaling efforts could become a self-improving test and analysis effort. This can only occur if state-of-the-art technology is constantly being included in the scaling effort.

The capabilities of combustion instrumentation are constantly being improved and enhanced. The current technology includes nonintrusive techniques for measuring flow velocity, flow rates, drop sizes, chemical species, and temperature distribution within the combustor. Development hardware needs to be designed to facilitate the application of advanced diagnostics to provide direct insights of the impact of design parameters on stability and performance. This would ideally include full-scale development hardware as well as subscale test articles. In this manner more information could be obtained from each test. This is an important consideration since test costs continue to rise. It is not an uncommon event for full-scale hardware to be manufactured with a fraction of the instrumentation (especially high-frequency pressure transducers) that the subscale development hardware had. A lack of instrumentation in the full-scale hardware makes it very difficult to cross reference the full-scale results to that of the subscale testing.

VII. Conclusions

Subscale test techniques can be used to gain substantial insight relative to full-scale hardware combustion stability characteristics. The purpose of the subscale hardware is to mitigate as much as possible the risk involved with unstable combustion in the full-scale design. Absolute stability characteristics of the full-scale hardware cannot be determined without testing the full-scale hardware, but subscale hardware can greatly reduce the cost of a full-scale stability development program.

In addition, subscale testing can serve to better anchor performance and stability analyses that can be utilized to assess the stability of the full-scale design before it is built and tested. In particular, where specific mechanisms such as injector element hydraulic coupling with the combustion process can be studied, key stability mechanism models can be anchored. Currently, the approach to achieving a successful full-scale design is to perform a portion of the development work at the subscale level and enter the full-scale design with as much information as possible. Some previous development programs started with full-scale hardware and backed into subscale stability tests to examine particular mechanisms as they arose. This practice has proven to be very costly. Careful planning of experiments and use of analysis techniques is required to produce the design information needed to develop a full-scale configuration within program schedules and budgets.

References

[1]Jensen, R. J., Dodson, H. C., and Claflin, S. E., "LOX/Hydrocarbon Combustion Instability Investigation," Final Report, NASA CR-182249, May 1989.

[2]Huzel, D. K., and Huang, D. H., *Design of Liquid Propellant Rocket Engine,* NASA SP-125, 2nd ed., 1971.

[3]Anon., *Guidelines for Combustion Stability Specifications and Verification Procedures for Liquid Propellant Rocket Engines,* CPIA Pub. 247, Chemical Propulsion Information Agency, Laurel, MD, Oct. 1973.

[4]Arbit, H. A., Tuegel, L. M., and Dodd, F. E., "Heavy Hydrocarbon Main Injector Technology Program Final Report," Rockwell International/Rocketdyne Div., RI/RD 91-118, Contract NAS8-36369, April 1991.

[5]Levine, R. S., "Experimental Status of High Frequency Liquid Rocket Combustion Instability," *Tenth Symposium (International) Combustion,* Combustion Inst., Pittsburgh, PA, 1965, pp. 1083–1099.

[6]LeCourt, R., and Foucaud, R., "Experiments on Stability of Liquid Rocket Motors," AIAA Paper 87-1772, June 1987.

[7]Crone, C., Stechman, C., and Woll, P., "Combustion Instability in Small Earth Storable Bipropellant Rocket Engines," AIAA/ASME/SAE/ASEE 29th Joint Propulsion Conf. and Exhibit, AIAA Paper 93-196, Monterey, CA, June 28–30, 1993.

[8]Hunt, K. W., Fisher, S. C., and Arbit, H. A., "Turbine Drive Combustor Technology," Rockwell International/Rocketdyne Div., RI/RD89-235, Contract NAS8-34928, Aug. 1989.

[9]Conrad, E., Bloomer, H. E., Wanhainen, J. P., and Vincent, D. W., "Interim Summary of Liquid Rocket Acoustic-Mode-Instability Studies at a Nominal Thrust of 20,000 Pounds," NASA TN D-4968, Dec. 1968.

[10]Levine, R. S., "Experimental Status of High Frequency Liquid Rocket Combustion Instability," *10th Symposium (International) on Combustion,* Combustion Inst., Pittsburgh, PA, 1965, pp. 1083–1099.

[11]Anon., "Thrust Chamber Assembly 40K Subscale Chamber and Injector Cycle Life," Rockwell International, Rocketdyne RSS-303-28, Rocketdyne Div., Canoga Park, CA, Sept. 1976.

[12]Bailey, C. R., "Test Evaluation of Oxygen-Methane Main Injectors," *JANNAF Propulsion Meeting,* CPIA Pub. 480, No. 1.5, Chemical Propulsion Information Agency, Laurel, MD, Dec. 1987, pp. 77–86.

[13]Price, H. G., and Masters, P. A., "Liquid Oxygen Cooling of High Pressure LOX/Hydrocarbon Rocket Thrust Chambers," NASA TM 88805, Aug. 1986.

[14]Anon., "Acoustic Cavity Technology for High Performance Injectors," Final Rept., NAS9-14232-3-1, Aerojet Liquid Rocket Co., Oct. 1976.

[15]Anon., "History, Project First, F-1 Combustion Stability Program," Rocketdyne Rept., R-5615-16, Contract NASW-16, Feb. 1967.

[16]Lawhead, R. B., and Combs, L. P., "Modeling Techniques for Liquid Propellant Rocket Combustion Processes," *Ninth Symposium (International) on Combustion,* Combustion Inst., Pittsburgh, PA, 1963.

[17]Crocco, L., and Cheng, S. I., *Theory of Combustion Instability in Liquid Propellant Rocket Motors,* Butterworths, London, 1956.

[18]Mitchell, C. E., and Eckert, K., "A Simplified Computer Program for the Prediction of the Linear Stability Behavior of Liquid Propellant Combustors," NASA CR 3169, Aug. 1979.

[19]Priem, R. J., and Guentert, D. C., "Combustion Instability Limits Determined by a Nonlinear Theory and a One-Dimensional Model," NASA TN D-1409, Oct. 1962.

[20]Johnson, J. R., Treinen, T. R., and McMillion, R. L., "Advanced Spacecraft Engine Components XLR-132 Breadboard Engine Program," Rockwell International/Rocketdyne Div., RI/RD 86-291, AFRPL TR-86-086, June 1987.

[21]Hampson, M. E., "Storable Upper Stage Engines, High Thrust Technology," Rocketdyne IR&D Tech. Rept. ITR-88-016, Rockwell International, Rocketdyne Div., Canoga Park, CA, Nov. 1988.

[22]Grenda, J. M., Venkateswaran, S., and Merkle, C. L., "Multi-Dimensional Analysis of Combustion Instabilities in Liquid Rocket Motors," AIAA/ASME/SAE/ASCE 28th Joint Propulsion Conf., AIAA Paper 92-3764, Nashville, TN, July 1992.

[23]Liang, P., and Ungewitter, R., "Multi-Phase Simulations of Coaxial Injector Combustion," AIAA 30th Aerospace Sciences Meeting, AIAA Paper 92-0345, Reno, NV, Jan. 1992.

Author Index